Spec Builder's Guide

by

Jack P. Jones

Craftsman Book Company
6058 Corte Del Cedro, Carlsbad, CA 92009

Acknowledgements

The author wishes to express his appreciation to the following companies and organizations for furnishing materials used in the preparation of various portions of this book.

American Plywood Association, 7011 South 19th Street, P. O. Box 11700, Tacoma, Washington 98411
Asphalt Roofing Manufacturers Assn. 1800 Massachusetts Ave., N.W., Ste. 702, Washington, D.C. 20036
Benjamin Moore & Co., Chestnut Ridge Road, Montvale, New Jersey 07645
The Celotex Corporation, P.O. Box 22602, Tampa, Florida 33622
First Federal Savings and Loan Association, 1570 Watson Boulevard, Warner Robins, Georgia 31099
Georgia-Pacific Corporation, 133 Peachtree St., N.E., Atlanta, Georgia 30303
Weyerhaeuser Company, Tacoma, Washington 98401
Woodmark Cabinets, American Woodmark Corporation, P.O. Box 514, Berryville, Virginia 22611

This book is dedicated to my wife, Tena

Library of Congress Cataloging in Publication Data

Jones, Jack Payne, 1928-
Spec builder's guide.

Includes index.
1. Building--Handbooks, manuals, etc. I. Title.
TH151.J66 1984 690'.68 84-1697
ISBN 0-910460-38-8

Fourth printing 1989

Edited by Sam Adrezin

Contents

Chapter 1

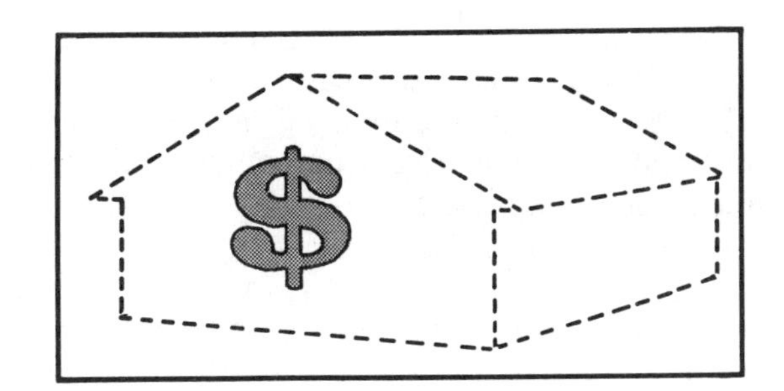

Opportunities for Spec Builders

I hope you've picked up this book because you want to become a home builder — a speculative builder who finds the land, gets the loan, builds the house and sells it at a good profit. That's exactly what I intend to explain.

If you're not ready to take the plunge into "spec" building, this book will serve equally well as a primer on custom home building — putting up a home at a price set by contract with the owner of the lot.

But my primary objective is to convince you to build and sell your first spec house. Every housing giant started with one house and one buyer. Maybe you can make it too. If you have some savings and some time, give it a try. Spec building is a profitable and challenging career — or sideline — for many professionals in the construction industry.

Home building takes skill, common sense, good judgment, hard work, money, and more paper shuffling than most tradesmen feel comfortable with. But the field is as wide open to eager young contractors today as it has ever been. If you're considering a career as a home builder, read on.

Advantages of the Small-Volume Builder

Home building is a risky and competitive business. That's always been true. But it offers rewards equal to the challenge. And, as a new home builder, you have some unique advantages.

First, you'll be working in one of the few major industries that aren't dominated by four or five big national companies. There's no General Motors or IBM of home building to put you out of business. You'll never be replaced by automation, and foreign imports will never wipe out your job. And there's another big plus: It doesn't take millions of dollars to get started as a home builder. Thousands do it every year with only limited savings and a few thousand dollars invested in tools and a truck.

There's yet another advantage every small-volume builder has. If the market for spec or custom houses is flat, switch to remodeling and additions. There's always remodeling work available for a reputable builder. Many small-volume builders handle these jobs on a regular basis because the markups are better. A major builder accustomed to selling several hundred homes a year can't shift smoothly into remodeling work. When a slump comes, he has to stop building. As a small volume builder, you don't have that problem.

A major home builder has some advantages, of course. For example, he builds the same basic home over and over again — maybe a hundred times or more in the same tract. By the time the tract is finished, he should be pretty good at it. Repetition tends to iron out the kinks and bring down costs.

But many small volume builders do nearly the same thing. They build the same basic house over and over again with only minor variations to meet the needs of the market, the lot, the owner or the neighborhood. Of course, these houses probably go up one at a time. But there's even an advantage to that. You have time to evaluate the problems and make corrections before the next unit is started.

Your best advantage over big 100-plus-homes-per-year builders is that you know your community. You know your customers and what they'll buy. You know the best subcontractors and tradesmen in the area. You know what parcels are available and what land prices are reasonable.

There isn't just one housing market. There are thousands of them. Every community has its own unique housing market. You have the opportunity to know yours best, and serve it better than anyone else can. If you're lucky, your community wouldn't support a 500-home tract thrown up by a major builder from out of state. That leaves you free to become the most active and professional home builder in the area. And that's exactly the aim of this book.

I'm not going to claim that small volume builders have all the advantages. They don't. And I'm not going to promise that you'll become a millionaire. I haven't, and I've been at it for over 25 years. But I've earned a good living as a contractor and spec builder every year during that period. I know many home builders that have done as well or better during those years. And some of the successful small builders I used to know are now successful big builders.

Be a Working Contractor

Over the last 25 years I've learned good ways to build quality houses and sell them at a profit. I know how to cut costs and waste so that my prices are competitive with builders who are less quality-conscious. I know good building practice when I see it. And I know what it takes to prevent a problem from becoming a major loss.

In my opinion, there's no substitute for this knowledge, hard work, and attention to detail. And the place where the most knowledge, hard work and attention to detail are needed is on the job site.

That's why I'm *not* going to explain how to become a "paper contractor." If your idea of home building is to run the whole show from your office with a couple of subcontractors, a telephone, and a typewriter, you're reading the wrong book.

In my company I'm my own best carpenter — and sometimes my own best roofer, cement mason, bricklayer, plumber, electrician, floor layer, painter and wallboard hanger. I don't ask any tradesman to do anything that I couldn't do myself and probably do just as well. If you want to call this one of the "secrets" of making a living as a home builder, then this is the secret that I intend to share with you.

It took time and trouble to learn all this. But I know how long every job should take, and I recognize when it's done right and when it's not. My tradesmen and subcontractors have learned that they can't pull the wool over my eyes — at least not very often and not for very long.

But don't get the impression that I spend more time building than I do running my company. I'm the boss first and a tradesman second. But there's no substitute for knowing every part of building houses — from foundations to roofing and everything in between. That's why this book explains exactly how all the work should be done.

Technical competence is an important part of every home builder's job. My customers know that I take personal responsibility for every nail and stud in every home I build. And your customers should know that you do too.

Opportunities — Where You Find Them

The United States is the land of opportunity for spec builders. In how many countries could nearly anyone go out and find a lot, get a loan, put up a house and sell it at a profit? There aren't many small-volume builders in Europe. And there are few or none in socialist countries where construction is handled under government authority. Most developing countries don't have lending institutions that can finance spec builders like banks and Savings & Loans can in the U.S. The truth is that spec builders probably have the best chance of success right here in the U.S.

It's true that the 500 largest home builders dominate spec building today. That wasn't true 25 years ago when I started building. But thousands of spec homes are built in the U.S. every year by hard-working, knowledgeable, professional small-volume builders. And many are making good money year after year.

And the rewards can be substantial. Tradesmen work for wages. The lender gets its fee and interest on the loan. Those are set by contract. The subcontractor who includes the smallest profit percentage in his bid probably gets the job. But you are the entrepreneur who takes the risk. You're entitled to the entrepreneur's profit — whatever the market will bear. And the profits can be major if you know your trade and judge your market correctly.

How to Avoid Losses

Spec building is a risky business. Thousands of home builders go broke or leave the industry during every housing recession. And you can expect a recession every 5 to 7 years, if past experience is a good guide to the future.

Most of these failures and nearly all of the bankruptcies are the result of over-eager expansion, misjudging the housing market, high overhead or debt that couldn't be repaid. In my opinion, you stand a better chance of success if you handle just one or two jobs at a time, keep your overhead at a bare minimum, and work along with your crew so you also earn a tradesman's wage.

Thousands of general contractors operate out of their homes. Don't worry about not having an office and receptionist. The customers that buy your houses don't care. The bank that lends you the money doesn't care. Get a business phone installed in your home. Maybe invest in a phone answering machine. Your pickup truck is the only office you'll need for several years — maybe the only office you'll ever need.

You can find jobs, plan them, compute the materials required, prepare the estimates, hire help, see that they get paid, and fill out forms right on your kitchen table if necessary. An office and receptionist just run up your overhead.

My advice is to subcontract less of the work and handle more yourself. The more work you and your crew can do, the more you can earn on each house. Move to a site, finish the work, and move on to another. You can make a good living as a spec or custom builder even if you handle only one job at a time. There are plenty of advantages to keeping your business as simple as possible.

Problem #1 — Finding Land

In many communities it's increasingly hard to find suitable raw land at a reasonable price. In many cities, few residential lots are available at any price. I can remember when the cost of land was usually about 20% of the selling price of a spec home. Now the land cost is closer to a third of the selling price on many homes, and can be close to a half in some communities. That makes it harder for spec builders. And you're not going anywhere until you either take title to the land, or have the cooperation of the landowner so that construction can begin.

But there are always opportunities available if you know where to look.

My advice on finding land is to map out an area within a 5-mile drive of your home. Make a list of 40 or 50 suitable parcels that are or may be offered for sale and the price asked for each. You'll spot some that seem to be offered below market value. Concentrate on lots in residential areas that seem to be increasing in value. Stay away from areas where too many homes are for sale. You don't want to just add one more "For Sale" sign on the street.

Stick with neighborhoods where water, electricity and phone service are available.

Eventually you'll narrow your choice down to perhaps a dozen parcels. Make an offer on some of these. If cash will be a problem, ask the owner to subordinate the lot to you and then split the profits when the house is sold. If this is impossible, find an owner who's willing to wait for payment until the house you build is sold. The longer the lot has been for sale, the better this is going to sound to the lot owner.

In some metropolitan areas there aren't any vacant lots. What then? That's easy! Make the same survey of parcels available, but look at old houses badly in need of repair and offered at bargain prices. You can make just as much money fixing up and reselling an old house as you can building on raw land in the suburbs. In fact, you have a big advantage. The utilities are already in and paid for.

Watch the Construction Cycle

There are times to be a spec builder and there are times to stay out of the business and concentrate on remodeling and add-ons. The years 1977 and 1978 were great years for home builders. A spec builder with an inventory of houses to sell could count on selling nearly everything before it was finished. Just the opposite was true in 1981 and 1982.

Recognize that there's a housing cycle. It rewards those who anticipate the cycle and punishes those who elect to ignore it. Don't get stuck with your life savings in a home that has to be sacrificed to satisfy an anxious lender.

The best time to build your first spec house is when construction activity is just picking up after a recession. You should be able to build and sell several homes during the two or three years of good times that builders usually enjoy after a housing bust.

Here's how to know where we are in the construction cycle at any given time: The U.S. Department of Commerce publishes two excellent reports (Figures 1-1 and 1-2) that make trends in construction activity very clear to anyone who takes the trouble to glance at them:

"New One-Family Houses Sold and For Sale" (Construction Reports publication C25) costs $20 per year. "Housing Starts" (Construction Reports publication C20) costs $21 per year. Subscribe by sending a check to Superintendent of Documents, U.S. Government Printing Office, Washington, D.C. 20402.

"Housing Starts" shows how many new, privately-owned housing units are started each month in the U.S. You'll note from the chart published in each issue that starts rise and fall fairly regularly. You wouldn't launch a new ship into the teeth of a gale. So don't start new projects when starts are falling. Instead, close out the work

CONSTRUCTION REPORTS

U.S. Department of Commerce
Bureau of the Census

U.S. Department of Housing
and Urban Development

New One-Family Houses Sold and For Sale

JUNE 1982

C25-82-6
Issued August 1982

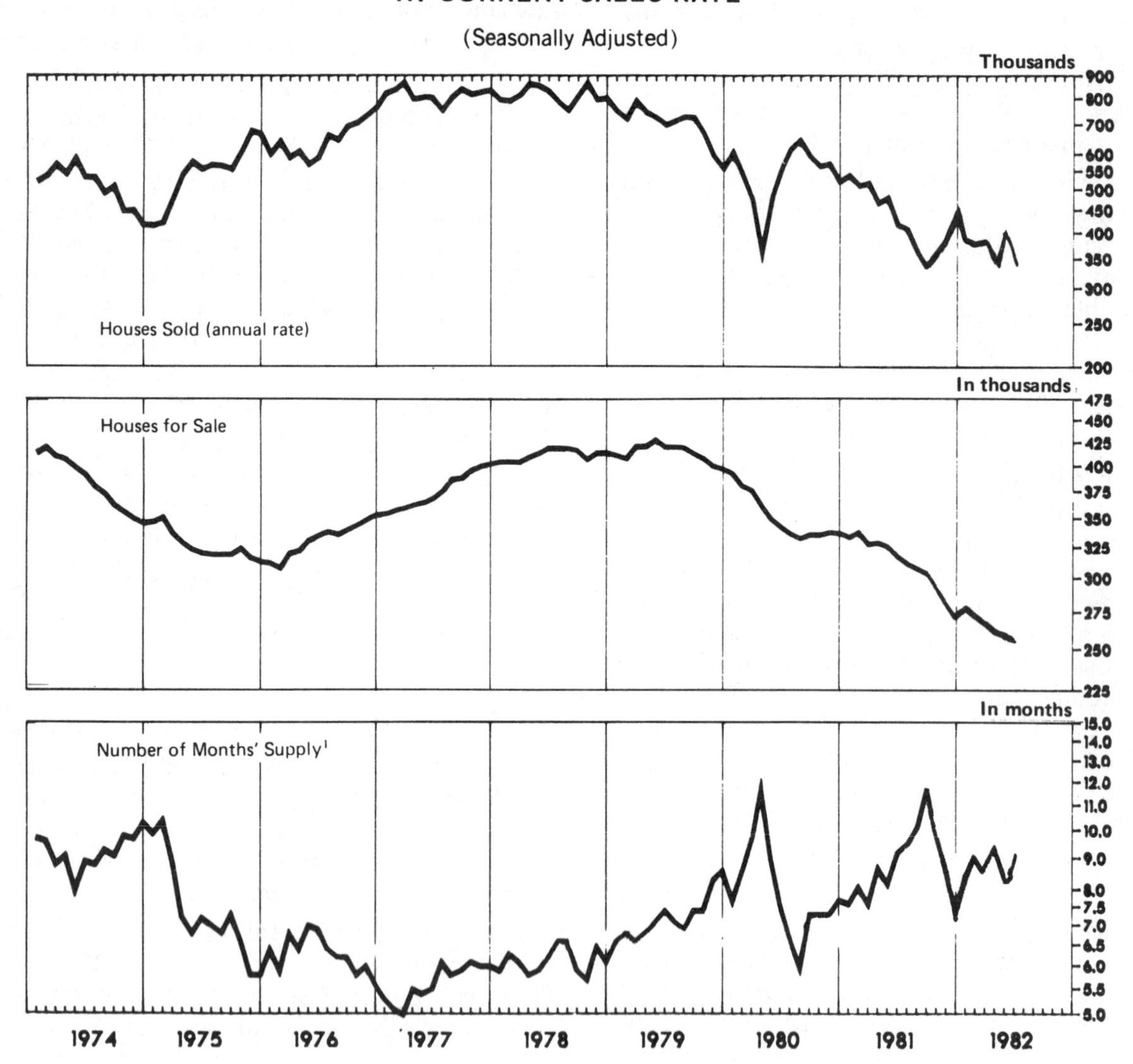

[1] Ratio of houses for sale to houses sold at current sales rate.
Source: Bureau of the Census.

Questions regarding these data may be directed to Barry A. Rappaport, Chief, Construction Starts Branch, Telephone 301-763-7842.

For sale by the Superintendent of Documents, U.S. Government Printing Office, Washington, D.C. 20402. Postage stamps not acceptable; currency submitted at sender's risk. Remittance from foreign countries must be by international money order or by draft on a U.S. bank. Annual subscription price - domestic $20.00, foreign $25.00. Single copy - domestic $2.00, foreign $2.50. Annual report - domestic $4.25, foreign $5.35.

Figure 1-1

CONSTRUCTION REPORTS

Housing Starts

FEBRUARY 1983

U.S. Department of Commerce
BUREAU OF THE CENSUS

C20-83-2
Issued April 1983

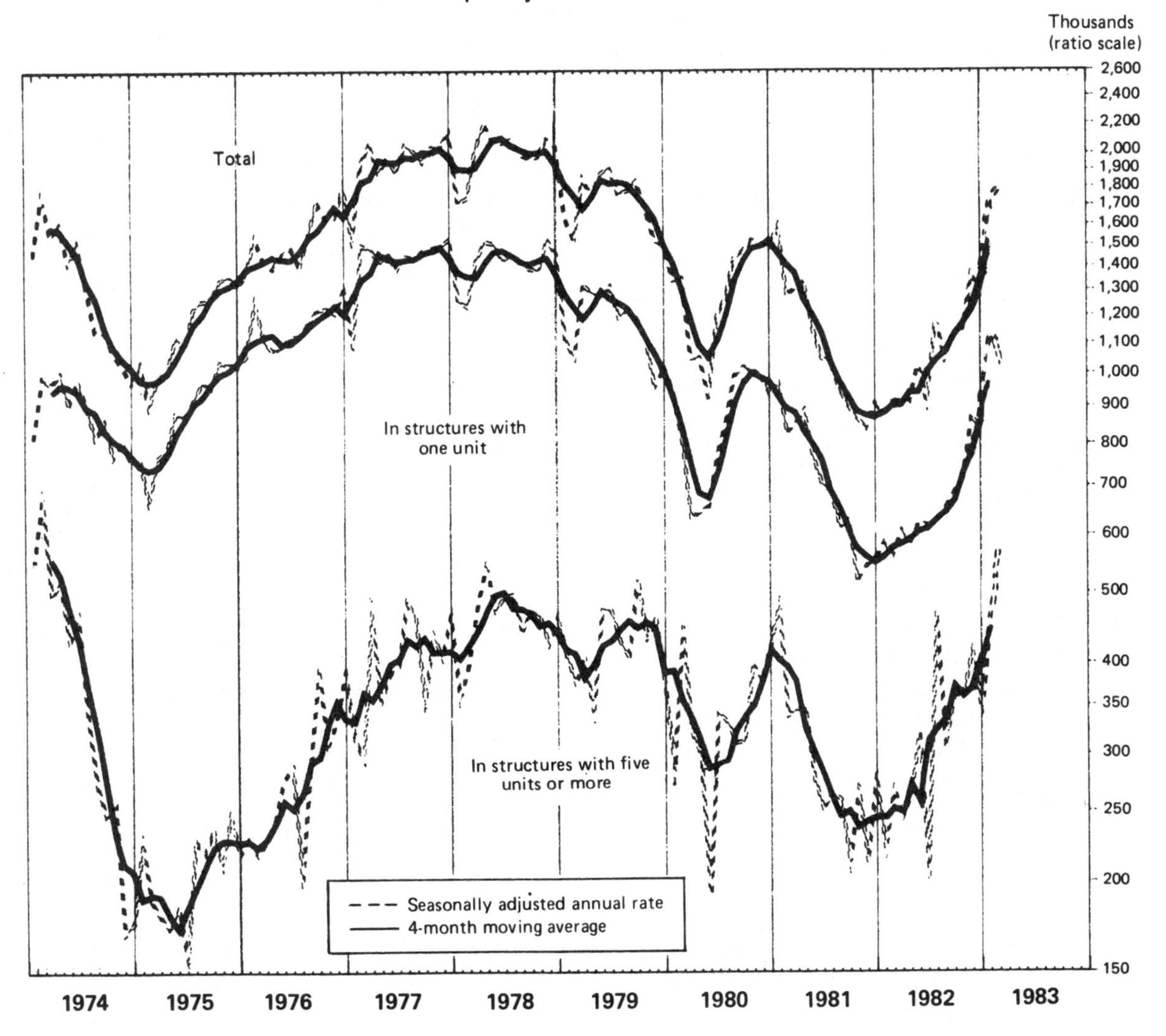

Note: Total includes units started in structures with two to four units.

Questions regarding these data may be directed to David Fondelier, Chief, Construction Starts Branch, Telephone (301) 763-5731.

For sale by the Superintendent of Documents, U.S. Government Printing Office, Washington, D.C. 20402. Postage stamps not acceptable; currency submitted at sender's risk. Remittances from foreign countries must be by international money order or by draft on a U.S. bank. Annual subscription price: domestic $21.00, foreign $26.25. Single copy: domestic $2.75, foreign $3.45.

Figure 1-2

you're doing and reduce the risk that you'll be stuck with a house that won't sell.

"Housing Starts" will help you visualize where we are in the construction cycle. But that's just part of the story. You can really stack the deck in your favor if you start building when new home sales are rising and the inventory of homes for sale is falling. That's where the other Department of Commerce report comes in.

"New One-Family Houses Sold and For Sale" shows the annual rate of home sales, how many houses are currently for sale, and the number of months of supply of homes at the current sales rates. Months of supply is the key figure. It's a builder's market when home sales are rising and the inventory of homes for sale is low. A 6-month inventory shows that there's a shortage of new homes for sale. When the unsold inventory reaches 12 months, that's panic time for most builders.

Plan to have your first home under construction (or expand your operation) when the inventory of new homes for sale is less than seven months. The "good times" will probably last two or three years, if the previous 20 years is a reliable guide to the future. In those two or three years you should make good money on every home you turn out.

Keep an Eye on the Competition

I should make one additional point before we get down to the business of building that first spec house. Study your competition like your profit depended on it — because it does. Your potential buyers will shop from home to home for the best value. Make sure your project measures up to standard in every way in its own price class.

Walk through other spec homes. Check the price against what you have to charge for your house. Check the features. Inquire about the financing that's available. Find out what makes one spec home sell faster than another. Decide for yourself what keeps a home from selling and why some homes have to be sacrificed. Look at each home as though you were a potential buyer. Then visualize how the home you plan to build will compare feature for feature.

You may decide that you can't match the other houses that are for sale at the prices being asked. But more likely, comparison will help you discover ways to make your house more appealing and a better value. And you may discover that the home you plan to build will sell for more than you expected.

If your community is like mine, you'll discover that low-priced homes seem to sell best in nearly any business climate. In general, you reduce the risk of an unsold house by keeping the price modest. True, there's less profit in low-priced homes. But there's also less risk.

Start with a home that's among the lowest-priced new homes available. Get a commitment from your lender that lets you offer an attractive loan to qualified buyers. That's a formula that's almost sure to succeed.

Using the Checklists

You'll note that each chapter after this one has a Checklist and Data Reference Sheet. This form is designed both to steer you through the construction process and to be a permanent labor and material cost record of each job completed. The Checklist and Data Reference Sheet is part of your second profit on every job. The first profit is the money that goes into your pocket. The second is what you've learned about home building that will make your next job easier, faster, and more profitable.

Now, let's get down to the business of making you a knowledgeable, professional home builder.

Chapter 2

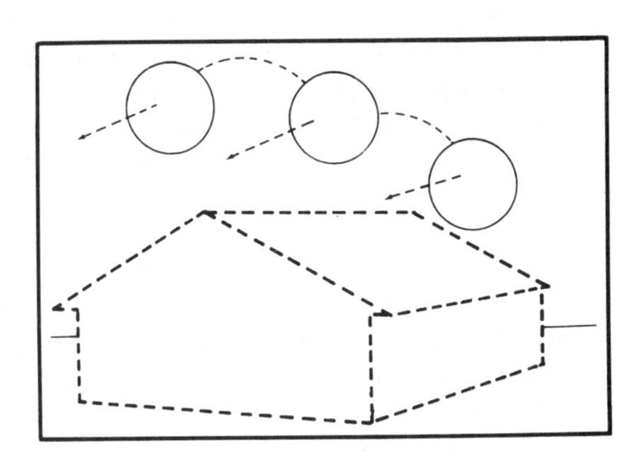

The Building Site

The building site deserves special attention, no matter what type of house you plan to build. If it's a spec house, you make the site selection. When you build for someone who already owns the land, your options are more limited. But in both cases, you need to know several things about the site. For example, don't plan a basement on a site with solid rock 3 feet below the surface. The same is true if the water table is 5 feet down. Make these discoveries long before actual construction begins.

Here are some of the things your site evaluation should include: How much excavating, tree removal, or grading will be required? Where does surface water drainage leave the site? Most states and counties won't let you divert drainage water across the adjoining property unless that's the natural flow path. Neither can you stop the natural flow from crossing your lot.

You or your owner can spend a lot of money on site preparation. It is literally sunk into the ground because it doesn't show up in the house. Avoid lots with site improvement problems if you are a spec builder, even if the purchase price seems to make it a bargain. The less money put into the ground, the better.

Hillsides make beautiful building sites. But every hillside comes with its own unique problems. Getting material to the site in wet weather will be difficult, expensive and maybe impossible in some weather. Consider potential mud slides and erosion. The cost of building on an incline will be higher. One rule of thumb is that every foot of elevation required above the natural ground level under the dwelling will add about $500 to construction costs in 1983 dollars. Finally, a house with different levels is not for everyone.

Be wary of low-lying areas. Heavy rains might flood the site even if it has elaborate drainage canals. Few houses can be sold if you can't get financing, and it's hard to get financing on any home that can't be insured.

Every builder should be familiar with soil conditions, drainage problems and the water table level where he plans to build. If you're not sure, check it out. Drill test holes if necessary. Talk to the city building department. Know your site. Avoid a costly surprise.

The Ideal Building Site

The ideal building site needs little preparation. It offers good drainage, has firm ground for footings, has easy access to water, phone, gas, sewer, electric and cable TV lines, and already has a well-paved street, curbs and sidewalks. It's close to schools, churches, hospitals, parks, playgrounds, stores and public transportation, and is served by police and fire departments. It's in a neighborhood with houses of similar size, style and price.

The ideal site has shade trees on the west side for late afternoon cooling on hot summer days, and is open on the south side so that windows can capture the warmth of the winter sun.

If you find that site, call me, because I've never seen one. Every building site I've worked on has compromises that make it less desirable or more expensive. Your job is to determine how much more expensive and how much less desirable.

Stay away from deteriorating neighborhoods. No one wants their dream castle in a neighborhood where hubcaps disappear nightly.

You can build a custom house anywhere your

client wants it if he has a loan. But lenders have guidelines that restrict construction loans on a questionable site.

Here's a trick you may not have thought of. Some larger builders and developers sell off lots from larger parcels to independent builders. They feel it helps sales to get a few houses up and sold. There will probably be restrictions on the style, size and price range of the house you build under an arrangement like this.

These building lots will be priced to include improvements such as streets, curbs, drains, and utilities. You pay a premium price for such lots, but it's a good investment for a builder not big enough to develop new areas on his own. But plan to delay your building until the streets are built and the utilities are available. Some development plans never materialize.

Houses located near interstate highways, airports, railroad tracks and large manufacturing plants are ill-suited to residential construction. These are major and permanent problems. You need as few of these as possible. Anything that detracts from a quiet atmosphere will reduce the value. A lot of people enjoy baseball, but few want to live next to the ball park.

Consider the view. Anything unsightly will keep the house from selling. Glare of headlights through windows at night is a source of discomfort and disturbance.

Don't over-build for the area. The house you build should be similar in size and style to the other houses. A mansion next to small, modest homes won't sell. Likewise, it's foolish to build a modest size house in an area of larger, more expensive homes. The land will cost too much.

The size of the house should be in proportion to the lot size. Local codes usually dictate setback from the street, side yard and back yard distances. Find out what the restrictions are on your lot.

The quicker a spec house sells, the more you make. An unsold house costs you plenty every month. Houses that blend in with the neighborhood seem more inviting and comfortable, and should sell more easily.

The important question is, will a particular size and style house in a given location sell within a reasonable period of time? If you can't be sure, talk to some real estate brokers in the area.

Summary

The better the site, the more appeal your spec house will have. Every community has good and bad locations. Choice lots cost more initially, but may save you money by insuring a quick sale.

A good lot puts you on the right path for making a profit. With a little luck, you'll sell a spec home before it's completed.

Don't bid on a custom job until you know the soil and conditions on the site. If in doubt, get a foundation soil test. It may cost $500, but it could prevent a big loss. The time you devote to site selection and analysis helps guarantee a fair profit on the job.

Figure 2-1 is a Site Evaluation Checklist and Data Reference Sheet. Don't buy or obligate yourself to build on a lot until you've checked each point on Figure 2-1.

Site Evaluation Checklist and Data Reference Sheet

Neighborhood:

Location	☐Urban	☐Suburban	☐Rural
Built-up	☐Over 75%	☐25% - 75%	☐Under 25%
Growth rate	☐Fully developed	☐Rapid	☐Slow
Property value	☐Increasing	☐Stable	☐Declining
Demand/Supply	☐Shortage	☐In balance	☐Oversupply
Marketing time	☐Under 3 months	☐4 - 6 months	☐Over 6 months
Present land use	________% 1 family	______% 2 - 4 family	____% apartments
	________% commercial	________% industrial	________% vacant
	________% condo		

Change in present land use ☐Not likely

☐Likely

☐Taking place (changing from ______________________________

______________________________ to ______________________________)

Predominant occupancy ☐Owner ☐Tenant ________________% vacant

Predominant value $________________

Single-family age ____________ up to ____________ predominant age ____________ years

	Good	Average	Fair	Poor
Employment stability	☐	☐	☐	☐
Convenience to employment	☐	☐	☐	☐
Convenience to shopping	☐	☐	☐	☐
Convenience to schools	☐	☐	☐	☐
Adequacy of public transportation	☐	☐	☐	☐
Recreational facilities	☐	☐	☐	☐
Adequacy of utilities	☐	☐	☐	☐
Property compatibility	☐	☐	☐	☐
Protection from detrimental conditions	☐	☐	☐	☐

Figure 2-1

	Good	**Average**	**Fair**	**Poor**
Police and fire protection	☐	☐	☐	☐
General appearance of properties	☐	☐	☐	☐
Appeal to market	☐	☐	☐	☐

Soil conditions ______________________________

Lot:

Dimensions ______________equals ______________square feet or acres

Corner lot ☐

Zoning classifications ______________________________

Present improvements ☐do ☐do not conform to zoning regulations

Best use: ☐Present use ☐Other (specify)______________________________

____

Lot offers an excellent site for ☐basement construction ☐crawl-space construction ☐slab-on-grade construction ☐split-level ☐1-story ☐2-story

Topography ______________________________

Shape ______________________________

View ______________________________

Drainage ______________________________

Soil condition ______________________________

Trees to remove ______________________________

Grading and excavation requirements ______________________________

____

Comments (favorable or unfavorable, including any apparent adverse easements, encroachments or other adverse conditions):

____

____

____

Utilities available:

	Public	Other (describe)
Electric	☐	________
Gas	☐	________
Water	☐	________
Sanitation/sewer	☐	________
Underground electric and telephone lines	☐	________

Offsite improvements:

Street surface ________

	Public	Private
Street maintenance	☐	☐
Street access	☐	☐
Storm sewer	☐	☐
Sidewalk	☐	☐
Curb and gutter	☐	☐
Street lights	☐	☐

Estimated cost to prepare lot for construction:

Clearing $________ Excavating $________ Retaining walls $________ Grading $________

Roads $________ Fill dirt $________ Other $________

Total preparation cost $________

Cost of lot $________

Total $________

A copy of this Checklist and Data Reference sheet has been provided:

☐Management

☐Accounting

☐Estimator

☐Foreman

☐File

☐Other: ________

Chapter 3

Staking Out the House

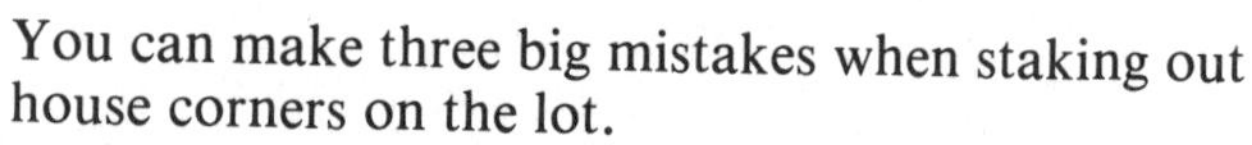

You can make three big mistakes when staking out house corners on the lot.

First, be sure you know where the lot is. You don't want to build on someone else's land. It's the owner's responsibility to locate the lot corners accurately.

Second, the location of the house on the lot must satisfy code requirements. This may be your problem. Every builder is assumed to know the code and regulations in the area where he builds. If there are no code restrictions, locate the house the same distance from the street as the adjacent houses. But in any case, the setback should provide for reasonable privacy and minimize noise, headlight glare and fumes. Know what the restrictions and setback requirements are.

Finally, don't overlook the requirements of the lender. For some types of loans and loan guarantees, you'll have to meet certain setback standards.

Even if you've satisfied yourself on these three points, there are other considerations to make before beginning work.

Location of the house on the lot should be in harmony with the site. Avoid the deep cuts, excessive foundation wall depth, steep inclines and unnecessary steps.

Plan for maximum protection from wind, sun, precipitation and temperature extremes. Make sure windows take advantage of attractive views. Windows in rooms used in the daytime should not face windows in adjacent houses. Avoid placing windows where they would be open to noise and headlight glare.

The location, arrangement and orientation of the building should give adequate space for outdoor living areas and service facilities. It should give natural inside light, air and privacy.

Consider yard depth and width, distances between houses, length and height of walls, and the location of the main entrance in relation to nearby dwellings.

Planning the Excavation

If a dirt contractor is going to dig a basement or prepare the site, have someone from that firm visit the site with you. Get approximate costs to:

1) clear the site of trees, rocks and stumps

2) strip and store top soil

3) excavate the basement

4) dig trenches and pits

5) backfill

6) spread topsoil for the finish grade

Make it clear that you won't allow any backfilling until the first-story floor framing is in place and all waterproofing is finished. Delaying backfill usually increases manhours, as workers must use a catwalk or ramps to get across the unfilled areas. But backfilling too soon can be even more expensive. Any load placed on green concrete walls can cause them to slump. And the inspector may want to see the waterproofing on basement walls.

Topsoil is spread when the building is completed. A small farm-type tractor with a landscape bucket is best for this work.

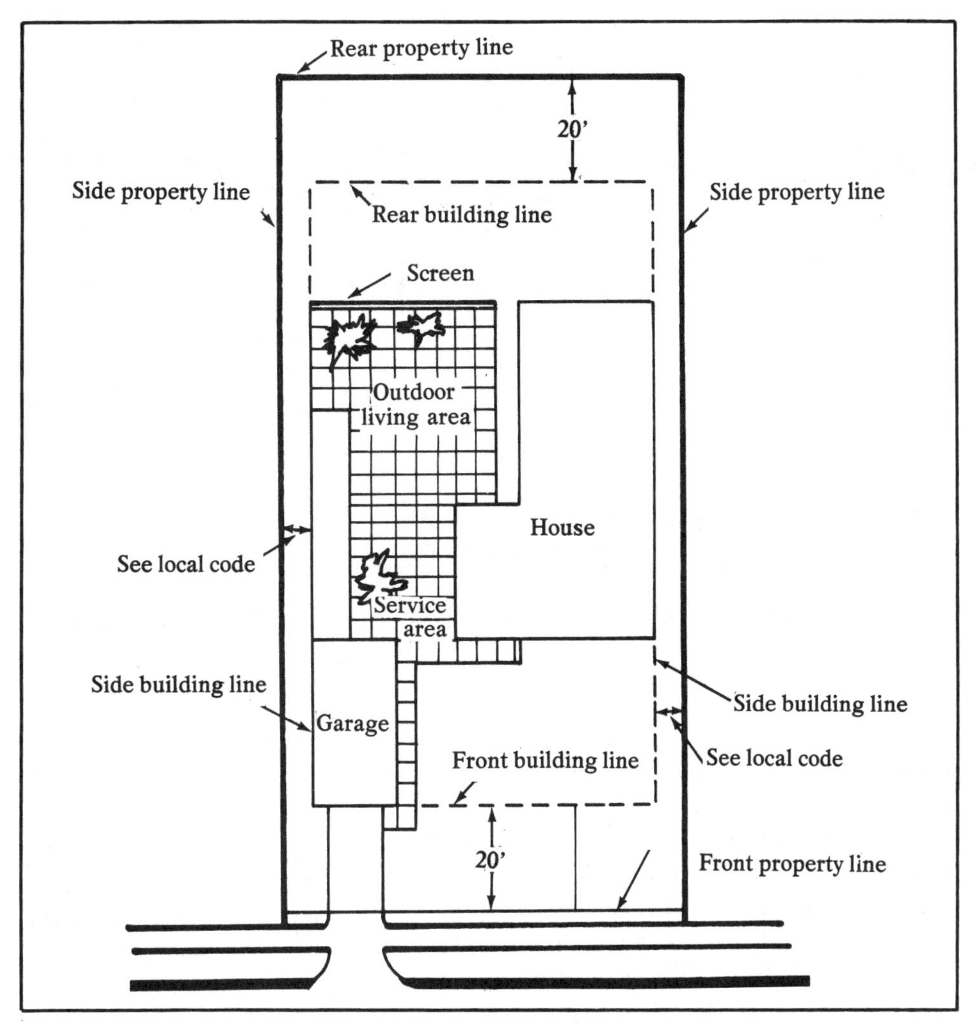

Plot plan
Figure 3-1

Beginning the Layout

A plot plan may be required by the building inspector before the permit can be issued. It shows the location on the lot, including the setback distance, and rear and side building lines. See Figure 3-1. Even if a plot plan isn't required, make up a simple sketch for your own use when doing the layout.

The front of the house is generally parallel to the street. Drive a small stake at each of the front corners. The house will be squared from these two stakes. One method of squaring a building uses a 6-8-10 triangle. See Figure 3-2. A triangle with sides that are any multiple of 3, 4, and 5 will always have one square corner.

The triangle in Figure 3-2 has sides 6, 8, and 10 feet long. Corner A is a right angle.

The building can also be squared by cross-measuring as illustrated in Figure 3-3. Stretch a tape from opposite corners — from A to C, and from B to D. When the two diagonal distances are exactly the same, the building is square. If there is a

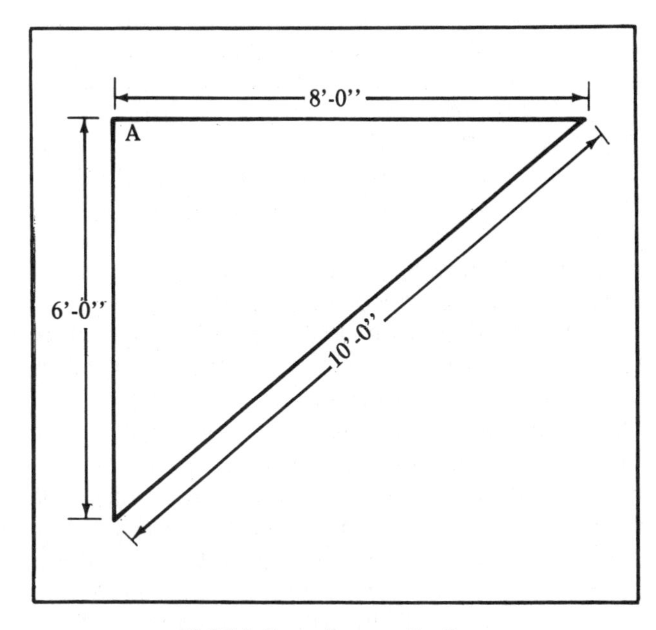

6-8-10 Squaring method
Figure 3-2

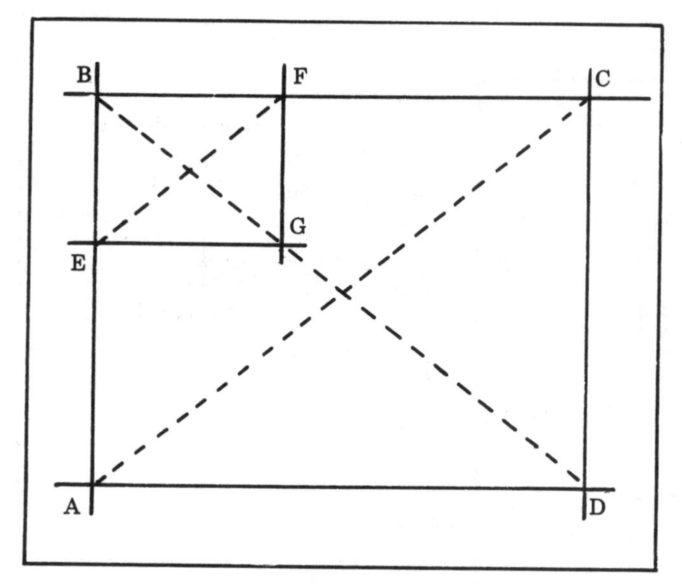

Squaring a large area by cross-measuring
Figure 3-3

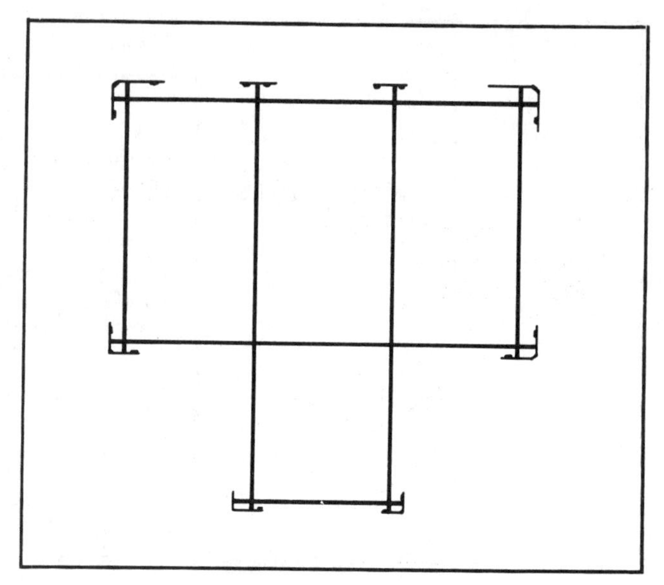

Squaring a T-shaped area
Figure 3-4

difference, adjust the corners. But don't adjust the two stakes that locate the front corners of the building. This would change the alignment of the house to the street.

On a large home or where obstructions make cross-measuring difficult, square smaller areas such as the square E-B-F-G (upper left of Figure 3-3). Then extend line B-G across the interior of the building to D.

Figure 3-4 shows how to stake out and square a T-shaped building.

Batterboards

Batterboards hold the string that marks the exterior wall line. Once this line is in place, footing lines, chimney lines and inside wall lines can be established. Saw kerfs (notches) on top of the boards mark the exact points where the string is attached, so that the string can be rerun after it is removed for excavating, pouring footings, and material handling. When rerunning the string, recheck measurements and squareness. Batterboards can get bumped out of place.

Batterboards should be located 4 to 5 feet back from the corners so there's room for digging, block laying or concrete pouring. Use 1 x 4's when making batterboards. They remain stable when driven into the ground, and are strong enough to withstand the pressure of taut nylon twine. Figure 3-5 shows the proper location of a batterboard.

Batterboards can be used as reference points when establishing the height of the foundation walls, the depth of the footing trenches and the height of the footing grade stakes.

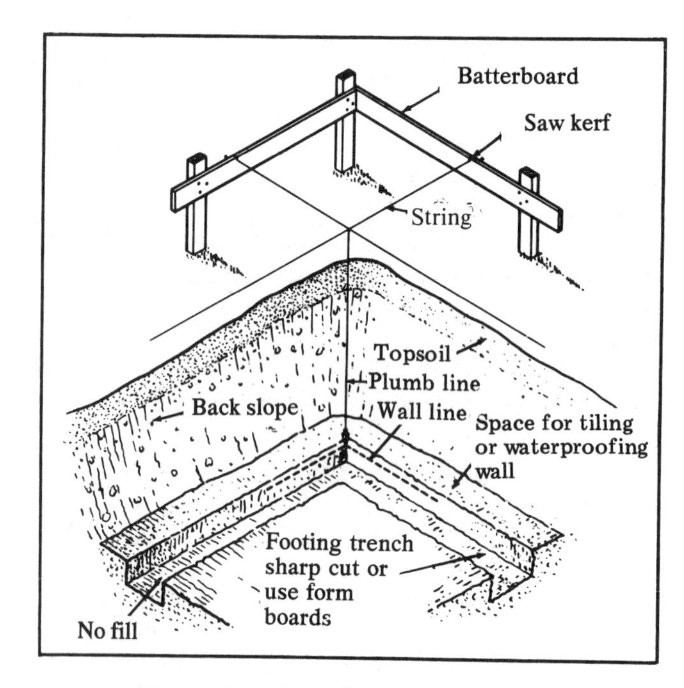

Proper location of a batterboard for squaring and excavation
Figure 3-5

Building on Sloping Lots

On a sloping lot, the highest point is your starting point because it controls the height of the foundation wall. See Figure 3-6. The foundation wall has to be high enough so that the finish grade slopes away from the building. Water pooling around the foundation is sure to cause problems. See Figure 3-7.

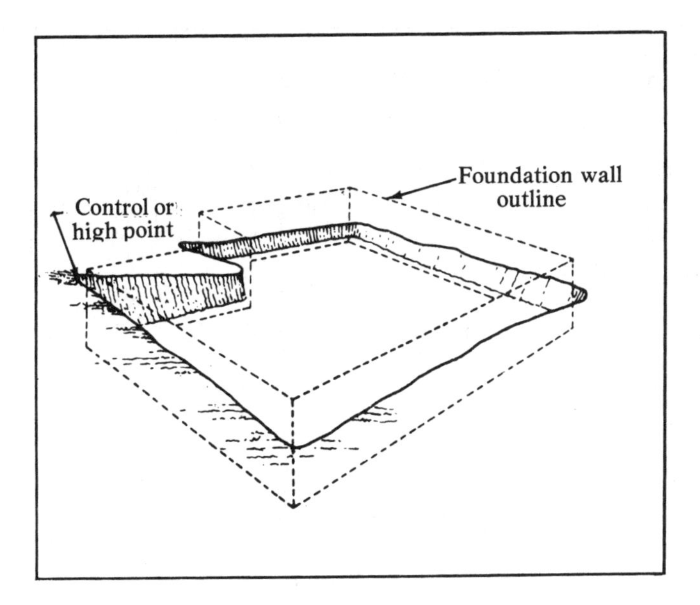

Establishing depth of excavation
Figure 3-6

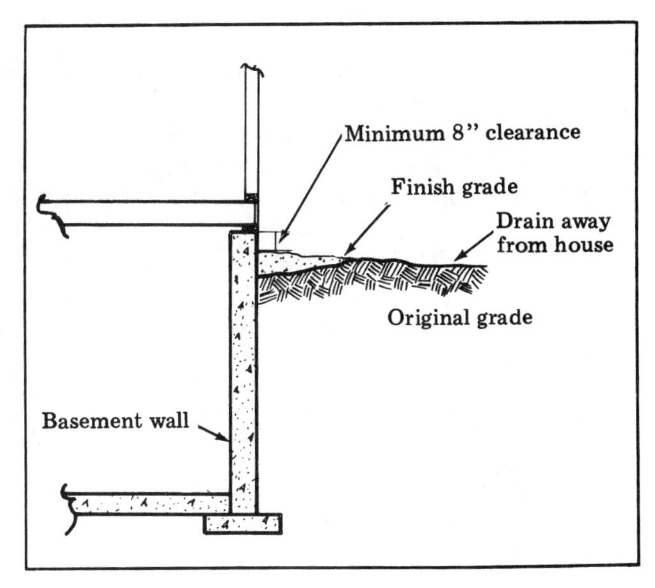

Finish grade sloped for drainage
Figure 3-7

The foundation wall for basements should allow a minimum finished ceiling height of 7'6''. The girders may be lower. A ceiling height of 6'8'' (6'4'' under girders) is allowed for basements without habitable space. A bedroom is considered a habitable space, but a bathroom or game room isn't. Basements that can later be converted into habitable space add value to the house. The 7'6'' ceiling offers opportunities that a lower ceiling can't.

Be sure to consider the height of the finish grade against basement walls and the size and placement of windows if you intend to use the basement as habitable space.

Summary

Start every job with level and square footings and foundation walls. A foundation out-of-square means that you have to adjust everything else in the house to fit. This is a waste of time and materials. The defect will be visible from both outside and inside. It's simpler, easier and cheaper to start the job right.

Figure 3-8 is your Checklist and Data Reference Sheet for staking out the house.

House Stake-Out Checklist and Data Reference Sheet

1) The lot corners are ☐staked ☐not staked.

2) Exact property lines are ☐known ☐unknown.

3) Setback distance is established by ☐code ☐adjacent houses.

4) Side and rear yard distances are __________ and __________ respectively and meet code requirements which are __________ and __________.

5) The orientation of the house takes advantage of:

☐Shade trees

☐Winter sun

☐Attractive view

☐Outdoor relaxation areas

☐Service facilities

☐Natural light

☐Other ______________________________

6) The orientation of the house protects against:

☐Winter winds

☐Heavy rains

☐Noise and fumes from street

☐Headlight glare

☐Invasion of privacy

☐Other ______________________________

7) Window locations in habitable rooms ☐ do ☐do not offer privacy.

8) Windows in ☐living room, ☐dining area and ☐family room are not face to face with those in adjacent houses.

9) Fencing or screen walls ☐are ☐are not necessary to preserve privacy.

10) The house ☐is ☐is not aligned with the street. If not, explain ______________________________

11) The house can be squared by the 6-8-10 method (or other proportionate numbers, such as 9-12-15), and checked by measuring diagonally, as shown in Figure 3-3. A diagonal measurement ☐was ☐was not made.

12) Batterboards were placed __________ feet from the building line to allow room for ☐excavating ☐pouring concrete ☐forming ☐laying block ☐handling material.

Figure 3-8

13) Saw kerfs ☐were ☐were not made in the batterboards to permit rerunning lines.

14) The squareness of the batterboards ☐was ☐was not checked by diagonal measurements when strings were removed and replaced.

15) Batterboards ☐were ☐were not placed so that they could be used as reference points for establishing:

☐Depth of footing trench

☐Height of footing grade stakes

☐Height of foundation wall

16) The foundation wall ☐was ☐was not started at the highest point on the lot.

17) The foundation wall ☐is ☐is not high enough to allow the finish grade to slope away from the building. If not, explain ______________________________

18) The basement foundation wall will allow a finished ceiling height of:

☐Minimum 6'-8''

☐7'-6'' (for later conversion to habitable living space)

☐Other (specify) ______________________________

19) Cost and/or manhours have been determined for:

☐Clearing the site of unwanted trees, rocks, stumps and litter $__________

☐Stripping and storing topsoil $__________

☐Basement excavating $__________

☐Digging trenches and pits $__________

☐Backfilling $__________

☐Spreading topsoil for finish grade $__________

Total $__________

20) The following materials were used to stake out the house:

______________________________ $__________

______________________________ $__________

______________________________ $__________

______________________________ $__________

Total $__________

21) A total of ________ manhours @ $________ hour were used to stake out the house.

Total $________

22) Total cost:

Materials $________________

Labor $________________

Other $________________

Total $________________

23) Other comments ____

____

____

24) A copy of this Checklist and Data Reference Sheet has been provided:

☐Management

☐Accounting

☐Estimator

☐Foreman

☐File

☐Other: ________________

Chapter 4

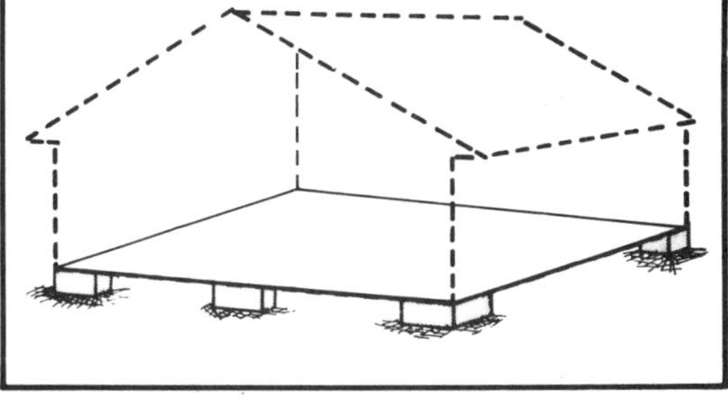

Footings and Foundations

A builder's reputation is only as good as the footings he pours. Unless you want to install taillights on the homes you build so the owners can follow them down the street, you had better be sure your foundations are first rate. Fortunately, it isn't hard to put a good footing under most buildings. You just have to follow a few basic principles.

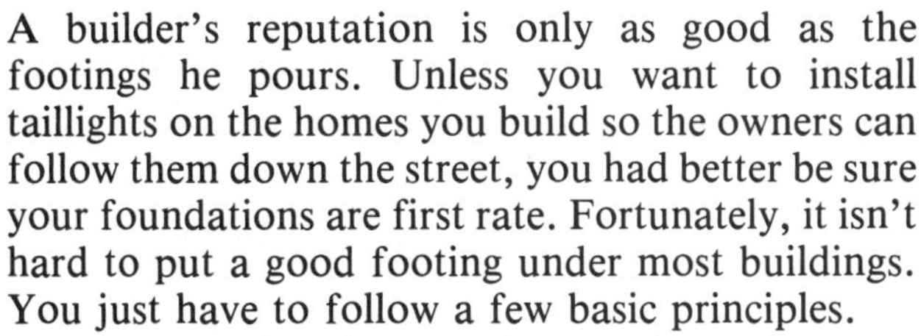

A poor footing shows up in excessive and uneven settlement, cracked foundation walls, doors and windows that won't close or open smoothly and many other problems. Poured concrete footings are still the most common in residential construction. Treated wood foundation systems, which permit all-weather construction of basement and crawl-space houses, are also used. We'll cover both poured concrete and treated wood foundations in this chapter.

Concrete Footing Sizes

The footing size varies with the type of soil and the weight of the structure. Figure 4-1 shows the sizes normally required for a conventionally loaded wall on soil of average bearing value (approximately 2,000 psf or better).

The excavation should extend at least 6 inches into natural undisturbed soil which will provide adequate bearing for the load. On fill, excavate through the fill to undisturbed soil.

Extend the footing below the frost line and at least 6 inches below the finish grade. The code in your area will usually specify how deep the footing must be. In some areas, the footing must go down at least 4 feet to get below the point where soil freezes in winter. Frozen soil "heaves" and moves and can fracture even major concrete foundations.

Here are some typical footing depths for various climates:

Anchorage, Alaska	-	42"
Albany, New York	-	42"
Bangor, Maine	-	48"
Charleston, West Virginia	-	24"
Helena, Montana	-	36"
Tampa, Florida	-	6"
Shreveport, Louisiana	-	18"

Four considerations have a bearing on the design of the footing:

1) The loading of the soil by the weight of the building and its intended use

2) The total settlement expected from the building

3) The differential in settlement expected between parts of the house

4) Local conditions that indicate a wider and thicker footing should be used

Correcting faulty footings after the house is built can be very expensive. Do it right the first time. Reputations are built on quality work, not on how well you fix the problem.

If a footing trench is dug too deep, *don't* refill with dirt. Use extra concrete. Use forms for footings where soil conditions prevent sharply cut trenches. Reinforce the footing with steel rods any

Number of Stories	Frame		Masonry or Masonry Veneer	
	Minimum Thickness (inches)	Projection Each Side of Wall (inches)	Minimum Thickness (inches)	Footing Projection Each Side of Foundation Wall (inches)
One story:				
No basement	6	2	6	3
Basement	6	3	6	4
Two story:				
No basement	6	3	6	4
Basement	6	4	8	5

Footing sizes
Figure 4-1

time the footing crosses a pipe trench. Protect freshly poured footings from freezing. Frozen concrete can't reach its maximum strength.

If the footing has "steps" because the lot isn't reasonably level, be sure the top on the footing is still level between steps. It's hard to lay block or brick on a footing that isn't level. A high spot will require cutting concrete block. A low point must have a course of brick under the block. These are the things that run up costs and dip into your profit.

Stepped Footings

Stepped footings are necessary whenever the ground slopes more than a foot or two. When you plan the step, be sure the height of the vertical step doesn't exceed 3/4 of the horizontal distance between steps. The horizontal distance between steps should be at least 2 feet. The vertical connection should be the same width as the footing and at least 6 inches thick. Figure 4-2 illustrates these points. The vertical and horizontal runs are poured at the same time so the two parts form a good bond.

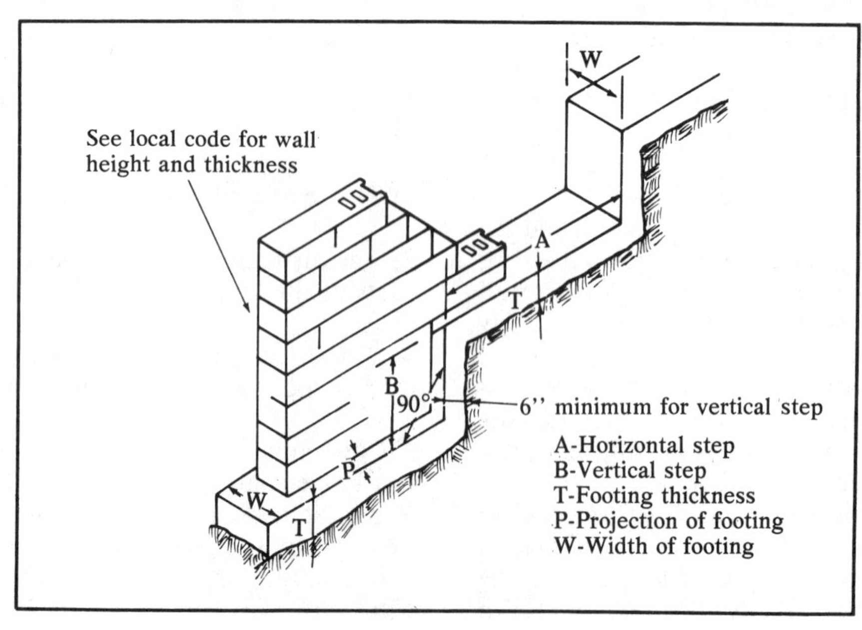

Stepped footings
Figure 4-2

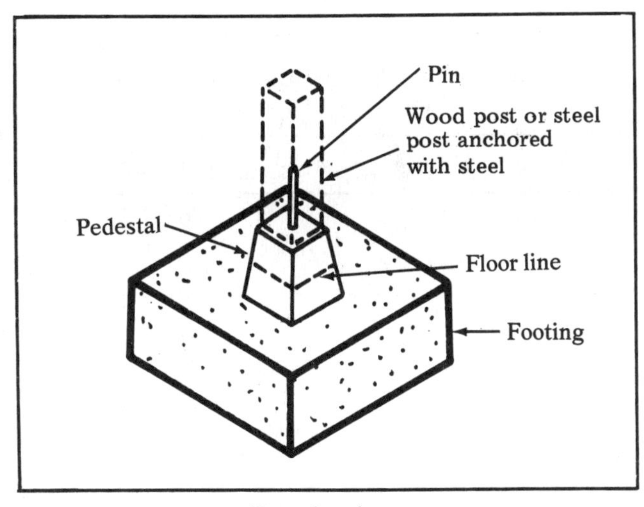

Post footings
Figure 4-3

Pier and Post Footings
Concrete block is the most common material for piers and foundation walls in crawl space construction. The pier footing should be a least 8 inches thick and large enough to support the total design load without excessive differential settlement. A 24 x 24-inch-square footing is adequate in most areas for a pier made of 8 x 8 x 16-inch concrete blocks spaced 8 feet on center. Local soil conditions may require a larger footing.

Figure 4-3 illustrates a typical post footing for basement posts. The pedestal may be eliminated when a steel post is set on the footing and the basement floor is poured around it.

Figure 4-4 shows another method of using a pier footing.

Footings for fireplaces and chimneys are normally poured at the same time as the foundation footings. We'll cover these footings in Chapter 21.

Concrete Mixes
Use 5 bag ready-mix concrete for most footings. A 5-bag mix has 5 bags of cement per cubic yard of concrete. According to the American Concrete Institute, the mixture must work readily into corners and angles of forms and around reinforcement without segregation of the materials or puddling of water on the surface.

The concrete should have a compressive strength, at 28 days, of at least 2,000 psi. Generally, the more water, the weaker the concrete is when set.

For small jobs and work in isolated areas, you might prefer to mix the concrete on the job. Figure 4-5 shows some acceptable mixes. Don't overwet the concrete. The proportions are for aggregates measured by volume and in a damp and loose condition.

Extremely hot weather dries the concrete too quickly and stops the hardening process too soon. Keep it moist for several days. And remember that freshly poured concrete must be protected against freezing.

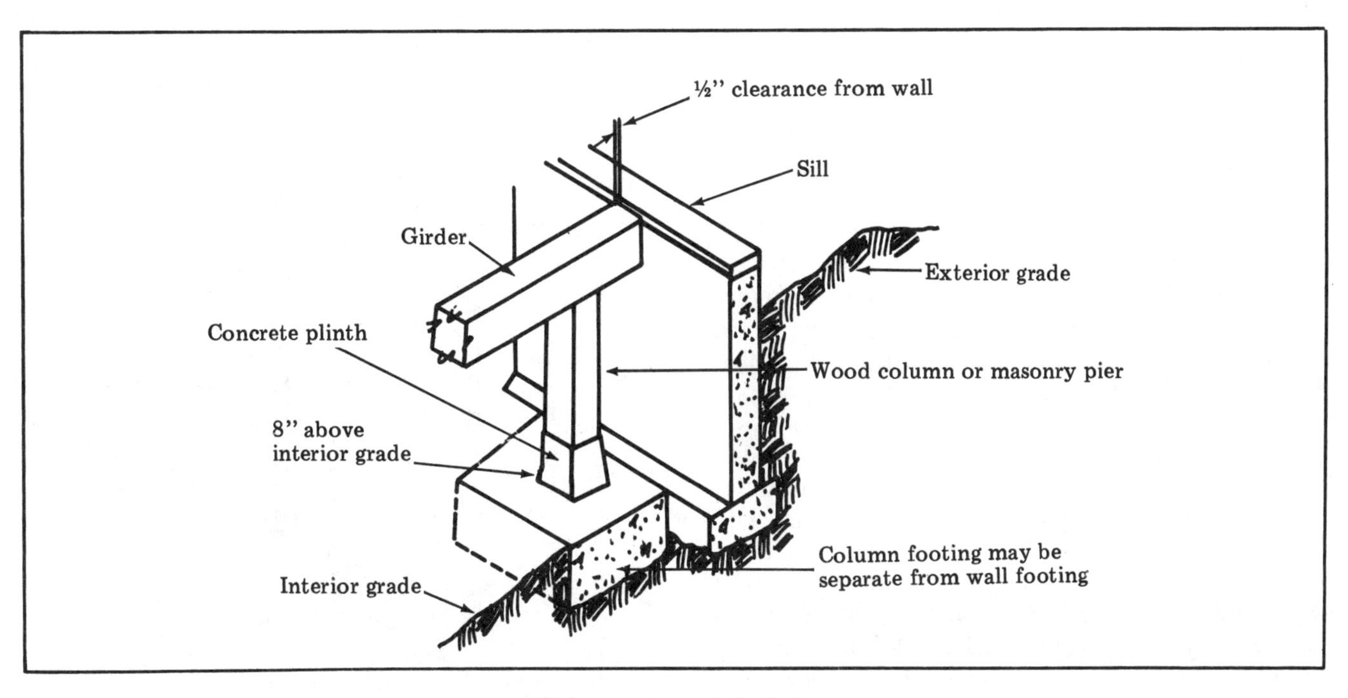

Column-supported girder
Figure 4-4

Maximum Size of Coarse Aggregate	Approximate Cement (Sacks per Cubic Yard)	Approximate Water (Gallons per Sack)	Approximate Proportions (by Volume) per Sack of Cement		
			Cement	Fine Aggregate	Coarse Aggregate
¾"	6.0	5	1	2½	2¾
1"	5.8	5	1	2½	3
1½"	5.4	5	1	2½	3½
2"*	5.2	5	1	2½	4

*Not recommended for slabs or other thin sections

Concrete proportions for field mixing
Figure 4-5

Estimating Concrete Requirements
When you've dug the footing trench and put in grade stakes, you're ready to pour the footing. How many yards of concrete are required? A good way to figure the concrete needed is:

Multiply the length in feet times the width in feet times the thickness in inches and divide by 314.

Thus, a building 40 x 24 feet would have 128 linear feet of perimeter or wall footing. If the footing is 18 inches wide (1½ feet) and 6 inches thick, your computation would be:

```
  128 length in feet
 x1.5 width in feet
  192 square feet
  x 6 thickness in inches
1,152
```

1,152 divided by 314 is 3.67 cubic yards. Round it out to 3.75 cubic yards to allow for waste.

Foundation Drains
Generally, any building with a basement or with habitable space below grade must have a sump and drain below the level to be protected. The drain should discharge into a drainage ditch or storm drainage system. See Figure 4-6.

Treated Wood Foundations

Treated wood basement construction has been accepted by the U.S. Department of Housing and Urban Development for FHA insured loans. Wiring, insulation and the finish wall can be installed in the same manner as stud-wall construction. Sections can be built in a shop and assembled on the site, reducing the on-site manhours. In wet or cold weather, this is a big advantage. See Figure 4-7.

Start by excavating the basement. Put in the plumbing lines below the basement floor. Then cover the basement floor with a layer of crushed stone or gravel at least 4 inches thick. The gravel must extend at least 6 inches beyond the footing line.

Tip the assembled, treated stud wall with treated plate and footing into place, and secure it with the other wall assemblies. Plywood wall panels are then installed over the studs and fastened in place. Seal the panel joints with a waterproof caulk.

Apply a continuous sheet of 6-mil polyethylene over the exterior side of the wall below grade level. Use 6-mil polyethylene over the entire stone or gravel floor bed. Then pour a concrete slab floor.

Don't do any backfilling until the floor of the first story is securely fastened to the top of the wood basement walls and the concrete floor slab in the basement has set up. Otherwise, they're likely to move under pressure from the soil.

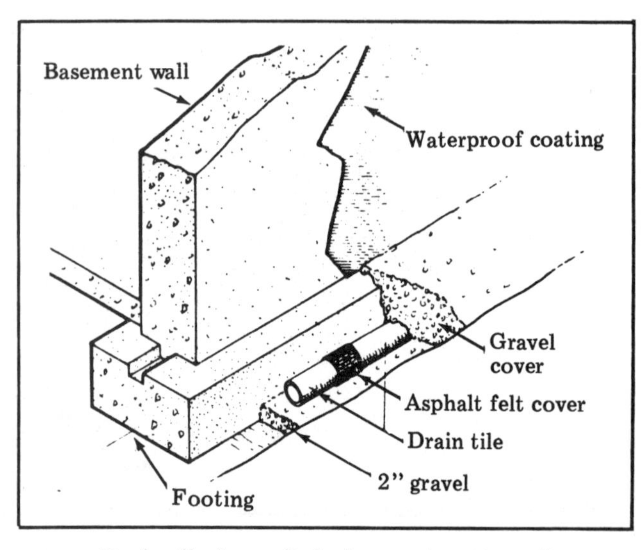

Drain tile for soil drainage at outer wall
Figure 4-6

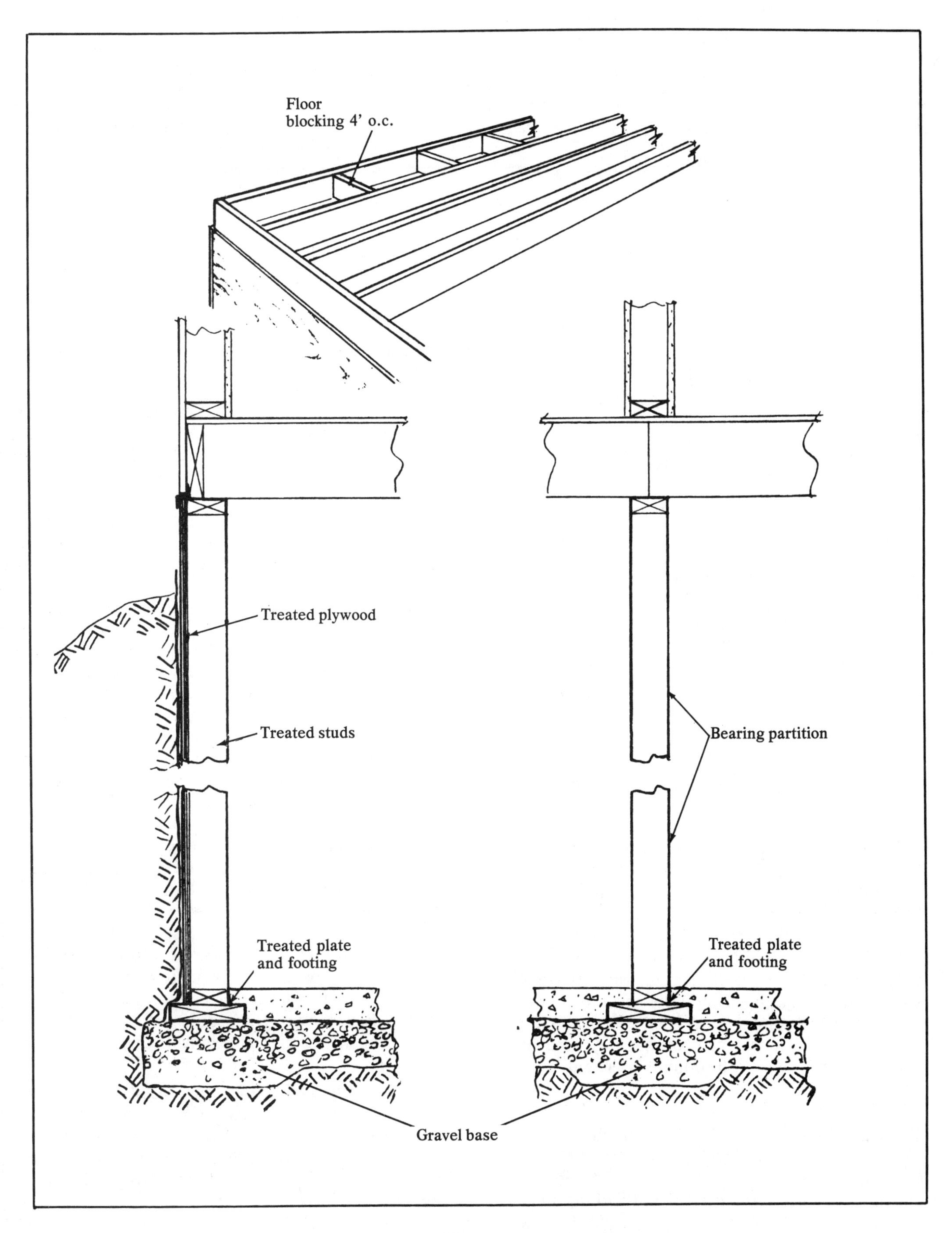

Treated wood basement construction
Figure 4-7

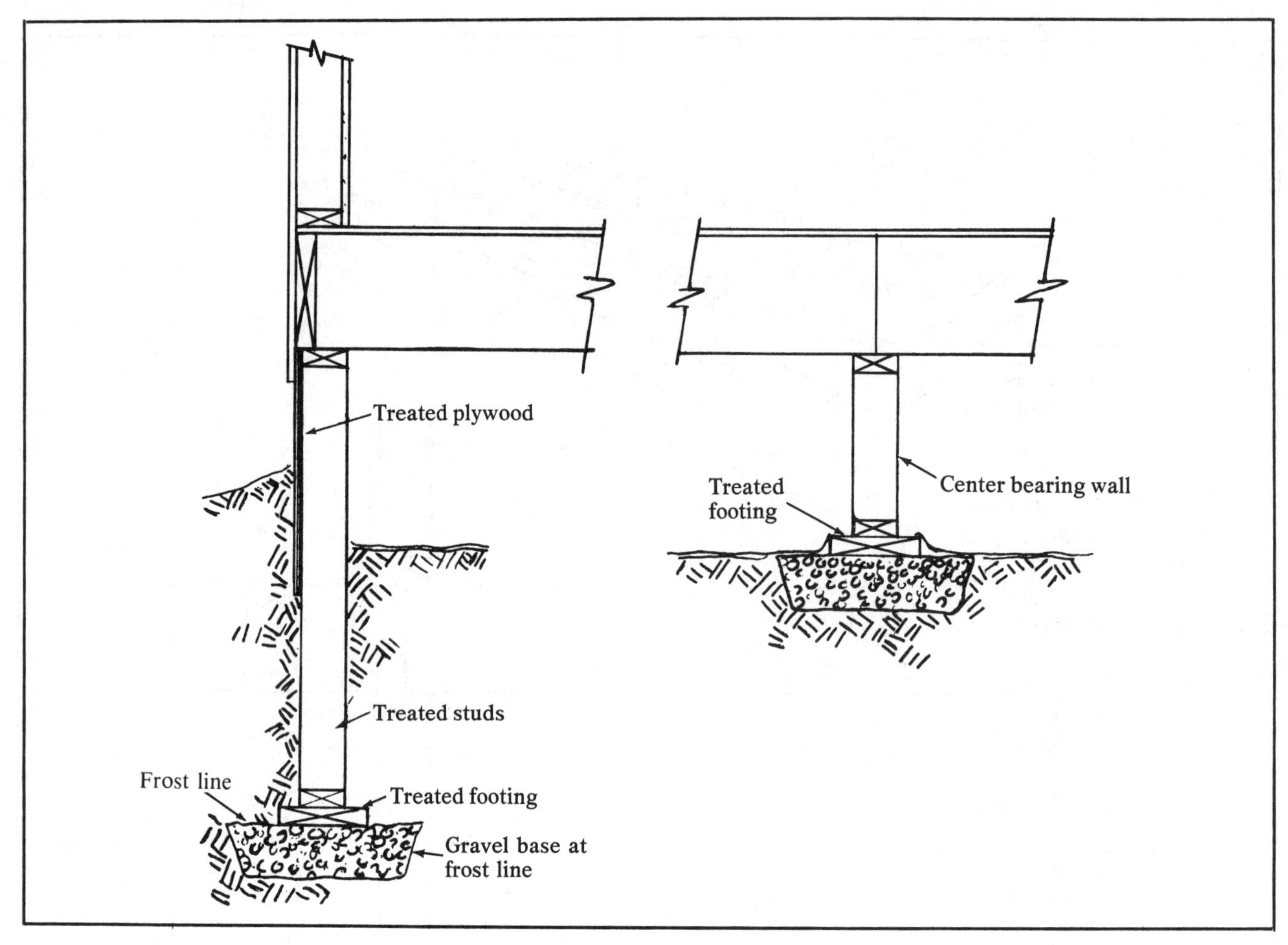

Treated wood crawl-space construction
Figure 4-8

Install solid blocking at end walls so the foundation wall loads are transmitted uniformly to the solid floor frame.

Treated wood basement walls can't be used in all soil conditions and at all backfill heights. If the backfill is over 48 inches deep, your code or engineering requirements will govern the stud size.

For the wall panels, use pressure-treated 1/2-inch-thick standard C-D (exterior glue) plywood. Be sure the face grain runs across the studs. Blocking at horizontal plywood joints is not required if the joints are at least 4 feet above the bottom plate. These specifications are based on a soil condition with 30 pounds per cubic foot equivalent fluid weight.

Treated Wood Crawl-Space Construction

Figure 4-8 shows a good pressure-treated wood crawl-space construction technique. Again, wall panels are fabricated with integral footing plates. Wall insulation can be installed between the studs. Use lumber and plywood treated as required by the American Wood Preservers Bureau Standard AWPB-FDN.

The difference with a crawl-space job is that the plywood panels are needed only for the first 2 feet below the foundation top plate. Crawl space construction needs no more than 24 inches of headroom between undisturbed soil and the floor level. Of course, the unfaced studs of the wall assembly continue down to the frost line. Treated 2 x 4 studs may be spaced 16 inches on center for single-story construction. Two-story construction requires 2 x 6 studs spaced 12 or 16 inches on center, depending on the height of the fill.

Masonry and Concrete Foundations

The most common foundation wall for basement and crawl-space construction is concrete block or poured concrete. Concrete and concrete block are highly durable, relatively inexpensive and resist moisture and insects.

Crawl-space construction is popular in mild climate areas. Plan on at least 18 inches of

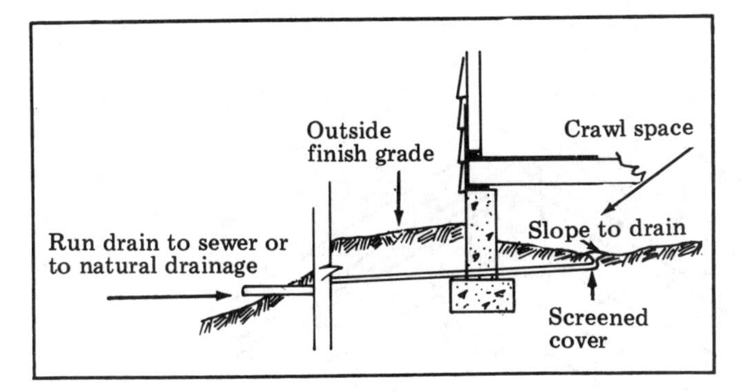

Crawl-space drainage
Figure 4-9

clearance from the ground level to the bottom of the floor joists, and 12 inches to the bottom of the wood girders. A minimum of 2 feet is required between the ground and the floor joists where mechanical equipment runs through the crawl space.

The ground level within the crawl space should generally be above the outside finish grade, unless drainage is provided or if water will not collect because of soil conditions. (See Figure 4-9.) A vapor barrier covering the ground under the house is required in some areas due to the soil and moisture conditions.

Walls or piers supporting wood-frame construction must extend at least 8 inches above the finish grade.

Hollow concrete block is the most common material used for foundation walls and piers of crawl-space construction in many parts of the country. The 8 x 8 x 16-inch block is generally used. The height of the wall or pier should not exceed 4 times its thickness, unless properly reinforced as prescribed by local code.

To reduce costs in brick veneer houses, free-standing exterior piers (Figure 4-10) may be used instead of a solid foundation wall.

In some areas, hollow block exterior piers supporting wood-frame walls must be filled with concrete or grout to prevent wind damage. Hollow block used for interior piers should have the cells of the top course filled with concrete or grout, unless a solid cap block is used.

The maximum height above grade of reinforced concrete, solid masonry or filled cell blocks is 10 times the smallest dimension of the block.

Exterior piers and interior girder piers should be spaced no more than 8 feet on center. Piers under exterior wall beams running parallel to the floor joists (at each end of the house) should be spaced no more than 12 feet on center.

Figure 4-11 shows a hollow block foundation wall with a properly anchored sill.

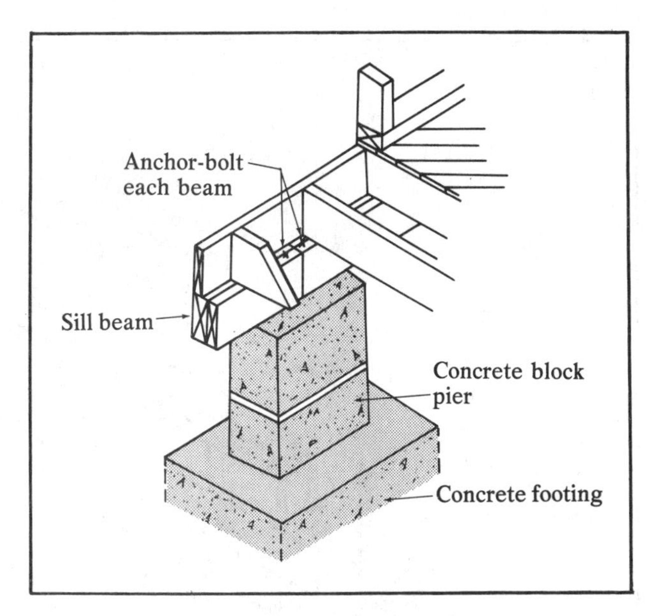

Free-standing exterior pier
Figure 4-10

Where sill anchorage is required, wood-frame floor and wall construction should be anchored with 1/2-inch bolts (with a washer at each end), embedded at least 15 inches in block piers and 6 inches in concrete piers.

Basement Walls

Basements make sense in colder climates. A deep foundation is needed to get below the freezing level of the soil. A deep foundation wall usually has to be extended only slightly to become a basement wall — providing extra interior space at little extra cost.

The wall can be either poured concrete or concrete block. Poured walls require forming. Block walls do not. Figure 4-12 shows forming for a poured concrete wall.

The form has to be both tight and able to withstand the pressure of liquid concrete. Wet concrete dumped into a form exerts terrific pressure. A poorly constructed form will collapse or distort. Most small builders prefer to build basement walls with blocks. That eliminates the cost of building and storing forms.

If poured concrete will be used, forms for windows, doors and other openings must be set in place along with the wall forms. Beam pockets or slots must also be framed. Anchor bolts are placed when the concrete has started to set.

Concrete blocks for basement walls are available in 8, 10 and 12-inch widths for residential construction. The 8-inch width is most common. The standard block allows for a 3/8-inch mortar joint.

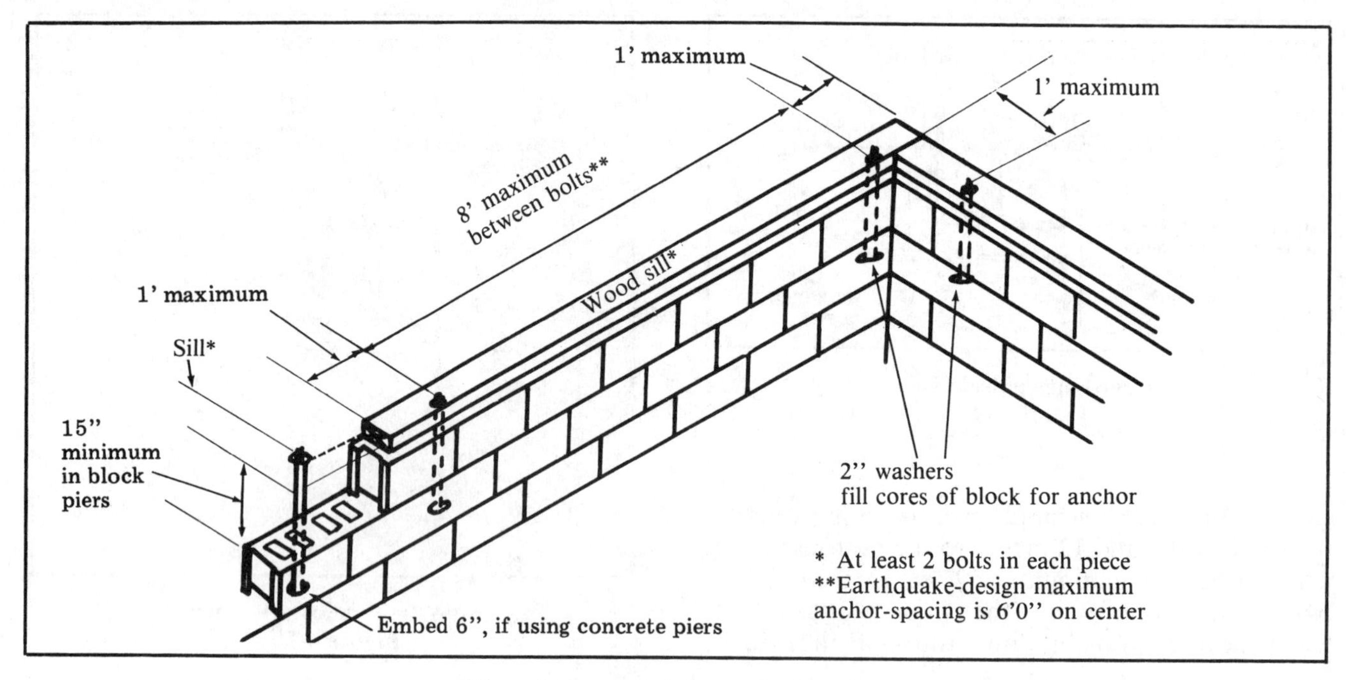

Sill anchorage on hollow block wall
Figure 4-11

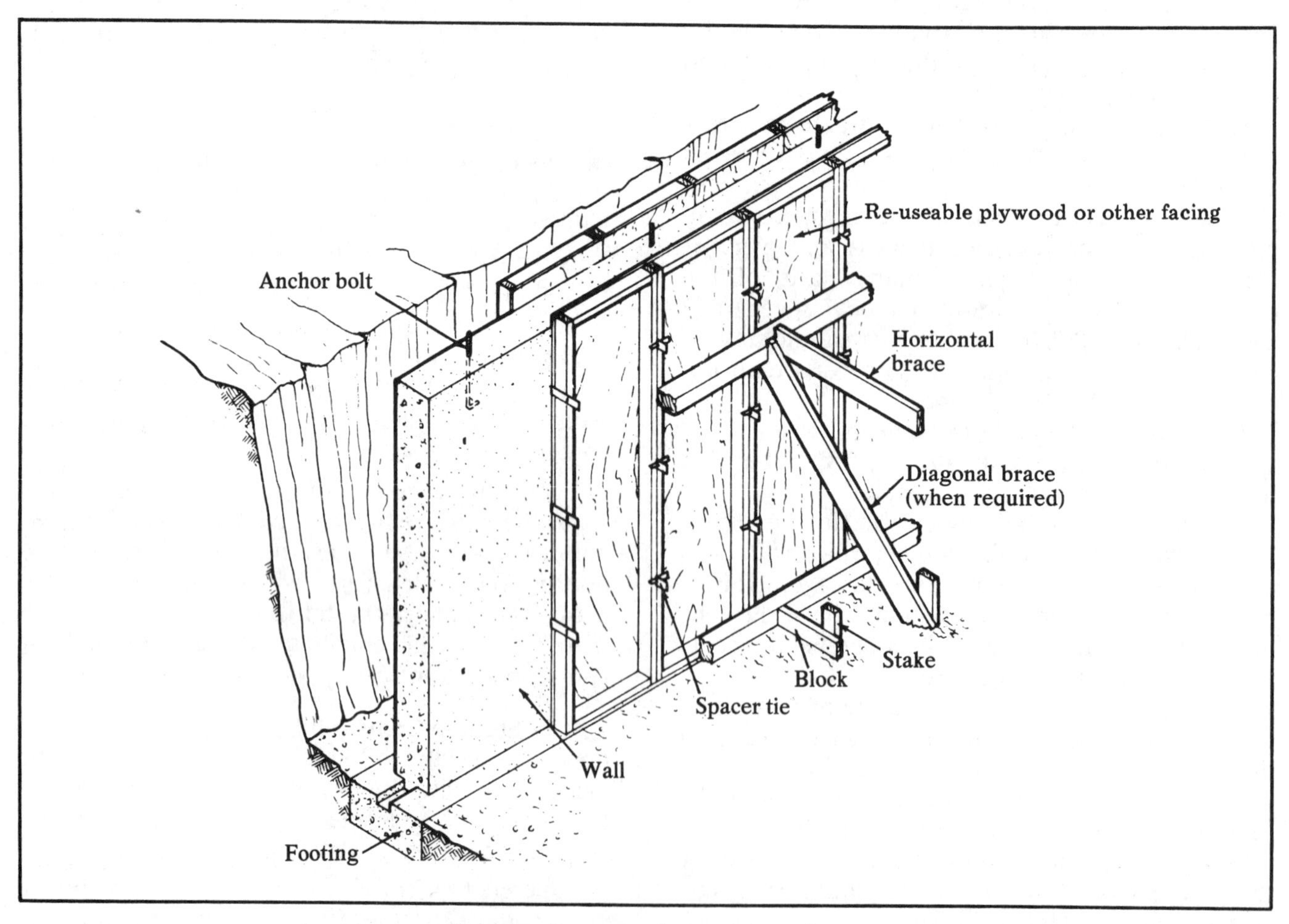

Forming for poured concrete wall
Figure 4-12

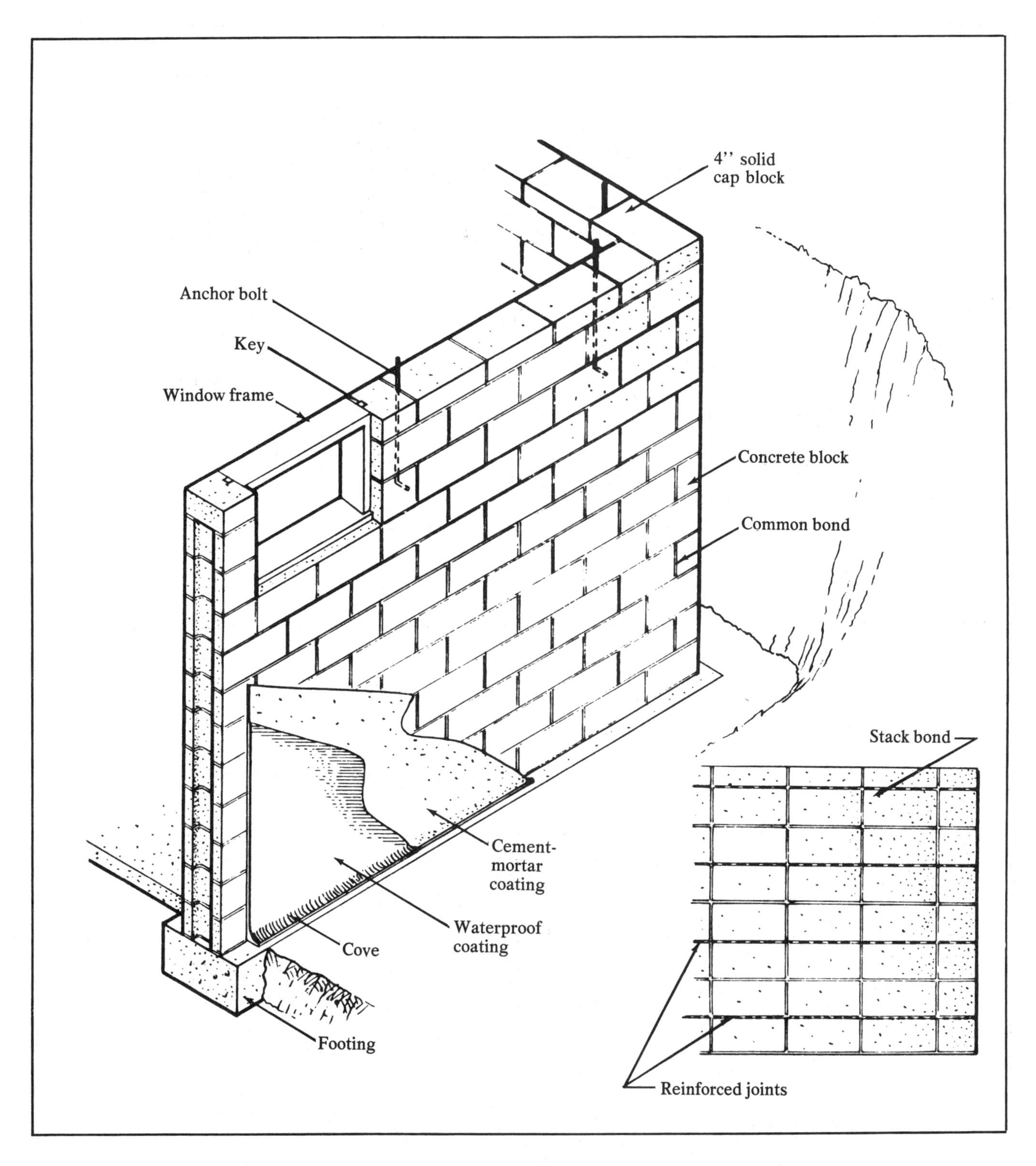

Common bond and stack bond concrete block walls
Figure 4-13

The common bond method of laying block is preferred. (See Figure 4-13.) It does not require steel reinforcement in every second course.

Keyed blocks are used at window openings for inserting metal windows. Concrete blocks are laid in full mortar joints. The joints are tooled smooth to resist water seepage.

In veneer construction, the upper portion of the foundation wall may be reduced to 4 inches. When the 4-inch portion is higher than 4 inches, reinforce-

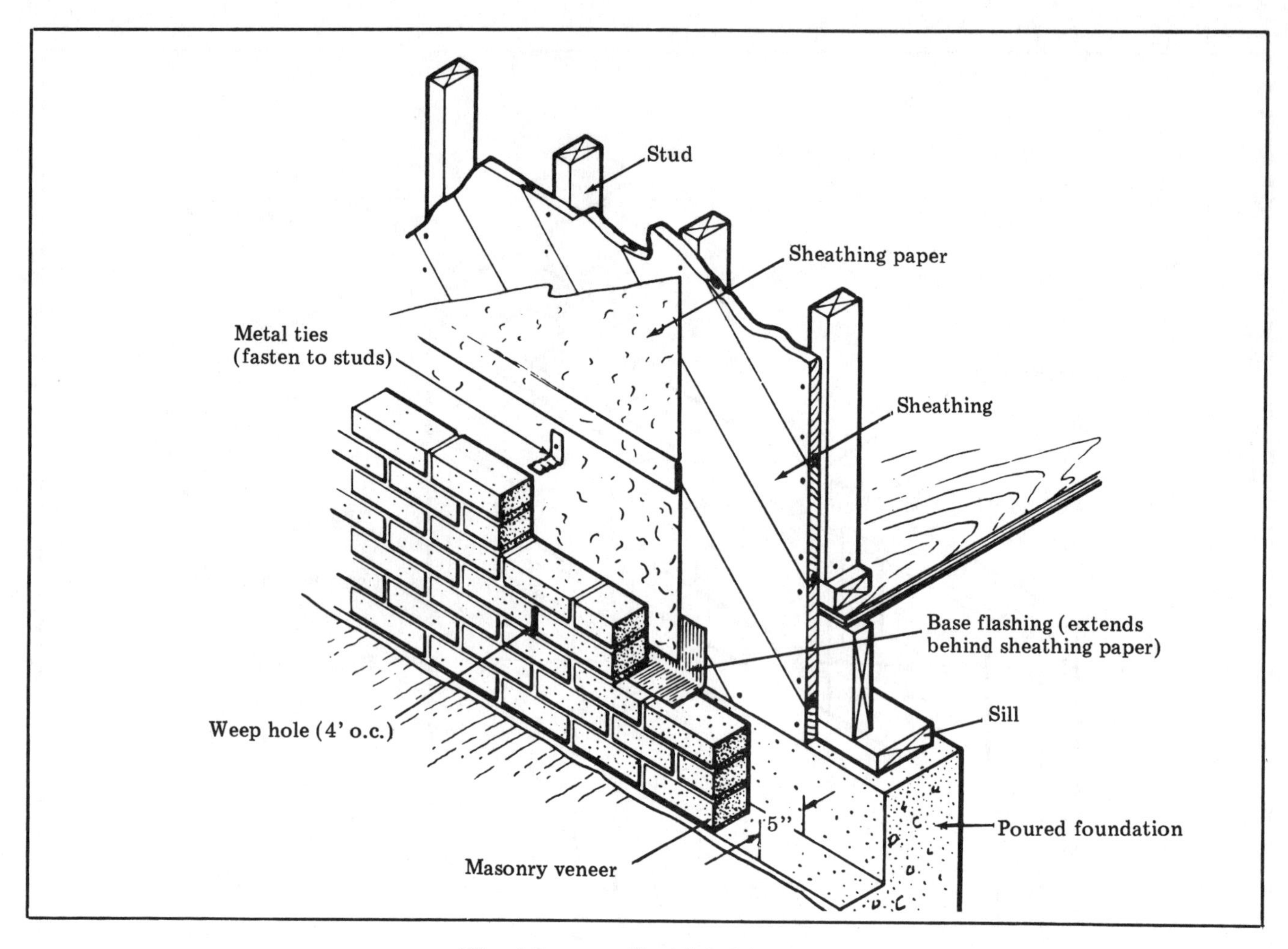

Wood-frame wall with brick veneer
Figure 4-14

ment should be provided. The maximum height of the 4-inch portion is 12 inches, unless the veneer is bonded to the wall. Figure 4-14 illustrates a wood frame wall with brick veneer on a poured foundation wall. Figure 4-15 shows a poured foundation wall with a wood-beam notch.

In many areas, termite shields are required. In heavy infestation areas, the soil should be treated chemically.

Estimating Concrete Block Requirements

Here's the way to estimate the number of standard-size (8 x 16-inch) concrete blocks required in a wall:

1) The height of the wall in feet times 1½ equals the number of courses (H)

2) The length of the wall in feet times 3/4 equals the number of blocks in each course (L)

3) H times L equals the number of blocks required

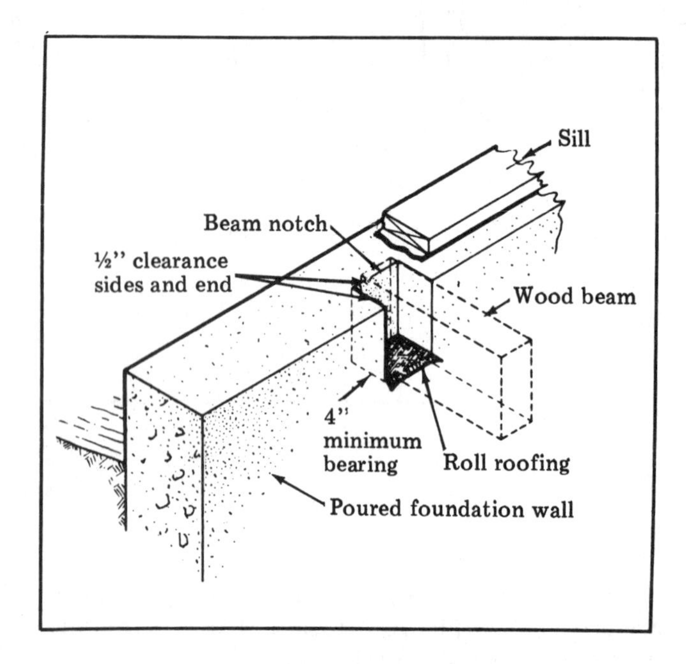

Wood-beam notch
Figure 4-15

Thus, for an 8-foot high wall 40 feet in length:

8 times 1.5 equals 12 (H)
40 times 0.75 equals 30 (L)
30 times 12 equals 360 blocks required

Note: This method counts corners twice, allowing approximately 2% per wall for waste.

To lay 100 blocks, a 70 lb. bag of masonry cement and 3 cubic feet of sand will be needed.

Waterproofing

Basements separate the professional from the amateur. It takes know-how to create a basement that will stay dry in any kind of weather. Waterproofing and drainage are critical. It has to be done during construction, before backfilling.

Seal the joint between the wall and the footing with an elastic caulking compound. Follow the manufacturer's instructions.

Block walls should be parged with a 3/8-inch coat of portland cement. Apply a waterproof membrane extending from the edge of the footing up to the finish grade. All laps should be sealed. Be sure the membrane is firmly affixed to the wall. The membrane material must provide the equivalent protection of 2-ply hot-mopped felt, 6-mil polyethylene or 55-pound roll roofing.

Drain tiles are essential in many areas to insure a waterproof basement. In arid areas, they may not be necessary. Your code will spell out what's needed. Where habitable rooms are below grade, most codes require drain tile and specific waterproofing procedures.

Summary

Most successful builders are cost-conscious builders. It's to your advantage to seize every opportunity to cut costs without sacrificing quality. If a concrete block basement wall costs less than poured concrete, stick with concrete block. Very few small builders can pour a wall that's 8 feet high as inexpensively as they can build one out of block.

Your goal is to make money. While there is no such thing as "cheap construction" any more, there *is* poor quality construction. *Avoid it, if you plan to succeed in the construction field.* The builders who make money on house after house are those who do quality work consistently. The small builder who builds quality homes is the builder who stays busy in slow periods as well as during boom times. There are acceptable ways to cut building costs without sacrificing quality. The following chapters will cover these ways in detail.

Figure 4-16 is a Footing and Foundation Checklist and Data Reference Sheet. Complete a sheet like this on every project so you have a permanent record of the job. This becomes an invaluable guide when estimating future work.

Footing and Foundation Checklist and Data Reference Sheet

1) The footing is ________ feet deep (to frost line), ________ feet wide, ________ inches thick, and meets ☐local code ☐engineering specifications.

2) The footing ☐does ☐does not extend at least 6 inches into natural undisturbed soil. If not, explain ____

____

3) The footing ☐does ☐does not comply with the requirements shown in Figure 4-1. If not, explain

____

4) The house ☐is ☐is not on a fill site. (If on a fill site, the footing must extend to rest on undisturbed soil unless otherwise permitted by local code or conditions.)

5) The footing ☐does ☐does not extend at least 6 inches below the finish grade.

6) The size and design of the footing will prevent:

☐Overloading the soil with excessive building weight.

☐Settlement of the building, to the extent that it will impair the usefulness of the house.

☐Differential settlement (between parts of the house) that will cause structural damage.

7) The footing ☐was ☐was not reinforced where pipe trenches run under the footing. If not, explain

____

8) All footing steps meet ☐local code ☐engineering specs or ☐following conditions:

☐Height of the vertical step does not exceed ¾ of the horizontal distance between steps.

☐Horizontal distance between steps is not less than 2 feet.

☐Vertical connections are the same width as the footing, and are not less than 6 inches thick.

☐Vertical and horizontal runs of the footing were poured at the same time to insure a proper bond.

9) Pier footing sizes meet ☐local code ☐engineering specs, and ☐are at least 8 inches thick by ________ inches by ________ inches. (This size must be large enough to support the total design load without excessive differential settlement.)

10) Post footings ☐do ☐do not have a pedestal because ☐wood ☐steel posts will be used. (Wood posts generally require a pedestal.)

11) Steel basement posts will be ☐bolted to ☐embedded in the concrete floor.

12) Footings for fireplaces and chimneys are ________ feet by ________ feet and ________ inches thick. The footings ☐were ☐were not poured at the same time as the foundation footings.

13) ☐Ready-mix ☐on-site-mix concrete was used. (The concrete should have a compressive strength, at 28 days, of not less than 2,000 psi.)

Figure 4-16

14) The concrete was protected from ☐freezing ☐excessive heat by ☐covering with ____________ ☐wetting for _________ days.

15) The footing (for walls and piers) is _________ linear feet in length, _________ feet wide and _________ inches thick. To determine how many cubic yards of concrete you need, multiply the length in feet times the width in feet times the thickness in inches, and divide by 314. Cubic yards of concrete required: _________ @ $_________per cubic yard. Total $__________.

16) Foundation drains ☐are ☐are not required.

17) Feet of drain tile installed: _________@ $ _________ per linear foot. Total $_________.

18) Treated wood members ☐were ☐were not considered for the footing and foundation walls because of:

☐Personal preference

☐Availability of materials

☐Local code requirements

☐Cost, compared to ☐poured concrete ☐concrete block

19) This job is ☐crawl-space ☐basement construction . Ceiling height is _________ inches/feet from ☐ground ☐floor level.

20) The crawl-space ground level ☐is ☐is not above the outside finish grade. If not, explain _______

____

21) A vapor barrier (covering the ground under the house) ☐is ☐is not required.

22) Walls or piers supporting wood-frame construction ☐are ☐are not at least 8 inches above finish grade. If not, explain ____

____.

23) Height of hollow block or poured concrete walls and piers for ☐crawl-space ☐basement construction meet ☐local code ☐engineering specs ☐other (specify)____________________________

____.

24) Exterior piers and interior girder piers are spaced ☐no greater than ☐greater than 8 feet on center as specified by ☐local code ☐engineering specs ☐other (explain) ______________________

____.

(Piers in house-end locations may be spaced 12 feet on center.)

25) Sill anchorage ☐is ☐is not required and is ☐installed ☐not installed _________ feet on center.

26) Pilasters ☐are ☐are not required by ☐local code ☐engineering specs.

27) A total of _________ hollow-core blocks @ $_________ each were required for the job. Total $_________. (Size of block: _________ inches.)

28) A total of ________ solid blocks @ $________ each were required for the job. Total $________.

29) A total of ________ bags (70-pound) masonry cement @ $________ each and ________ cubic yards of sand @ $________ per cubic yard were required for the job. Total $__________.

30) Waterproofing of the basement walls consisted of:

☐ ⅜-inch coat of portland cement (parged)

☐ Caulked joint between wall and footing

☐ Waterproof membrane ☐ 6-mil polyethylene ☐ 2-ply hot-mopped felt ☐ 55-lb. roll roofing

☐ All joints and overlaps sealed

☐ Other (specify) ______________________________

31) Material costs: Specify all materials used, giving quantity and price. This information will be of great value when estimating material and labor costs for future jobs.

______________________	@ $________	each =	$__________
______________________	@ $________	each =	$__________
______________________	@ $________	each =	$__________
______________________	@ $________	each =	$__________
______________________	@ $________	each =	$__________
		Total	**$**__________

32) Labor costs:

Carpentry (forms)	________hours	@ $________	hour =	$________
Masonry:				
Pouring concrete	________hours	@ $________	hour =	$________
Laying block	________hours	@ $________	hour =	$________
Waterproofing	________hours	@ $________	hour =	$________
Other (specify)				
______________	________hours	@ $________	hour =	$________
			Total	**$**________

33) Total cost:

Materials $____________________

Labor $____________________

Other $____________________

Total $____________________

34) Problems that could have been avoided ______________________________

Remember, by listing problems and mistakes that occurred on this job, you will have an accurate, detailed record which will help you reduce your costs and increase your profits on future jobs. Was the footing as level as it should have been? Are you really satisfied with the waterproofing of the basement? Concrete blocks were left over? Bags of mortar mix? Sand? How can your crew improve production? How can you cut costs on the next job?

Comments ______________________________

35) Footing and Foundation costs per house square foot $________________

36) A copy of this Checklist and Data Reference Sheet has been provided:

☐ Management

☐ Accounting

☐ Estimator

☐ Foreman

☐ File

☐ Other: ____________________

Chapter 5

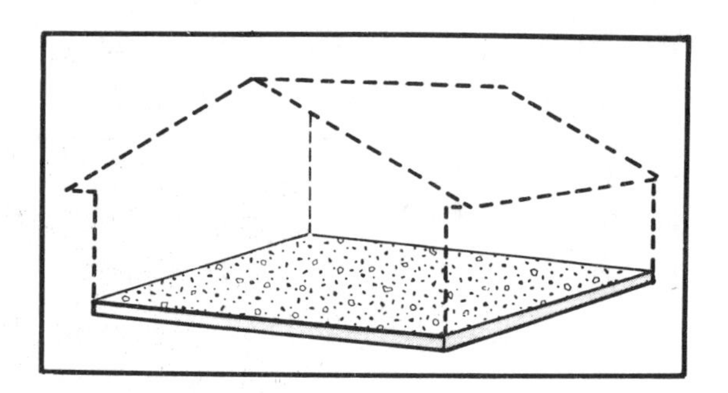

Slab-on-grade Construction

Houses with ground-supported concrete slab floors are popular in some areas. There are both advantages and disadvantages to slab construction. One big advantage is that it's cheaper. Some contractors claim a savings of a dollar a square foot of floor, compared to conventional foundation homes.

But slabs are not suited for wet sites. Even in dry areas, site grading must be planned to keep surface and ground water from collecting under the slab.

It's important that all vegetation, topsoil and foreign material be removed from the slab area. The fill material must be free of vegetation and trash.

When residential on-slab construction was first introduced, the floors tended to be cold and damp. Today we know how to handle insulation and waterproofing better. On early slab jobs, asphalt tile laid in mastic was the finish floor. Now carpeting, sheet vinyl and wood coverings are used. But a lot of homeowners object to having their heating and cooling ducts, sewer drain lines and water pipes buried in concrete and therefore inaccessible.

A properly built on-slab house may offer some savings over crawl-space construction in some areas. But in other areas, it may cost more. Let your area and market dictate which type of floor system to build.

Dampproofing Slabs

Concrete slab-on-grade construction requires special precautions to keep moisture from damaging the flooring. What you need depends on the location of the slab in relation to the finish grade, the height of the ground water table and the type of subsoil. The top of the slab should be at least 8 inches above the exterior finish grade. The bottom of the wood sills or sleepers should also be at least 8 inches above grade.

Once the topsoil has been removed and the utility lines run, 4 to 6 inches of gravel or crushed stone are tamped into place.

Rigid insulation (permanent, waterproof, nonabsorbent type) should be installed around the perimeter of the wall. The insulation may extend down on the inside of the wall vertically (the preferred method in many areas) or under the slab edge horizontally for at least 24 inches. Figure 5-1 illustrates one method of slab-on-ground construction in an area with a deep frost line.

Concrete slabs must be reinforced with 6 x 6-inch No. 10 wire mesh. The slab should be at least 4 inches thick. Figure 5-2 illustrates a monolithic slab. In warm climates where termite infestation is a problem, the monolithic slab offers the best protection. In monolithic slabs, the footing and slab are poured at the same time. The vapor barrier extends under the footing. The bottom of the footing should be at least 1 foot below the natural grade line. Solid, well-drained soil is required for this method.

In areas with a deep frost line, the walls of the house should be supported by foundations or piers which extend below the frost line to solid bearing on unfilled soil. In such construction, the slab and foundation wall are usually separate. Figure 5-1 shows one method of independent (separate) construction. Figures 5-3 and 5-4 show two other methods.

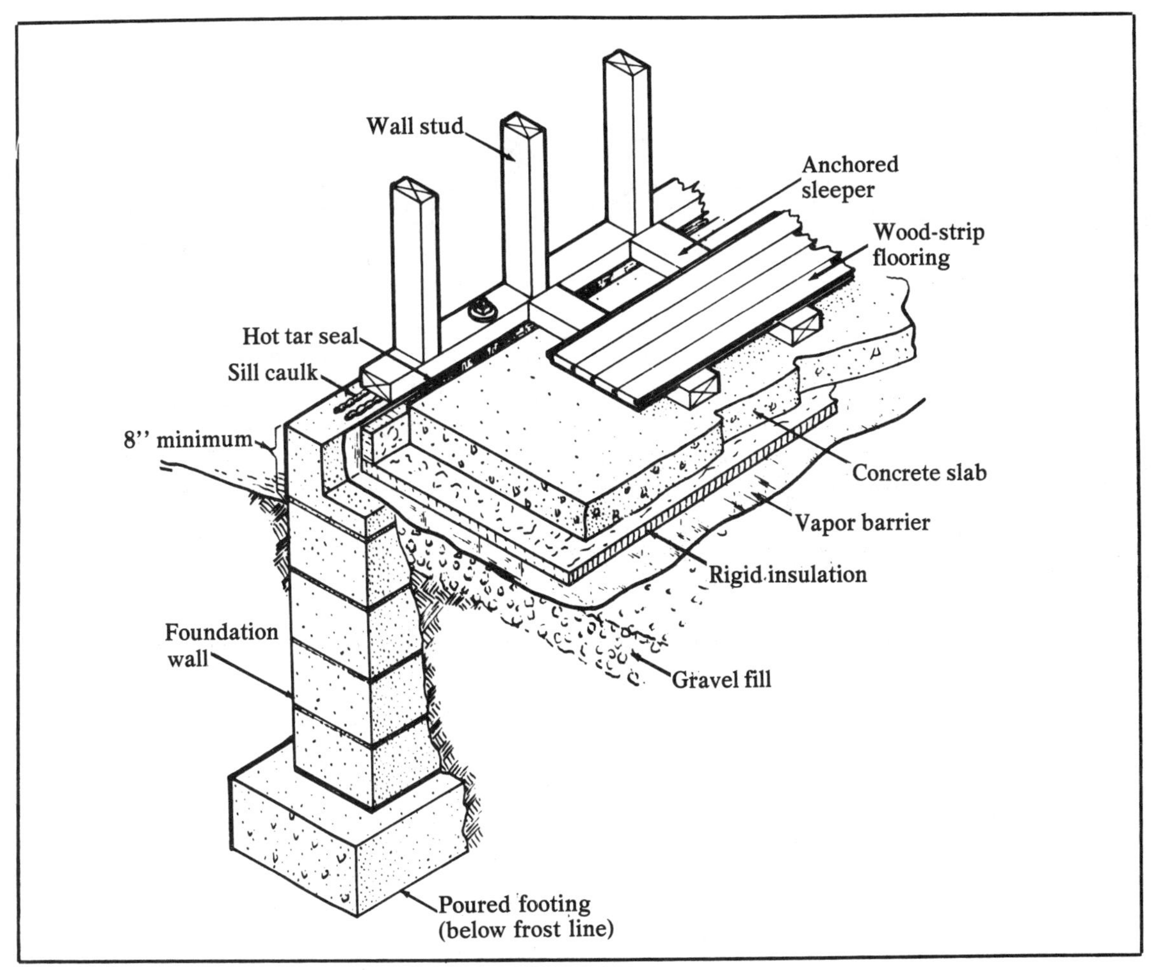

Independent concrete floor slab and wall
Figure 5-1

Vapor Barriers

Vapor barriers are essential under concrete slabs. Your building code probably is very specific on the vapor barrier to use. Usually the vapor barrier must have a vapor transmission rating of less than 0.5 perm, have the ability to withstand rupturing, and have good resistance to moisture damage and rot. The following materials usually satisfy code requirements:

- Heavy plastic film, such as 6-mil or heavier polyethylene or similar plastic film laminated to a duplex-treated paper
- 55-pound roll roofing or heavy asphalt-laminated duplex-treated paper
- Heavy asphalt-impregnated and vapor-resistant rigid sheet material with sealed joints
- Three layers of 15-pound roofing felt mopped with hot asphalt

All seams in the vapor barrier should be sealed.

Floor Coverings

The following finish flooring materials may be used on slabs with a 4-inch base course of gravel or stone and vapor barrier:

Group A*	Group B	Group C**
Asphalt tile	Vinyl sheet	Wood block
Rubber tile	Rubber sheet	Wood strip
Vinyl tile	Vinyl-cork tile	
Vinyl-asbestos	Cork tile	
Terrazzo	Linoleum tile	
Ceramic tile	Concrete, stained or painted	
	Carpet	

* A base course is not required except when the capillarity of the subsoil is such that the liquid rising from the ground water table will permit water to reach the bottom of the slab.

** A vapor barrier may be installed either under or on top of the slab.

In arid regions where no irrigation or heavy sprinkling is expected on the site, and where no

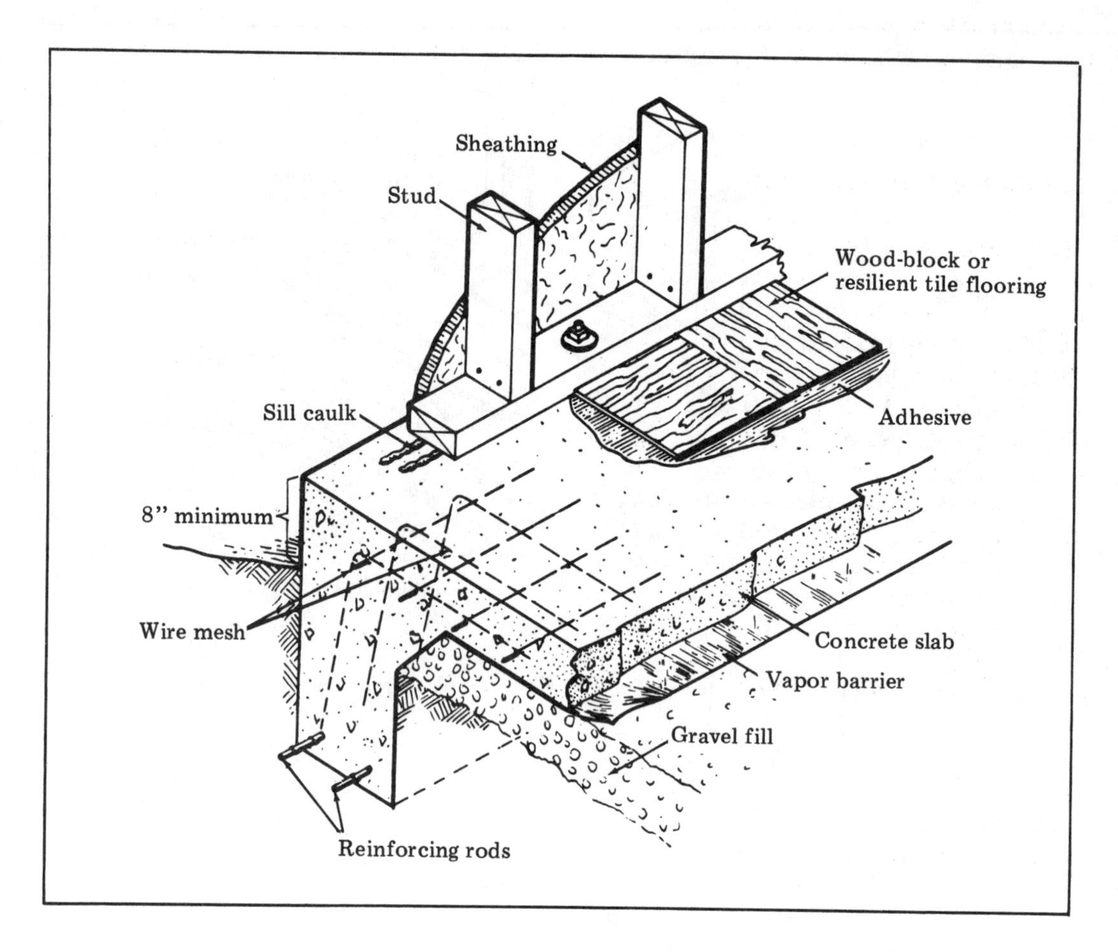

Monolithic slab
Figure 5-2

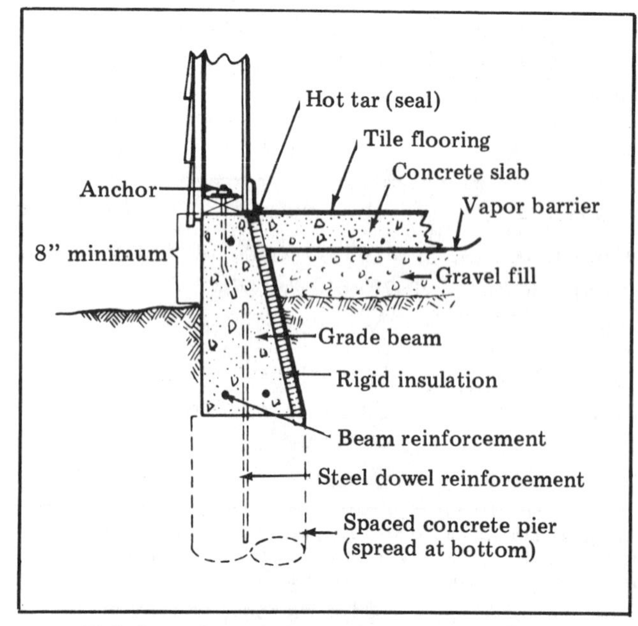

Reinforced grade beam for concrete slab
Figure 5-3

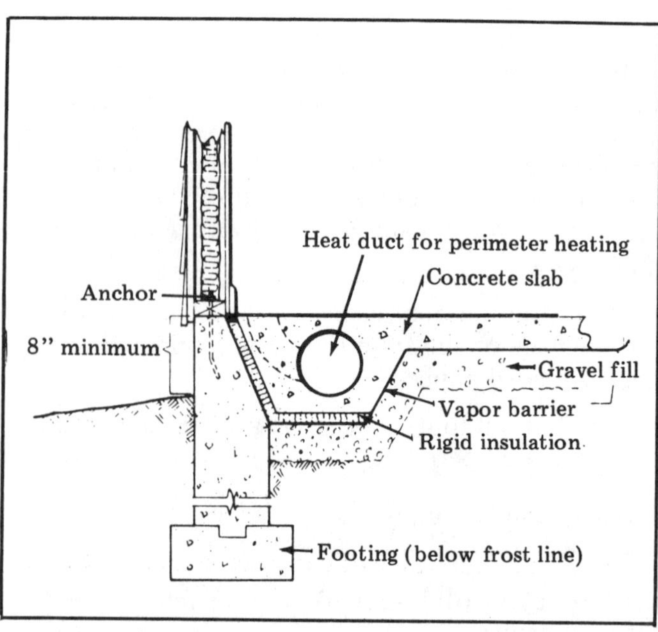

Full foundation wall for cold climates
Figure 5-4

drainage or soil problem exists, vapor barriers are generally not required with Group A floor coverings.

Insulating Concrete Floors

Heat is lost from slab floors mainly around the perimeter. Heat loss from the center is minimal. Perimeter insulation is used to reduce heat loss. Figures 5-1, 5-3 and 5-4 show the various methods of slab insulation. In Figure 5-1, the slab's weight on the foundation lip is not supported by the insulation. The insulation is omitted for 6 inches (horizontally) at approximately 3-foot intervals to provide solid bearing for the slab.

Insulation thickness is determined by the climate and insulation material. Local codes generally specify the R-value of perimeter insulation. Insulation material in contact with the ground must resist damage from soil, vermin and water.

Perimeter insulation meeting the following specifications satisfies most local codes for on-slab insulation:

Winter Degree Days (65 F Base)	Minimum R Values	
	Heated Slab	Unheated Slab
500 or less	2.8	---
1,000	3.5	---
2,000	4.0	---
2,500	4.4	2.5
3,000	4.8	2.8
4,000	5.5	3.5
5,000	6.3	4.2
6,000	7.0	4.8
7,000	7.8	5.5
8,000	8.5	6.2
9,000	9.2	6.8
10,000 or greater	10.0	7.5

Foam plastic insulation such as polystyrene and polyurethane is good for on-slab construction. It's available in 1/2-, 1-, 1½- and 2-inch thicknesses. Polystyrene offers 3.7 R-value per inch; polyurethane offers about 6.0. These materials have a low water vapor transmission rate. They have, however, a low crushing strength.

The insulation value of concrete can be increased to about 1.1 R-value per inch of thickness by using mica aggregate (1 part cement to 6 parts aggregate). It should not be used where moisture might exist.

Porches and Entrance Slabs

Slabs should be at least 4 inches thick and should be anchored to or supported at the foundation wall by anchors, piers or corbels built into the wall. The outer edge of the slab should be supported on a foundation wall or grade beam and piers if the soil under the slab is uncompacted fill or is susceptible to excessive expansion or frost action. Always slope the surface of slabs *away* from the foundation.

Concrete Slab Finish

A smooth finish is best on concrete floors. The surface should be steel troweled for a dense, smooth surface. Once the concrete has hardened, the sole plates for exterior and partition walls can be installed (Figure 5-5). Then you can begin framing.

Sole plates installed on smooth-finish slab
Figure 5-5

Termite Protection

In termite areas, treat the ground under the footing and slab with a soil poison before dumping the base course of gravel or stone. Be sure to leave a space 1 inch wide and 1 inch deep around plumbing pipes where they pass through the floor within walls or under cabinets. Fill the opening with hot tar.

Where insulation is used between the slab and the foundation wall, keep the insulation 1 inch below the top of the slab and fill the space with hot tar, as shown in Figure 5-3.

Summary

A concrete slab is about as nearly trouble-free as you can get in construction. It should last for years with no deterioration or maintenance. But it is possible to pour a bad slab. A slab poured on fill that has not been properly compacted can sink, shift, tilt or crack. The only remedy may be to remove the slab and repour it.

Don't pour a slab in areas with poor drainage. Instead, use a conventional foundation after you have graded the site properly.

Some areas are ideal for on-slab houses. In other areas, the climate will dictate a basement or crawl-space structure. The slab job is objectionable only when built on ill-suited lots or where climatic conditions are unfavorable.

Figure 5-6 is your On-slab Construction Checklist and Data Reference Sheet.

On-slab Construction Checklist and Data Reference Sheet

1) The site is acceptable for slab-on-grade construction because:

☐Building site is level

☐Site grading and drainage system will prevent surface and ground water from collecting under the slab

☐Site is not in a low-lying area

☐Site is not a fill site

☐Site is a compacted-fill site approved for slab-on-grade construction

☐Other (specify) ______________________________

2) All vegetation, topsoil and foreign material have been removed from the site.

3) A form was constructed and required the following materials and manhours:

Materials:

______________________	@ $________	each =	$____________
______________________	@ $________	each =	$____________
______________________	@ $________	each =	$____________
______________________	@ $________	each =	$____________
		Total	$____________

Manhours: ________hours @ $________ hour. **Total** $____________

4) A base course ☐was ☐was not required.

5) The base course is ______inches thick and is ☐ crushed stone ☐gravel ☐other (specify) ________. A total of ________ tons were used @ $________ ton. **Total** $________.

6) A vapor barrier ☐was ☐was not required.

7) The vapor barrier consists of:

☐Heavy plastic film, 6-mil or heavier

☐55-pound roll roofing

☐Heavy asphalt-laminated duplex barrier

☐Heavy asphalt-impregnated and vapor-resistant rigid sheet material

☐Three layers of 15-pound roofing felt mopped with hot asphalt

8) All seams in the vapor barrier ☐were ☐ were not sealed.

Figure 5-6

9) Total material used for vapor barrier was ________square feet @ $________square foot. **Total** $________

10) Labor for installing vapor barrier:________hours @ $________hour. **Total** $________

11) The vapor barrier ☐does ☐does not extend under the footing. If not, explain ________

12) The slab is ☐monolithic ☐independent.

13) The bottom of the ☐footing ☐foundation ☐piers extends ________feet below the natural grade line. (It should extend at least 1 foot.)

14) The frost line is ________ inches deep.

15) The slab is reinforced by:

☐6 x 6-inch No. 10 wire mesh

☐Other (specify) ________

Cost: ________square feet @ $________square foot. **Total** $________

16) The ground under the footing and slab ☐was ☐was not chemically treated for termites. Cost of treatment: $________

17) The slab is insulated at perimeters with ________-inch-thick ________installed ☐vertically at foundation wall ☐horizontally under slab edge. R-value of insulation:________. Total ________square feet of insulation @ $________square foot. **Total** $________

18) Expanded mica aggregate mix ☐was ☐was not used. Total ________cubic yards @ $________cubic yard. **Total** $________

19) The concrete was protected from ☐freezing ☐excessive heat by ☐covering with________ ☐wetting for ________days.

20) The top of the slab ☐is ☐is not at least 8 inches above the finish grade. If not, explain ________

____

____

21) The size of the slab is ________feet by ________feet and ______inches thick for a total of ________cubic yards of concrete @ $________cubic yard. **Total** $________

(Length in feet times width in feet times thickness in inches divided by 314 equals number of cubic yards required.)

22) A ☐porch ☐entrance slab ☐carport slab ☐was ☐was not anchored to or supported at foundation wall by ☐anchors ☐piers ☐corbels built into the wall. The porch/slab required a total of ________cubic yards of concrete @ $________cubic yard. **Total** $________

23) Material costs:

____________________ @ $______ each = $________

____________________ @ $______ each = $________

____________________ @ $______ each = $________

____________________ @ $______ each = $________

____________________ @ $______ each = $________

Total $________

24) Labor costs:

Pouring slab ______hours @ $______ hour = $________

____________ ______hours @ $______ hour = $________

____________ ______hours @ $______ hour = $________

Total $________

25) Total Cost:

Materials $____________

Labor $____________

Other $____________

Total $____________

26) Slab-on-grade cost per square foot $__________

27) Problems that could have been avoided ________________________

____

____

____

28) Comments __________________________________

____

____

____

29) A copy of this Checklist and Data Reference Sheet has been provided:

☐Management

☐Accounting

☐Estimator

☐Foreman

☐File

☐Other: ____________________

Chapter 6

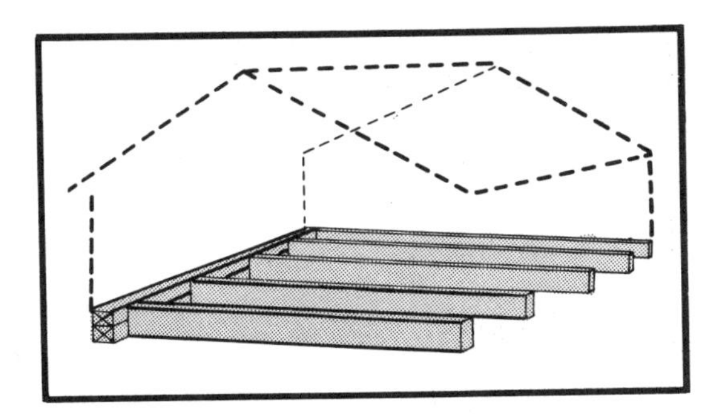

Framing the Floor

Knowing how to build and how to make money at it are not necessarily the same thing. Keep that in mind as we go through this chapter. The purpose of this book is to explain both. Don't let the technical details obscure your reason for being in construction: You want to make a good living as a builder — and to do that, you need to know how to turn your technical competence into profits.

The four figures that follow have a lot of good basic information that you'll want to refer to occasionally. The book wouldn't be complete without them, so we'll present these figures together at the beginning of this chapter to make reference easy.

Figure 6-1 shows typical lumber product classifications.

Figure 6-2 shows nominal and actual sizes for softwood lumber.

Figure 6-3 is a nail size chart.

Figure 6-4 is a nailing schedule for a wood-frame house.

Joist Grades and Sizes for Different Spans

You should know what size material to use for floor joists over different spans. Figure 6-5 gives this information by wood species. It lists the minimum grades and sizes of joists required for house depths from 20 to 32 feet at 1-foot intervals. The calculation is based on 40 pounds per square foot uniform live load, framing with center girder, nominal 2-inch thick band (header) joists, and 2 x 4 sill and center bearing plates.

	Thickness	Width
Board lumber	1"	2" or more
Light framing	2" to 4"	2" to 4"
Studs	2" to 4"	2" to 6", 10' and shorter
Structural light framing	2" to 4"	2" to 4"
Structural joists and planks	2" to 4"	5" and wider
Beams and stringers	5" and thicker	more than 2" greater than thickness
Posts and timbers	5" x 5" and larger	not more than 2" greater than thickness
Decking	2" to 4"	4" to 12" wide
Siding	thickness expressed by dimension of butt edge	
Moldings	size at thickest and widest points	

Lengths of lumber generally are 6 feet and longer in multiples of 2'.
Source: Western Wood Products Association

Typical lumber product classifications
Figure 6-1

The thicknesses apply to all widths.

Thicknesses			Face Widths		
Nominal	Minimum Dressed		Nominal	Minimum Dressed	
	Dry* Inches	Green* Inches		Dry* Inches	Green* Inches
BOARDS**					
1	¾	25/32	2	1½	1 9/16
			3	2½	2 9/16
			4	3½	3 9/16
			5	4½	4⅝
1¼	1	1 1/32	6	5½	5⅝
			7	6½	6⅝
			8	7¼	7½
			9	8¼	8½
1½	1¼	1 9/32	10	9¼	9½
			11	10¼	10½
			12	11¼	11½
			14	13¼	13½
			16	15¼	15½
DIMENSION					
2	1½	1 9/16	2	1½	1 9/16
			3	2½	2 9/16
2½	2	2 1/16	4	3½	3 9/16
			5	4½	4⅝
			6	5½	5⅝
3	2½	2 9/16	8	7¼	7½
			10	9¼	9½
3½	3	3 1/16	12	11¼	11½
			14	13¼	13½
			16	15¼	15½
DIMENSION					
4	3½	3 9/16	2	1½	1 9/16
			3	2½	2 9/16
			4	3½	3 9/16
			5	4½	4⅝
			6	5½	5⅝
4½	4	4 1/16	8	7¼	7½
			10	9¼	9½
			12	11¼	11½
			14		13½
			16		15½
TIMBERS					
5 & thicker	—	½ off	5 & wider	—	½ off

*These are minimum dressed sizes. Source: American Softwood Lumber Standard, PS 20-70.

**Boards less than the minimum thickness for 1 inch nominal but ⅝ inch or greater thickness dry (11/16 inch green) may be regarded as American Standard Lumber, but such boards shall be marked to show the size and condition of seasoning at the time of dressing. They shall also be distinguished from 1-inch boards on invoices and certificates.

Nominal size chart for softwood lumber
Figure 6-2

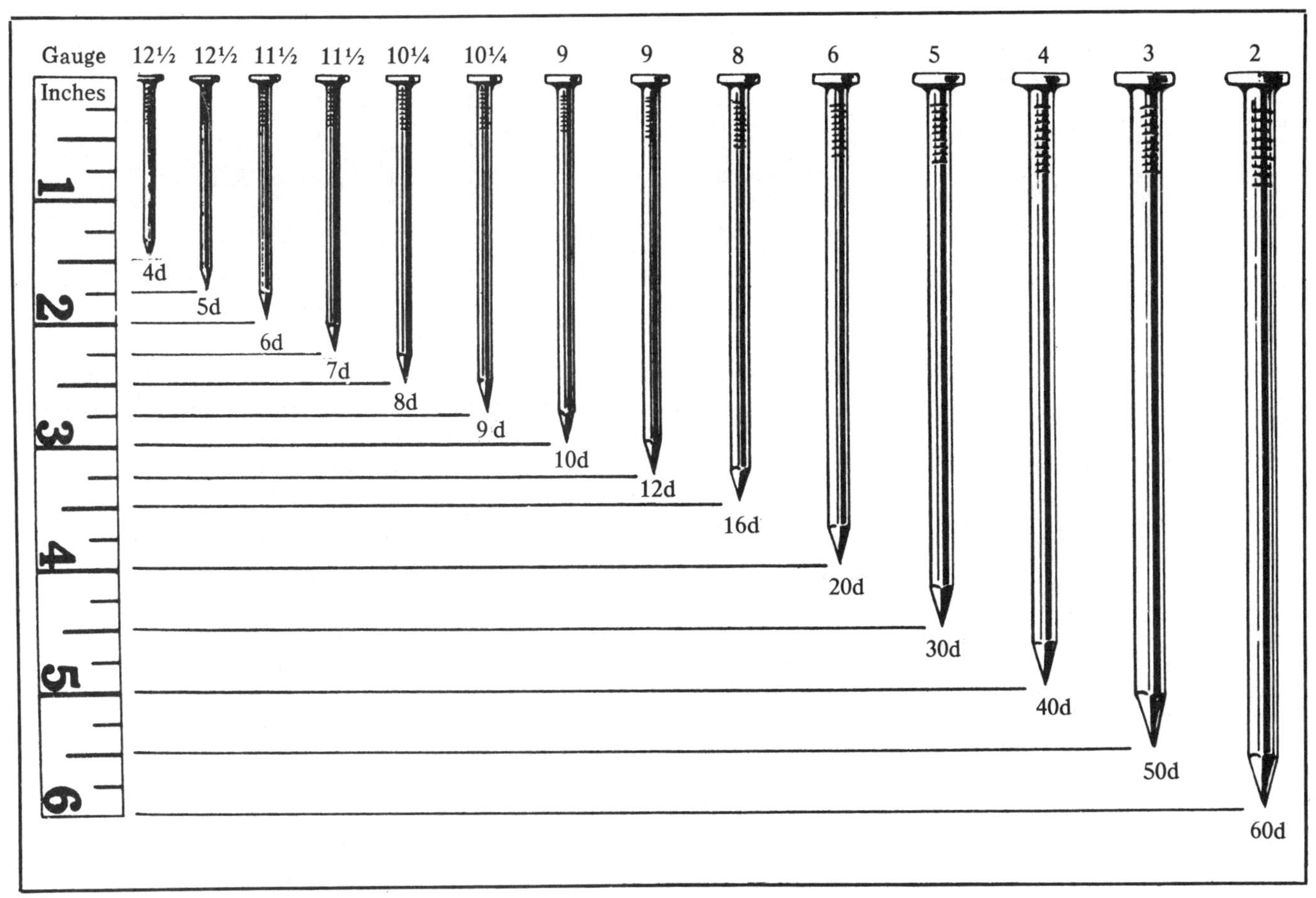

Sizes of common wire nails
Figure 6-3

Where use of a 2 x 6 or 2 x 8 sill or bearing plate (as in Figures 6-6A and 6-6B) would permit a particular size and grade of joist to be used for a larger house depth, it's noted in the chart.

You will note in these two figures that the wider the sill plate, the shorter the span. Where joists bear on a center girder, the clear span of each joist is equal to 1/2 the house depth (measured between outside surfaces of the exterior studs), minus the thickness of the band header joist, minus the length of joist bearing on the foundation wall sill plate, minus 1/2 the width of the bearing plate on the center support.

Where 2 x 4 sill plates and a 2 x 4 center bearing plate are used, the clear span of each joist is 1/2 the house depth minus 5¼ inches. For practical purposes, the joist span is the *unsupported* area of the joist run.

The Floor Frame

Figure 6-7 shows a typical platform floor frame. Let's analyze it.

We begin with the *sill plate*. This is the bearing member placed directly on the foundation wall. See Figure 6-8. (A sill beam is placed directly on the exterior free-standing piers. Refer back to Figure 4-10 in Chapter 4 for a picture of a sill beam and pier.) Where required, sill sealer is used between the foundation wall and the sill. In termite-infested areas, a metal shield or barrier is installed between the sill and the pier or foundation wall.

The sill must be square and level. This should be no problem if the foundation is built properly.

Sills treated with a wood preservative are required in some areas. Southern yellow pine, Douglas fir, redwood and cypress make good sills. Cost will generally dictate which species to use.

Platform and balloon framing are the two general types of wood-sill construction. The box method is commonly used in platform construction. It consists of a plate anchored (when required) to the foundation wall or piers. This provides support and fastening for the joists and header. See Figure 6-9.

This method is good because it provides a fire stop and solid bearing for nailing the joists and subfloor. The disadvantage is that the wall shrinks more, due to the thick horizontal grain.

Joining	Nailing Method	Number	Nails: Size	Nails: Placement
Header to joist	End-nail	3	16d	
Joist to sill or girder	Toenail	2	10d or	
		3	8d	
Header and stringer joist to sill	Toenail		10d	16-in. on center
Bridging to joist	Toenail ea. end	2	6d	
Ledger strip to beam, 2-in. thick	Face-nail	3	16d	At each joist
Subfloor, boards:				
1 x 6 and smaller	Face-nail	2	8d	To each joist
1 x 8	Face-nail	3	8d	To each joist
Subfloor, plywood:				
At edges	Face-nail		8d	6-in. on center
At intermediate joists	Face-nail		8d	8-in. on center
Subfloor (2 x 6-in., T&G) to joist or girder	Blind-nail (casing) & face-nail	2	16d	
Sole plate to stud, horizontal assembly	End-nail	2	16d	At each stud
Top plate to stud	End-nail	2	16d	
Stud to sole plate, upright assembly	Toenail	4	8d	
Sole plate to joist or blocking	Face-nail		16d	16-in. on center
Double studs	Face-nail, stagger		10d	16 in. on center
End stud of intersecting wall to ext. wall stud	Face-nail		16d	16-in. on center
Upper top plate to lower top plate	Face-nail		16d	16-in. on center
Upper top plate, laps and intersections	Face-nail	2	16d	
Continuous header, two pieces	Each edge		12d	12-in. on center
Ceiling joist to top wall plates	Toenail	3	8d	
Ceiling joist laps at partition	Face-nail	4	16d	
Rafter to top plate	Toenail	2	8d	
Rafter to ceiling joist	Face-nail	5	10d	
Rafter to valley or hip rafter	Toenail	3	10d	
Ridge board to rafter	End-nail	3	10d	
Rafter to ridge board	Toenail	3	10d	
Collar beam to rafter:				
2-in. member	Face-nail	2	12d	
1-in. member	Face-nail	3	8d	
1-in. diagonal let-in brace to each stud and plate (4 nails at top)		2	8d	
Built-up corner studs:				
Studs to blocking	Face-nail	2	10d	Each side
Intersecting stud to corner studs	Face-nail		16d	12-in. on center
Built-up girders and beams, 3 or more members	Face-nail		20d	32-in. on center, each side
Wall sheathing:				
1 x 8 or less, horizontal	Face-nail	2	8d	At each stud
1 x 6 or greater, diagonal	Face-nail	3	8d	At each stud
Wall sheathing, vertically applied plywood:				
⅜-in. and less thick	Face-nail		6d	6-in. edge
½-in. and over thick	Face-nail		8d	12-in. intermediate
Wall sheathing, vertically applied fiberboard:				
½-in. thick	Face-nail			1½-in. roofing nail } 3-in. edge & 6-in. intermediate
25/32-in. thick	Face-nail			1¾-in. roofing nail } 3-in. edge & 6-in. intermediate
Roof sheathing, boards, 4-, 6-, 8-in. width	Face-nail	2	8d	At each rafter
Roof sheathing, plywood:				
⅜-in. and less thick	Face-nail		6d	6-in. edge and 12-in. intermediate
½-in. and over thick	Face-nail		8d	6-in. edge and 12-in. intermediate

Recommended schedule for nailing framing and sheathing of wood-frame house
Figure 6-4

Species	Joist spacing	Joist size	House depth (measured between outside surfaces of exterior studs), in feet													Grading rule agency*
			20	21	22	23	24	25	26	27	28	29	30	31	32	
Balsam fir	16"	2x8	No.2	No.2	No.2	No.2	No.2[c]	No.1[c]								NeLMA NH&PMA
		2x10	No.3	No.3	No.3	No.3[a]	No.2	No.2	No.2	No.2	No.2	No.2	No.2[c]	No.1[a]		
	24"	2x8	No.2[c]	No.1[a]												
		2x10	No.2	No.2	No.2	No.2	No.2	No.2[c]	No.1	No.1[c]						
California redwood (open grain)	16"	2x8	No.3[c]	No.2	No.2	No.2[a]	No.1[b]									RIS
		2x10	No.3	No.3	No.3	No.3	No.3	No.3[c]	No.2	No.2	No.2	No.2[a]	No.1[a]			
		2x12						No.3	No.3	No.3	No.3	No.3	No.3[b]	No.2	No.2	
	24"	2x8	No.2	No.1[a]												
		2x10	No.3	No.3[c]	No.2	No.2	No.2	No.2	No.2[c]	No.1[c]						
		2x12		No.3	No.3	No.3	No.3	No.3[c]	No.2	No.2	No.2	No.2	No.2	No.2[a]	No.1[b]	
Douglas-fir-larch	16"	2x8	No.3	No.3	No.3	No.2	No.2	No.2	No.2	No.2	No.1[b]					NLGA WCLIB WWPA
		2x10				No.3	No.3	No.3	No.3	No.3	No.3[a]	No.2	No.2	No.2	No.2	
		2x12									No.3	No.3	No.3	No.3	No.3	
	24"	2x8	No.2	No.2	No.2	No.2	No.2[c]	No.1[c]								
		2x10	No.3	No.3	No.3	No.3[a]	No.2	No.2	No.2	No.2	No.2	No.2	No.2[b]	No.1[a]	No.1[c] DENSE	
		2x12				No.3	No.3	No.3	No.3	No.3	No.3[b]		No.2	No.2	No.2	
Douglas-fir-south	16"	2x8	No.3	No.3	No.3[c]	No.2	No.2	No.2[a]	No.1[c]							WWPA
		2x10			No.3	No.3	No.3	No.3	No.3	No.3	No.2	No.2	No.2	No.2	No.2[c]	
		2x12									No.3	No.3	No.3	No.3	No.3	
	24"	2x8	No.2	No.2	No.2[c]	No.1[c]										
		2x10	No.3	No.3	No.3	No.3[c]	No.2	No.2	No.2	No.2	No.2[b]	No.1[c]				
		2x12				No.3	No.3	No.3	No.3	No.3[a]	No.2	No.2	No.2	No.2	No.2	
Eastern hemlock-tamarack	16"	2x8	No.3	No.3[c]	No.2	No.2	No.2[b]	No.1[a]								NeLMA NLGA
		2x10			No.3	No.3	No.3	No.3	No.3[a]	No.2	No.2	No.2	No.2[a]	No.1	No.1[c]	
	24"	2x8	No.2	No.2[a]	No.1[a]											
		2x10	No.3	No.3	No.3[c]	No.2	No.2	No.2	No.2	No.2[c]	No.1[b]					
Eastern spruce	16"	2x8	No.2	No.2	No.2	No.2	No.2[a]	No.1	No.1[c]							NeLMA NH&PMA
		2x10	No.3	No.3	No.3	No.3	No.3[b]	No.2	No.2	No.2	No.2	No.2	No.2	No.2[c]	No.1	
	24"	2x8	No.2[b]	No.1	No.1[a]											
		2x10	No.3[c]	No.2	No.2	No.2	No.2	No.2[a]	No.1	No.1	No.1[b]					
Englemann spruce-alpine fir or (Englemann spruce-lodgepole pine)	16"	2x8	No.2	No.2	No.2	No.2	No.2[c]	No.1[a]								WWPA
		2x10	No.3	No.3	No.3	No.3[a]	No.2	No.2	No.2	No.2	No.2	No.2	No.2[c]	No.1	No.1[c]	
		2x12				No.3	No.3	No.3	No.3	No.3			No.2	No.2	No.2	
	24"	2x8	No.2[c]	No.1[a]												
		2x10	No.2	No.2	No.2	No.2	No.2	No.2[c]	No.1	No.1[c]						
		2x12	No.3	No.3	No.3	No.3[c]		No.2	No.2	No.2	No.2	No.2	No.2[c]	No.1	No.1	
Hem-fir	16"	2x8	No.3[a]	No.2	No.2	No.2	No.2	No.2	No.2[c]							NLGA WCLIB WWPA
		2x10		No.3	No.3	No.3	No.3	No.3[a]	No.2	No.2	No.2	No.2	No.2	No.2	No.2	
		2x12						No.3	No.3	No.3	No.3	No.3	No.3	No.3[c]		
	24"	2x8	No.2	No.2	No.1	No.1										
		2x10	No.3	No.3[c]	No.2	No.2	No.2	No.2	No.2	No.2[c]	No.1	No.1[a]				
		2x12		No.3	No.3	No.3	No.3	No.3[a]		No.2	No.2	No.2	No.2	No.2	No.2	
Idaho white pine or Western white pine	16"	2x8	No.2	No.2	No.2	No.2[a]	No.1	No.1[a]								WWPA
		2x10	No.3	No.3	No.3	No.3[a]	No.2	No.2	No.2	No.2	No.2	No.2	No.1	No.1	No.1[b]	
		2x12				No.3	No.3	No.3	No.3	No.3	No.3[c]		No.2	No.2	No.2	
	24	2x8	No.1	No.1[b]												NLGA
		2x10	No.2	No.2	No.2	No.2	No.2[a]	No.1	No.1[b]							
		2x12	No.3	No.3	No.3	No.3[c]	No.2	No.2	No.2	No.2	No.2	No.2[a]	No.1	No.1	No.1[c]	

Based on construction with center girder, nominal 2-inch-thick band (header) joists and, except where footnoted, nominal 2x4 sill and 2x4 center bearing plates

Minimum joist grades for different house depths 40 psf uniform live load
Figure 6-5

Species	Joist spacing	Joist size	20	21	22	23	24	25	26	27	28	29	30	31	32	Grading rule agency*
			House depth (measured between outside surfaces of exterior studs), in feet													
Lodgepole pine	16"	2x8	No.2	No.2	No.2	No.2	No.2	No.1[a] No.2[c]								WWPA
		2x10	No.3	No.3	No.3	No.3	No.3	No.2	No.2	No.2	No.2	No.2	No.2	No.2[a]	No.1[c]	
		2x12						No.3	No.3	No.3	No.3	No.3[a]		No.2	No.2	
	24"	2x8	No.2	No.1	No.1[a]											
		2x10	No.3[a]	No.2	No.2	No.2	No.2	No.2	No.2[c]	No.1	No.1[b]					
		2x12		No.3	No.3	No.3	No.3[a]		No.2	No.2	No.2	No.2	No.2	No.2[b]	No.1	
Northern pine	16"	2x8	No.3[c]	No.2	No.2	No.2	No.2	No.2[a]	No.1[c]							NeLMA NH&PMA
		2x10	No.3	No.3	No.3	No.3	No.3	No.3[c]	No.2	No.2	No.2	No.2	No.2	No.2	No.2[c]	
	24"	2x8	No.2	No.2[b]	No.1	No.1[c]										
		2x10	No.3	No.3[c]	No.2	No.2	No.2	No.2	No.2[a]	No.1	No.1	No.1[c]				
Ponderosa pine-sugar pine	16"	2x8	No.2	No.2	No.2	No.2	No.2[b]	No.1[c]								WWPA
		2x10	No.3	No.3	No.3	No.3	No.2	No.2	No.2	No.2	No.2	No.2	No.2[a]	No.1[a]		
		2x12					No.3	No.3	No.3	No.3	No.3[a]		No.2	No.2	No.2	
	24"	2x8	No.2[c]	No.1	No.1[c]											
		2x10	No.2	No.2	No.2	No.2	No.2	No.2[c]	No.1	No.1[a]						
		2x12	No.3	No.3	No.3	No.3[a]		No.2	No.2	No.2	No.2	No.2	No.2[b]	No.1	No.1	
Southern pine	16"	2x8	No.3	No.3	No.3[a]	No.2	No.2	No.2	No.2[c]	MG No.2[b]	No.1[b]					SPIB
		2x10			No.3	No.3	No.3	No.3	No.3	No.3	No.3[c]	No.2	No.2	No.2	No.2	
		2x12									No.3	No.3	No.3	No.3	No.3	
	24"	2x8	No.2	No.2	No.2[c]	No.2	MG No.2[c]	No.1[c]								
		2x10	No.3	No.3	No.3	No.3[b]	No.2	No.2	No.2	No.2	MG No.2	MG No.2	MG No.2[b]	No.1[a]	No.1[c] DENSE	
		2x12				No.3	No.3	No.3	No.3	No.3	No.3[c]	No.2	No.2	No.2	No.2	
Southern pine KD (15% mc)	16"	2x8	No.3	No.3	No.3	No.3[b]	No.2	No.2	No.2	MG No.2	No.1	No.1[b] DENSE				SPIB
		2x10				No.3	No.3	No.3	No.3	No.3	No.3	No.3[b]	No.2	No.2	No.2	
		2x12										No.3	No.3	No.3	No.3	
	24"	2x8	No.2	No.2	No.2	No.2[c]	MG No.2[a]	No.1[c]								
		2x10	No.3	No.3	No.3	No.3	No.3[b]	No.2	No.2	No.2	No.2	No.2[c]	MG No.2	No.1	No.1[c]	
		2x12					No.3	No.3	No.3	No.3	No.3	No.3	No.3[a]	No.2	No.2	
Spruce-pine fir Coast Sitka spruce or Sitka spruce	16"	2x8	No.2	No.2	No.2	No.2	No.2[a]	No.1	No.1[a]							NLGA
		2x10	No.3	No.3	No.3	No.3	No.2	No.2	No.2	No.2	No.2	No.2	No.2	No.2[c]	No.1	
		2x12					No.3	No.3	No.3	No.3	No.3[a]			No.2	No.2	
	24"	2x8	No.2[b]	No.1	No.1[c]											WCLIB
		2x10	No.2	No.2	No.2	No.2	No.2	No.2[c]	No.1	No.1						
		2x12	No.3	No.3	No.3	No.3[c]		No.2	No.2	No.2	No.2	No.2	No.2[a]	No.1	No.1	
Western hemlock	16"	2x8	No.3	No.3[a]	No.2	No.2	No.2	No.2	No.2[c]	No.1[b]						WWPA
		2x10		No.3	No.3	No.3	No.3	No.3	No.3	No.3[c]	No.2	No.2	No.2	No.2	No.2	
		2x12								No.3	No.3	No.3	No.3	No.3	No.3[a]	
	24"	2x8	No.2	No.2	No.2[a]	No.1	No.1[c]									
		2x10	No.3	No.3	No.3[b]	No.2	No.2	No.2	No.2	No.2	No.2[b]	No.1	No.1[b]			
		2x12			No.3	No.3	No.3	No.3	No.3		No.2	No.2	No.2	No.2	No.2	
White woods (western woods)	16"	2x8	No.2	No.2	No.2	No.2[a]	No.1[b]									WWPA
		2x10	No.3	No.3	No.3	No.3[a]	No.2	No.2	No.2	No.2	No.2	No.2[a]	No.1[a]			
		2x12				No.3	No.3	No.3	No.3	No.3	No.3[c]	No.2	No.2	No.2	No.2	
	24"	2x8	No.1	No.1[b]												
		2x10	No.2	No.2	No.2	No.2	No.2[a]	No.1	No.1[a]							
		2x12	No.3	No.3	No.3	No.3[c]	No.2	No.2	No.2	No.2	No.2	No.2[a]	No.1	No.1	No.1[c]	

[a]Nominal 2 x 6 sill plate and 2 x 4 center bearing plate; or width of sill plate plus one-half width of center bearing equal to 7¼" or more.
[b]Nominal 2 x 6 sill plate and 2 x 6 center bearing plate; or width of sill plate plus one-half width of center bearing equal to 8¼" or more.
[c]Nominal 2 x 8 sill plate and 2 x 8 center bearing plate; or width of sill plate plus one-half width of center bearing equal to 10⅞" or more.
*NeLMA—Northeastern Lumber Manufacturers Association.
NH&PMA—Northern Hardwood and Pine Manufacturers Association.
RIS—Redwood Inspection Service.
NLGA—National Lumber Grades Authority, a Canadian Agency.
WCLIB—West Coast Lumber Inspection Bureau.
WWPA—Western Wood Products Association.
SPIB—Southern Pine Inspection Bureau.

Minimum joist grades for different house depths 40 psf uniform live load
Figure 6-5 (continued)

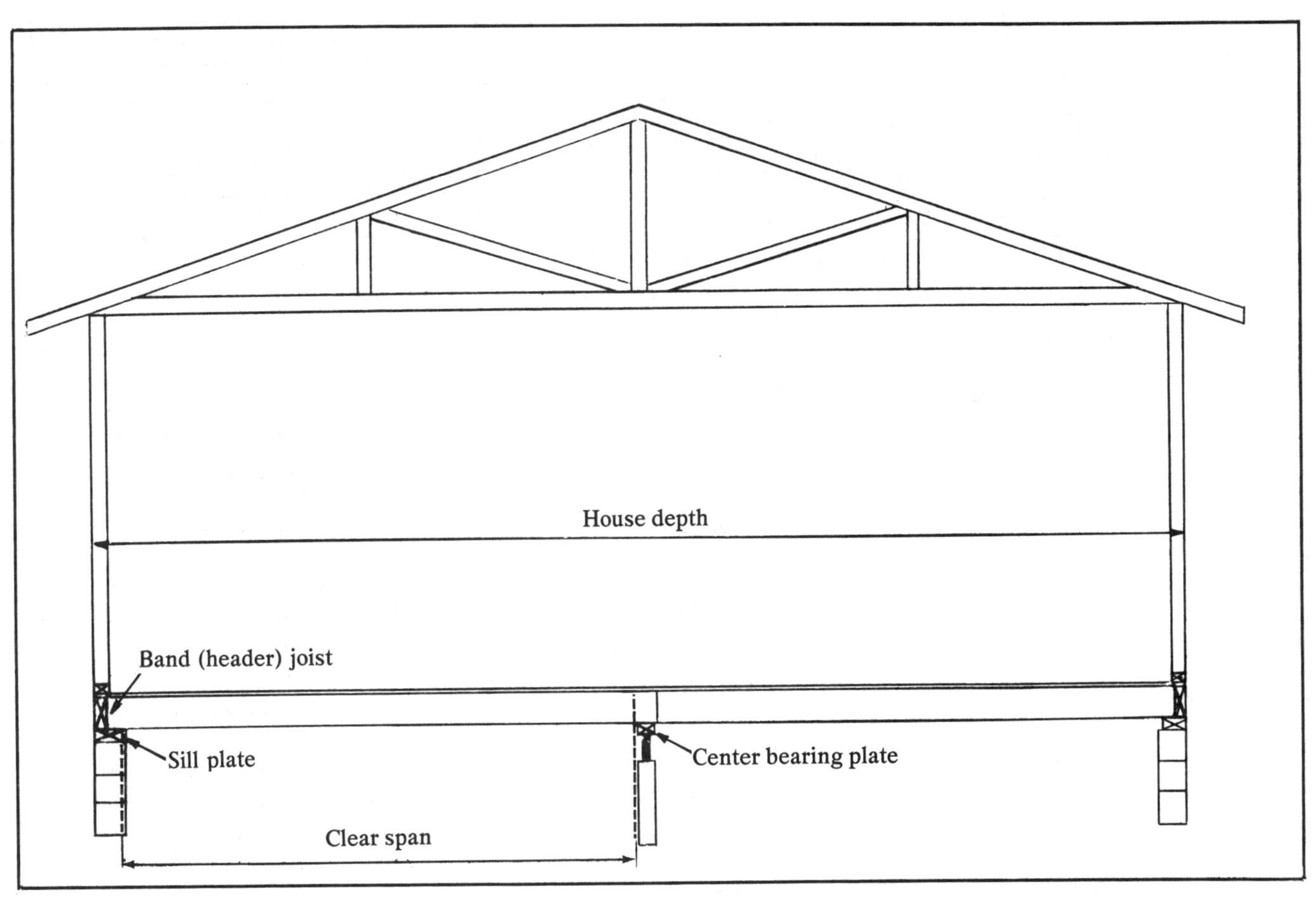

Clear span of joist
Figure 6-6A

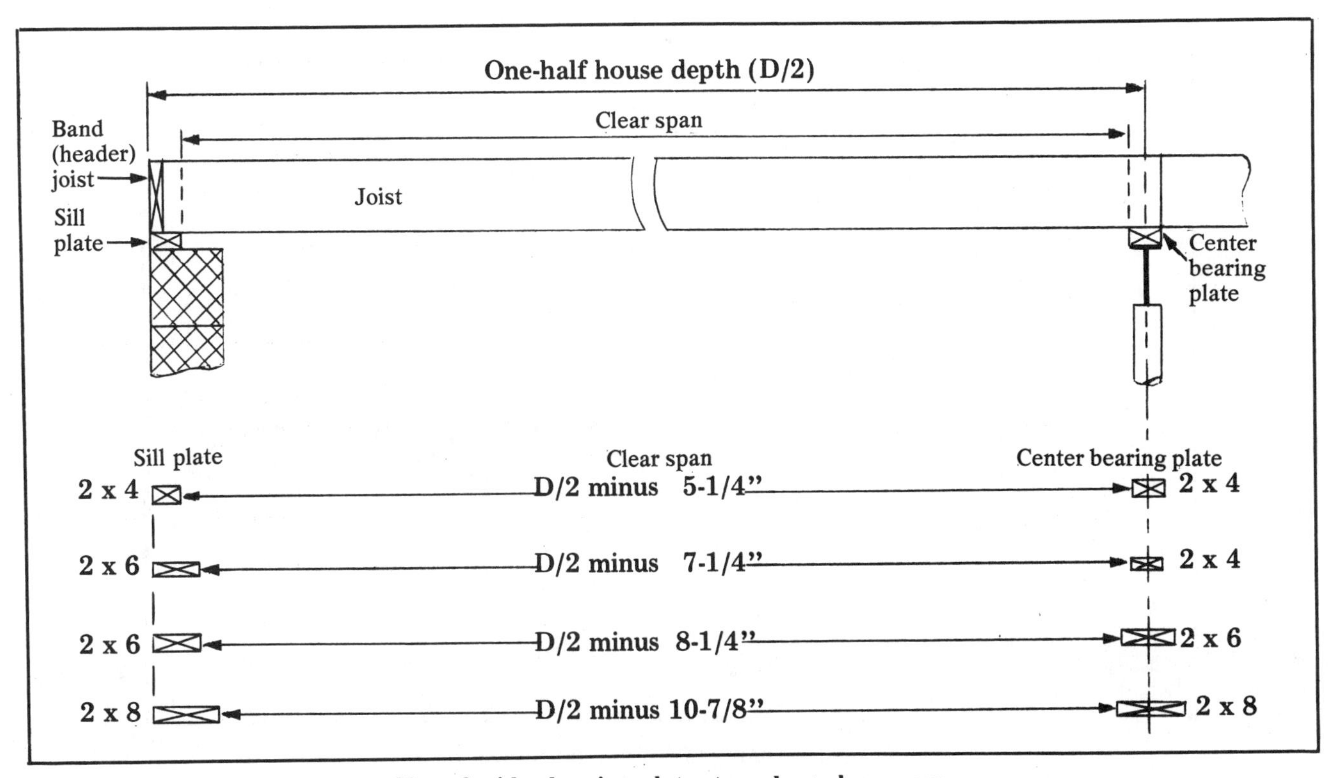

Use of wider bearing plates to reduce clear spans
Figure 6-6B

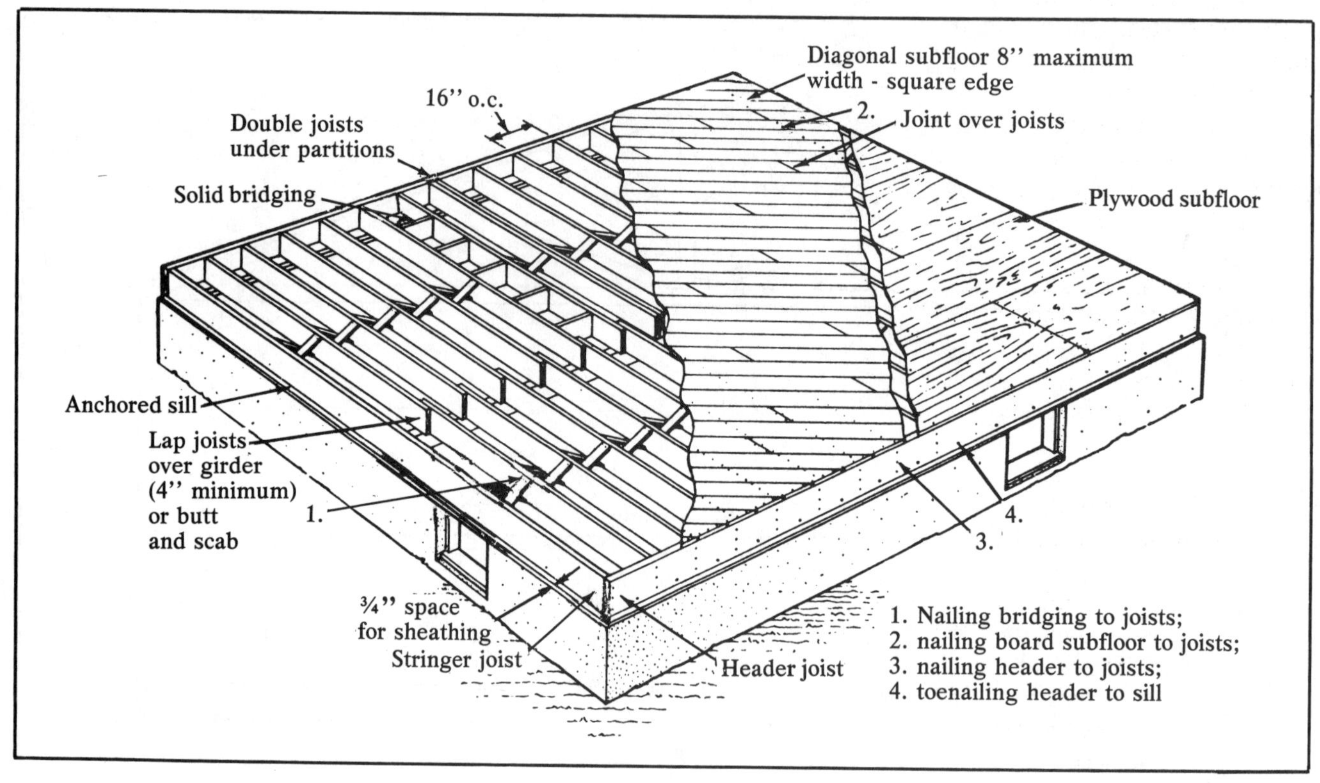

Floor framing
Figure 6-7

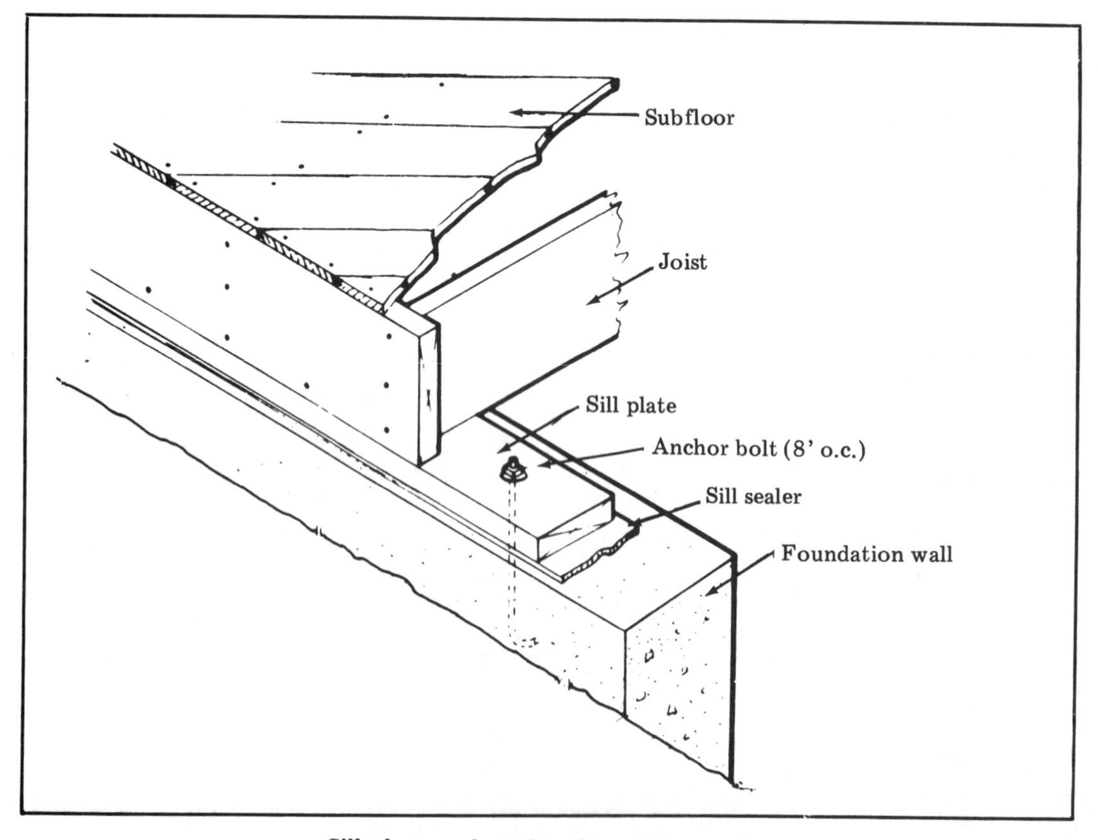

Sill plate anchored to foundation wall
Figure 6-8

Lumber, laid horizontally, will shrink slightly as it dries and ages. The "greener" the lumber, the more it will shrink. The moisture content of material used for floor framing should not exceed 19%. A lower moisture content is preferred. The moisture content will eventually stabilize. But before that happens, some problems can occur. Finish walls and trim can crack as the floor framing shrinks.

To avoid this, use the same total depth of horizontal wood at the outside walls as at the center beams and girders. This keeps the shrinkage equal. Figure 6-10 shows several ways to do this. The ledgers shown are the same thickness as the sill plate.

Figure 6-11 shows a single type box sill plate. The header rests directly on the foundation. The plate upon which the side-wall studs rest is let into the top of the joists.

In Figure 6-12, a double sill plate method is used. This laps one plate over the other at the corners of the building. Note Figure 6-13.

Obviously, these methods use more labor and material and are not recommended where a less costly method will do the job as well. Avoid all unnecessary costs.

Balloon Framing

Balloon framing allows less shrinkage in exterior walls. In many areas, it's the preferred method for full two-story brick or stone veneer houses.

This method uses a 2-inch-thick wood sill on the foundation wall. The joists rest directly on the sill. The studs also bear on the sill and are nailed both to it and the floor joists. The subfloor is laid diagonally (board) or at right angles (plywood or board) to the joists. A fire stop is added between the studs at the floor line. See Figure 6-14. When a diagonal subfloor is used, a nailing member is required between the joists at the wall line.

Horizontal shrinkage is limited to the sill. In two-story framing, the studs extend to the top plate of the second story.

Girder and Post Supports

An open basement requires post-supported girders. A crawl space requires piers and girders. Wood or steel posts are generally used in the basement. Masonry piers are commonly used in the crawl space.

The round steel post can be used to support wood girders or steel beams. The steel post should have a bearing plate at both ends. Figure 6-15 shows anchoring methods for the post.

Wood posts should be pressure treated and at least 6 inches square. A 4 x 6-inch post can be used with a frame wall to conform to the depth of the wall framing. Figure 6-16 shows the proper method of wood-post anchorage.

Both steel beams and wood girders are used in house construction. I-shape and wide-flange beams are the most common steel shapes used.

Wood girders can be built-up on the job, but glue-laminated beams are now available in most areas. They come in various dimensions and lengths. The longer lengths are fabricated with a crown to absorb sag. The crown is always placed up. The laminated beam works well over long spans where a center post is objectionable.

The built-up girder is popular with most builders. It's more economical and can be built in place. The girder is made up of two or more pieces joined over a supporting post. While a 2-piece girder may be nailed from one side with 10d nails, many carpenters prefer to nail it from both sides. Drive two nails at the ends of each piece and stagger the other nails 16 inches apart. On 3-piece girders, use 20d nails and drive from both sides. Stagger them 24 to 32 inches apart.

The ends of girders should bear at least 4 inches on the masonry walls or pilasters. When you use untreated lumber, provide a 1/2-inch air space at each end and side of the wood girder within the masonry. See Figure 6-17. In termite-infested areas, line the girder pockets with metal.

Girders support a lot of weight. They must be strong enough to handle the job. If a girder sags too much, the floor and ceiling will sag and doors will bind.

Use joist hangers to control shrinkage at the inner beam. They also provide greater headroom in the basement. See Figure 6-18. Joist hangers can replace a supporting ledger.

Create a continuous horizontal tie between exterior walls by nailing notched joists together. Or use either a connecting scab at each pair of joists or a steel strap, as shown in Figure 6-10. Leave a small space above the girder for joist shrinkage.

The subfloor can also be used as a tie-between. A sheet of plywood centered over the girder or boards laid diagonally provide an excellent tie. The starting course of plywood subfloor is centered over the girder and laid to the outsides.

Floor joist framing on steel beams is illustrated in Figure 6-19. A 2 x 4 is sometimes bolted on top of the beam as a base for joists. Each joist is then nailed in place at the girder. The 2 x 4 provides the same thickness of horizontal wood at the girder as at the wall. Keep the top of the steel beam even with the top of the foundation wall.

Figure 6-20 shows two other methods for running joists to steel. Where joists bear directly on the flange of the beam, as in 6-20 B, wood blocking between the joists at the beam flange will prevent joist twisting.

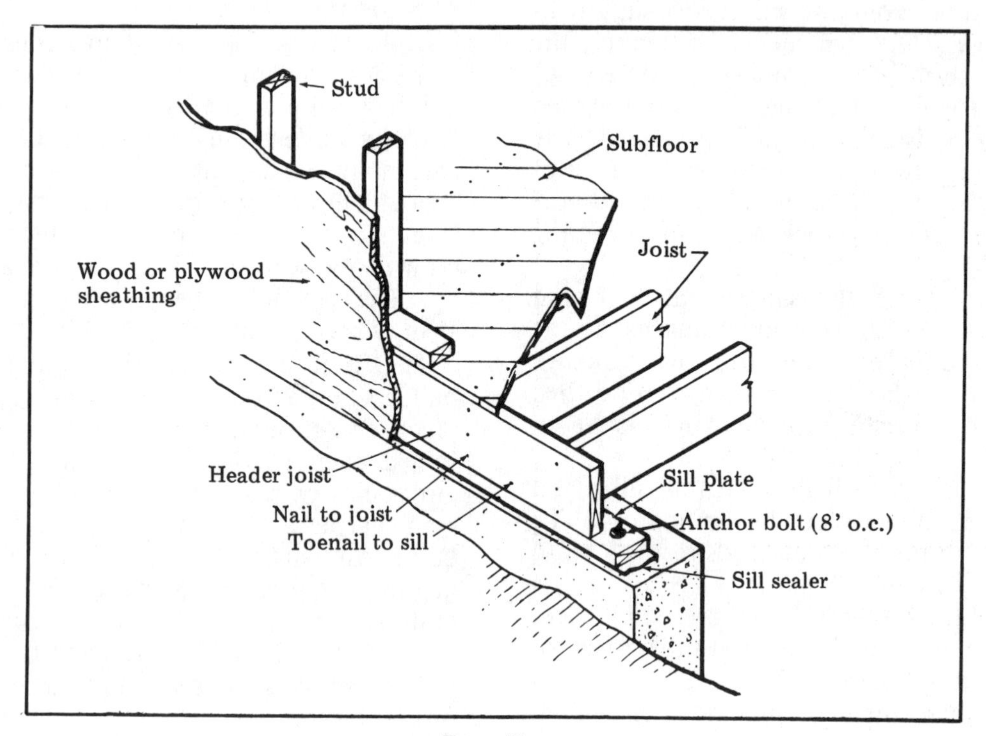

Box sill
Figure 6-9

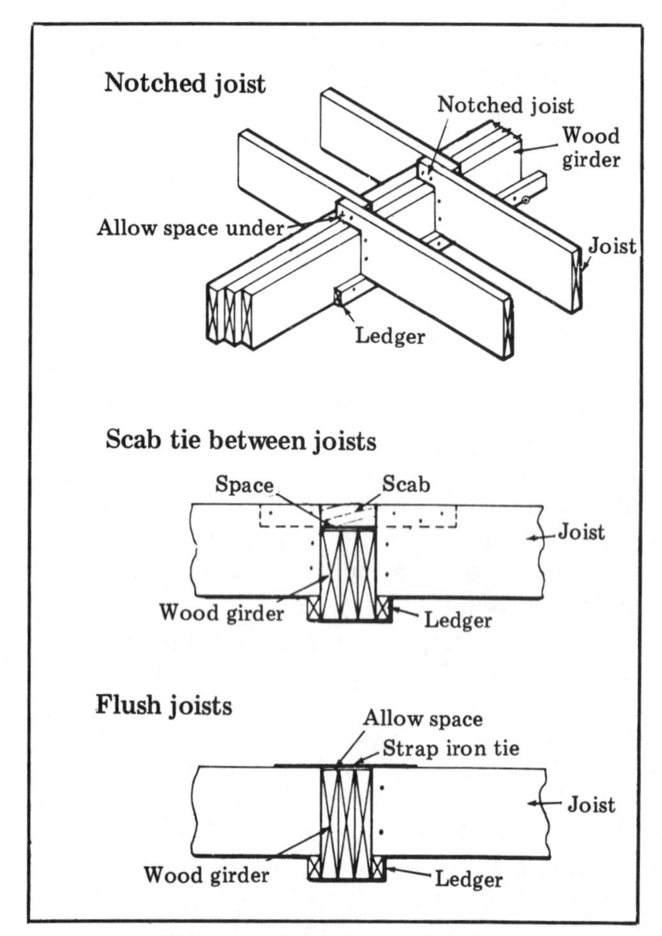

Ledger on center wood girder
Figure 6-10

Floor Joists

Floor joists must be big enough to provide a rigid floor. They must carry normal loads without excess deflection. Don't use lumber with warps and bows as floor joists. Be sure the bearing surfaces of the joists lie squarely on the sill and girder. The tops of the joists must form a straight and level floor line.

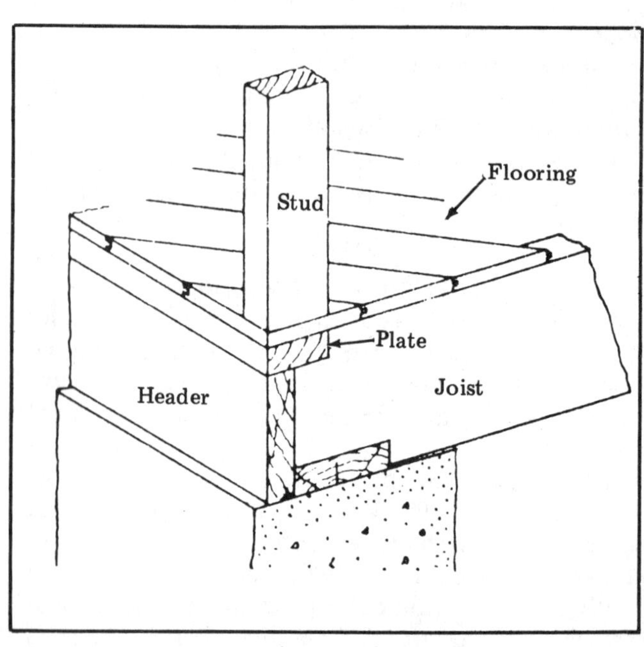

Alternate box sill
Figure 6-11

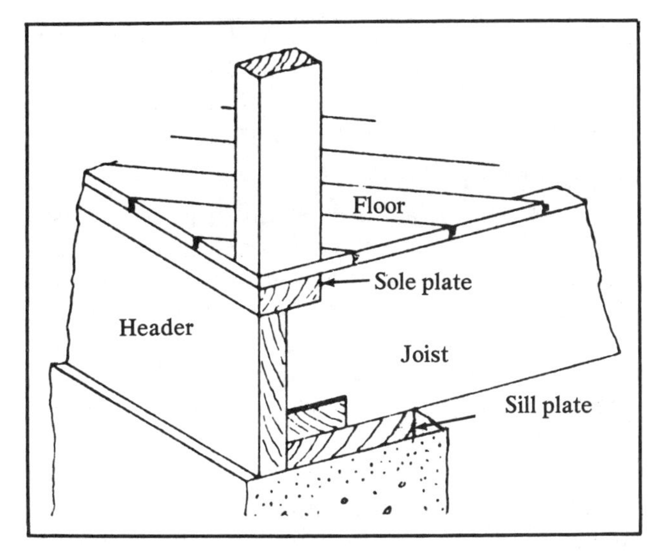

Double sill plate
Figure 6-12

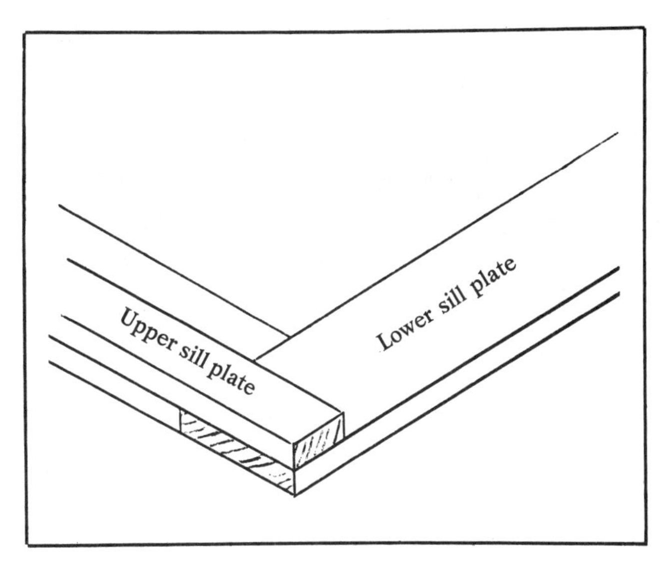

Lap of double sill plate
Figure 6-13

Floor joists can be 2 x 8, 2 x 10, or 2 x 12 lumber. The size to use depends on the species, span, spacing and grade of lumber. Douglas fir "standard" grade and Southern yellow pine (Number 2 or Number 2KD) are good choices (See Figure 6-5.)

Lumber with excessive edgewise bow should not be used. A 1/4-inch bow over a 10-foot span is acceptable. Always put the crown up. But watch out for large edge knots which would weaken the joists. Place any knots on the top side of the joists.

The header joist is nailed to each floor joist with three 16d nails and toenailed to the sill with 10d nails 16 inches apart (See Figure 6-9). The stringer joists are also toenailed to the sill in a like manner. Each joist is lapped at the girder (See Figure 6-21) and toenailed to this girder with two 10d nails (or three 8d nails). Then nail the joists to each other at the overlap with three or four 16d nails to lock the floor system in.

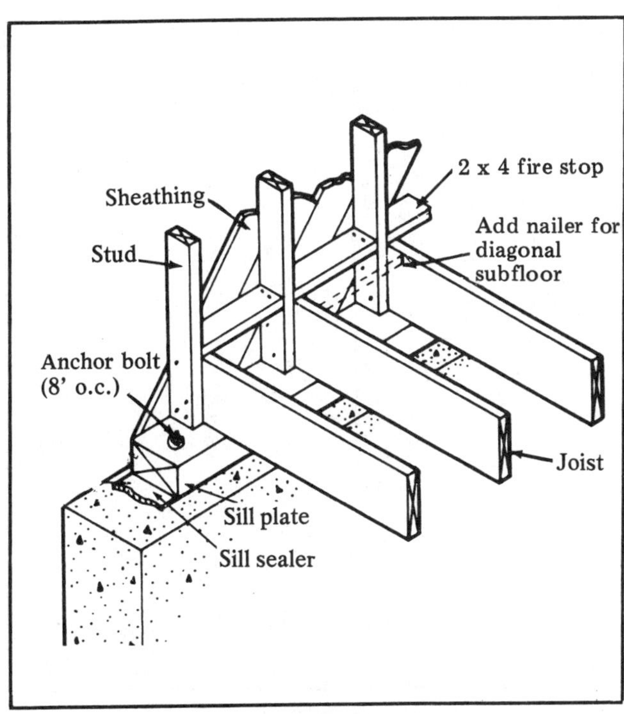

Sill for balloon framing
Figure 6-14

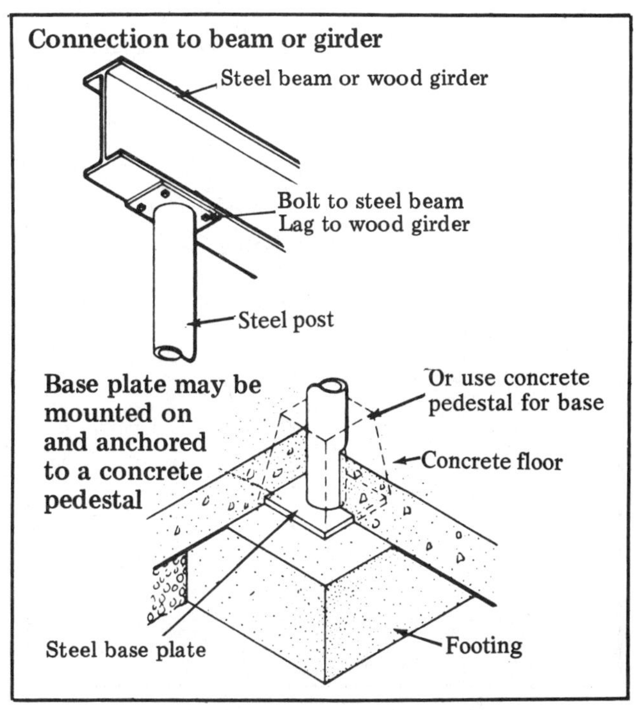

Steel post for wood or steel girder
Figure 6-15

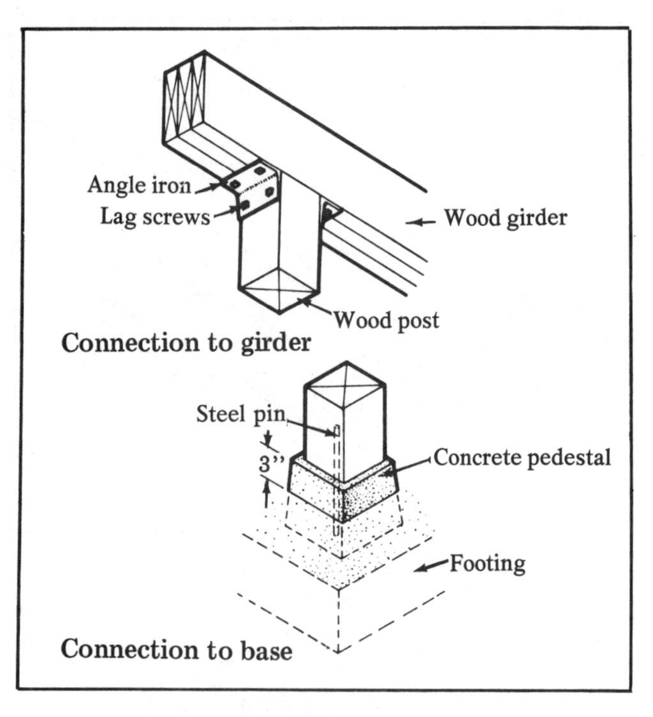

Wood post for wood girder
Figure 6-16

Floor joist spacing is generally 16 inches from center to center, with double joists under partitions that are parallel with the joist run. Figure 6-22 shows a floor plan for joist layout.

In-line Floor Systems

Figure 6-23 shows in-line floor joists. This is recommended for modular floor sheathing applications. There is less cutting and waste of plywood.

Preassembled in-line joists are available at some lumber yards. They're available in different dimensions and lengths. On-site assembly may be cheaper, however. Figure 6-23 shows in-line construction.

Bridging

Bridging between wood joists provides rigidity. It's questionable whether the gain is worth the expense. Some building codes require cross-bridging or solid bridging, as in Figure 6-24A. Solid bridging is often used under partition walls which are parallel to the joist run. (See Figure 6-7.)

Unless bridging is required, leave it out. It doesn't add that much to floor strength. It's an unnecessary cost for most floors.

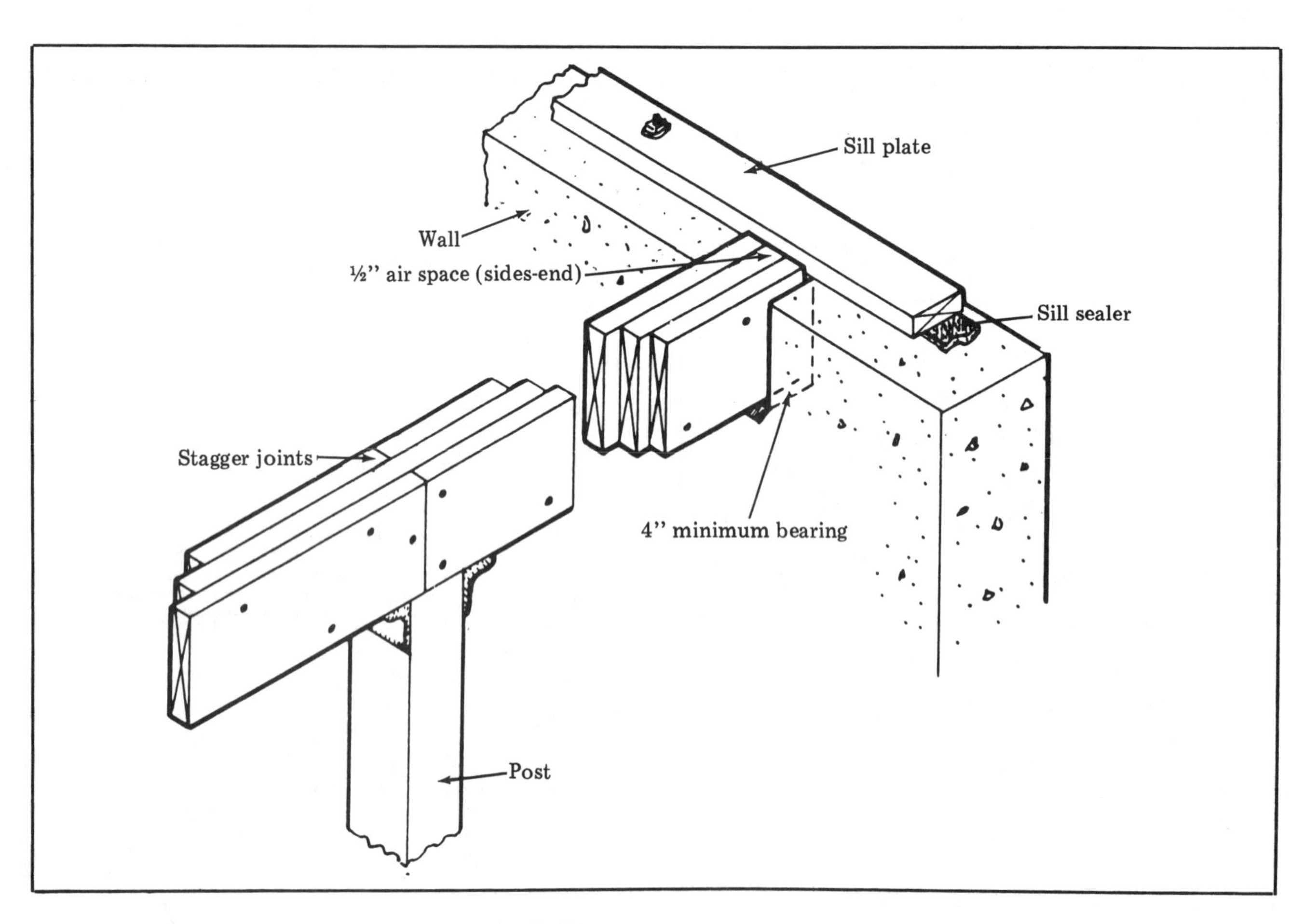

Built-up wood girder
Figure 6-17

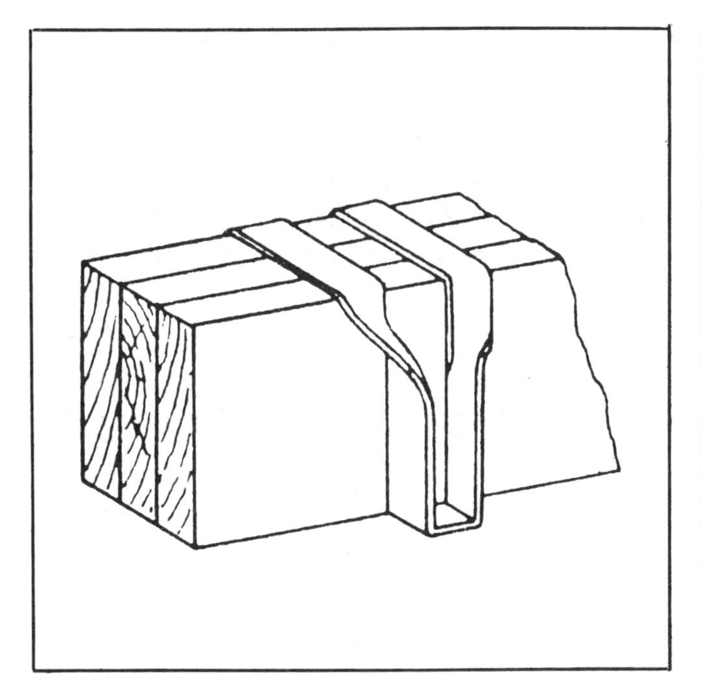

Girder with joist hanger
Figure 6-18

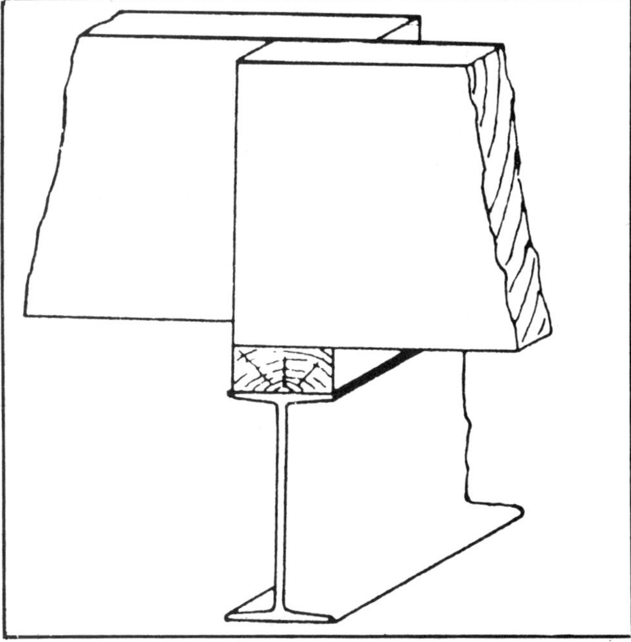

Steel girder
Figure 6-19

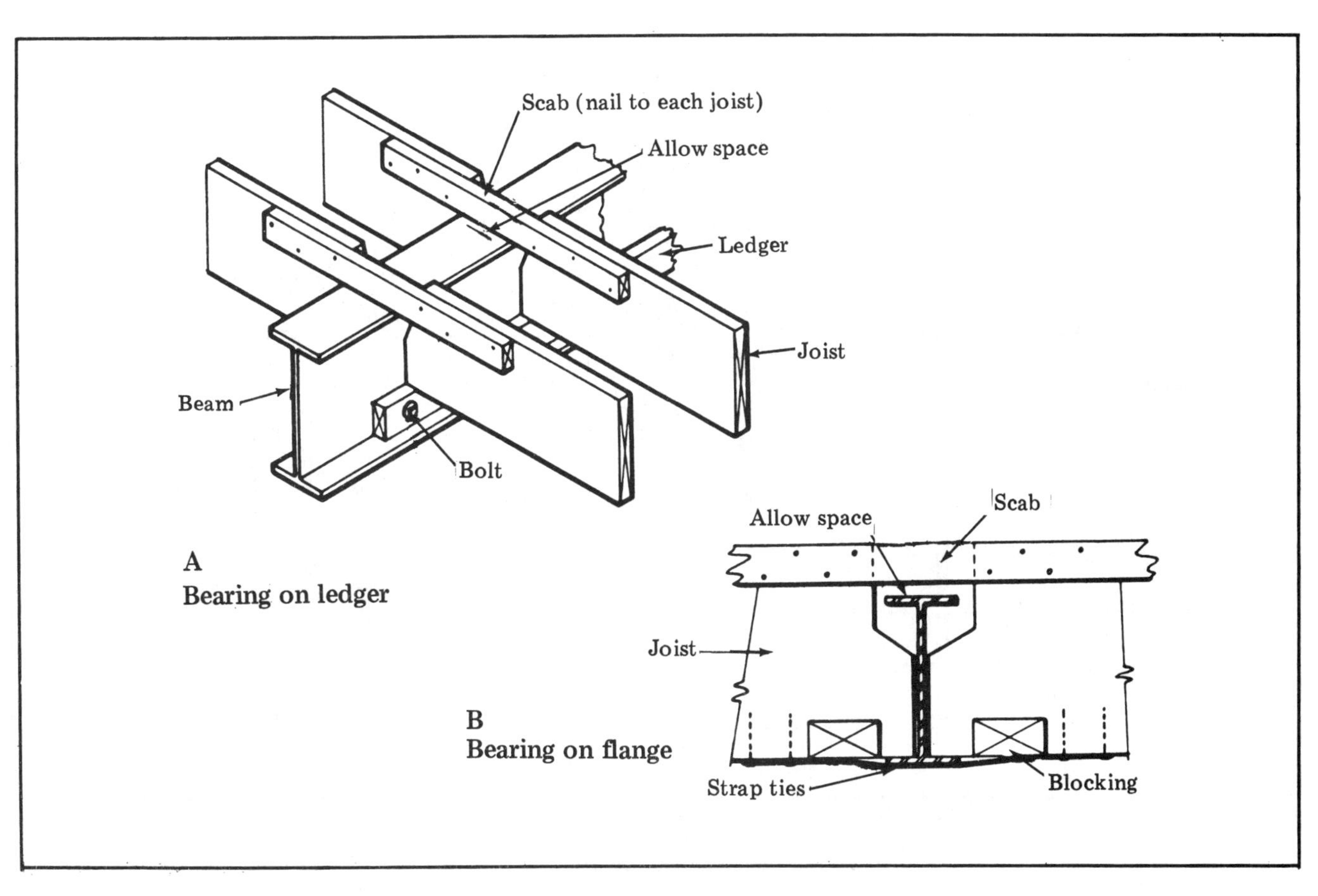

Steel beam and joists
Figure 6-20

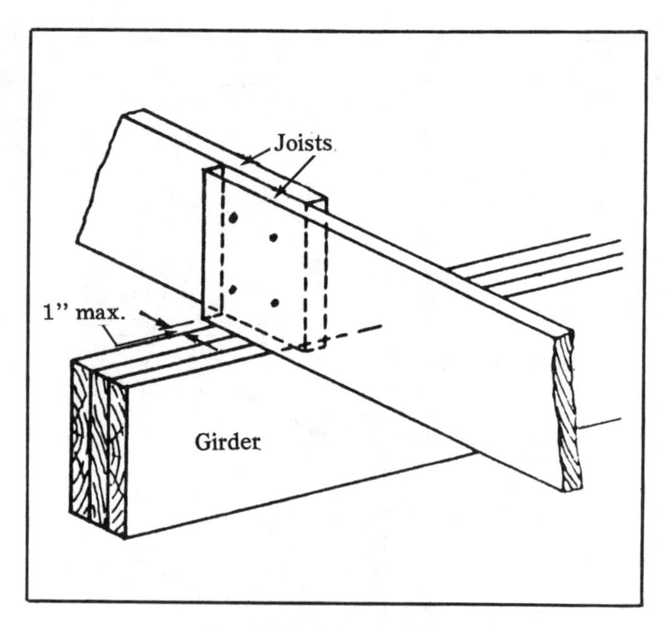

Lap of joists at girder
Figure 6-21

If bridging is needed, consider metal cross-bridging. It can save on manhours and isn't much more expensive than wood.

With cross-bridging, you usually nail the bottom of each bridge after installing the subfloor, as shown in Figure 6-24B. This allows the joists to be "pulled" into place. Bridging is generally installed at intervals of not more than 8 feet along the joist's span. Figure 6-25 shows metal cross-bridging in place.

Subfloor

The subfloor may be either boards or plywood. Plywood is a labor saver and makes a smooth, strong floor. It does not have the shrinkage problem of boards.

Boards may be applied diagonally or at right angles to the joists. Diagonal application permits the installation of finish strip flooring at right angles to the joists. This cuts down on damage due to shrinkage of the subfloor.

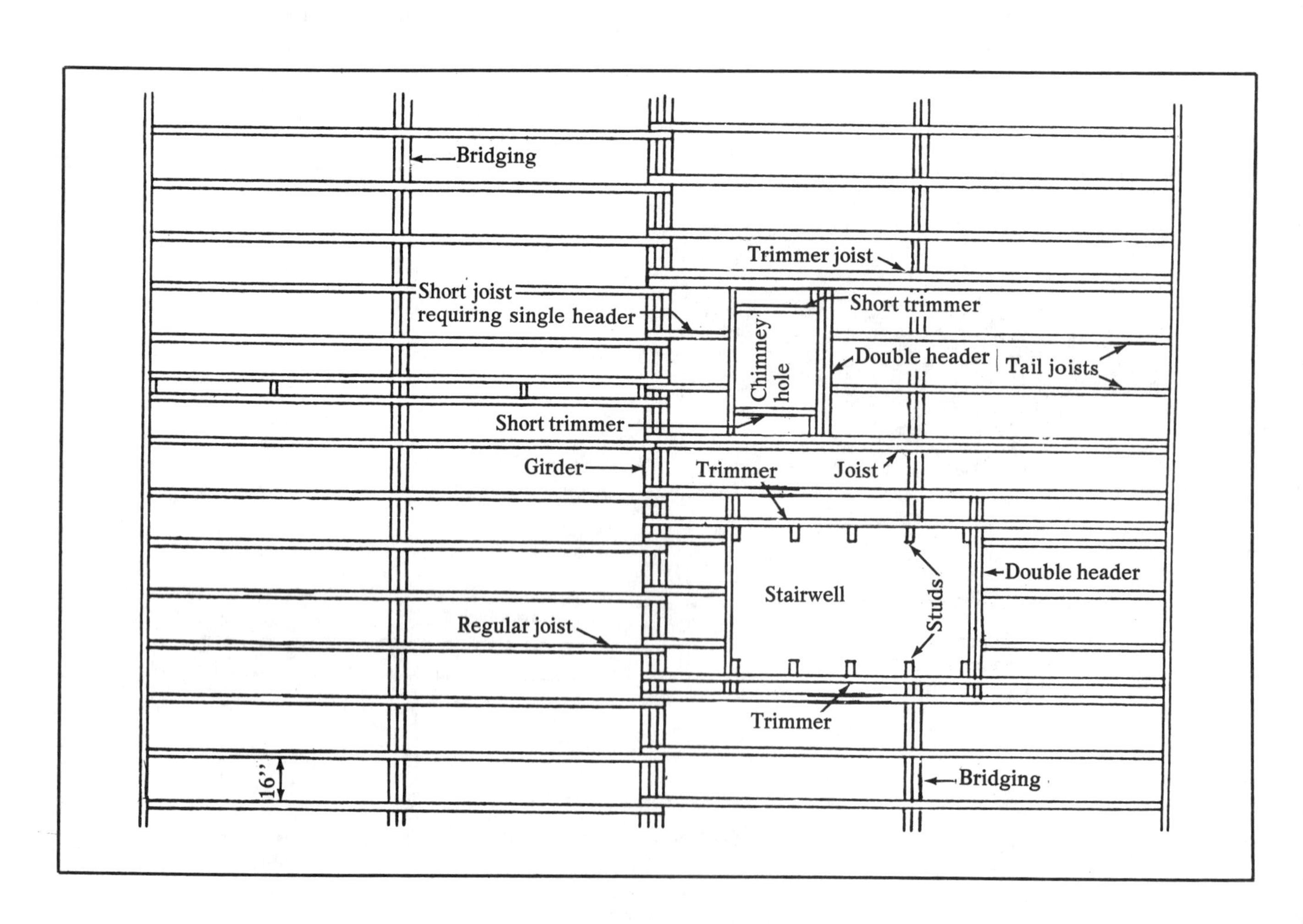

Floor plan for joist layout
Figure 6-22

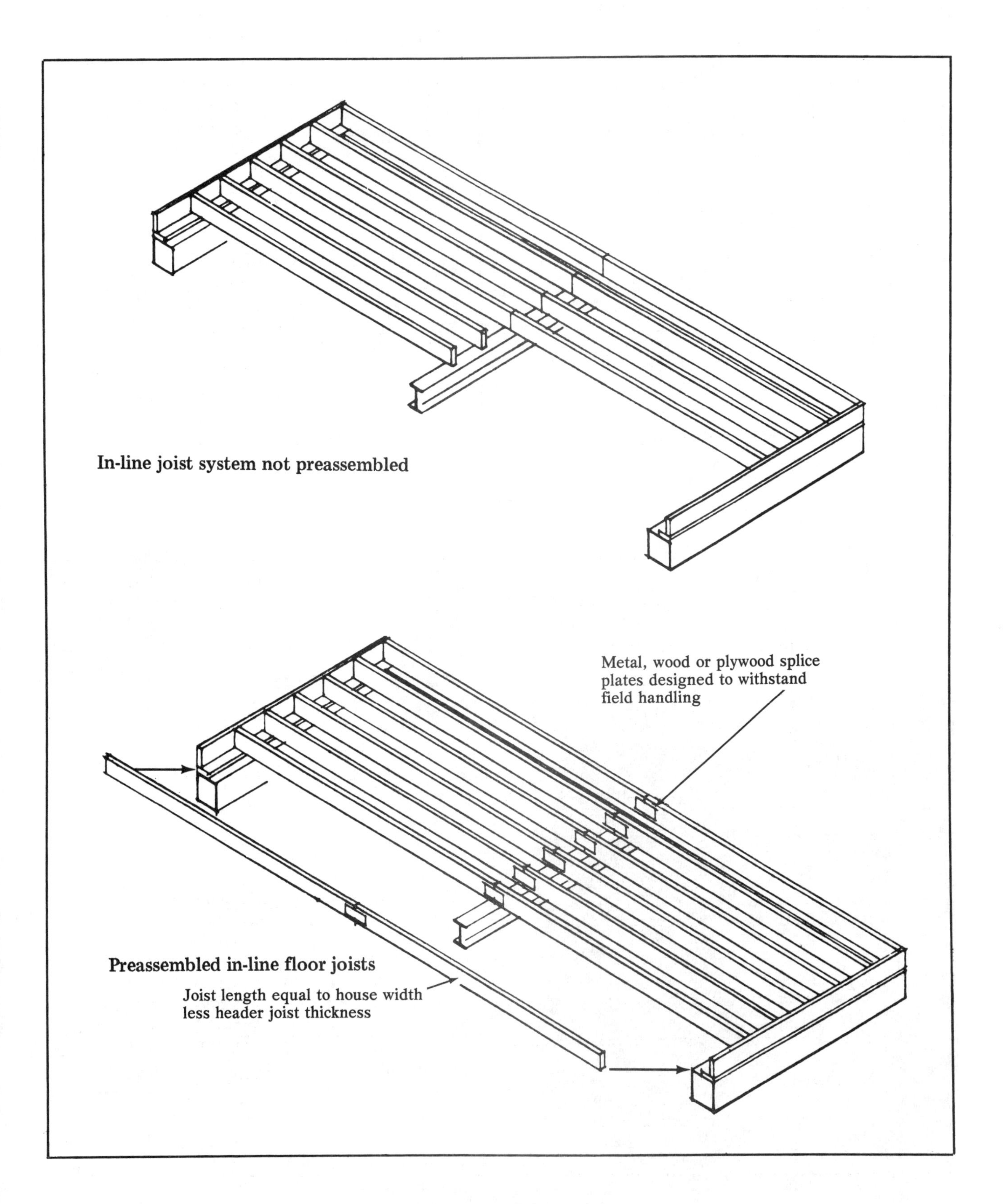

In-line floor joist systems
Figure 6-23

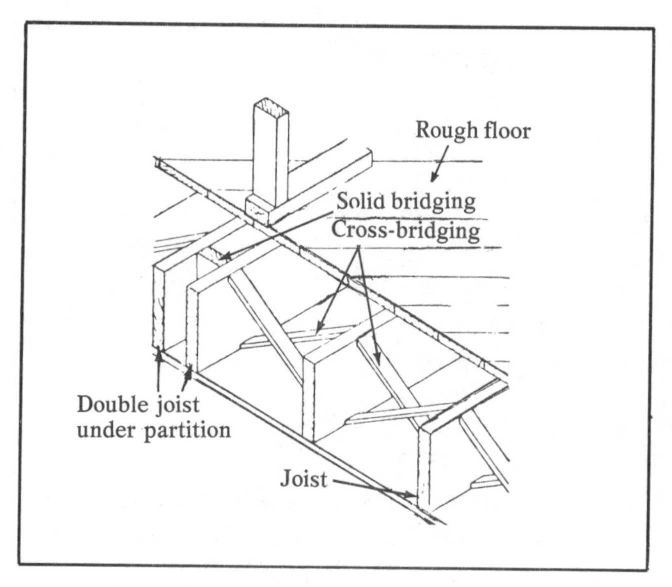

Solid and cross-bridging
Figure 6-24A

How bridging is nailed
Figure 6-24B

Figure 6-26 shows strip flooring being laid over a diagonal board subfloor.

Board subflooring is nailed to each joist with two 8d nails for widths under 8 inches and three 8d nails for 8-inch widths.

Metal cross-bridging
Figure 6-25

Laying strip flooring over diagonal board subfloor
Figure 6-26

Estimating board subflooring is a little more difficult. For 1 x 6-inch widths laid at right angles to the joists, add about 20% to the board footage. For 1 x 8-inch boards, add about 16%. Unless floor openings are very large, don't deduct for the opening. If the floor is laid diagonally, add about 25% for 6-inch boards and 22% for 8-inch boards.

Plywood subflooring reduces waste and saves manhours. Figure 6-27 describes the panels used for subflooring.

Figure 6-28 explains what the American Plywood Association (APA) Registered Trademarks mean. The span rating "32/16"

Grade Designation	Description & Common Uses	Typical Trademarks	Most Common Thicknesses (ins.)			
			3/8	1/2	5/8	3/4
APA rated sheathing EXP 1 or 2	Specially designed for subflooring and wall and roof sheathing, but can also be used for a broad range of other construction and applications. Can be manufactured as conventional veneered plywood, as a composite, or as a nonveneered panel for special engineered applications, including high load requirements and certain industrial uses, veneered panels conforming to PS1 may be required. Specify Exposure 1 when construction delays are anticipated	APA RATED SHEATHING 32/16 1/2 INCH SIZED FOR SPACING EXPOSURE 1 000 PS 1-74 C-D INT/EXT GLUE NRB-108	•	•	•	•
APA rated Sturd-I-Floor EXP 1 or 2	For combination subfloor-underlayment. Provides smooth surface for application of resilient floor covering and possesses high concentrated and impact load resistance. Can be manufactured as conventional veneered plywood, as a composite. Available square edge or tongue-and-groove. Specify Exposure 1 when construction delays are anticipated	APA RATED STURD-I-FLOOR 24 oc 23/32 INCH SIZED FOR SPACING T&G NET WIDTH 47-1/2 EXPOSURE 1 000 INT/EXT GLUE NRB-108 FHA-UM-66			• 19/32	• 23/32

Performance-rated panels for subflooring
Figure 6-27

means the rafter span should not exceed 32 inches center to center, and the floor joist span should not exceed 16 inches center to center.

Plywood should be installed with the grain direction of the outer plies at a right angle to the joists. Nail 1/2- and 3/4-inch plywood to the joists at each bearing with 8d common or 7d threaded nails. Space the nails 6 inches apart along all edges and 10 inches along intermediate members. When plywood serves as both subfloor and underlayment, space nails 6 to 7 inches apart at all joists and blocking. Use 8d or 9d common nails or 7d or 8d threaded nails.

Don't jam plywood subfloor panels together when they are laid. Leave a 1/8-inch gap at the edges and a 1/16-inch gap at the ends.

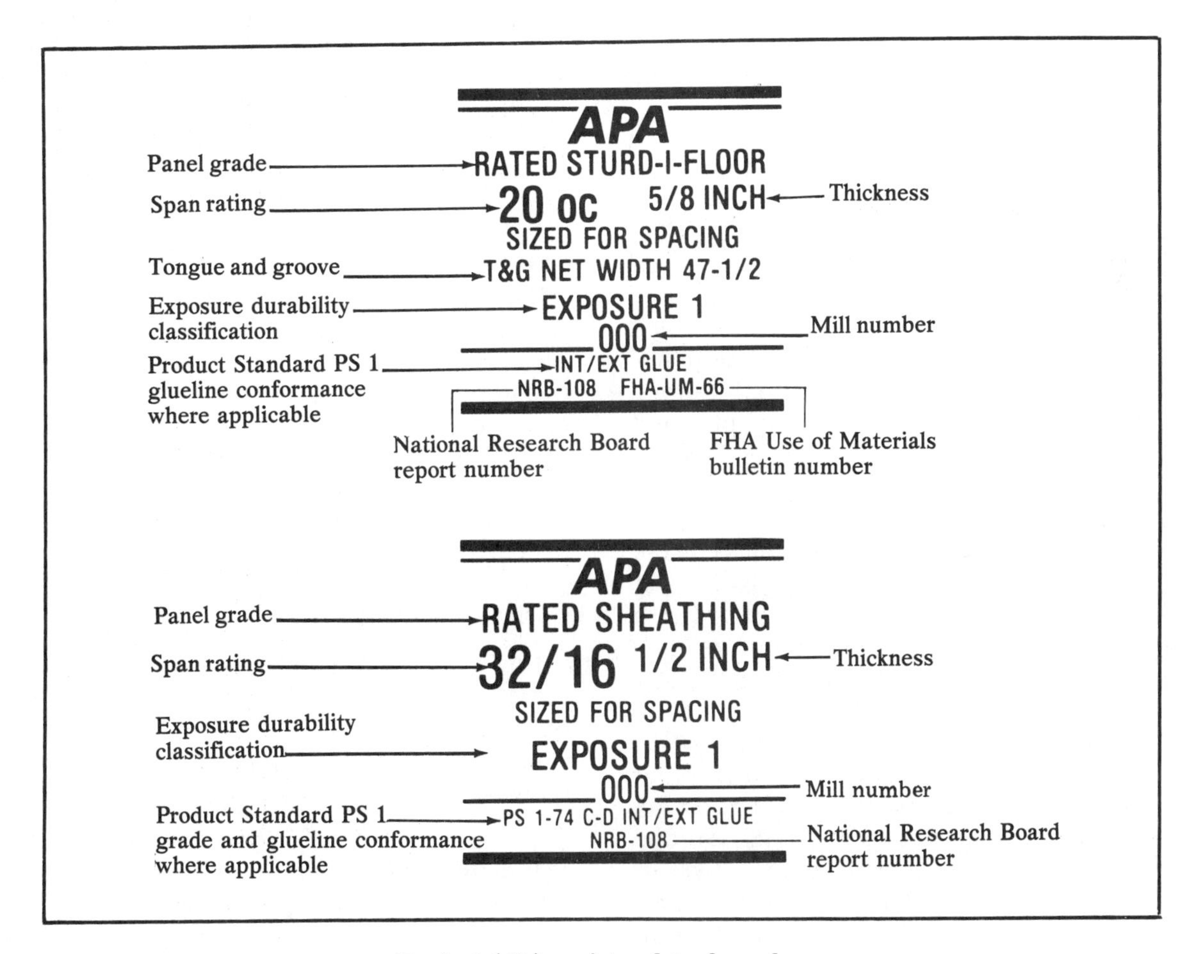

Typical APA registered trademarks
Figure 6-28

Plywood subflooring laid with staggered end joints
Figure 6-29

Plywood glued to floor joists makes a stronger floor. The plywood and joists become a solid structural member and offer greater resistance to deflection. This makes longer spans possible for a given size floor joist.

When glue is used, the plywood is still nailed firmly in place. This glue-nailed installation improves the structural quality of the floor. Figure 6-29 shows one method of installing plywood subflooring. Note the staggering of the end joints.

APA-Rated Sturd-I-Floor

APA-rated Sturd-I-Floor is a span-rated APA floor designed specifically for use in single-layer floor construction beneath nonstructural finish flooring. See Figure 6-30. Sturd-I-Floor reduces costs because it combines subfloor and underlayment. It's accepted by the FHA and most codes. And it's easy to use because the span rating (the maximum recommended spacing between floor joists) is stamped on each panel back. Panels are manufactured with span ratings of 16, 20, 24, and 48 inches. These measurements assume two or more spans with the long dimension of the panel across the supports. APA-rated Sturd-I-Floor panels will support at least 75 psf live load at maximum span. Sturd-I-Floor 48 oc (2-4-2) will support only 65 psf live load.

Sturd-I-Floor 48 oc (2-4-1)

APA-rated Sturd-I-Floor 48 oc incorporates all the features of panels formerly designated 2-4-1. Application recommendations remain the same. You install panels over 2-inch joists spaced 32 inches on center (Figure 6-31), or over 4-inch girders spaced 48 inches on center (Figure 6-32). For the 48 oc method, supports may be 2-inch joists spiked together, 4-inch lumber, lightweight steel beams, or wood-steel floor trusses. Girders of doubled 2-inch members should have their top edges flush so the panels butt together smoothly.

For a low profile with supports 48 inches on center, set the beams in foundation pockets or on posts supported by footings. The panels should bear directly on the sill. The 4-inch-thick wood girders should be air dried or set higher than the sill to allow for shrinkage.

In hallways and other heavy traffic areas, greater stiffness in the floor may be needed. Straight or diagonal blocking will increase stiffness considerably.

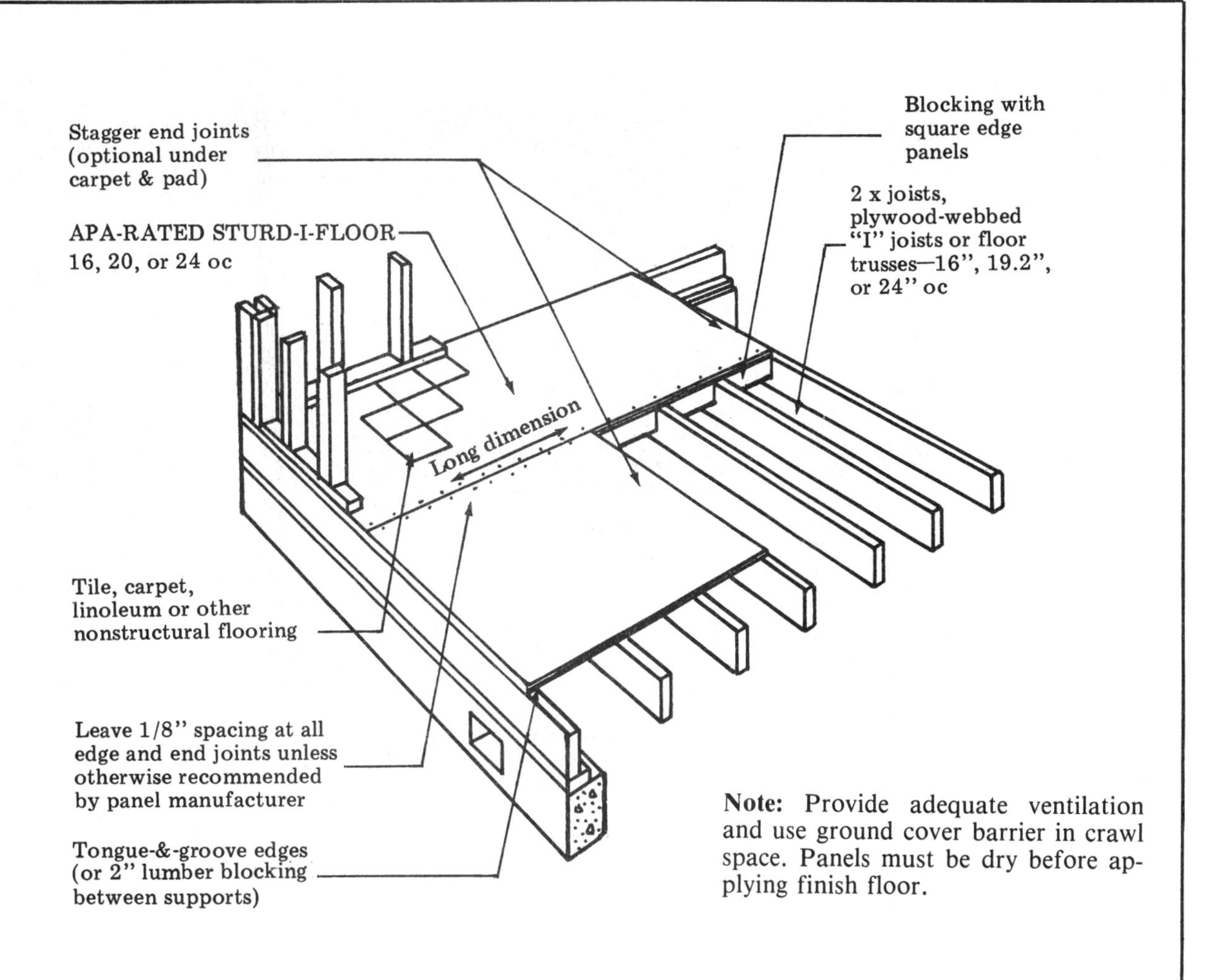

Note: Provide adequate ventilation and use ground cover barrier in crawl space. Panels must be dry before applying finish floor.

A) APA-Rated Sturd-I-Floor panels include plywood manufactured under the general provisions of U.S. Product Standard PS-1/ANSI A199.1 for Construction and Industrial Plywood. However, for panels with Span Ratings of 24 or less, wood-based materials other than veneer may be used either as inner layers or for the entire panel, provided the panels satisfy APA performance standards for applicable mechanical and physical properties and for durability.

B) The National Research Board is sponsored jointly by the three model code organizations — the Building Officials and Code Administrators International, promulgators of the Basic Building Code; the International Conference of Building Officials, promulgators of the Uniform Building Code; and the Southern Building Code Congress International, promulgators of the Standard Building Code. See National Research Board Report No. NRB-108. The NRB report acknowledges that Sturd-I-Floor panels may be used over the spans indicated in the trademark. It also recognizes certain Sturd-I-Floor panels in lieu of the same thickness of Underlayment grade plywood in over 30 floor-ceiling (or roof-ceiling) assemblies listed in the Underwriters Laboratories Fire Resistance Index. These assemblies are generally accepted by building code officials. FHA recognition of Sturd-I-Floor is contained in its Use of Material Bulletin FHA-UM-66. Copies of the NRB and FHA acceptances are available upon request from APA.

APA-rated Sturd-I-Floor 16, 20, and 24 oc
Figure 6-30

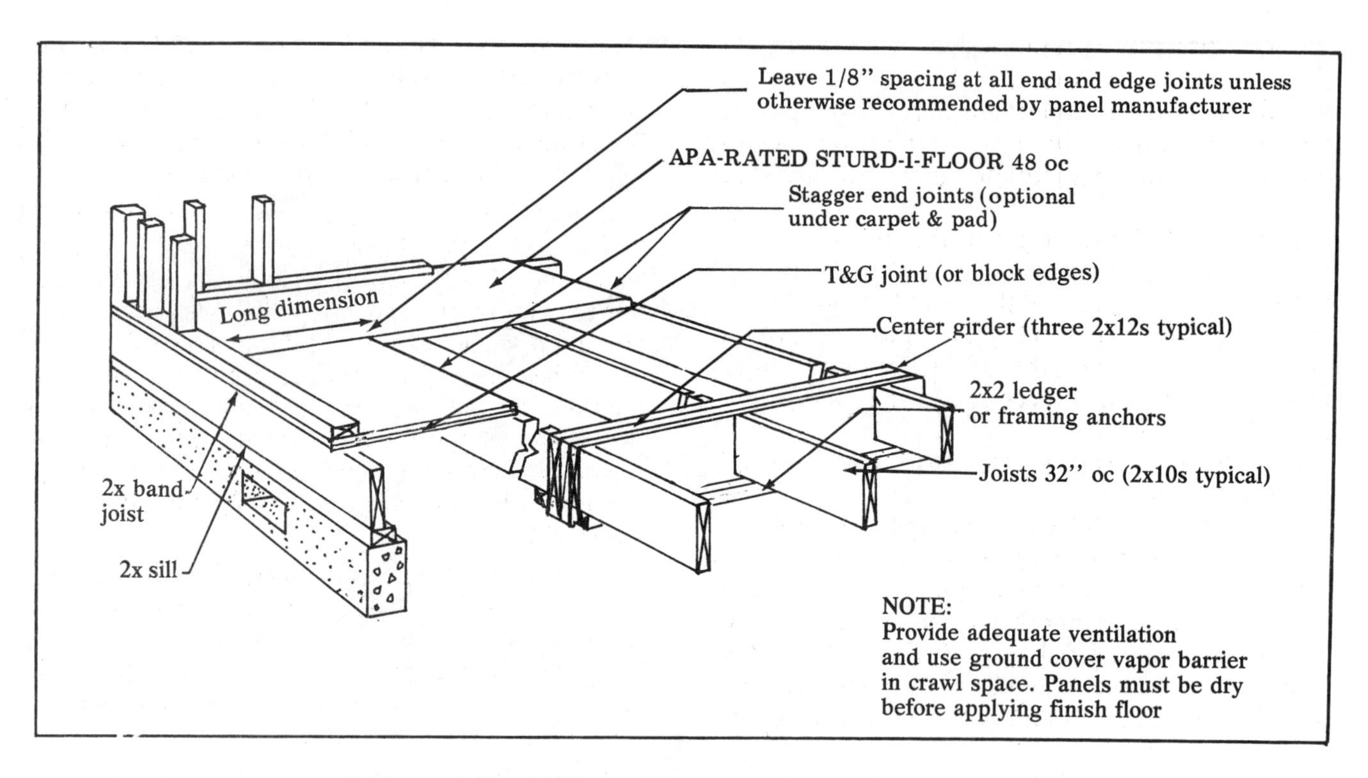

APA-rated Sturd-I-Floor 48 oc (2-4-1) (over supports 32" oc)
Figure 6-31

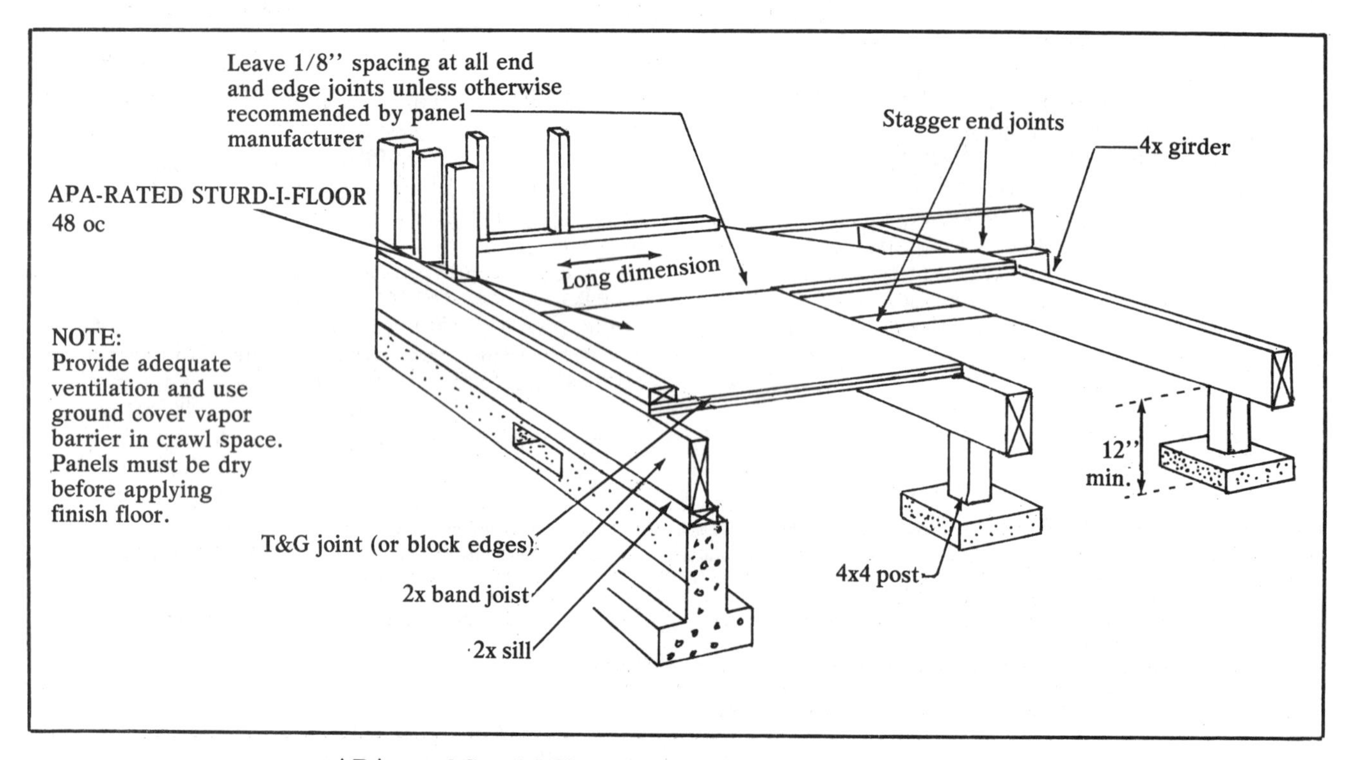

APA-rated Sturd-I-Floor 48 oc (2-4-1) (over supports 48" oc)
Figure 6-32

Glue-nailing is recommended for plywood and composite Sturd-I-Floor panels. See Figure 6-33. When the floor members are dry, make sure the fasteners are flush with or below the floor surface, especially if you plan to put down tile or sheet flooring.

Nails should be set if green framing will present nail popping problems on drying. *Do not* fill nail holes. Fill edge joints and thoroughly sand any surface roughness, particularly at joints and around nails, before applying the finish flooring. Always protect smooth panel faces and tongue and groove edges from damage.

If permanent exposure to moisture is expected, use exterior panels.

Although Sturd-I-Floor is suitable for direct application of finish flooring, an additional underlayment is often used under tile or linoleum. This added layer restores a smooth surface over panels that may have been scuffed or roughened. Underlayment runs up the cost. So install the panels correctly and guard against damage during construction.

Underlayment

Underlayment is installed later in the construction process. Particleboard, tempered hardboard or plywood can be used as underlayment. Plywood underlayment grades have a solid touch-sanded surface for direct application of nonstructural finish flooring. They also have special inner-ply construction for resistance to dents and punctures from concentrated loads. Plywood underlayment is dimensionally stable and eliminates excessive swelling and buckling when applied correctly.

Install plywood underlayment immediately before laying the finish floor. Protect it against water and physical damage before its application. For maximum stiffness, place the face grain across the supports and place the end joints over the framing. When the floor members are dry, make sure the fasteners are flush with or below the floor surface. Fill and thoroughly sand the edge joints of the panels before applying resilient floor covering. Fill any other damaged or open areas and then sand lightly.

The plywood underlayment needed to bridge an uneven floor will depend on roughness and the loads applied. Although a minimum 11/32-inch thickness is recommended, 1/4-inch plywood underlayment may also be enough over smooth subfloors. See Figure 6-34.

Where floors may be subject to unusual moisture, use panels with exterior glue or APA C-C Plugged Exterior. APA C-D Plugged Interior is not an adequate substitute since it doesn't have equivalent dent resistance. Figure 6-35 shows subfloor and underlayment application techniques.

Span rating (maximum joist spacing) (inch)	Panel thickness[b] (inch)	Fastening: Glue-Nailed[c] Nail size and type	Fastening: Glue-Nailed[c] Spacing (inch) Panel edges	Fastening: Glue-Nailed[c] Spacing (inch) Intermediate	Fastening: Nailed-Only Nail size and type	Fastening: Nailed-Only Spacing (inch) Panel edges	Fastening: Nailed-Only Spacing (inch) Intermediate
16	19/32, 5/8	6d deformed-shank[d]	12	12	6d deformed-shank	6	10
20	19/32, 5/8, 23/32, 3/4	6d deformed-shank[d]	12	12	6d deformed-shank	6	10
24	23/32, 3/4	6d deformed-shank[d]	12	12	6d deformed-shank	6	10
24	7/8	8d deformed-shank[d]	12	12	8d deformed-shank	6	10
48 (2-4-1)	1 1/8	8d deformed-shank[e]	6	(f)	8d deformed-shank[e]	6	10

(a) Special conditions may impose heavy traffic and concentrated loads that require construction in excess of the minimums shown.
(b) As indicated above, panels in a given thickness may be manufactured in more than one span rating. Panels with a span rating greater than the actual joist spacing may be substituted for panels of the same thickness with a span rating matching the actual joist spacing. For example, 19/32-inch-thick Sturd-I-Floor 20 oc may be substituted for 19/32-inch-thick Sturd-I-Floor 16 oc over joists 16 inches on center.
(c) Use only adhesives conforming to APA Specification AFG-01 with plywood and composite panels, applied in accordance with the manufacturer's recommendation.
(d) 8d common nails may be substituted if deformed-shank nails are not available.
(e) 10d common nails may be substituted with 1 1/8-inch panels if supports are well seasoned.
(f) Space nails 6 inches for 48-inch spans and 10 inches for 32-inch spans.

APA-Rated Sturd-I-Floor[a]
Figure 6-33

Plywood grades[a] and species group	Application	Minimum plywood thickness (inch)	Fastener size and type	Fastener Spacing (in)[b] Panel edges	Intermediate
Groups 1, 2, 3, 4, 5 APA Underlayment Int (with interior or exterior glue)	Over smooth subfloor	¼	18 ga. staples or 3d ring-shank nails[c][d]	3	6 each way
APA Underlayment Ext APA C-C plugged Ext	Over lumber subfloor or other uneven surfaces	11/32 11/32	16 ga. staples[c]	3	6 each way
			3d ring-shank nails[d]	6	8 each way
Same grades as above, but Group 1 only	Over lumber floor up to 4" wide. Face grain must be perpendicular to boards.	¼	18 ga. staples or 3d ring-shank nails[c][d]	3	6 each way

(a) When thicker underlayment is desired, APA-RATED STURD-I-FLOOR may be specified. For ¼-inch thickness, A-C EXT may be substituted.

(b) If green framing is used, space fasteners so they do not penetrate framing.

(c) Use 16 gauge staples for 11/32-inch and thicker plywood. Crown width ⅜-inch for 16 gauge, 3/16-inch for 18 gauge staples, length sufficient to penetrate completely through, or at least ⅝-inch into, subflooring.

(d) Use 3d ring-shank nail also for ½-inch panels and 4d ring-shank nail for ⅝-inch or ¾-inch panels.

APA plywood underlayment
Figure 6-34

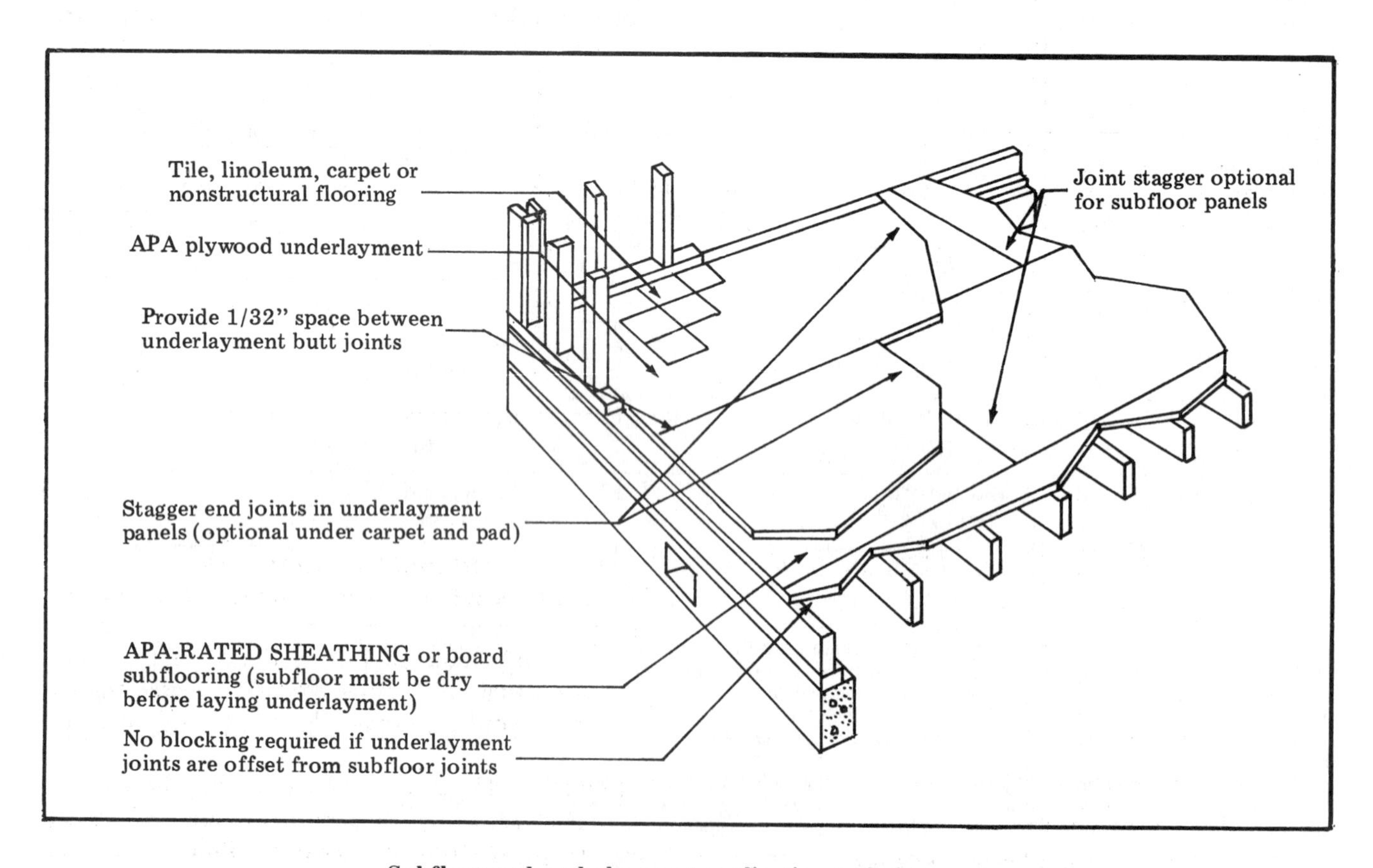

Subfloor and underlayment application techniques
Figure 6-35

You can also use 5/8-inch particleboard as underlayment. When laid over a 1/2-inch plywood subfloor on joists spaced 16 inches on center, it makes a good floor. Don't use particleboard in rooms like laundry rooms and bathrooms where spills or dampness are a hazard. Use exterior plywood in these areas.

Particleboard edges should not be jammed tight. Leave about a 1/16-inch gap at the sides and ends.

Estimating Material Requirements

Figure 6-36 shows the number of floor joists required for any floor length and spacing. Or you can find the number of joists required by dividing the length of the house in feet by the joist spacing center to center in feet. Add one joist for the end. Add for double joists used under wall partitions.

Assume a house 24 feet wide and 40 feet long with joists installed at 16 inches on center. You are using 12-foot-long 2 x 10's for the joists. Your computation would be:

40 divided by 4/3	=	30
Add 1 for end		1
Total		31

There are 31 joists required for one side or bay. Since there is one on each side of the center girder, 62 floor joists are needed.

A 2 x 10 which is 12 feet long has 20 board feet, and 62 pieces have a total of 1,240 board feet.

The board footage is found by multiplying the width times the thickness in inches times the length in feet, and dividing by 12. A board foot is one square foot of wood 1 inch thick (based on nominal dimensions).

Figure 6-37 shows the board footage of lumber in a built-up girder.

Length of span	Spacing of joists 12"	16"	20"	24"	30"	36"	42"	48"	54"	60"
6	7	6	5	4	3	3	3	3	2	2
7	8	6	5	5	4	4	3	3	3	2
8	9	7	6	5	4	4	3	3	3	3
9	10	8	6	6	5	4	4	3	3	3
10	11	9	7	6	5	4	4	4	3	3
11	12	9	8	7	5	5	4	4	3	3
12	13	10	8	7	6	5	4	4	4	3
13	14	11	9	8	6	5	5	4	4	4
14	15	12	9	8	7	6	5	5	4	4
15	16	12	10	9	7	6	5	5	4	4
16	17	13	11	9	7	6	6	5	5	4
17	18	14	11	10	8	7	6	5	5	4
18	19	15	12	10	8	7	6	6	5	4
19	20	15	12	11	9	7	6	6	5	5
20	21	16	13	11	9	8	7	6	5	5
21	22	17	14	12	9	8	7	6	6	5
22	23	18	14	12	10	8	7	7	6	5
23	24	18	15	13	10	9	8	7	6	6
24	25	19	15	13	11	9	8	7	6	6
25	26	20	16	14	11	9	8	7	7	6
26	27	21	17	14	11	10	8	8	7	6
27	28	21	17	15	12	10	9	8	7	6
28	29	22	18	15	12	10	9	8	7	7
29	30	23	18	16	13	11	9	8	7	7
30	31	24	19	16	13	11	10	9	8	7
31	32	24	20	17	13	11	10	9	8	7
32	33	25	20	17	14	12	10	9	8	7
33	34	26	21	18	14	12	10	9	8	8
34	35	27	21	18	15	12	11	10	9	8
35	36	27	22	19	15	13	11	10	9	8
36	37	28	23	19	15	13	11	10	9	8
37	38	29	23	20	16	13	12	10	9	8
38	39	30	24	20	16	14	12	11	9	9
39	40	30	24	21	17	14	12	11	10	9
40	41	31	25	21	17	14	12	11	10	9

One joist has been added to each of the above quantities to take care of extra joist required at end of span.
Add for doubling joists under all partitions.

Number of wood joists required for any floor and spacing
Figure 6-36

Size of girder	Board feet per linear foot	Nails per 1000 board feet
4 x 6	2.15	53
4 x 8	2.85	40
4 x 10	3.58	32
4 x 12	4.28	26
6 x 6	3.21	43
6 x 8	4.28	32
6 x 10	5.35	26
6 x 12	6.42	22
8 x 8	5.71	30
8 x 10	7.13	24
8 x 12	8.56	20

Built-up girders
Figure 6-37

Estimating Manhours

There is no single rule for estimating the labor required for framing. No two men work alike and no two jobs are identical. But there are some general guidelines that will help until you've accumulated figures for your crews and for the type of work you do.

Figure 6-38 gives estimated manhours for various tasks. These will apply on many jobs when modern power tools are used by skilled craftsmen with reasonable motivation. Times are in hours per 1,000 board feet of lumber. In the example above,

Item	Manhours Skilled	Manhours Unskilled	Item	Manhours Skilled	Manhours Unskilled
Sill, bolted and grouted per 100 L.F.	3	2	Wall sheathing, 1 x 6 or 1 x 8, diagonal, per 1,000 B.F.	15	5
Joists, per 1,000 B.F.			Wall sheathing, 1 x 6 or 1 x 8, horizontal, per 1,000 B.F.	12	5
2 x 4	20	8	Plywood or composition wall sheathing, per 1,000 S.F.	9	4
2 x 6	18	7	Roof sheathing, board, per 1,000 B.F.		
2 x 8	16	6	Plain gables	13	6
2 x 10	15	5	Hips, valleys or cut-up	16	7
Built-up girders, per 1,000 B.F.	13	5	Spaced plain gables	14	7
Rafters, per 1,000 B.F.			Roof sheathing, plywood, per 1,000 S.F.		
Plain gable, 2 x 6 or 2 x 8	23	6	Plain gables	9	4
Plain gable, 2 x 10	21	6	Hips, valleys or cut-up	12	5
Hip, no dormers or gables	26	7	Spaced plain gables	10	5
Hip, with dormers or gables	38	8	Building paper, side wall or roof, per 500 S.F. roll	2	--
Install roof trusses, 20' to 40', per 1,000 B.F.	23	7	Siding, per 1,000 S.F. of wall		
Joist bridging, 1 x 3 or 1 x 4, per 100 pieces	4	--	Drop, 6"	21	--
Exterior wall studding and plates, per 1,000 B.F.	21	6	Drop, 8"	19	--
Partition studding and plates, per 1,000 B.F.	20	5	Drop, 10" or 12"	17	--
Rafter bracing and ties, per 1,000 B.F.	20	5	Bevel, 6"	22	--
Wall bridging (solid blocking) per 1,000 B.F.	50	--	Bevel, 8"	20	--
Subflooring, diagonal, per 1,000 B.F.	12	5	Bevel, 10" or 12"	18	--
Subflooring, right angle, per 1,000 B.F.	10	5	Vertical patterns		
			6" or 8"	30	--
			10" or 12"	26	--
			Plywood, 4' x 8' panels	14	--

Manhours for rough carpentry
Figure 6-38

we needed 1,240 board feet of floor joist. According to Figure 6-38, it takes 20 manhours, 15 carpenter and 5 laborer, to install 1,000 board feet. The 1,240 board foot job would need 18.6 carpenter and 6.2 laborer hours.

From Figure 6-37 we know that there are 4.28 board feet of lumber per linear foot of built-up 4 x 12 girder. In a 40-foot house, the length of the center girder would be 38'8" inches (inside to inside of an 8-inch-thick concrete block foundation wall) for a total of about 166 board feet. Figure 6-38 shows that 13 carpenter hours and 5 laborer hours are required for each 1,000 board feet of built-up girder. That's 2.15 carpenter and 0.83 laborer hours for the girder on this job.

Sills are computed by the linear foot at 3 carpenter hours and 2 laborer hours per 100 linear feet. A 24 x 40-foot house has 128 linear feet of sill, and would require 3.8 carpenter and 2.5 laborer hours.

Plywood subflooring goes down at the rate of about 9 carpenter and 4 laborer hours per 1,000 square feet. A 24 x 40-foot house has 960 square feet. Thus, 8.64 carpenter (or skilled) and 3.84 laborer (or unskilled) hours are required to lay the plywood subfloor.

Framing the floor and laying the subfloor would require 33.19 carpenter hours and 13.3 laborer hours.

Description of Material	Unit of Measure	Size and Kind of Nail	Number of Nails Required	Pounds of Nails Required
Wood shingles	1,000 BF	3d common	2,560	4 lbs.
Individual asphalt shingles	100 SF	⅞" roofing	848	4 lbs.
Three in one asphalt shingles	100 SF	⅞" roofing	320	1 lb.
Wood lath	1,000 BF	3d fine	4,000	6 lbs.
Wood lath	1,000 BF	2d fine	4,000	4 lbs.
Bevel or lap siding, ½" x 4"	1,000 BF	6d coated	2,250	*15 lbs.
Bevel or lap siding, ½" x 6"	1,000 BF	6d coated	1,500	*10 lbs.
Byrkit lath, 1" x 6"	1,000 BF	6d common	2,400	15 lbs.
Drop siding, 1" x 6"	1,000 BF	8d common	3,000	25 lbs.
⅜" hardwood flooring	1,000 BF	4d finish	9,300	16 lbs.
13/16" hardwood flooring	1,000 BF	8d casing	9,300	64 lbs.
Softwood flooring, 1" x 3"	1,000 BF	8d casing	3,350	23 lbs.
Softwood flooring, 1" x 4"	1,000 BF	8d casing	2,500	17 lbs.
Softwood flooring, 1" x 6"	1,000 BF	8d casing	2,600	18 lbs.
Ceiling, ⅝" x 4"	1,000 BF	6d casing	2,250	10 lbs.
Sheathing boards, 1" x 4"	1,000 BF	8d common	4,500	40 lbs.
Sheathing boards, 1" x 6"	1,000 BF	8d common	3,000	25 lbs.
Sheathing boards, 1" x 8"	1,000 BF	8d common	2,250	20 lbs.
Sheathing boards, 1" x 10"	1,000 BF	8d common	1,800	15 lbs.
Sheathing boards, 1" x 12"	1,000 BF	8d common	1,500	12½ lbs.
Studding, 2" x 4"	1,000 BF	16d common	500	10 lbs.
Joist, 2" x 6"	1,000 BF	16d common	332	7 lbs.
Joist, 2" x 8"	1,000 BF	16d common	252	5 lbs.
Joist, 2" x 10"	1,000 BF	16d common	200	4 lbs.
Joist, 2" x 12"	1,000 BF	16d common	168	3½ lbs.
Interior trim, ⅝" thick	1,000 BF	6d finish	2,250	7 lbs.
Interior trim, ¾" thick	1,000 BF	8d finish	3,000	14 lbs.
⅝" trim where nailed to jamb	1,000 BF	4d finish	2,250	3 lbs.
1" x 2" furring or bridging	1,000 BF	6d common	2,400	15 lbs.
1" x 1" grounds	1,000 BF	6d common	4,800	30 lbs.

***Note:** Cement coated nails sold as two-thirds of pound equals 1 pound of common nails.

Standard nail requirements
Figure 6-39

Size of nails	Length of nails inches	Gauge number	Approximate number to pound	Size of nails	Length of nails inches	Gauge number	Approximate number to pound
2d	1	15	876	10d	3	9	69
3d	1¼	14	568	12d	3¼	9	63
4d	1½	12½	316	16d	3½	8	49
5d	1¾	12½	271	20d	4	6	31
6d	2	11½	181	30d	4½	5	24
7d	2¼	11½	161	40d	5	4	18
8d	2½	10¼	106	50d	5½	3	14
9d	2¾	10¼	96	60d	6	2	11

Penny conversions
Figure 6-40

A good rule of thumb for finding the manhour requirements for floor framing and subflooring with plywood is:

Divide the square feet of the area by 18. Use 2/3 of the sum for carpenter (skilled) hours and 1/3 for laborer (unskilled) hours. Add 18% for installing wood bridging.

Figure 6-39 shows nail requirements for various framing tasks.

Nailing practices vary by area and from builder to builder. For example, sheathing may be nailed either with 8d, 9d or 10d nails. When 10d nails are used, more pounds will be required. This costs more. The "d" behind the number stands for "penny." That's an English term which originally indicated the price per hundred. It now signifies length only. Figure 6-40 converts penny designations into sizes and weights.

Summary

No manhour estimate will apply on every job. There are too many variables. Your judgment is always necessary. Your best estimates will always be based on your own cost records. That's why this book places so much emphasis on keeping good notes on every job.

Figure 6-41, Floor Framing Checklist and Data Reference Sheet, will help you collect facts about each floor framing job.

Floor Framing Checklist and Data Reference Sheet

1) The ☐sill plate ☐sill beam, size __________, ☐is ☐is not anchored to the ☐foundation wall ☐free-standing piers.

2) The ☐sill plate ☐sill beam is ☐treated ☐untreated:

☐Southern yellow pine

☐Douglas fir

☐Other (specify) ______________________________

3) A ☐sill sealer ☐termite shield ☐was ☐was not required and ☐was ☐was not installed.

4) The house is ☐1-story ☐2-story and the floor framing is :

☐Platform

☐Balloon

5) The ☐platform ☐balloon framing has __________ inches (thickness) of horizontally-laid wood at exterior walls.

6) The house ☐does ☐does not have a basement and has girders supported by:

☐Steel posts (number and size) ______________________________

☐Wood posts (number and size) ______________________________

☐Masonry piers for crawl-space

7) The girder is:

☐Built-up with ______pieces of 2 x __________

☐Glue-laminated (size) ____________________

☐Solid wood (size) ____________________

☐Steel I-beam (size) ____________________

☐Steel wide-flange (size) ____________________

8) The ☐wood ☐steel girder bears at least 4 inches on the masonry walls or pilasters. (If you use untreated lumber, a ½-inch air space should be provided at each end and each side of the girder within the masonry. In termite-infested areas, line the pocket with metal.)

9) The girder top ☐is ☐is not level with the top of the sill plate. (If level with top of sill plate, joists will bear on top of girder.)

10) The girder has __________ inches (thickness) of horizontally-laid wood and ☐does ☐does not equal the horizontally-laid wood in item 5, above.

11) A continuous horizontal tie between exterior walls was made by ☐nailing notched joists together ☐connecting scab ☐steel strap ☐subflooring material.

Figure 6-41

12) If steel girder is used, joists are:

☐Nailed to a 2 x 4 bolted to the steel

☐Bearing on ledger

☐Bearing on flange, ☐with ☐without blocking between joists

☐Scab-tied

13) The floor joists are ☐Southern yellow pine ☐Douglas fir ☐other (specify) ________________ and are:

☐2 x 8 x __________ feet long

☐2 x 10 x __________ feet long

☐2 x 12 x __________ feet long

14) Floor joist clear span is ______ feet ______ inches.

15) Floor joists are spaced ______ inches on center.

16) Floor joists ☐are ☐are not in-line.

17) Floor joists ☐are ☐are not double under partitions that are parallel with the joist run.

18) A total of __________ floor joists were required. (To find the number of joists required, divide the house length (in feet) by the joist spacing (in feet). Add 1 extra joist for the end, and add for double joists under wall partitions.)

19) Bridging ☐was ☐was not required. Type of bridging:

☐Wood cross-bridging

☐Metal cross-bridging

☐Solid bridging

20) Subfloor is:

☐Boards (size)__________laid diagonally

☐Boards (size)__________laid right angles to the joists

☐Plywood (thickness)__________applied at right angles to the joists

☐Sturd-I-Floor with span rating of ☐16'' ☐20'' ☐24'' ☐48'' on center:

☐Thickness __________

☐Square-edge

☐T&G

☐Nailed

☐Glue-nailed

21) The floor is ______feet long and ______feet wide for a total area of ______square feet.

22) Board subflooring required a total of ______BF of boards. (For right-angle subflooring, add to total board footage: 20% for 1 x 6's or 16% for 1 x 8's. For diagonal subflooring, add to total board footage: 25% for 1 x 6's or 22% for 1 x 8's.)

23) Board subflooring required a total of ______manhours, __________ skilled and __________ unskilled. (1,000 BF right-angle subflooring require 10 skilled and 5 unskilled labor hours. 1,000 BF diagonal subflooring require 12 skilled and 5 unskilled labor hours.)

24) Plywood subflooring required a total of __________ sheets of plywood, and a total of ______ manhours, __________ skilled and __________ unskilled. (1,000 square feet plywood subflooring require 9 skilled and 4 unskilled labor hours.)

25) Material costs:

Sill plate	__________LF	@	$________LF	=	$__________
Sill sealer	__________LF	@	$________LF	=	$__________
Termite shield	__________LF	@	$________LF	=	$__________
Joist headers & stringers	__________LF	@	$________LF	=	$__________
Steel posts	__________pcs	@	$________ea	=	$__________
Wood posts	__________pcs	@	$________ea	=	$__________
Built-up girder	__________LF	@	$________LF	=	$__________
Steel girder	__________LF	@	$________LF	=	$__________
Glue-laminated girder	__________LF	@	$________LF	=	$__________
Ledger	__________LF	@	$________LF	=	$__________
Floor joists:					
2 x 8	__________pcs	@	$________ea	=	$__________
2 x 10	__________pcs	@	$________ea	=	$__________
2 x 12	__________pcs	@	$________ea	=	$__________
Bridging:					
Wood cross-bridging					
1 x 3	__________LF	@	$________LF	=	$__________
1 x 4	__________LF	@	$________LF	=	$__________
Metal cross-bridging	__________pcs	@	$________ea	=	$__________
Solid wood bridging	__________LF	@	$________LF	=	$__________

Subflooring:

1 x 6 boards	________BF	@	$______M BF	=	$________
1 x 8 boards	________BF	@	$______M BF	=	$________
Plywood	________pcs	@	$______ea	=	$________
Other (specify)					
________________	________	@	$________	=	$________

Nails:

16d	________lbs	@	$________lb	=	$________
12d	________lbs	@	$________lb	=	$________
10d	________lbs	@	$________lb	=	$________
8d	________lbs	@	$________lb	=	$________
6d	________lbs	@	$________lb	=	$________
Other (specify)					
________________	________lbs	@	$________lb	=	$________

Other materials, not listed above:

________________	________	@	$________	=	$________
________________	________	@	$________	=	$________
________________	________	@	$________	=	$________
			Total for all materials		$________

26) Labor costs:

Skilled	________hours	@	$______hour	=	$________
Unskilled	________hours	@	$______hour	=	$________
Other (specify)					
________________	________hours	@	$______hour	=	$________
			Total		$________

27) Total cost:

Materials	$________
Labor	$________
Other	$________
Total	$________

28) Cost of floor framing per square foot $__________

29) Problems that could have been avoided __________

30) Comments __________

31) A copy of this Checklist and Data Reference Sheet has been provided:

☐ Management

☐ Accounting

☐ Estimator

☐ Foreman

☐ File

☐ Other: __________

Chapter 7

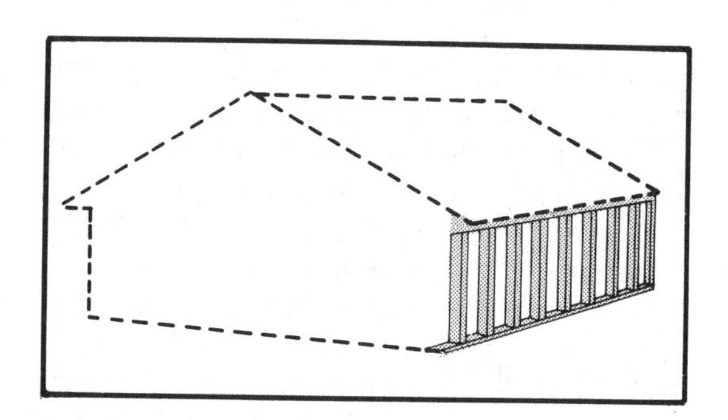

Framing the Wall

Many experienced builders have learned that it's good business to build the same type of house over and over. And that's even better advice for beginners in the industry. Minor changes in roof line, exterior treatment and ornamentation can make otherwise identical houses look very different. Only you and your carpentry crew will know that they're basically the same structure.

The point is that switching construction methods can be costly. Making money at building is our goal. Stack the cards in your favor: Master a particular design or type of construction and then keep improving on it. Naturally, you have to compete on the jobs that are available. But you have a big advantage if what your client wants built is what you specialize in.

This chapter will explain the most commonly used framing systems, how they differ, and how to handle each to the best advantage. But before we get into the different types of framing, we'll hit the high points on the selection of framing lumber.

The American Softwood Lumber Standard, PS20-70

The American Softwood Lumber Standard, PS20-70, establishes a uniform set of rules for grading framing lumber. Before PS20-70, every regional species had its own set of regional rules and standards. These standards took into consideration the unique characteristics of the species and the expected uses in that particular region. This didn't create any problem, as long as you bought your lumber from the local mill. But the lumber from different mills and areas was not comparable. You needed a set of span tables for each grade on each job.

PS20-70 changed that. It established a National Grading Rule Committee "to maintain and make fully and fairly available grade strength ratios, nomenclature and descriptions of grades for dimension lumber." These grades and grade requirements, as developed, are now used by all regional grade-writing agencies. Now nearly all framing lumber is graded by grade-writing agencies that are certified by the American Lumber Standards Committee.

The National Grading Rule separates dimension lumber into two *width* categories. Pieces up to 4 inches wide are graded as *Structural Light Framing, Light Framing* and *Studs.* Pieces 6 inches and wider are graded as *Structural Joists* and *Planks.* (See Figure 7-1.) Where a fine appearance and high bending strength are required, the National Grading Rule also provides an *Appearance Framing* grade.

Structural Light Framing grades are for those uses where higher bending strengths are required. The four grades included in this category are *Sel Str* (Select Structural), *No. 1, No. 2,* and *No. 3.*

Light Framing grades are used where good appearance at a lower design level is satisfactory. Grades in this category are called *Const* (Construction), *Std* (Standard), and *Util* (Utility).

A single Stud grade is also provided under the National Grading Rule. It's intended specifically for use as a vertical bearing member in walls and partitions to a maximum height of 10 feet.

Structural Joist and Plank grades are available in widths 6 inches and wider for use as joists, rafters, headers, built-up beams, etc. Grades in this category are *Sel Str* (Select Structural), *No. 1, No. 2,* and *No. 3.*

Not all grades that are described in the National Grading Rule and listed in Figure 7-1 will be available in all species or regions. The Sel Str and No. 1 grades are frequently used for truss construction when high strength is required. For general construction, you will use No. 2, or No. 2 and Btr, or Std and Btr. The No. 3 and Util grades are used where less strength is needed.

Lumber that is imbedded in the ground or often exposed to damp conditions either must be cut from naturally durable species or pressure treated to stop deterioration. This also includes lumber used for sills resting on a concrete slab which is in direct contact with the earth, and joists which are closer than 18 inches to the ground.

The durable species most frequently used are California redwood, Western red cedar and tidewater red cypress. But not all redwood and red cedar are adequate for these purposes. Be sure to specify foundation grades of redwood and red cedar. In cypress, select a heart structural grade.

Using Lumber Grades

Your job in selecting lumber is to buy material appropriate for its intended use. In structural lumber, that means getting lumber rated for the load anticipated. Lumber grading rules determine the maximum number, size, and position of the defects each piece may have in each grade.

In Appearance Framing grades, the principle is the same, but the standards are different. In siding, for example, a reasonable number of knots are covered when the siding is in place. In flooring, some knots and other defects on the underside are allowable, since they will not show. Sheathing and subflooring may have a considerable number of defects since both are entirely covered by finish materials.

The condition of each defect may also influence the grade. Tight knots in certain grades of siding or ceiling material may be allowed. Loose knots would be objectionable.

The grading rules describe the defects in lumber, not the quality of the wood itself. If two boards of the same kind of wood are "clear" or have similar defects, they fall into the same grade, regardless of the wood species. The wood in one board may be dense, heavy and strong; the other may be light and weak. In siding, ceiling or finish material, these differences may not matter. But for structural timber or flooring, where strength is essential, the wood must be dense to give satisfactory service.

2"-4" Thick, 2"-4" Wide	
Structural Light Framing	Sel Str (Select Structural) No. 1 No. 2 No. 3
Studs	Stud
Light Framing	Const (Construction) Std (Standard) Util (Utility)
2"-4" Thick, 6" and Wider	
Structural Joists and Planks	Sel Str (Select Structural) No.1 No. 2 No. 3
2"-4" Thick, 2" and Wider	
Appearance Framing	A (Appearance)

Dimension lumber grades
(National Grading Rule)
Figure 7-1

Grades Recommended for Framing

For your jobs, you want to use the lowest grade of lumber suitable for the application. It's a waste of money to use Select Structural or No. 1 studs when No. 3 will do the job just as well. Follow the recommendations in Figure 7-2 to reduce the cost of framing lumber and still provide quality construction that meets code requirements.

Every lumber grade exists to meet certain requirements. In every home, there are places where you can use all three framing grades.

No. 1 and No. 2 Douglas fir, Southern pine and West Coast hemlock are used for horizontal, load-bearing members, such as joists and rafters, because of their superior strength. No. 2 Douglas fir joists and rafters can be used on spans up to 3/4 the maximum allowable span for No. 1 joists and rafters of the same thickness and depth. For example, in a dwelling with a live load of 40 pounds per square foot, 2-inch No. 1 joists, 16 inches on center, can be used safely for a span (in feet) equal to 1½ times their nominal depth (in inches). 2-inch No. 2 joists, 16 inches on center, can be used safely on a span (in feet) slightly more than their nominal depth (in inches). For example, a 2 x 10 No. 1 can be used on a span of 15 feet; a 2 x 10 No. 2 can be used on a span of 11'3".

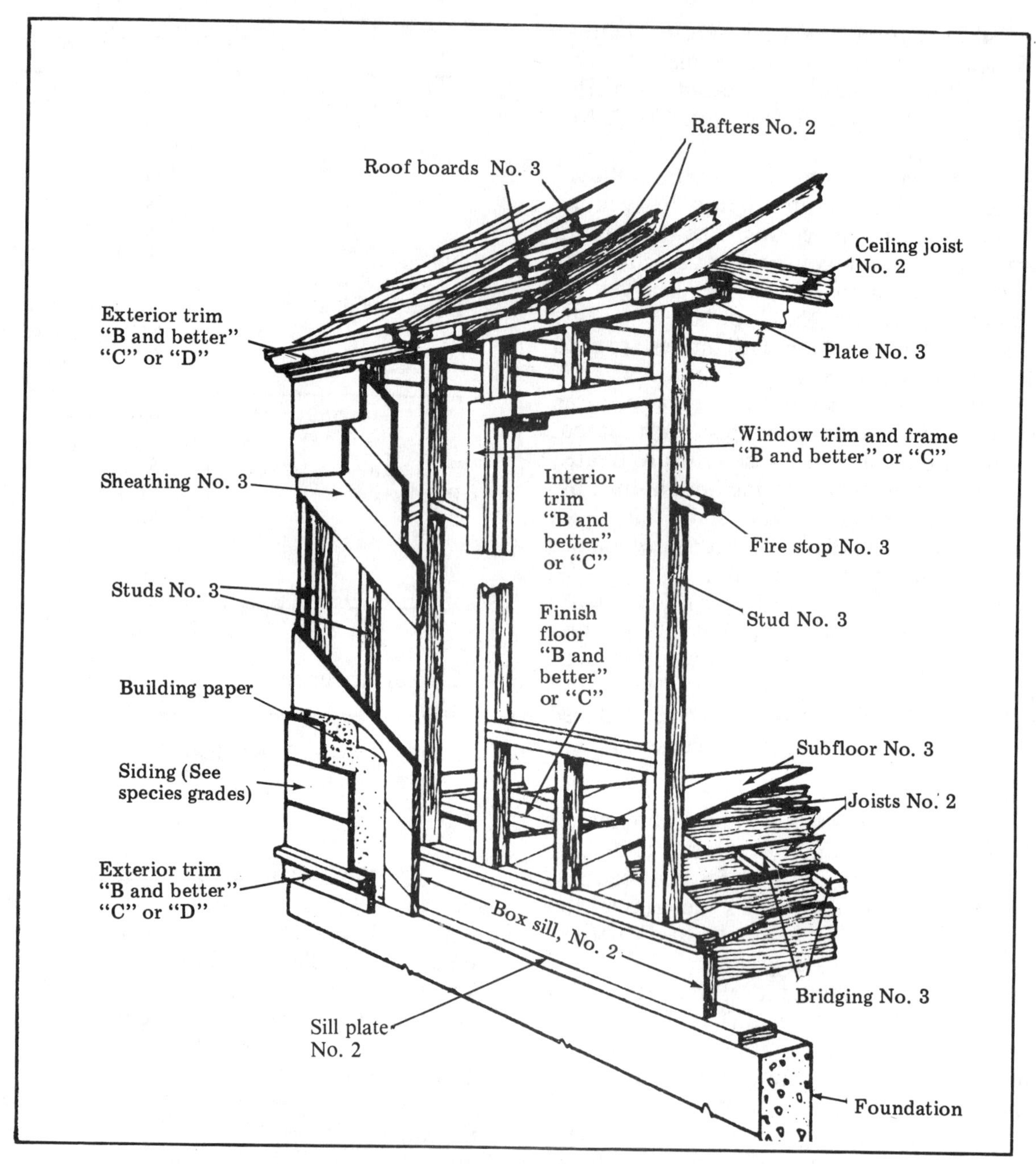

Grade use key for low-cost one-story construction
Figure 7-2

Use No. 3 studs on partitions in one-story homes and on nonbearing or minor partitions in two-story structures. In two-story construction, use No. 2 studs for the first floor and No. 3 for the second floor. This provides enough strength and a reasonable margin of safety.

No. 3 studs don't look as good as No. 1. They may have large knots, small knot holes, short strips of bark along one edge, and white specks or pitch pockets. No. 3 studs still provide enough strength with an adequate margin of safety. Studs are for strength; appearance is irrelevant. Don't buy better lumber than you need. See Figure 7-3.

You're going to get some trash lumber with most orders. Examine the material when it's delivered to the job site. If you've ordered No. 2 lumber, be sure the grade mark says No. 2. Don't accept No. 3!

Basic Framing Concept

The frame provides the "skeleton" on which all exterior and interior finishes are hung. Two popular variations on conventional framing are shown in Figures 7-4 and 7-5. Figure 7-4 shows a second story which cantilevers over the first floor. Figure 7-5 shows a cathedral-type ceiling.

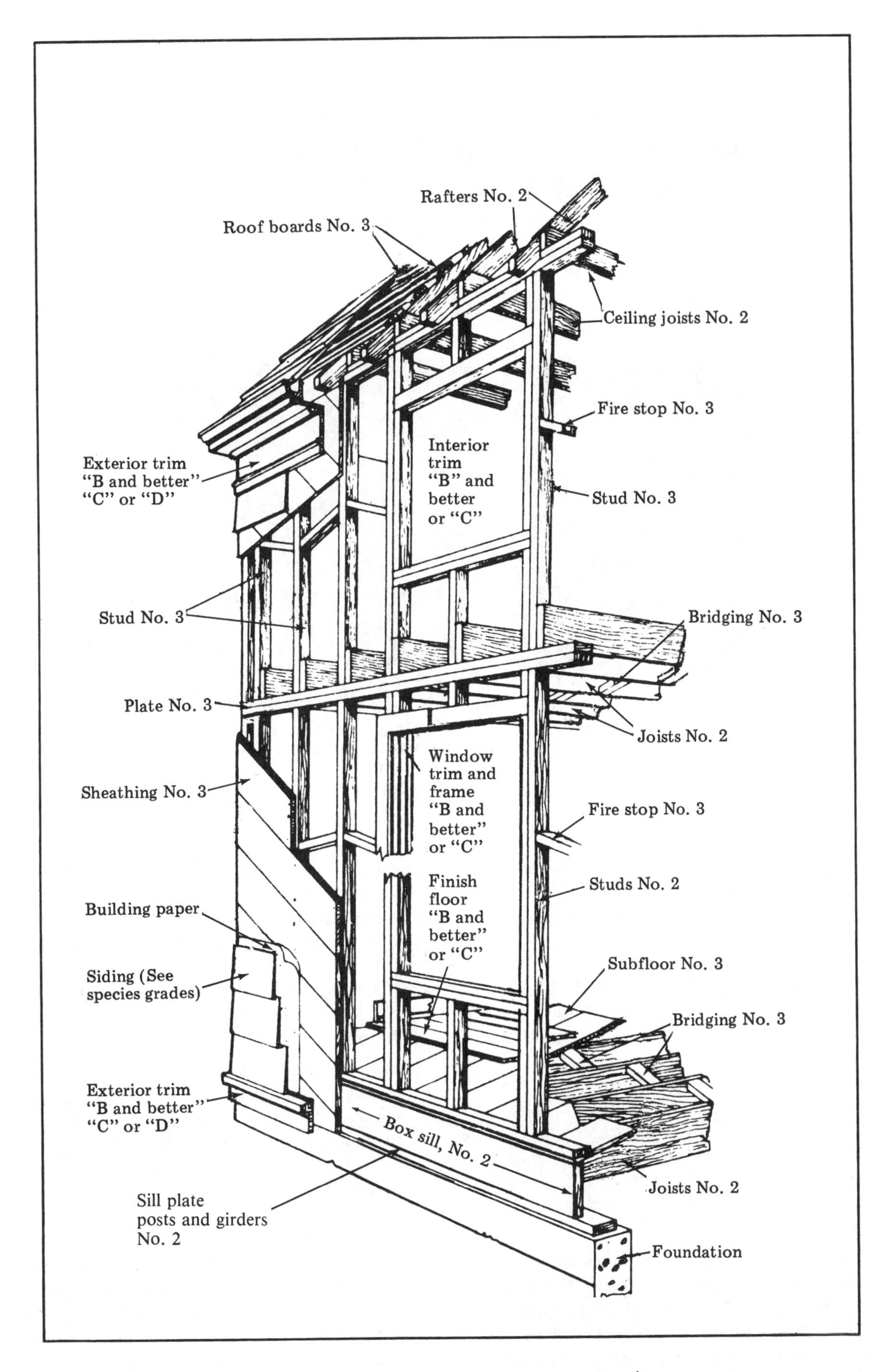

Grade use key for low-cost two-story construction
Figure 7-3

Second floor cantilever
Figure 7-4

The building frame includes the floor joists, the rafters, the studs, and the girders which support the ends of the joists and which are supported by posts or columns.

Framing practice varies from area to area. Each of the principal regions of the United States — the Atlantic Coast, the South, the Mid-West, the West, and the Pacific Coast — has some unique characteristics. But the same basic principles of framing and bracing are used throughout the U.S.

Buildings constructed entirely of wood above the foundation may be classified as *early braced frame, modern braced frame, Western* (or *platform) frame, balloon frame,* and *plank-and-beam frame.* We'll explore the differences in each system next.

Early Braced Frame

This system is no longer used in modern construction. Each post and beam was mortised and tenoned together. The angles formed by these posts or beams were made rigid by diagonal braces. These in turn were held in place by wooden pins or dowels through the joints. This construction did not rely on the sheathing for rigidity; in fact, it could run vertically up the sides of the structure. At one time, most barns were built this way.

A more modern method of bracing the framework has replaced this type of construction. It's used where sheathing is not the primary rigid support for the building. It's called modern braced frame construction.

Modern Braced Frame

Figure 7-6 shows the details of a modern braced frame. The studs of the side walls and center partitions are cut to exactly the same length. The side-wall studs are attached to the sill or plate. The partition studs are attached to the girder or sill. This permits uniform shrinkage in the outside and inside walls when a steel girder is used as shown.

Diagonal braces are let into the studs at the corners of the building to provide rigidity. Several systems of bracing may be used at the corners, depending on the placement of the openings in the side or end walls. The joists are lapped and spiked to the side wall studs. They are either continuous

Framing a cathedral-type ceiling
Figure 7-5

between the two opposite side walls, or are side-lapped and spiked at the center partition. In either case, these joists form a tie for the two side walls. The joists are bridged (where required) and are supported by the full width of the sill and girder. Fire stops are provided at the ends of all joists.

The flooring is laid diagonally in opposite directions on the upper and lower floors to prevent twisting of the structure.

Western (or Platform) Frame

Figure 7-7 shows the Western frame. Each story of the building is built as a separate unit. The floor is laid before the side walls are raised. This provides a safe working platform. Western framing is similar to the modern braced frame except that the ends of the joists form a bearing for the side and center wall studs. This allows equal shrinkage at the side and center walls.

This type of framing became more popular when kiln-dried lumber became generally available. It's probably the fastest and safest form of good construction. It also lets you use more short material for studs. The studding extends only one story and rests on a sole plate nailed on top of the flooring. The second floor joists are carried on a plate placed on top of the studding for the story below. Since the rough floor extends to the outside edge of the building, no wooden fire stops are required.

In Western framing, a continuous header is used at the first floor. It's the same size as the joists. The bottom edge of the header rests on the sill. Its outside face is flush with the outside edge of the sill to form a box-sill. At the second floor, instead of using a continuous header, the joists are extended flush to the outside of the plate. Blocking is placed between the joists.

Partitions are built and supported just like partitions in any other framing system. They require girders, double joists, sole pieces and plates for support.

In Western frame construction, outside wall members are framed the same way as interior partitions. This makes the shrinkage uniform in both outside walls and partitions. Floors and ceilings should be level regardless of shrinkage. Where steel beams are used, wood of the same cross-section size as the sills should be used on the steel to give equal shrinkage.

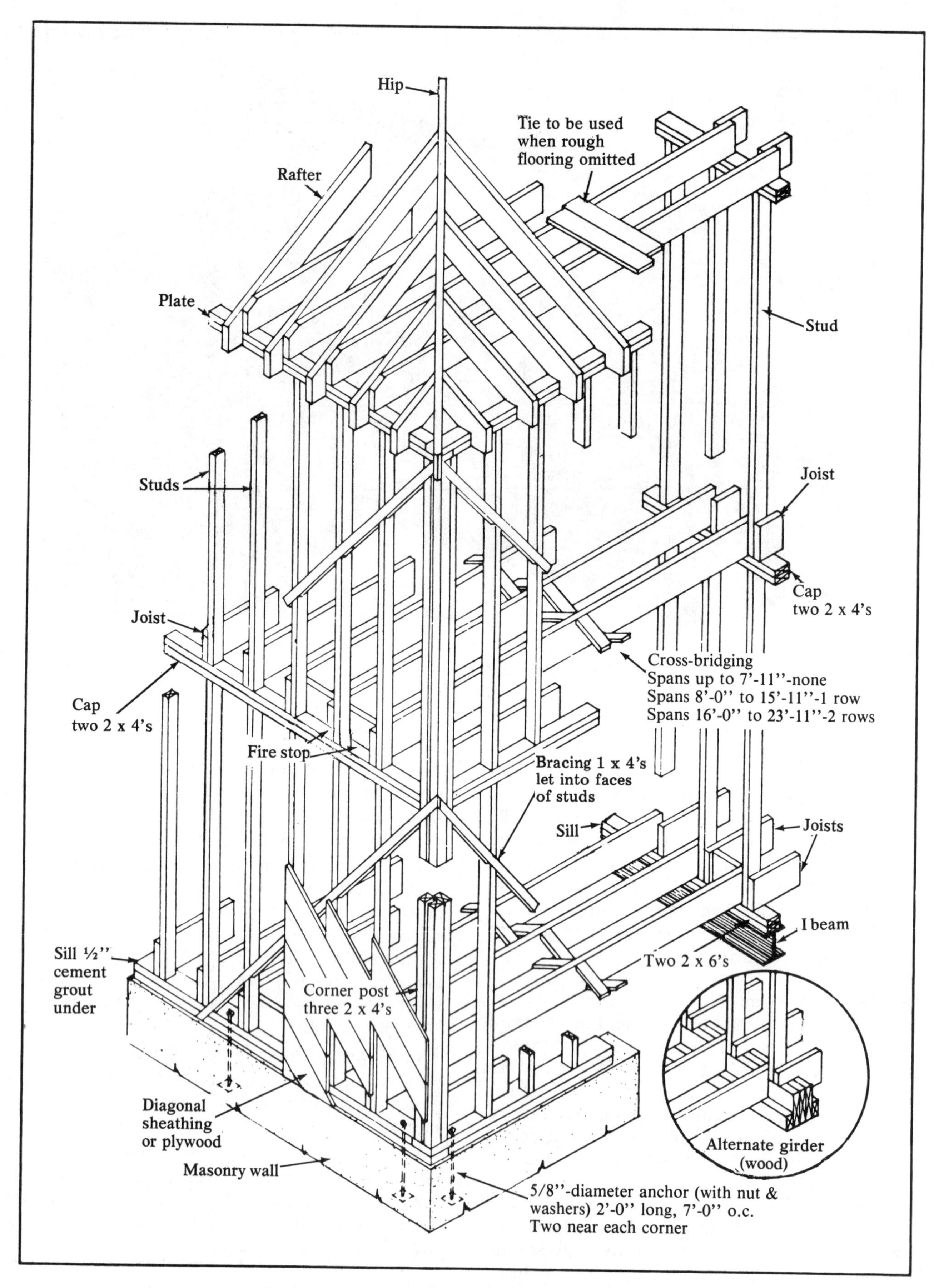

Modern braced framing
Figure 7-6

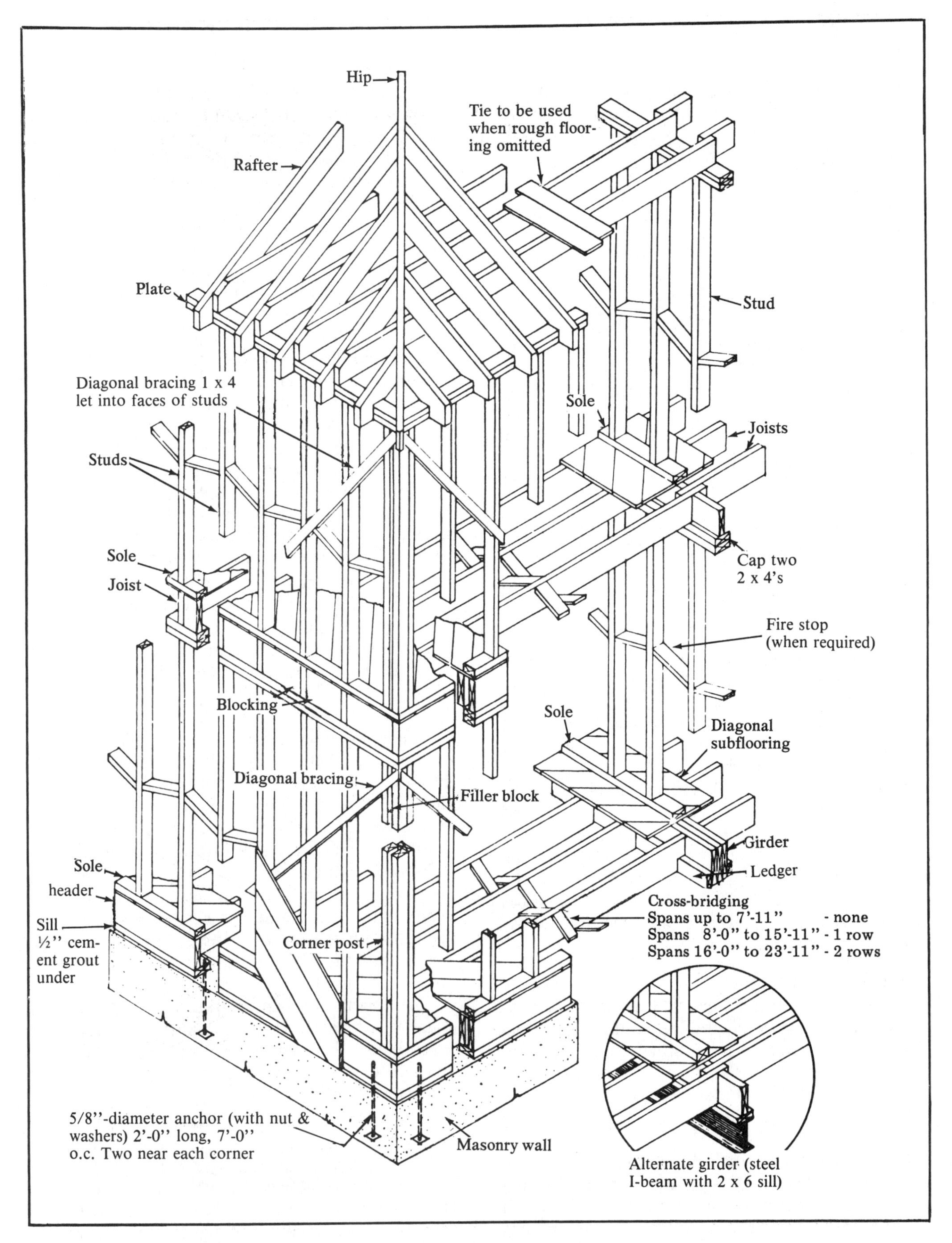

Western framing
Figure 7-7

In Western framing, you can use shorter lengths of lumber. This will generally reduce lumber costs. Lengths can be readily determined from the plans and then precut. Walls are generally built horizontally, and then raised into place in a single operation.

Balloon Frame

Figure 7-8 shows the balloon frame. The second-floor joists are carried by a ribbon let into the studs. Balloon framing uses long side-wall and end-wall studs, but requires fire stops between the studs.

Balloon framing replaced the old braced frame, but is now less common than the modern braced frame and Western frame in most parts of the country.

The balloon frame has many features to recommend it. Outside studs go up quickly because they extend the full two stories, from the sill to the roof plate. The load-bearing partitions can also be built this way. However, the preferred method is shown in Figure 7-8. At the second floor level, joists rest on a 1 x 6 ribbon or ledger board and are nailed against the studs. Attic or ceiling joists rest on the double top plate.

The flue effect in openings between the studs is much stronger. That makes the fire stops much more important. Without fire stops, flames from the basement could sweep to the top plate or into the attic. There must be blocking, both at the sills and at half-story heights.

The balloon frame is not as rigid as braced types of framing until after the outside sheathing has been applied. After sheathing, a balloon frame building is as strong and stiff as the other types. But costs may be lower because it can be erected faster. When the sheathing is run diagonally or if plywood is used, you can omit the 1 x 4 diagonal braces. In many cases, plywood panels provide the only bracing needed. The balloon frame reduces shrinkage by keeping the amount of diagonally-laid lumber to a minimum.

Plank-and-Beam Frame

The plank-and beam frame (also called a *post-and-beam frame*) is very adaptable to modern building styles. It works well with one-story structures, large glass areas, modular set-up, open-span planning, and natural finish materials. Plank-and-beam can make good use of glue-laminated beams where large spans are required.

The plank-and-beam frame concentrates the structural load on fewer and larger framing members than conventional construction. This can cut the manhours spent in carpentry. Using planks continuously over two or more spans, plus the elimination of most interior and exterior finishes, can cut costs further.

Plank-and-beam framing was adapted from heavy timber construction. It can be used for framing floors and roofs in combination with ordinary wood stud or masonry walls, or as the skeleton frame in curtain wall construction.

Figure 7-9 shows the plank-and-beam frame roof applied to masonry walls. Assuming that Figure 7-9 represents a 24-foot-wide house with either a flat or a low-pitch roof, 14-foot planks automatically provide an overhang of approximately 2 feet. The overhang is virtually completed when the roof is sheathed.

It's a wise designer who lays out the house on a 2-foot module. If every dimension is a multiple of 2 feet, there will be a minimum of cutting, fitting and waste. The plank-and-beam frame lends itself to modular concepts. The key is in deciding what the beam spacing will be. Your decision should be based on efficient use of material dimensions. For example, you might decide on an 8-foot spacing for beams and columns so that 16-foot planks will be continuous over two spans. 4 x 8 sheets of drywall and plywood will fit the walls without cutting.

The plank-and-beam frame can save time, material, and money over conventional residential construction, if you know how to use it. These savings come from the dual function of materials, the fewer pieces to handle, and the high structural efficiency.

The wide beam spacing and 2 x 6, or 2 x 8 tongue-and-groove dual-function planking can eliminate conventional ceiling construction. You save on ceiling joists, bridging, lath and plaster, or drywall. Plan to stain and seal the planks and beams on the ground before erection. The ceiling is complete when the planks are nailed into place.

Figures 7-10A and 7-10B illustrate the basic differences between conventional framing and the plank-and-beam frame for a typical 40-foot wall section with the same openings, the same ceiling height, and the same height of the rough floor over the outside grade.

The plank-and-beam frame has a lot less material and labor. But top quality workmanship is essential in this type of construction. All joints must fit exactly since there are fewer contacts between members.

Beginning the Wall Framing

In the last chapter, we covered floor framing. So far in this chapter we've only discussed the various types of frames. Generally, it's a good idea to use the framing method that is most commonly used in your area. Too many innovations can make any house hard to sell.

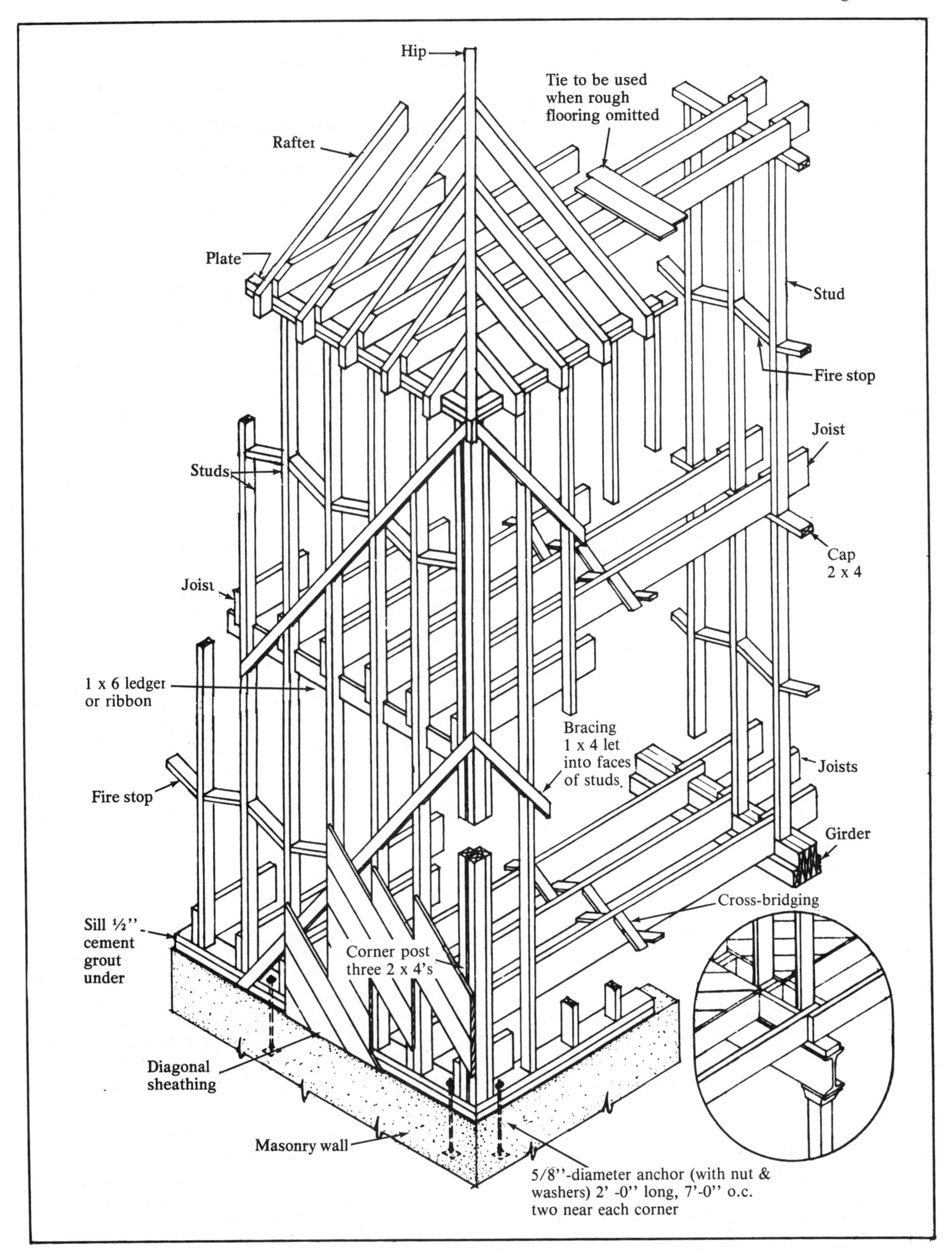

Balloon framing
Figure 7-8

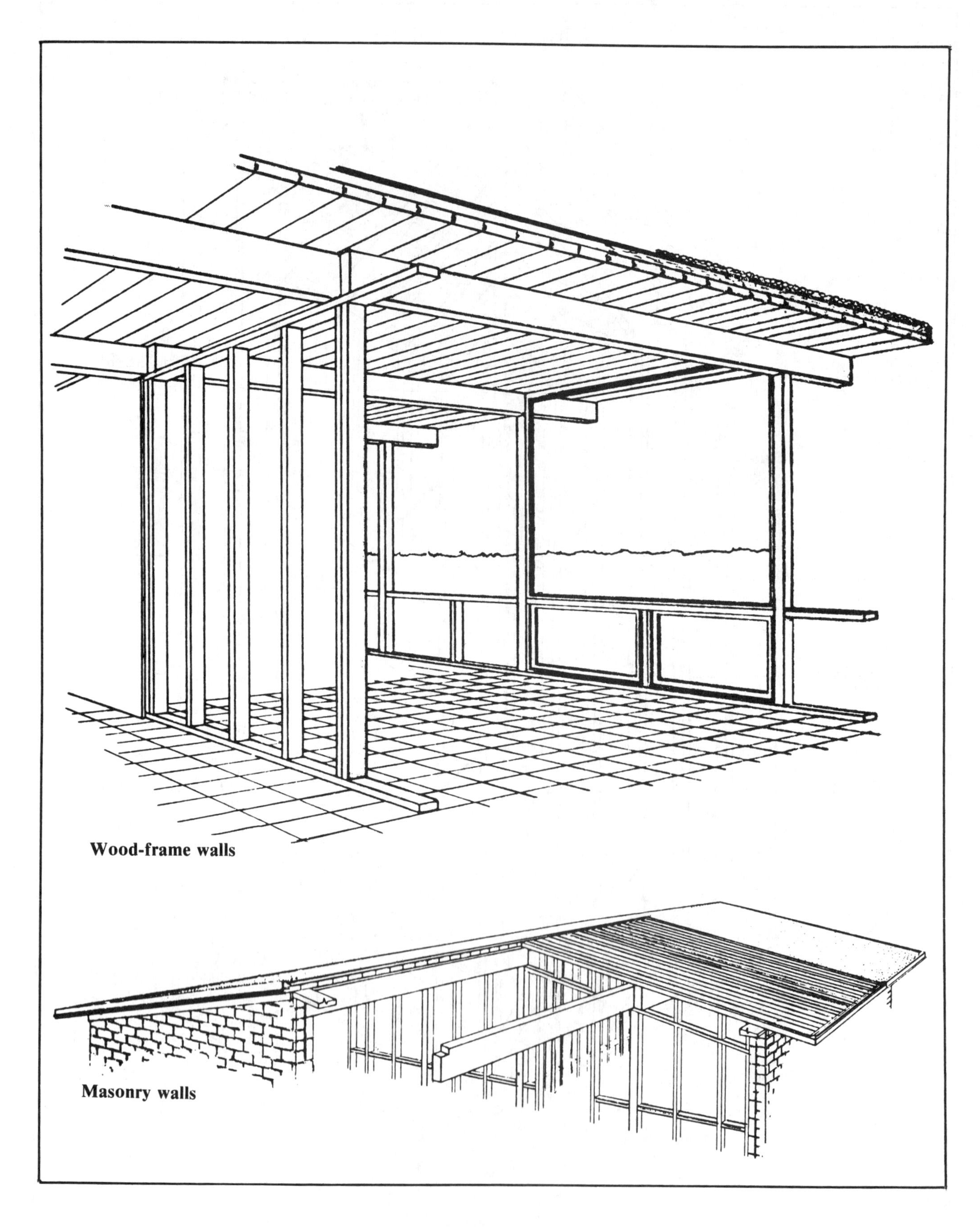

Plank-and-beam framing
Figure 7-9

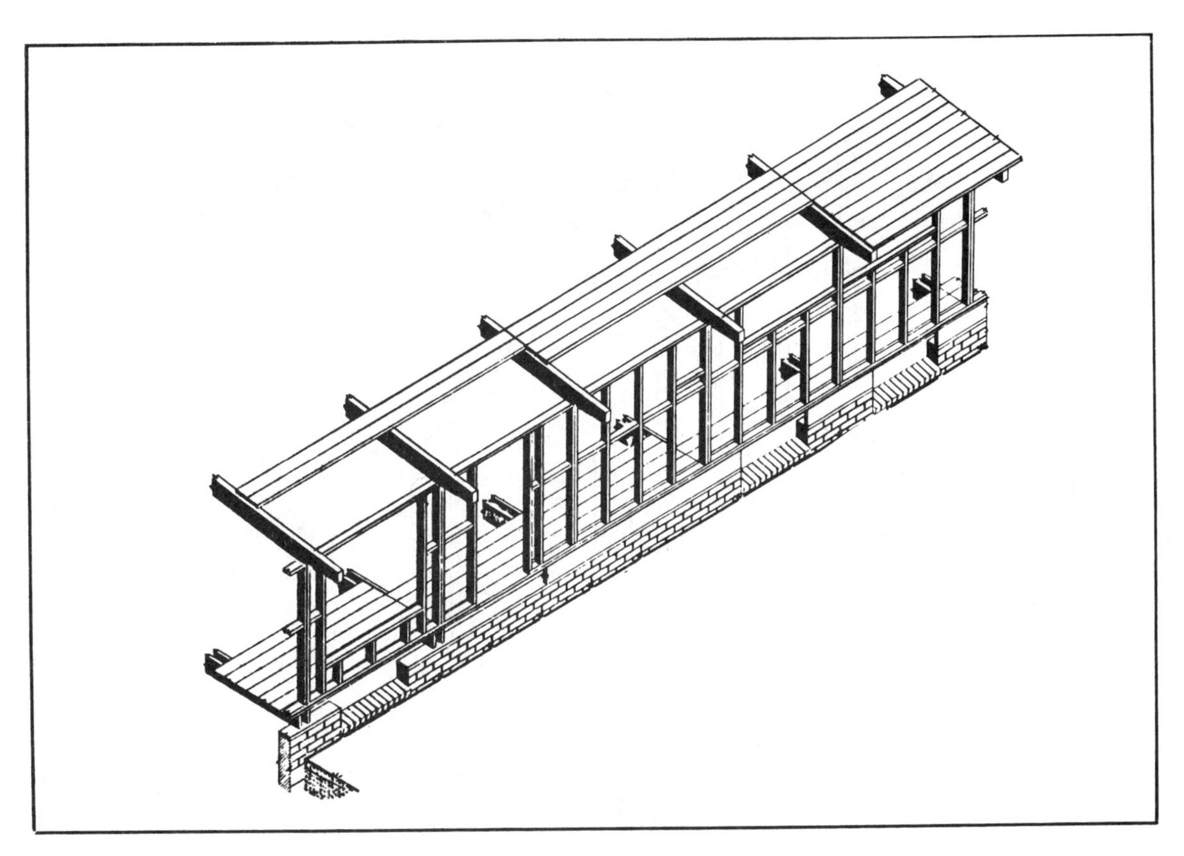

Plank-and-beam framing
Figure 7-10A

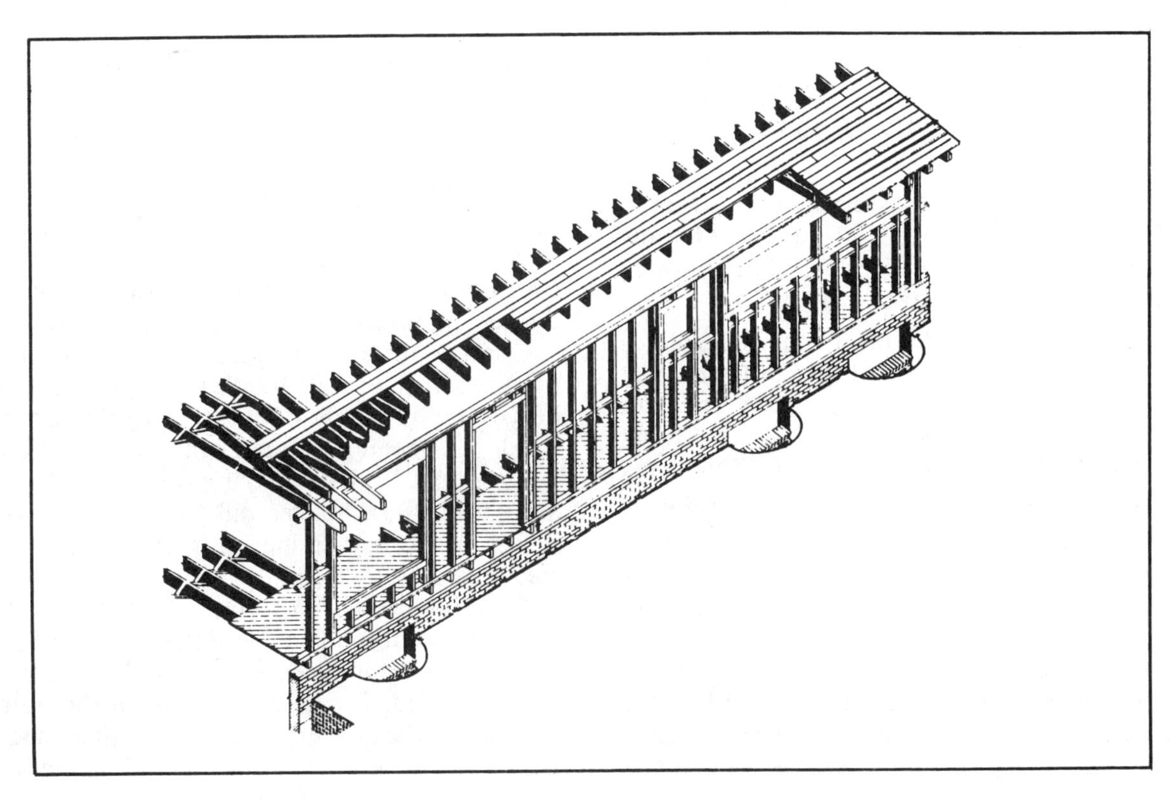

Conventional framing
Figure 7-10B

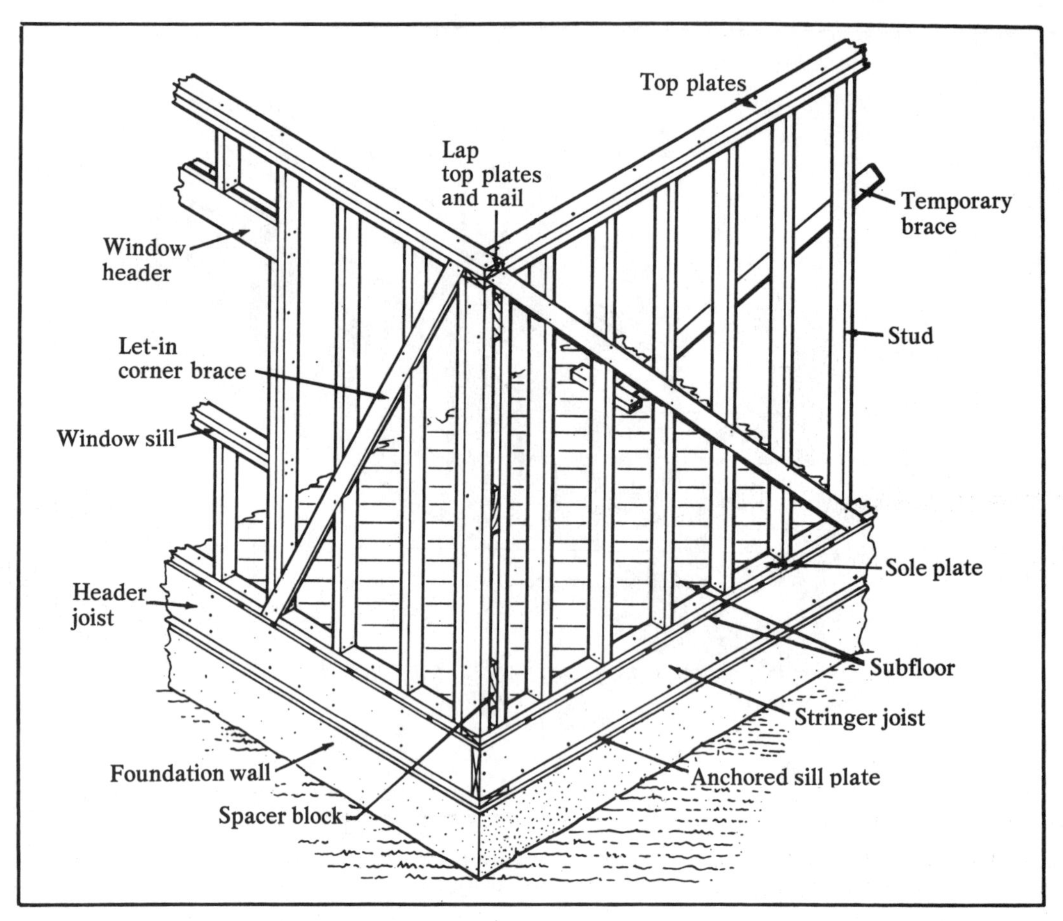

Wall framing with platform construction
Figure 7-11

Western (Platform) construction is the method I prefer and have used for many years. I put the wall together on the deck (subfloor), and raise it into place.

This is called horizontal assembly or "tilt-up." You can fabricate whole walls or sections of walls in this way. Only the availability of manpower limits the wall size that can be "tilted-up."

Some builders prefer to install the sole plate, nailing it in place on the subfloor. Then they assemble the studs to the top plate (bottom member) on the deck. Finally, they raise the frame and toenail the studs to the sole plate. This takes more time, but can still produce a sound wall.

The sole plate and top plate should always be marked for studs, window and door openings, and partition intersections at the same time. Lay the sole plate out where it's to be installed. Temporarily tack it in place at both ends. Lay the top plate on the floor alongside the sole plate. Using a square, measure and mark both top plate and sole plate for all studs, openings, and intersections. When the studs and joists are on the same spacing, align the side wall studs directly over the floor joists.

Lay out precut studs, window and door headers, cripple studs, jamb studs (studs supporting headers), corner posts, intersection T's, and window sills.

Top and sole plates are then nailed to all vertical members (studs). Use 16d nails. Then install window and door framing and cut for let-in corner bracing, if required. The entire wall frame is then erected, plumbed, and braced. See Figure 7-11.

Corner posts, T's, and headers are usually preassembled so that they can be installed during horizontal assembly of the wall frame. Complete window opening units can be preassembled horizontally by nailing the double stud (full stud and jamb stud), header, sill and cripples together. Assemble door opening units on the deck in the same manner.

Some builders also install sheathing before the wall is raised into place. See Figure 7-12. You could even produce complete finished walls, with window and door units in place, and most of the siding installed. In the winter, raise one of these sheathed frame walls to block the wind and create a more comfortable working area.

In high wind areas, wall anchorage is essential.

Walls raised with sheathing installed
Figure 7-12

Sheathing can provide the necessary anchorage when it extends over and is nailed to the sill plate. Figure 7-13 shows another means of anchorage.

There are several ways to assemble studs to make up the corner posts. Figure 7-14 shows three methods of stud assembly. Figure 7-11 shows the standard assembly. This method uses three pieces of 2 x 4 blocking.

Interior walls must be firmly secured to all intersecting walls. There are several ways to provide a nailing surface at these corners. The T shown in Figure 7-15 A is commonly used because it's assembled quickly and provides a strong wall tie-in. In Figure 7-15 B, short pieces of 2 x 4 blocking are used between the studs to support and provide backing for a 1 x 6 nailer.

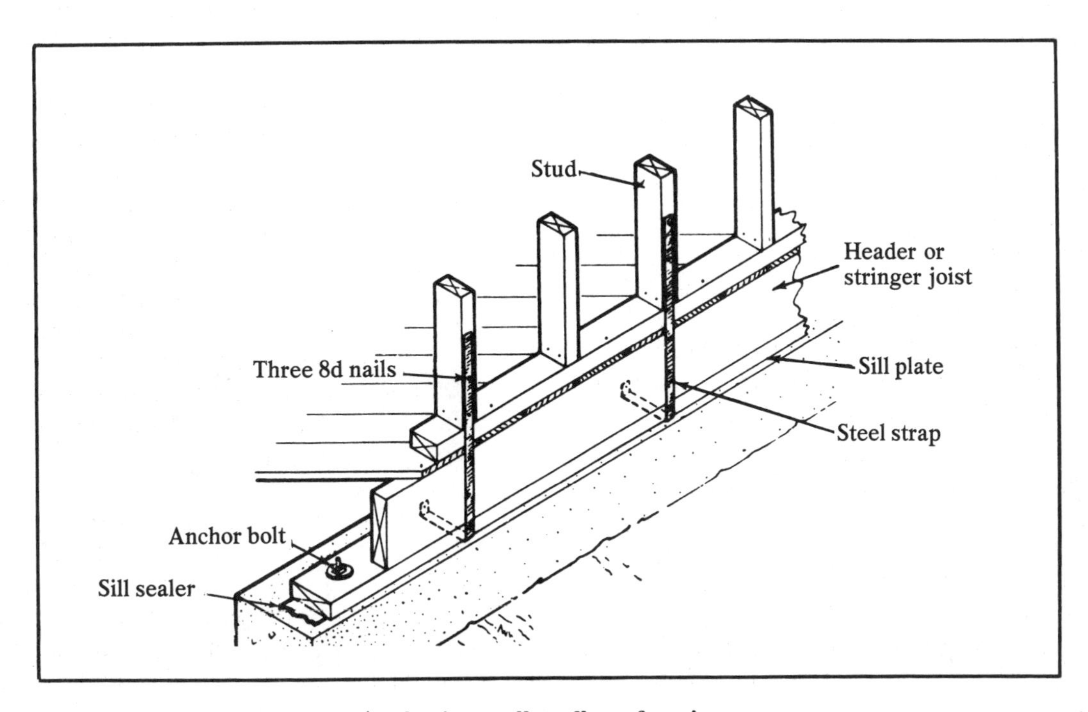

Anchoring wall to floor framing
Figure 7-13

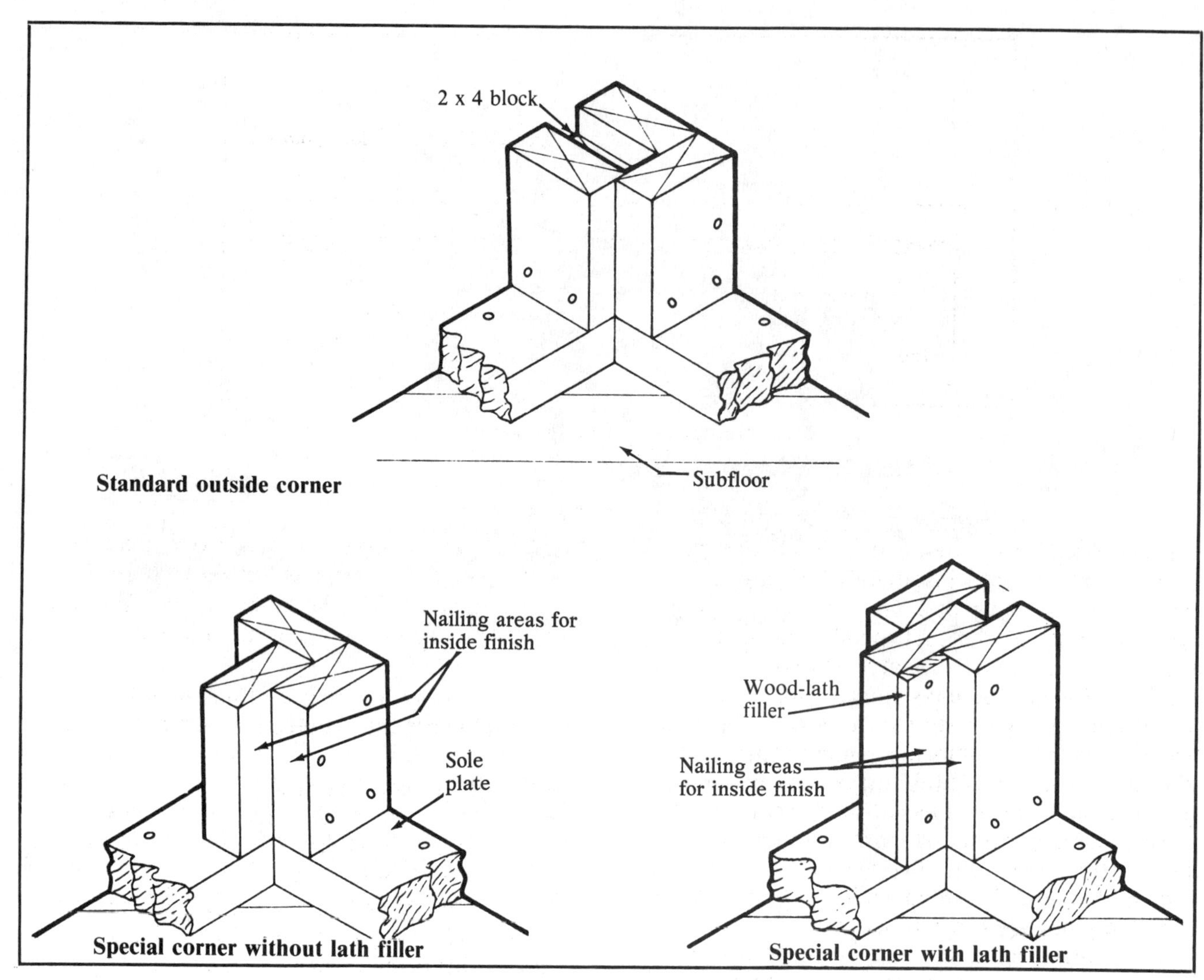

Examples of corner stud assembly
Figure 7-14

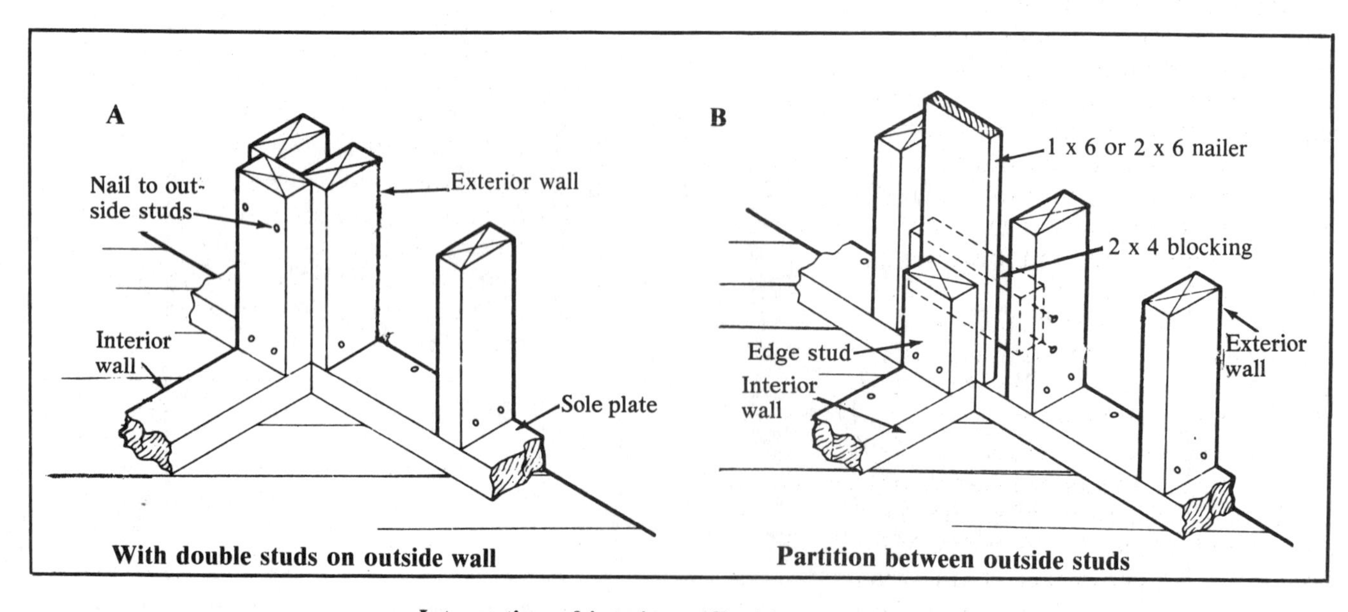

Intersection of interior wall with exterior wall
Figure 7-15

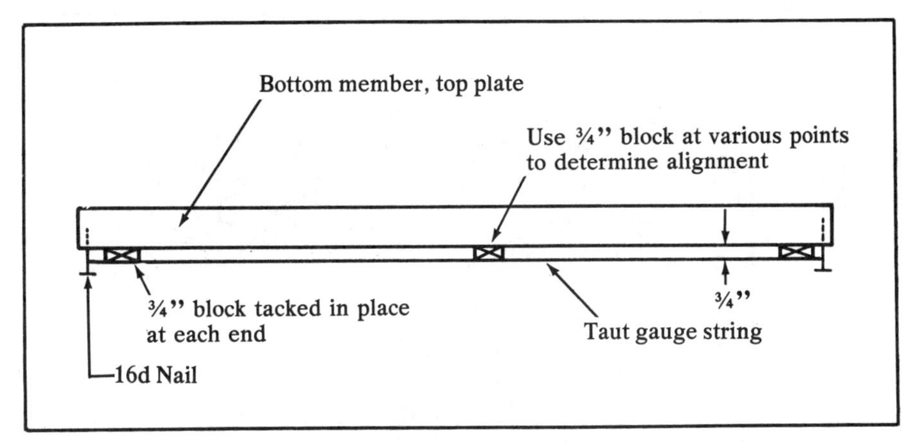

Determining wall alignment
Figure 7-16

The key to a solid tie-in of all walls and partitions is the proper installation of the second member of the top plate. The top plate is lapped over the joints of the first member at all corners and wall intersections. Use at least two 16d nails at overlaps. Stagger 16d nails about 16 inches apart to firmly join the two plate members. It's important that the walls be plumbed and aligned before the second member is added. Use a taut gauge string to align the outside walls, as shown in Figure 7-16.

Accurate alignment of outside walls is important. Rafters won't fit when the side walls are out of alignment. Gable or end walls won't be straight. Use plenty of temporary braces and get the plumb and alignment right. It prevents problems before they happen. Quality construction demands vertical walls and square corners. Leave all temporary bracing in place until the roof is on and the sheathing is applied to the outside walls.

In balloon framing, both the wall studs and the floor joists rest on the sill, as shown in Figure 7-17. The studs and joists are toenailed to the sill with 8d nails and nailed to each other with three 10d nails.

The ends of the second-floor joists bear on a 1 x 4 or 1 x 6 ribbon let into the studs. The joists are also nailed to the studs at this point with four 10d nails. In addition, the end joists that are parallel to the exterior end wall on both the first and second floors are nailed to each stud.

Fire stops in balloon framing are extremely important, and are required by most codes. Use 2 x 4 blocking between the studs. It adds strength as well as serving as a good fire stop. See Figure 7-17.

Diagonal braces can be inserted in the corners of the wall of each story after the walls have been erected and plumbed. These braces are usually 1 x 4's. A notch is cut in each stud at the point where the brace crosses it. This allows the brace to fit flush with the outside of the studs. If an opening near the corner makes it impossible to use diagonal braces, use knee braces as shown in Figure 7-18. These are installed in the same manner. This bracing system can also be used in other types of frames.

Figure 7-19 shows how to make the joist tie-in at the corners in a balloon frame.

Window and Door Framing

Figure 7-20 shows typical window framing. Notice that the builder was careful not to damage trees already existing on the site.

The members used over window and door openings are called headers. See Figure 7-21. As the span of the opening increases, the depth of the header must increase to support the ceiling and roof loads. A header is built-up from 2-inch members, usually spaced with 3/8 inch lath or wood strips. Also, 3/8-inch-thick plywood the same size as the header can be sandwiched between the 2-inch members. This gives added strength. Be sure to nail through both the 2-inch pieces and the spacer when nailing up the header.

The header is supported at the ends by the inner (jamb) studs of the double-stud joint at the exterior walls and interior load-bearing walls. The species used for floor joists is generally strong enough to use for header material in houses. Here is a good guide for headers:

Maximum Span (Feet)	Header Size (Inches)
3½	2 x 6
5	2 x 8
6½	2 x 10
8	2 x 12

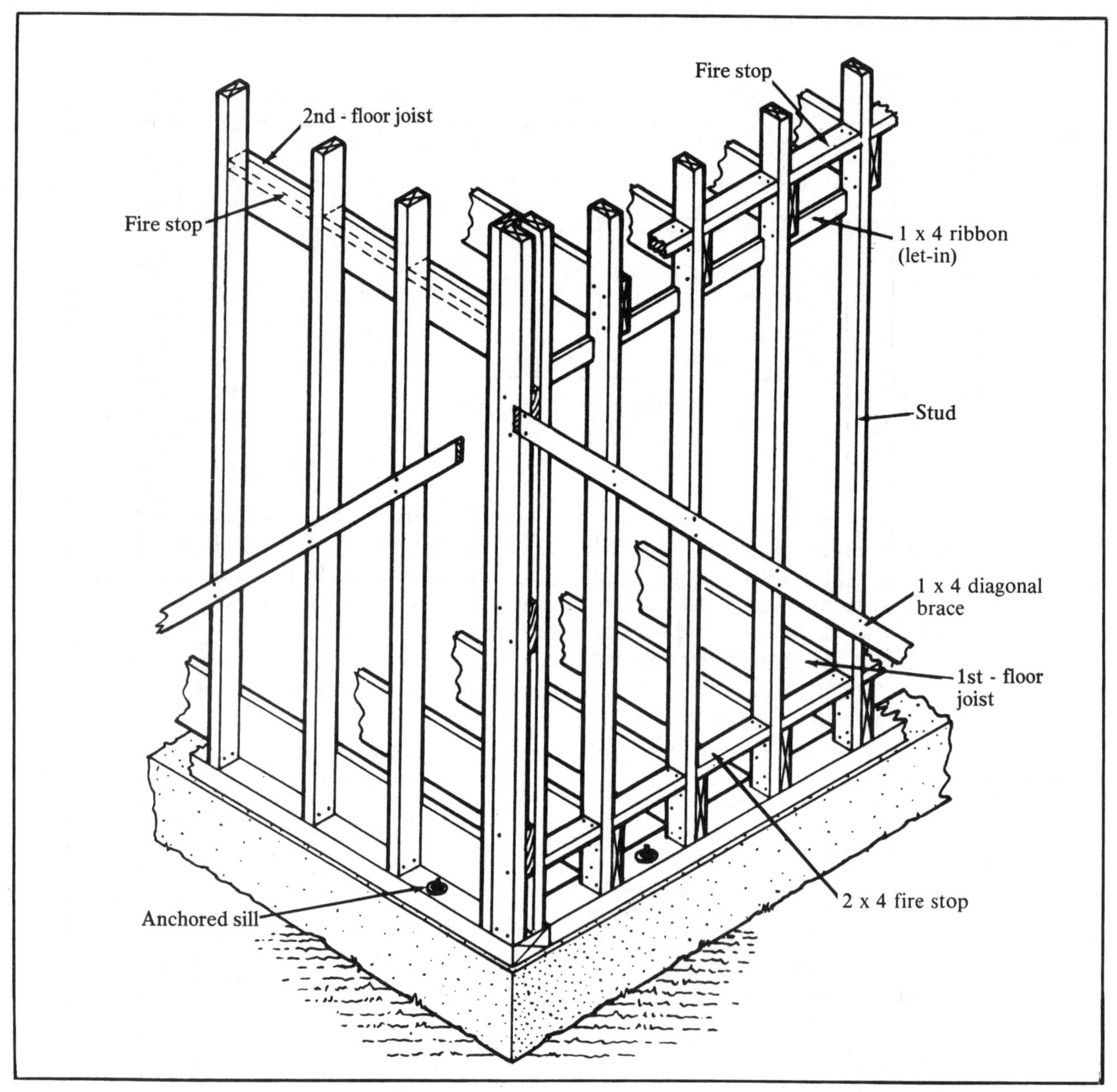

Wall framing used in balloon construction
Figure 7-17

Wider openings might require specially-designed trussed headers. See Figure 7-22.

In trussed headers, the diagonal members of the truss transmit the load directly to the sides of the opening. It's then carried down by the studs to the sill or load-bearing members.

Make sure that the location of studs, headers, and sills around window openings conforms to the rough opening sizes recommended by the window manufacturers. Rough openings (RO) for some styles and makes of windows can be tricky. When framing the window openings, you may want to have sample window sizes on the site. This should guarantee proper window fit.

The framing height to the bottom of the window and door headers is normally based on door heights of 6'8''. To allow for the thickness and clearance of the head jambs and the finish floor, the bottom of the headers are usually located 6'10'' inches to 6'11'' inches above the subfloor, depending on the thickness of the finish floor used.

End Walls
Compare Figures 7-23 and 7-24. They highlight the differences between the framing of the end walls in Western (platform) and balloon framing. Figure 7-23 shows a commonly used method of wall and ceiling framing for platform construction in a one and one-half or two-story house with finished

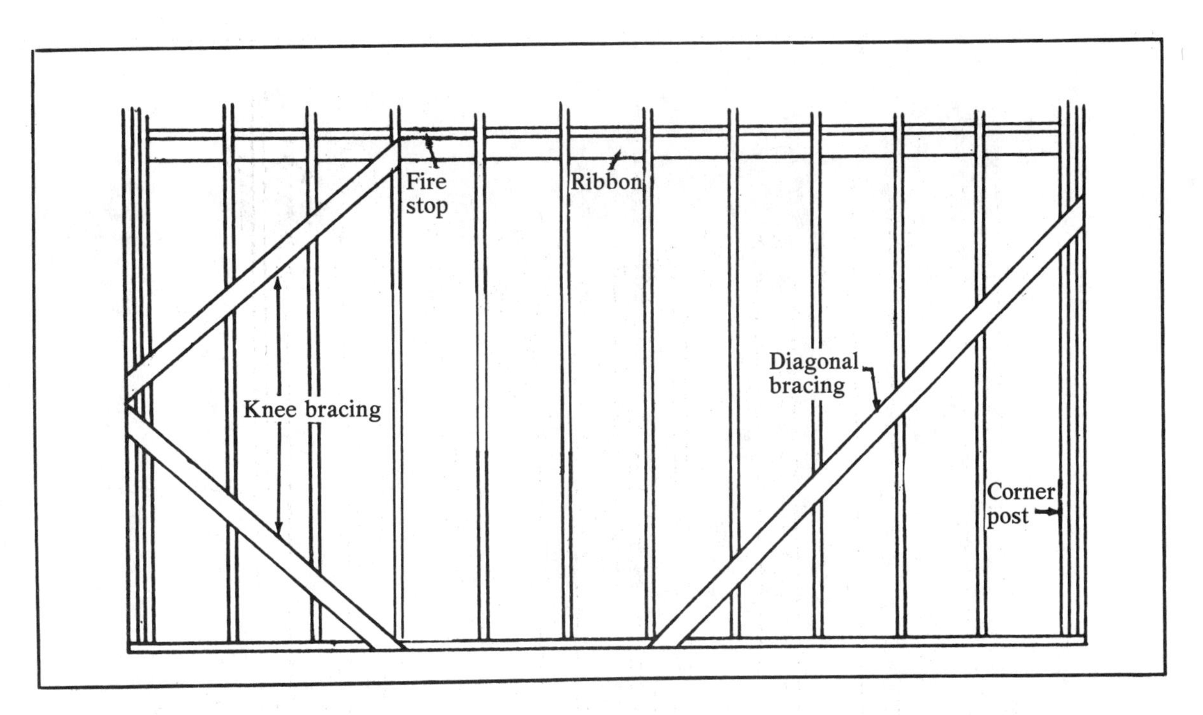

Bracing a wall section
Figure 7-18

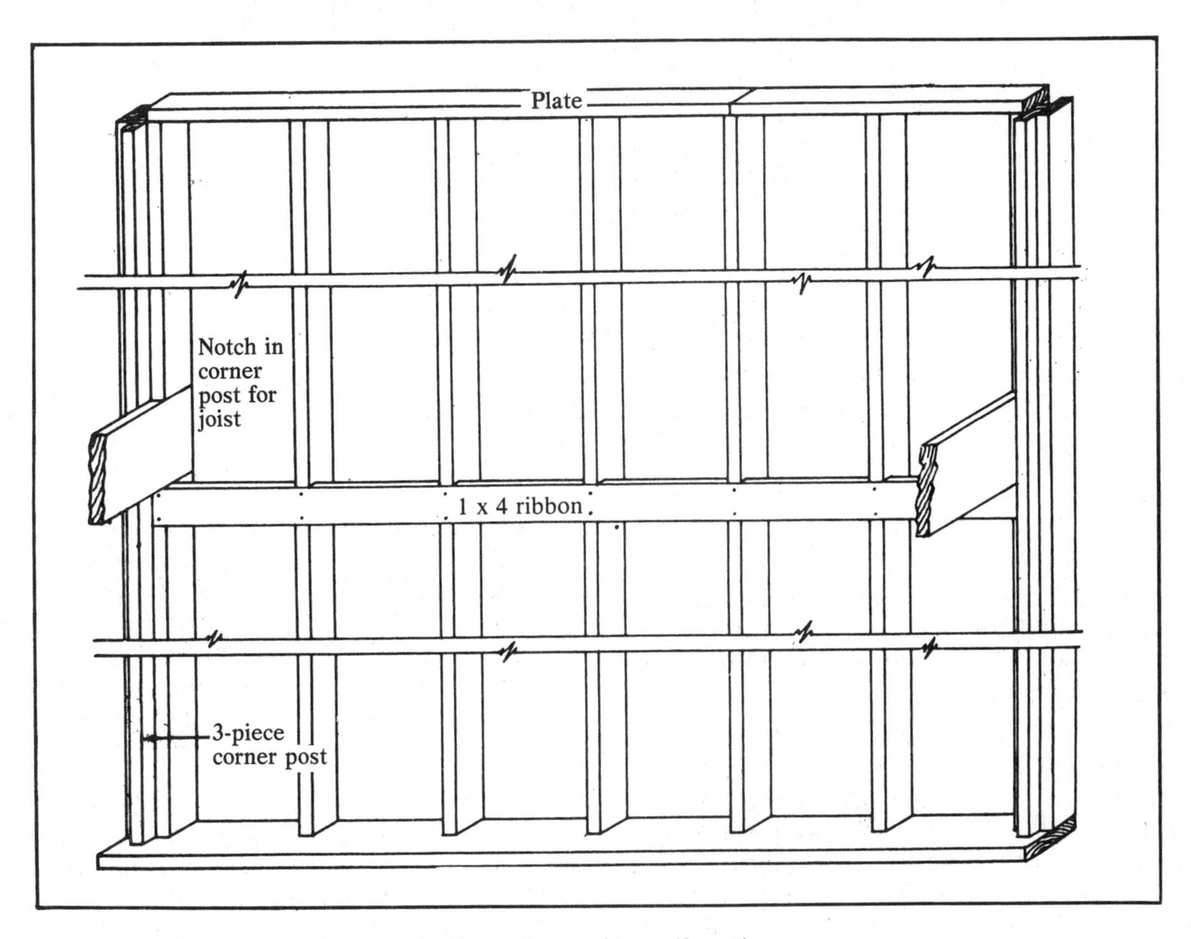

Balloon frame side-wall section
Figure 7-19

Typical window (RO) framing
Figure 7-20

rooms above the first floor. The edge floor joist is toenailed to the top wall plate with 8d nails spaced 16 inches on center. The subfloor, sole plate, and wall framing are then installed in the same way as for the first floor.

In balloon framing, the studs continue through the first and second floors. The edge joist can be nailed to each stud with three 10d nails. At the first and second floor levels, 2 x 4 fire stops are installed between each stud spacing.

Interior Walls

The interior walls in a house with conventional joist and rafter construction are usually both loadbearing walls and partitions, if they run at right angles to the floor joists. Walls that run parallel to the joists are usually non-loadbearing. Use 2 x 3 studs in a wall like this. Walls that help hold up the roof should be framed with 2 x 4 studs. Some builders use 2 x 4 studs throughout.

The interior walls have a sole plate and may have a double top plate just like exterior walls. The upper member of the top plate is used to tie intersecting and crossing walls together by overlapping. A single stud can be used to frame a door opening in non-loadbearing partitions. In loadbearing walls, use double studding as shown in Figure 7-25.

When trussed rafters (roof trusses) are used, no loadbearing interior partitions are required. The location of walls is determined by room size. The bottom chord of the trusses is the anchor for crossing partitions. When partition walls are parallel to and located between trusses, use 2 x 4 blocks nailed between the lower chords to anchor the partition plates.

The studs may be spaced either 16 or 24 inches on center. Many builders use 24-inch spacing to reduce framing materials and manhours. Studs spaced 24 inches on center provide a structurally sound wall.

Lath Nailers

Wall and ceiling framing must provide both vertical and horizontal fastening points for plaster-

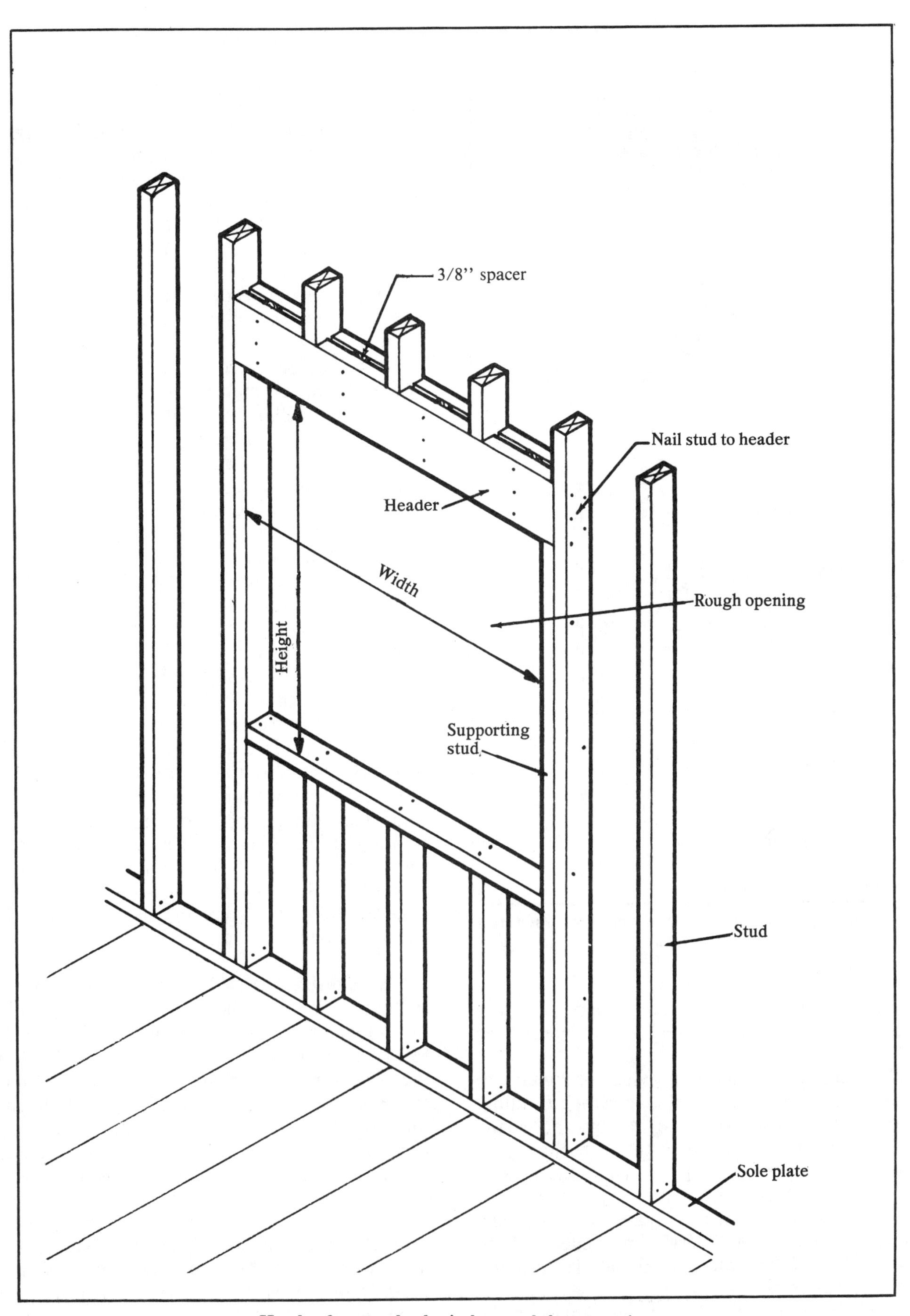

Header for standard window and door openings
Figure 7-21

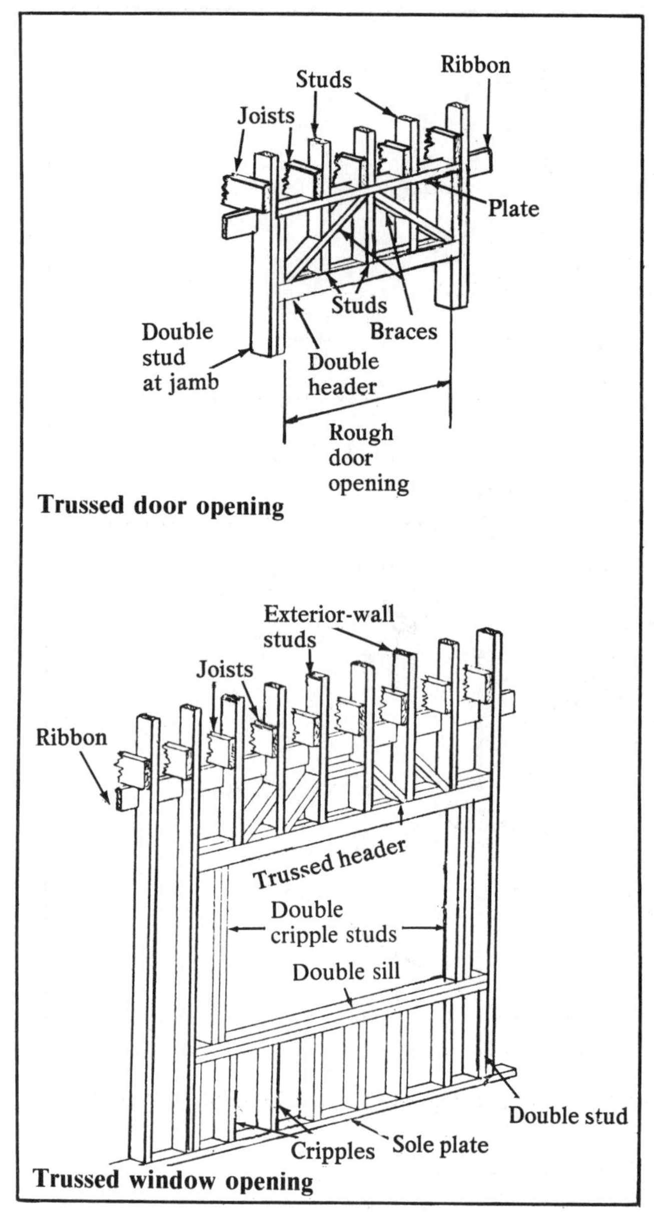

Special headers for wider window and door openings
Figure 7-22

base lath or drywall at all inside corners. In Figure 7-15, we saw two corner methods at intersecting walls.

There are several ways to provide horizontal lath nailers at the junction of wall and ceiling framing. Figure 7-26 A shows double ceiling joists above the wall. The joists are spaced so that each joist provides a nailing point. In Figure 7-26 B, the parallel wall is located between two ceiling joists. A 1 x 6 board is nailed to the top plates with backing blocks spaced 3 to 4 feet on center. Many carpenters use 2-inch thick scrap material fastened with 16d box (wax-coated) nails.

When the partition wall is at right angles to the ceiling joists, let in 2 x 6 blocks between the joists as shown in Figure 7-26 C. They are nailed to the top plate and toenailed to the ceiling joists. Use 2-inch scrap lumber here also.

Estimating Studs for Western Framing

Here's how to figure the lumber required for stud framing. Start with the basement plan, and measure the number of linear feet of inside stud partitions. Then move to the first floor. Begin at one corner of the building. Measure the linear feet all the way around the outside walls. Measure the linear feet of the inside stud partitions running in one direction. Then measure the partitions running at right angles.

Follow the same procedure for all exterior walls and partitions on higher floors.

For a wall without openings, find the number of studs needed by multiplying the length of the wall in feet by 3/4 and add 1 piece. This gives the number of studs spaced at 16 inches on center. Figure 7-27 may be used to estimate partition studs. Figure 7-28 gives coverage figures for all standard stud walls.

Here's the rule for walls with openings: For 16 inches o.c. framing, estimate one stud for each linear foot of wall. This allows for doubling studs at corners and at window or door openings.

The rule for top and bottom plates is: Multiply the linear feet of all walls and partitions by 2. If a double plate consisting of two top members and a single bottom plate are used, multiply the linear feet of walls and partitions by 3.

Here's a rule of thumb that can prevent some errors. The number of studs required in Western framing 16 inches o.c. will be about equal to *the total square feet of floor (omitting basements) divided by 3*. Thus, a 2,400 square-foot house would require approximately 800 studs for all exterior walls and interior partitions.

Estimating Studs for Balloon Framing

In balloon framing, the studding extends from the sill to the top plate of the second story. The board feet of lumber will be essentially the same as for Western framing, but the lengths will be greater and you will need only half as many. Full-length studs from the sill to the top plate of the second story are required for exterior walls and for all double studding at openings. Shorter lengths can be used under and over openings, such as from the top of a first floor door or window opening to the top plate of the second floor. The inside partitions are figured like the partitions in Western framing.

Many builders add 10% when estimating stud requirements. This margin allows for waste and rejects.

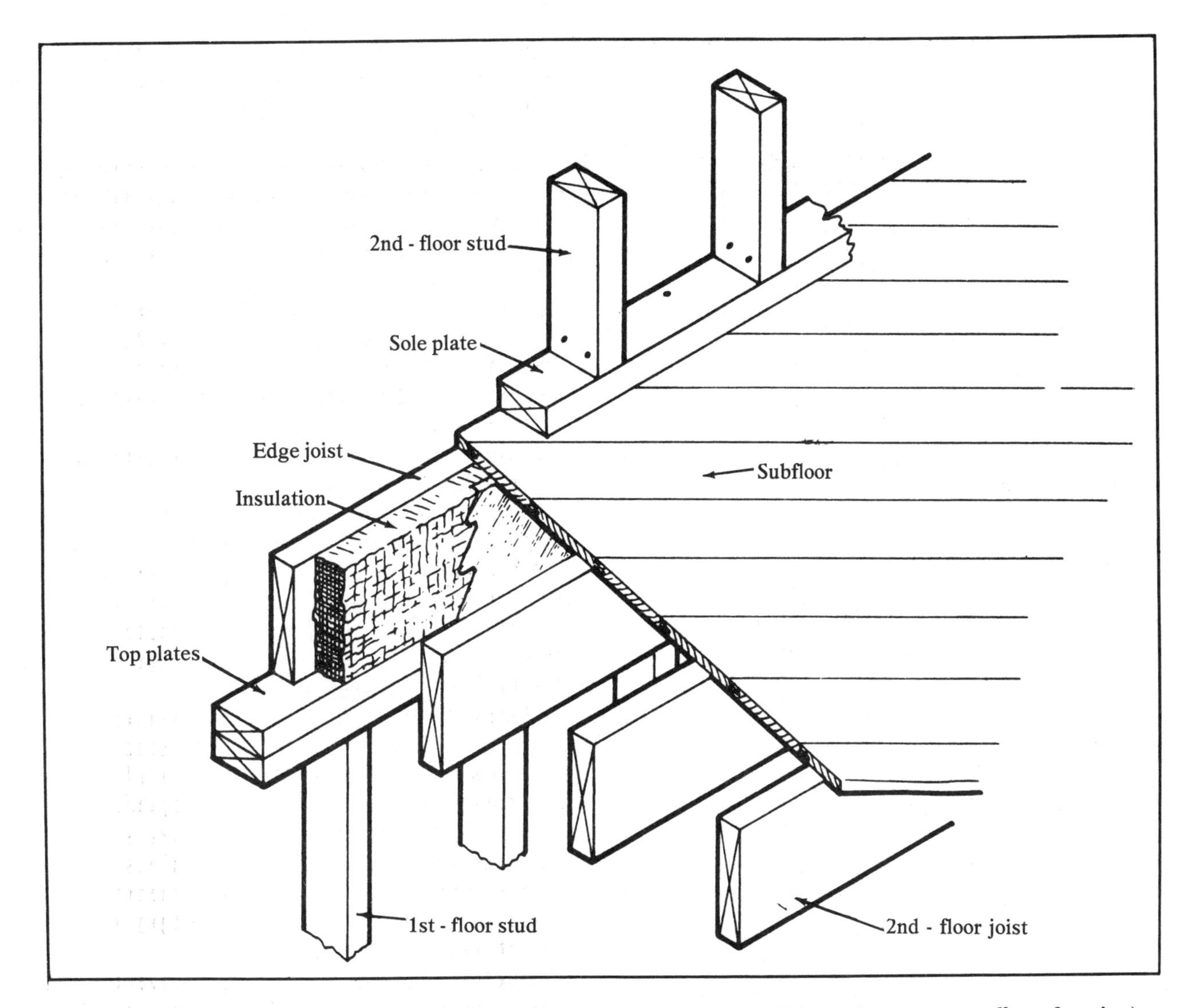

End-wall framing for platform construction (junction of first-floor ceiling and upper-story floor framing)
Figure 7-23

To figure the amount of ribbon required in balloon framing, measure the linear feet of the outside supporting walls (walls supporting second-floor joists). Any wall parallel to the joists does not require a ribbon. See Figure 7-24.

Estimating Studding Manhours

Exterior wall studding and plates require about 21 skilled and 6 unskilled labor hours per 1,000 BF. (See Figure 6-38 in Chapter 6.) Partition studding and plates require about 20 skilled and 5 unskilled hours per 1,000 BF.

Using our previous example, a building 24 x 40 feet, a 40-foot exterior wall with an 8-foot ceiling height and stud spacing at 16-inches on center has 1.05 board feet per square foot of wall area. See Figure 7-28.

Our 40-foot wall has a 320 square feet of wall area and would need 336 board feet of lumber. Exterior wall studding requiring 21 skilled and 6 unskilled hours per 1,000 board feet will require 2.1 skilled and 0.6 unskilled hours per 100 board feet.

Thus:

2.1 (skilled hours per 100 BF) times 336 BF equals approximately 7 hours.

.60 (unskilled hours per 100 BF) times 336 BF equals approximately 2 hours.

A total of 7 skilled and 2 unskilled hours are required to frame the 40-foot exterior wall.

Partition studding requires 2 skilled and 0.5 unskilled hours per 100 BF.

From Figure 7-28, a partition with 8-foot ceiling height and studs placed 16 inches o.c. has 1.12 BF per square foot of wall area. When using 2 x

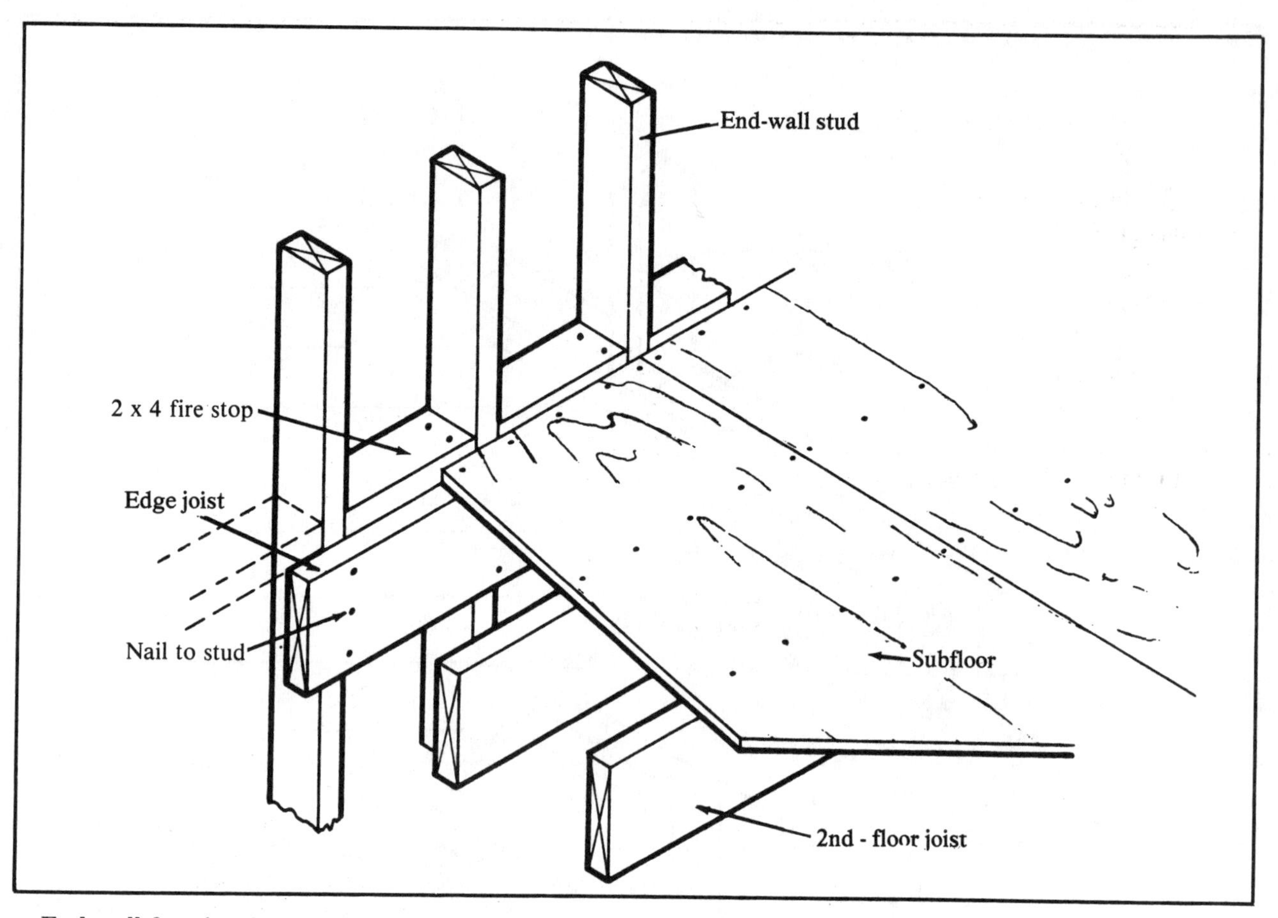

End-wall framing for balloon construction (junction of first-floor ceiling and upper-story floor framing)
Figure 7-24

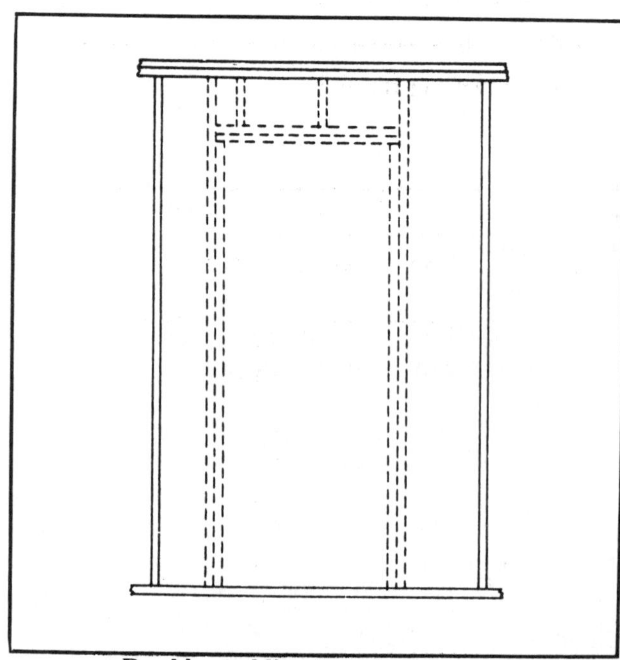

Double studding in a door opening in a load-bearing wall
Figure 7-25

4 studs, a 20-foot partition has 160 square feet and needs 179 BF of lumber and 3.5 skilled hours and 1 unskilled hour for framing.

Summary

Learn to order the right grades of lumber and the right header sizes for the work you do. Find the best method of framing for your jobs. It will earn extra profits for your company. There are several ways to frame every job. Find the method that saves both material and time. Sometimes what you save in lumber will be paid out in extra framing manhours.

Here's a good rule of thumb for manhours needed to erect walls (Western framing) with studs 16 inches o.c.:

Exterior wall framing requires 15 minutes per linear foot (2/3 skilled, 1/3 unskilled).

Partition wall framing requires 13 minutes per linear foot (2/3 skilled, 1/3 unskilled).

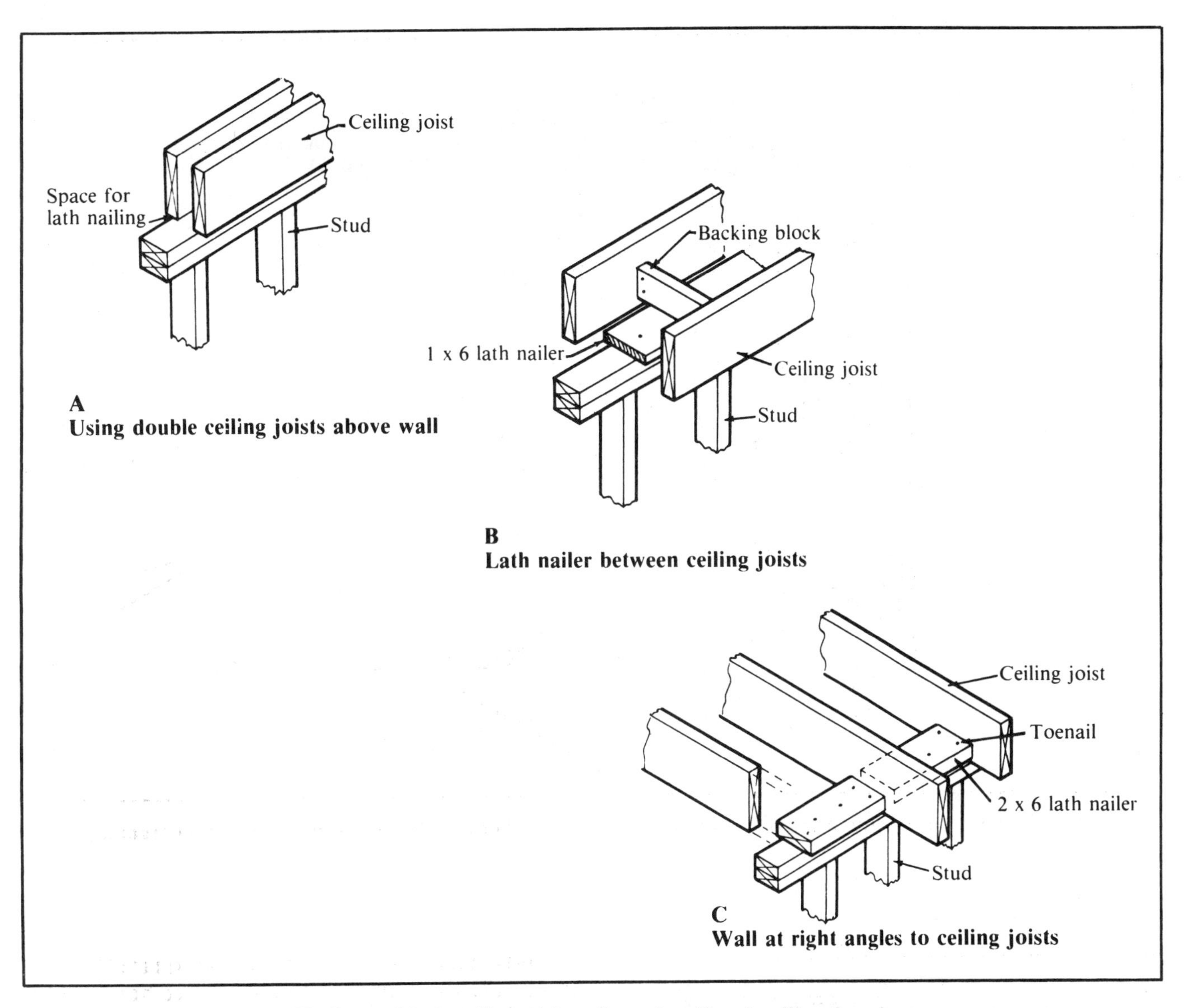

Horizontal lath nailers at junction of wall and ceiling framing
Figure 7-26

Length of partition in feet	Number of studs required	Length of partition in feet	Number of studs required
		11	9
2	3	12	10
3	3	13	11
4	4	14	12
5	5	15	12
6	6	16	13
7	6	17	14
8	7	18	15
9	8	19	15
10	9	20	16

Partition studs required 16'' on center
Figure 7-27

Exterior-Wall Studs			
Size of studs	Spacing on center	Bd. ft. per sq. ft. of area	Lbs. nails per 1000 bd. ft.
2 x 3	12"	.83	30
	16"	.78	
	20"	.74	
	24"	.71	
2 x 4	12"	1.09	22
	16"	1.05	
	20"	.98	
	24"	.94	
2 x 6	12"	1.66	15
	16"	1.51	
	20"	1.44	
	24"	1.38	

Includes an allowance for corner bracing,

Partition Studs			
Size of studs	Spacing on center	Bd. ft. per sq. ft. of area	Lbs. nails per 1000 bd. ft.
2 x 3	12"	.91	25
	16"	.83	
	24"	.76	
2 x 4	12"	1.22	19
	16"	1.12	
	24"	1.02	
2 x 6	16"	1.48	16
	24"	1.22	

Includes an allowance for top and bottom plates, end studs, blocks, backing, framing around openings and normal waste.

Board feet per square foot for wall studs
Figure 7-28

Figure 7-29 is your Wall Framing Checklist and Data Reference Sheet. Use it to maintain accurate wall framing records and as a guide on your next construction project. This list also includes materials and labor for framing and subflooring the second floor.

Wall Framing Checklist and Data Reference Sheet

1) The house is ☐1-story ☐2-story and ☐does ☐does not have an open basement.

2) The house is ________ feet by ________ feet and has ________ square feet of floor space.

3) The following lumber grades were used in all framing:

First Floor	**Grade Recommended**	**Grade Used**
☐Sill plates	No. 2	No.________
☐Box sill	No. 2	No.________
☐Floor joists	No. 2	No.________
☐Bridging	No. 3	No.________
☐Subfloor	No. 3	No.________
☐Sole and top plates	No. 3	No.________
☐Studs:		
First story of two-story house	No. 2	No.________
One-story house	No. 3	No.________
☐Headers	No. 2	No.________
☐Fire stop	No. 3	No.________
☐Ceiling joists	No. 2	No.________
☐Rafters	No. 2	No.________
☐Roof boards	No. 3	No.________

☐Other (specify) ____

____

Second Floor		
☐Floor joists	No. 2	No.________
☐Bridging	No. 3	No.________
☐Floor boards	No. 3	No.________
☐Studs	No. 3	No.________
☐Headers	No. 2	No.________
☐Fire stop	No. 3	No.________

Figure 7-29

☐Ribbon	No. 2	No.________
☐Ceiling joists	No. 2	No.________
☐Rafters	No. 2	No.________

Other (specify) ____

____

4) No. 2 grade lumber ☐was ☐was not used throughout. If it was used throughout, explain ________

____

5) No. 3 grade lumber was used as indicated in item 3, above. The waste (due to unuseable pieces) ☐was ☐was not excessive.

6) The framing method used was:

☐Modern braced

☐Western (Platform)

☐Balloon

☐Plank-and-beam

7) The framing method used ☐is ☐is not the predominant framing method in this area.

8) ☐The framing was assembled horizontally and raised into place. The following items were installed prior to tilt-up:

☐Window framing

☐Door framing

☐Bracing

☐Sheathing

☐Ribbon (for balloon frame)

☐Other (specify)____

9) ☐The framing was only partially assembled horizontally. Explain ________________

____

10) ☐The framing was not assembled horizontally. Explain ________________

____

11) The sole plates and top plates ☐were ☐were not marked simultaneously for stud spacing, partition intersections, and window and door openings.

12) The joist spacing ☐is ☐is not the same as the stud spacing, and the studs ☐do ☐do not rest directly over the joists.

13) The following were preassembled prior to beginning wall framing:

☐Corner posts

☐Intersection T's

☐Headers

☐Complete window opening units

☐Complete door opening units

14) Wall anchorage to sill plate ☐was ☐was not required.

15) Wall anchorage to sill plate was accomplished by ☐metal strap ☐sheathing overlap on sill plate.

16) The top member of the top plate ☐does ☐does not lap the joints of first member at all corners and wall intersections. If not, explain (Most codes require overlapping) ____________________

____

17) A line gauge ☐was ☐was not used to align the exterior walls before installing the top member of the top plate.

18) Temporary braces ☐were ☐were not used to plumb and align walls. (Temporary braces should remain in place until roof and wall sheathing are applied.)

19) All required fire stops ☐were ☐were not installed. If not, explain ____________________

20) All required bracing ☐was ☐was not installed. If not, explain ____________________
____.

Bracing consists of:

☐1 x 4 diagonal let-in braces

☐4 x 8 foot, ½-inch plywood installed at corners

☐Fiberboard, size ____________________

☐Other (specify) ____________________

21) Headers in exterior and load-bearing partition walls consist of double members of:

☐2 x 6 for a __________-foot span

☐2 x 8 for a __________-foot span

☐2 x 10 for a __________-foot span

☐2 x 12 for a __________-foot span

☐Other (specify) ____________________

22) Openings in nonbearing walls ☐are ☐are not double-studded with ☐2 x 4 ☐2 x 6 ☐2 x 8 headers.

23) Window and door RO's are based on ☐sample sizes ☐manufacturer's recommendations ☐house plans.

24) Height of door openings from bottom of header to subfloor ☐is ☐is not based on first-class door heights. The height of door opening is:

☐6 feet 10 inches

☐6 feet 11 inches

☐Other (specify) ____________________

25) Height of door openings allows for a __________-inch-thick underlayment and a __________-inch-thick finish floor material.

26) Interior wall studs are ☐2 x 4 ☐2 x 3.

27) Studs are spaced ☐16 inches ☐24 inches on center in ☐exterior ☐interior walls.

28) Lath nailers are provided at all corners and intersecting walls by:

☐Standard corner posts

☐Intersection T's

☐Other (specify) ____________________

29) Stud requirements were determined by measuring the linear footage of outside and inside walls, and applying the following formula:

☐Walls with openings: For studs spaced 16 inches o.c., use 1 stud per linear foot of wall.

☐Walls without openings: For studs spaced 16 inches o.c., multiply length of wall (in feet) by ¾ or 0.75 and add 1 stud.

☐Other (specify) ____________________

____

(A rule of thumb for determining number of studs required for Western framing spaced 16 inches o.c. is: The total square footage of floor space divided by 3.)

☐Ten percent added for waste

30) Sole and top plate requirements were determined by:

☐Linear footage of all walls, multiplied by 2. (For a sole plate and single-member top plate.)

☐Linear footage of all walls, multiplied by 3. (For a sole plate and double-member top plate.)

31) Material costs:

Sole plates (2 x ____)	________LF	@	$_______LF	=	$__________
Top plates (2 x ____)	________LF	@	$_______LF	=	$__________
Studs:					
Precut	________pcs	@	$________ea	=	$__________
8-foot	________pcs	@	$________ea	=	$__________
16-foot	________pcs	@	$________ea	=	$__________
2-story	________pcs	@	$________ea	=	$__________
Fire stop	________LF	@	$_______LF	=	$__________
Ribbon	________LF	@	$_______LF	=	$__________
Floor joists:					
2nd floor (2 x ____ x ____)	________pcs	@	$________ea	=	$__________
Subflooring (2nd-floor):					
Boards (1 x ____ x ____)	________LF	@	$_______LF	=	$__________
Plywood (4 x 8 x ____)	________pcs	@	$________ea	=	$__________
Other (specify)					
______________________	__________	@	$__________	=	$__________
Headers:					
2 x 6	________LF	@	$_______LF	=	$__________
2 x 8	________LF	@	$_______LF	=	$__________
2 x 10	________LF	@	$_______LF	=	$__________
2 x 12	________LF	@	$_______LF	=	$__________
Other (specify)					
______________________	__________	@	$__________	=	$__________
Nails:					
16d	________lbs	@	$________lb	=	$__________
10d	________lbs	@	$________lb	=	$__________
8d	________lbs	@	$________lb	=	$__________

Other materials, not listed above:

__________ __________ @ $__________ = $__________

__________ __________ @ $__________ = $__________

__________ __________ @ $__________ = $__________

__________ __________ @ $__________ = $__________

Total cost of all materials $__________

32) Manhour requirements were determined as follows:

☐ Exterior wall studding and plates require about 21 skilled and 6 unskilled labor hours per 1,000 BF when stud spacing is 16 inches o.c.

☐ Partition studding and plates require about 20 skilled and 5 unskilled hours per 1,000 BF when spaced 16 inches o.c.

☐ A rule of thumb for figuring manhours required for wall framing having 16-inch stud spacing is: Exterior wall requires 15 minutes per linear foot (⅔ skilled, ⅓ unskilled). Partition wall requires 13 minutes per linear foot (⅔ skilled, ⅓ unskilled).

33) A total of __________ linear feet of exterior walls were erected with:

☐ __________ skilled hours

☐ __________ unskilled hours

34) A total of __________ linear feet of partition walls were erected with:

☐ __________ skilled hours

☐ __________ unskilled hours

35) Labor costs:

Skilled ________ hours @ $______ hour = $__________

Unskilled ________ hours @ $______ hour = $__________

Other (specify)

__________ ________ hours @ $______ hour = $__________

Total $__________

36) Total cost:

Materials $__________

Labor $__________

Other $__________

Total $__________

37) Cost of wall framing per square foot of floor space is $__________ for this house with a ceiling height of __________.

38) Problems that could have been avoided __________

39) Comments __________

40) A copy of this Checklist and Data Reference Sheet has been provided:

☐Management

☐Accounting

☐Estimator

☐Foreman

☐File

☐Other: __________

Chapter 8

Framing the Ceiling and Roof

Even before you frame the roof on your spec house, put up a "For Sale" sign out front. The rooms are partitioned, and potential buyers can get a feel for the layout. A lot of houses are sold before the roof is framed. If, at this point, you sell the house yourself, you save the hefty commission that a realtor takes off the top.

Your "For Sale" sign should look sturdy and professional. Have it done by a good sign company. Emphasize the quality of the house. Use something like "*Another Quality Home By Jones Construction*." Or, if you're fully insulating and installing insulated windows and doors, use "*The Energy Saver*" or "*The Energy Miser*." Many buyers want to select carpet, room colors and paneling. Emphasize that buying early lets them make these changes at little added cost.

If you plan to make a living building houses, spend some time learning to sell houses. You don't need to be a slick salesman. Many people would prefer to buy directly from the builder. But you should learn the basics of a salesman's trade. Leave the paperwork and legal matters to an attorney or escrow company. Your bank will be able to recommend several for your consideration.

A realtor gets a commission of about 6%. That's $6,000 on a $100,000 house. You have to drive a lot of nails to earn that much. So, put that sign up! Later on, put an ad in the local newspaper. But plan to sell your own houses if you can. The man who raised the hog deserves a slice of the ham.

Now, while your sign is drawing inquiries from prospective buyers, let's get back to framing the roof.

Roof framing can get pretty complex, and can increase the construction cost by more than a few hundred dollars. A straight-line, plain gable roof has never kept many houses from selling. Why get involved with a complicated roof when it's not necessary? The idea is to keep costs down. A simple roof does just that. Avoid complex or cut-up roof designs.

Ceiling Joists

When wall framing has been plumbed and aligned and temporary braces are in place, it's time to begin with the ceiling joists. Ceiling joists and rafters are normally placed across the shortest dimension of the house. Walls should be laid out so that you can use ceiling joists of even lengths: 10, 12, 14, and 16 feet or longer. This avoids waste.

Ceiling joists run from exterior walls to load-bearing interior walls. The size of the joists depends on the span, wood species, spacing between joists, and the load on the second floor or attic. The house plan will usually give the size of joist to use. Local building codes generally stipulate joist size, span and spacing. Most lumber yards will have joist tables which also provide this information on the species that they sell. Where preassembled trussed rafters (roof trusses) are used the lower chords act as the ceiling joists. The truss also eliminates the need for load-bearing partitions.

Ceiling joists support the ceiling finishes. They often act as floor joists for second-story or attic floors. They also act as ties between exterior walls and interior partitions. In short, they lock the

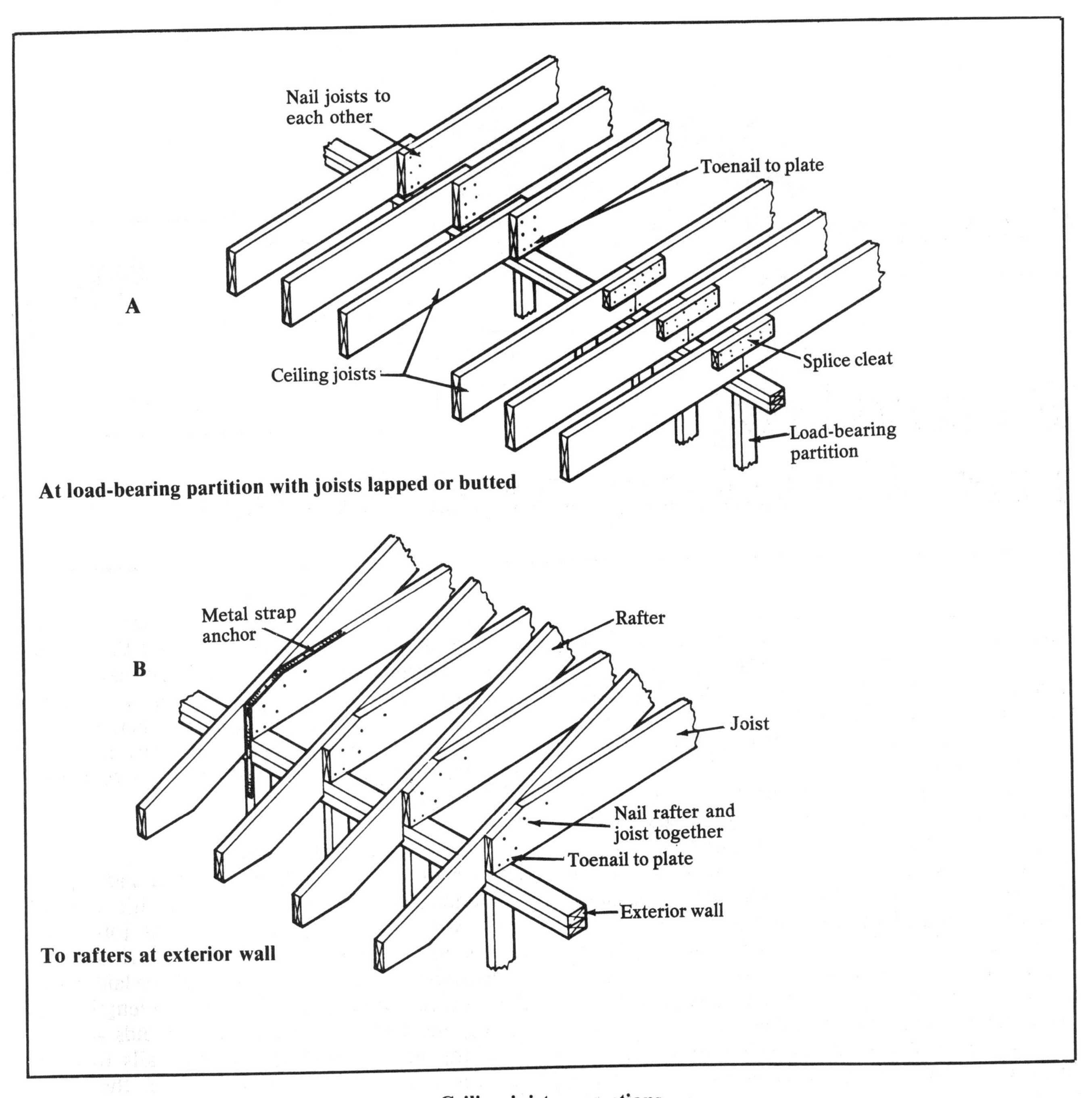

Ceiling joist connections
Figure 8-1

house together and resist the outward thrust of the rafters on pitched roofs.

Joists are nailed to the plate at outer and inner walls. They are also nailed together, directly or with wood or metal cleats, where they cross or join at the load-bearing partition (Figure 8-1 A) and to the rafters at the exterior walls (Figure 8-1 B). Toenail the joists at each wall with 8d nails. Don't try to use 16d nails. The wood will split.

In high wind areas, you'll need metal straps for anchoring the joists and rafters to the wall framing. When ceiling joists are perpendicular to the rafter run, collar beams and cross ties should be used on the rafters to resist thrust.

Flush Ceiling Framing Method

Where it is desirable to have a continuous flush ceiling over a large span, such as an open basement, a flush beam is used. The flush beam replaces the load-bearing partition for that particular area. A nail-laminated beam with sufficient load-carrying capacity supports the ends of the

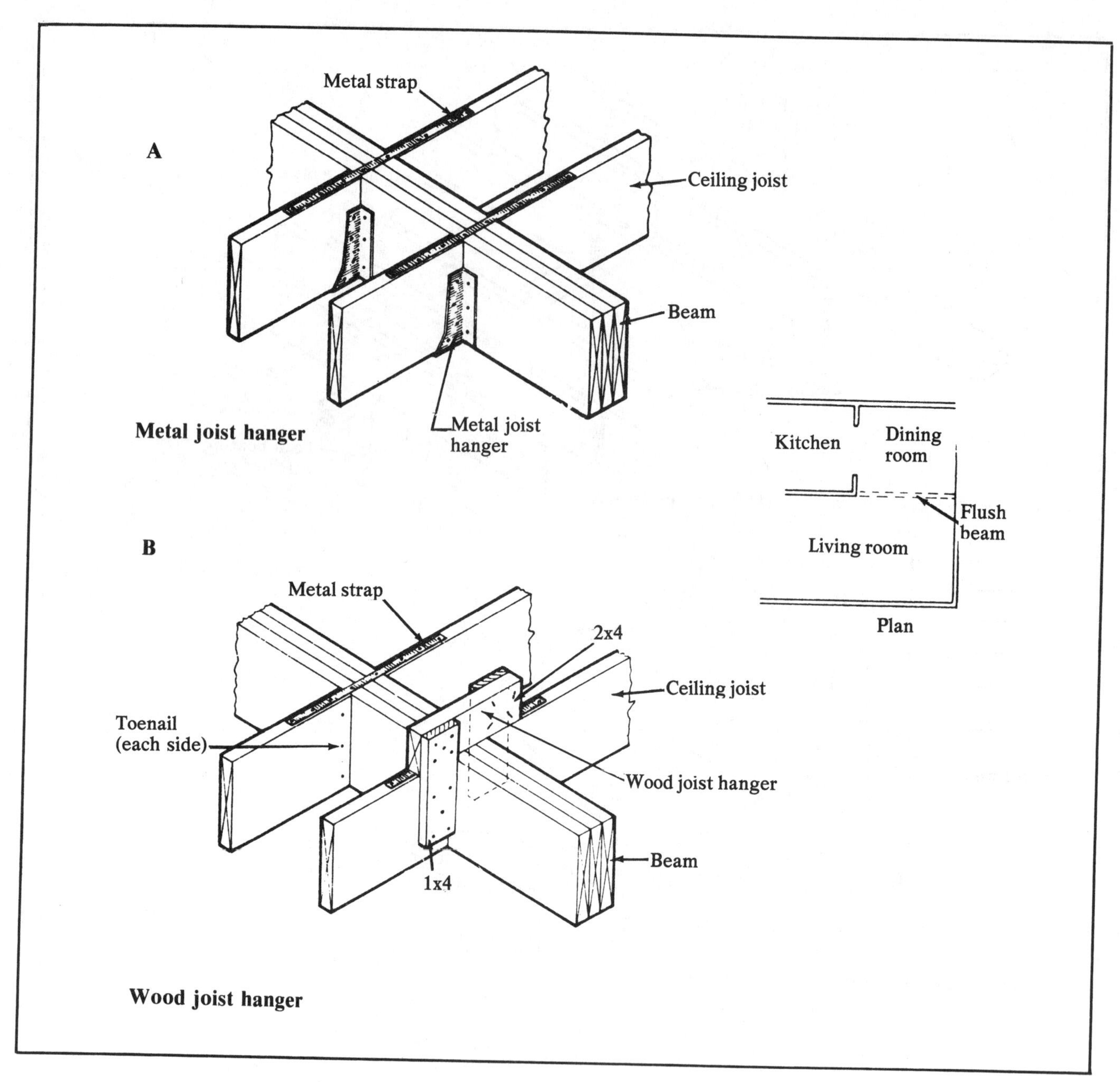

Flush ceiling framing
Figure 8-2

joists. Joists are toenailed into the beam, and are supported by metal joist hangers (Figure 8-2 A) or wood hangers (Figure 8-2 B), or a ledger.

Rafter thrust must still be taken into account. Metal strapping is used to tie in the system and resist the thrust. Nail the strapping to each opposite joist with four 8d nails.

Post-and-Beam Framing Method

Look at Figure 8-3. We discussed plank-and-beam framing in the preceding chapter. The post and beam are a part of that system. Contemporary houses often accent exposed beams as part of the interior design. These beams may also replace interior and exterior load-bearing walls. Exterior walls can consist of fully glazed panels between posts, requiring no additional support. The areas below interior beams can remain open or can be closed with wardrobes, cabinets, or light curtain walls.

Post-and-beam construction is seldom used in multi-unit housing.

Shear and racking resistance in the exterior walls is usually provided by solid masonry walls or fully sheathed frame walls between open glazed areas.

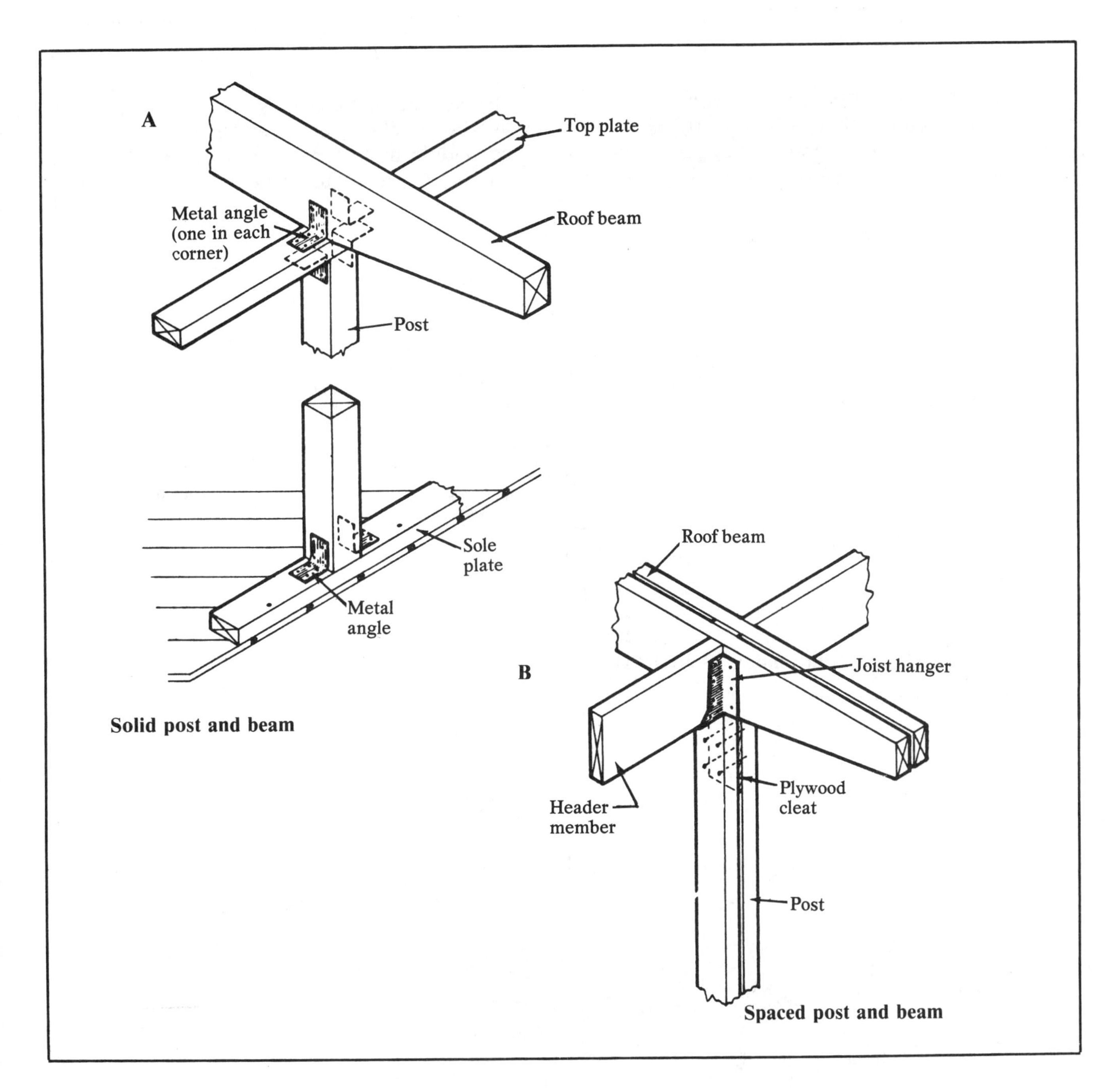

Post and beam connections
Figure 8-3

The roof in a post-and-beam house will usually be low-pitched or flat. It may have a conventional rafter-joist combination, or consist of thick wood decking spanning between beams. Insulation requirements dictate the type of roof decking to use.

Figure 8-3 shows, for both solid and spaced members, the connection of the supporting posts at the sole plate and roof beam. This connection is important, because it provides uplift resistance. The solid post and beam are fastened together with metal angles nailed to the top plate, the sole plate, and the roof beam, as illustrated in Figure 8-3 A. The spaced beam and post are fastened together with 3/8-inch or thicker plywood cleats extending between and nailed to the spaced members. See Figure 8-3 B. A wall-header member between beams can be fastened with joist hangers.

Where a continuous header is used between spaced posts, the beams should be fastened well and reinforced at the corners with lag screws or metal straps. Figure 8-4 A shows how to do this. This makes possible large openings for glass or panels.

Wood or fiberboard decking is a common roof cover. Thick wood decking is frequently used for beam spacings up to 10 feet. For longer spans, a special application is required. Depending on the type of fiberboard, 2- to 3-inch thick fiberboard decking is limited to a beam or purlin spacing of 4 feet.

The best technique is to use tongue-and-grooved solid wood decking, in either 3 x 6 or 4 x 6 sizes. These should be toenailed and face-nailed directly to the beams, and should be edge-nailed to each

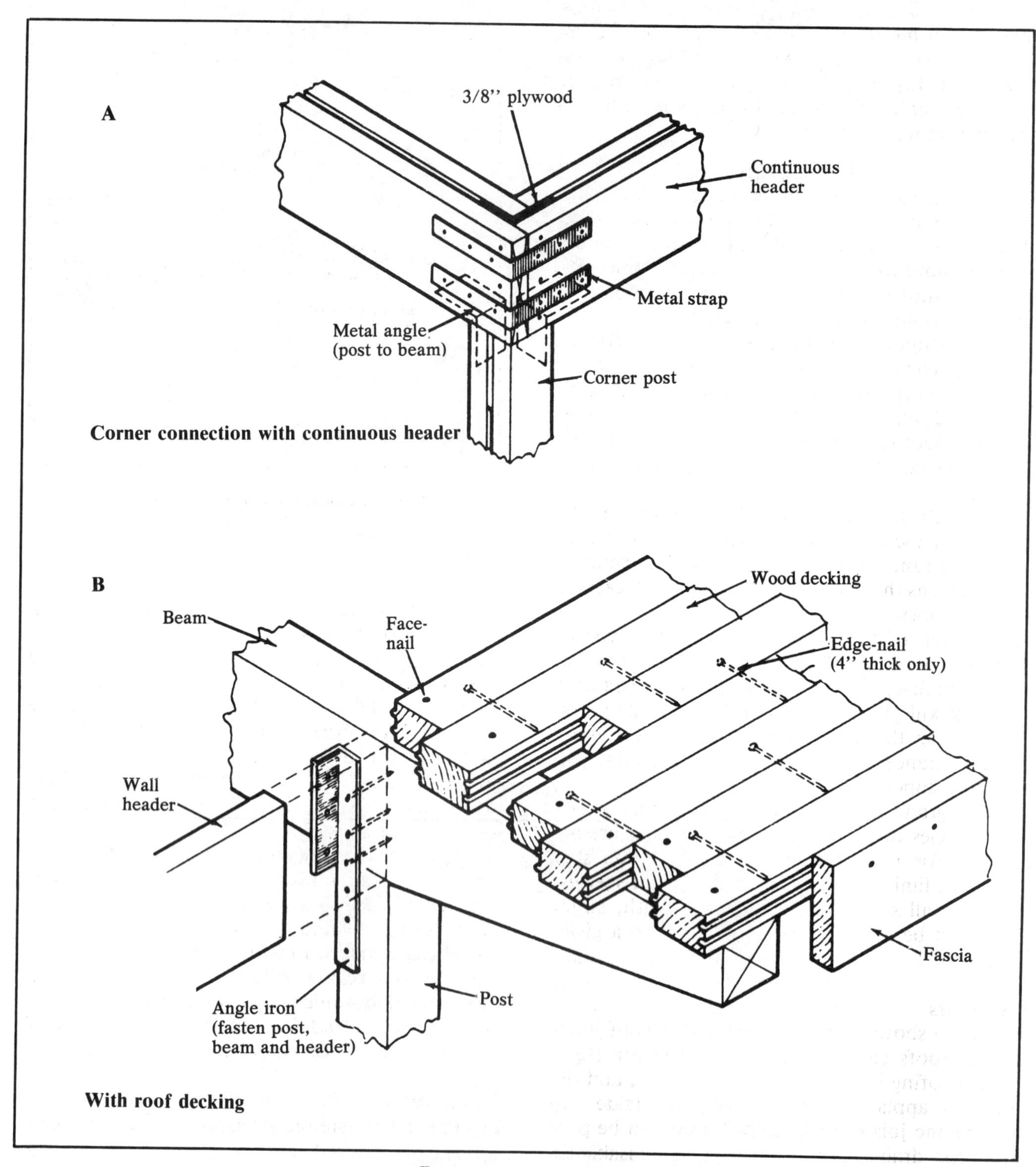

Post and beam details
Figure 8-4

other with long nails driven into predrilled holes. See Figure 8-4 B. Thinner decking is usually face-nailed to the beams. The decking must be square-end-trimmed to provide a good fit.

If additional insulation is required, apply fiberboard or an expanded foamed plastic, in sheet form, to the decking before the roof is installed. Watch the moisture content of the decking material. It should be near its in-service condition (about 15% moisture content) to keep joints from opening later. The more moisture the wood contains when installed, the wider the joints will be when the lumber has dried. Wide joints will ruin the appearance of the ceiling.

Roof Slopes

Roof slopes vary with climate, materials used and changes in buyer tastes. Roofs of straw, wood shingles and similar materials require steep, sloping roofs to move the water quickly. Steep slopes prevent build-ups of snow and ice. A flat roof in a snow belt is an invitation to disaster.

A contemporary house may require a flat or slightly pitched roof, an intermediate slope such as 5/12, or a steep slope. The two basic types are called *flat* and *pitched*. The flat type has one member acting as both roof and ceiling support. In pitched roofs, both ceiling joists and rafters (or trusses) are required.

The pitch of a roof is usually expressed as the number of inches of vertical rise in 12 inches of horizontal run. The rise is given first. For example, 4 in 12 means that the roof rises 4 inches in each 12 inches of horizontal run.

The pitch of the roof depends on the type of roofing to be used. Shingles don't work on a flat roof. A built-up roof that is properly applied on a flat deck will give good service for 20 to 25 years. By doubling the underlay and decreasing the exposure distance of wood or asphalt shingles, you can have slopes as low as 3/12.

No. 2 grade wood is normally used for rafters. Most species of softwood framing lumber are acceptable for roof framing, as long as you stay within the limits of the span table for that species. Since not all species are equal in strength, larger sizes must be used for weaker species on a given span.

Flat Roofs

Figure 8-5 shows two shed-type roofs. Roof joists for flat roofs can be dead level or have a slight pitch. Roofing is laid on top of the joists, and the ceiling is applied directly to the underside. By tapering the joists, a slight roof slope can be provided for drainage. Some slope is advisable in order to avoid the puddling of water on the roof.

Roofs using single-roof construction
Figure 8-5

The roof design usually includes an overhang beyond the walls. When insulation is installed, airways are provided just under the roof sheathing so that moisture can escape and winter condensation problems are minimized. Roofs of this type require larger-size members than roofs with a steeper pitch, because these larger members must carry both roof and ceiling loads. To further reduce heat loss in cold climates, rigid insulating materials are used over the decking.

To provide overhang on all sides of the flat roof, lookout rafters are used. See Figure 8-6. Lookout rafters are nailed to a double header and toenailed to the wall plate with 8d nails. The distance from the double header to the wall line is usually twice the overhang. Rafter ends may be finished with a nailing header, which also serves as a fastening point for soffit and fascia boards. Regardless of the type of soffit, be sure to provide ventilation.

Gable Roofs

Figure 8-7 A is a simple gable roof. It's the simplest form of the pitched roof. Both rafters and ceiling joists are required because of the attic space form-

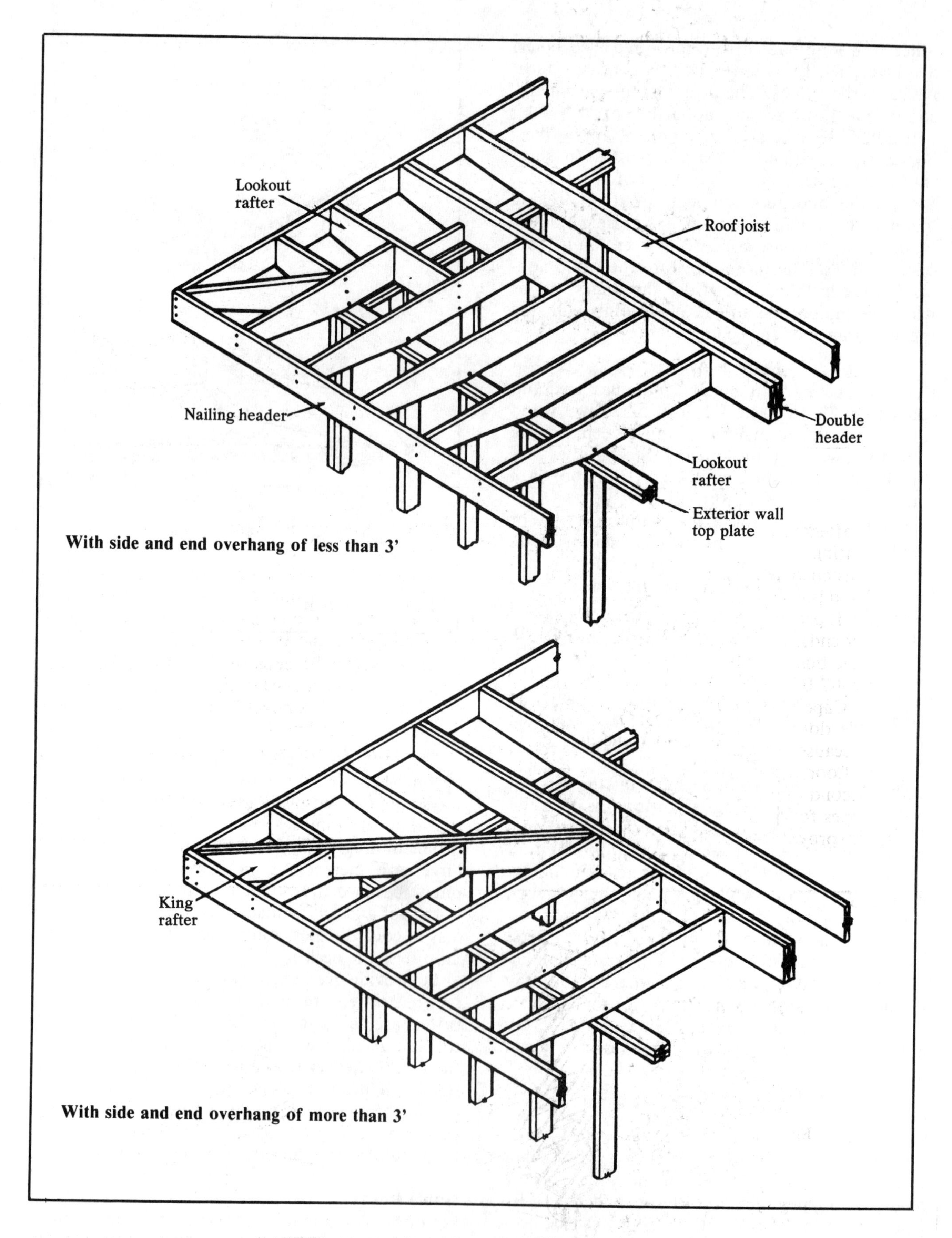

Typical construction of flat or low-pitched roof
Figure 8-6

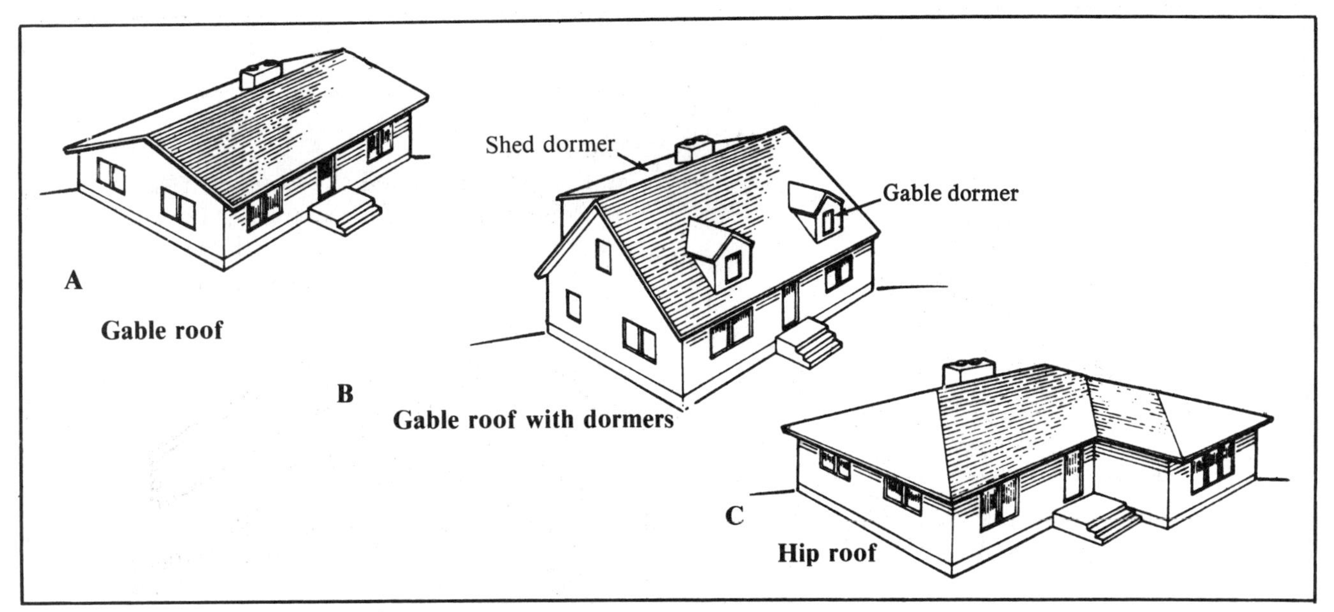

Types of pitched roofs
Figure 8-7

ed. All rafters are cut to the same length and pattern. Putting them up is relatively simple. Each pair is fastened at the top to a ridge board. The ridge board is usually a 1 x 8 member, if you use 2 x 6 rafters. It provides support and a nailing area for the rafter ends. Some builders prefer to use a 2 x 8 for a ridge board.

Figure 8-7 B is a variation of the gable roof. It's used for Cape Cod styles and often includes shed and gable dormers. This is basically a one-story house, because the majority of the rafters rest on the first-floor plate. Space and light are provided on the second floor by the shed and gable dormers. Roof slopes for this style have to be from 9/12 to 12/12 to provide headroom. This is a practical design, because it makes maximum use of materials.

Figure 8-7 C shows a hip roof. Center rafters are tied to the ridge board. Hip rafters support the shorter jack rafters. Cornice lines are carried around the perimeter of the building.

In pitched-roof construction, the ceiling joists are nailed in place after the interior and exterior wall framing is complete. Rafters should not be erected until ceiling joists are fastened in place. Otherwise, the thrust of the rafters will push out the exterior walls.

Rafters are usually precut to length, with angles cut at the ridge and eave, and with notches cut for the top plate. See Figure 8-8. Rafters are erected in

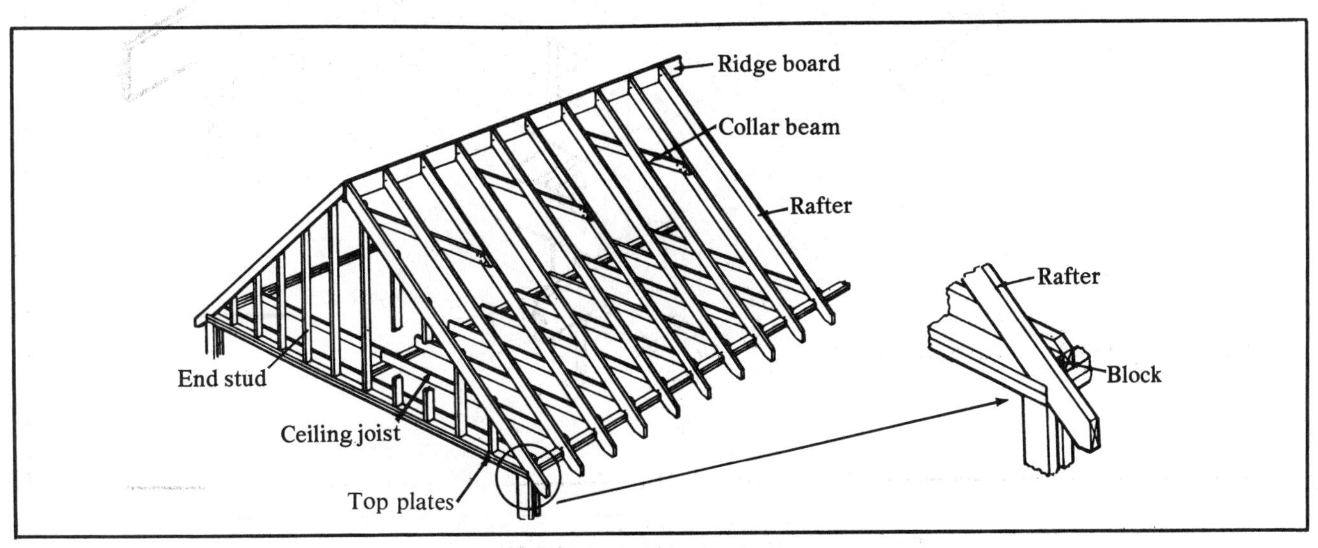

Overall view of gable-roof framing
Figure 8-8

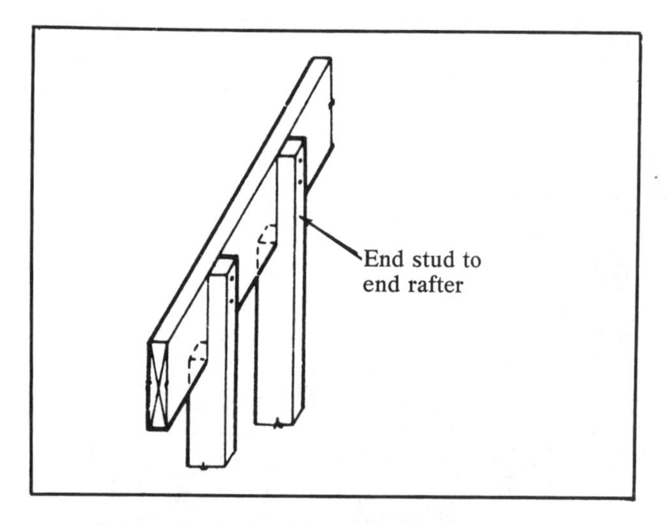

Connection of gable end studs to end rafter
Figure 8-9

pairs or in pair groups. Studs for gable end walls are cut to fit, and are nailed to the end rafters and the top plate of the end wall. See Figure 8-9. With a gable overhang, a fly (or floating) rafter is used beyond the end rafter. It's fastened with blocking and sheathing. (See Figure 8-10.)

Figure 8-11 shows an economical way to handle rafter overhangs. The ridge board and the bottom member of the side-wall top plate extend out to fit flush with the outside edge of the fly rafter. The extension of the bottom member does not interfere with the overlapping of the top member at the corners. A short top member is added at the extension.

Hip Roofs

Hip roofs are framed the same as a gable roof at the center section of a rectangular house. The ends are framed with hip rafters which extend from each outside corner of the wall to the ridge board at a 45-degree angle. Jack rafters extend from the top plates to the hip rafters. See Figure 8-12.

When roof spans are long and slopes are flat, use collar boards between opposing rafters. For steep slopes and shorter spans, collar boards may be 1 x 6 material nailed with 8d nails on every third pair of rafters. In 1½ story houses, use 2 x 4's as collars at each pair of rafters. These also serve as ceiling joists for the finished rooms. Follow the nailing schedule in Figure 6-4 in Chapter 6 when installing rafters and ceiling joists.

Valleys

The valley is the internal angle formed where two sloping roof surfaces meet. The valley rafter forms the bottom of the valley. When two equal-size roof sections come together, use double valley rafters and use rafter material that is two inches deeper than the common rafters. These two extra inches are needed to provide full contact with the jack

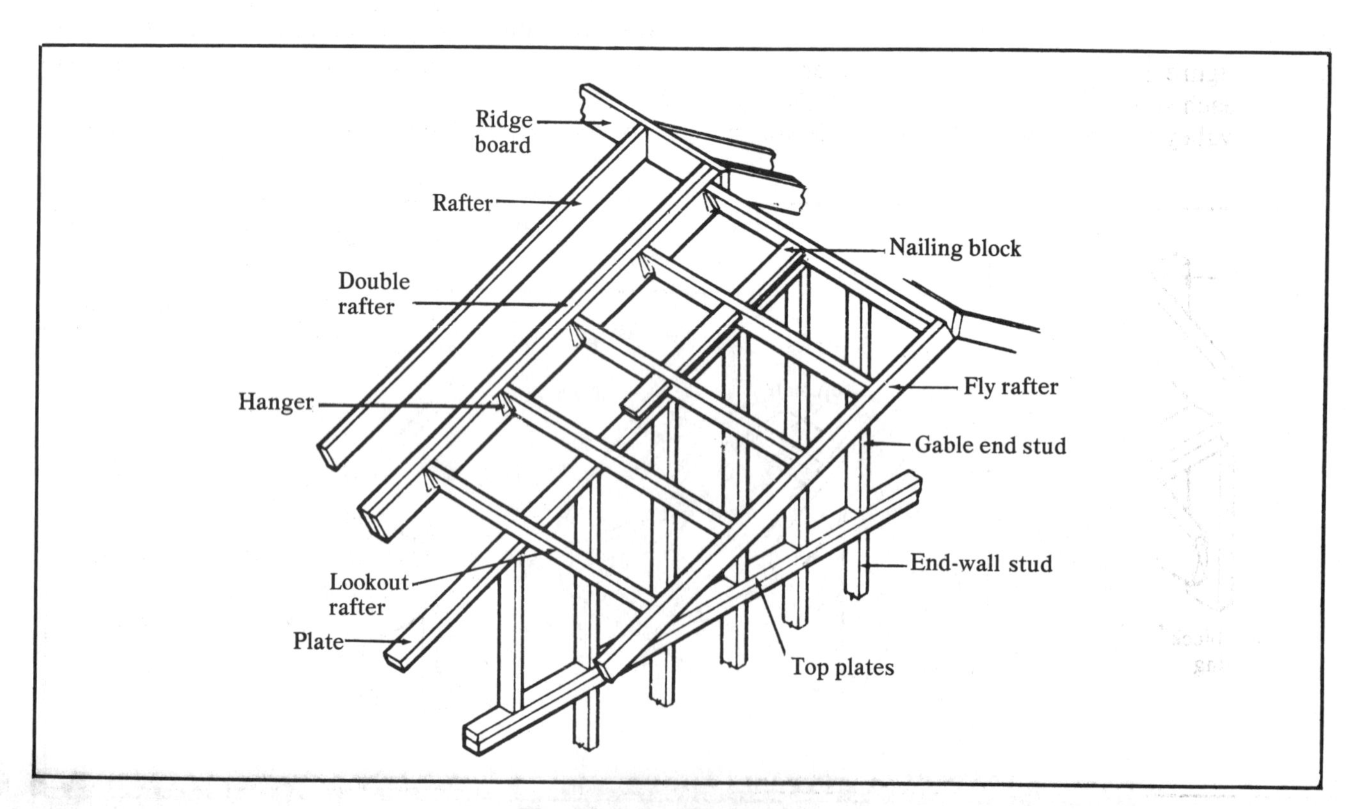

Fly rafter in gable overhang
Figure 8-10

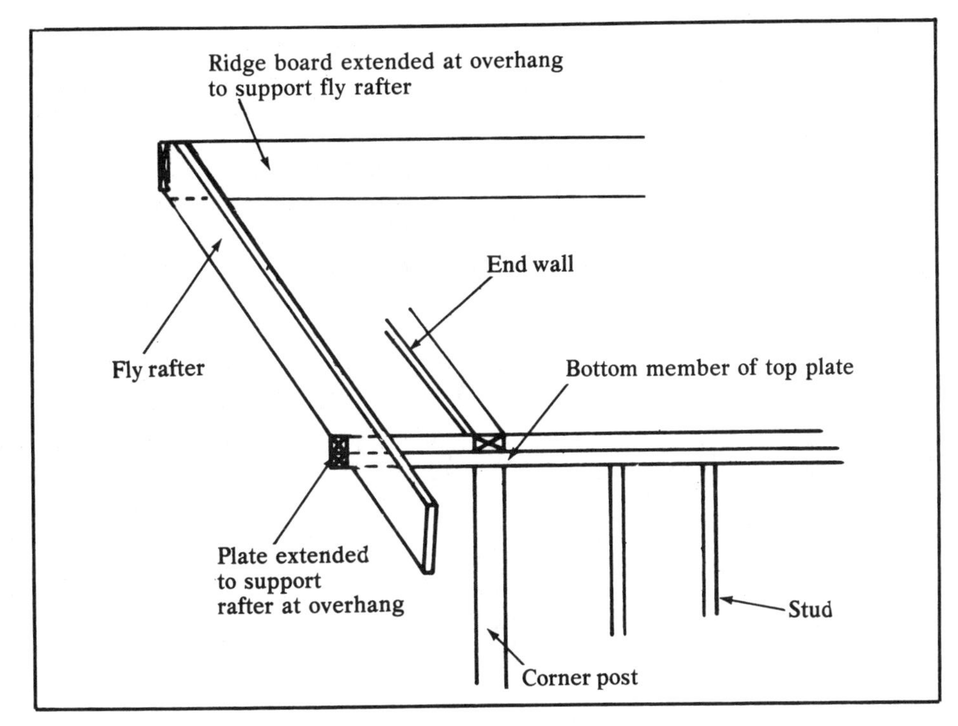

Rafter overhang supported by extended ridge board and top plate
Figure 8-11

rafters. See Figure 8-13. Jack rafters are nailed to the ridge and toenailed to the valley rafter with three 10d nails.

Dormers
Look at Figure 8-14. This is a small dormer. The rafters at each side are doubled. The side studs and the short valley rest on these members. Side studs may also be carried past the doubled rafter, and may bear on a sole plate nailed to the floor framing and the subfloor. This type of framing may also be used for the side walls of shed dormers. The valley rafter is tied to the roof framing by a header. Use previously described methods of fastening at the top plates.

If the style of the houses you build permits

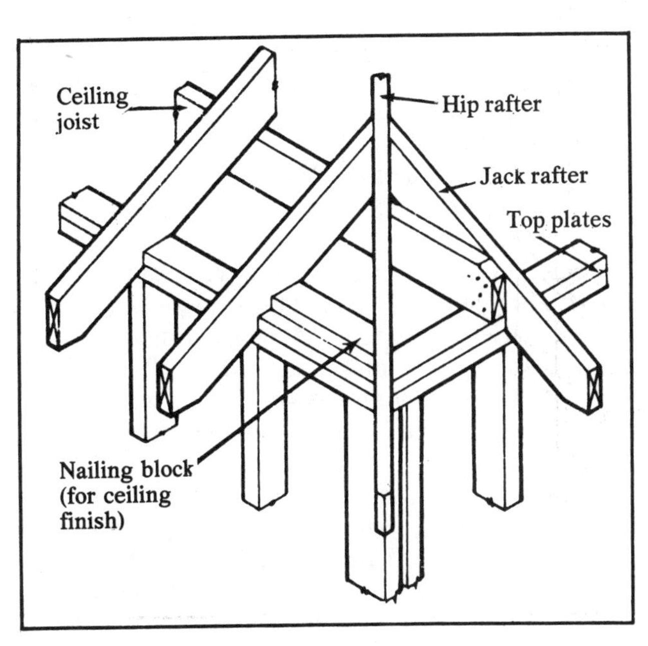

Detail of corner of hip roof
Figure 8-12

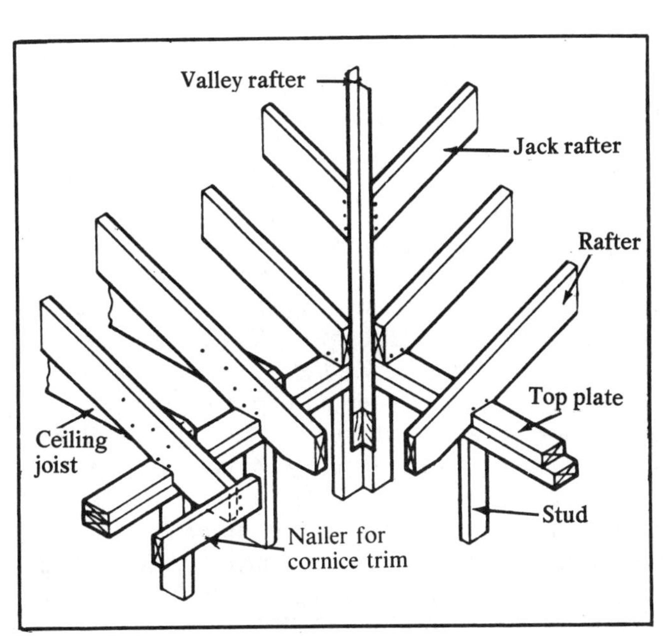

Framing at a valley
Figure 8-13

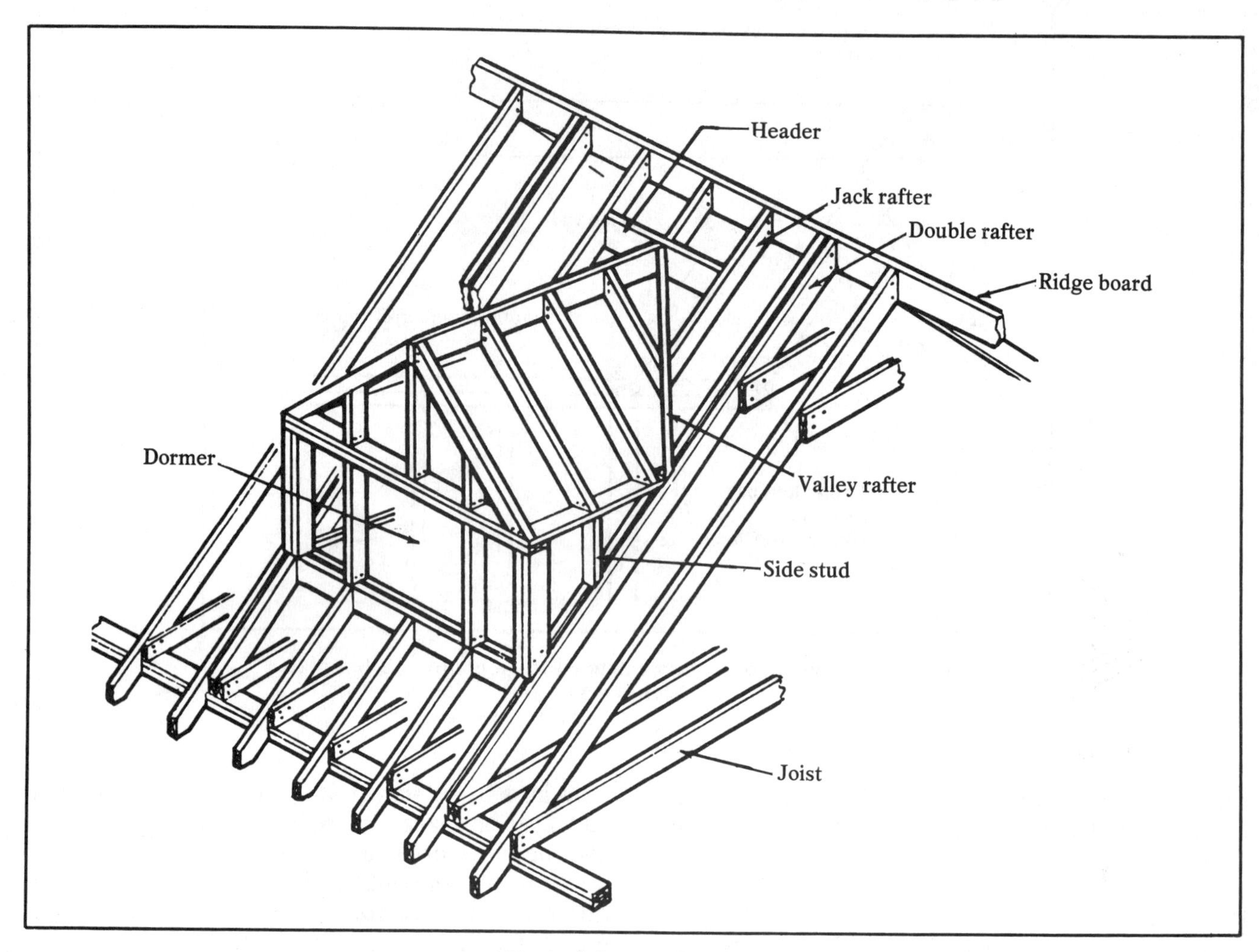

Typical dormer framing
Figure 8-14

dormers, do consider including them in the plan. Dormer space can be turned into living space at a relatively low cost. It's an economical way to create space for a growing family.

Ridge Beams

Low-sloped roof designs often require a glue-laminated ridge beam to span the open area. They are usually supported by an exterior wall, and an interior partition wall or post. The beam must be adequate for the roof load and the span selected.

Wood decking can serve as roof support and as sheathing. Spaced rafters, placed over the ridge beam or hung on metal joist hangers, are an alternate framing method. When a ridge beam and wood decking are used, as in Figure 8-15 A, good anchoring is essential. Use long ringshank nails supplemented with metal strapping or angle iron at both bearing areas.

A combination of large space (purlin) rafters and structural fiberboard decking is another acceptable system. Rafters can be supported by metal hangers at the ridge beam. See Figure 8-15 B. Rafters extend beyond the outer walls to form an overhang.

Roof Trusses

Mill-manufactured roof trusses are popular with many builders. Less roof-framing material is required, and there is a substantial reduction of job-site framing labor. Trusses can be manufactured in a wide variety of sizes and types. Look into the possibility of using mill-made trusses. The money you save goes into your pocket. That's what we are concerned with in this book — increasing your profits.

The simple truss is a rigid framework of triangular shapes, and is capable of supporting loads over long spans without intermediate support. It saves on material, and can be erected quickly. In addition, the house can be "dried-in" in a shorter time.

Trusses are usually designed to span from one

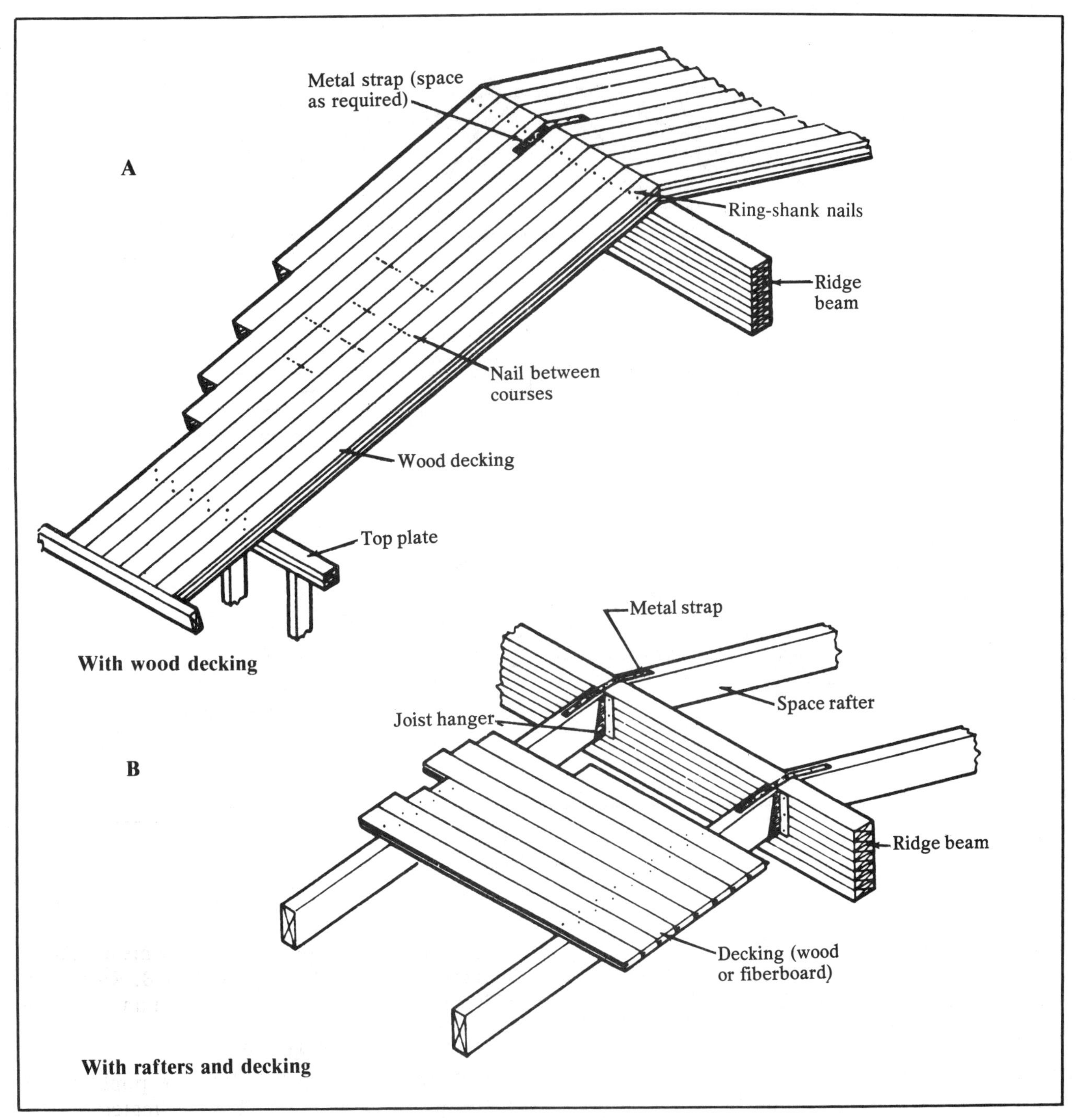

Ridge beam for roof
Figure 8-15

exterior wall to the other. Partitions can be placed without regard to structural requirements.

The wood trusses most commonly used for houses are the: *W-type, King-post,* and *Scissors.* See Figure 8-16. These work best on rectangular houses. A constant width requires only one type of truss. However, they can also be used for L plans and for hip roofs. Special hip trusses can be built for each end and for valley areas.

Trusses are commonly designed for 2-foot spacing, rather than 16-inch spacing. The extra 8-inch width requires somewhat thicker interior and exterior sheathing.

W-type Truss

The W-type truss (Figure 8-16 A) is the most popular light wood truss. This design uses three more members, but the distance between connections is less. This allows the use of lower grade lumber and greater spans.

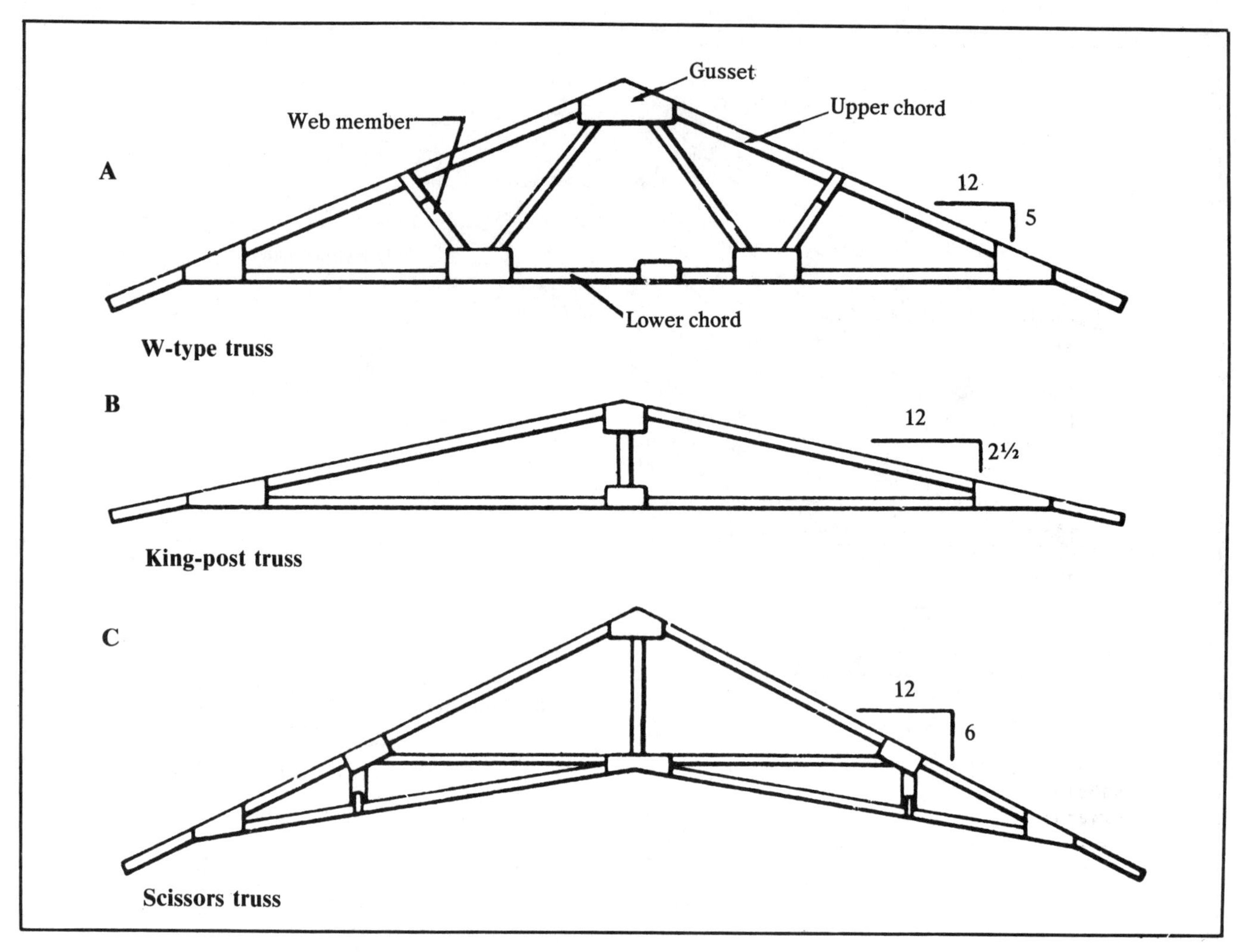

Light wood trusses
Figure 8-16

King-post Truss

The king-post truss is composed only of upper and lower chords and a center vertical post. See Figure 8-16 B. Allowable spans are somewhat less than for the W-type truss, even when the same-size members are used. For example, a plywood gusset king-post truss, with 4/12 pitch and 2-foot spacing, is limited to a 26-foot span for 2 x 4 members. The W-type truss, with the same size members and spacing, could be used for a 32-foot span. The unsupported length of the upper chord reduces the acceptable span in king-post trusses. For short and medium spans, the king-post truss is probably more economical. It has fewer pieces, and can be fabricated faster.

Local prices, design load requirements, and span will govern the type of truss to be used.

Scissors Truss

The scissors truss (Figure 8-16 C) is intended for houses with a sloping ceiling. It provides good roof construction for a cathedral-type ceiling, and requires less material than conventional framing methods.

Truss Design and Fabrication

The design of a truss has to consider snow, wind force, slope, and the weight of the roof itself. Generally, the lesser the slope, the greater the stress. You need larger members and stronger connections.

Most trusses are fabricated with gussets of plywood (nailed, glued, or bolted in place) or metal. Others are assembled with split-ring connectors. Designs for standard W-type and king-post trusses with plywood gussets are available from many lumber dealers. Some will provide you with completed trusses ready for erection.

Here's how to design and build a typical wood W-type truss:

The span for the glue-nailed gusset truss (Figure 8-17) is 26 feet, the slope is 4/12, and the spacing is

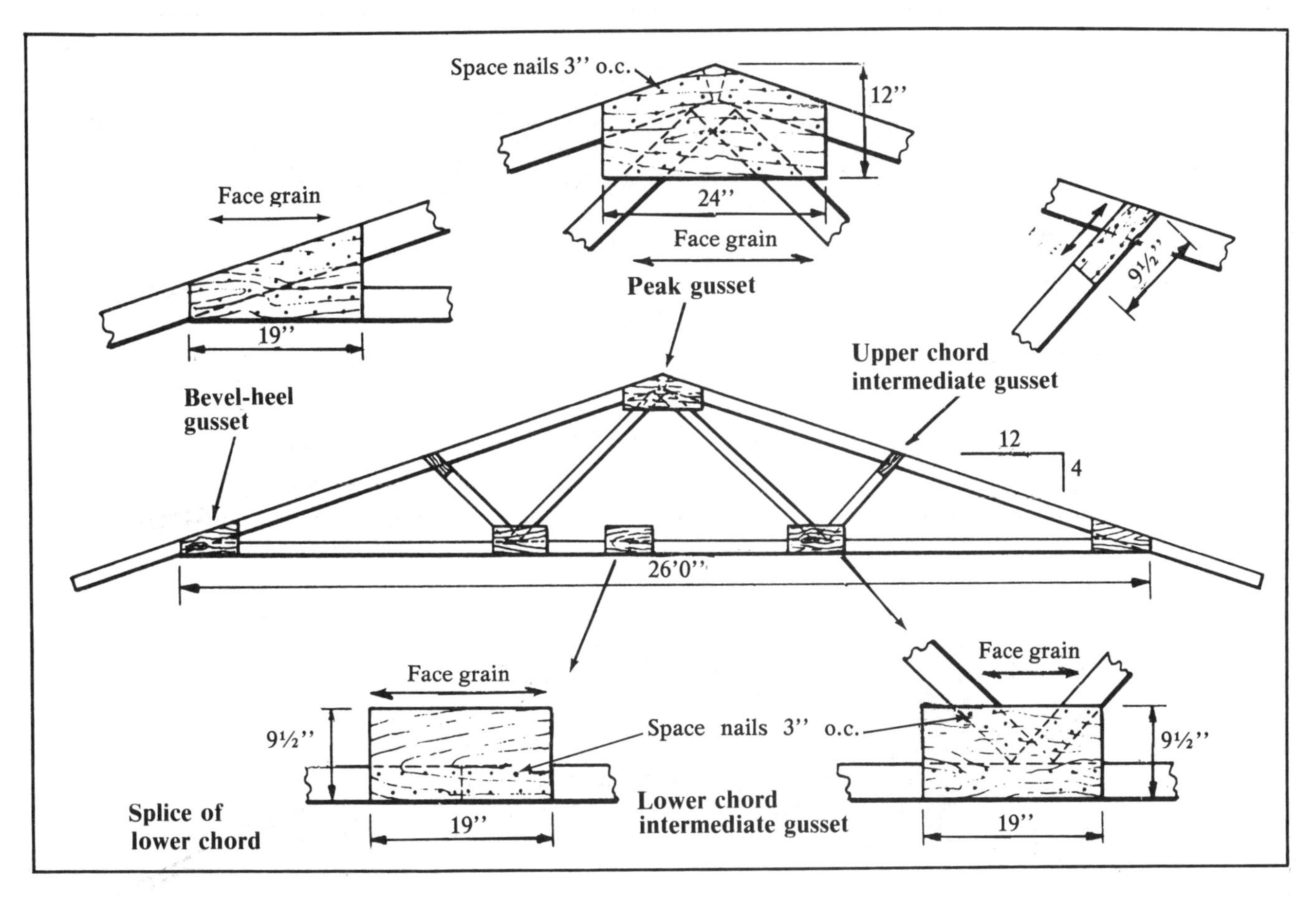

Construction of 26' W-type truss
Figure 8-17

24 inches. The total roof load assumed by the building code is 40 pounds per square foot. This is sufficient for moderate snow belt areas. The upper and lower chords are often 2 x 4 members. The upper chord requires a slightly higher grade of material. It's best to have a moisture content between 15% and 19%.

Plywood gussets can be made from 3/8- or 1/2-inch standard plywood with exterior glue, or from exterior-sheathing grade plywood. The cutout size of the gussets, and the general nailing pattern for glue-nailing are shown in Figure 8-17. Use 4d nails for plywood gussets up to 3/8-inch thick, and 6d nails for plywood 1/2- to 7/8-inch thick. When the plywood is no more than 3/8-inch thick, 3-inch nail-spacing should be used. Use 4 inches for thicker plywood.

When wood truss members are nominally 4 inches wide, use two rows of nails with a 3/4-inch edge distance. Use three rows when truss members are 6 inches wide. Gussets are used on both sides of the truss.

For normal and high relative-humidity conditions, use a resorcinol glue for the gussets. In dry areas, a casein glue might be considered.

Spread glue on the clean surfaces of the gusset and truss members. Use either nails or staples to hold the pieces together until the glue has set. For plywood 1/2-inch and thicker, nails (not staples) are recommended. Use the nail-spacing previously outlined. Closer spacing or intermediate nails may be needed to insure "squeeze-out" of the glue at visible edges. Follow the manufacturer's recommendations on temperature conditions. This is especially true when using resorcinol.

Truss Handling

Trusses are very strong when secured in place but very weak if handled improperly. They're designed to carry roof loads in a vertical position. Be sure your crews store and lift them in an upright position. If they must be handled in a flat position, provide some support in the middle to minimize bending deflections. Never support them only at the center or at each end.

Truss Erection

You can start either at an end section of the building or at the center, as shown in Figure 8-18. Raise the trusses by hand, and nail them into place.

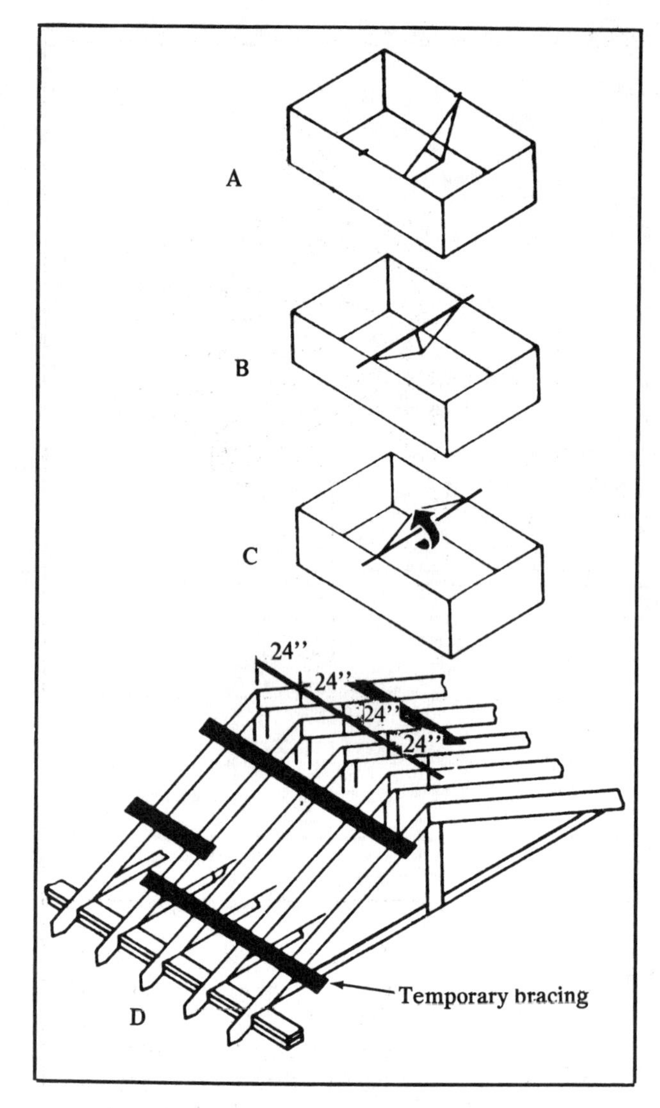

Erection of trusses
Figure 8-18

Temporarily brace the truss to the end section of the building until the sheathing is applied. Temporary benches may be needed for the crew to stand on. Knee braces are not used on a truss, unless required.

The trusses are installed as follows:

1) Mark on the top plates the proper position of all trusses. Mark the exact position on the face of all rafters.

2) Rest one end of the assembly, peak down, on a mark on the top plate (Figure 8-18 A).

3) Rest the other end on the opposing mark (Figure 8-18 B).

4) Rotate the assembly into position using a pole or rope (Figure 8-18 C).

5) Flush the faces against the marks and secure them.

6) Raise and nail three assemblies into position. Nail temporary 1 x 6 braces across these and the other assemblies as they are placed into position (Figure 8-18 D). As the braces are nailed on, check the rafter spacing at the peak.

7) The temporary braces may be used as a platform for the crew to stand on. A three-man crew can handle most house trusses, but a small crane will save time if you have to place trusses for several houses on one site.

Anchoring is important. It's best to use a metal connector to supplement toenailing. Plate anchors are available commercially, or they can be fabricated from sheet metal. Remember that you have to secure the trusses against uplift stress as well as downward thrust. Many dealers supply trusses with a 2 x 4 soffit return at the end of each upper chord. This makes it easy to nail at the soffit.

Trusses can be toenailed to the top wall plates. But the heel gusset is located at the wall plates, and this makes toenailing the trusses difficult. When nailing the lower chord to the plate, use two 10d nails at each side of the truss. See Figure 8-19 A. Predrilling may be necessary to prevent splitting. A simple metal connector or plate anchor is better. See Figure 8-19 B. Nail plate anchors at the sides and top with 8d nails. Use 6d or 1½ inch roofing nails to nail the anchor to the lower chords of the truss.

Rafter Lengths and Cuts

There are several ways to find the lengths and cuts of roof rafters. The basis of each method is trigonometry. Use the method that seems easiest for you.

For many carpenters, the step-off method of laying out rafters is the most logical. It's the method I prefer and the one described here. Figure 8-20 will help you determine the rough length of the lumber you need for the rafters.

Rafter layout is complicated. Each step of the work must be done in order. Get started right by laying out the pattern rafter correctly. The crowned (or top edge) should be toward you as you are laying out the rafter.

A carpenter should hold the tongue of the steel square in his left hand and the body in his right hand when laying out rafters. In this position, the tongue forms the vertical or top cut of the rafter. The body is the level or seat cut. See Figure 8-21.

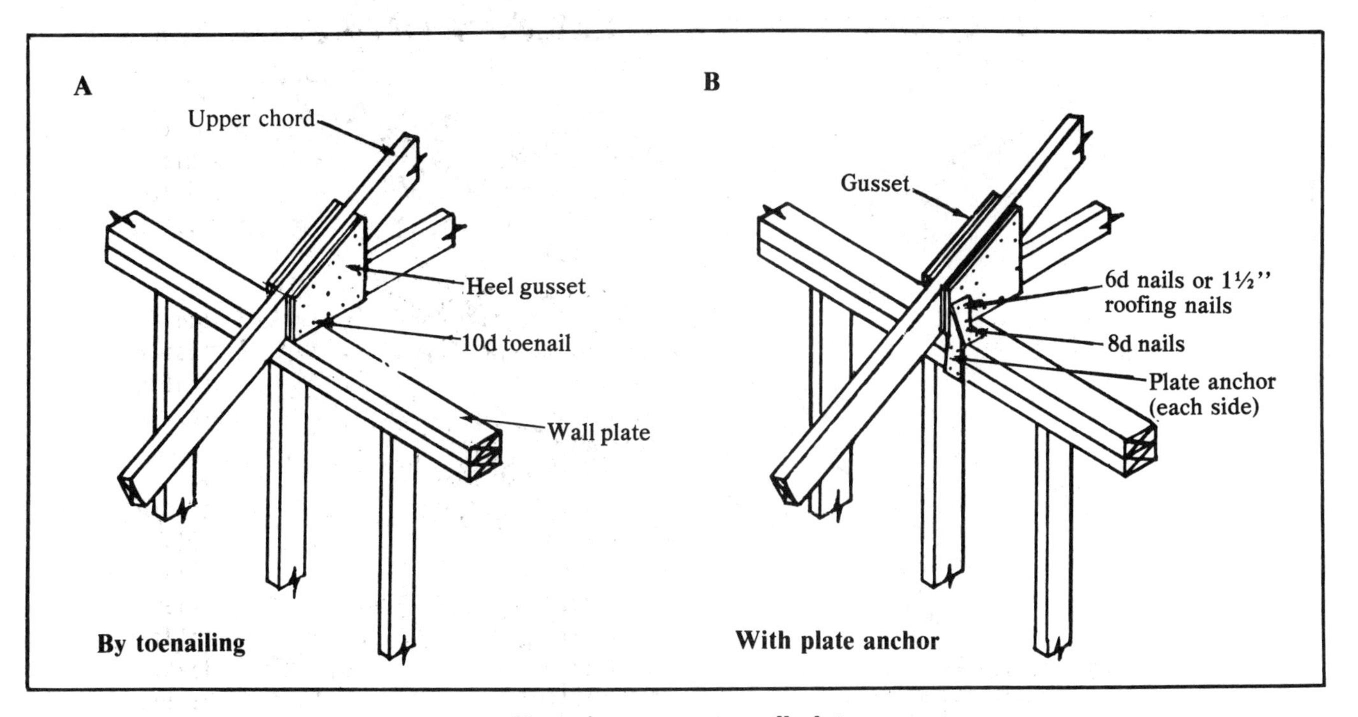

Fastening trusses to wall plate
Figure 8-19

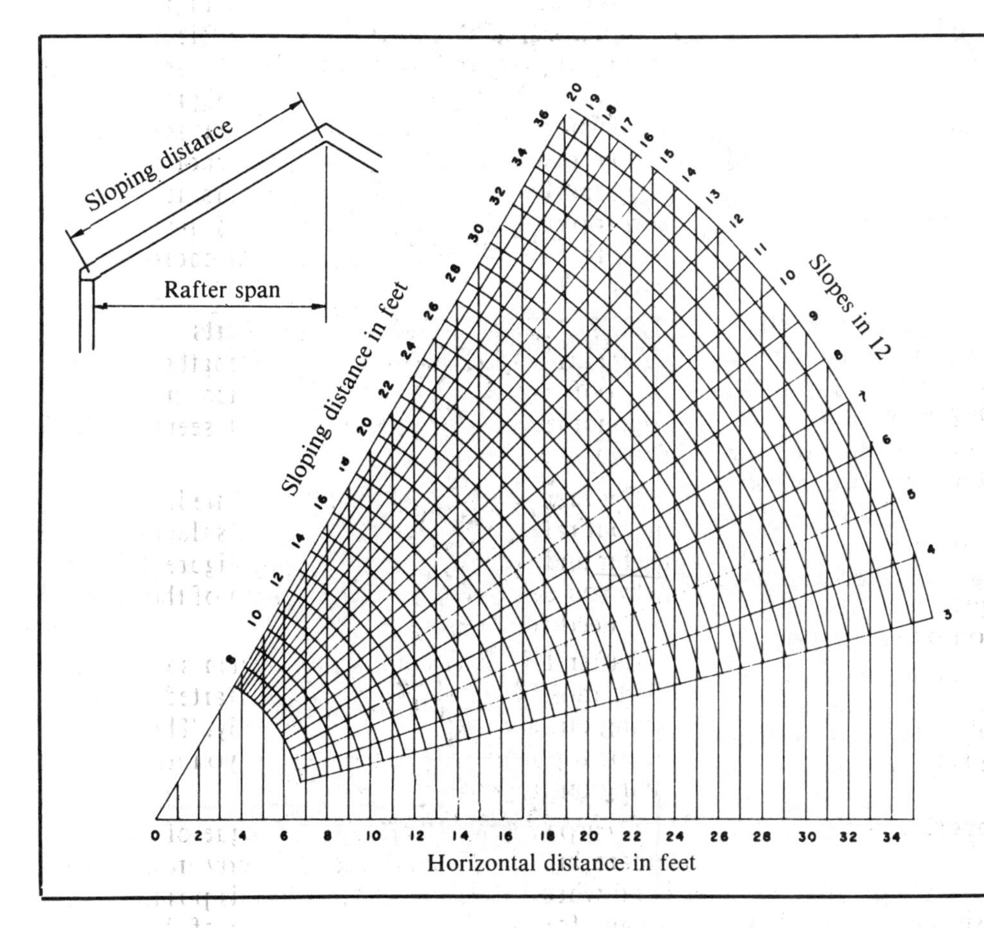

To use the diagram select the known horizontal distance and follow the vertical line to its intersection with the radial line of the specified slope, then proceed along the arc to read the sloping distance. In some cases it may be desirable to interpolate between the one foot separations. The diagram also may be used to find the horizontal distance corresponding to a given sloping distance or to find the slope when the horizontal and sloping distances are known.

Example: With a roof slope of 8 in 12 and a horizontal distance of 20 feet the sloping distance may be read as 24 feet.

Conversion diagram for rafters
Figure 8-20

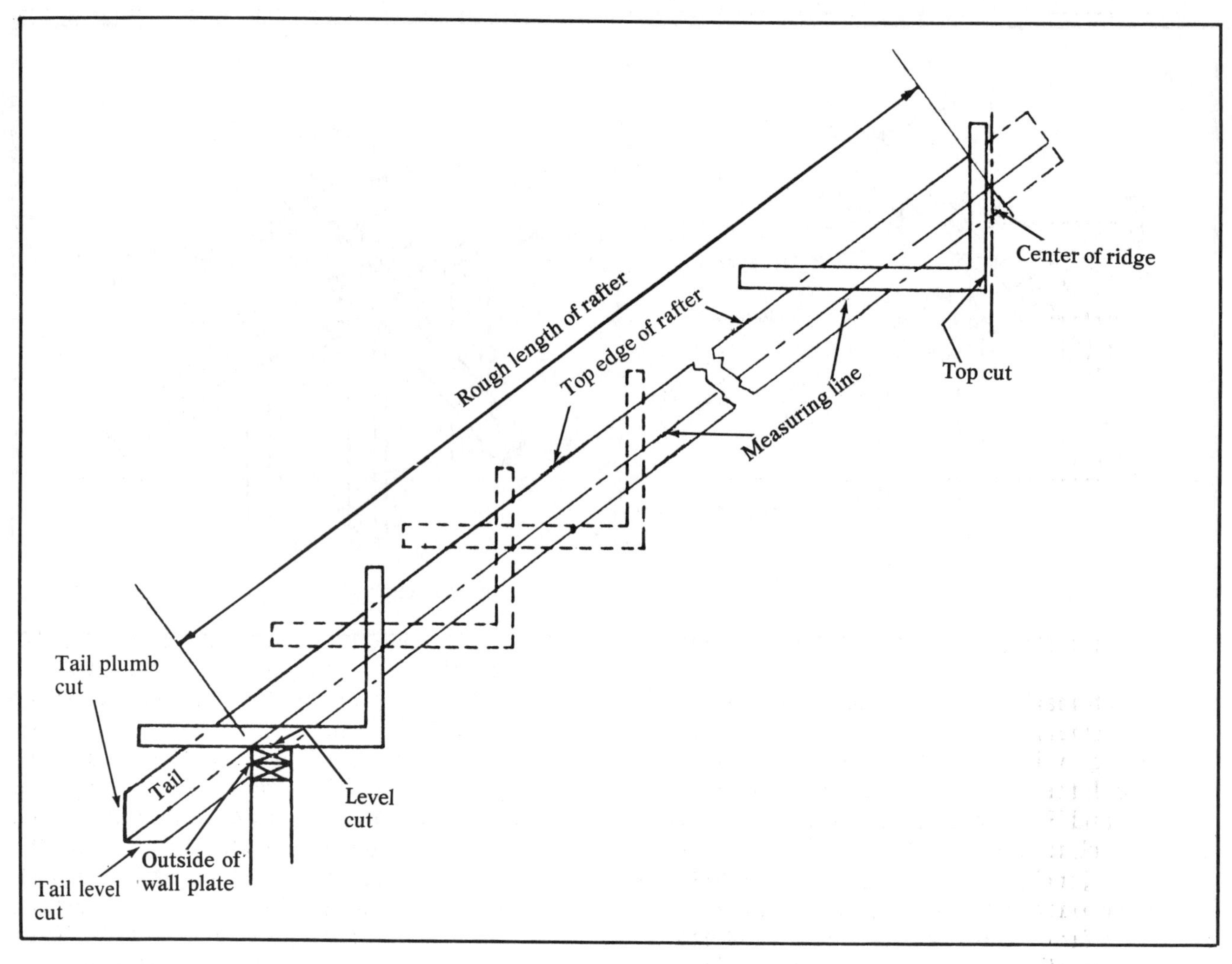

Common rafter
Figure 8-21

Finding the Rough Length of a Common Rafter
You can find the approximate length of a common rafter by using your square as a calculator. Let the tongue of the square stand for the rise in feet. The body then shows the run in feet. Measure the length of the diagonal between these two points. This, when expressed in feet, is the rough length of the rafter.

Remember to add the length of the overhang when making this calculation.

Here's an example. Assume that the total rise of a rafter is 9 feet and the run is 12 feet. Locate 9 and 12 on the square. See Figure 8-22. Measure the diagonal. This is 15. Therefore, the rough length of the rafter will be 15 feet plus the overhang of 1 foot. The rafters, then, will have to be at least 16 feet long.

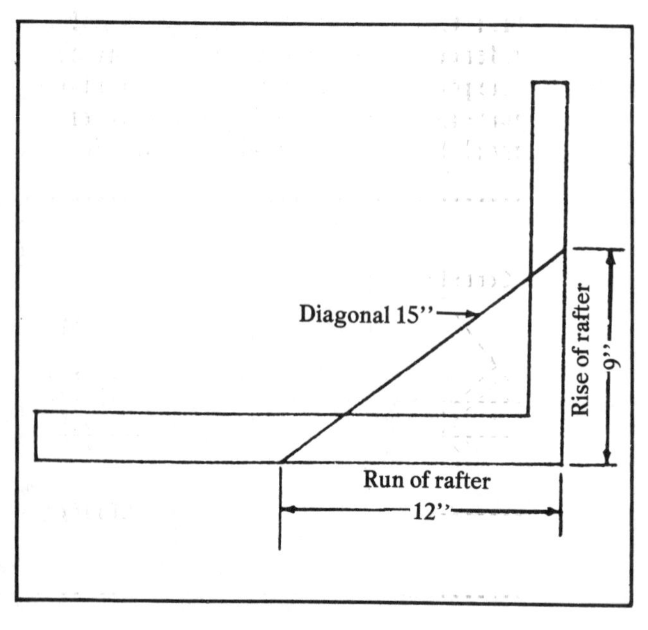

Rough length of rafter
Figure 8-22

Locating the Measuring Line on a Rafter
The stock for the first pattern rafter must be straight. Lay the board flat across two saw horses.

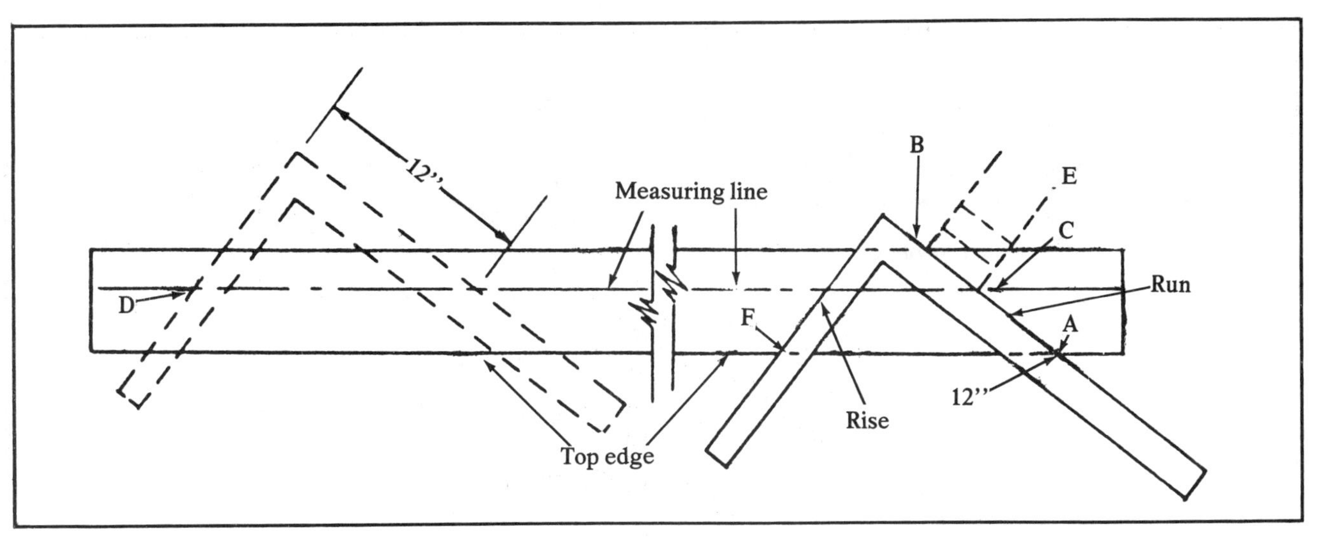

Locating measuring line on rafter
Figure 8-23

The square is placed near the right end. See Figure 8-23.

The inch mark corresponding to the run of the roof (12 inches in all cases), and the inch mark corresponding to the rise of the roof, should both intersect with the top edge of the rafter, as shown in points A and F of Figure 8-23. Then draw line A-B on the plank to represent the top of the wall plate. Measure 3⅝ inches along this line, from B to C, to locate the outside top corner of the plate. This should be far enough from the right-hand end of the rafter to allow for the tail. The measuring line (C-D) is then gauged parallel to the edge of the rafter.

Laying Out a Common Rafter (Step-off Method)

A wooden fence or small metal clamps may be attached at the proper points on the tongue and body of the square as an aid in laying out a rafter. Of course, any skilled carpenter will tell you that the rafter can be laid out expertly without these aids.

Assume that a building is 24 feet wide and the pitch of the rafter is to be 3/8. This is 9 inches of rise on 12 inches of run. You have to find the exact length of the rafter. Lay the square on the stock so the 12-inch mark on the outside of the body is at point C, as shown in Figure 8-24, and the 9-inch mark on the outside edge of the tongue is on the measuring line at E. The square is kept in this position. Adjust the fence until its edge lies against the top edge of the rafter. (See Figure 8-24.) Then tighten the fence so it won't move on the square.

It's important that the 12-inch mark on the body of the square and the 9-inch mark on the tongue be exactly on the measuring line when the fence is against the edge of the rafter.

Place the fence against the rafter stock as shown in the extreme right of Figure 8-24. Mark along the outside edge of the body and tongue. This locates point E on the measuring line. Slide the square to

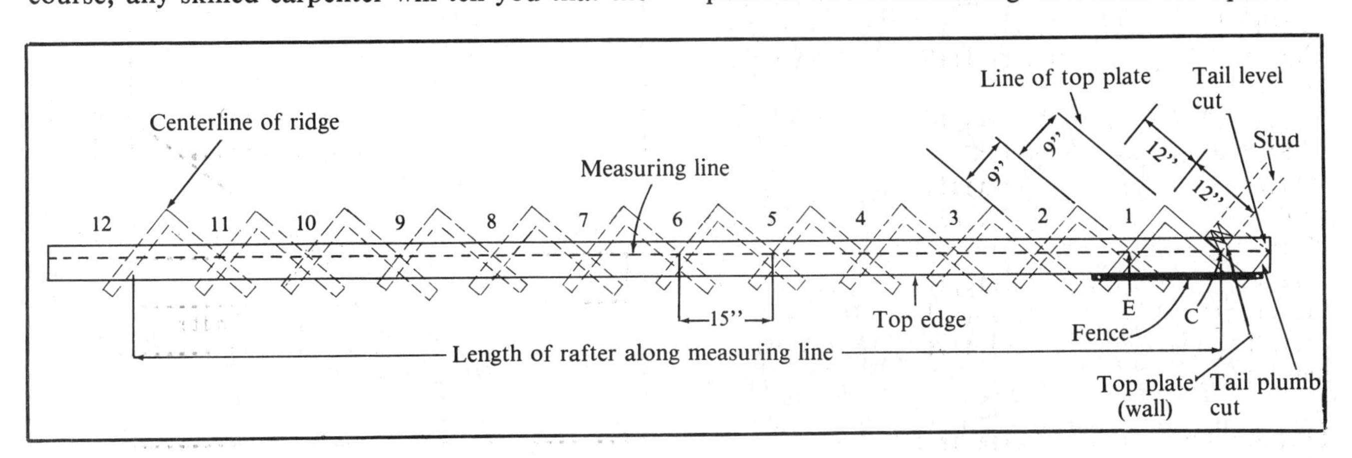

Laying out a common rafter (step-off method)
Figure 8-24

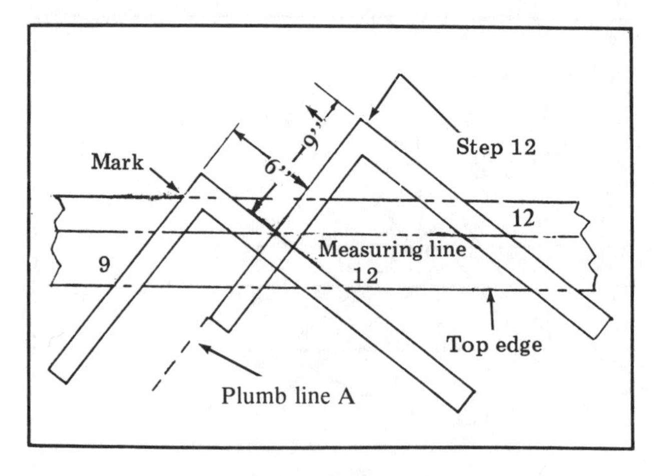

Additional half step
Figure 8-25

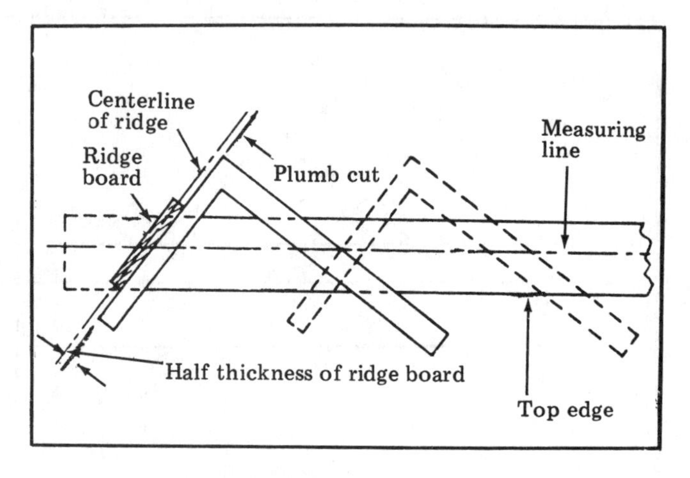

Allowance for ridge board
Figure 8-26

the left until the 12-inch mark is over E (position 2) and again mark along the tongue and body. Continue this operation as many times as there are feet of run in the rafter (12 in this case).

The successive positions of the square are shown by positions 1 through 12, Figure 8-24. When the last position has been reached, draw a line along the tongue, across the rafter, to indicate the centerline of the ridge. Each step in this whole process should be clearly marked and performed very carefully. Even a slight error will change the fit and length of the rafter.

Layout of a Rafter with an Odd Span

Let's assume that a building is 25 feet wide and the pitch of the rafter is 3/8. This is 9 inches of rise to 12 inches of run. The run of the rafter would be 1/2 the width of the building, or 12 feet 6 inches.

The layout of the rafter is very similar to the one just described, except that 12½ steps will be taken. This additional 1/2 step is added for the extra 6 inches of run in the rafter. After the 12th step has been taken and marked as shown in Figure 8-25, lay the square on the rafter so 9 and 12 will intersect with the top edge of the rafter as shown. Move the square until 6 on the outside of the body is directly over the 12th step plumb line A. Note in Figure 8-25. Mark a line along the outside edge of the tongue to indicate the centerline of the ridge.

Allowance for Ridge Board

The last line marked shows where the rafter would be cut off, if there were no ridge board. Since a ridge board is to be used, the rafter will be a little too long. Cut a piece off the end of the rafter. This piece should be equal to 1/2 the thickness of the ridge board. (See Figure 8-26.)

To lay out this line, slide the square back away from the last line (position 12 in Figure 8-24). Keep the fence tight against the top edge of the rafter. When the square has been moved back 1/2 the thickness of the ridge board, mark the plumb cut along the edge of the tongue. This line should be inside of, and parallel to, the original line. Take measurements from this original line and at right angles to it. See Figure 8-26.

Bottom or Seat Cut

The bottom or seat cut (sometimes called the "bird mouth") of the rafter is a combination of the level and plumb cuts. The level cut (B-C, in Figure 8-23) rests on the top face of the side wall plate. The plumb cut (C-E) fits against the outside edge of the wall plate. The plumb cut is laid out by squaring a line from line A-B, Figure 8-23, through point C. C-E then represents the plumb cut.

Tail of the Rafter

If the rafter tail is to be as shown in Figure 8-24, make the plumb cut along the line as shown. The level cut at the end of the tail (see Figure 8-21) is found by sliding the square toward the tail until the body intersects with the plumb and measuring line.

The tail of the rafter must be cut differently for the box cornice shown in Figure 8-27. In this case, the level and plumb cuts are laid out as before. Make an allowance for the thickness of the sheathing. To do this, lay out line A-B parallel to line C-D, and the thicknesss of the sheathing away from C-D. The line of the level cut at D is then extended to meet A-B. When the rafter is cut, it will fit over the plate and sheathing.

Let's assume that a 12-inch piece is to be used at E. Continue the line of the level cut through B to F on top of the rafter. Lay the body of the square along this line with the 12-inch mark at the outside

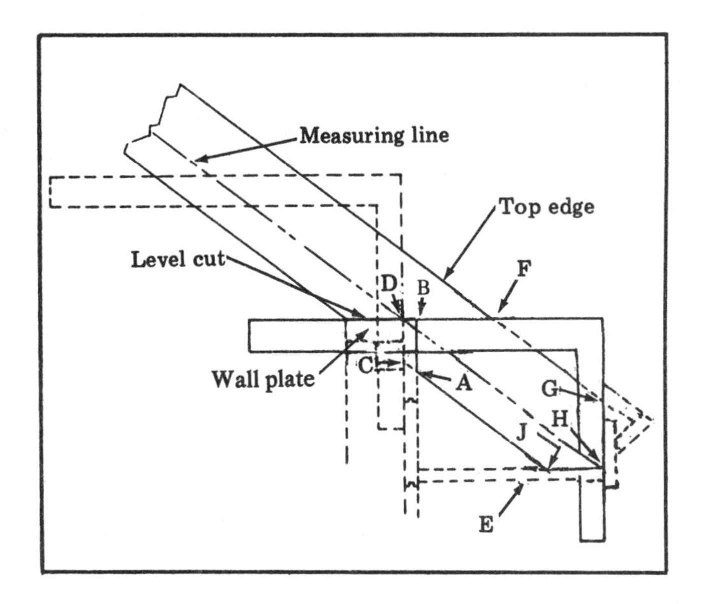

Tail cut
Figure 8-27

edge directly over point B. Mark a line G-H along the outer edge of the tongue across the rafter. The point H where this line meets the measuring line represents the tip of the rafter. Square the line H-J across the rafter from line G-H to locate J.

In a different treatment of the cornice framing, the cut on line J is omitted. Many carpenters prefer to delay cutting off the end of the rafter (line G-H) until after all the rafters are installed. At this time, a taut line is run from the outside rafter at each end to establish the tail end of all the rafters. Line G-H is then marked on each rafter with the aid of a bevel square at the string. Saw each rafter carefully with a power saw on the same side of the line. This insures a straight-line cut of the rafter tail, even if there is a slight variance in the alignment of the top plate.

Rafter Tables

Many steel squares have rafter tables stamped on them. The manufacturer can probably supply an instruction booklet that explains how to use the tables. A book titled *Rafter Length Manual* published by Craftsman Book Company, Box 6500, Carlsbad, California 92008, gives the exact rafter lengths for every roof span and rise. But you can calculate these lengths with the table on most framing squares.

Here's how to use the rafter length table. (See Figure 8-28.) The inch marks on the outside edge of the square indicate the rise per foot of run. For example, the figure 8 means 8 inches of rise per foot of run. Directly below each of these figures is the length of the main rafters per foot of run, 14.42 in this case. If the run of the rafter is 10 feet, the 14.42 inches should be multiplied by 10. This gives 144.2 inches or 12.01 feet as the length of the rafter.

Estimating Manhours for Ceiling-Joist Framing

Labor for installing ceiling joists is usually based on the board feet of lumber used. Note Figure 6-38 in Chapter 6.

Assume you have a building 24 x 40 feet with a load-bearing partition in the center. A 12-foot ceiling joist is required for each side of the partition. A 40-foot building will require 31 ceiling joists (40 multiplied by 0.75, plus 1) for one side, or 62 joists for both sides of the partition. A 2 x 6 joist 12 feet long has 12 board feet. The 62 pieces have a total of 744 BF. It takes about 18 skilled and 7 unskilled manhours to erect 1,000 BF of 2 x 6 ceiling joists. This is 1.8 skilled and 0.7 unskilled hours for 100 BF. Your job with 744 BF will take 13.39 skilled and 5.2 unskilled hours or a total of approximately 18.5 manhours.

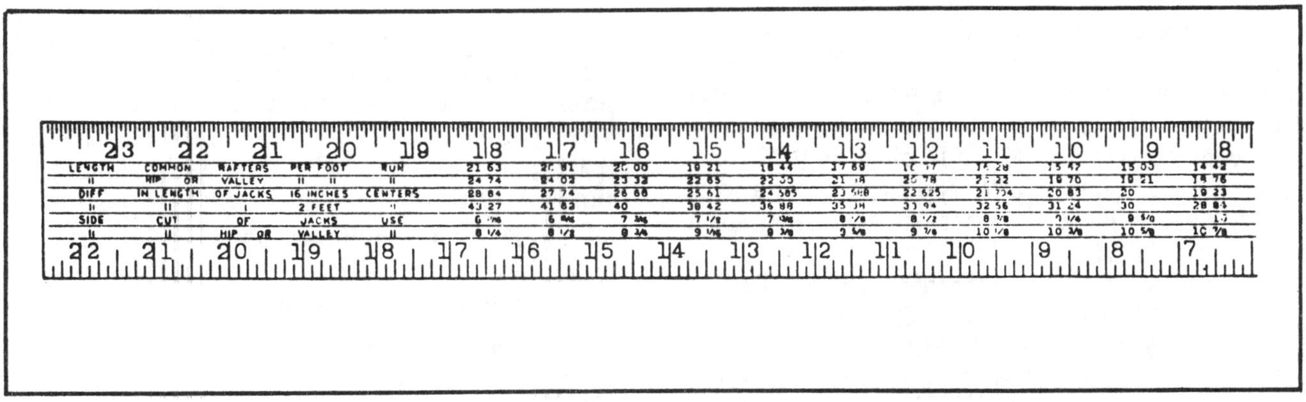

Rafter length table
Figure 8-28

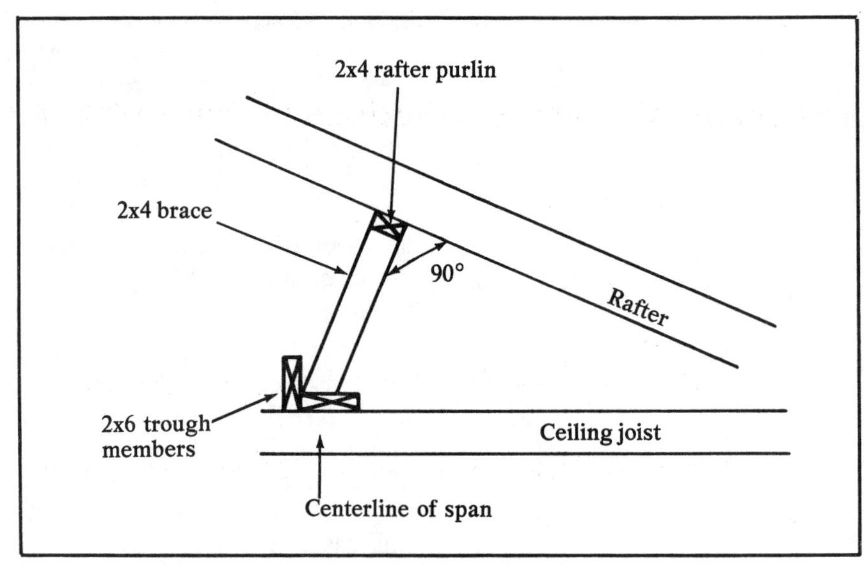

Trough nailed to ceiling joist and braced to rafter purlin
Figure 8-29

Here's a rule of thumb for determining manhour requirements for ceiling-joist framing:

Multiply the square feet of the ceiling area by 1.2 minutes to determine manhours (2/3 skilled, 1/3 unskilled).

To allow manhours for installing "troughs" on ceiling joists, purlins on rafters, and bracing from "trough" to purlin (as in Figure 8-29), add 1 minute per square foot of ceiling area (2/3 skilled and 1/3 unskilled). For a 960-square-foot structure, this adds 16 manhours.

Estimating Manhours for Rafter Framing

Labor for installing rafters is also figured on a board-foot basis. See Figure 6-38 in Chapter 6. In a building 24 x 40 feet with a 5/12 pitch and plain gable roof, the rafter horizontal distance is 12 feet. Look at Figure 8-20 for a quick estimate of the rafter length. The rafter span is 13 feet. Assume that there is a 1-foot overhang. A 14-foot rafter is required. A 40-foot building having a 1-foot overhang at the gable would result in a 42-foot roof. The number of rafters required for spacing 16 inches on center is 42, multiplied by 0.75, or a subtotal of 31. Add 1 rafter for the end, for a total of 32. There will be 32 rafters on each side, for a grand total of 64.

A 2 x 6 rafter 14 feet long has 14 board feet of lumber. In 64 rafters you have 896 BF.

It takes approximately 23 skilled and 6 unskilled hours to erect 1,000 BF of 2 x 6 rafters. See Figure 6-38 in Chapter 6. It takes 2.3 skilled and 0.6 unskilled hours for 100 BF. Thus, 20.6 skilled and 5.37 unskilled hours are required to erect 896 BF or 64 rafters.

Here's a rule of thumb for estimating manhours on plain gable rafter framing 16 inches o.c.:

Multiply the square feet of roof area by 1.5 minutes to determine manhours (2/3 skilled, 1/3 unskilled).

Estimating Manhours for Erecting Trusses

A 24 x 40-foot structure will take 21 roof trusses for 24-inch spacing on center. A rough estimate for installing roof trusses is 23 skilled and 7 unskilled manhours per 1,000 BF. A better guess on most small jobs would be 1.5 manhours per truss. Your 21 trusses would require 31.5 manhours, (2/3 skilled, 1/3 unskilled). Carpenters experienced in installing trusses could cut this time considerably.

Summary

Roof trusses save money on most jobs. Get the house dried-in as quickly as possible so that work can continue during bad weather. Stay with the simple gable roof to cut time and material costs. When you have extra money in the bank and experience under your belt, then tackle the more complicated roofs.

Making money building houses is a series of money-saving steps which add up to major savings. This is simply good management.

Figure 8-30 is your Ceiling Joist and Roof Framing Checklist and Data Reference Sheet. It will help you stay on the road to more profit per house.

Ceiling Joist and Roof Framing Checklist and Data Reference Sheet

1) The house is ☐1-story ☐1½-story ☐2-story.

2) The house is ☐rectangular ☐L-shape ☐T-shape.

3) Its size is _________ feet by _________ feet, for a total of _________ square feet.

4) The house ☐does ☐does not have an attic space suitable for conversion to habitable space.

5) The roof is:

☐Gable

☐Low-pitch (____/12)

☐Medium-pitch (____/12)

☐High-pitch (____/12)

☐Straight-line

☐Broken-line

☐With gable dormers (number)______

☐With shed-type dormers (number)______

☐Overhang of ______ inches

☐Other (explain)______________________________

☐Post-and-Beam construction

☐Beams exposed to interior

☐Post and beam joined with metal angles at top and at sole plate

☐Spaced beams and post joined with ⅜-inch or thicker plywood cleats

☐Decking tied with ☐metal straps ☐other (specify) _________

☐Flat

☐Has one member acting as roof and ceiling support

☐Uses 2 x ______ material

☐Has ______-inch overhang

Figure 8-30

6) Flush ceiling framing method ☐was ☐was not used in this house. Also used were:

☐Nail-laminated beam

☐Glue-laminated beam

☐Beam anchored at exterior wall

☐Beam anchored at interior post or partition

7) Ceiling joists are No. 2 grade (species)________________________________

☐2 x 6

☐2 x 8

☐2 x 10

8) Ceiling joists are securely fastened at the load-bearing partition. The joists are:

☐Toe-nailed to the top plate with 8d nails

☐Nailed together at the joint (overlapped at center partition)

☐Secured with metal cleats at the joint

☐Secured with wood cleats at the joint

9) Ceiling joists are securely fastened at exterior wall by:

☐8d nails toe-nailed to plate

☐Metal straps anchored to plate

☐16d nails (face-nailed to rafters)

☐Other (specify)____________________________________

10) Ceiling joists are spaced ☐16 inches ☐24 inches on center.

11) Joists are anchored to laminated beam with:

☐Metal joist hangers

☐Wood hangers

☐Metal straps

12) A total of __________ ceiling joists were used. (To determine number of ceiling joists required for 16-inch spacing, multiply the length of the house by 0.75 and add 1.)

13) Rafter layout was done by:

☐Step-off method

☐Rafter table method

14) The rafter length is ______ feet ______ inches and requires ______ foot-long lumber.

15) A total of ______ rafters were used. (To determine number of rafters required for 16-inch spacing, multiply the total roof length (including overhang) by 0.75 and add 1. Double the sum to obtain the number of rafters required for both sides.)

16) The ridge board used was ☐1 x 8 ☐2 x 8

17) Trusses ☐were ☐were not used on this job. Truss data:

☐W-type truss ☐King-post truss ☐Scissors truss ☐Metal gusset ☐Plywood gusset ☐Nailed ☐Glued

☐Anchorage: ☐Soffit return ☐Toe-nailed ☐Metal connectors ☐Plate anchors

18) Material costs:

Ceiling joists:

2 x 6	________pcs	@	$______ea	=	$________
2 x 8	________pcs	@	$______ea	=	$________
2 x 10	________pcs	@	$______ea	=	$________

Bracing:

2 x 4	________LF	@	$______LF	=	$________
2 x 6	________LF	@	$______LF	=	$________

Bridging:

1 x 3	________LF	@	$______LF	=	$________
1 x 4	________LF	@	$______LF	=	$________
Metal crosses	________pcs	@	$______ea	=	$________

Rafters:

2 x 6	________pcs	@	$______ea	=	$________
2 x 8	________pcs	@	$______ea	=	$________

Ridge board:

1 x 8	________LF	@	$______LF	=	$________
2 x 8	________LF	@	$______LF	=	$________

Tail header:

2 x 6	________LF	@	$______LF	=	$________
2 x 8	________LF	@	$______LF	=	$________

Nails:

16d ________lbs @ $________lb = $__________

10d ________lbs @ $________lb = $__________

8d ________lbs @ $________lb = $__________

6d ________lbs @ $________lb = $__________

Other (specify)

____________________ ________lbs @ $________lb = $__________

Trusses ________pcs @ $________ea = $__________

Other materials, not listed above:

____________________ ________ @ $__________ = $__________

____________________ ________ @ $__________ = $__________

____________________ ________ @ $__________ = $__________

____________________ ________ @ $__________ = $__________

Total cost of materials $__________

19) Manhours required to erect ceiling joists:

☐________ Skilled hours @ $________ hour = $________

☐________ Unskilled hours @ $________ hour = $________

It takes about 18 skilled and 7 unskilled manhours to erect 1,000 BF of 2 x 6 ceiling joists. A rule of thumb for determining hours for erecting ceiling joists spaced 16 inches o.c. is:

Multiply the square feet of ceiling area by 1.2 minutes for total manhours (⅔ skilled, ⅓ unskilled).

20) Manhours required to erect rafters:

☐________ Skilled hours @ $________ hour = $________

☐________ Unskilled hours @ $________ hour = $________

It takes about 23 skilled and 6 unskilled hours to erect 1,000 BF of 2 x 6 rafters. A rule of thumb for determining hours for erecting rafters spaced 16 inches o.c. is:

Multiply the square feet of roof area by 1.5 minutes for total manhours (⅔ skilled, ⅓ unskilled).

21) Manhours required to install trough, purlin and bracing:

☐________ Skilled hours @ $________ hour = $________

☐________ Unskilled hours @ $________ hour = $________

A rule of thumb for determing manhours required to install trough, purlin and bracing is:

Add 1 minute per square foot of ceiling area.

22) Manhours required to erect trusses:

☐__________ Skilled hours @ $__________ hour = $__________

☐__________ Unskilled hours @ $__________ hour = $__________

A rule of thumb for erecting trusses is:

Multiply the number of trusses by 1.5 hours for total manhours (⅔ skilled, ⅓ unskilled).

23) Labor costs:

Skilled	________hours	@	$______hour	=	$__________
Unskilled	________hours	@	$______hour	=	$__________
Other (specify)					
____________________	________hours	@	$______hour	=	$__________
				Total	$__________

24) Total cost:

Materials $__________

Labor $__________

Other $__________

Total $__________

25) Cost of ceiling joist and rafter framing (including bracing) per square foot of floor space $__________

26) Problems that could have been avoided ______________________________

____

____

____

____

27) Comments ______________________________

____

____

____

____

28) A copy of this Checklist and Data Reference Sheet has been provided:

☐Management

☐Accounting

☐Estimator

☐Foreman

☐File

☐Other: ___________

Chapter 9

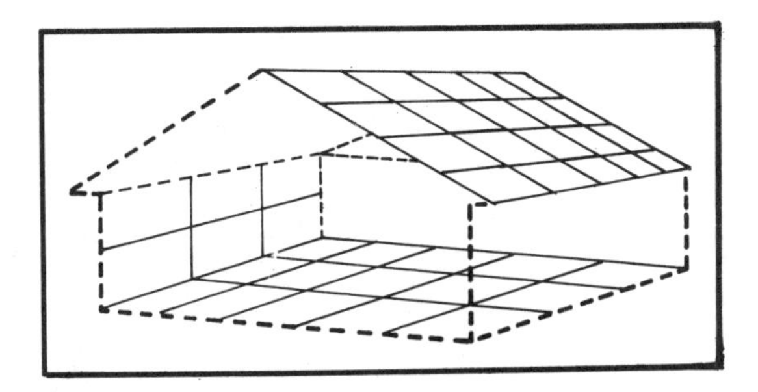

Modular Construction

Most carpenters who have worked on more than a few jobs have their favorite story about the board feet of lumber that were wasted on some job because plan dimensions were poorly coordinated with material dimensions.

They'll tell you about rafters that had to be cut from 20-foot stock because the length needed was 18'2''. They'll estimate how many extra studs were needed in some homes because the wall length selected by the designer wasn't a multiple of 16 or 24 inches. And if you insist on a guess, you might hear a claim that nearly 10% of the framing lumber delivered to some jobs becomes cutting waste and has to be hauled away again.

And it isn't just lumber that's wasted due to lack of modular coordination. Remember that vinyl comes in rolls 8 feet wide, roofing comes in rolls 3 feet wide, wallboard comes in 4 x 8 sheets. In fact, most construction materials that are delivered to the site in rigid form are sold in a fairly limited number of standard sizes. Material cut off to provide adequate fit in a given area often can't be used and must be discarded.

And here's the clincher. It isn't only the material you waste. Your biggest expense may be the added cost of the time spent cutting and fitting small pieces or adding extra sections.

If, due to lack of modular coordination, the waste of materials on your jobs is close to 10%, you could be building houses with close to 10% more space at no extra cost! That's why profitable builders do their level best to plan houses that make full use of standard-size materials.

Make the commitment to yourself right now that your houses will use modular dimensions whenever possible. That's an excellent way to provide more house for your buyer's dollar, and put more profit in your pocket.

This chapter explains how to put modular planning principles to work for you. Every modular plan involves three dimensions: length, width and height. The modular length and width dimensions form the base. Your modular plan should be based on a planning grid like Figure 9-1.

There's no rigid formula for modular planning, but you always begin by dividing the horizontal plane or footprint of the house into units of 4, 16, 24 or 48 inches. You'll probably always want overall dimensions to be in multiples of 4 inches.

The 16-inch unit offers flexibility in spacing windows and doors. Use increments of 24 and 48 inches for exterior dimensions of the house. Floor, ceiling and roof construction also use these dimensions.

You'll want to draw modular plans on the grid you select. Grid layout gives you a quick way of planning a modular set-up with accuracy.

Thickness and Tolerance Variables

Divide the house into horizontal and vertical elements at regular intervals, as shown in Figure 9-2A. Although the elements are shown as planes without thickness, allow for wall thickness and tolerance based on fixed module lines at the outside face of the studs, as shown in Figure 9-2B.

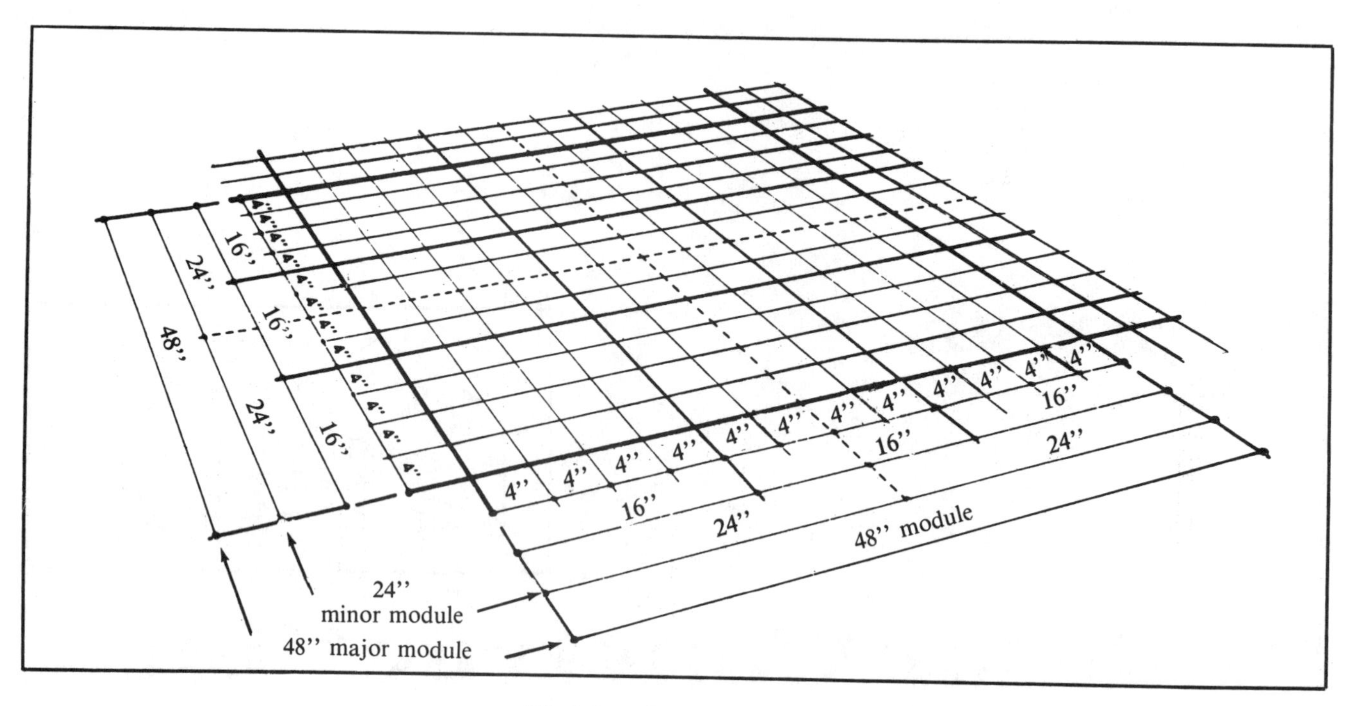

Modular planning grid
Figure 9-1

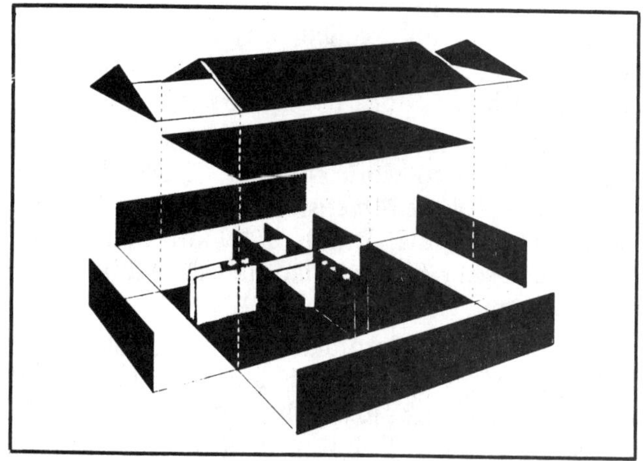

Division of house into horizontal and vertical planes
Figure 9-2A

Exterior walls and partitions may have many thickness variables. Floor and roof construction can also vary in thickness, of course, depending on structural requirements and types of framing and finishing.

Exterior Walls, Doors and Windows

An important feature of preplanning is to separate exterior wall elements at natural division points. In Figure 9-3, overall house dimensions are based on the 48-inch and 24-inch modules. However, maximum flexibility in placing door and window openings requires the 16-inch module. Also, the precise location of wall openings on the 16-inch module eliminates the extra wall framing that is frequently required in nonmodular house construction.

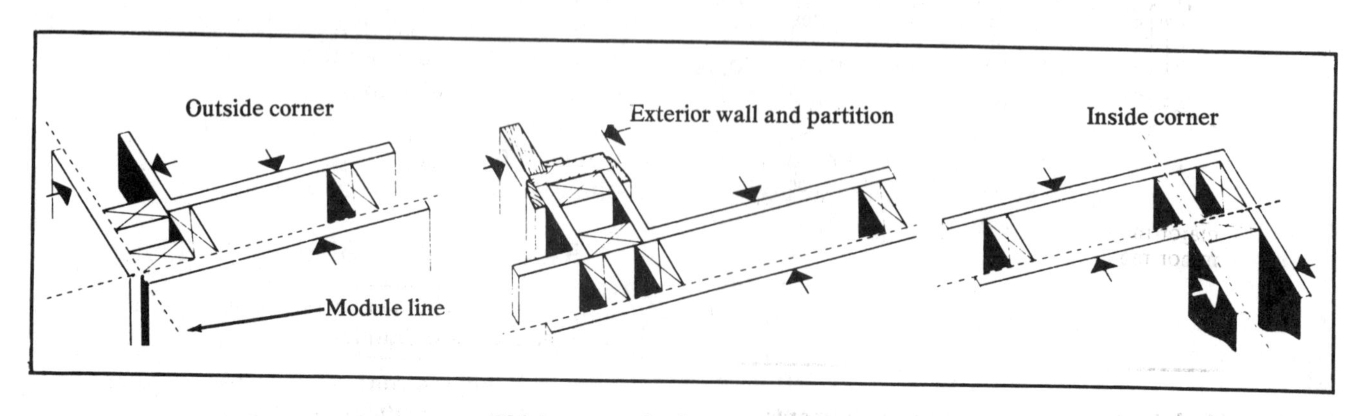

Thickness and tolerance variables
Figure 9-2B

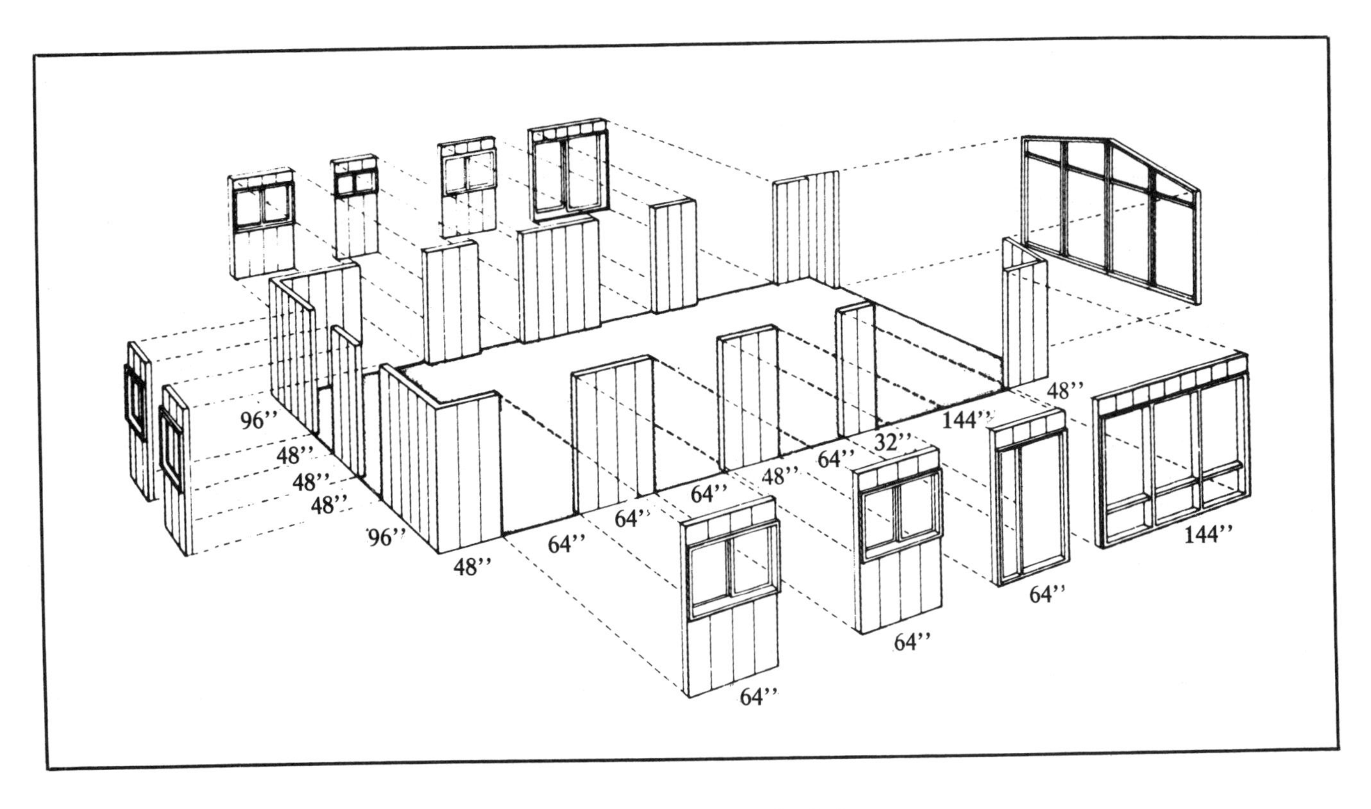

Exterior walls, doors and windows
Figure 9-3

Roof Dimensioning

Increments for house depths are in 24-inch multiples. Six 48-inch module depths and five 24-inch module widths fulfill most roof-span requirements. (See Figure 9-4A.) Standard roof slopes combined with modular house depths provide all the dimensions necessary to design rafter, truss, and panel roofs (Figure 9-4B).

The pivotal point shown in Figure 9-5 is the fixed point of reference in the modular line of the exterior wall. Modular roof design and construction dimensions are determined from this point.

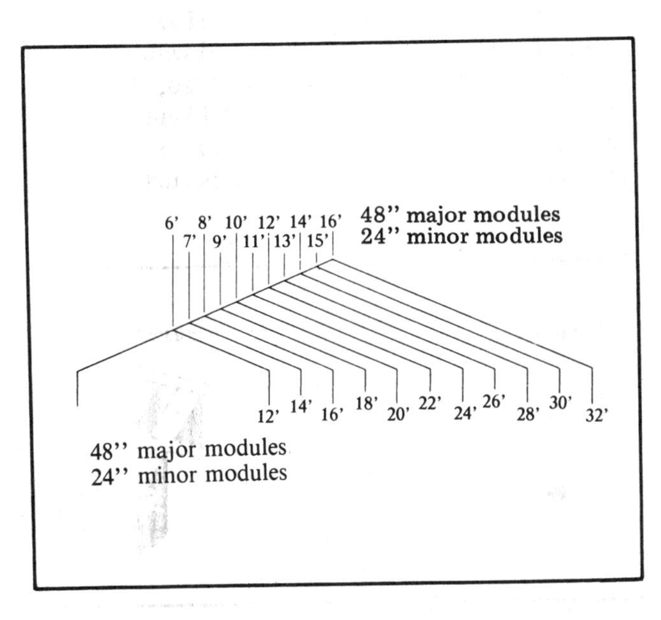

Modular house depths in 24'' increments
Figure 9-4A

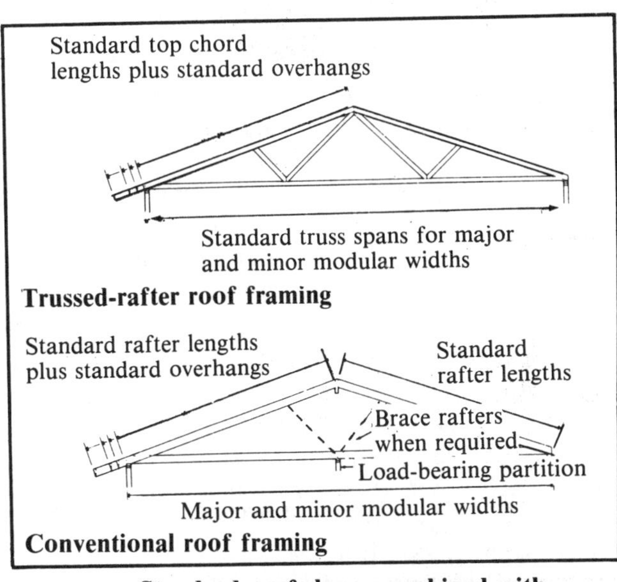

Standard roof slopes combined with modular house depths
Figure 9-4B

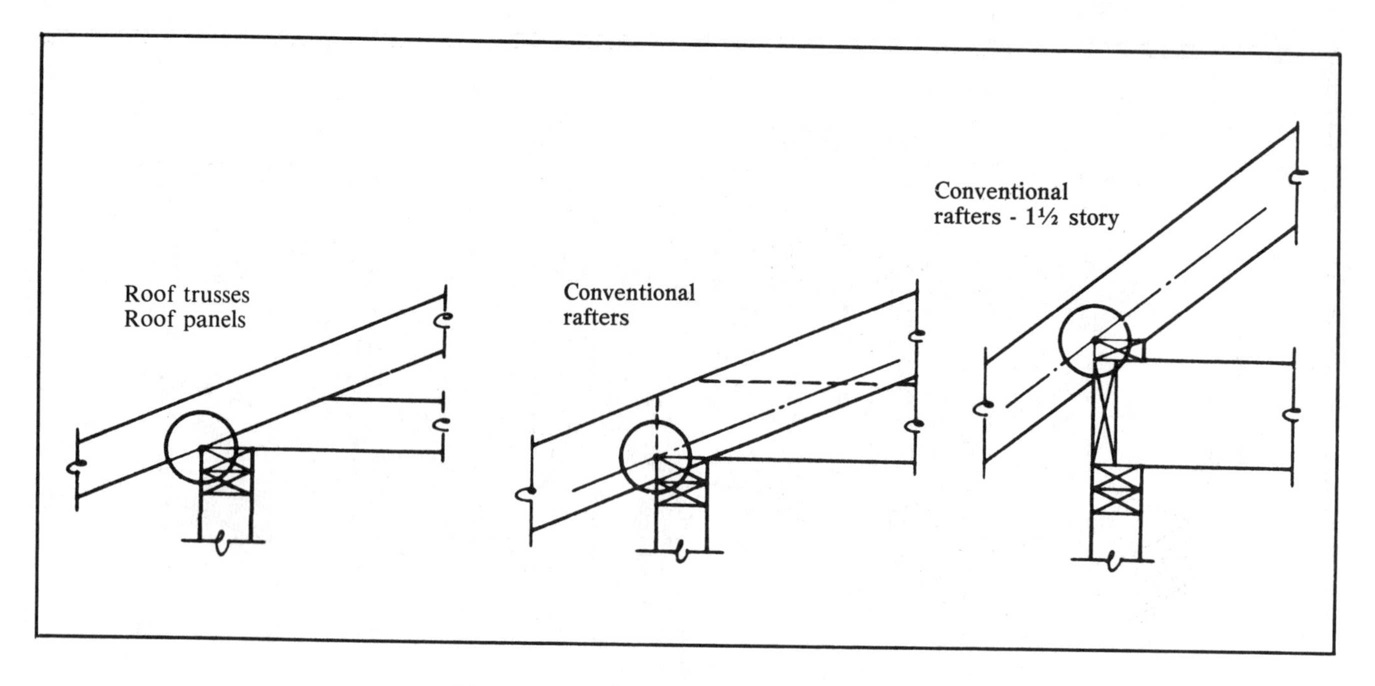

Pivotal point for modular roof planning
Figure 9-5

Floor Framing

In many new homes, 5 to 15% of floor framing is wasted. If the depth of the floor is not evenly divisible by 4, you are losing the maximum length of your floor joists. (See Figure 9-6.) Check and determine the efficiency of your floor framing system. You may save money!

Space floor joists at 12, 16, or 24 inches on center, depending on the floor load. They may be placed "lapped" or "in-line." Joist spacings of 13.7 inches and 19.2 inches are alternatives. They also divide the 8-foot length of a plywood subfloor panel into seven and five equal spaces, respectively.

Use of the 48-inch module on house depths permits greater use of full 4 x 8 plywood subflooring and minimizes cutting and waste.

Joist material is sized in 2-foot increments, with a tolerance of minus 0 and plus 3 inches. Joists 12 feet and longer usually are an extra ½-inch or more longer to allow for end-squaring.

In normal Platform construction with lapped joists bearing on top of a center girder, the length of joist required is half the house depth. Note also that the thickness of the band joist (1½ inches) and half the required overlap at the support (1½ inches) cancel each other. (See Figure 9-7.)

House Depths vs. Materials

Joist lengths correspond to standard lumber sizes when the house depth is on the 4-foot module of 24, 28 or 32 feet. Look at Figure 9-8. The same total linear footage of standard joist lengths is required to frame the floor of a 25-foot house depth as a 28-foot house depth. The same is true for a 29-foot house depth and a 32-footer. This holds for any joist spacing, and for in-line as well as lapped joists.

Changing to the 4-Foot Depth Module— If joist spans are increased when you change to the 4-foot module, the allowable span for the species and grade being used should be rechecked. It will be satisfactory in most cases.

Where spans are reduced to achieve the 4-foot module, a smaller size joist may be possible.

Subfloor Panel Layout— You're going to save even more with a 4-foot house depth module if the layout of the subfloor panels is preplanned. 48-inch wide panels can be used, without ripping, for the 24-, 28- and 32-foot house depths when joists are either lapped or trimmed for in-line placement. A panel layout option for the 28-foot house depth using lapped joists is shown in Figure 9-9. Similar layouts can be used with other lapped joist designs.

Exterior Wall Framing

Exterior wall framing can be cut as much as 25% with a modular layout. The key to eliminating unnecessary materials and labor is pre-planning.

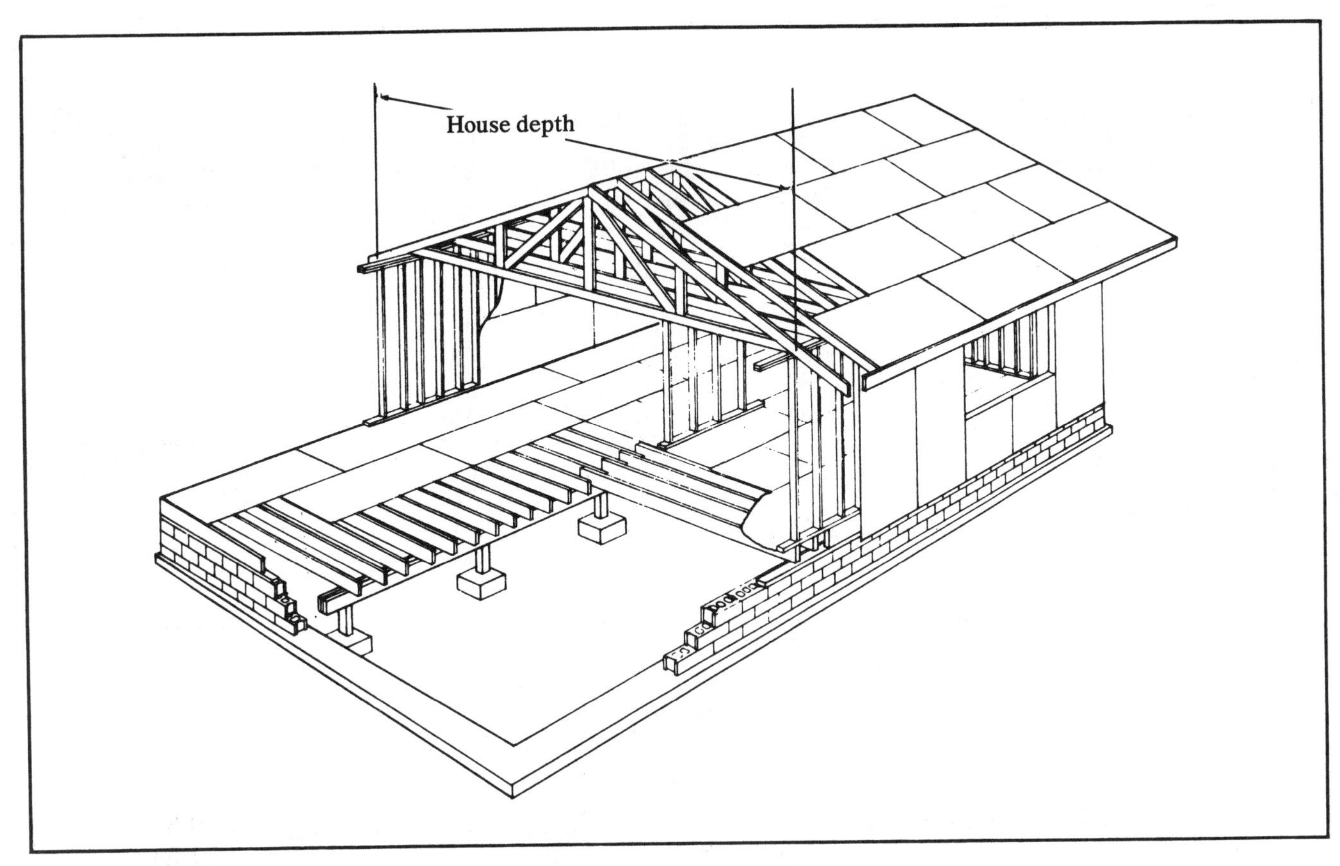

House depth measurement
Figure 9-6

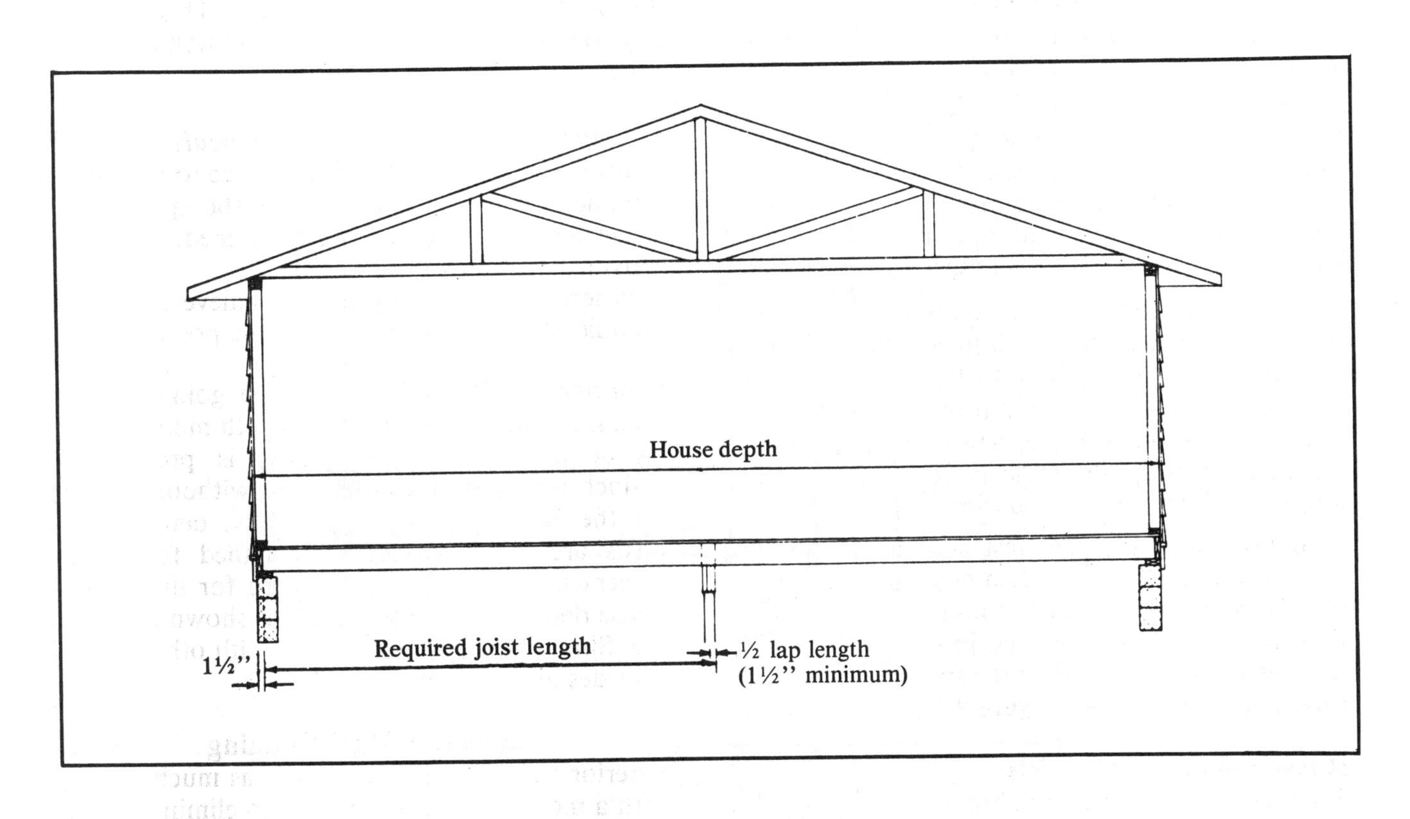

Required joist length
Figure 9-7

House Depth Feet	Joist Required Feet	Standard Length Feet	Linear feet 16" joist Spacing	Linear feet 24" joist Spacing	Bd. ft. per sq. ft. of floor area 16" spacing 2 x 8	16" spacing 2 x 10	24" spacing 2 x 10	24" spacing 2 x 12
21	10½	12	72	48	1.14	1.43	.95	1.14
22	11	12	72	48	1.09	1.36	.91	1.09
23	11½	12	72	48	1.04	1.30	.87	1.04
24	12	12	72	48	1.00	1.25	.83	1.00
25	12½	14	84	56	1.12	1.40	.93	1.12
26	13	14	84	56	1.08	1.35	.89	1.08
27	13½	14	84	56	1.04	1.30	.86	1.04
28	14	14	84	56	1.00	1.25	.83	1.00
29	14½	16	96	64	1.10	1.38	.92	1.10
30	15	16	96	64	1.07	1.33	.89	1.07
31	15½	16	96	64	1.03	1.29	.86	1.03
32	16	16	96	64	1.00	1.25	.84	1.00

Required footage of joists per 4' of house length based on standard joist length
Figure 9-8

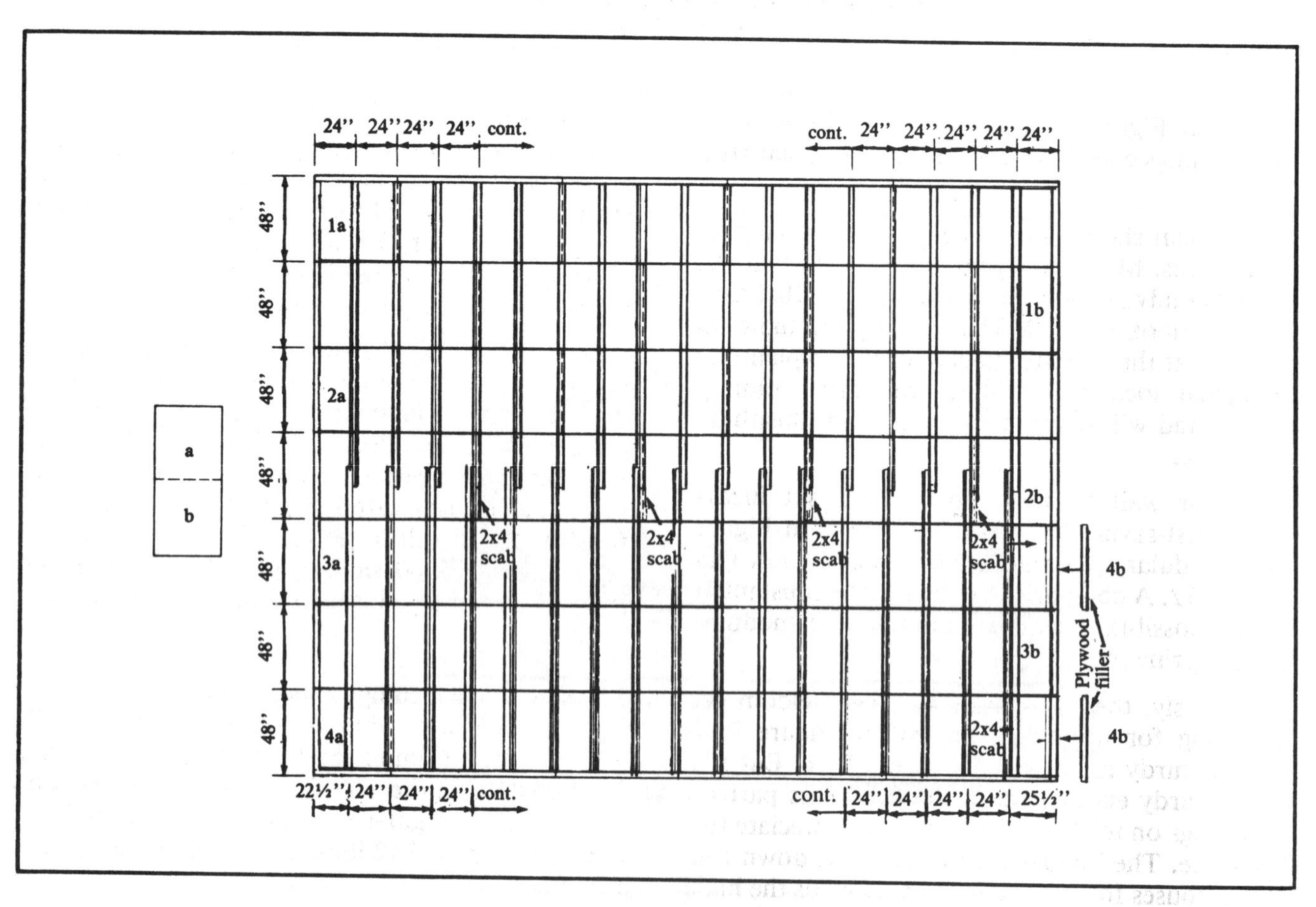

Subfloor panel layout for 28' house depth when joists are lapped over center support
Figure 9-9

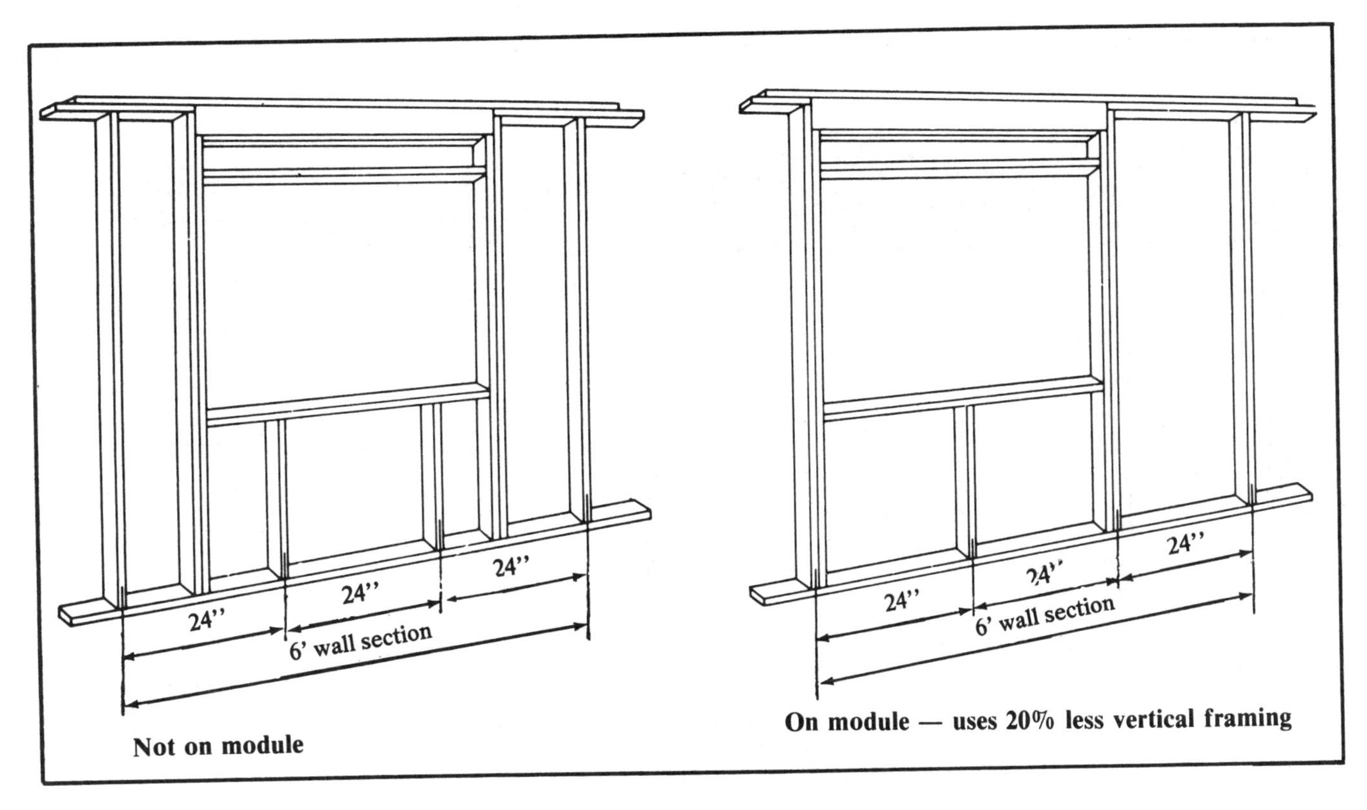

Windows located on module can save framing
Figure 9-10

Look at Figure 9-10. The size and location of wall openings coincide with standard modular stud spacings.

Your plan should show the location of all wall components. Make sure your crew knows that the plan takes advantage of modular sizes, so that they don't overlook possibilities for labor and material savings that the designer anticipated. Pre-planning the actual location of studs will save framing lumber, and will reduce cutting on both sheathing and siding.

Exterior wall framing which does not incorporate cost-saving techniques is shown in Figure 9-11. Modular, pre-planned framing is shown in Figure 9-12. A comparison indicates the substantial savings possible if you pay attention to modular framing principles.

Obviously, there is a trade-off. You seldom get something for nothing. The wall in Figure 9-12 isn't as sturdy as the one in Figure 9-11. But it's plenty sturdy enough to do the job. Pass part of the savings on to the buyers. They'll appreciate the difference. The builder who keeps costs down can sell his houses for less money. That avoids the high cost of carrying houses that don't sell quickly.

Let's take a closer look at some of the ways modular framing can reduce costs.

Stud Spacing

You can use 24-inch stud spacing in virtually any style house where exterior and interior facing materials can span 24 inches. Tests and structural analyses show that, for many installations, 24-inch spacing can even be used for walls supporting the upper floor and roof of two-story houses.

Using 24-inch stud spacing will save you money. For a 1,440 square foot one-story house, with standard door and window areas, a savings of $150 is possible when exterior wall framing is 24 inches rather than 16 inches on center. In many cases, the same thickness of exterior sheathing and exterior finish material will be acceptable. There could be an additional $100 savings in reduced labor for drywall, sheathing, insulation and wiring installations.

Placement of Openings and Intersections

Poor placement of window and door locations will increase wall framing costs. Whenever possible, have one side of the window and door rough openings fall on a regular stud position. Details E, F and M of Figure 9-12 illustrate this modular layout principle.

Windows and doors are available that have a rough-opening width that is a multiple of the stud spacing. Then both sides can be on the stud

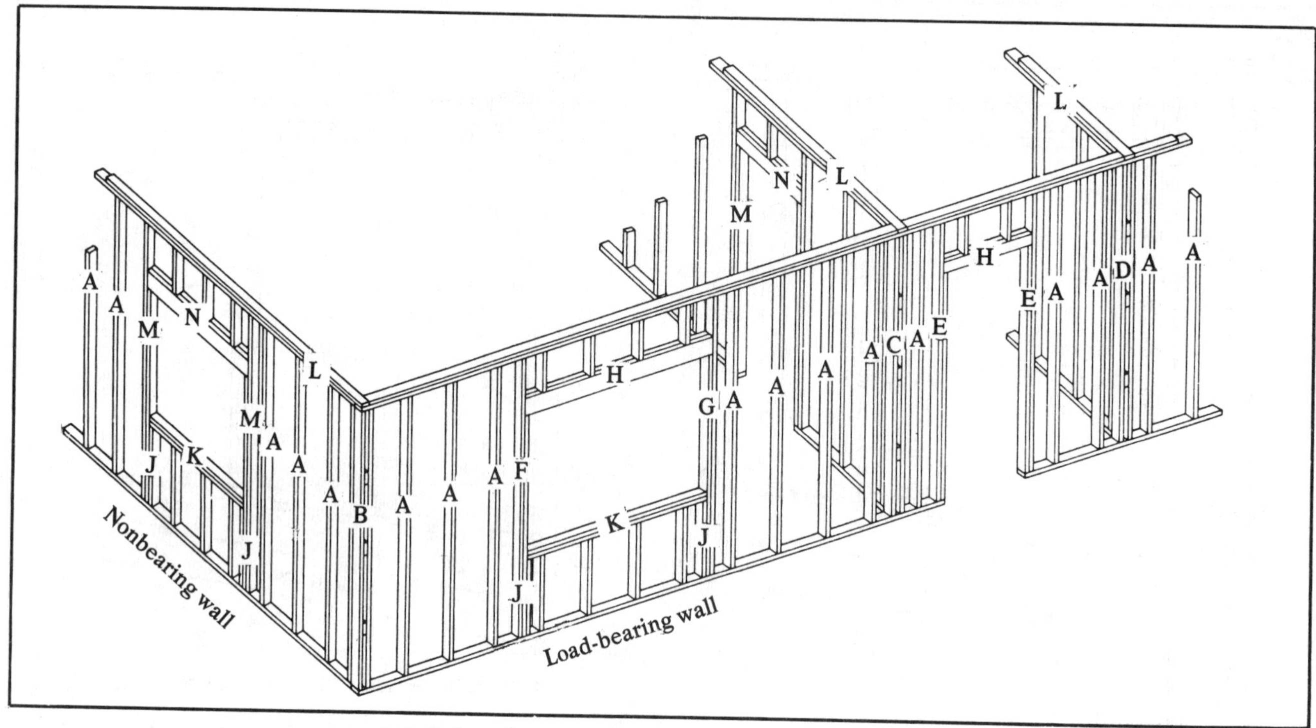

Wall framing: cost-saving principles not applied
Figure 9-11

module. This reduces the number of studs required, and permits the installation of sheathing and siding with a minimum of waste. Figure 9-12 shows how this can be done. The result is additional savings in your pocket.

Now look at Figure 9-12, Detail D. Here the partition is located at a normal wall stud, eliminating the need for a corner nailer on one side of the partition. Scrap plywood or boards can be used for back-up cleats for opposite corner nailing.

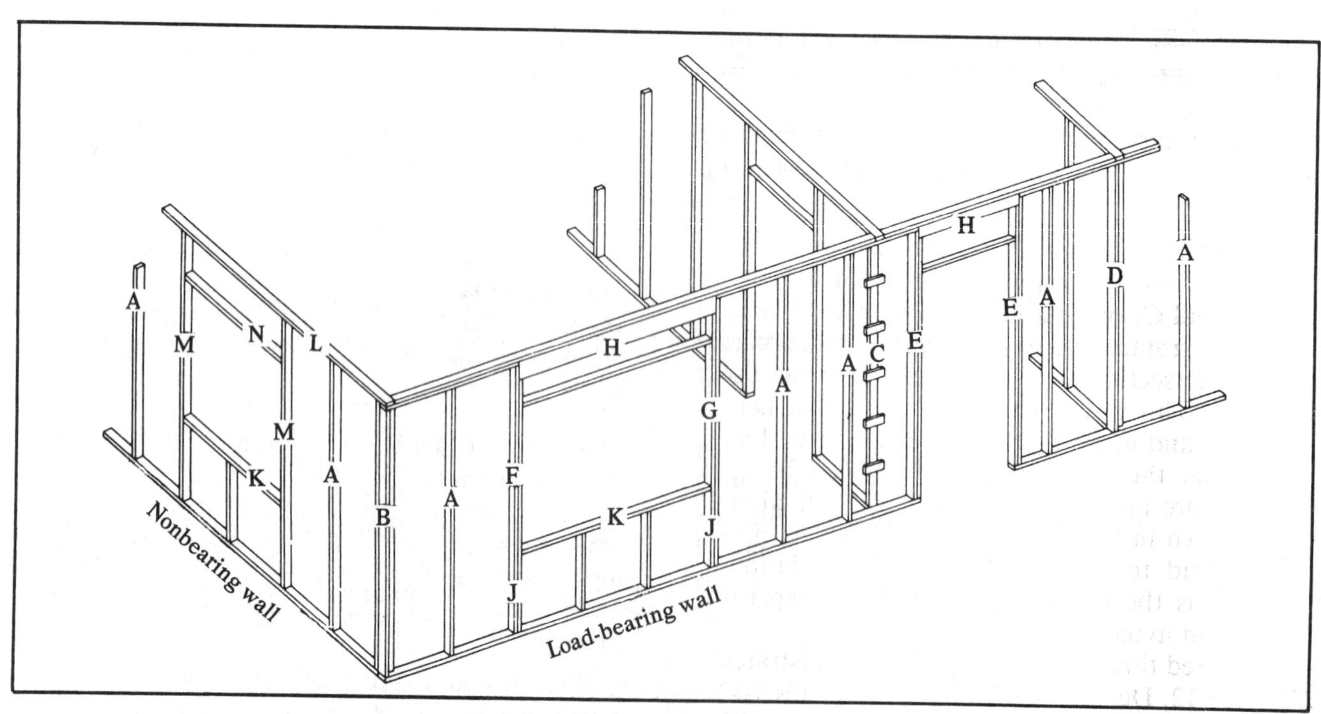

Wall framing: cost-saving principles applied
Figure 9-12

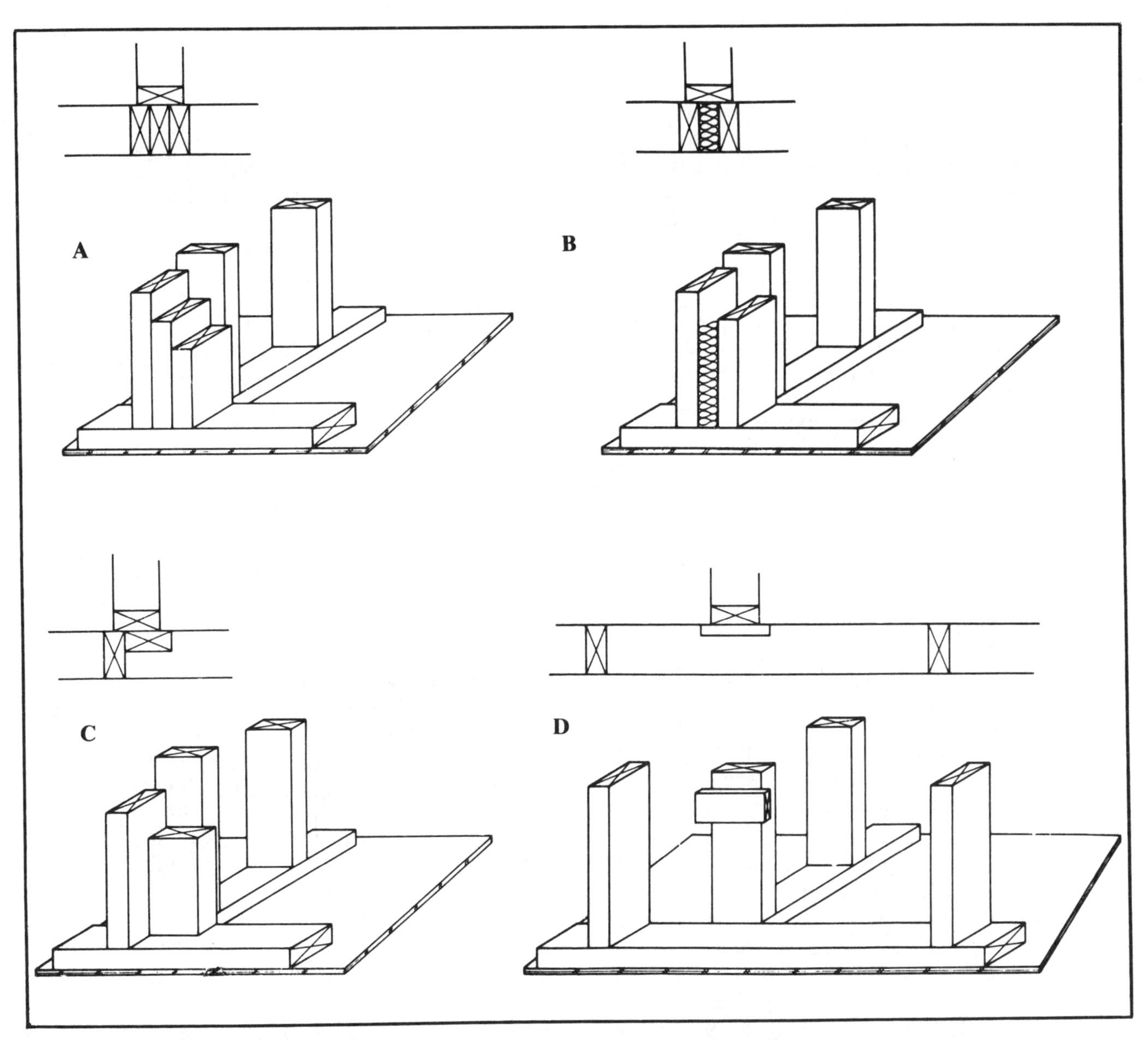

Framing alternatives at wall and partition intersections
Figure 9-13

Studs at Wall Corners and Intersections
Traditional framing methods for exterior corners and for intersections of partitions and walls use three studs. This is shown in Details B, C and D of Figure 9-11 and in Figures 9-13 B and 9-14 B. This requires that the space between the studs be insulated before applying the sheathing. One alternative, shown in Figures 9-13 A and 9-14 A, uses an extra stud to provide the desired insulation. While this is thermally effective, it's an expense that you can avoid.

A preferred three-stud arrangement is illustrated in Figure 9-12, Details B and D, and in Figures 9-13 C and 9-14 C. This stud placement eliminates spacer blocks. It provides the same back-up support for wall linings as the more traditional arrangement. Insulation can be installed after the sheathing is applied.

Another alternative is shown in Figure 9-12, Detail C, and in Figure 9-13 D. Studs, serving only as a back-up for interior corner-nailing surfaces for finish materials, are eliminated by attaching back-up cleats to the partition stud. These cleats can be 3/8-inch plywood or 1-inch lumber or any suitable scrap material. A similar arrangement of cleats can be used to eliminate the back-up nailer stud at interior corners. (See Figure 9-14 C.)

You'll save about $60 per house in labor and materials by using the three-stud exterior corner arrangement and back-up cleats at all intersections of

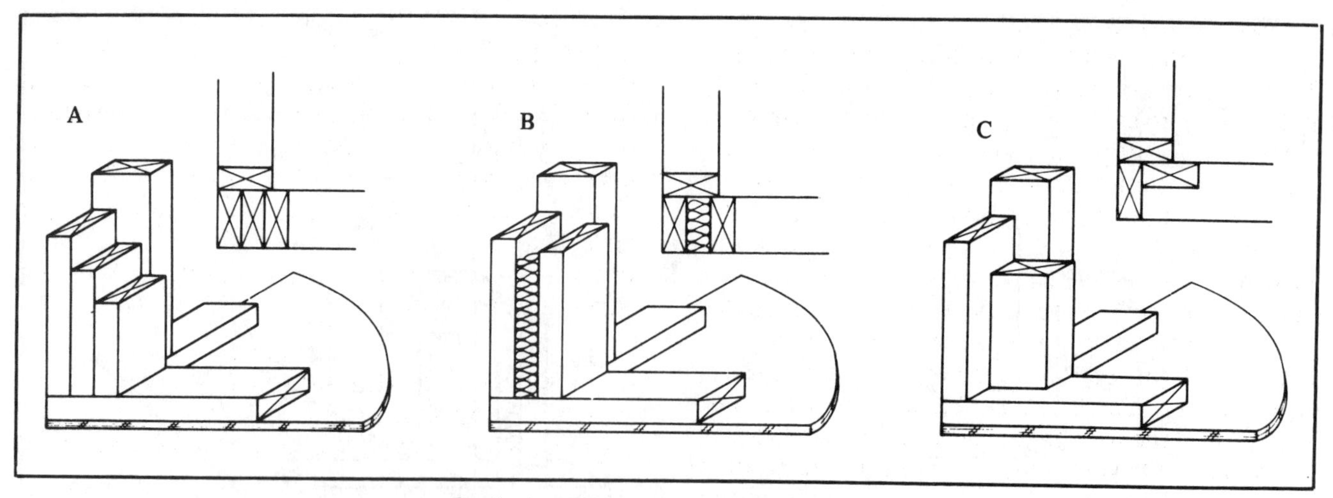

Stud arrangements at exterior corners
Figure 9-14

partitions and exterior walls. (Compare Details B and C, Figures 9-11 and 9-12.)

Framing Around Windows and Doors

Most carpenters have long suspected that too much wood is used in framing window openings. A window a few inches too wide or too narrow or located a few inches too far left or right usually means that you need at least one extra stud. Some carpenters put enough material under and over a window to support a truck.

Details J and K of Figure 9-11 show doubled sills under windows, plus short support studs at the ends of the sill. Figure 9-15 shows that vertical loads at windows are transferred downward by the studs that support the headers. The sill and wall beneath the window are non-bearing. Details J and K of Figure 9-12, and Figures 9-10 and 9-15 show how to eliminate the second sill and the two sill-support studs. The ends of the sill are supported by end-nailing through the adjacent stud. Use of a single sill member at the bottom of all window openings, and eliminating all window sill support studs will save you about $45 per house.

You can save even more by moving the header beam up to replace the lower top plate. See Figures 9-11 and 9-12, Detail H, and Figures 9-10 and 9-15. This requires longer support studs. But, it can eliminate most of the short fill-in studs between the top plate and the header. Vertical loads are not transferred to the top of the window opening. Short studs below the header are needed only when required to support the sheathing and finish materials.

If the sheathing or finish materials can span the distance between the header beam and the rough opening top plate, no short studs are needed. Savings in framing material and labor are only part of the advantage. You also save on cutting and fitting insulation.

The same principle applies to header beams across doorways. The objection to a single member top plate over the header, as shown in Figures 9-10, 9-12 and 9-15, is that it can slow down the horizontal wall framing. As you will recall, horizontal wall framing on the deck depends on the bottom member of the top plate being assembled continuously with the wall. Some adjustment has to be made when using this technique, and this might offset material savings.

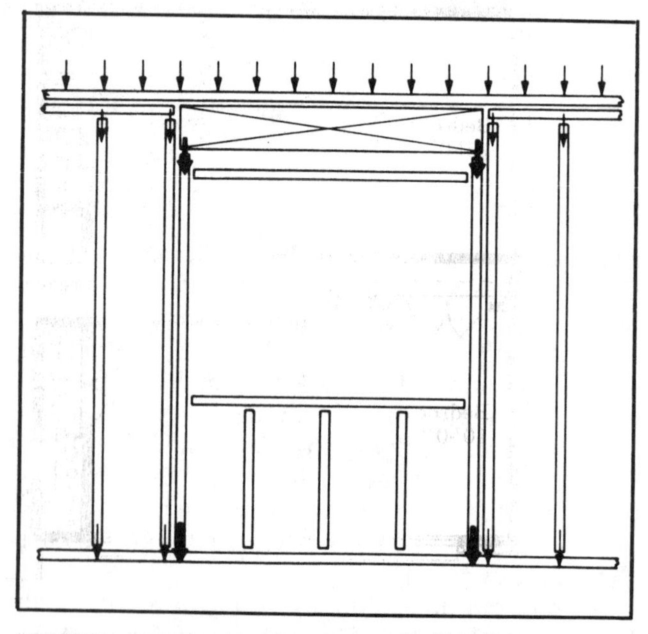

Load distribution through header and support studs at opening in load-bearing wall
Figure 9-15

Framing Non-bearing Walls
Not all exterior walls are load-bearing. Generally, there are two or more walls, parallel to the joists and roof trusses, which support only their own weight. Such non-bearing exterior walls can be framed the same way as interior partitions, as shown in Details K, L, M and N of Figure 9-12.

Since there are no loads to transfer, doors and windows can be framed with single members. A single top plate is enough if metal straps or plates are used to tie them together at joints. Of course, when single top plates are used, studs must be 1½ inches longer. This is no problem, if studs are cut to length on the job. If precut studs are used, order two lengths. Or you can use standard precut studs for the load-bearing walls, and cut the studs on the job for the non-bearing walls.

Mid-height Wall Blocking
Mid-height blocking between wall studs is generally unnecessary and may be eliminated. Top and bottom plates provide adequate fire stopping, and 1/2-inch sheet rock applied horizontally does not require blocking for support when the studs are 24 inches on center or less. Eliminating mid-height blocking will save you about $75 per house.

Wall Bracing
Most codes require that exterior walls have minimum "racking" strength for stability under wind loads. There's enough "racking" strength in panel-type sheathing or siding. No additional let-in corner bracing is needed. If a nonstructural sheathing is used, such as low-density fiberboard or gypsumboard, additional bracing may be necessary. You can use structural sheathing or siding panels at the corners of the structure, or use let-in 1 x 4 bracing. But 1 x 4 bracing requires additional skilled craftsman time. It's cheaper to use panels for bracing.

Estimating Studding Requirements

Modular construction with framing 24 inches on center will reduce your building costs considerably.

The drawings in this chapter illustrate the money-saving techniques available with a little pre-planning.

Here's a rule of thumb to use when estimating the number of studs required for 24-inch spaced modular framing:

For load-bearing walls: 1 stud for every 2 feet, plus 2 studs for each window opening, and 2 studs for each door opening. Add 2 studs for each corner.

For non-bearing walls: 1 stud for every 2 feet, plus 1 stud for each window opening, and 1 stud for each door opening. Add 2 studs for each intersection.

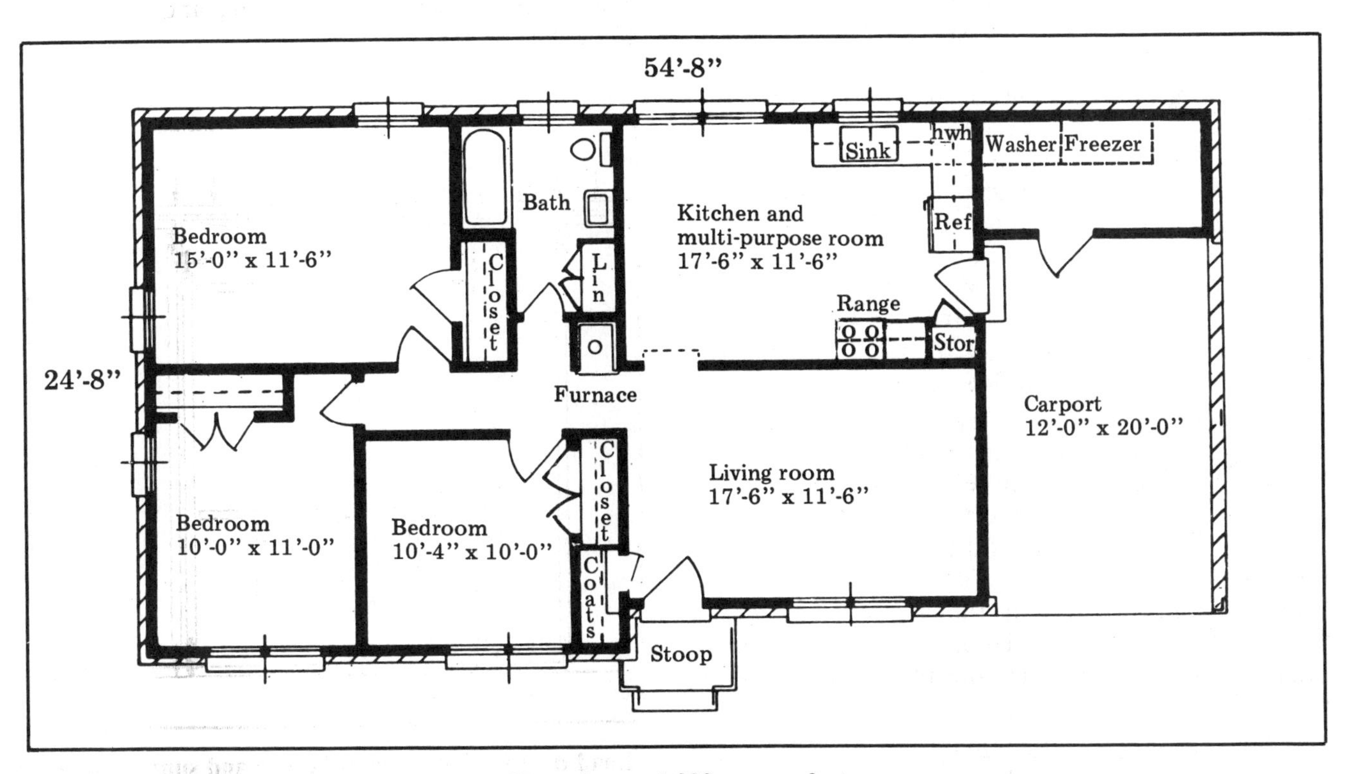

Heated area 1,092 square feet
Figure 9-16

Figure 9-16 shows a home with 1,092 square feet of floor space (heated area). Our rule of thumb for modular wall framing calls for about 200 studs. The rule of thumb for conventional wall framing 16 inches o.c., requiring 1 stud for every 3 feet of floor space, would require about 364 studs.

At today's prices, saving 164 studs and the labor to install those 164 studs is a very healthy chunk of change. Modular framing can give you a competitive edge, and will put additional profit in your pocket.

Figure 9-17 is your Modular Construction Checklist and Data Reference Sheet. Use it to create a record of your work on modular building jobs.

Summary

The concepts explored in this chapter will give you some powerful tools for making and saving money on every home you build. But you don't need to know *everything* to succeed in building houses. Indeed, no one knows it all! But this book is intended to give you all the essential know-how for building marketable, speculative custom homes.

Modular Construction Checklist and Data Reference Sheet

1) The house is _______ feet wide and _______ feet in length, for a total of _______ square feet.

2) The house is: ☐1-story ☐1½ story ☐2-story with ☐basement ☐crawl space ☐slab.

3) Modular framing is ☐16 inches o.c. ☐24 inches o.c. ☐other (specify) ____________________________

____

4) Floor joists are:

2 x 8 x __________

2 x 10 x __________

2 x 12 x __________

Total number of floor joists: __________

(To determine number of joists required for 24-inch spacing divide the length of the house by 2 and add 1 joist. Use additional joists under partitions as required.)

5) Floor joists are ☐lapped at girder ☐in-line.

6) Subfloor plywood panels ☐did ☐did not require cutting prior to installation. For example, panels don't require cutting when applied to a 28-foot by 56-foot house (measured from outside stud surfaces). A total of 49 sheets of 4 x 8-foot plywood will cover the floor without using any labor for measuring and ripping.

7) Studs are ☐2 x 4 precut ☐2 x 4 cut on job site.

Total number of studs: __________

(A rule of thumb for determining stud requirements for modular construction with 24-inch spacing is:

For load-bearing walls, use 1 stud for every 2 feet plus 2 studs for each window opening and 2 studs for each door opening. Add 2 studs for each corner.

For example a 48-foot wall with 1 door and two windows requires 34 studs.

For non-bearing walls, use 1 stud for every 2 feet plus 1 stud for each window opening and 1 stud for each door opening. Add 2 studs for each intersection.)

8) Openings were preplanned to allow ☐one side ☐two sides to fall on a regular spaced stud at ☐windows ☐doors.

9) Top plate is:

☐Single-member in non-bearing walls

☐Single-member over headers

☐Double-member in load-bearing walls

☐Single-member in partitions.

Figure 9-17

10) Ceiling joists and rafter framing are:

☐Conventional "stick" method

☐Trusses

11) Materials used in conventional method:

☐Ceiling joists

☐2 x 6 x ________

☐2 x 8 x ________

Total number of ceiling joists: ________

(Ceiling joists required for 24-inch spacing are determined the same way as floor joists.)

☐Rafters:

☐2 x 4 x ________

☐2 x 6 x ________

☐2 x 8 x ________

Total number of rafters: ________

(To determine rafter requirements for 24-inch spacing, divide the length of the structure (measured from outside stud surfaces) by 2 and add 1. Where overhang fly (floating) rafters are used, add 2 rafters to each side of roof.)

☐Rafter bracing is ☐2 x 4 ☐2 x 6.

☐Collar beams are ☐1 x 6 ☐2 x 4 ☐2 x 6.

12) Truss data:

☐W-type (length) ________ feet

☐King-post (length) ________ feet

☐Scissors (length) ________ feet

☐Studded gable ends

☐Soffit return

☐No soffit return

☐Plywood gussets

☐Metal gussets

☐Pitch (______/12)

Total number of trusses: __________

(Trusses are commonly spaced 24 inches o.c. To determine number of trusses required. divide length of structure by 2 and add 1.)

13) Wall bracing is:

☐4 x 8-foot structural panel bracing at corners

☐Other (specify) ______________________________

14) Material costs:

Sills, girders, bridging:

1 x 3	______LF	@	$______LF	=	$________
2 x 2	______LF	@	$______LF	=	$________
2 x 4	______LF	@	$______LF	=	$________
2 x 6	______LF	@	$______LF	=	$________
2 x 8	______LF	@	$______LF	=	$________
2 x 10	______LF	@	$______LF	=	$________
2 x 12	______LF	@	$______LF	=	$________
Other (specify)					
____________	________	@	$________	=	$________

Floor joists:

2 x 8 x ______	______pcs	@	$______ea	=	$________
2 x 10 x ______	______pcs	@	$______ea	=	$________
2 x 12 x ______	______pcs	@	$______ea	=	$________
Subfloor panels	______pcs	@	$______ea	=	$________
Studs	______pcs	@	$______ea	=	$________
Sole plates (2 x ____)	______LF	@	$______LF	=	$________
Top plates (2 x ____)	______LF	@	$______LF	=	$________

Ceiling joists:					
2 x 6 x ______	______pcs	@	$______ea	=	$______
2 x 8 x ______	______pcs	@	$______ea	=	$______
Rafters:					
2 x 4 x ______	______pcs	@	$______ea	=	$______
2 x 6 x ______	______pcs	@	$______ea	=	$______
2 x 8 x ______	______pcs	@	$______ea	=	$______
Rafter bracing:					
2 x 4	______LF	@	$______LF	=	$______
2 x 6	______LF	@	$______LF	=	$______
Collar beams:					
1 x 6	______LF	@	$______LF	=	$______
2 x 4	______LF	@	$______LF	=	$______
2 x 6	______LF	@	$______LF	=	$______
Trusses	______pcs	@	$______ea	=	$______
Wall bracing:					
4 x 8 panels	______pcs	@	$______ea	=	$______
1 x 4 let-in	______LF	@	$______LF	=	$______
Other (specify)					
______________	______	@	$______	=	$______
Nails:					
16d	______lbs	@	$______lb	=	$______
10d	______lbs	@	$______lb	=	$______
8d	______lbs	@	$______lb	=	$______
6d	______lbs	@	$______lb	=	$______
Other (specify)					
______________	______	@	$______	=	$______

Other materials, not listed above:

____________________ _________ @ $_________ = $_________

____________________ _________ @ $_________ = $_________

____________________ _________ @ $_________ = $_________

____________________ _________ @ $_________ = $_________

Total cost of all materials $_________

15) Labor costs:

Sills, bolted and grouted:
(Require 3 skilled, 2 unskilled hours per 100 LF.)

Skilled _______hours @ $_______hour = $_________

Unskilled _______hours @ $_______hour = $_________

Built-up girders:
(Require 13 skilled, 5 unskilled hours per 1,000 BF.)

Skilled _______hours @ $_______hour = $_________

Unskilled _______hours @ $_______hour = $_________

Floor joists:
(2 x 8 joists require 15 skilled, 6 unskilled hours per 1,000 BF.)

Skilled _______hours @ $_______hour = $_________

Unskilled _______hours @ $_______hour = $_________

Joist bridging:
(1 x 3 bridging requires 4 skilled hours per 100 pieces.)

Skilled _______hours @ $_______hour = $_________

Subflooring (plywood):
(Modular subflooring requires 8 skilled, 4 unskilled hours per 1,000 SF.)

Skilled _______hours @ $_______hour = $_________

Unskilled _______hours @ $_______hour = $_________

Studs (24 inches on center):
Exterior walls and plates:
(Require 21 skilled, 6 unskilled hours per 1,000 BF.)

Skilled _______hours @ $_______hour = $_________

Unskilled _______hours @ $_______hour = $_________

Partition walls and plates:
(Require 20 skilled, 5 unskilled hours per 1,000 BF.)

Skilled	________hours	@	$________hour	=	$____________
Unskilled	________hours	@	$________hour	=	$____________

Ceiling joists:
(2 x 6 joists require 18 skilled, 7 unskilled hours per 1,000 BF.)

Skilled	________hours	@	$________hour	=	$____________
Unskilled	________hours	@	$________hour	=	$____________

Rafters:
(2 x 6 material for plain gable construction requires 23 skilled, 6 unskilled hours per 1,000 BF.)

Skilled	________hours	@	$________hour	=	$____________
Unskilled	________hours	@	$________hour	=	$____________

Rafter bracing and ties:
(Require 20 skilled, 5 unskilled hours per 1,000 BF.)

Skilled	________hours	@	$________hour	=	$____________
Unskilled	________hours	@	$________hour	=	$____________

Truss installation:
(Each truss requires 1.5 hours, ⅔ skilled, ⅓ unskilled.)

Skilled	________hours	@	$________hour	=	$____________
Unskilled	________hours	@	$________hour	=	$____________

Total labor costs:

Skilled	________hours	@	$________hour	=	$____________
Unskilled	________hours	@	$________hour	=	$____________
Other (specify)					
________________________	________hours	@	________hour	=	$____________
				Total	$____________

16) Total cost:

Materials	$____________
Labor	$____________
Other	$____________
Total	$____________

17) Cost of modular framing per square foot of floor space $__________.

18) Problems that could have been avoided ______________________________

____

____

____

____

19) Comments ______________________________

____

____

____

____

20) A copy of this Checklist and Data Reference Sheet has been provided:

☐Management

☐Accounting

☐Estimator

☐Foreman

☐File

☐Other: ____________________

Chapter 10

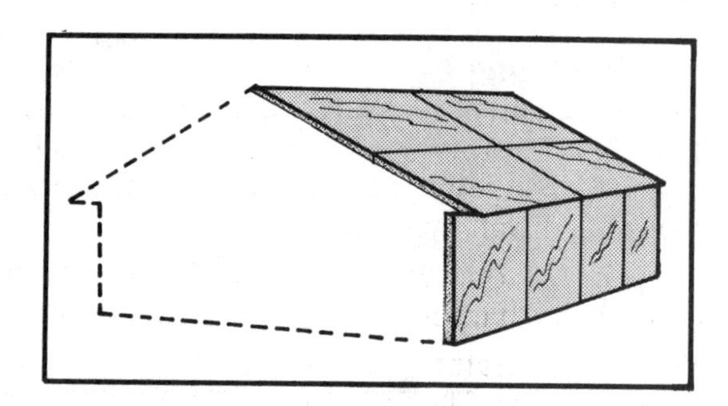

Insulated Sheathing

Every builder has the obligation to build energy efficiency into the homes he constructs. Buyers demand it, the building code requires it, and your good sense and professionalism should assume it. Remember that the cost of fuel burned to heat and cool a home during its useful life expectancy will nearly always exceed the cost of all materials needed to build the house. Build houses that reflect your wisdom and professionalism as an energy-conscious builder.

Many types of insulated sheathing are available now to make your job easier. This chapter will cover foam plastic sheathing sold under names such as Thermax.

But the place to start is with a few of the basic principles you need to know about insulation.

FTC Rules on Home Insulation

The Federal Trade Commission Rule "Labeling and Advertising of Home Insulation" requires that certain information be provided about home insulation.

Manufacturers and builders must provide information to potential buyers and users to help them evaluate competing products.

The home buyer must be advised of the R-value and thickness of the insulation used in the home prior to the sale. This information must be included as part of the contract.

Here's what the rule requires: All R-values must be listed for each material individually (air spaces or reflectivity not included) at 75 degrees F mean temperature. If reflective R-values are included, they must be listed separately, and details must be given as to how they were determined.

Written advertisements mentioning R-values must include the statement:

The higher the R-value, the greater the insulating power. Ask your seller for the fact sheet on R-values.

Written advertisements or promotional materials stating or implying fuel or dollar savings must include the statement:

Savings vary. Find out why in the seller's fact sheet on R-values. Higher R-values mean greater insulating power.

The following is from the FTC rule:

If you say or imply that a combination of products can cut fuel bills or fuel use, you must have a reasonable basis for the claim. You must make the statement about savings in subsection (b). Also, you must list the combination of products used. They may be two or more types of insulation; one or more types of insulation and one or more other insulating products, like storm windows or siding; or insulation for two or more parts of the house, like the attic and walls. You must say how much of the savings came from each product or location. If you cannot give exact or appropriate figures, you must give a ranking. For instance, if your ad says that insulation and storm doors combined to cut fuel use by 50%, you must say which one saved more.

The FTC rule has the effect of law. Compliance is mandatory. Protect yourself. Get a current copy of the rule on home insulation from the FTC at:

Public Reference Branch
Federal Trade Commission Headquarters
6th Street and Pennsylvania Ave., N.W.
Washington, D.C. 20580

Insulating Foam Plastic Sheathing

Insulating foam plastic sheathing (FPS) is efficient. Efficiency measures how much insulating thickness provides what resistance to heat transfer through an insulated wall or ceiling.

Thermax is one of the most efficient FPS products on the market today. (See Figure 10-1.)

Thermax sheathing is an insulation board consisting of glass fiber-reinforced polyiscolyanurate foam plastic core with aluminum foil faces. The sheathing's *aged* R-values are shown in Figure 10-2. It is available for conventional 2 x 4-inch (nominal dimension) framing, without using more costly window and door trim or unconventional construction techniques. It comes in standard insulation board sizes of 4 x 8 and 4 x 9 feet. The thickness range is from 1/2-inch to 3 inches.

Thermax and similar foam plastic sheathing are not structural panels. You must provide diagonal bracing with either 1 x 4 let-ins or with metal straps (as permitted by your building code).

What are aged R-values? You know that R-value is a measure of insulation. The higher the number, the better the insulating value. But R-value can change with age. Closed-cell foams using halocarbon blowing agents tend to lose their thermal resistance (R-value) when atmospheric gas moves into the cells, or when the blowing agent escapes from the cells. This process generally occurs over a period of time.

The California Energy Commission Conservation Division, Title 20, Chapter 2, Sub-chapter 4, Article 3, "Standards for Insulation Material" defines an aged R-value as the thermal resistance of "test samples certified by an approved testing laboratory to have been aged while exposed to free air in a ventilated room for at least two years."

For ASHRAE winter design purposes: If one side of Thermax is exposed to a 3/4-inch air space in the wall, add 3 "R's" to the value shown in Figure 10-2 per ASHRAE method (ASHRAE 1977 Fundamentals Handbook). For consumer comparison purposes, add 2.8 "R's" instead.

R-20 Wall Components

In many parts of the U.S. it's good practice to provide a minimum wall R-value of 20. This is possible even with conventional 2 x 4 framing, if you substitute Thermax or a similar sheathing for ordinary exterior sheathing. No major changes in house design, framing lumber, or trim are needed.

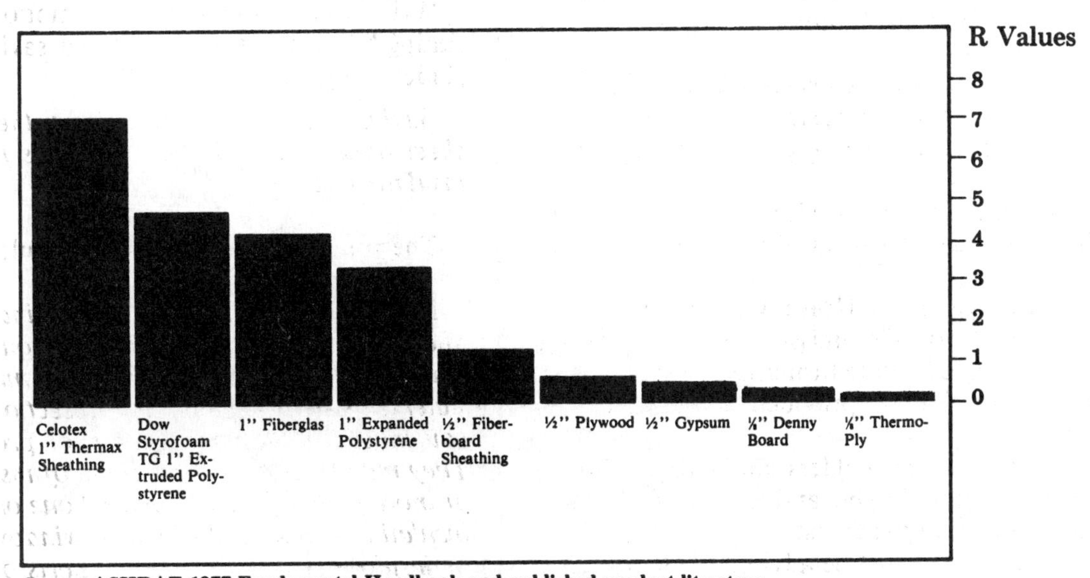

Source ASHRAE 1977 Fundamental Handbook and published product literature

Thermo-Ply—Reg. trademark of Simplex Industries, Adrian, MI
Denny Board—Reg. trademark of Denny-Corp., Calwell, OH
Expanded Polystyrene is sold by many different manufacturers
Styrofoam—Reg. trademark of Dow Chemical manufacturers
Fiberglas—Reg. trademark of Owens-Corning Fiberglas Corporation, Toledo, OH

Comparative R-values (at 75°F mean temperature) of sheathings in available thicknesses
Figure 10-1

THERMAX - AGED R VALUES*

Nominal Thickness	½"	⅝"	¾"	⅞"	1"	1¼"	1½"	1¾"	2"	2¼"	2½"	2¾"	3"
Information for consumer comparison purposes													
At 15 Months Aged 75°F Mean Temp.	3.6	4.5	5.4	6.3	7.2	9	10.8	12.6	14.4	16.2	18.0	19.8	21.6
At 68 Months Aged 75°F Mean Temp.	3.6	4.5	5.4	6.3	7.2	9	10.8	12.6	14.4	16.2	18.0	19.8	21.6
Additional information for design professionals													
At 15 Months Aged 40°F Mean Temp.	4	5	6	7	8	10	12	14	16	18	20	22	24
At 68 Months Aged 40°F Mean Temp.	4	5	6	7	8	10	12	14	16	18	20	22	24
At 15 Months Aged 110°F† Mean Temp.	3.3	4.1	4.9	5.7	6.5	8.2	9.8	11.5	13.1	14.7	16.4	18.0	19.6
At 68 Months Aged 110°F Mean Temp.	3.3	4.1	4.9	5.7	6.5	8.2	9.8	11.5	13.1	14.7	16.4	18.0	19.6

**Aged R values are based on Guarded Hot Box Tests conducted in accordance with ASTM C236 on full sized product, and reported without air space effects. Also applicable to 16" and 24" widths. For "R" values at other mean temperatures, consult Celotex.*

Thermax sheathing: aged R-values
Figure 10-2

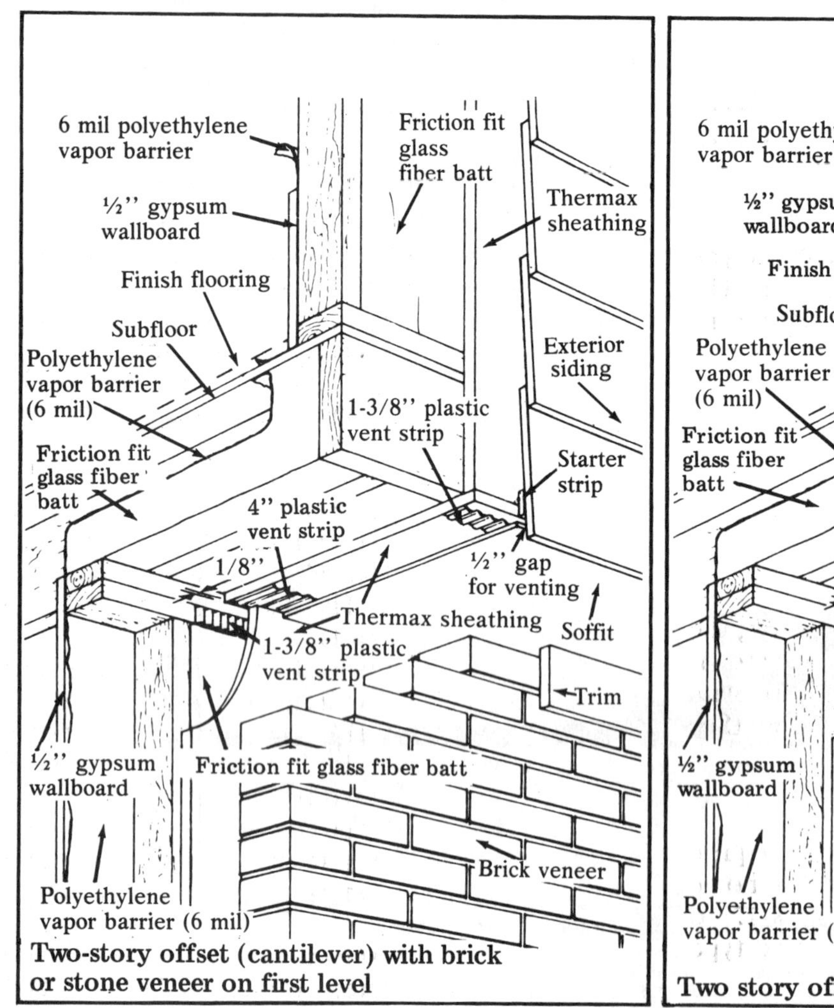

Two-story offset (cantilever) with brick or stone veneer on first level

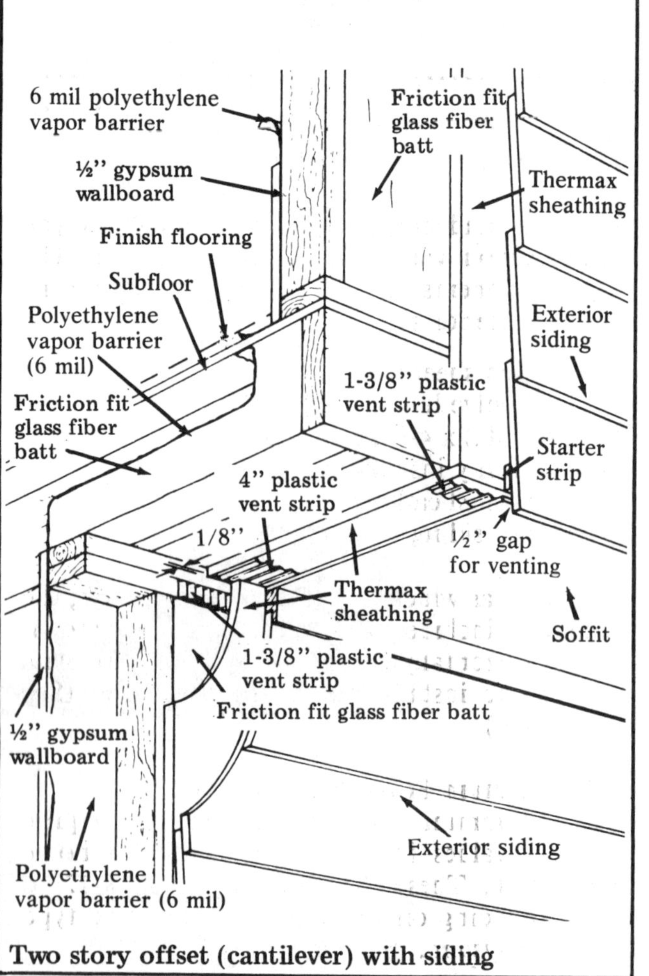

Two story offset (cantilever) with siding

Plastic vent strip installation
Figure 10-3

The Minumum Property Standards contain provisions for "trade-offs" between opaque wall R-values and window types and area. For example, by increasing the R-value of the wall over the minimums shown, greater areas of glass may be permitted. Consult the HUD Standards for complete details.

HUD Minimum Property Standards require specific R-values for exterior walls in new home construction. These R-values vary from R-12 to R-20 depending on geographic location and type of heating system. (See map.)

An exterior wall system using Thermax Sheathing in the proper thickness can meet these federal requirements using conventional 2 x 4 construction. See Minimum Property Standards for specific requirements.

HUD MINIMUM PROPERTY STANDARD WALL REQUIREMENTS

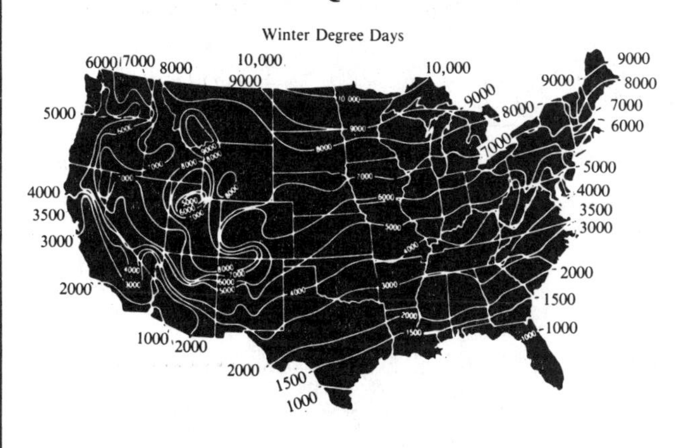

Note: This map shows approximate degree days based on 65°F. For more accurate data consult local weather bureau, energy office or HUD office.

Values for "R's" pertain to area between studs.

R-value	Degree days
	9000
R-20 (For all heating systems)	8000
	7000
R-20 (HP +R, ER) R-14 (HP, FF)	6000
	5000
R-20 (ER only) R-14 (HP +R, HP, FF)	4000
	3500
R-20 (ER only) R-12.5 (HP +R, HP, FF)	3000
	2500
R-14 (ER only) R-12.5 (HP +R, HP, FF)	2000
	1000
R-12.5 (For all heating systems)	below 1000

KEY:

HP + R:	Heat pump with electric Resistance heat
HP:	Heat pump only
ER:	Electric resistance
FF:	Fossil fuel (gas, oil, coal)

HUD Minimum Property Standards
Figure 10-4

Thermax and similar exterior sheathing give you design freedom with a wide range of materials. The basic components of an R-20 wall system, starting from the interior are:

- 1/2-inch sheet rock
- 6-mil polyethylene vapor barrier
- nominal 2 x 4-inch framing
- stud cavity insulation
- Thermax sheathing
- exterior siding material

Note: In areas which exceed 4,000-degree days, the system may include plastic vent strips which are installed horizontally along the exterior top plate prior to the installation of the sheathing. (See Figure 10-3.)

HUD Minimum Property Standards

HUD Minimum Property Standards require specific R-values for exterior walls in new home construction. These R-values vary from R-12 to R-20, depending on geographic location and type of heating system. (See Figure 10-4.)

These standards and many building codes limit the amount of glass area you can put in most walls. As a home builder, you may find it hard to sell homes that have limited window area. But insulated siding can help solve this problem.

The HUD Minimum Property Standard allows a trade-off between wall R-values and window types and area. For example, by increasing the R-value of the wall so that it exceeds the minimums shown in Figure 10-4, more glass area is permitted. The HUD Standard has complete details.

Opaque Wall Systems: Comparative Data— Figure 10-5 shows calculated R- and U-values at the wall cavity and the framing. (This excludes windows and doors.) The values shown represent different thicknesses of Thermax sheathing, and two types of glass fiber batts, R-11 and R-13.

Figure 10-5 is intended to show comparisons with other systems or insulation requirements as defined by HUD/FHA and by various energy codes. The data may also be used for evaluation of alternatives as permitted by such authorities.

Thermax aged R-values incorporated in Figure 10-5 are at 75 degrees F mean temperature because data is for comparison rather than for design.

Opaque Wall Systems: Design Data— Figure 10-6 shows calculated winter design R- and U-values for the opaque section of walls (excluding windows and doors but including framing).

Values shown represent different thicknesses of Thermax and both R-11 (3½ inches) and R-13 (3⅝ inches) glass fiber batts.

CALCULATED AT 75°F MEAN TEMPERATURE

2 x 4 Wood Frame Wall Systems 16″ o.c., 19% Framing
Calculated R and U values for Thermax Sheathing Thicknesses shown

Exterior Finish	Glass Fiber Batt Thickness		½″ (R3.6) R	½″ (R3.6) U	⅝″ (R4.5) R	⅝″ (R4.5) U	¾″ (R5.4) R	¾″ (R5.4) U	⅞″ (R6.3) R	⅞″ (R6.3) U	1″ (R7.2) R	1″ (R7.2) U
¾" (19.1 mm) wood siding	R11 (3½″)	Frame	10.30	.097	11.20	.089	12.10	.083	13.00	.077	13.90	.072
		Cavity	16.95	.059	17.85	.056	18.75	.053	19.65	.051	20.55	.049
		Wall	15	.066	16	.062	17	.059	18	.056	19	.053
	R13 (3⅝″)	Frame	10.30	.097	11.20	.089	12.10	.083	13.00	.077	13.90	.072
		Cavity	18.63	.054	19.53	.051	20.43	.049	21.33	.047	22.23	.045
		Wall	16	.062	17	.058	18	.055	19	.053	20	.050
7/16" (11.1 mm) hardboard siding	R11 (3½″)	Frame	9.69	.103	10.59	.094	11.49	.087	12.39	.081	13.29	.075
		Cavity	16.34	.061	17.24	.058	18.14	.055	19.04	.053	19.94	.050
		Wall	14	.069	15	.065	16	.061	17	.058	18	.055
	R13 (3⅝″)	Frame	9.69	.103	10.59	.094	11.49	.087	12.39	.081	13.29	.075
		Cavity	18.02	.056	18.92	.053	19.82	.051	20.72	.048	21.62	.046
		Wall	15	.065	16	.061	17	.057	18	.054	19	.052
4" (101.6 mm) face brick[1] 1" (25.4 mm) reflective air space	R11 (3½″)	Frame	12.46	.080	13.36	.075	14.26	.070	15.16	.066	16.06	.062
		Cavity	19.11	.052	20.01	.050	20.91	.048	21.81	.046	22.71	.044
		Wall	17	.058	18	.055	19	.052	20	.050	21	.048
	R13 (3⅝″)	Frame	12.46	.080	13.36	.075	14.26	.070	15.16	.066	16.06	.062
		Cavity	20.79	.048	21.69	.046	22.59	.044	23.49	.043	24.39	.041
		Wall	18	.054	19	.052	20	.049	21	.047	22	.045
Aluminum siding* (no backerboard)	R11 (3½″)	Frame	9.86	.101	10.76	.093	11.66	.086	12.56	.080	13.46	.074
		Cavity	16.51	.061	17.41	.057	18.31	.055	19.21	.052	20.11	.050
		Wall	15	.068	16	.064	17	.061	17	.057	18	.054
	R13 (3⅝″)	Frame	9.86	.101	10.76	.093	11.66	.086	12.56	.080	13.46	.074
		Cavity	18.19	.055	19.09	.052	19.99	.050	20.89	.048	21.79	.046
		Wall	16	.064	17	.060	18	.057	19	.054	19	.051
Vinyl siding** (no backerboard)	R11 (3½″)	Frame	9.72	.103	10.62	.094	11.52	.087	12.42	.081	13.32	.075
		Cavity	16.37	.061	17.27	.058	18.17	.055	19.07	.052	19.97	.050
		Wall	14	.069	15	.065	16	.061	17	.058	18	.055
	R13 (3⅝″)	Frame	9.72	.103	10.62	.094	11.52	.087	12.42	.081	13.32	.075
		Cavity	18.05	.055	18.95	.053	19.85	.050	20.75	.048	21.65	.046
		Wall	16	.064	16	.061	17	.057	18	.054	19	.052
Stucco	R11 (3½″)	Frame	9.45	.106	10.35	.097	11.25	.089	12.15	.082	13.05	.077
		Cavity	16.10	.062	17.00	.059	17.90	.056	18.80	.053	19.70	.051
		Wall	14	.070	15	.066	16	.062	17	.059	18	.056
	R13 (3⅝″)	Frame	9.45	.106	10.35	.097	11.25	.089	12.15	.082	13.05	.077
		Cavity	17.78	.056	18.68	.054	19.58	.051	20.48	.049	21.38	.047
		Wall	15	.066	16	.062	17	.058	18	.055	19	.052

*[1]Air space R-value between brick veneer and Thermax based on 50°F mean temp., 30°F temperature difference. *R-values for aluminum siding based on tests of a wall assembly per ASTM C 236 at 40°F mean temperature with this siding applied over non reflective substrate. **R-value for vinyl siding based on same conditions as above.*

SAMPLE CALCULATION FOR COMPARISON WALL R VALUES USING ¾″ THERMAX SHEATHING R5.4 AT 75°F MEAN TEMPERATURE

Illustrated below are the calculations for one design based on a 19% framing factor.

	Winter Conditions: "R" Values Thru Frame	Winter Conditions: "R" Values Thru Cavity
Inside surface film	0.68	0.68
½″ (12.7 mm) Gypsum wallboard	0.45	0.45
6 mil (0.15 mm) Poly film	negl.	negl.
Wood framing 2 x 4's 16″ (406.4 mm) o.c.	4 35	—
3⅝″ (92.1 mm) Friction fit glass batt compressed	—	12.68
¾″ (19.1 mm) Thermax Sheathing (75°F mean)	5.4	5.4
¾″x10″ (19.1 mm x 254 mm) Lapped wood siding	1.05	1.05
Outside surface film 15 mph wind	0.17	0.17
"R"s at Section	12.10	20.43
"U"s at Section	.0826	.0489

$$\text{Total Design "U"} = \frac{.19}{12.10} + \frac{.81}{20.43} = .055$$

$$\text{Total Design "R" Value} = \frac{1}{.055} = 18$$

Per ASHRAE 1977 "Handbook of Fundamentals" Methods. Note: Standard rules for rounding numbers applied.

Opaque wall systems — comparative data
Figure 10-5

CALCULATED AT 40°F MEAN TEMPERATURE
2 x 4 Wood Frame Wall Systems 16″ o.c., 19% Framing
Calculated R and U values for Thermax Sheathing Thicknesses shown.

Exterior Finish	Glass Fiber Batt Thickness		½″ (R4)		⅝″ (R5)		¾″ (R6)		⅞″ (R7)		1″ (R8)	
			R	U	R	U	R	U	R	U	R	U
¾" (19.1 mm) wood siding	R11 (3½″)	Frame	10.70	.093	11.70	.085	12.70	.079	13.70	.073	14.70	.068
		Cavity	17.35	.058	18.35	.054	19.35	.052	20.35	.049	21.35	.047
		Wall	16	.064	17	.060	18	.057	19	.054	20	.051
	R13 (3⅝″)	Frame	10.70	.093	11.70	.085	12.70	.079	13.70	.073	14.70	.068
		Cavity	19.03	.053	20.03	.050	21.03	.048	22.03	.045	23.03	.043
		Wall	17	.060	18	.057	19	.053	20	.051	21	.048
7/16" (11.1 mm) hardboard siding	R11 (3½″)	Frame	10.09	.099	11.09	.090	12.09	.083	13.09	.076	14.09	.071
		Cavity	16.74	.060	17.74	.056	18.74	.053	19.74	.051	20.74	.048
		Wall	15	.067	16	.063	17	.059	18	.056	19	.053
	R13 (3⅝″)	Frame	10.09	.099	11.09	.090	12.09	.083	13.09	.076	14.09	.071
		Cavity	18.42	.054	19.42	.051	20.42	.049	21.42	.047	22.42	.045
		Wall	16	.063	17	.059	18	.055	19	.052	20	.050
4" (101.6 mm) face brick[1] 1" (25.4 mm) reflective air space	R11 (3½″)	Frame	13.11	.076	14.11	.071	15.11	.066	16.11	.062	17.11	.058
		Cavity	19.76	.051	20.76	.048	21.76	.046	22.76	.044	23.76	.042
		Wall	18	.055	19	.052	20	.050	21	.047	22	.045
	R13 (3⅝″)	Frame	13.11	.076	14.11	.071	15.11	.066	16.11	.062	17.11	.058
		Cavity	21.44	.047	22.44	.045	23.44	.043	24.44	.041	25.44	.039
		Wall	19	.052	20	.050	21	.047	22	.045	23	.043
Aluminum siding* (no backerboard)	R11 (3½″)	Frame	10.26	.097	11.26	.089	12.26	.082	13.26	.075	14.26	.070
		Cavity	16.91	.059	17.91	.056	18.91	.053	19.91	.050	20.91	.048
		Wall	15	.066	16	.062	17	.058	18	.055	19	.052
	R13 (3⅝″)	Frame	10.26	.097	11.26	.089	12.26	.082	13.26	.075	14.26	.070
		Cavity	18.59	.054	19.59	.051	20.59	.049	21.59	.046	22.59	.044
		Wall	16	.062	17	.058	18	.055	19	.052	20	.049
Vinyl siding** (no backerboard)	R11 (3½″)	Frame	10.12	.099	11.12	.090	12.12	.083	13.12	.076	14.12	.071
		Cavity	16.77	.060	17.77	.056	18.77	.053	19.77	.051	20.77	.048
		Wall	15	.067	16	.063	17	.059	18	.055	19	.052
	R13 (3⅝″)	Frame	10.12	.099	11.12	.090	12.12	.083	13.12	.076	14.12	.071
		Cavity	18.45	.054	19.45	.051	20.45	.049	21.45	.047	22.45	.045
		Wall	16	.063	17	.059	18	.059	19	.052	20	.050
Stucco	R11 (3½″)	Frame	9.85	.102	10.85	.092	11.85	.084	12.85	.078	13.85	.072
		Cavity	16.50	.061	17.50	.057	18.50	.054	19.50	.051	20.50	.049
		Wall	15	.068	16	.064	17	.060	18	.056	19	.053
	R13 (3⅝″)	Frame	9.85	.102	10.85	.092	11.85	.084	12.85	.078	13.85	.072
		Cavity	18.18	.055	19.18	.052	20.18	.050	21.18	.047	22.18	.045
		Wall	16	.064	17	.060	18	.056	19	.053	20	.050

[1] ***Air space R-value between brick veneer and Thermax based on 50°F mean temp., 30°F temperature difference. *R-values for aluminum siding based on tests of a wall assembly per ASTM C 236 at 40°F mean temperature with this siding applied over non reflective substrate. **R-value for vinyl siding based on same conditions as above.***

PER ASHRAE 1977 "HANDBOOK OF FUNDAMENTALS" METHODS. NOTE: STANDARD RULES FOR ROUNDING NUMBERS APPLIED.

Opaque wall systems — design data
Figure 10-6

Figure 10-6 is based on winter conditions. Aged R-values for Thermax at 40 degrees F mean temperature was included in these calculations.

U-Values

U-value expresses the overall heat transmission of a material. It's the amount of heat expressed in Btu's, transmitted through 1 square foot of a given material in 1 hour at a temperature differential of 1 degree F, from the air on one side to the air on the other side. The lower the U-value, the better. It looks like this:

$$U = \frac{\text{Btu}}{\text{hour sq. ft. } ^\circ\text{F}}$$

FPS Recommended Uses

Foam plastic sheathing is generally concealed in the wall in residential construction. You then provide an interior finish acceptable to your building code. Examples of suggested concealed uses are:

- High performance insulation sheathing in new frame wall construction
- Thin profile cavity wall insulation in new masonry construction
- High performance insulating vapor barrier undercourse behind new interior wall or ceiling finish material
- Thin profile insulating underlayment under roof shingles in vaulted, cathedral-type ceilings and "A-frame" construction
- Under slab or perimeter thermal insulation
- Thin profile insulating underlayment installed behind new exterior siding

• Thin profile insulating undercourse behind new interior wallboard

Foam plastic sheathing is not intended for exposed use in living quarters. When it is used in other exposed areas of the home, you'll want to observe the following points:

• In areas immediately adjacent to or above combustion equipment (such as furnaces or chimneys) or other surfaces which could reach high temperatures, foam plastic sheathing and other combustibles must be shielded. Your building code will have more on this requirement.

• In crawl spaces, attics, basements and other storage areas, foam panel siding should be protected against physical impact damage.

• In attached garages, the insulated walls and ceiling separating the garage from the living area must comply with the building code criteria for fire resistance. As a minimum, 1/2-inch sheet rock should be applied to both sides of the frame wall. A minimum of 1/2-inch type-X sheet rock should be installed on ceilings.

When handling and installing foam plastic sheathing, use care to avoid puncturing the foil face of the board. After installation, the aluminum foil facers protect the insulation from ultraviolet light which would harm the foam.

If the sheathing is accidentally damaged, small tears can be repaired with aluminized tape. Larger holes or breaks can be plugged or patched with appropriately sized pieces of similar sheathing held in place with aluminized tape.

The insulation must be installed the right way to get the given R-value. Follow the manufacturer's instructions.

Vent Strip Systems for Moisture Relief

Thermax sheathing has a closed-cell structure and aluminum foil facers that make it an efficient barrier to both moisture and heat. In colder climates, that creates the potential problem of trapped water vapor within stud spaces. Condensation could accumulate in the wall insulation. Water is an excellent conductor of heat and cold, so any insulation material that has absorbed moisture has lost much of its insulation value.

In older houses, there wasn't much problem with condensation. Moisture escapes very quickly from a poorly-insulated house. But the lack of a good vapor barrier in a tighter, better-insulated house will result in:

• Loss of insulating efficiency
• Paint blistering, peeling and cracking
• Siding and sheathing warping and buckling
• Corrosion and staining
• Mildew and odor
• Deterioration in framing, sheathing and siding
• Wet interior wall surfaces

Vapor Barriers

You can see why vapor barriers are more important in a well-insulated house. Without a vapor barrier on the inside (warm in winter) side of the wall, moisture from the interior will condense on the colder exterior wall and saturate the insulation. A 6-mil polyethylene vapor barrier properly installed under the interior wall finish will stop this accumulation of moisture. It should be installed as follows:

1) Use the polyethylene to cover the entire height of the wall, from the bottom to the top wall plates. Staple it about every 16 inches on center at the top and bottom plates, around all framing openings and at lap joints. Some additional staples may be needed to hold down the film. The vapor barrier is applied to the interior stud face.

2) No horizontal joints should be permitted.

3) Vertical joints must overlap at least two wall studs. You could also use a lap joint, if the lap is sealed at a wall stud with cloth-backed tape.

4) The polyethylene should be cut tight around holes made for pipes, ducts and electrical boxes, and then taped or caulked.

5) Windows and all other openings should be totally covered at the time of application. Make the cut undersize so the vapor barrier film can be folded over during finishing.

6) Except for windows, don't allow any penetration in the polyethylene in exterior bathroom walls.

7) Inspect the vapor barrier and make any necessary repairs before installing the interior wall finish. Repair all tears and accidental punctures with cloth-backed tape or another polyethylene film. Keep in mind that the vapor barrier will protect the home and its occupants for many years if it is installed correctly.

In warmer areas (below 4,000-winter-degree days), there's less potential for condensation in the wall cavity. Vapor barriers rated at 1 perm or less, such as 2-mil polyethylene film, aluminum foil,

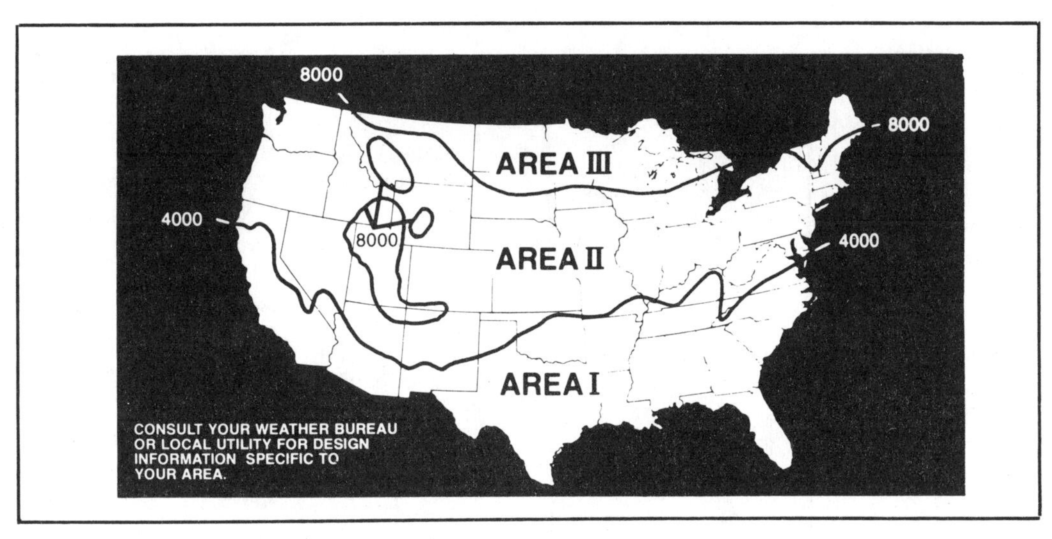

Winter-degree days
Figure 10-7

and vapor barrier paint may be all the barrier you need.

A moisture vent strip may be needed in very cold climates to vent moisture from the wall cavity. The vent strip allows moisture to escape, but doesn't reduce the effectiveness of the insulation. See Figure 10-3. Moisture vent strips make sense in colder climates or wherever special precautions against moisture accumulation are needed.

Moisture vent strips are corrugated plastic strips. They're available in various widths and are easily nailed or stapled at the top plates.

Here are some standards to follow for the areas shown in Figure 10-7:

Area I (To 4,000-winter-degree days)— Use a vapor barrier rated at 1 perm or less on the interior side of the wall.

Area II (4,000 to 8,000-winter-degree days)— Use a 6-mil polyethylene vapor barrier on the interior side of the wall. Although 4-mil polyethylene vapor barrier film has an acceptable perm rating, it is thinner, weaker, and requires too much repairing of tears, punctures, etc.

Area III (Over 8,000-winter-degree days)— Use moisture vent strips to supplement 6-mil polyethylene vapor barrier.

Installation of Foam Sheathing

1) Install diagonal bracing using either 1 x 4 let-ins or metal strapping. (See Figure 10-8.)

2) If required, fasten plastic vent strips 12 inches o.c., as shown in Figures 10-3 and 10-12, to vent wall moisture into the attic soffit airway.

3) Install foam plastic sheathing vertically, with the long joints in moderate contact with one another and bearing directly on the framing members. Avoid horizontal joints unless the joints bear on a framing member or fall over vent strips.

4) Use galvanized roofing nails with 3/8-inch heads and a length sufficient to penetrate the framing a minimum of 3/4-inch. Use 16-gauge wire staples with a minimum of 3/4-inch crown, and leg length necessary to penetrate the framing a minimum of 1/2-inch.

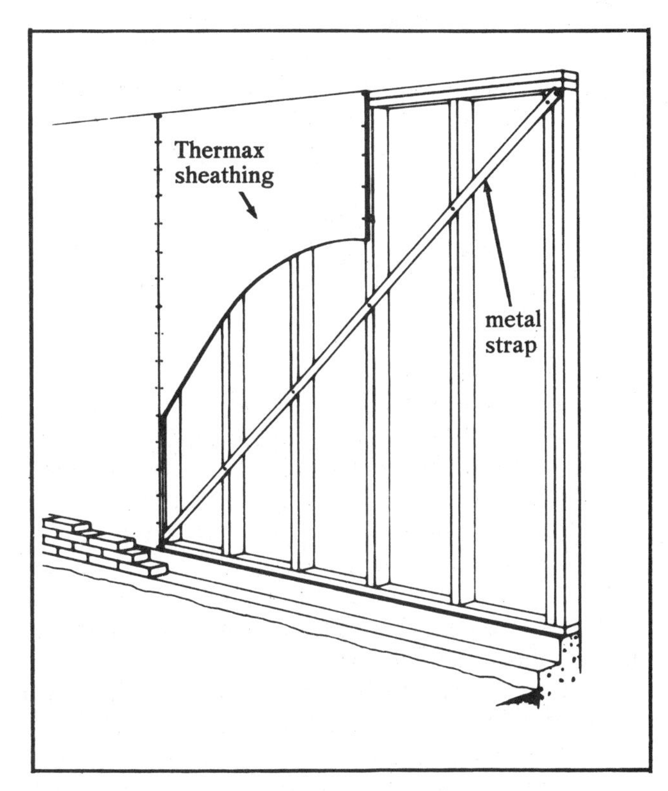

Corner bracing
Figure 10-8

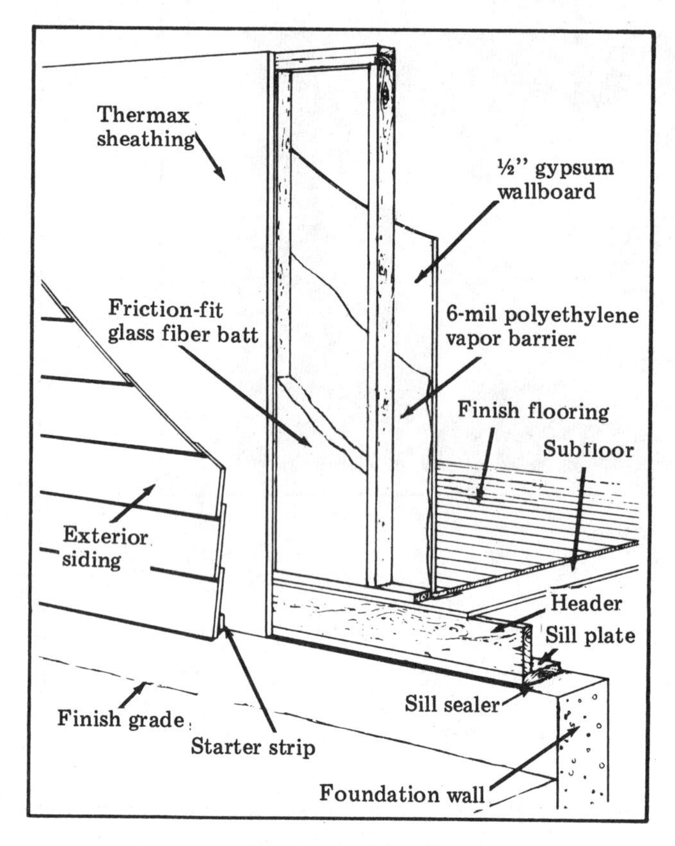

Basic wall with siding
Figure 10-9

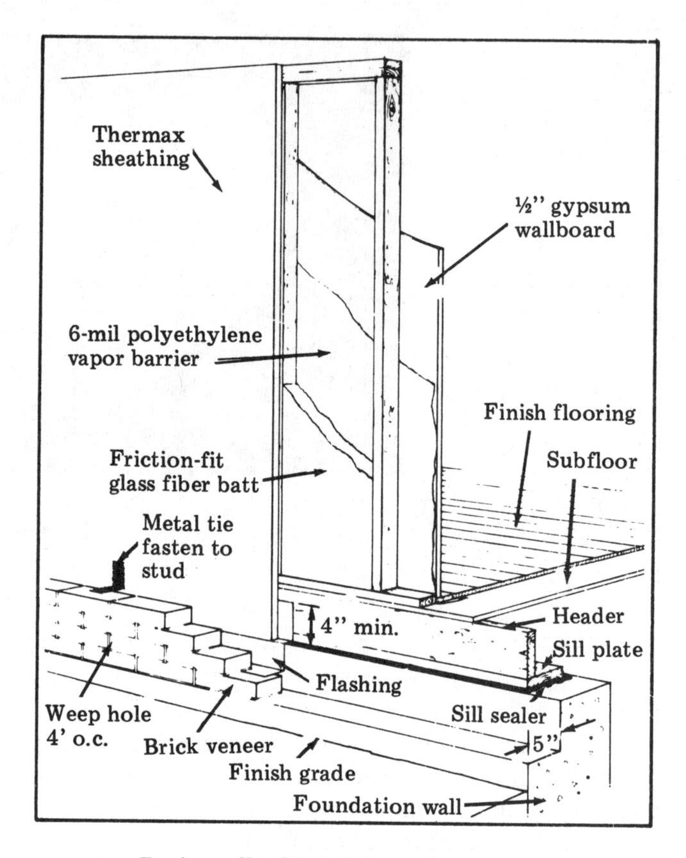

Basic wall with brick or stone veneer
Figure 10-10

Staples should be applied with the crowns running parallel to the direction of the framing. Don't over-drive the heads. That would tear the facer sheets.

Use a utility knife and straightedge to trim the insulation board to shape.

Figure 10-9 shows a basic wall with siding. The thermal performance ranking is:

• 3⅝-inch glass fiber batt, R-13

• 7/8-inch foam plastic sheathing, R-7.2

• Siding and wallboard don't contribute very much to the overall R-value of this wall.

Figure 10-10 shows a basic wall with brick veneer. Stone veneer may also be used.

Figure 10-11 illustrates the technique of venting under windows.

Figure 10-12 shows plastic vent strips applied to the top plate, venting into the attic.

Figure 10-13 illustrates application at a gable end with top-floor end ceiling-joist.

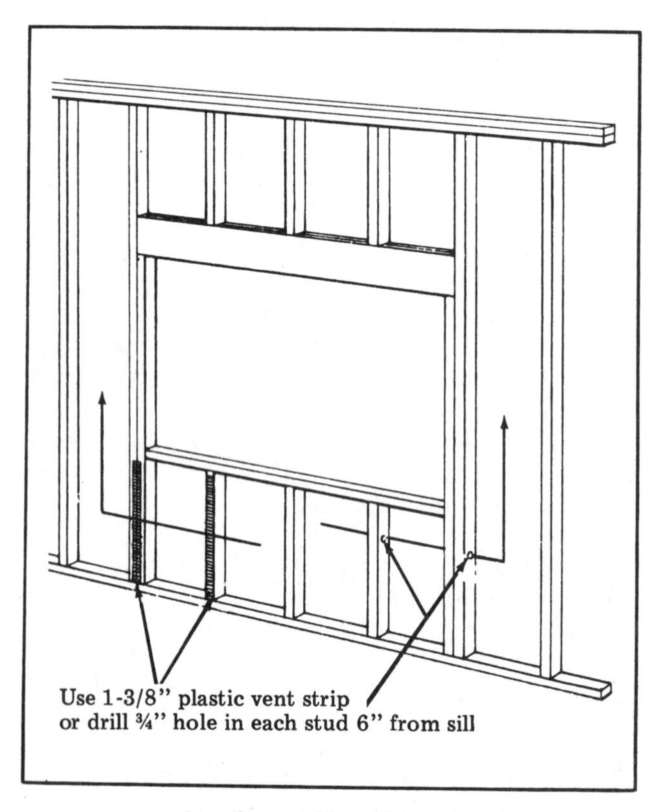

Venting under windows
Figure 10-11

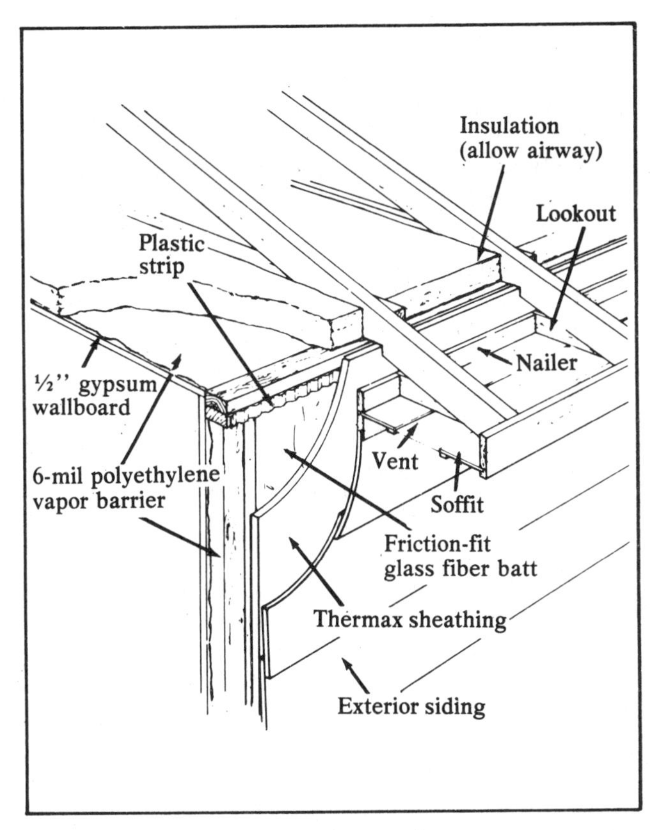

Venting into attic
Figure 10-12

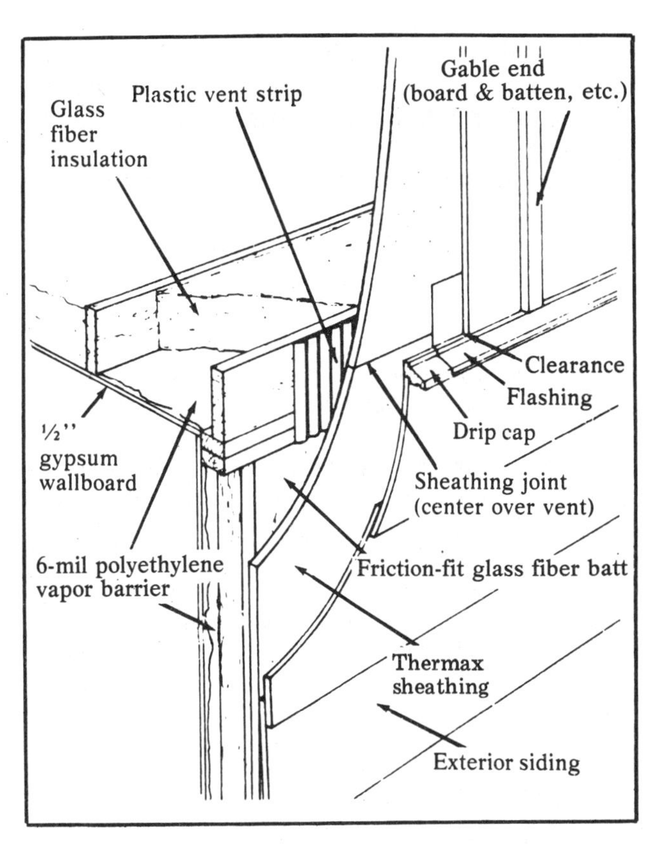

Gable end with top floor end ceiling joist
Figure 10-13

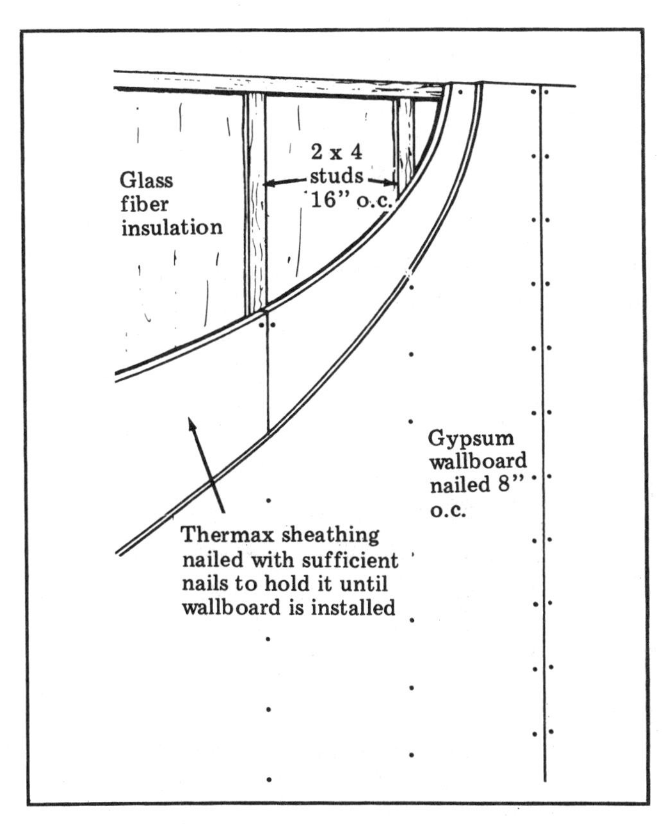

Underlayment frame wall system
Figure 10-14

Underlayment Frame Wall Systems

Foam plastic sheathing with taped joints and sealed penetration makes an effective vapor barrier. The sheathing is applied directly to the interior stud surface, and is then covered with 1/2-inch gypsum wallboard staggered over the joints.

Glass fiber batts are then installed between the framing members from the exterior side of the wall. This saves the expense of a polyethylene vapor barrier.

Installation

1) Install foam plastic sheathing vertically with the long joints touching and meeting over studs. (See Figure 10-14.)

2) Nail the foam plastic sheathing with galvanized roofing nails with 3/8-inch heads. The nails should be long enough to go at least 3/4-inch into the framing. You could also use 16-gauge wire staples with at least a 3/4-inch crown, and a leg that will go at least 1/2-inch into the framing.

3) Tape all sheathing joints and penetrations with aluminized tape.

4) Apply gypsum wallboard over foam plastic sheathing with the joints staggered.

5) Nail the gypsum wallboard 8 inches o.c. on all studs and top and bottom plates. Caulk the top and bottom joints where necessary.

6) Install glass fiber batts in the wall cavity.

7) Apply exterior fiberboard sheathing or No. 15 asphalt-saturated felt on the exterior of the wall.

Wood Roof Deck Systems

You can use foam plastic sheathing over 2-inch T & G wood decks or plywood decks. Foam plastic sheathing by itself is not a suitable nail or staple base for shingles. But you can use FPS under shingles if you install it according to the method described below.

Installation

1) Start with a deck that is smooth, dry and free of debris.

2) At the eave and rake, install a wood nailer strip at least 2½ inches wide and the same thickness as the insulation panel. (See Figure 10-15.)

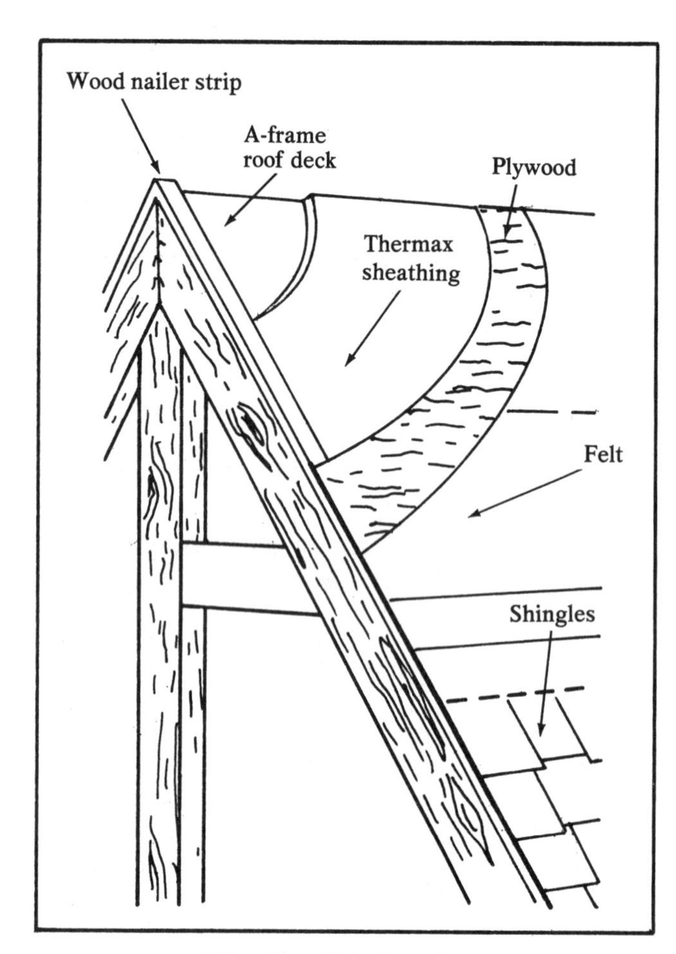

Wood roof deck system
Figure 10-15

3) Install foam plastic sheathing with enough nails to hold it in place. Fit panels snugly at the perimeter and butt all joints.

4) Install exterior grade plywood of the thickness recommended by the shingle manufacturer (1/2-inch-thick plywood is generally sufficient) over the foam plastic sheathing and eave and rake strips. Stagger plywood joints so they don't coincide with joints in the FPS. Nail 12 inches o.c. in each direction using galvanized nails with 3/16-inch heads. The nails should be long enough to penetrate into the structural deck at least 1 inch.

5) Install the roofing material according to the manufacturer's instructions.

Conventional Ceiling Systems

Sometimes you'll want to install panel insulation on the room side of a ceiling because there isn't enough space to do it on the attic side of the ceiling. (Scissor-truss construction, low-pitched roofs, and floored attics are prime examples.) You can also increase the R-value in a conventional ceiling by installing the panel insulation on the room side of the ceiling.

Installation

1) Nail the sheathing directly to the underside of the ceiling joists with large headed (3/8-inch-diameter) galvanized nails spaced 12 inches o.c. Penetrate the joists at least 1/2-inch. (See Figure 10-16.)

2) Tape sheathing joints and penetrations with aluminized tape.

3) Install 1/2-inch gypsum wallboard, with joints staggered, by nailing through the sheathing into the ceiling joists. Nails should be 7 inches o.c. You may want to use screws if the foam plastic sheathing is very thick.

4) Apply glass fiber insulation over the sheathing between the joists.

Masonry Wall Systems

FPS is an ideal insulation for masonry or concrete wall construction. The material's thin profile makes it easy to apply on a masonry surface.

Installation

1) Apply foam plastic sheathing directly to the wall with an appropriate adhesive. See Figure 10-17 A.

2) Apply metal channels (or treated wood furring strips) 24 inches o.c. vertically over the insulation.

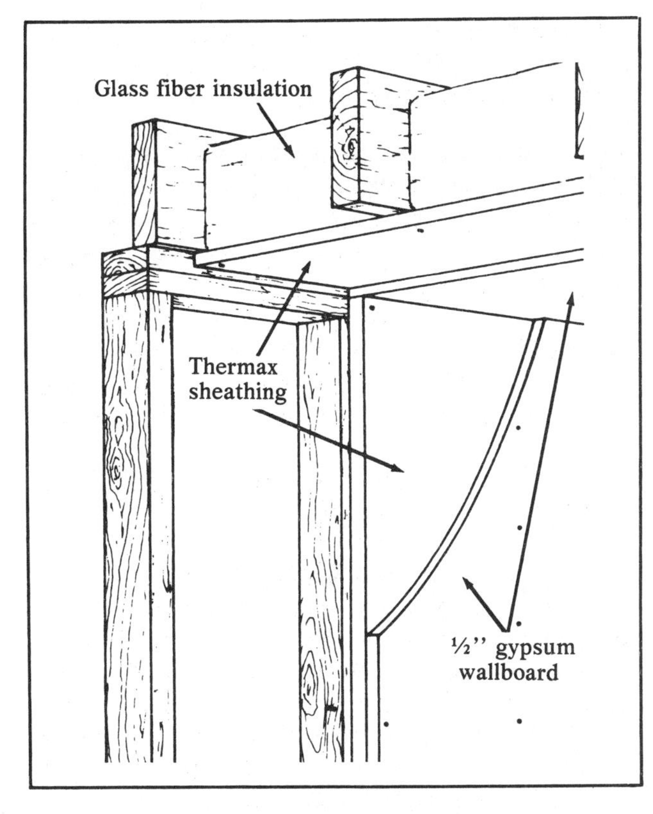

Conventional ceiling system
Figure 10-16

You can use masonry nails, Tapcon screws, or pneumatically-activated fasteners spaced 12 inches o.c. on alternate sides of the channels.

3) Apply gypsum wallboard either vertically or horizontally on the channels. Use gypsum wallboard screws, Type S, spaced 12 inches o.c. that are long enough to penetrate the furring channel at least 3/8-inch but no more than 5/8-inch. Conventional drywall nails are used with wood furring.

You can also put furring strips right on the masonry wall, and apply foam plastic insulation directly to the furring strips. Apply the furring strips vertically at 24 inches o.c. on the masonry wall. Then put the desired thickness of foam plastic sheathing and 1/2-inch gypsum wallboard over the furring. (See Figure 10-17 B.)

Metal Furred Systems
FPS can also be applied over the interior of a masonry wall. Furring strips are then applied over the sheathing, and the wall is finished with 1/2-inch gypsum wallboard or a similar interior finish. This provides both good insulation and space to run piping and electrical lines behind the wall surface.

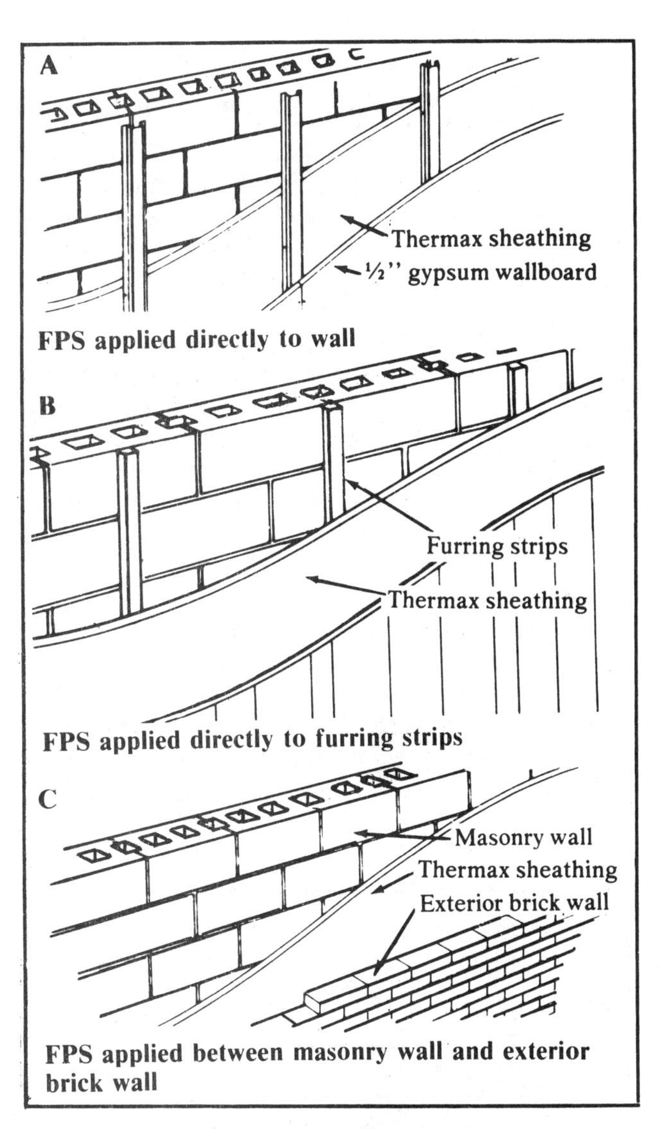

Masonry wall system
Figure 10-17

Cavity Wall Systems
In this case, foam plastic sheathing is applied between the masonry wall and the exterior brick wall, as shown in Figure 10-17 C:

1) Install the sheathing horizontally within the cavity space and against the inner masonry wall.

2) Use adjustable wall ties secured through the insulating sheathing to the masonry. If there are ties in the masonry wall, stick the insulation on the ties.

System R-Values

In masonry wall systems with foam plastic sheathing applied on the interior side of the masonry or concrete wall construction, the application is:

- 1/2-inch gypsum wallboard
- "hat channel" (or wood furring)

- 1-inch FPS
- air space
- exterior finish (brick for this example)

The R-value for such a wall system would be calculated as shown in Figure 10-18 ("hat channel" at 24 inches o.c. at an assumed 15% framing factor).

In cavity wall systems, foam plastic sheathing is applied between the exterior finish and the concrete block wall. In Figure 10-19, the exterior finish is 4-inch face brick and the typical calculation is shown.

Foam plastic sheathing (such as Thermax) of greater thickness will increase the wall system R-value, as indicated in Figure 10-20.

	R-Value thru furring	R-Values between furring
Inside surface film	0.68	0.68
½" (12.7 mm) gypsum wallboard	0.45	0.45
Reflective air space (hat channel)	2.77	--
Reflective air space	--	2.77
1" (25.4 mm) (nominal thickness) Thermax sheathing (75° F mean)	7.2	7.2
8" (203.2 mm) concrete block	1.11	1.11
Non-reflective air space	0.94	0.94
4" (100.8 mm) face brick	0.44	0.44
Outside surface film	0.17	0.17
"R's" at sections	13.76	13.76
"U's" at sections	.0727	.0727

Total design R-Value $= \frac{1}{U} = \frac{1}{.0727} = 14$

System R-Value (Masonry Wall)
Figure 10-18

	R-Value
Inside surface film	0.68
8" (203.2 mm) concrete block	1.11
1" (25.4 mm) Thermax sheathing (75° F mean)	7.2
¾" reflective air space (50° F mean 30° TD)	2.77
4" (101.6 mm) face brick	0.44
Outside surface film	0.17
Total design R-Value =	12.37

Total design U-Value $= \frac{1}{12.37} = .081$

System R-Value (Cavity Wall)
Figure 10-19

Calculated cavity wall system comparative values with selected Thermax sheathing thicknesses at 75° mean temperature

	1" (R-7.2)	1¼" (R-9)	1½" (R-10.8)	1¾" (R-12.6)	2" (R-14.4)	2¼" (R-16.2)
4" (101.6) face brick exterior	R-12	R-14	R-16	R-18	R-20	R-21
8" (203.2 mm) concrete block interior, Thermax sheathing in wall cavity	U-.081	.071	.063	.056	.051	.047

Calculated cavity wall system winter design values with selected Thermax sheathing thicknesses at 40° mean temperature

	1" (R-8)	1¼" (R-10)	1½" (R-12)	1¾" (R-14)	2" (R-16)	2¼" (R-18)
4" (101.6) face brick exterior	R-13	R-15	R-17	R-19	R-21	R-23
8" (203.2 mm) concrete block interior, Thermax sheathing in wall cavity	U-.075	.065	.057	.051	.047	.043

R-value increases with greater thickness FPS
Figure 10-20

	R-Value thru frame	R-Values thru cavity
Top surface film	0.61	0.61
6" (152.8 mm) R-19 glass batt insulation	--	19
Wood joist nominal 2" x 6"	6.84	--
7/8" Thermax sheathing (75° F mean)	6.3	6.3
½" (12.7 mm) gypsum wallboard	0.45	0.45
Bottom surface film	0.61	0.61
R's at Sections	14.81	26.97

$$\text{Total Design U} = \frac{.10}{14.81} + \frac{.90}{26.97} = 0.0401$$

$$\text{Total Design R} = 1/U = 1/0.0401 = 25$$

System R-Value (vented attic)
Figure 10-21

The system R-value in a vented attic is calculated as shown in Figure 10-21. Framing members are assumed to be 10% of the ceiling area.

Estimated Manhours

Insulation sheathing panels require about 9 skilled and 4 unskilled hours per 1,000 SF. A 24 x 56-foot house will require 40 four-foot-wide panels. Assuming the panels are 9 feet long, there are 1,440 square feet, which will require approximately 12 skilled and 6 unskilled manhours for installation.

Summary

Foam plastic sheathing (FPS) gives you maximum insulation in a minimum of space. It's easy to install. One 3/4-inch-thick 4 x 8-foot panel weighs about 4½ pounds. One person can handle and install it quite easily.

In frame construction, no matter how much or what type of insulation is installed between the studs, the studs themselves transfer heat and cold quite well between the inside and outside surfaces of the wall. This framing area, known as the framing factor, varies between 18% and 27% of the total opaque exterior wall area, depending on the construction method.

FPS covers these insulation short circuits by providing an insulating envelope over the entire exterior frame wall.

Wood, hardboard, brick, aluminum, and vinyl can all be applied quite easily over FPS. They are fastened to the wood frame through the sheathing.

Wood and asbestos siding shingles can also be applied by installing furring strips or a plywood nailer base over the sheathing.

You can use stucco over FPS when an expanded metal lath is fastened over the sheathing to the wood frame.

In 1983 dollars, it costs only about $350 extra to substitute foil-faced foam plastic sheathing for conventional fiberboard sheathing.

Figure 10-22 is your Insulated Sheathing Checklist and Data Reference Sheet.

Insulated Sheathing Checklist and Data Reference Sheet

1) The house is ________ feet wide and ________ feet in length, for a total of ________ square feet.

2) Corner bracing is ☐1 x 4 let-in ☐metal strapping ☐other (specify)________________

3) Size of foam plastic sheathing (FPS) is ☐4 x 8 ☐4 x 9 feet ☐Other (specify) ________________

4) Nominal thickness of Thermax or equivalent FPS is ☐½" (3.6 R) ☐⅝" (4.5 R) ☐⅞" (6.3 R) ☐1" (7.2 R) ☐1¼" (9.0 R) ☐1½" (10.8 R) ☐1¾" (12.6 R) ☐2" (14.4 R) ☐2¼" (16.2 R) ☐2½" (18.0 R) ☐2¾" (19.8 R) ☐3" (21.6 R).

(Note: R-value shown is for consumer comparison purposes.)

5) Total number of FPS sheets used: ________________

6) Plastic vent strips ☐were ☐were not installed because the house is located in:

☐Area I (to 4,000-winter-degree days)

☐Area II (4,000 to 8,000-winter-degree days)

☐Area III (over 8,000-winter-degree days)

7) The house ☐does ☐does not contain the basic components for an R-20 wall system:

☐½-inch sheet rock

☐6-mil polyethylene vapor barrier

☐Nominal 2 x 4 framing

☐Stud cavity insulation

☐FPS sheathing

☐Exterior siding material

8) FPS was installed with:

☐Galvanized roofing nails with ⅜ inch heads and lengths sufficient to penetrate framing a minimum of ¾-inch.

☐16-gauge wire staples having (minimum) ¾-inch crown, and leg length sufficient to penetrate framing a minimum of ½ inch.

9) The house ☐does ☐does not have an energy-saving FPS system A-frame roof construction with exposed wood deck.

10) The house ☐does ☐does not have an energy-saving FPS conventional ceiling system.

11) The house ☐does ☐does not have an energy-saving FPS masonry wall system using ☐wood furred system ☐metal furred system ☐cavity wall system.

Figure 10-22

12) The house has ☐plain gable roof ☐flat or low-pitch roof ☐A-frame ☐cathedral-type ceiling ☐basement ☐crawl space ☐on-slab.

13) Material costs:

FPS	________pcs	@	$________ea	=	$________
Nails or staples	________lbs	@	$________lb	=	$________
Tape	________rolls	@	$________ea	=	$________
Vent strips	________LF	@	$________LF	=	$________

Other materials, not listed above:

________________	________	@	$________	=	$________
________________	________	@	$________	=	$________
________________	________	@	$________	=	$________

Total cost of all materials $________

14) Labor costs:

Skilled	________hours	@	$________hour	=	$________
Unskilled	________hours	@	$________hour	=	$________

Other (specify)

________________	________hours	@	$________hour	=	$________

Total $________

15) Total cost:

Materials	$________
Labor	$________
Other	$________
Total	$________

16) Cost of insulated sheathing per SF of floor space $________________

17) Problems that could have been avoided ________________________________

____

____

____

____

18) Comments ______________________________

____

____

____

____

19) A copy of this Checklist and Data Reference Sheet has been provided:

☐Management

☐Accounting

☐Estimator

☐Foreman

☐File

☐Other: ______________

Chapter 11

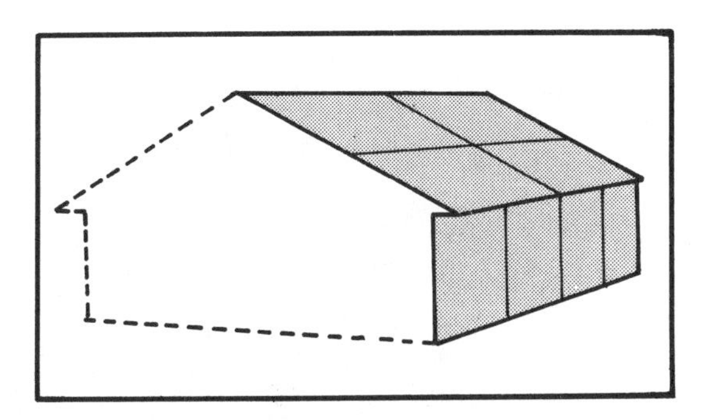

Sheathing

This chapter deals with wall and roof sheathing. In Chapter 10, we examined the benefits of foam plastic sheathing. There are other wall-sheathing materials, and each has its purpose. Fiberboard is the wall sheathing most commonly used in masonry-veneer construction. Plywood is the material most builders use as roof decking or sheathing. Some sheet materials serve both as sheathing and siding.

In some of your homes, all wall and roof sheathing will be plywood. Wall sheathing is generally installed vertically, as shown in Figure 11-1, though it may also be installed horizontally, as shown in Figure 11-2. Plywood roof sheathing is always installed with the face grain perpendicular to the rafters. The long side goes across the rafters and the joints are staggered.

Types of Wall Sheathing

Board Wall Sheathing

Board sheathing is usually made of nominal 1-inch boards in a shiplap, a tongue-and-groove, or a square-edge pattern. Resawed 11/16-inch boards are also used under certain conditions. The boards can be 6, 8 or 10 inches wide. They may be applied horizontally or diagonally. See Figure 11-3 A.

Wall sheathing is sometimes carried only to the subfloor, as shown in Figure 11-3 B. But you'll get a stronger wall when diagonal sheathing or sheet materials are applied as shown in Figure 11-3 C. This method ties the wall to the floor system. To minimize shrinkage, the moisture content of the wood should not exceed 15%.

Random-length side- and end-matched boards are sometimes used for sheathing. Most of the softwoods are suitable. These include pine, spruce, Douglas fir and hemlock. Grades vary between species, but No. 3 is commonly used.

The boards should be nailed at each stud with two nails for 6- and 8-inch widths, and three nails for wider boards. In diagonal applications, one more nail can be used at each stud. Use box nails whenever possible. These have better holding power than uncoated nails.

Joints should fall on a stud unless the boards are T & G end-matched. T & G end-matched boards are applied continuously, either diagonally or horizontally. The end joints fall where they may. However, no two adjoining boards should have end joints in the same location. Each board should bear on at least two studs.

Board sheathing applied horizontally, as shown in Figure 11-3 A, is easy to install, and there is less waste than with diagonally applied sheathing. But the horizontal method requires corner bracing.

Diagonal sheathing should be installed at a 45-degree angle to the foundation for maximum wall strength and rigidity. No corner bracing is required.

Plywood sheathing has replaced board sheathing in most areas because of the higher labor costs for applying board sheathing. Plywood sheathing is

Wall sheathing installed vertically
Figure 11-1

Wall sheathing installed horizontally
Figure 11-2

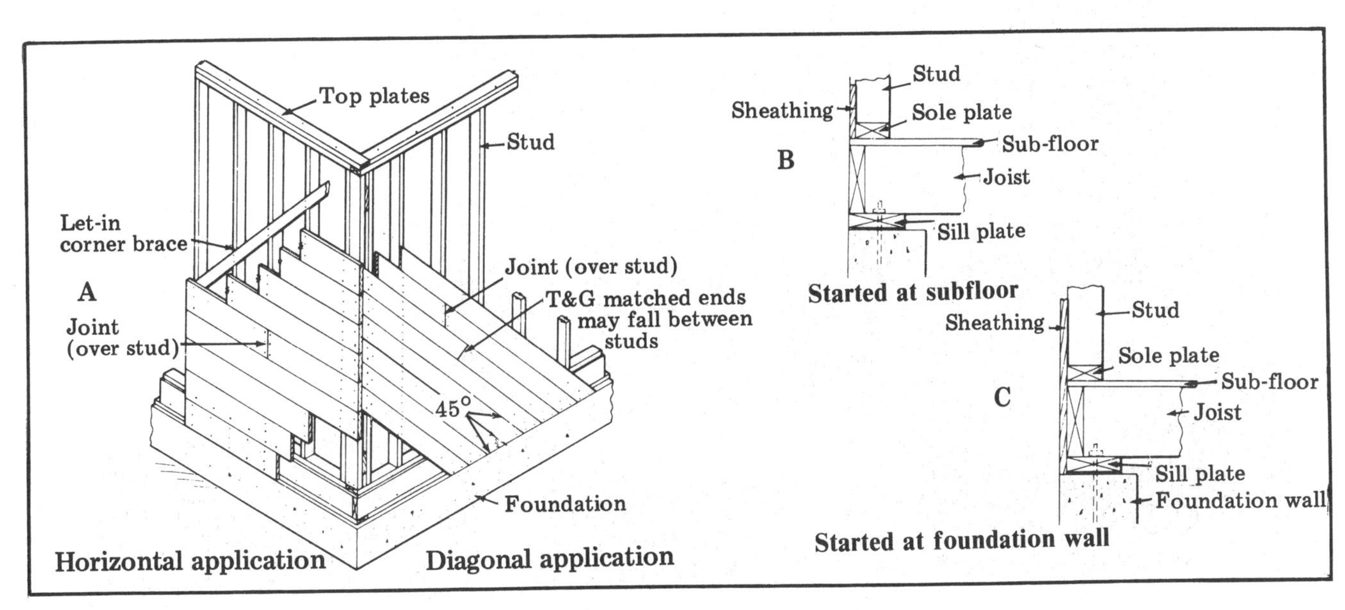

Application of board wall sheathing
Figure 11-3

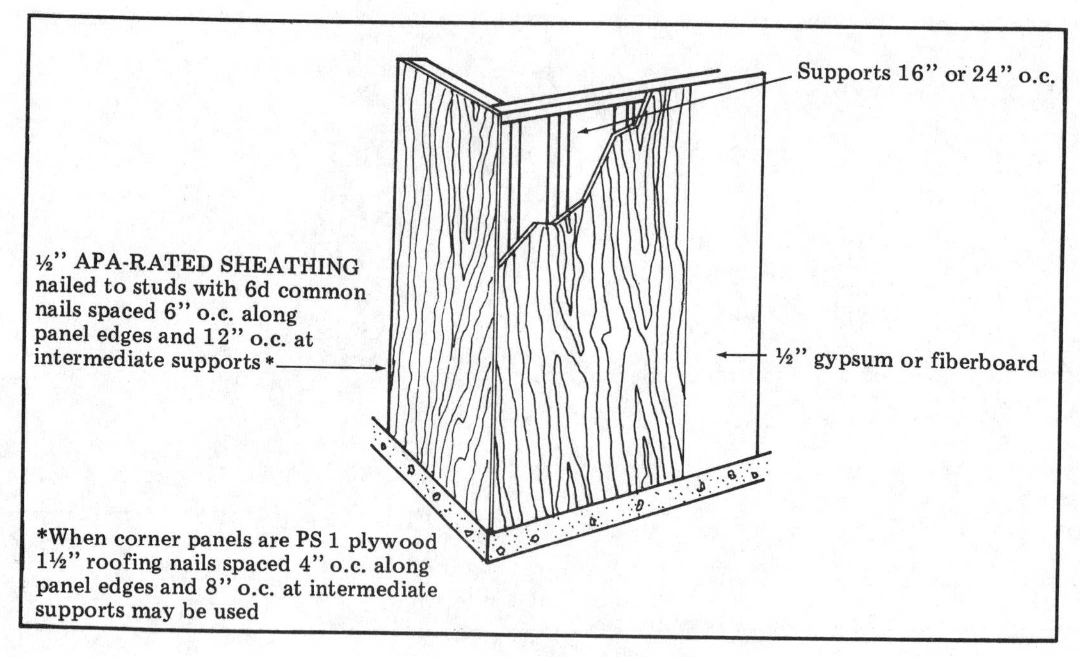

APA sheathing corner panels
Figure 11-4

generally more economical, easier to handle and goes up faster. The percentage of waste is also less.

Plywood Wall Sheathing

APA (American Plywood Association) -rated plywood sheathing meets building code requirements for bending and racking strength without let-in corner bracing. And APA-rated sheathing corner panels can be used to eliminate the costly let-in bracing often required by fiberboard sheathing. See Figure 11-4. Installation recommendations are given in Figure 11-5.

Building paper is not required over plywood wall sheathing with brick veneer or masonry, if you provide an air space. Building paper (15-pound saturated asphalt felt) is required under stucco.

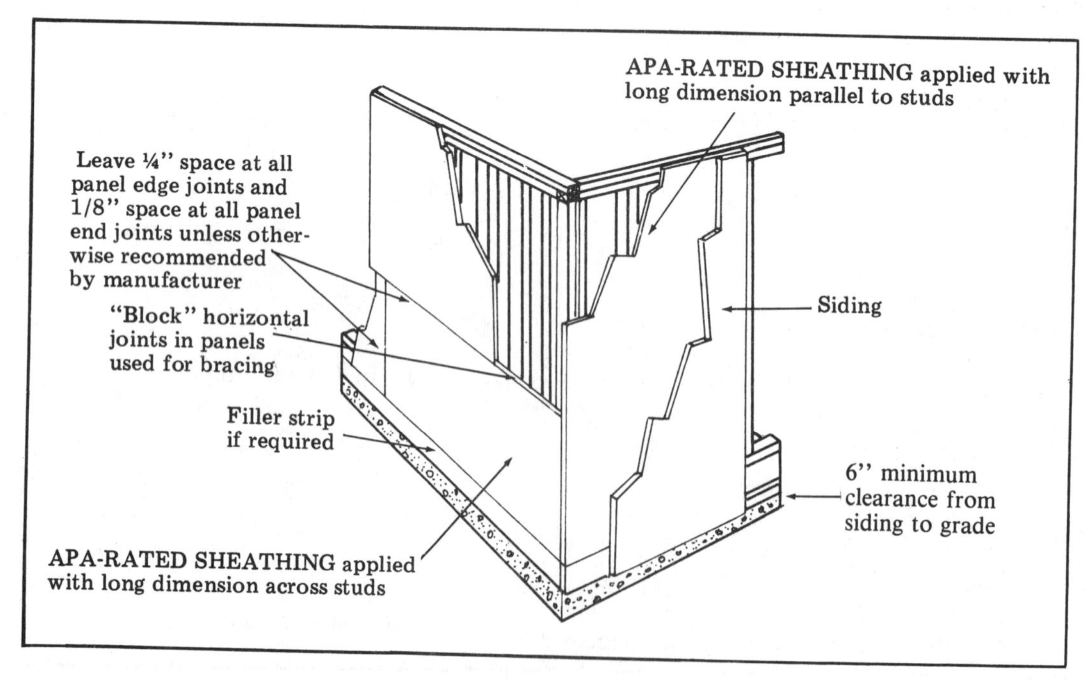

APA panel wall-sheathing installation recommendations
Figure 11-5

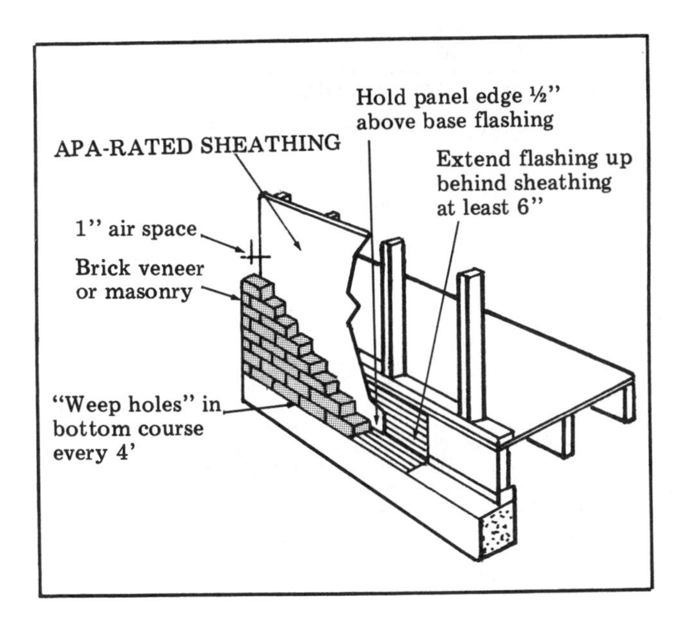

Brick veneer over APA panel wall sheathing
Figure 11-6

(See Figures 11-6 and 11-7.) Recommended wall-sheathing spans with brick veneer, masonry or stucco are the same as those for siding over sheathing.

Figure 11-8 gives specifications for APA panel wall sheathing.

Figure 11-9 shows stapling recommendations for APA panel wall sheathing.

Standard sheathing-grade plywood is commonly used for wall sheathing. For more severe exposures, this same plywood is available with an exterior glue-line. For 16-inch stud spacing, the minimum plywood thickness should be 3/8 of an inch or thicker, especially if the exterior finish is nailed to the sheathing. The plywood thickness may also be influenced by the standard jamb widths in windows and exterior doors. They may require sheathing 1/2 inch thick or thicker.

Structural Fiberboard Wall Sheathing

Structural fiberboard (also called structural insulating board) is the cheapest sheathing material. Fiberboard is coated or impregnated with asphalt to make it water-resistant. During construction, you don't have to worry about the material getting rain-soaked. It's a good "dry-in" material.

The three common types of fiberboard used for sheathing include: regular density, intermediate density and nail base. Regular-density sheathing is made in 1/2- and 25/32-inch thicknesses and 2 x 8-, 4 x 8- and 4 x 9-foot panels. Intermediate-density and nail-base sheathing are denser. They're manufactured only in a 1/2-inch thickness and in 4 x 8- and 4 x 9-foot panels. While 2 x 8-foot panels with matched edges are applied horizontally, 4 x 8- and 4 x 9-foot panels are usually installed vertically.

Corner bracing is required on horizontally applied panels and on 1/2-inch regular-density sheathing applied vertically. Additional corner bracing is usually not required for regular-density fiberboard sheathing that is 25/32 of an inch thick. Neither is it required for intermediate-density and nail-base sheathing when applied with the long edge vertical, as shown in Figure 11-10. The fastenings must be adequate around the perimeter and at intermediate studs.

Nail-base sheathing also permits the direct application of shingles as siding if you use special annular-grooved nails. Galvanized or other corrosion-resistant fasteners are recommended for structural fiberboard sheathing.

Many builders who use 1/2-inch regular-density fiberboard sheathing will use a sheet of 1/2-inch-thick sheathing-grade plywood at each corner for bracing instead of let-in, diagonal bracing.

Use 1¾-inch galvanized roofing nails for the 25/32-inch structural fiberboard sheathing and 1½-inch nails for the 1/2-inch sheathing. The manufacturers recommend 1/8-inch spacing between panels. Joints are centered on the studs.

Gypsum Wall Sheathing

Gypsum sheathing has a treated gypsum filler faced on two sides with water-resistant paper. Usually one edge is grooved and the other has a matched

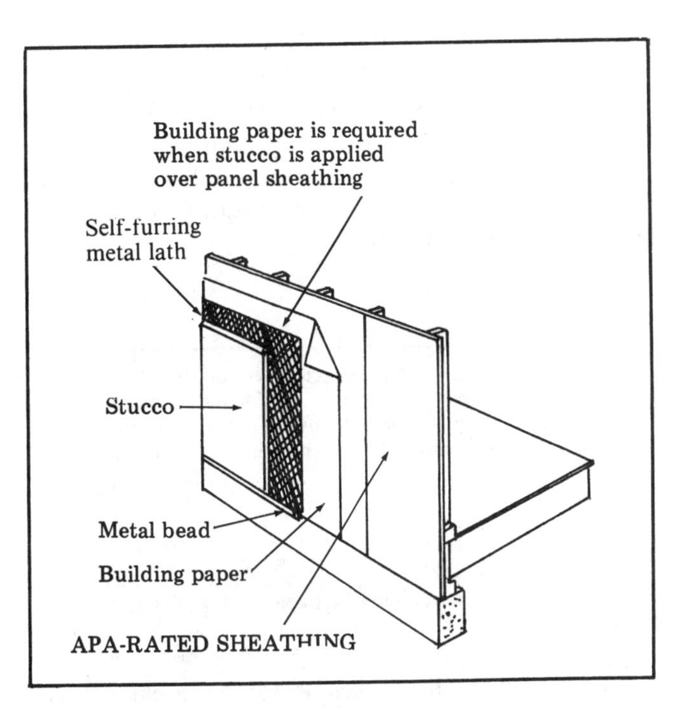

Stucco over APA panel wall sheathing
Figure 11-7

Panel span rating	Maximum stud spacing (inches)	Nail size***	Nail spacing***	
			Panel edges	Intermediate
12/0, 16/0, 20/0 or wall-16 o.c.	16*	6d for panels ½" thick or less, 8d for thicker panels	6"	12"
24/0, 24/16, 32/16	24**			

APA-rated sheathing panels continuous over two or more spans

*Apply plywood panels less than ⅜" thick with face grain across studs when exterior covering is nailed to sheathing
**Apply 3-ply plywood panels with face grain across studs 24" o.c. when exterior covering is nailed to sheathing
***Common, smooth, annular, spiral-thread, or galvanized box; or T-nails of the same diameter as common nails (0.113" diameter for 6d, 0.131" for 8d) may be used. Staples also permitted at reduced spacing.

APA panel wall sheathing specifications
Figure 11-8

"V". The 1/2-inch-thick 2 x 8-foot panels are applied horizontally for stud spacings of 24 inches or less. Figure 11-11 shows details for horizontal application of 2 x 8-foot sheathing materials.

Applying 2 x 8-Foot Panel Wall Sheathing

Gypsum and structural fiberboard sheathing in 2 x 8-foot panels require corner bracing. You can use 1 x 4 let-in bracing, as shown in Figure 11-11. Vertical joints should be staggered. Nail thc 25/32-inch structural fiberboard to each stud with 1¾-inch galvanized roofing nails spaced about 4½ inches apart.

Nail 1/2-inch gypsum and fiberboard sheathing to the framing members with 1½-inch galvanized roofing nails spaced about 3½ inches apart.

When wood bevel or similar sidings are used over plywood sheathing less than 5/8 of an inch thick, or over fiberboard or gypsum board, the nails must penetrate the stud.

Panel Thickness (inches)	Staple Leg length (inches)	Spacing around entire perimeter of sheet (inches)	Spacing at intermediate members (inches)
5/16	1-1/4	4	8
3/8	1-3/8	4	8
1/2	1-1/2	4	8

Values are for 16 ga. galvanized wire staples with a minimum crown width of 3/8".

Recommended stapling schedule for APA panel wall sheathing
Figure 11-9

If you want to use wood shingles over gypsum or regular-density fiberboard sheathing, first apply horizontal rows of 1 x 3-inch nailing strips spaced to conform to the shingle exposure. Nail these wood strips to each stud with two 8d or 10d threaded nails, depending on the sheathing thickness. (See Figure 11-11.) Nail-base fiberboard sheathing usually does not require wood strips when threaded nails are used.

Saving on Wall-Sheathing Material and Labor Costs

Every home builder should be interested in cutting unnecessary material and labor costs. On many houses, the wall sheathing can be eliminated entirely. In mild climates, sheathing can be eliminated by using an appropriate finish siding. (See Figure 11-12.) You can apply 4 x 8-foot structural siding of plywood, hardboard, or high-density fiberboard directly to 2 x 4-stud framing, even if the studs are spaced 24 inches on center. The siding provides enough racking strength to resist horizontal wind and earthquake loads.

Plywood used for sheathing can be the minimum thickness — 5/16-inch-thick for studs spaced 16 inches o.c. and 3/8-inch-thick for studs 24 inches o.c.

Plywood siding placed over sheathing should be no less than 1/4 of an inch thick. If no sheathing is used, the plywood siding should be 3/8-inch-thick for studs spaced 16 inches o.c. or 1/2-inch-thick for studs 24 inches o.c.

Wall Sheathing Paper

Sheathing paper for walls should be water-resistant but not vapor-resistant. It should allow the move-

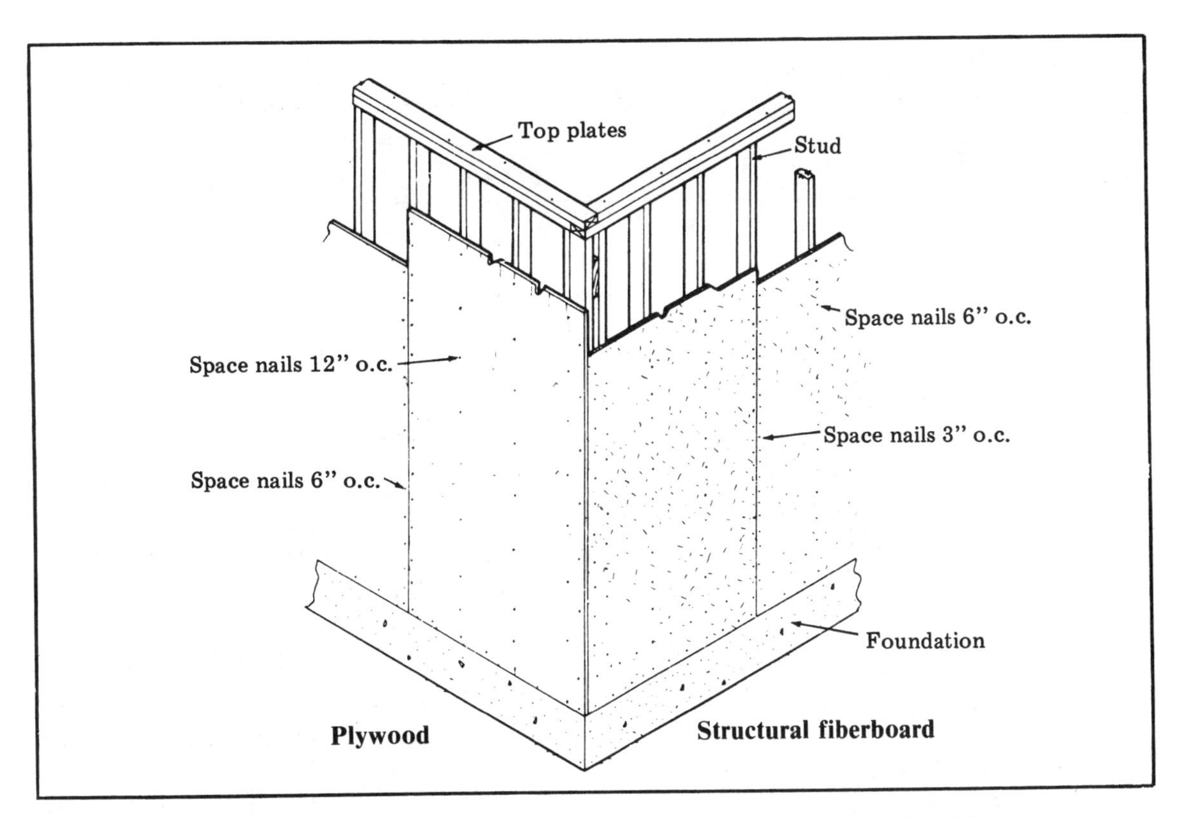

Vertical application of plywood or structural fiberboard wall sheathing
Figure 11-10

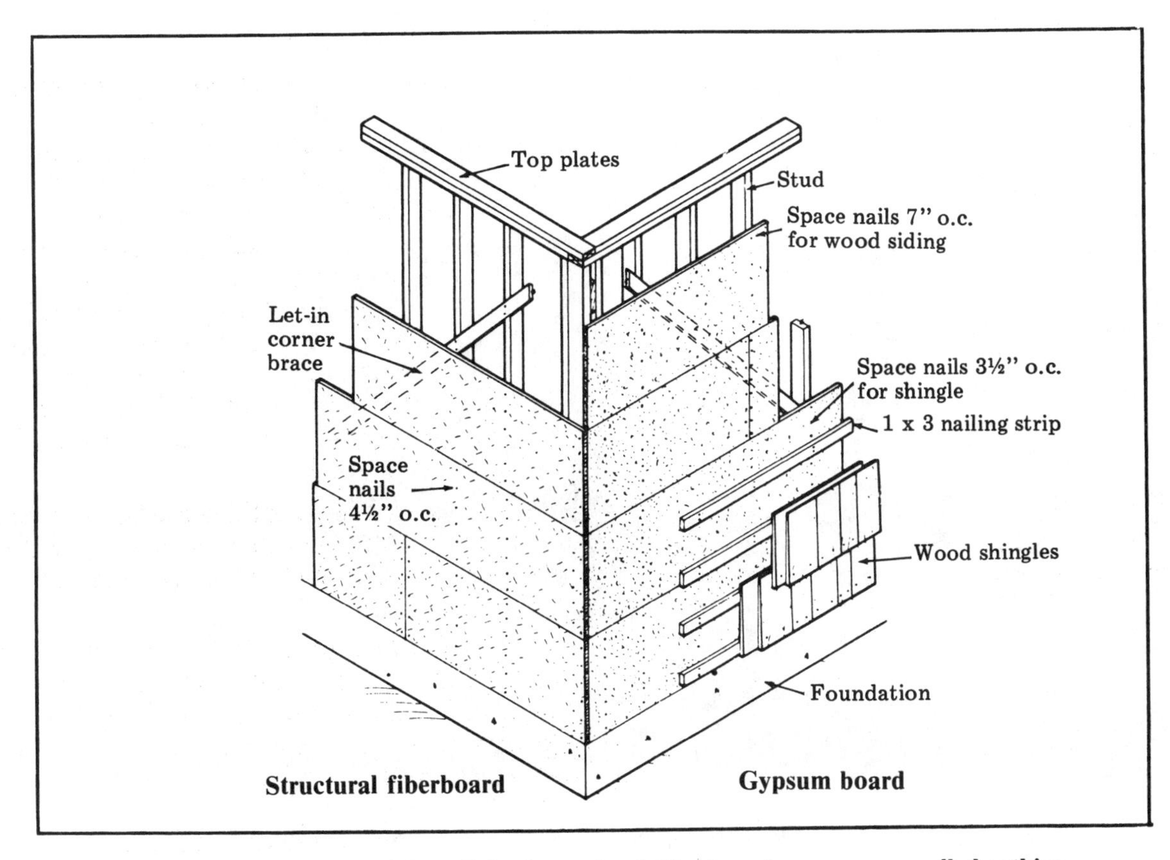

Horizontal application of 2 by 8-foot structural fiberboard or gypsum wall sheathing
Figure 11-11

Finish siding applied directly to studs
Figure 11-12

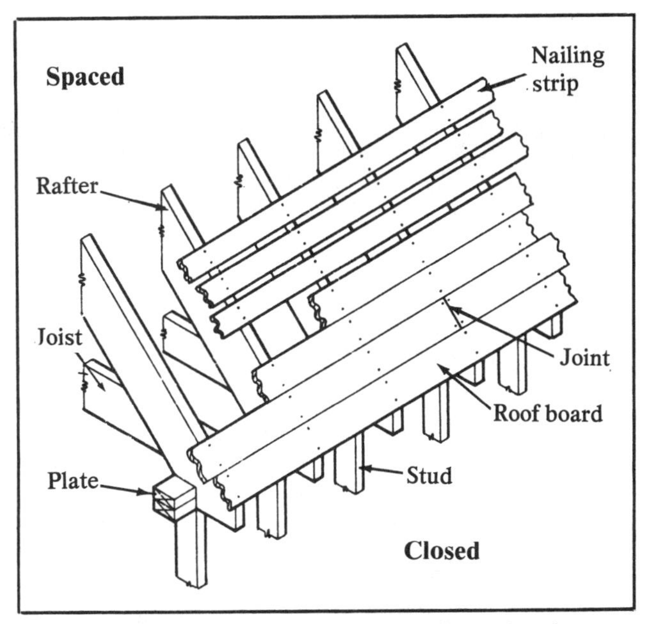

Installation of board roof sheathing, showing both closed and spaced types
Figure 11-13

ment of water vapor but resist the entry of direct moisture, such as blowing rain. Use either 15-pound asphalt felt or rosin paper for most applications. Sheathing paper should have a "perm" value of 6 or more. The paper also helps resist air infiltration.

Always use sheathing paper behind a stucco or masonry veneer finish and over board sheathing. Install it horizontally, starting at the bottom of the wall. Succeeding layers should lap about 4 inches. It's not generally used over plywood, fiberboard, or other sheet materials that are water-resistant. However, 8-inch or wider strips of the paper should be used around window and door openings to minimize air infiltration.

Types of Roof Sheathing

Board Roof Sheathing

If you use roof sheathing boards, use No. 3 pine, redwood, hemlock, Western larch, fir or spruce. Only thoroughly seasoned material should be used with asphalt shingles. Unseasoned wood will dry out and shrink. This causes buckling or lifting of the shingles along the length of the board.

In some cases, board sheathing is preferred. An example is where wood shingles or shakes are used in damp climates. Board sheathing should be laid closed (without spacing) when used under asphalt shingles, metal-sheet roofing, or other materials that require continuous support. (See Figure 11-13.) Wood shingles can also be used over such sheathing. Boards should be matched, shiplapped, or square-edged with joints staggered and placed over the center of the rafters.

To minimize shrinkage, use boards no wider than 8 inches. Boards that are 6 inches wide are preferred. They should have a minimum thickness of 3/4 of an inch for rafter spacing of 16 to 24 inches o.c., and should be nailed with two 8d nails for each board at each bearing.

Use long sheathing boards at roof ends to get good framing anchorage, especially in gable roofs where there is a rake overhang or an unsupported fly rafter. (See Figure 11-14.)

When wood shingles or shakes are used in damp climates, space the roof boards as shown in Figure 11-13. This allows free air circulation under the shingles. That's essential for good drying. Wood nailing strips in nominal 1 x 3- or 1 x 4-inch sizes are spaced the same distance on center as the shingles are to be laid to the weather. For example, if shingles are laid 5 inches to the weather and nominal 1 x 4-inch strips are used, there would be spaces of 1⅜ to 1½ inches between each board to provide the necessary ventilation.

Plywood Roof Sheathing

Most roof sheathing is plywood. It goes on fast, so labor costs are less than they are with board sheathing. Plywood makes a smooth, solid deck that works well with nearly any roofing material. Plywood roof sheathing can be standard sheathing-grade plywood.

Plywood roof sheathing is laid with the face grain perpendicular to the rafters. See Figure 11-15. Standard sheathing-grade plywood is generally specified. But in damp climates, use a standard sheathing grade with exterior glue. End joints are made over the center of the rafters and are staggered by at least one rafter. Plywood should be nailed at each bearing, 6 inches o.c. along all edges and 12 inches o.c. along intermediate members. Unless plywood has an exterior glue-line, raw edges should not be exposed to the weather at the gable end or at the cornice. Instead, protect it with a trim piece.

Some manufacturers recommend spacing panels at least 1/8 of an inch at ends and 1/4 of an inch at edge joints. That creates a problem. An 8-foot plywood panel is a full 96 inches long. When you end-space the sheathing that much, the panels will very quickly run off the rafter centers. So you would have to saw about 1/2 of an inch off the end of every third or fourth sheet. A lot of builders don't take the time to do this. They butt the panels snug. But if snug sheathing gets wet, you can expect some buckling.

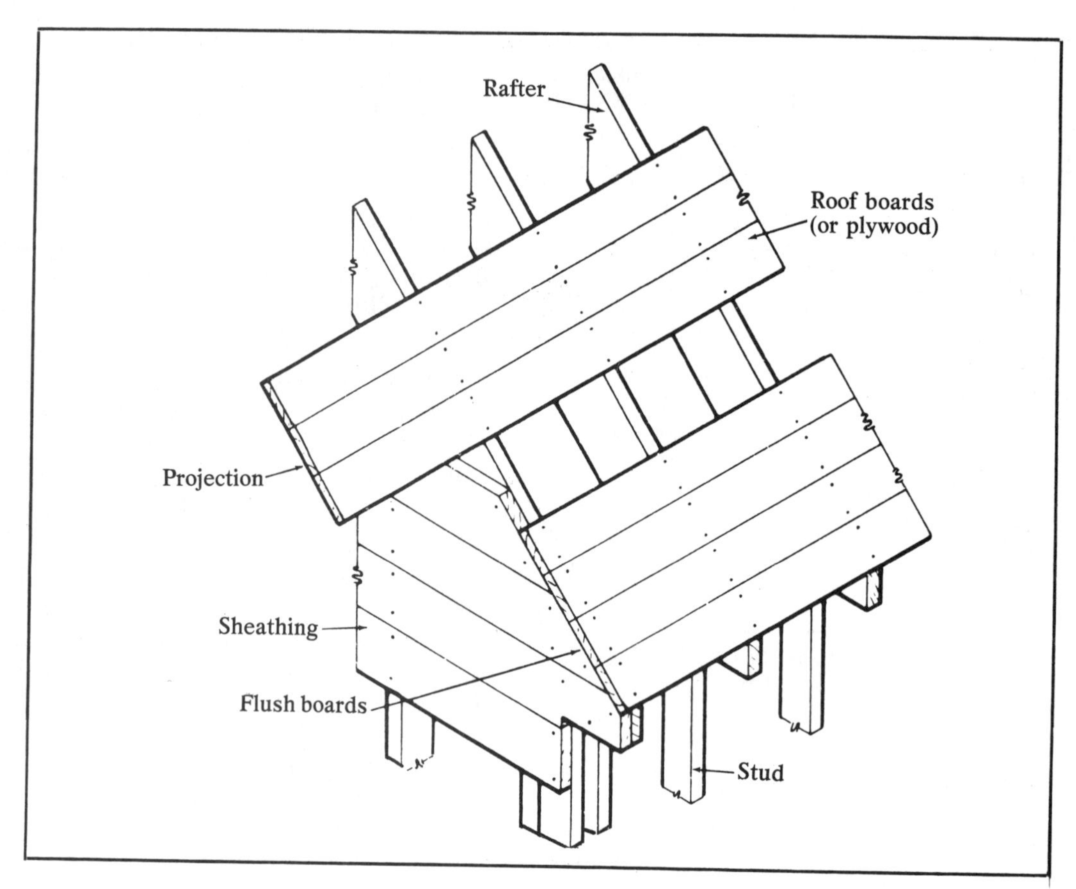

Board roof sheathing at ends of gable
Figure 11-14

APA Panel Roof Sheathing

The recommendations for roof sheathing in Figure 11-16 apply to APA-rated sheathing labeled Exposure 1, Exposure 2 or Exterior, and APA Structural-rated sheathing 1 and 2, Exposure 1 or Exterior. Uniform load deflection limits are 1/180 of span under live load plus dead load, and 1/240 under live load only. Panels are assumed continuous over two or more spans with the long dimension across supports.

Special conditions, such as heavy, concentrated loads, may require thicker panels than shown in Figure 11-17. Also, allowable live loads may have to be decreased for tile roofs with dead loads greater than 5 psf. Figure 11-16 shows the typical APA panel roof sheathing procedure.

APA-rated panel sheathing makes a good base under built-up roofing; asphalt, fiberglass or asbestos shingles; tile roofing; or wood shingles or shakes (except in damp areas).

When roof trusses spaced 24 inches o.c. are used, you can use 3/8-inch 24/0 plywood sheathing applied with panel clips. But it may be cheaper to use fewer supports with thicker panels: 3/4-inch 48/24 panels over framing 48 inches o.c. are also a good choice for long-span flat or sloped roofs.

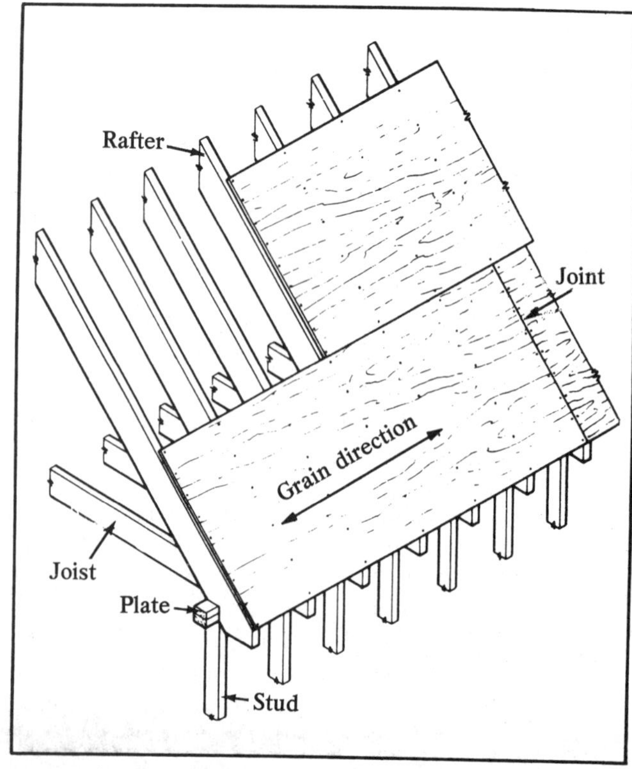

Application of plywood roof sheathing
Figure 11-15

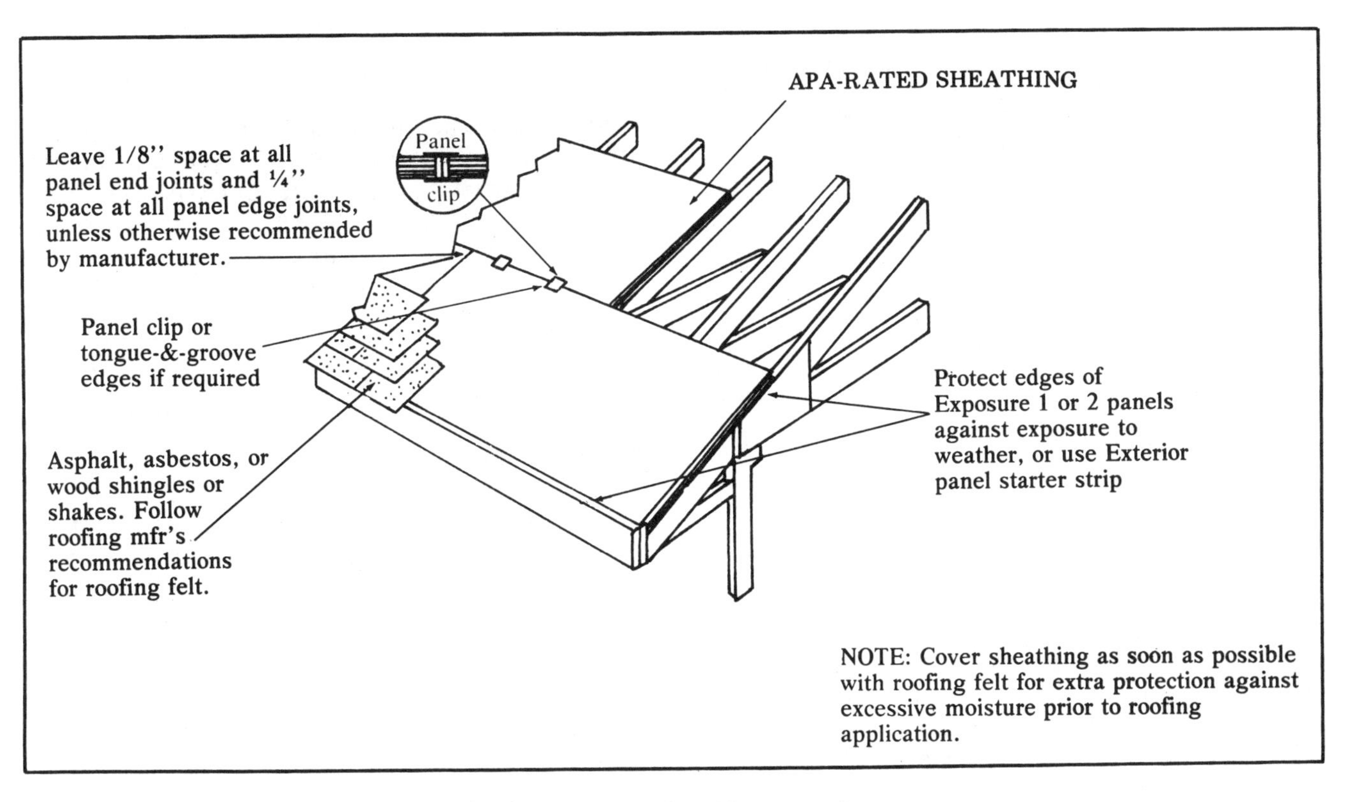

APA panel roof sheathing procedure
Figure 11-16

Nailing recommendations are given in Figure 11-17. Stapling recommendations are in Figure 11-18.

When support spacing exceeds the maximum length of an unsupported edge (see Figure 11-17), provide blocking, tongue-and-groove edges, or other edge-support, such as panel clips. Panel clips provide edge-support and assure panel spacing. When required, use one panel clip for spans less than 48 inches and two for 48-inch or longer spans.

Panel span rating	Panel thickness (inches)	Maximum span (inches) With edge support*	Without edge support	Nail size and type	Nail spacing (inches) Panel edges	Intermediate
12/0	5/16	12	12	6d common	6	12
16/0	5/16, 3/8	16	16			
20/0	5/16, 3/8	20	20			
24/0	3/8, 7/16, 1/2	24	20***			
24/16	7/16, 1/2	24	24			
32/16	1/2	32	28			
32/16	5/8	32	28	8d common		
42/20**	5/8, 3/4, 7/8	42	32			
48/24**	3/4, 7/8	48	36		6	6

All panels will support at least 40 psf live load plus 5 psf dead load at maximum span, except as noted.

*Tongue-and-groove edges, panel edge clips (one between each support, except two between supports 48 inches o.c.) lumber blocking or other.
**PS 1 plywood panels with Span Ratings of 42/20 and 48/24 will support 35 psf live load plus 5 psf dead load at maximum span. For 40 psf live load, specify Structural I.
***24 inches for 1/2-inch panels.

APA panel roof sheathing nailing recommendations
Figure 11-17

Panel thickness (inches)	Staple leg length (inches)	Staple spacing (inches)	
		Panel edges	Intermediate
5/16	1-¼		
3/8	1-⅜	4	8
1/2	1-½		

Values are for 16-ga. galvanized wire staples with a minimum crown width of 3/8-inch.

For stapling asphalt shingles to 5/16-inch and thicker panels, use staples with a 3/4-inch minimum crown width and a 3/4-inch leg length. Space according to shingle manufacturer's recommendations.

APA panel roof sheathing recommended minimum stapling schedule
Figure 11-18

APA Panel Soffits

An open soffit is shown in Figure 11-19, and a closed soffit is shown in Figure 11-20. Recommended spans for open and closed soffits are given in Figures 11-21 and 11-22. Panels are assumed to be continuous over two or more spans, with the long dimension across the supports. For spans of 32 and 48 inches in open-soffit construction, provide blocking or use tongue-and-groove edges or edge-support, such as panel clips. Minimum loads are at least 40 psf live load, plus 5 psf dead load, with the exception of the 1⅛-inch panels of Group 2, 3, or 4 species, which support 35 psf live load.

For open-soffit construction, panels designated Exposure 1 or Interior with exterior glue are recommended as a minimum where appearance is not a major consideration. But it may pay to check your building code on this.

Only panels identified as *Exterior* should be used for closed soffits. For open soffits and in closed-soffit construction where Interior (Exposure 1 or 2) sheathing is used for roof decking, use fascia trim to protect the panel edges against direct exposure to the weather.

The recommendations in Figure 11-21 for open soffits also apply to combined roof-ceiling construction.

Plank Roof Decking

Plank roof decking is commonly used for flat or low-pitched roofs in post-and-beam construction. It's usually nominal 2-inch tongue-and-groove wood planking. Common sizes are nominal 2 x 6-, 3 x 6-, and 4 x 6-inch V-grooved plank. The thicker planking is suitable for spans up to 10 to 12 feet. The maximum span for 2-inch planking in most grades and species is 8 feet when continuous over two supports, and 6 feet over single spans.

Special load requirements may reduce the allowable span. Roof decking can serve both as an interior ceiling finish and as a base for roofing. But plan to provide fiberboard or other rigid insulation over the wood decking to reduce heat loss and gain.

The decking is blind-nailed through the tongue and face-nailed at each support. In the 4 x 6-inch size, it's predrilled for edge-nailing. For thinner decking, a vapor barrier is ordinarily installed between the top of the plank and the roof insulation.

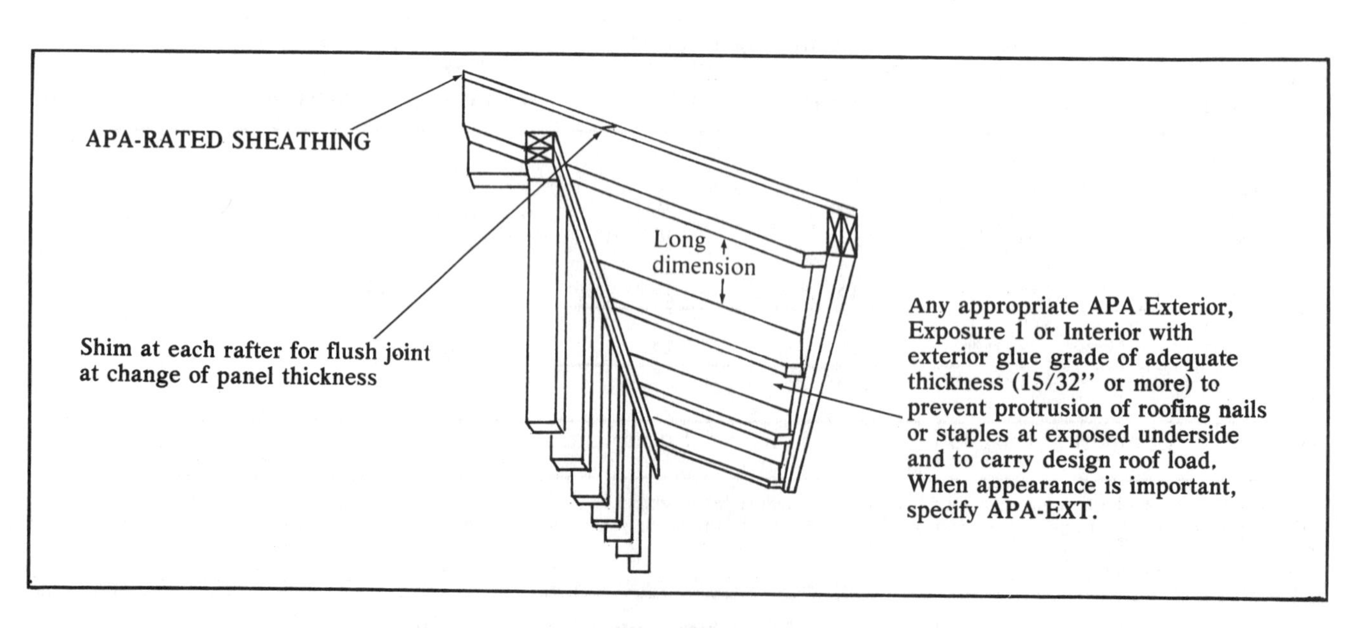

Open soffit
Figure 11-19

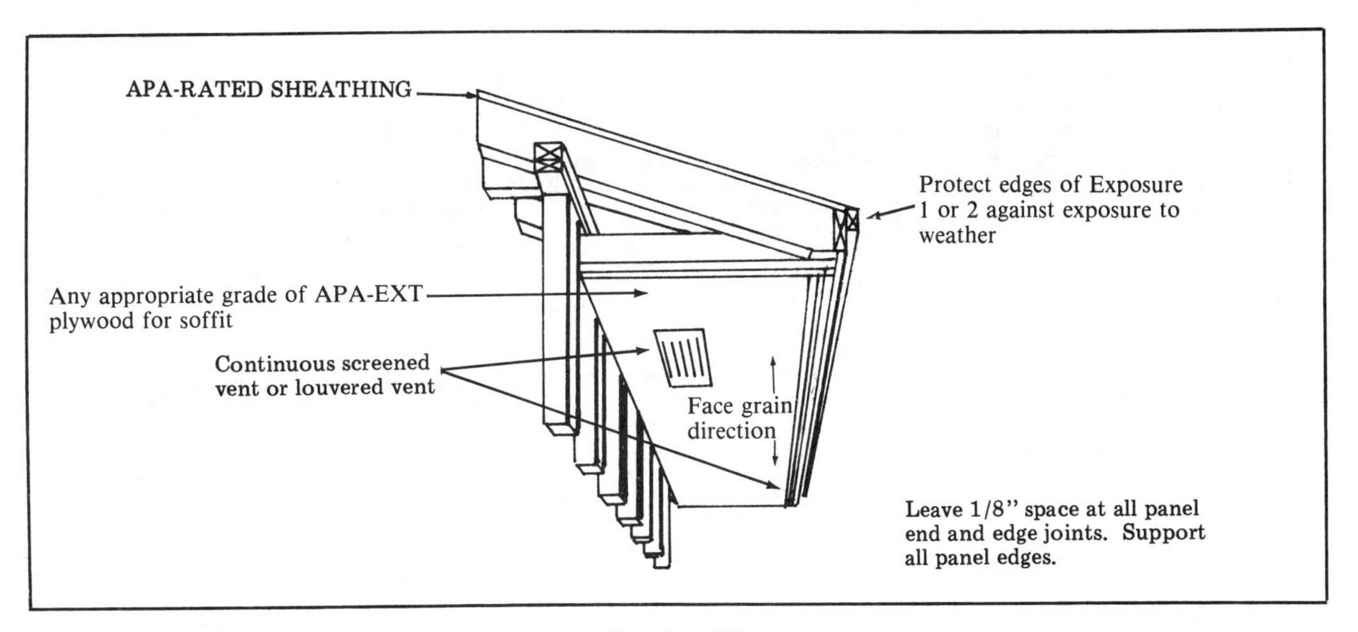

Closed soffit
Figure 11-20

Maximum span (inches)	Panel description (all panels Exterior, Exposure 1 or Interior with exterior glue)	Species group for plywood
16	15/32" APA 303 siding 1/2" APA sanded plywood APA Rated Sturd-I-Floor 16 o.c.	1,2,3,4 1,2,3,4 ----
24	15/32" APA 303 siding 1/2" APA sanded plywood 19/32" APA 303 siding 5/8" APA sanded plywood APA Rated Sturd-I-Floor 20 o.c.	1 1,2,3,4 1,2,3,4 1,2,3,4 ----
32*	5/8" APA sanded plywood 23/32" APA 303 siding 3/4" APA sanded plywood APA Rated Sturd-I-Floor 24 o.c.	1 1,2,3,4 1,2,3,4 ----
48*	1-1/8" APA textured plywood** APA Rated Sturd-I-Floor 48 o.c.	1,2,3,4 ----

All panels will support at least 40 psf live load plus 5 psf dead load at maximum span, except as noted

*Provide adequate blocking, tongue-and-groove edges or other suitable edge-supports, such as panel clips.
**1-⅛"panels of Group 2,3, or 4 species will support 35 psf live load plus 5 psf dead load.

APA panels for open soffits
Figure 11-21

Maximum span (inches) All edges supported	Nominal panel thickness	Species group	Nail size and type*
24	11/32" APA**		6d nonstaining
32	15/32" APA**	All	box or casing
		species	
48	19/32"APA**	groups	8d nonstaining
			box or casing

*Space nails 6 inches at panel edges and 12 inches at intermediate supports for spans less than 48 inches; 6 inches at all supports for 48-inch spans.
**Any suitable grade of Exterior panel which meets appearance requirements

APA panels for closed soffits
Figure 11-22

Fiberboard Roof Decking

Fiberboard roof decking is used the same way as plank roof decking. Of course, the supports must be spaced much closer together. Fiberboard decking is usually supplied in 2 x 8-foot sheets with tongue-and-groove edges. The thickness of the plank and the spacing of supports should comply with the following schedule:

Minimum Thickness (Inches)	Maximum Joist Spacing (Inches)
1½	24
2	32
3	48

Nails used to fasten fiberboard to the framing should be corrosion-resistant and spaced not more than 5 inches o.c. Use nails long enough to penetrate the joist or beam at least 1½ inches.

Sheathing at Chimney Openings

Where a chimney penetrates the roof, the sheathing should be 3/4 of an inch from the finished masonry on all sides. See Figure 11-23A. Rafters and headers around the opening need a clearance of 2 inches from the masonry for fire prevention. Check your local code for the exact clearance needed.

Sheathing at Valleys and Hips

Wood or plywood sheathing at the valleys and hips should be installed so it provides a tight joint. Make sure that panels are nailed securely to hip and valley rafters. See Figure 11-23B. This helps provide a solid and smooth base for flashing.

Manhour Estimates for Roof and Wall Sheathing

Board roof sheathing for a plain gable roof takes about 13 skilled and 6 unskilled hours per 1,000 BF. (See Figure 6-38 in Chapter 6.) The steeper the roof, the more manhours required. Remember that 1,000 BF of lumber will not cover 1,000 square feet of roof. Board footage is based on nominal or name sizes. You get the actual size. If you're using square-edge 1 x 6-inch boards, find the number of square feet in the roof and add 18%. This also allows for waste from end-cutting.

A roof 40 feet long and 14 feet wide (from ridge to end of rafter) has a total of 1,120 square feet. Adding 18% (202 square feet) gives a total of 1,322 square feet, which will require about 17 skilled and 8 unskilled manhours.

Plywood roof sheathing for a plain gable roof takes about 9 skilled and 4 unskilled hours per 1,000 square feet (Figure 6-38 in Chapter 6). Covering 1,120 square feet of roof will require about 10 skilled and 4.5 unskilled hours.

A good rule of thumb for estimating manhour requirements for installing plywood sheathing on a plain gable roof is:

Multiply the number of 4 x 8-foot panels required to cover the roof by 30 minutes. 2/3 of the sum will be for skilled labor and 1/3 will be for unskilled labor.

A rule of thumb for estimating manhours to install plywood or fiberboard wall sheathing is:

Multiply the number of 4 x 8-foot panels required to cover the wall by 25 minutes. 2/3 of the sum will be for skilled labor and 1/3 will be for unskilled labor.

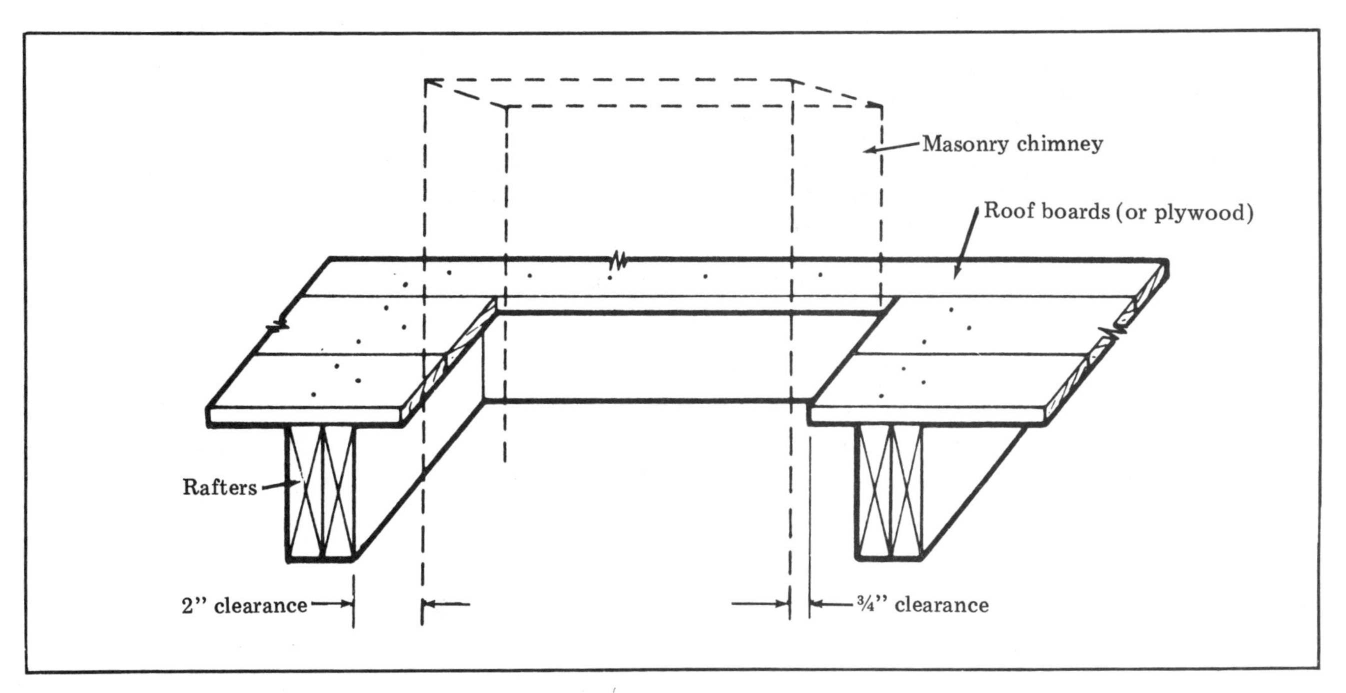

Sheathing at a chimney opening
Figure 11-23A

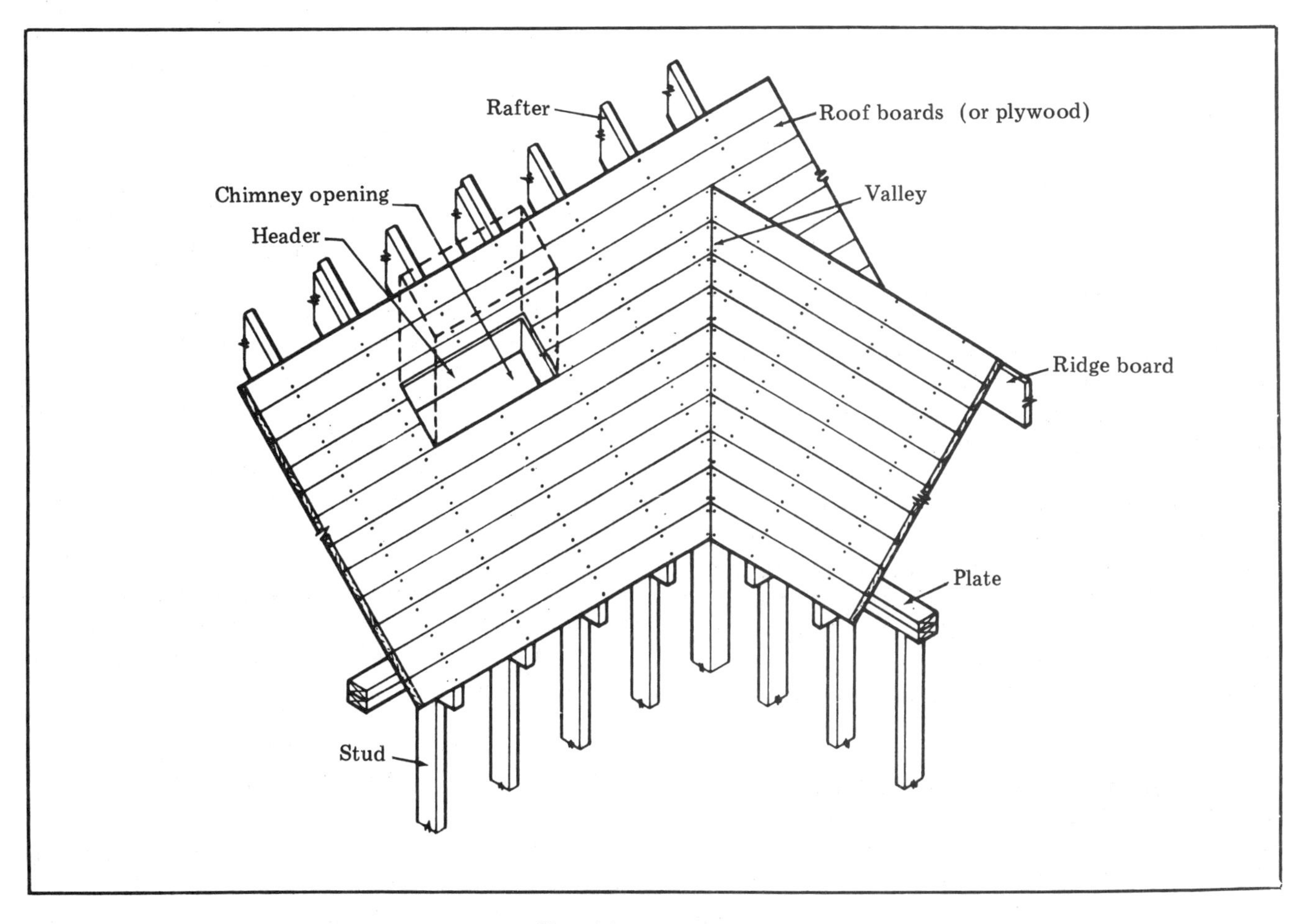

Sheathing at valley
Figure 11-23B

Summary

Wall and roof sheathing are high priority components in your construction schedule. They "dry-in" the structure, protecting it from weather. That permits you to move work inside on rainy days.

Once the roof sheathing is applied, cover it with building paper or the finish roofing material as quickly as possible. Building paper contracts and expands a great deal as the temperature changes, working loose from the roofing nails. It will not stay nailed down longer than a few days unless you use square cap roofing nails or felt nails with large tin washers.

The house with a simple gable roof needs only minimum thickness plywood sheathing for the span. Don't go to the next greater thickness. It's an unnecessary cost. If wall sheathing is not necessary in your area, don't install it. Every unnecessary dollar you save adds an extra dollar to your profit.

For an accurate record of materials and labor, complete the Sheathing Checklist and Data Reference Sheet (Figure 11-24) on each of your houses. It will be a valuable guide when estimating future jobs.

Wall and Roof Sheathing Checklist and Data Reference Sheet

1) The house is _______ feet in length and _______ feet wide, for a total of _______ square feet.

2) The roof is:

☐Plain gable,____/12 pitch

☐Gable with dormers,____/12 pitch

☐Gable dormers. Number ______

☐Shed-type dormers. Number ________

☐Variable gable

☐Hips, valleys, cut-up

☐Other (specify)________________________________

☐Hip, ____/12 pitch

☐Flat

☐Other (specify)________________________________

3) The roof has a ______-inch overhang

4) The roof has a total of ______ square feet.

5) The wall framing is ☐16 inches o.c. ☐24 inches o.c. ☐Other (specify) ____________

6) The rafters are ☐16 inches o.c. ☐24 inches o.c. ☐beam construction ☐trusses ☐other specify_______

7) Wall sheathing is:

☐Plywood, size________

☐Boards, size________

☐Fiberboard, size________

☐Gypsum board, size________

☐Other (specify) ________________________________

8) Wall sheathing is ________ inches thick.

9) Wall sheathing is installed ☐horizontally ☐vertically ☐diagonally.

10) Roof sheathing is:

☐Plywood, size________

☐Boards, size________

Figure 11-24

☐Fiberboard, size________

☐Other (specify) ____

11) Roof sheathing is ________ inches thick.

12) Material costs:

Wall sheathing	______BF/SF	@	$______BF/SF	=	$__________
Roof sheathing	______BF/SF	@	$______BF/SF	=	$__________
Nails:					
10d	__________lbs	@	$__________lb	=	$__________
10d threaded	__________lbs	@	$__________lb	=	$__________
8d	__________lbs	@	$__________lb	=	$__________
8d threaded	__________lbs	@	$__________lb	=	$__________
6d	__________lbs	@	$__________lb	=	$__________
⅞'' galvanized	__________lbs	@	$__________lb	=	$__________
1½'' galvanized	__________lbs	@	$__________lb	=	$__________
1¾'' galvanized	__________lbs	@	$__________lb	=	$__________
Staples (size ____)	__________lbs	@	$__________lb	=	$__________
Building paper	________rolls	@	$__________ea	=	$__________
Other materials, not listed above:					
________________________	____________	@	$____________	=	$__________
________________________	____________	@	$____________	=	$__________
________________________	____________	@	$____________	=	$__________
________________________	____________	@	$____________	=	$__________

Total cost of all materials $__________

13) Labor costs:

Wall sheathing:

Skilled	________hours	@	$________hour	=	$__________
Unskilled	________hours	@	$________hour	=	$__________

Roof sheathing:

Skilled	________hours	@	$________hour	=	$__________
Unskilled	________hours	@	$________hour	=	$__________

Building paper:

Skilled	________hours	@	$________hour	=	$__________
Unskilled	________hours	@	$________hour	=	$__________

Total labor costs:

Skilled	________hours	@	$________hour	=	$__________
Unskilled	________hours	@	$________hour	=	$__________
Other (specify)					
______________________	________hours	@	$________hour	=	$__________
				Total	$__________

14) Total cost:

Materials $__________

Labor $__________

Other $__________

Total $__________

15) Total cost of wall and roof sheathing per SF of floor space $__________.

16) Problems that could have been avoided ____

____

____

____

____

____

17) Comments ____

____

____

____

18) A copy of this Checklist and Data Reference Sheet has been provided:

☐Mangement

☐Accounting

☐Estimator

☐Foreman

☐File

☐Other: ___________

Chapter 12

Horizontal and Rake Cornices

You'll probably want to give cornice trim work (often called "boxing") special attention. Exterior trim requires exact fitting and installation if you want your houses to make a good first impression. Poor trim work can keep a house from selling.

Exterior trim includes door and window trim, cornice moldings, fascia boards and soffits, and porch trim and moldings. Contemporary houses have simple cornices and moldings. Traditional style houses have considerably more trim. Much of the exterior trim is cut and fitted at the job site. Trim items such as shutters, louvers, railings and posts are shop-fabricated and arrive on the job ready to install.

I prefer to install the fascia board *before* nailing on the roof sheathing. The fascia board is extended (above the rafters) the thickness of the sheathing and acts as a stop for the sheathing.

Selecting the Right Materials

The material used for exterior trim needs both good painting and weathering characteristics. It should be easy to work with and warp-resistant. Decay-resistance is also essential in the fascia boards, rails, shutters, and caps and bases of porch columns. Heartwood of cedar, redwood and cypress have a high resistance to decay. But heartwood is expensive. You might prefer to use a less durable species which can be treated to make it decay-resistant.

It's important to use the right nails to fasten exterior trim. Use rust-resistant nails such as galvanized, stainless steel, aluminum, or cadmium-plated. When a natural finish is used, the nails should be stainless steel or aluminum to prevent staining and discoloration.

Siding and trim are fastened in place with a standard siding nail — the kind that has a small flat head. Finish or casing nails can also be used. If the nails are not rust-resistant, they should be set below the surface and puttied after the prime coat of paint has been applied. This, of course, adds to your labor costs. Using the right rust-resistant nail eliminates this added cost.

Cornices

Look at Figure 12-1. The cornice of a building is where the lower edge of the roof meets the wall line. It is usually built up of several plain or molded members. The design and type of cornice depends on the architectural style of the house. There's no standard cornice treatment that can be used as a model. Cornices fall into two broad categories: open cornices and closed cornices. The open cornice has exposed rafter tails. In a closed cornice, the rafter tails are enclosed. Each type has many variations.

In a hip roof structure, the cornice is usually continuous around the building. In a gable roof, it may extend along two sides of the building and be returned a short distance around the ends of the building. This is called a *cornice return,* shown in Figure 12-1. The cornice may also be extended up the slope of the roof. This is the rake cornice. A cornice is also required on the dormer. The two sections of this cornice are called the *dormer*

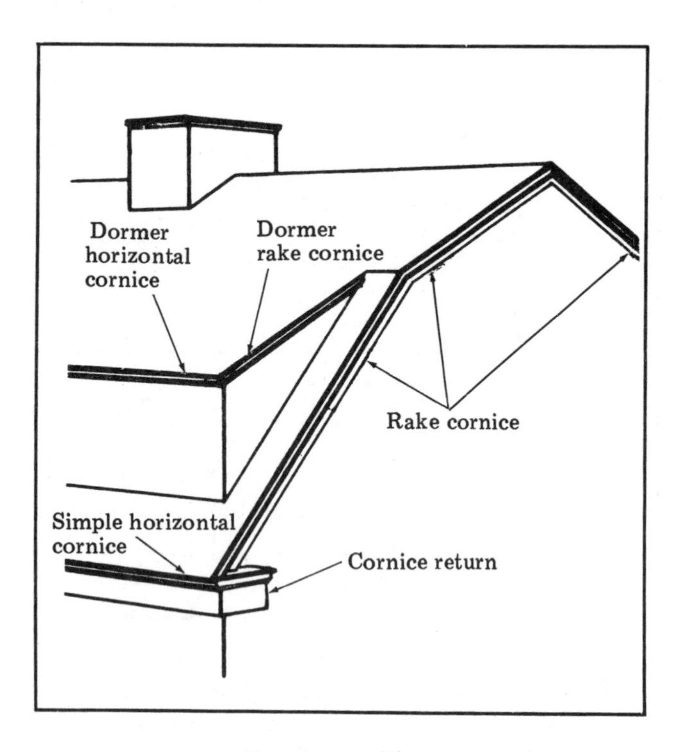

Cornice positions
Figure 12-1

horizontal cornice and the *dormer rake cornice*.

The cornice provides a means of decorating the section where the roof and walls of a building meet. It's often used to carry out the roof lines so that they are in harmony with the general style of the building. A projecting cornice protects the wall of the building. This "overhang" sheds water away from the wall.

Simple Cornice

Figure 12-2 shows a simple cornice. It consists of a single strip called a frieze board. The frieze board may be beveled on the upper edge to fit closely under the overhang of the eaves. It may be rabbeted on the lower edge to overlap the upper edge of the top course of siding. A crown or shingle molding is used to give the eave line a finished appearance.

A simple cornice doesn't add much to the appearance of a house, and it doesn't add any protection to the walls, windows or doors either.

Open Cornice

Figure 12-3 shows an open cornice. The rafters extend (beyond the wall line) the distance of the overhang. A fascia board is sometimes nailed to the plumb or square cut of the rafter to "kill" the rafter ends or to form a solid base for the gutter.

In climates where snow may accumulate during winter months, the open cornice can present a problem. So can a closed cornice with no soffit ventilation. If the building is not adequately insulated, heat will escape through the roof and melt the snow. The water then drains from the roof surface until it reaches the overhang of the roof beyond the wall. The overhang is exposed on the underside. This causes the draining water to freeze. The accumulating ice causes water to back up underneath the shingles to a point inside the wall line. From there, it usually seeps into the building. When the ceiling is adequately insulated, heat from the interior doesn't melt snow on the roof, and the ice dam problem doesn't occur.

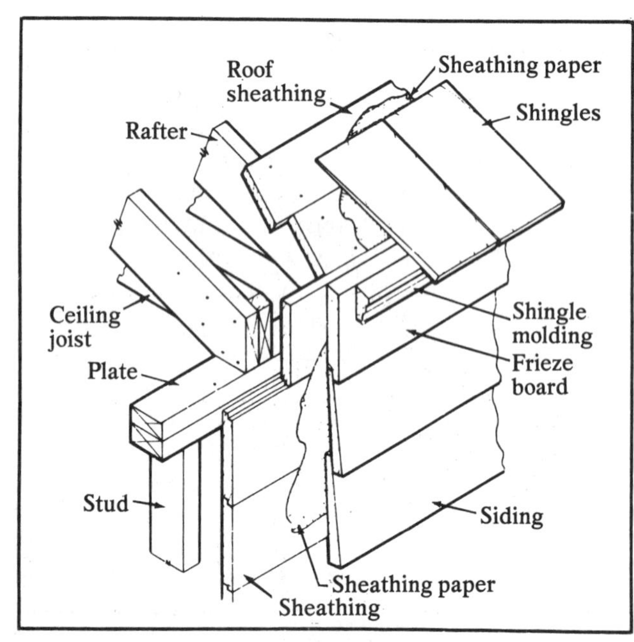

Simple cornice
Figure 12-2

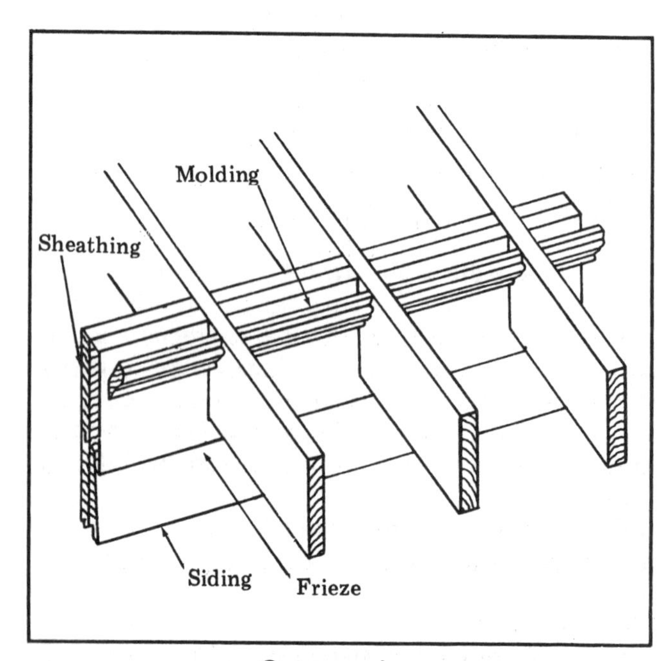

Open cornice
Figure 12-3

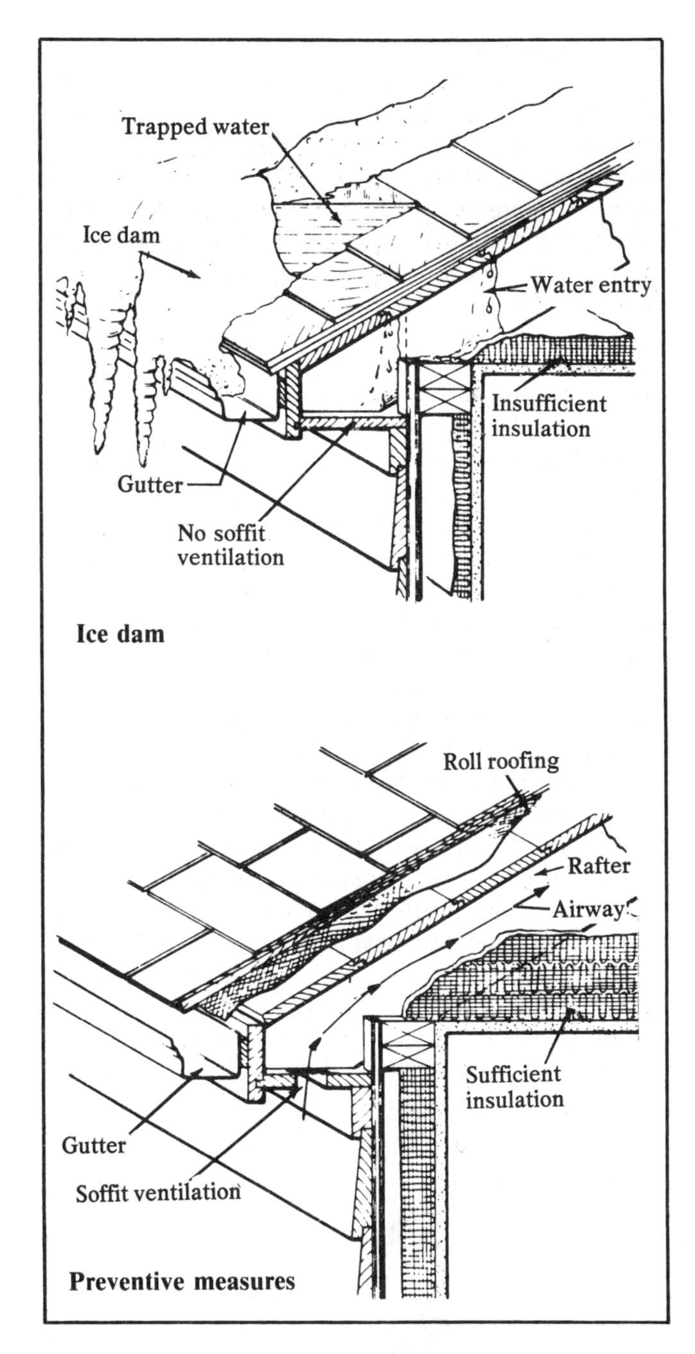

Winterizing the cornice
Figure 12-4

Smooth roll roofing, 50 lbs. or heavier, can reduce or eliminate ice dam problems. Lay the roofing over the overhang area and up the roof for 36 inches. That should eliminate seepage, even if snow on the roof melts and freezes again. See Figure 12-4.

Frieze ventilators are usually installed in open-cornice construction, as shown in Figure 12-5.

Closed Cornice

Figure 12-6 shows a closed cornice. This cornice is built on extended common rafters. The fascia is

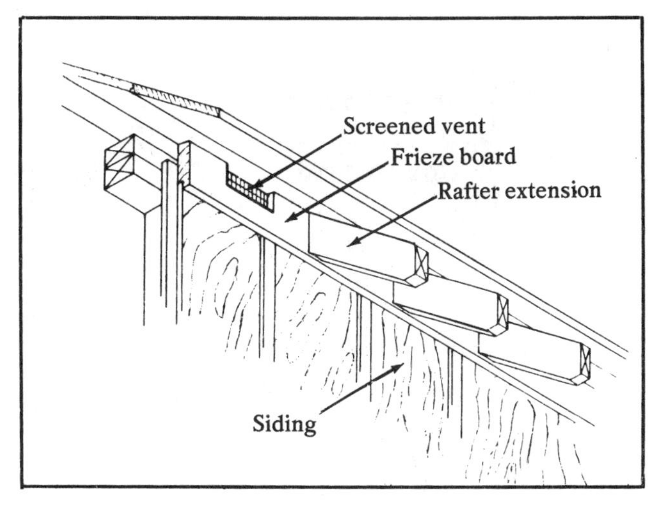

Frieze ventilator for open cornice
Figure 12-5

nailed to the plumb cut at the end of the rafter tail. The soffit is fitted to the back of the fascia and nailed to the bottom edge of the rafter. It extends up the rafter to the top edge of the frieze. A molding is fitted at the intersection of the soffit and the frieze to cover the joint there. (Note that the same materials are also used for the return up the gable rake of this type of cornice.)

Allow the roof sheathing to project (over the outside common rafter) the combined distance of the thickness of the frieze, the width of the soffit, and the thickness of the fascia. It can extend beyond the fascia, depending on the type of crown mold used. (See Figure 12-7, line A-B.)

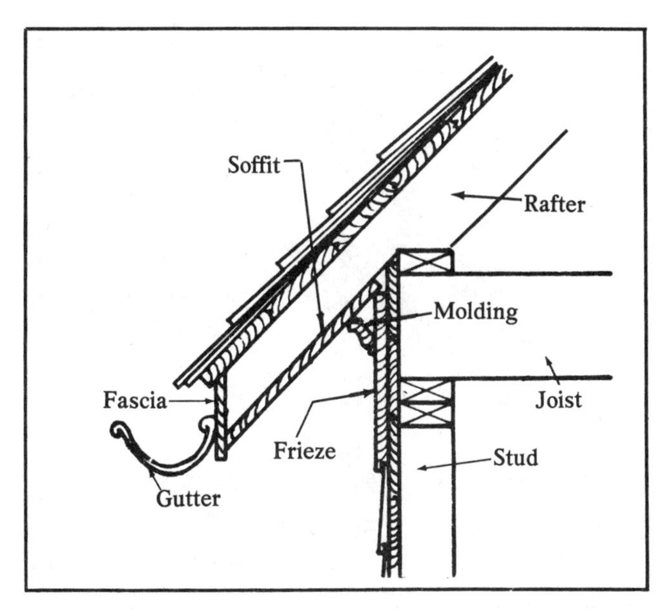

Closed cornice
Figure 12-6

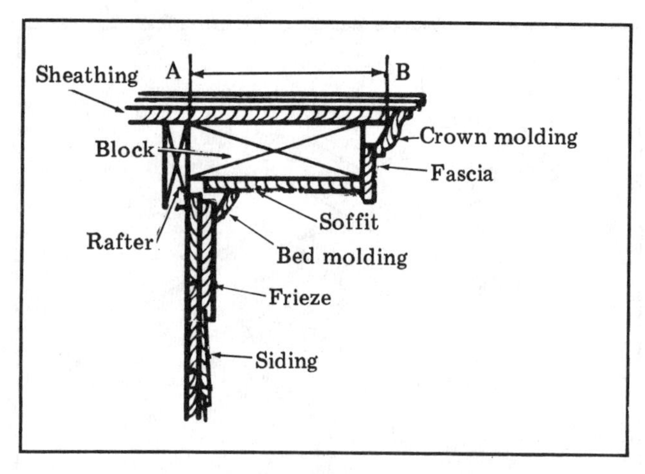

Section of rake with closed cornice
Figure 12-7

Every 4 feet, nail blocks to the underside of the sheathing, along the rake. These blocks should be the proper thickness to bring the bottom face of the rake soffit in line with the bottom face of the horizontal soffit, as shown in Figures 12-7 and 12-8.

Box Cornice

The box cornice is probably the one you see most often. It offers a more finished look and more weather protection for walls, windows and doors. In a narrow box cornice, the rafter serves as a nailing surface for the soffit as well as for the fascia (Figure 12-9). Depending on the roof slope and the size of the rafters, the rafter extension may vary between 6 and 12 inches. A wide box cornice normally requires additional framing for fastening the soffit. This is done with "lookouts" which are toenailed to the wall and nailing header, and face-nailed to the ends of the rafter extensions. See Figure 12-10.

Soffit material can be lumber, plywood, paper-overlaid plywood, hardboard, or medium-density fiberboard. Thicknesses should be based on the distance between supports. Use 3/8-inch plywood or 1/2-inch fiberboard for 16-inch rafter spacing. A nailing header (straightener) at the end of the rafters will provide a nailing base for the soffit and fascia. The nailing header is sometimes eliminated in a moderate cornice extension when a rabbeted fascia is used.

For best results in conventional roof framing, use the nailing header. It helps align the rafter tails, both horizontally and vertically. Inlet ventilators, often narrow continuous slots covered with screen wire, can be installed in the soffit area. This type of cornice is often used on a hip roof.

The projection of the cornice beyond the wall should not prevent the use of a narrow frieze board or a frieze molding above the top of the windows. The combination of a steeper slope and a wide projection will bring the soffit in the wide box cornice too low. Instead, use a narrow box cornice.

Prefabricated soffit systems are available in aluminum and fiberboard. These come in sections

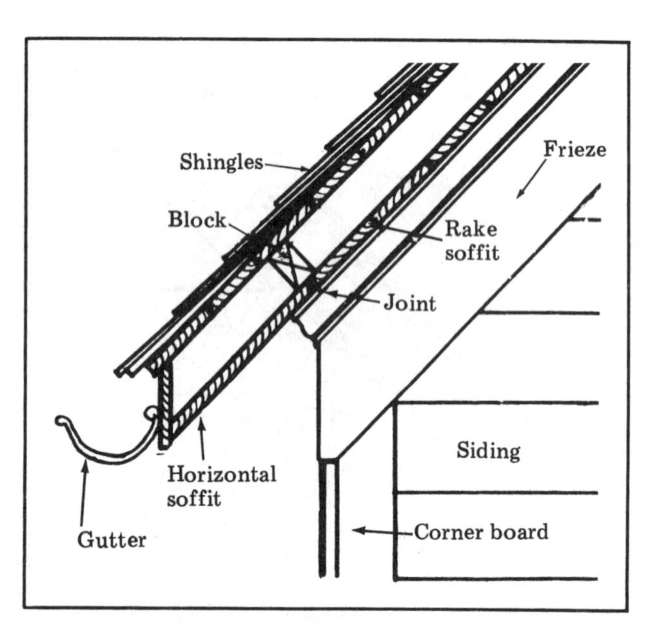

Horizontal and rake soffit joint
Figure 12-8

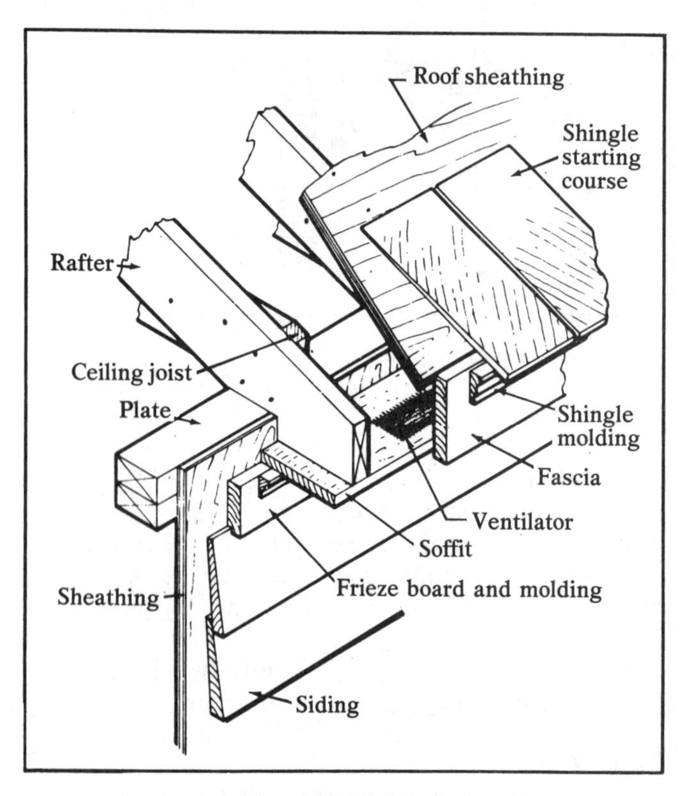

Narrow box cornice
Figure 12-9

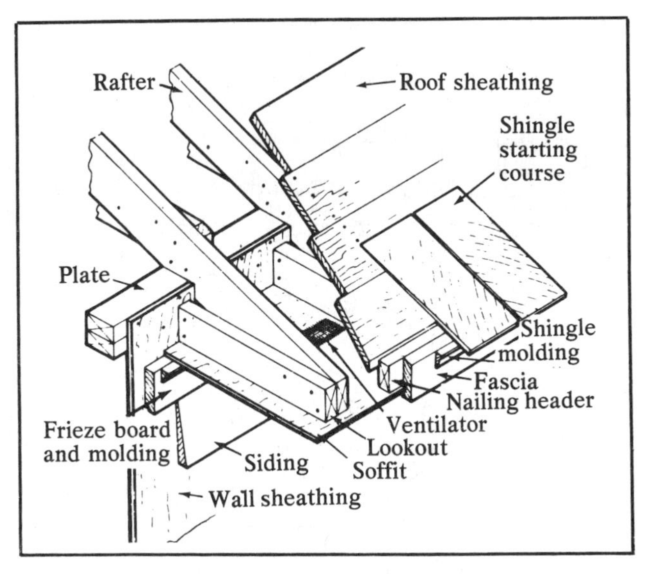

Wide box cornice
Figure 12-10

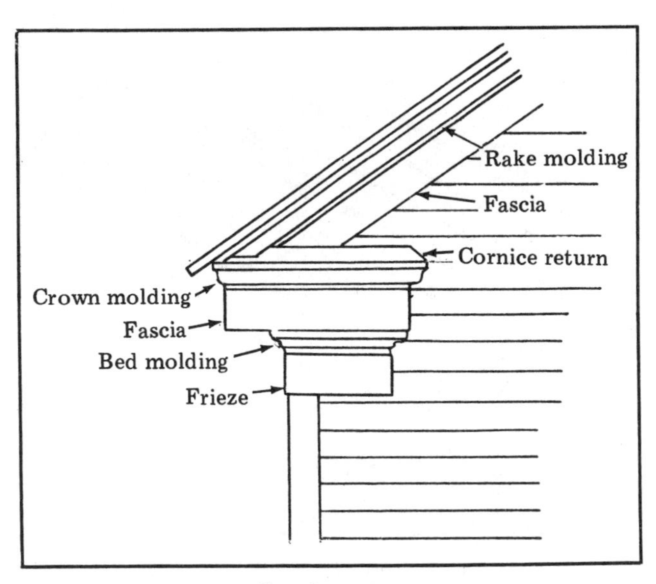

Cornice return
Figure 12-11

that save cutting and fitting time and have ready-made vents. To give a finished appearance, most soffit systems have complete supports, trim and brackets. But compare the cost of these prefabricated systems against the cost of conventional methods before deciding which system to use.

Cornice Return

The cornice return is the end-finish of the cornice on a gable roof. On hip and flat roofs, the cornice is usually continuous around the entire house. On a gable house, it must be terminated or joined with the gable ends. The type of detail selected depends on the type of cornice and the projection of the gable roof beyond the end wall. Figure 12-11 shows a cornice return. When the soffit is in a level position, the cornice is returned against the building on the gable end. The rake cornice at the roof line of the gable has the same design as the cornice at the eaves. It does not miter with the eave cornice members but ends on top of the eave cornice return.

A narrow box cornice, often used in houses with Cape Cod or colonial details, has a boxed return when the rake section projects. See Figure 12-12. The fascia board of the horizontal cornice is carried around the corner of the rake projection.

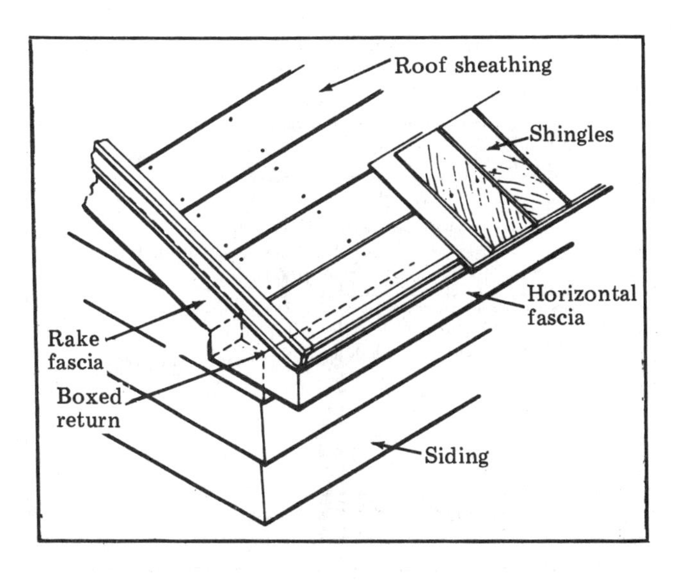

Narrow box cornice with boxed cornice return
Figure 12-12

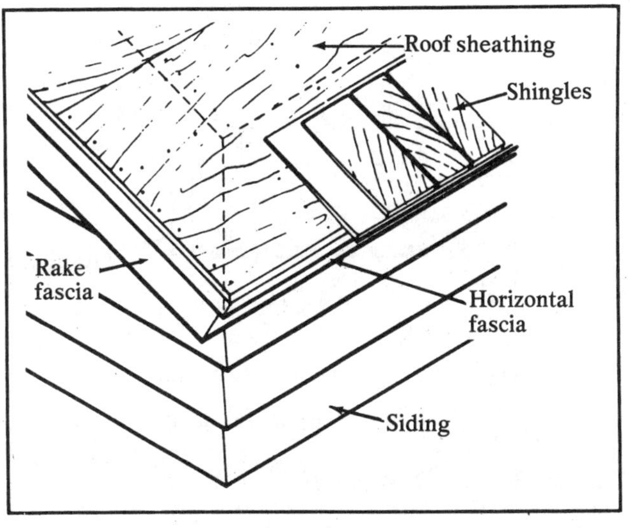

Wide box cornice with equally sloping rake and horizontal soffits
Figure 12-13

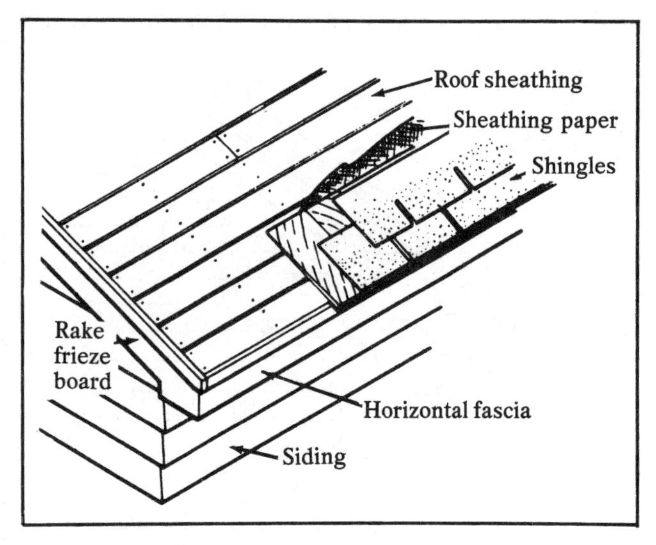

Narrow box cornice with close rake
12-14

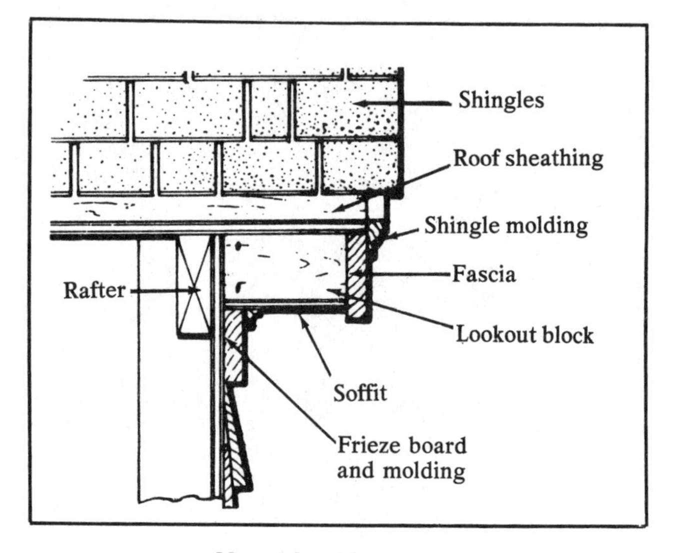

Normal gable overhang
Figure 12-15

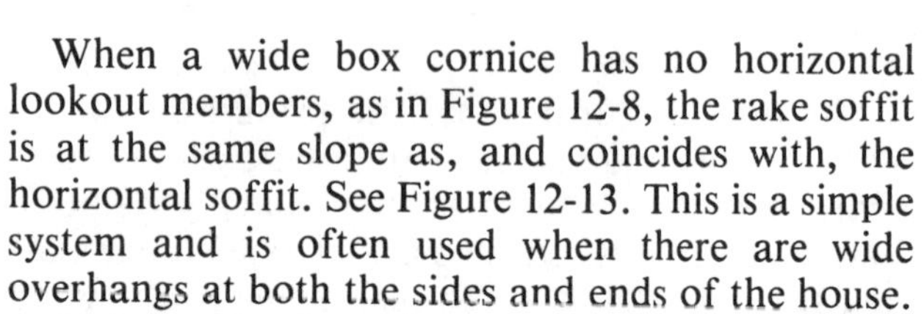

When a wide box cornice has no horizontal lookout members, as in Figure 12-8, the rake soffit is at the same slope as, and coincides with, the horizontal soffit. See Figure 12-13. This is a simple system and is often used when there are wide overhangs at both the sides and ends of the house.

A close rake (a gable end with little projection) may be used with a narrow box cornice or a simple or closed cornice. The rake frieze board, into which the siding butts, joins the frieze board or fascia of the horizontal cornice. See Figure 12-14.

Rake Cornice

The rake section is the extension of a gable roof beyond the end wall of the house. This overhang might vary from 6 inches to 2 feet or more.

In a normal overhang, when the rake extension is only 6 to 8 inches, the fascia and soffit can be nailed to a series of short lookout blocks, as shown in Figure 12-15. The fascia is further secured by nailing through the projected roof sheathing. A frieze board and appropriate moldings complete the construction.

In a moderate overhang of up to 20 inches, both the extended roof sheathing and a fly rafter support the rake section (Figure 12-16). The fly rafter extends from the ridge board to the nailing header that connects the ends of the rafters. The roof sheathing should extend from inner rafters to the end of the gable projection to provide rigidity and strength.

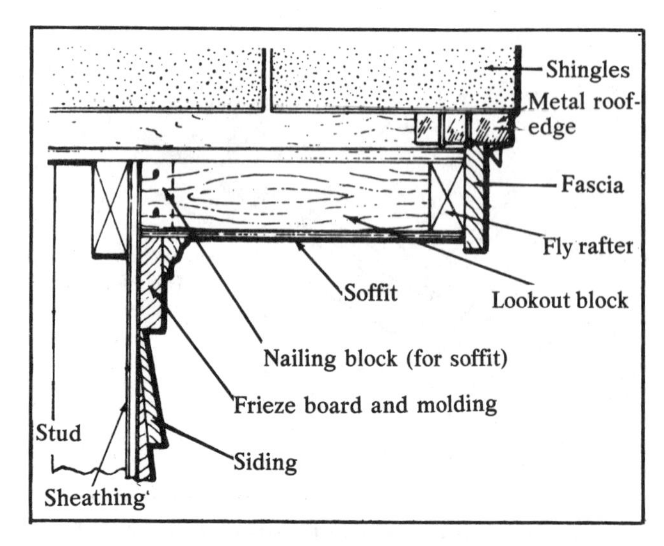

Moderate gable overhang
Figure 12-16

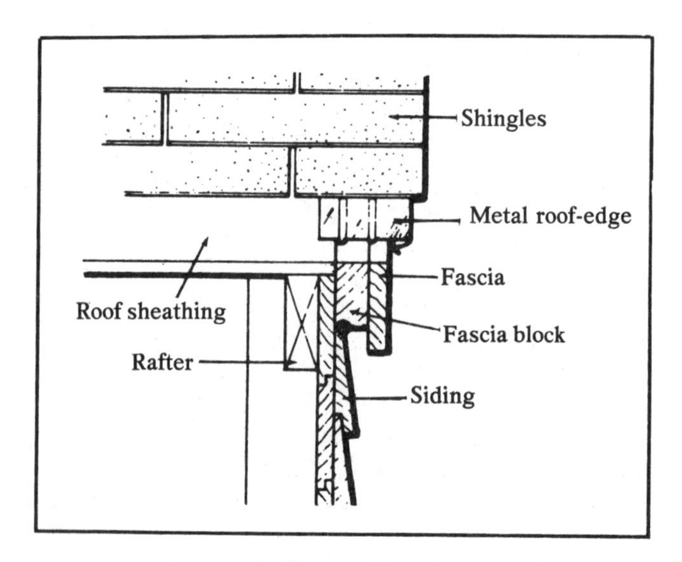

Close rake
Figure 12-17

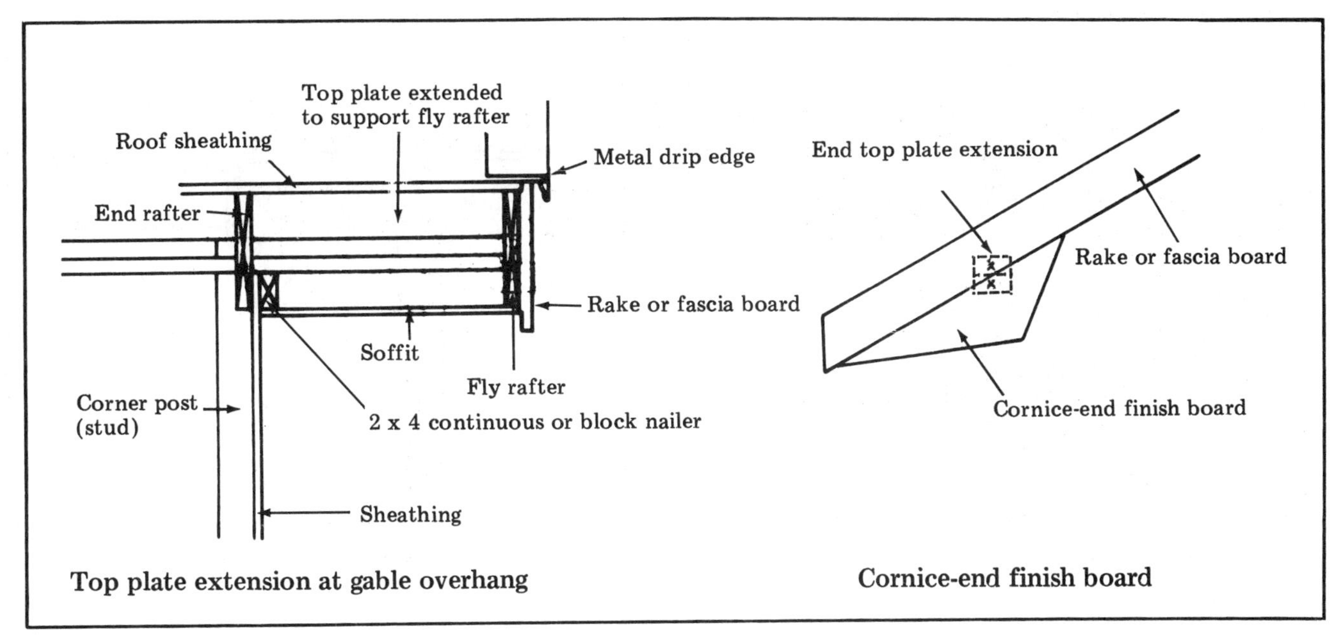

Boxed return for supported fly rafter
Figure 12-18

The roof sheathing is nailed to the fly rafter and to the lookout blocks which aid in supporting the rake section.

A close rake usually has no extension beyond the end wall other than the frieze board and moldings. Additional protection and overhang can be provided by using a 2 x 3- or 2 x 4-inch fascia block over the sheathing instead of a frieze board. See Figure 12-17. The fascia block acts as a frieze board against which the siding can be butted. The fascia, often 1 x 6 inches, serves as a trim member. Metal roof-edging is recommended for use as flashing along the rake section.

Where the rake overhang uses a supported fly rafter (as in Figure 8-11), the extension or overhang can be 30 inches or more. The sheathing is nailed to the rafter in the conventional manner. The soffit is fastened at the gable end to a continuous or block nailer. See Figure 12-18. The cornice-end finish board finishes the joint where the eave box cornice joins the rake box cornice.

Estimating Manhours

Time required for cornice trim work depends on the type of horizontal cornice and rake cornice treatments. For example, Figure 12-10 shows a box cornice with a frieze board on the face of the house, a soffit underneath the cornice and a fascia on the cornice face. The soffit and fascia are nailed to a nailing header. Additional trim, such as shingle molding, may be used on the cornice face. Molding may also be used on the frieze board.

Lookout members (as in Figures 12-10 and 12-16) are estimated on the same basis as wall bridging (solid blocking). See Figure 6-38 in Chapter 6. Figure 50 skilled hours per 1,000 BF.

When a fascia is used to finish off the gable end of a building, as shown in Figure 12-11, allow 4 hours per 100 LF.

When estimating manhours, list the linear feet of each item and establish a unit cost per 100 linear feet for each member. Figure 12-10 shows a cornice with three members (i.e., frieze board, soffit, and fascia). In addition, this cornice treatment has two separate runs of molding. The approximate manhours to install 100 linear feet of cornice trim work are shown in Figure 12-19.

A two-member closed cornice (including molding and frieze board) on a 40 x 24-foot house will take about 18 skilled manhours. Approximately 23 skilled hours will be required for a

Trim	Skilled Manhours Per 100 LF
2-member closed cornice	8
3-member boxed cornice	12
Molding	3
Barge board (verge) or separate frieze board	4

Cornice manhour requirements
Figure 12-19

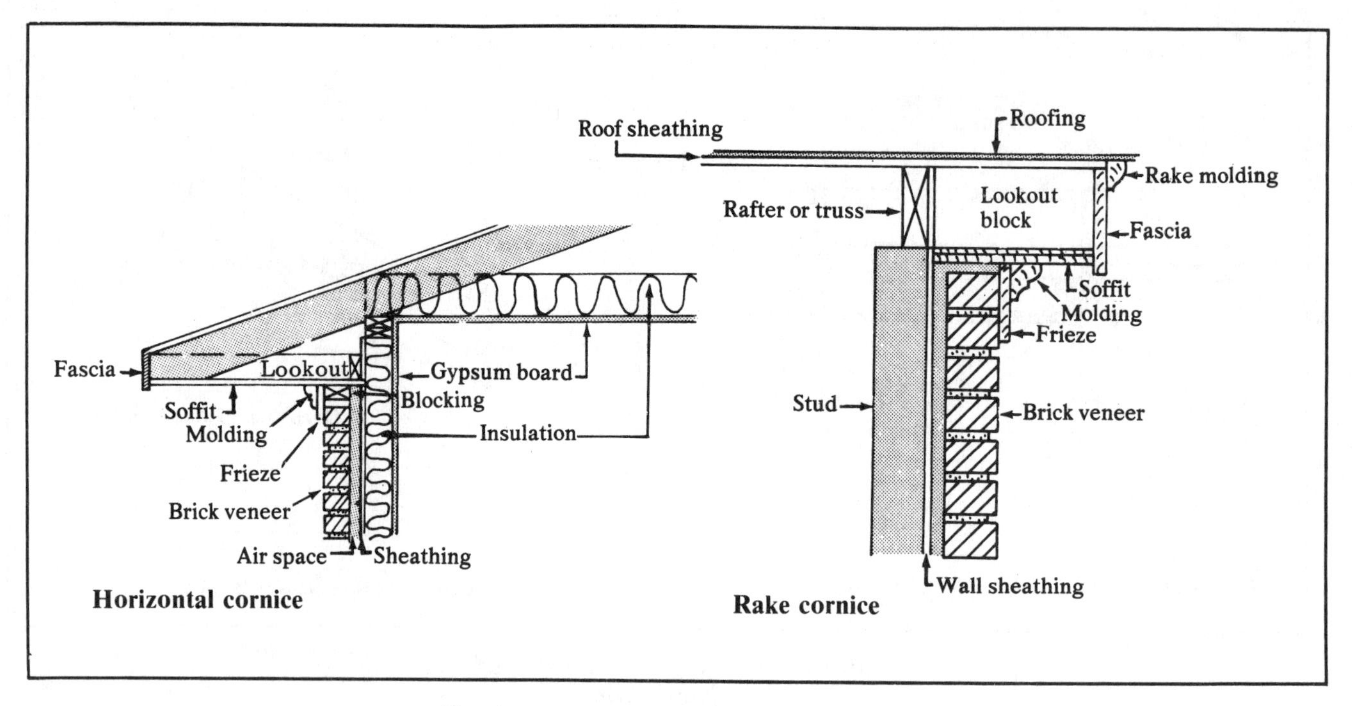

Cornice treatment on brick veneer house
Figure 12-20

three-member box cornice. The installation of lookout members in the horizontal and rake cornices will require about 5 skilled hours.

Cornice trim work must be done from ladders or scaffolding. Take this into consideration when estimating manhour requirements.

Remember that the weather can affect production. Also, no two crews will perform at the same rate. The estimates given here will serve as a general guide. Your manhour requirements will vary widely from job to job.

Summary

The style of the house determines the cornice treatment. The closed cornice offers you the opportunity to cut costs. The open cornice costs less, but buyers tend to prefer a closed or box cornice. The box cornice requires more materials and labor. The wide box cornice offers good protection for walls, windows and doors.

Exterior trim work for brick and masonry veneer houses is done the same way as for houses with siding. Many builders prefer to finish most of the cornice work on brick veneer houses before the brick is laid. The mason "lays to" the finished carpentry. Figure 12-20 shows a cornice treatment on a brick veneer house.

All exterior trim work is *finish* work. This is no place to get careless. Sloppy work is magnified in exterior trim. Take the time to do it right. A slight wave or curve can be spotted a block away. After the prime coat of paint has been applied, take the trouble to caulk all cracks and nail holes.

Figure 12-21 is your Checklist and Data Reference Sheet for exterior trim on horizontal and rake cornices. Use it the same way as previous checklists. It's your basis for accurate estimates on future jobs.

Horizontal and Rake Cornice Checklist and Data Reference Sheet

1) The house is ________ feet in length and ________ feet in width for a total of ________ square feet.

2) The house is ☐1 story ☐1½ story ☐2 story.

3) The roof pitch is ______/12 and the type of roof is:

☐Hip

☐Plain gable

☐Gable with dormers

☐Gable dormers. Number:________

☐Shed-type dormers. Number:________

☐Variable gable

☐Flat

☐Other (specify)_______________

4) The house has:

☐Masonry veneer

☐Siding

☐Combination masonry and siding

☐Other (specify)_______________

5) The cornice is ☐simple ☐open ☐closed ☐narrow box cornice (width: ________ inches) ☐wide box cornice (width: ________ inches) ☐cornice return.

6) The cornice is ☐2-member ☐3-member ☐with molding (number of runs: ________) ☐without molding.

7) The rake is ☐close ☐normal (______ inches) ☐moderate (________ inches) ☐extended overhang (________ inches) ☐open ☐closed.

8) A rafter nailing header (straightener) ☐was ☐was not installed.

9) Lookout members ☐were ☐were not installed at:

☐Eaves with spacing ☐16 inches o.c. ☐24 inches o.c. ☐32 inches o.c. and using ☐2 x 2 material ☐2 x 4 material ☐other (specify) ____.

☐ Rakes with spacing ☐24 inches o.c. ☐32 inches o.c. ☐48 inches o.c. and using ☐2 x 4 material ☐2 x 6 material ☐other (specify) ____.

10) Fascia board material is ☐cedar ☐redwood ☐fir ☐white pine ☐other (specify)_______________

Figure 12-21

11) Prefabricated soffit system ☐was ☐was not used. The prefab system is ☐aluminum ☐fiberboard ☐other (specify)______________.

12) Nails used in trim were ☐galvanized ☐stainless steel ☐aluminum ☐cadmium-plated ☐other (specify)______________.

13) Frieze ventilators ☐were ☐were not installed in the open cornice.

14) Soffit ventilators ☐were ☐were not installed in the box cornice.

15) Soffit material used:

☐⅜-inch exterior plywood

☐½-inch fiberboard

☐Lumber (size and species ______________)

☐Paper-overlaid plywood

☐Hardboard

☐Other (specify)______________

16) Material costs:

Lookouts	________BF	@	$________BF	=	$__________
Fascias	________LF	@	$________LF	=	$__________
Soffits:					
Prefab	________LF	@	$________LF	=	$__________
On-site	______LF/SF	@	$______LF/SF	=	$__________
Molding	________LF	@	$________LF	=	$__________
Nails	________lbs	@	$________lb	=	$__________
Ventilators	________pcs	@	$________ea	=	$__________

Other materials, not listed above:

____________________	__________	@	$__________	=	$__________
____________________	__________	@	$__________	=	$__________
____________________	__________	@	$__________	=	$__________
____________________	________	@	$__________	=	$__________

Total cost of all materials $__________

17) Labor costs:

Skilled ________hours @ ________hour = $____________

Unskilled ________hours @ ________hour = $__________

Other (specify)

____________________________ ________hours @ ________hour = $____________

Total $____________

18) Total cost:

Materials $____________

Labor $____________

Other $____________

Total $____________

19) Cost of trim per square foot of floor space $__________.

20) Problems that could have been avoided ____

____

____

____

21) Comments ____

____

____

____

22) A copy of this Checklist and Data Reference Sheet has been provided:

☐Management

☐Accounting

☐Estimator

☐Foreman

☐File

☐Other: ____________

Chapter 13

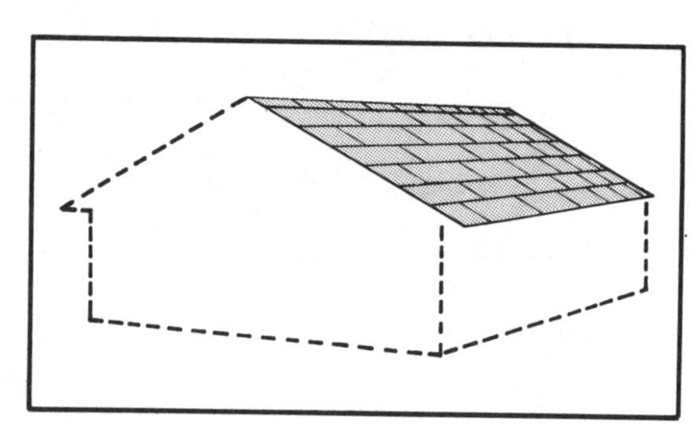

Roof Coverings

Roofing separates the professionals from the amateurs. Nothing puts a home builder in disrepute faster than a leaky roof. Roofing is one of the most important and yet most neglected parts of house construction.

Even some experienced roofers don't understand basic roofing principles. A few are downright ignorant on the subject. Even if you plan to subcontract all your roofing jobs, take the time to learn what makes a good roof and what creates problems.

The vast majority of houses are covered with asphalt roofing. So we'll give most of our attention to asphalt roofing, and then touch on other popular roofing materials.

The traditional supporting membrane for asphalt roofing is "felt." Thicker and more absorbent than conventional paper, felt is composed primarily of cellulose fibers made from recycled waste paper or converted wood chips. At one time, cotton or wool fibers from rags made up one third of the felt content, giving rise to the term "rag felt." These rag fibers haven't been used since 1942. However, you'll still hear the term "rag felt" used by some roofers.

In the late 1950's, inorganic base materials were introduced. Inorganic bases are made entirely of glass fibers. Today's improved technology makes these glass-fiber mat bases competitive with organic felt.

The thickness and weight of a glass-fiber mat is usually much less than that of organic felt. For example, glass fiber may be 0.03 inches thick, versus 0.055 inches for organic felt. And glass fiber may weigh 2 to 3 lbs. per 100 square feet, versus 12 lbs. per 100 square feet for organic felt.

Asphalt shingles with a glass-fiber base will probably replace organic fiber shingles entirely within the next 10 to 20 years.

Advantages of Asphalt Roofing

Asphalt roofing has been used for many years. Characteristics that make it so popular include:

1) Weather-resistance. Asphalt roofing resists heat, cold, water and ice.

2) Fire-resistance. Roofs are particularly vulnerable to fire. Many types of asphalt roofing resist fire better than wood shingle or shake roofing.

3) Wind-resistance. Asphalt roofing that bears the Underwriters Laboratories "wind-resistant" label has been tested to withstand gale-force winds. Wind-resistant shingles were originally developed for use only in high wind areas, but they are now available throughout the U.S.

4) Economy. High-volume production and the relatively low cost of application make asphalt shingles an economical choice for most homes.

5) Ease of application. Of all standard roofing materials, asphalt roofing is the easiest to apply.

6) Adaptability. Because of its flexibility and strength, asphalt roofing is appropriate for a wide variety of roof styles.

7) Beauty. It is available in many colors and depths which provide bold roof textures.

Figure 13-1 is a listing of typical asphalt shingles. Figure 13-2 is a listing of typical asphalt rolls.

Classifications A, B and C of the Underwriters Laboratories Listing (shown in Figures 13-1 and 13-2) indicate the ability of the roofing material to withstand external fires. The intensity of the test fire establishes the classification rating:

Class A - Severe exposure to fire
Class B - Moderate exposure to fire
Class C - Light exposure to fire

Figure 13-3 shows how a professional asphalt roofing job looks. This chapter explains how to do a professional roofing application like the one shown in Figure 13-3.

Selecting the Right Materials

No one product is the best for every job. On many jobs, there will be several alternatives.

Your choice of roofing material will depend on roof slope, coverage and exposure, wind conditions, color and texture requirements, and budget.

Roof Slope

Of all factors to consider, the most critical is the slope of the roof. It affects the surface drainage of water. Free drainage is essential for asphalt roofing. Good drainage makes the difference between a weathertight roof and one that leaks.

The slope of a roof is usually a function of style. A number of common roof styles are illustrated in Figure 13-4.

Use asphalt shingles for roof slopes that are between 4 inches and 21 inches per foot. Beyond this maximum slope, you'll have to follow steep-slope application procedures. Square-tab strip shingles can be used on slopes that are between 2 inches and 4 inches per foot if you observe special low-slope application rules.

Roll roofing is often used on nearly flat roofs. The minimum slope depends on the application method and the type of roll roofing. Most roll roofing can be used on roof slopes down to 2 inches per foot if applied by the exposed-nail method. With the concealed-nail method and at least 3 inches of top lap, you can go down to slopes of 1 inch per foot. Double-coverage roll roofing applied with a top lap of 19 inches can also be used on slopes down to 1 inch per foot.

Slope limitations for asphalt roofing materials are summarized in Figure 13-5.

Coverage and Exposure

Coverage— Describes the number of layers of material that lie between the deck and the exposed surface of the roof. For example, you may have single, double or triple coverage. Where the number of layers varies, coverage is based on the number of layers covering most of the roof. For example, if most of the roof is covered with two thicknesses of material, it would be considered double coverage.

Most asphalt roll roofing products are single-coverage materials. An exception is roll roofing applied with a 19-inch lap and 17-inch exposure. That qualifies as double-coverage material. Asphalt strip shingles are also considered double-coverage materials because their top lap is at least 2 inches greater than their exposure.

Exposure— Describes that portion of the roofing exposed to the weather after installation. The exposures for asphalt roofing products are specified by the manufacturer. Typical exposures are listed in Figures 13-1 and 13-2.

Wind Conditions

Self-sealing shingles or interlocking shingles are required wherever high winds are anticipated. Free-tab shingles don't have a factory-applied adhesive, but they can be made more wind-resistant by cementing each tab in place during installation.

Roof Colors and Textures

Be creative in your selection of shingle colors and textures. From down the block, most of what your potential buyers see is the roof line of the house. Make the roof color and texture important selling points by selecting one of the many attractive shingles now available.

Asphalt roofing shingles in blends of red, brown and green cost little more than the standard greys and off-whites. These newer colors help relate the building to a natural environment. They also complement the natural colors of brick walls and wood siding. Figure 13-6 is a convenient guide for choosing shingle colors compatible with siding, trim, shutters and doors.

Roof colors can create certain psychological effects. A small house looks bigger with a light-colored roof. This directs the eye upward and creates a sense of airiness. Dark colors on a tall or steeply sloped building tend to create the opposite effect. They bring the structure down to scale.

PRODUCT	Configuration	Per Square: Approximate Shipping Weight	Per Square: Shingles	Per Square: Bundles	Size: Width	Size: Length	Exposure	Underwriters Laboratories Listing
Self-sealing random-tab strip shingle 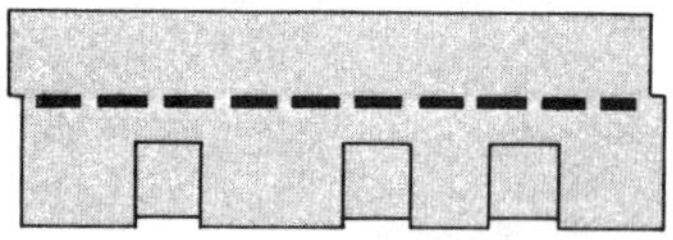 Multi-thickness	Various edge, surface texture and application treatments	285# to 390#	66 to 90	4 or 5	11½" to 14"	36" to 40"	4" to 6"	A or C - Many wind resistant
Self-sealing random-tab strip shingle 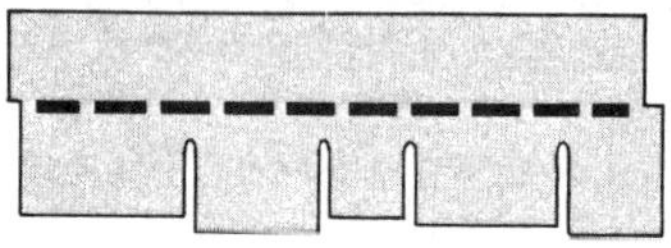 Single-thickness	Various edge, surface texture and application treatments	250# to 300#	66 to 80	3 or 4	12" to 13¼"	36" to 40"	5" to 5⅝"	A or C - Many wind resistant
Self-sealing square-tab strip shingle 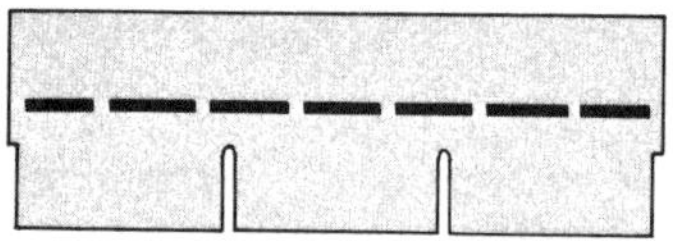 Three-tab	Two-tab or Four-tab	215# to 325#	66 to 80	3 or 4	12" to 13¼"	36" to 40"	5" to 5⅝"	A or C - All wind resistant
	Three-tab	215# to 300#	66 to 80	3 or 4	12" to 13¼"	36" to 40"	5" to 5⅝"	
Self-sealing square-tab strip shingle 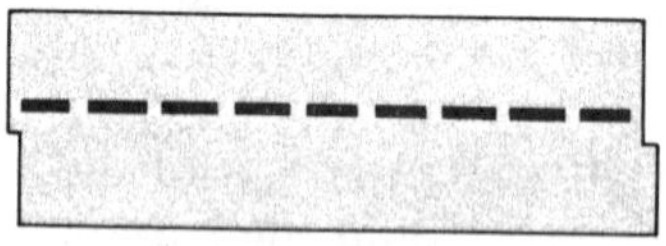 No-cutout	Various edge and surface texture treatments	215# to 290#	66 to 81	3 or 4	12" to 13¼"	36" to 40"	5" to 5⅝"	A or C - All wind resistant
Individual interlocking shingle 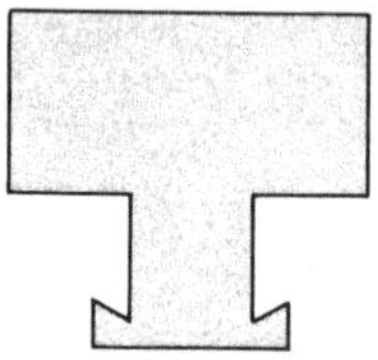 Basic design	Several design variations	180# to 250#	72 to 120	3 or 4	18" to 22¼"	20" to 22½"	—	C - Many wind resistant

Typical asphalt shingles
Figure 13-1

PRODUCT	Approximate Shipping Weight		Squares Per Package	Length	Width	Side or End Lap	Top Lap	Exposure	Underwriters Laboratories Listing
	Per Roll	Per Square							
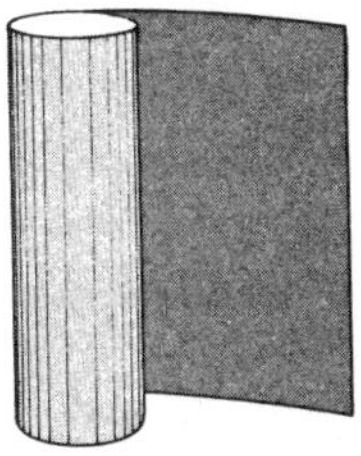 Mineral surface roll	75# to 90#	75# to 90#	1	36' to 38'	36"	6"	2" to 4"	32" to 34"	C
		Available in some areas in 9/10 or 3/4 square rolls.							
 Mineral surface roll (double coverage)	55# to 70#	110 # to 140 #	½	36'	36"	6"	19"	17"	C
 Smooth surface roll	40# to 65#	40# to 65#	1	36'	36"	6"	2"	34"	None
 Saturated felt (non-perforated)	60#	15# to 30#	2 to 4	72' to 144'	36"	4" to 6"	2" to 19"	17" to 34"	None

Typical asphalt rolls
Figure 13-2

Professional asphalt roofing job using multi-thickness self-sealing random-tab strip shingles
Figure 13-3

The use of laminated asphalt shingles can also contribute to the overall architectural effect of a building. Many of these shingles offer the "look" of wood but have the long-term wear and fire-resistance of asphalt. Others offer the "look" of slate and tile. All create interesting visual effects of light and shadow.

How Much Roofing is Required?

Shingles or roll roofing, starter strips, drip edges, hip and ridge shingles, and valley flashing are needed on most roofing jobs. Quantities are based on the dimensions and surface area of the roof. To find the surface area and dimensions of the roof, you will need to do a few fairly simple calculations.

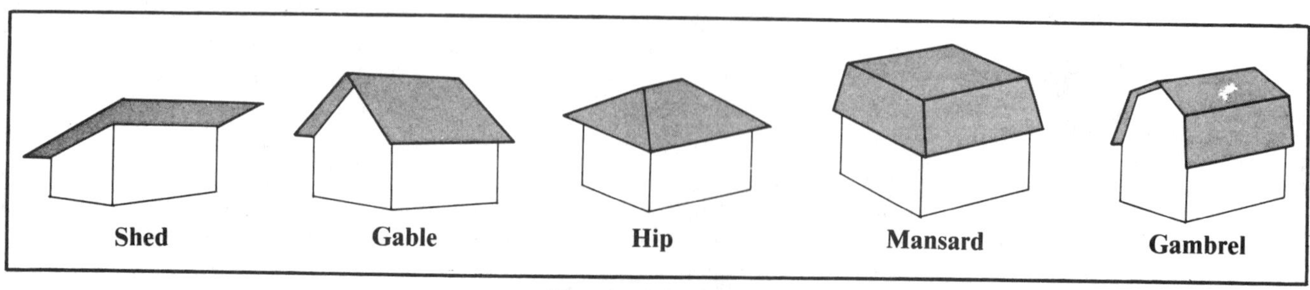

Common roof styles
Figure 13-4

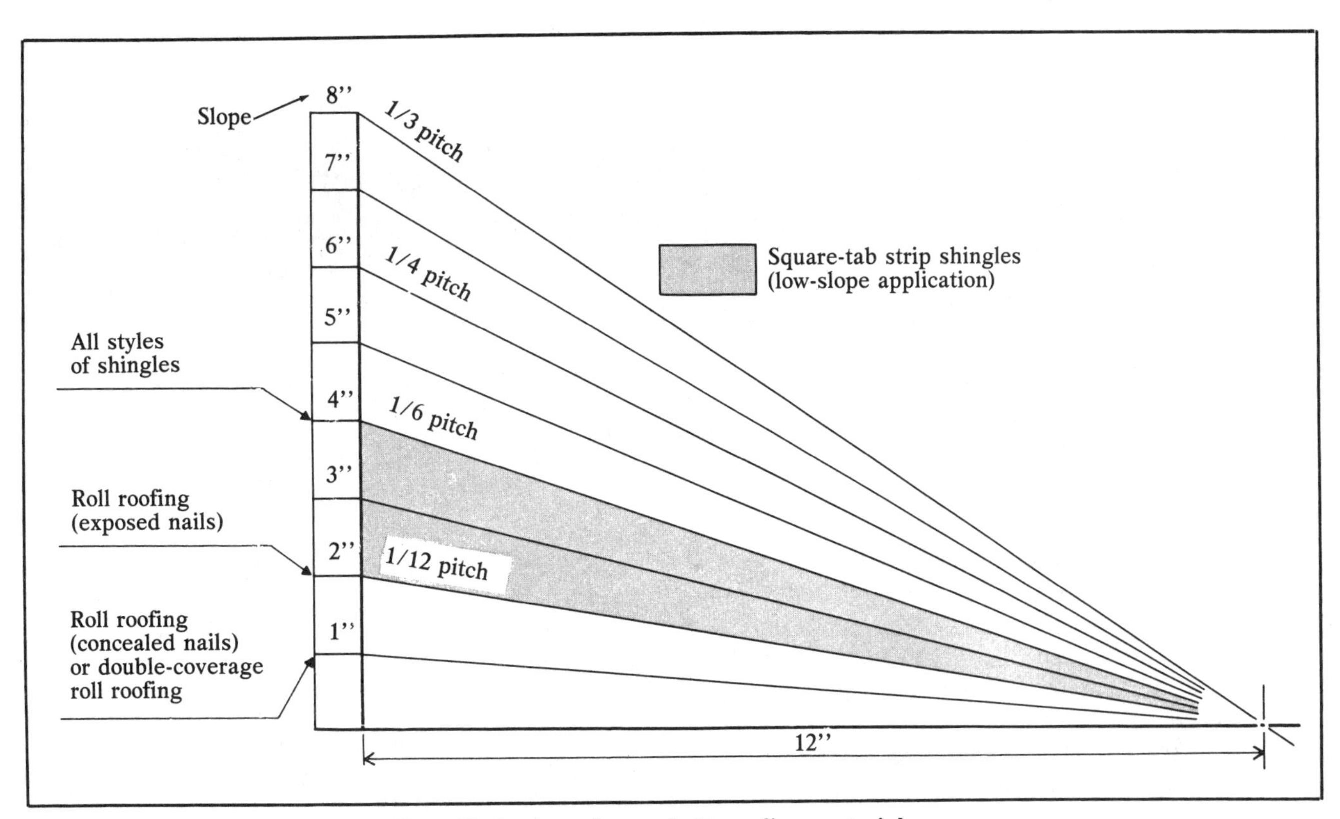

Slope limitations for asphalt roofing materials
Figure 13-5

Roof Shingles	Siding	Trim	Shutters and/or Doors
White	White	White	Deep Gold, Maroon
	White	Grey	Charcoal
	Green	White	Dark Brown, Dark Green
Black	White	White	Black, Maroon
	Yellow	White	Black, Deep Olive Green
	Gold	White	Black, Deep Olive Green
Grey	Red	White	Black, White
	Yellow	White	Grey, Charcoal, Green
	Coral Pink	Light Grey	Charcoal
Red	White	Grey	Charcoal
	White	White	Red
	Beige	White	Dark Brown
Brown	White	White	Dark Brown, Terra Cotta
	Green	White	Dark Brown, Dark Green
	Yellow	White	Dark Brown, White
Green	White	White	Dark Green, Black
	Yellow,	White	Dark Green
	Light Green	White	Dark Green, Terra Cotta

Asphalt shingle color guide
Figure 13-6

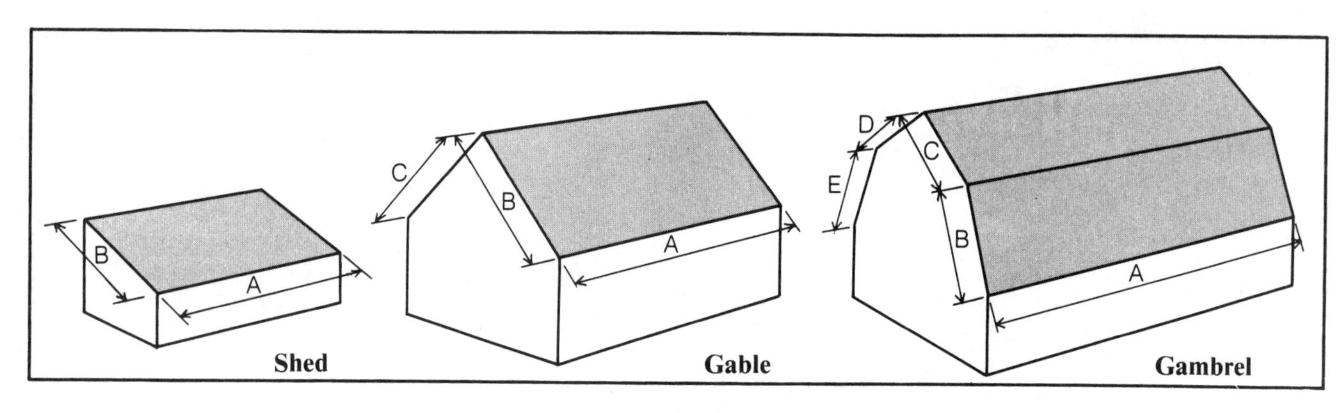

Simple roofs
Figure 13-7

Estimating Surface Area for Simple Roofs
Although roofs come in many sizes, shapes and styles, every roof is composed of flat surfaces. To figure the total roof surface area, first divide the roof into simple geometric planes: squares, rectangles, trapezoids and triangles.

The simplest roof has no projecting dormers or intersecting wings. Examples of this roof type are shown in Figure 13-7. Each of these roofs is made up of rectangles. In each case, the total roof area is equal to the sum of the areas of the rectangles included in the roof.

The shed roof has only one rectangle. The area is found by multiplying the eave line (A) times the rake line (B).

The gable roof has two rectangular planes. The area of the gable roof is found by multiplying the eave line times the sum of the two rake lines, or A times the sum of B plus C.

For the gambrel roof, four rake lines are involved. The area is found by multiplying the eave line times the sum of the four rake lines, or A times the sum of B plus C plus D plus E.

Estimating Surface Area for Complex Roofs
Complex roofs have projecting dormers or intersecting wings. Area calculations for these roofs use the same basic method as for simple roofs. The work is just a little more complicated because there are more roof surfaces to add up. Each surface is calculated separately. Then all are added together to find the total roof surface area.

If the plans of the building are available, use them. Otherwise, you can measure the distances right on the roof.

There's still another way to calculate the surface area for complex roofs. It allows you to measure the area indirectly by calculating the projected horizontal area of the roof. Then you use a roof slope conversion table to convert the horizontal area to the true surface area. Both the roof slope and the projected horizontal area can be determined indirectly as described below. Tables are included for converting indirect measurements to true surface area.

Roof pitch and slope— The degree of incline of a roof is usually expressed either as "pitch" or "slope." Pitch is the ratio of rise to span. For example, if the rise of a roof is 8 feet and the span is 24 feet, the pitch is 8/24 or 1/3. See Figure 13-8. Slope is the ratio of rise (expressed in inches) per foot of horizontal run. The run equals half the span. Using the same roof as an example, the slope would be 8 inches per foot of horizontal run. If the rise of the same roof span were 6 feet, the pitch would be 1/4 and the slope would be 6 inches per foot of run.

The pitch can also be expressed in inches of rise per foot of run. For example, 5/12 means that the rise is 5 inches per 12-inch run. Regardless of whether a particular roof incline is expressed in pitch or slope, the area is the same.

You don't have to get up on the roof to measure pitch or slope. With the aid of a carpenter's folding rule, it can be approximated from the ground.

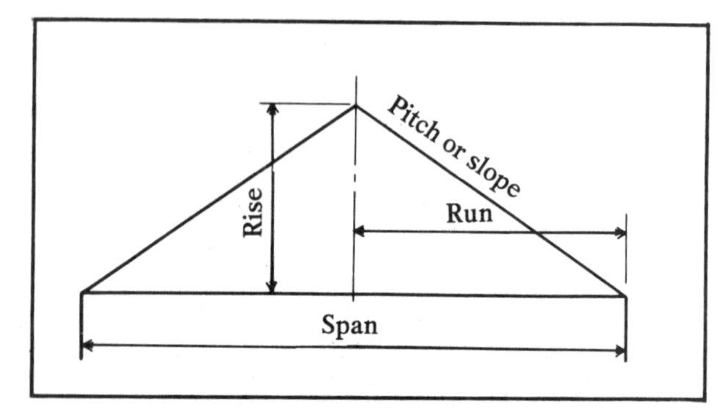

Pitch and slope relationships
Figure 13-8

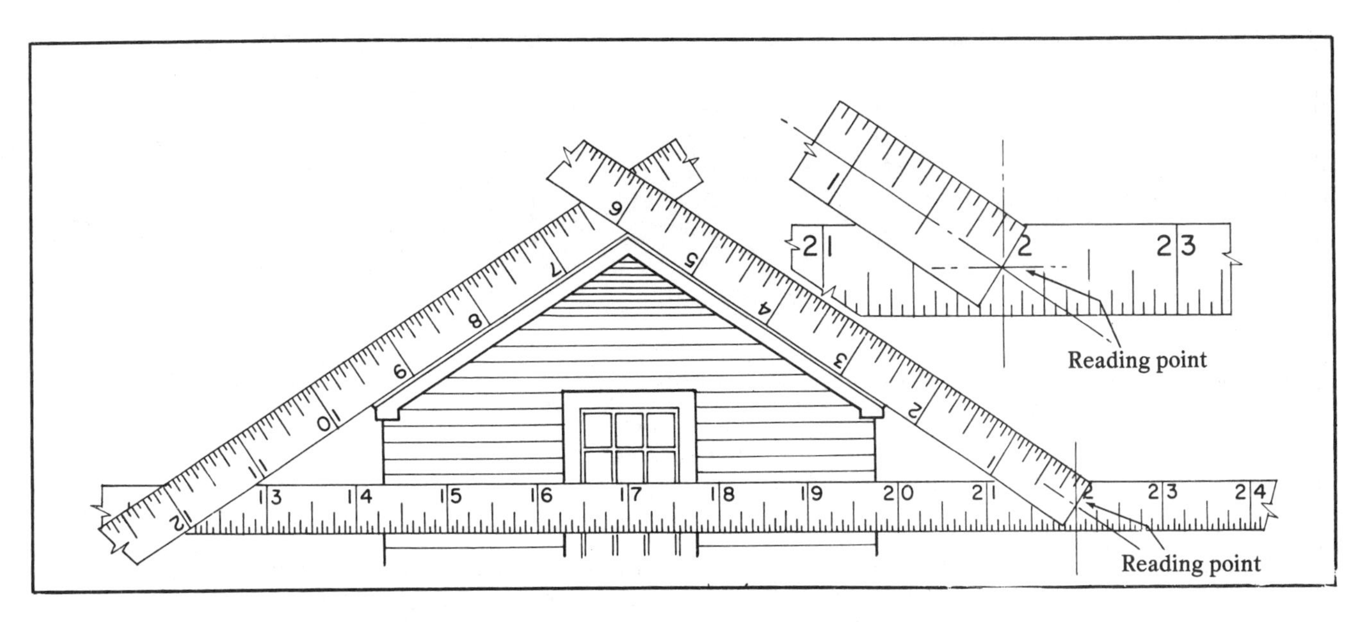

Use of carpenter's folding rule to determine pitch and slope
Figure 13-9

Start by standing away from the building. Form the rule into a triangle with the 6-inch joint at the apex and the 12-inch joint at the left side of the horizontal base line. See Figure 13-9. Holding the rule at arm's length, line up the sides of the triangle with the roof, as shown in Figure 13-9. Be sure to keep the base of the triangle horizontal.

Then, align the center of the zero-inch mark with the center of the intersecting mark on the right-hand side of the horizontal base line. In Figure 13-9, this occurs at the 22-inch mark. Next, locate the nearest "rule reading," as shown in Figure 13-10. Read the pitch and slope of the roof. In this example, the pitch is 1/3 and the slope is 8 inches per foot.

Projected horizontal area— No matter how complicated a roof is, its projection onto a horizontal plane will show you the total horizontal surface covered by the roof. Figure 13-11 shows a roof complicated by valleys, dormers and ridges at different elevations. The lower half of the figure shows the projection of the roof onto a horizontal plane. In the projection, inclined surfaces appear flat and intersecting surfaces appear as lines.

Measurements for the horizontal projection of the roof can be made from the plans, from the ground or from inside the attic. Once the measurements are made, the horizontal area covered by the roof can be drawn to scale and calculated.

Because surface area is a function of slope, calculations must be grouped by roof slope. Don't combine calculations for different slopes until the

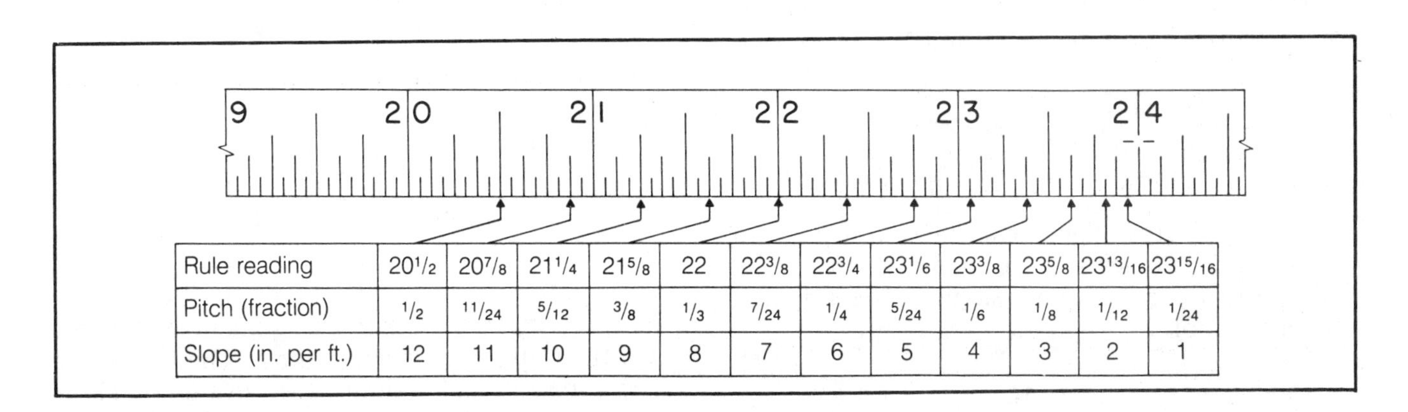

Rule reading	20 1/2	20 7/8	21 1/4	21 5/8	22	22 3/8	22 3/4	23 1/6	23 3/8	23 5/8	23 13/16	23 15/16
Pitch (fraction)	1/2	11/24	5/12	3/8	1/3	7/24	1/4	5/24	1/6	1/8	1/12	1/24
Slope (in. per ft.)	12	11	10	9	8	7	6	5	4	3	2	1

Rule reading: conversions to pitch and slope
Figure 13-10

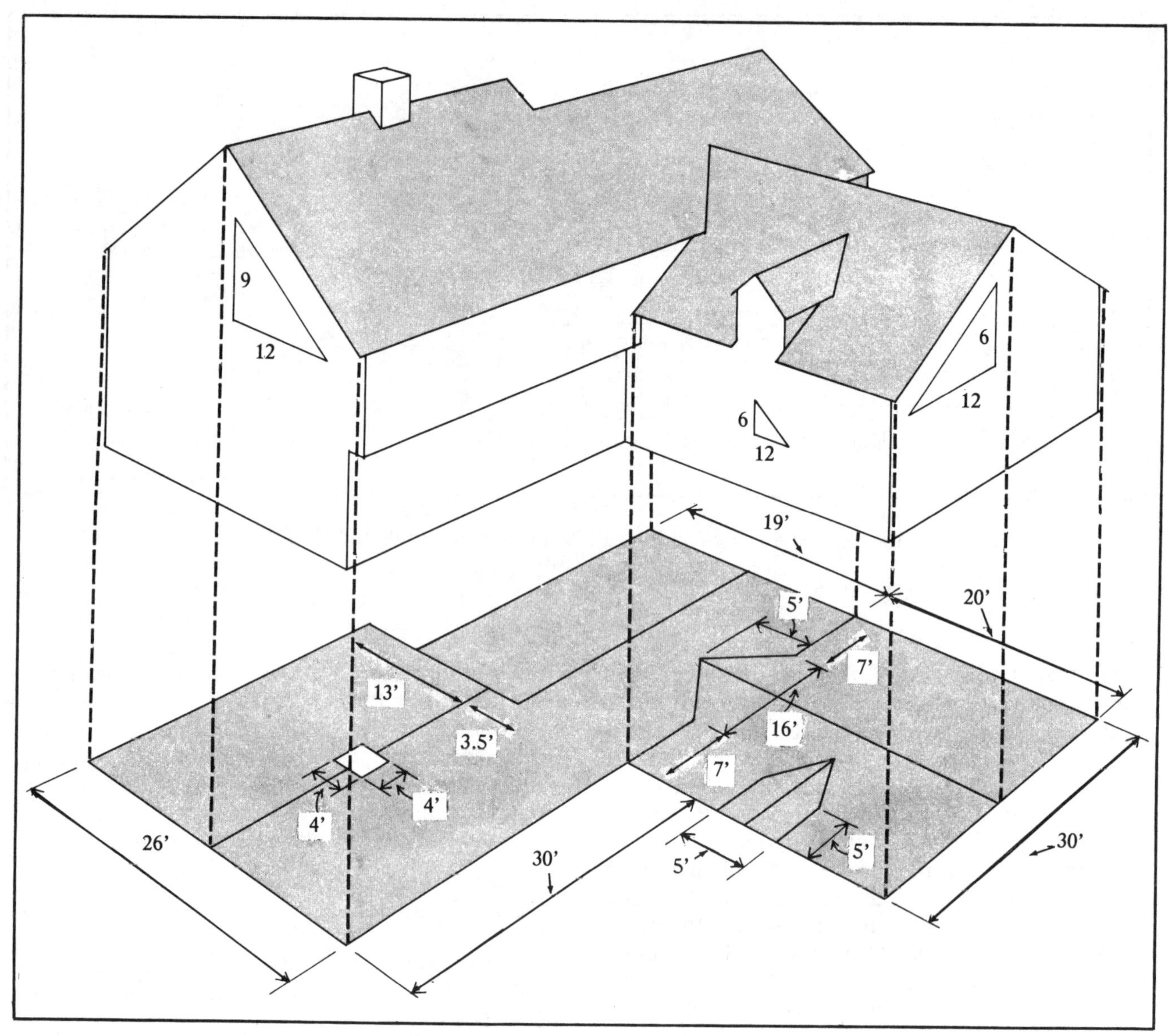

Horizontal projection of complex roof
Figure 13-11

area of each has been determined. Sample calculations for the roof in Figure 13-11 are given below. The horizontal area under the 9-inch-slope roof is:

26	x 30	=	780	
19	x 30	=	570	
Total		=	1350	square feet

From this gross figure you have to deduct the area of the chimney and the triangular area of the ell roof that overlaps and is sloped differently than the main roof:

Chimney	=	16	
Ell roof ½ (16 x 5)	=	40	(triangular area)
		56	square feet

The net projected area of the main roof is:

1,350 - 56 = 1,294 square feet

The horizontal area under the 6-inch-slope roof is:

20 x 30	=	600	
½ (16 x 5)	=	40	
Total	=	640	square feet

Don't forget to include the duplications. Portions of higher roof surfaces sometimes project over the roof surfaces below them. The horizontal projection doesn't show the overlap. These duplicated areas must be added to the total horizontal area. In Figure 13-11, there are three overlaps:

1) On the 6-inch-slope roof where the two dormer eaves overhang the ell roof.

2) On the 9-inch-slope roof where the main-roof eave overhangs the ell section.

3) On the 9-inch-slope roof where the main-roof rake overhangs the smaller section of the main roof in the rear of the building.

In each case, if the overhang extends 4 inches beyond the structure, the duplications are calculated as follows:

1) 2(5 x 4/12) = 3⅓ square feet. Add this to the horizontal area of the 6-inch-slope roof.

2) 2(7 x 4/12) = 4⅔ square feet. (The ell section is 30' wide. Sixteen feet of the ell roof connects with the main roof, leaving a 7' eave overhang on each side of the ell ridge.) Add the 4⅔ to the horizontal area of the 9-inch-slope roof.

3) 9.5 x 4/12 = 3-1/6 square feet. (Overhang covers only half of the 19-foot-wide section.) Add this to the horizontal area of the 9-inch-slope roof.

For the 6-inch-slope roof, the adjusted total is 640 plus 3, equals 643 square feet. For the 9-inch-slope roof, the adjusted total is 1,294 plus 8, equals 1,302 square feet. (Fractions are rounded off to the nearest foot.)

Conversion to true surface area— Once you know the projected horizontal area for each roof slope, the next step is to convert the results to true surface area. The conversion table in Figure 13-12 handles this for you.

Slope (Inches Per Foot)	Area/Rake Factor
4	1.054
5	1.083
6	1.118
7	1.157
8	1.202
9	1.250
10	1.302
11	1.356
12	1.414

Area/rake conversion table
Figure 13-12

To use the table, simply multiply the projected horizontal area by the conversion factor for the appropriate roof slope. The result is the true surface area of the roof.

For example, for the 9-inch-slope roof:

Horizontal area	x	Conversion factor	=	Actual area
1,302 SF	x	1.250	=	1,627.5 SF

For the 6-inch slope roof:

Horizontal area	x	Conversion factor	=	Actual area
643 SF	x	1.118	=	718.8 SF

When all horizontal areas have been converted to true surface areas, total the surface areas. The sum is your total roof area: 1,628 plus 719, equals 2,347 square feet.

You'll want to make an allowance for waste when ordering materials. In this case, assume 10% waste. Thus, the roofing material required is: 2,347 plus 235, equals 2,582 square feet.

The same roof slope and same horizontal area will always result in the same true surface area, regardless of roof style. In other words, if a shed roof, gable roof or hip roof with or without dormers each had the same slope and covered the same horizontal area, they would each require the same amount of roofing material to cover them.

Additional Material Estimates

To complete the estimate, find the necessary quantities of starter strips, drip edges, hip and ridge shingles, and valley flashing. Each of these quantities depends on the length of the eaves, ridges, rakes, hips, and valleys.

Eaves and ridges are horizontal. Their length can be scaled directly from the horizontal projection drawing or from your plans. Rakes, hips and valleys are sloped. Thus, their lengths must be calculated using a procedure similar to the one you used for calculating sloped surface areas.

To find the true length of a rake, first measure its projected horizontal length. Then use Figure 13-12 to convert projected horizontal length to true length. To use the table, multiply the rake's projected horizontal length by the conversion factor for the appropriate roof slope. The result is the true length of the rake.

For the house in Figure 13-11, the rakes at the ends of the main house have horizontal distances of 26 and 19 feet. There's another rake in the middle of the main house where the higher roof section meets the lower. Its horizontal distance is: 13 plus 3.5 for the short rake, equals 16.5 feet. Adding all these horizontal distances together gives a total of 61.5 feet. Then use the conversion table in Figure 13-12:

Horizontal length	x	Conversion factor	=	Actual length
61.5 feet	x	1.250	=	76.9 feet

Follow the same procedure for the ell section with its 6-inch-slope roof and dormer. The horizontal length of rakes is found to be 35 feet.

To find the quantity of drip edge required to do the job, add these rake lengths to the sum of the lengths of the eaves. The eave lengths are true horizontal distances, so no conversion is necessary.

The quantity of ridge shingles required is taken directly from the drawings, since ridge lines are true horizontal distances.

Hips and valleys again involve sloped distances. Convert their projected horizontal lengths to true lengths with the aid of Figure 13-13.

Slope (Inches per Foot)	Hip/Valley Factor
4	1.452
5	1.474
6	1.500
7	1.524
8	1.564
9	1.600
10	1.642
11	1.684
12	1.732

Hip/valley conversion table
Figure 13-13

First, measure the length of the hip or valley on the horizontal projection drawing. Multiply that figure by the conversion factor for the appropriate roof slope. The result is the true length of the hip or valley. The estimate for hip shingles or valley flashing material can be made from the sum of the true lengths of hips or valleys on the house.

Now we'll find the sum of the lengths of the valleys for the house in Figure 13-11.

There's a valley formed on both sides of the ell-roof intersection with the main roof. The total measured distance of these valleys on the horizontal projection is 16 feet.

The fact that two different slopes are involved makes the procedure a little more complicated. If there was only one roof slope, the true length could be calculated directly from Figure 13-13. But in this case, you need to calculate both slopes and then average them together to approximate the true length of the valleys. Thus:

Horizontal length	x	Conversion factor	=	Actual length
16 feet	x	1.600 (for 9" slope)	=	25.6 ft.
16 feet	x	1.500 (for 6" slope)	=	24.0 ft.
		Average: (24.0 + 25.6)/2	=	24.8 feet

The approximate true length of the two valleys is 24.8 feet, or 12.4 feet each.

The projected horizontal length of the dormer valleys in Figure 13-11 is 5 feet. Since both the ell roof and the dormer roof have slopes of 6 inches, the actual length of the valleys will be 7.5 feet, using Figure 13-13. The projected horizontal length of valleys is the distance at the widest point between the valleys. In the case of dormer valleys, it's the width of the dormer.

Total true valley length for the house is: 25 plus 8, equals 33 feet.

Preparing for a Roofing Job

Every roofing job requires more than just shingles and flashing. You'll need tools, cements and coatings, and fasteners. And now's a good time to plan for ventilation and the storage of materials at the job site.

Tools

All roofing work requires a few basic tools. These include:

1) Ladders and scaffolding for access to the roof, for carrying up materials, and for safe footing, especially when applying the starter strip and first course

2) Folding and tape rules for making measurements

3) Chalk reel for snapping the chalk lines used to align roofing materials

4) Roofing knife for cutting, shaping and fitting materials

5) Hammer or roofer's hatchet. The hatchet doubles as a tool for aligning shingles

6) Putty knife, pointed trowel or brush for applying asphalt cement

7) Caulking gun for applying continuous beads of asphalt cement

8) Broom for cleaning the deck before applying roofing materials and for cleaning up after the job is done

9) Chisel and saw for repairing or replacing damaged decking

Cements and Coatings

These asphalt-based materials are generally used as sealants and adhesives. All of these cements and coatings are highly flammable. They should never be heated over an open fire or placed in direct contact with a hot surface. If you need to soften the cement before application, put the unopened containers in hot water or store them in a warm place until ready for use.

Apply asphalt cements and coatings only on clean, dry surfaces. To eliminate air bubbles and force the material into all cracks and openings, trowel or brush the material onto the surface.

Any time you use asphalt cement, follow the manufacturer's recommendations. And don't use too much cement. Thick layers will crack and blister in the sun. Use just enough to do a neat, clean job.

Asphalt plastic cement— Also known as flashing cement, this material is generally applied to flashings where the roof meets a wall, chimney, vent pipe or other vertical surface. It won't flow at high surface temperatures during the summer. Asphalt cement remains soft and workable even at low temperatures.

Lap cement— Use this material to create a watertight bond between overlapping layers of roll roofing. When driving an exposed nail through felt bonded with lap cement, the nail should pass through the cement so that the shank is sealed where it penetrates the deck.

Quick-setting asphalt cement— This material can be applied by brush, trowel or gun. It's made with a solvent that evaporates quickly in the air. That makes it set up very quickly. Use quick-setting cement to bond the free tabs of strip shingles and roll-roofing laps that are applied by the concealed-nail method.

Masonry primer— Use this asphalt primer to prepare masonry surfaces for bonding with other asphalt products, such as built-up roofing components, asphalt plastic cements or roof coatings. The primer penetrates the masonry surface pores and forms a bond with both the masonry and the material applied on top of it.

Roofing tapes— Roofing tapes are made from asphalt-saturated cotton, glass fiber or other porous fabric. They're used in conjunction with asphalt cements and coatings for flashings and for patching seams, breaks and holes in metal and asphalt roofs. The tapes are available in rolls up to 50 yards long and 4 inches to 36 inches wide.

Fasteners

Both nails and staples are used to secure asphalt roofing to the deck. Nails can be used in most situations. Staples are acceptable *only* on new construction, and only if they meet the requirements discussed below.

Nails— Roofing nails are made of steel or aluminum. Steel roofing nails are zinc-coated for corrosion protection.

Roofing nails usually have barbed or deformed shanks, and are 11- or 12-gauge with large heads 3/8-inch to 7/16-inch in diameter. Be sure that the nails are long enough to penetrate through the roofing material and at least 3/4-inch into the deck. Figure 13-14 gives recommended nail lengths for various applications.

Application	Nail Length
Roll roofing on new deck	1"
Strip or individual shingles on new deck	1¼"
Reroofing over old asphalt roofing	1½ to 2"
Reroofing over old wood shingles	2"

Recommended nail lengths
Figure 13-14

The number and location of nails will be specified by the roofing manufacturer.

Staples— Staples can be used on wind-resistant shingles with factory-applied adhesives. You can use them on new construction, or on reroofing if the old roof has been torn off down to the wood deck.

Roofing staples are usually zinc-coated and are at least 16-gauge with a minimum crown width of 15/16-inch. Be sure that the staple is long enough to penetrate 3/4-inch into the deck lumber.

The staples should be driven so that the entire crown bears tightly against the shingle but doesn't cut into the shingle surface. This is critical. A roofer with an improperly adjusted stapling gun can do hundreds of dollars damage in a very short period of time. Sealing failure, raised tabs, buckling, leaks and blow-offs can result from incorrect stapling.

Ventilation

Good ventilation is an important part of every roof. Poor ventilation will cause blistering and buckling.

Good insulation, weatherstripping and caulking tend to make a home more airtight. An airtight home will keep the water vapor confined to the inside of the house. During winter, when warm moist air comes in contact with the cold underside of the roof deck, water vapor may accumulate under the deck. Vapor barriers reduce the passage of water vapor to the roof, but they don't eliminate it entirely.

With proper ventilation, air circulates freely under the roof deck, carrying away the water vapor before it can condense.

Some builders use a combination of ridge vents and eave or soffit vents to create a natural draft from the bottom to the top of the attic space. Another good choice is a combination of gable vents and eave or soffit vents. If you don't want to use eave or soffit vents, properly-sized gable vents alone will usually be enough. An electrically powered vent, controlled by a thermostat and humidistat switch, can also do a good job of reducing accumulated moisture.

If you use eave and soffit vents, be sure that they are not blocked by insulation.

Here are some standards to follow when sizing vents. If the attic has no vapor barrier, use a minimum of 1 square foot of free vent opening for each 150 square feet of attic surface. If there is a vapor barrier, and half of the vent openings are along the ridge and the other half are at the eave or soffit, use 1 square foot of free vent opening for each 300 square feet of attic space. On mansard roofs, triple these ventilation ratios.

When calculating the net area for the free vent opening, be sure to take screening into account. Screens reduce the air circulation through the vent. A No. 8 coarse screen will reduce free vent area by about 1/3. Ventilation is covered in more detail in a later chapter.

Storage of Materials at the Job Site

Be sure to protect your asphalt roofing materials from pre-installation damage from bad weather or improper handling. Here are some cautions to observe:

1) Never stack shingle bundles more than 4 feet high.

2) Don't store asphalt roofing products in direct contact with the ground. Put them on a raised flat platform.

3) Cover asphalt roofing materials with plastic or tarps that are vented to provide free air circulation.

4) Never store shingles in the direct hot sun.

Preparing for Asphalt Strip Shingle Installation

Roofing materials are only as good as the quality of their installation. And of all roofing materials, asphalt shingles are probably the easiest to install. But there are some critical points that need special attention: drip edges, underlayment, chalk lines, eave flashing and valley flashing. Let's look at a shingle roofing job in detail.

Drip Edges

Drip edges have to be applied before the roofing is put down. They help shed water at the eaves and rakes, and protect the underlying wood from rotting. Drip edges should be made of a corrosion-resistant material that extends approximately 3 inches back from the roof edge and bends downward over the roof edge. Along the eaves, apply drip edges directly to the deck. Along the rakes, apply drip edges over the underlayment. Figure 13-15 shows drip-edge application details.

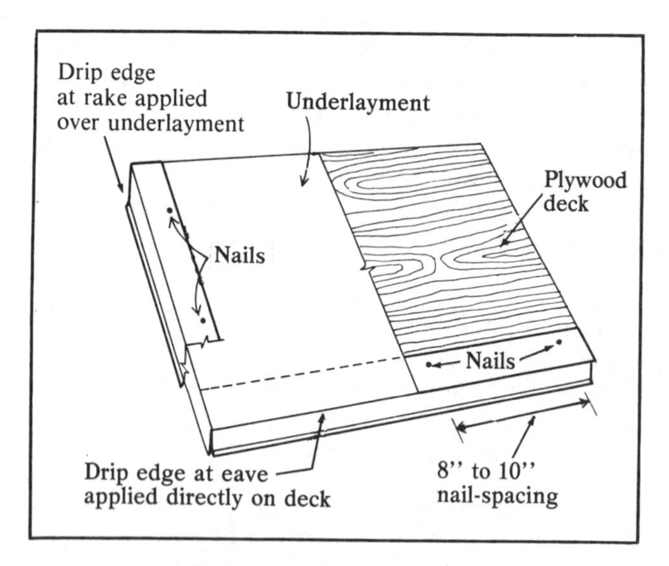

Drip-edge application
Figure 13-15

Underlayment

After the deck has been properly prepared, cover it with an asphalt-saturated felt underlayment. Apply the felt when the deck is dry.

On decks with a slope of 4 or more inches per foot, the underlayment should be one layer of nonperforated 15-lb. asphalt-saturated felt. Some codes require 30-lb. asphalt-saturated felt. Don't

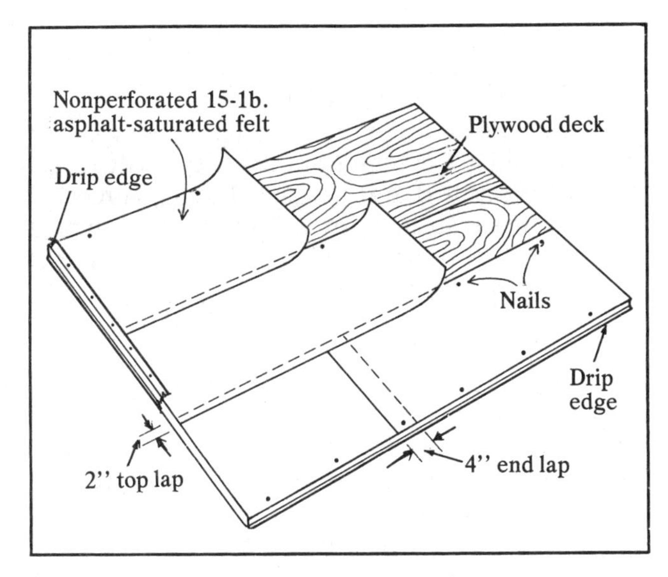

Underlayment application
Figure 13-16

use coated felts, tar-saturated materials, or laminated waterproof papers. These could act as vapor barriers and trap moisture or frost between the roof covering and the roof deck.

Lay the felt parallel to the eave, and top-lap each course at least 2 inches over the underlying course. Use the minimum number of nails necessary to hold it in place. See Figure 13-16. If two or more pieces are required in one course, lap the ends at least 4 inches. End laps in higher courses should be at least 6 feet from end laps in lower courses. Lap the felt 6 inches from both sides over all hips and ridges. Where the roof meets a vertical surface, carry the underlayment 3 or 4 inches up the surface.

Asphalt-saturated felt underlayment should always be used on new construction. It serves two important functions:

1) It keeps the deck dry until shingles are installed. This avoids the problems you get when shingles are laid on wet lumber or plywood.

2) If shingles are lifted, damaged or torn by the wind, the felt provides a secondary protection against wind-driven rain.

Chalk Lines
You'll find that it's pretty hard to lay a straight line of shingles across the roof unless you snap a chalk line as a guide. The shingles vary slightly in size and shape. On anything but a narrow roof, use a chalk line to keep the rows in good horizontal and vertical alignment.

Start with a horizontal chalk line parallel to the eave. Measure the appropriate distance up the roof and make a mark at three locations: at each end and in the middle. Drive a nail on the mark at each end. Stretch the chalk line between the nails and pull it taut. Check alignment at the middle mark, then snap the line from the center.

Vertical chalk lines are important for aligning cutouts from eave to ridge. They're also important for aligning shingles on each side of a dormer. As the line of shingles passes above the dormer, the shingles and cutouts should meet above the dormer without any gaps or overlaps.

On long runs, snap a vertical chalk line in the center of the run and begin each row of shingles to the left and right of that line. As shingle application approaches the ridge, check your horizontal chalk lines to make sure that the upper courses will be parallel to the ridge.

Eave Flashing
Ice dams form when snow melts and backs up under the roofing. It can do serious damage to ceilings, walls and insulation. Eave flashing helps prevent ice dams.

Install eave flashing wherever there is a good chance of icing along the eaves. The type of flashing material and the width of the flashing strip depend on the climate.

On roofs with a slope of 4 or more inches per foot, install a course of smooth surface roll roofing parallel to the eave and overhanging the drip edge by 1/4-inch to 3/8-inch. Use at least 50-lb. roofing material. Apply this flashing strip up the roof to a point at least 12 inches beyond the interior wall line. If a second flashing strip is required to reach that point, make sure that the lap is in front of the exterior wall line. Overlap the flashing strips at least 2 inches, and cement the horizontal joint the entire length of the roof. See Figure 13-17. Any end laps should overlap by 12 inches and should be sealed with cement.

If you anticipate severe icing, extend the flashing strip up the roof to a point at least 36 inches beyond the interior wall line, and seal it to the underlayment with a smooth coat of plastic cement applied in a ratio of two gallons per 100 square feet. Press the flashing strip firmly into the cement.

Valley Flashing
The valley is where two sloping roof surfaces meet to form an interior angle. During a heavy rain, water from the two sloping roof surfaces meets and runs off at the valley joint. This makes the valley especially vulnerable to leaks. Good roof performance depends on proper valley flashing.

First, install the underlayment in the valley. Center a 36-inch-wide strip of 15-lb. asphalt-saturated felt in the valley, and secure it with the

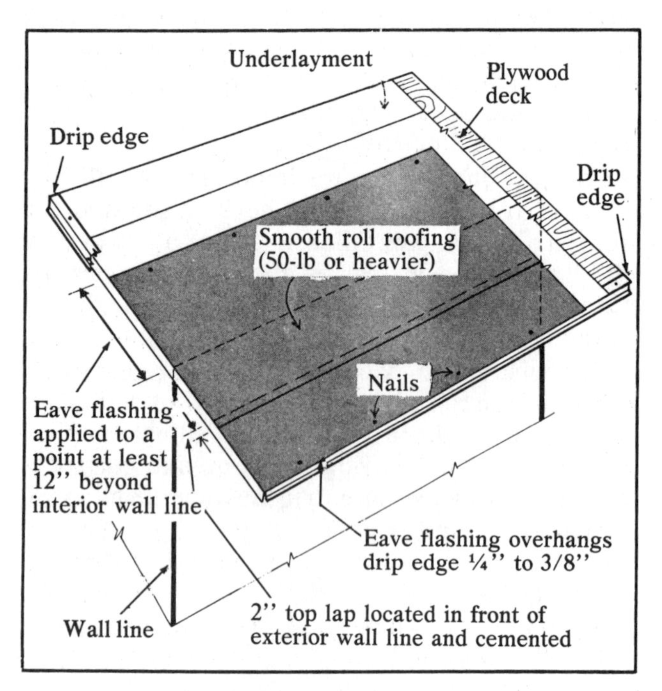

Eave flashing application
Figure 13-17

minimum number of nails necessary to hold it in place. Then trim the horizontal courses of felt underlayment already applied to the roof so that they overlap the valley strip at least 6 inches. See Figure 13-18.

Next comes the construction of the valley. There are three types of valleys: open, woven, and closed-cut. Open valleys can be used with most types of shingles and roll roofing. Woven and closed-cut valleys are used with strip shingles. Regardless of the type of valley you use, make it smooth, unobstructed, and large enough to carry away the water from the heaviest rainfall.

Open valleys— An open valley does not have shingles laid in the center of the valley. Instead, roll roofing is exposed in the valley. The recommended flashing material in open valleys is 90-lb. mineral surface roll roofing. Use a color that either matches or blends with the roofing shingles you are going to install. Apply the valley flashing in two layers. See Figure 13-19.

Center the first layer of 18-inch-wide 90-lb. mineral surface roll roofing in the valley. The mineral surface should be *down*. Trim the lower edge flush with the drip edge at the eave. This first layer goes up the entire length of the valley. If two or more strips of roll roofing are required, lap the upper piece over the lower piece so that drainage is carried over the joint, not into it. The overlap should be 12 inches and fully bonded with asphalt plastic cement.

Use the minimum number of nails necessary to hold the strip in place. Nail along a line one inch from each edge. Start at one edge and work all the way up. Then return to nail the other edge, pressing the flashing strip firmly into the valley.

When the first 18-inch strip has been secured, begin with the second layer. It should be of the same 90-lb. material, but 36 inches wide. Center

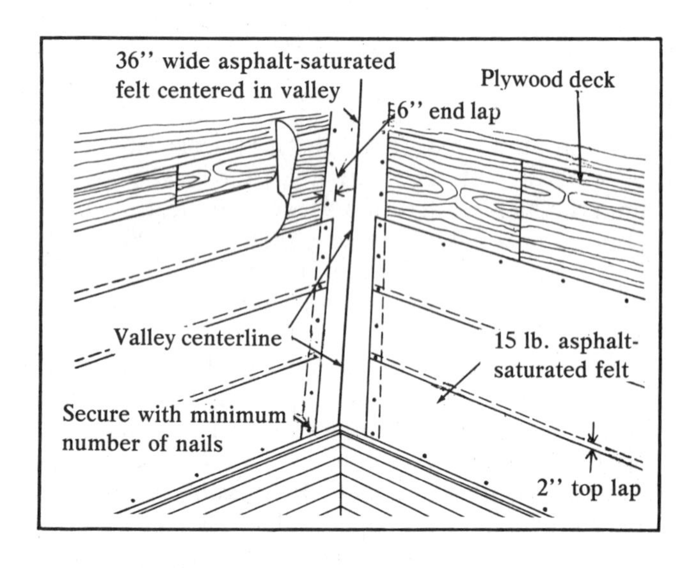

Valley underlayment application
Figure 13-18

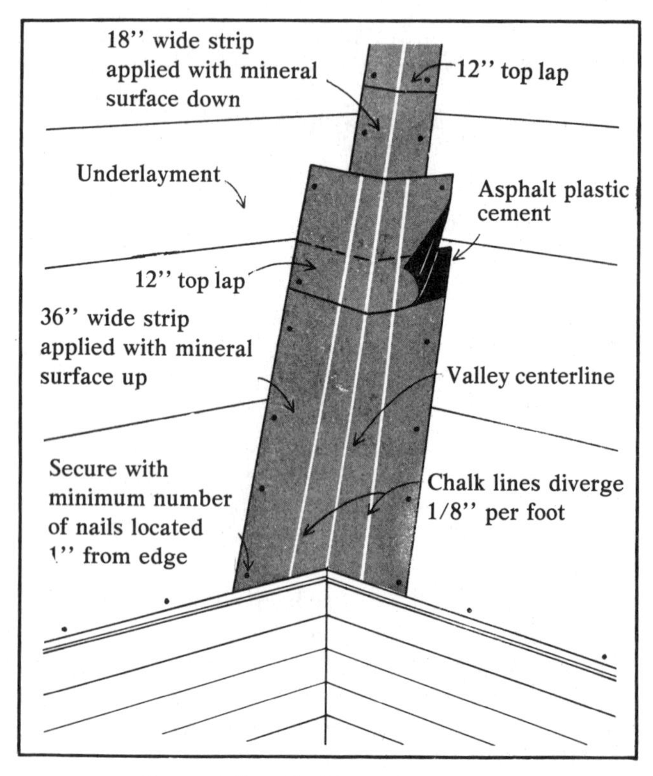

Open-valley flashing application
Figure 13-19

this second strip (over the first strip) in the valley. The mineral surface should face *up* this time. Nail this strip in place the same way as the underlying strip. Overlaps should be 12 inches and cemented. This completes the valley flashing. Shingles are laid over the edges of the roll roofing. Shingle installation will be explained later in this chapter.

Woven and closed-cut valleys— You'll seldom use an open valley on better quality homes. Woven and closed-cut valleys make a more attractive and professional roof. For both types of valley, start with 36-inch-wide mineral surface or smooth surface roll roofing. Use 50-lb. or heavier material. Center the strip in the valley, securing it with the minimum number of nails necessary to hold it in place. Drive nails along a line one inch from each edge, first on one edge and all the way up, then on the other edge and all the way up. While nailing, press the flashing strip into the valley so that it's smooth and tight. Overlaps should be 12 inches and cemented.

Asphalt Strip Shingle Installation

Installation procedures for all asphalt strip shingles are about the same. But there are different methods of application for the first shingle in each course. There are also different methods of shingle fastening. Be sure to follow the shingle manufacturer's directions.

Don't apply shingles until all chimneys are completed and all vent pipes, soil stacks and ventilators are in place. Be sure that all flashing is in place around chimneys and stacks and at vertical wall abutments.

If a roof surface is broken by a dormer or valley, start applying the shingles from a rake and work toward the dormer or valley. If the roof surface is unbroken, start at the rake that is most visible from the front of the house. If both rakes are equally visible, start at the center and work in both directions.

No matter where you start, apply the shingles across and diagonally up the roof. This practically guarantees that each shingle will be nailed properly. Straight-up application is sometimes called "racking," and it usually results in using less than the recommended number of nails. "Racking" requires that part of the shingle (in some courses) be placed under a shingle already applied. Because part of the shingle is hidden, it may be overlooked when nailed. With a diagonal application up the roof, each shingle is completely visible until covered by the course above.

Starter Strip

The starter strip can be either a row of shingles trimmed to the shingle manufacturer's recommendations or a strip of mineral surface roll roofing at least 7 inches wide. The starter strip protects the roof by covering the spaces under the cutouts and joints of the first course of shingles. It should overhang the rake edges and eaves by 1/4-inch to 3/8-inch.

If self-sealing shingles are used, cut off the tab portion of each shingle and position the remaining strip with the factory-applied adhesive face up along the eave. Trim at least 3 inches from the end of the first shingle in the starter strip. This insures that the cutouts of the first course of shingles won't fall over the starter-strip joints. Nail along a line parallel to the eave and 3 to 4 inches above it. Drive nails so that they will not be exposed under the cutouts in the first course. See Figure 13-20.

If you use shingles without self-sealing adhesive, remove the tab portion of each shingle and position the remaining strip along the eave. Complete the starter strip by following the instructions above.

If roll roofing is used, nail along a line 3 to 4 inches above the eave. Space the nails 12 inches apart. If more than one piece of roll roofing is needed, lap the end joint 2 inches and cement the lap.

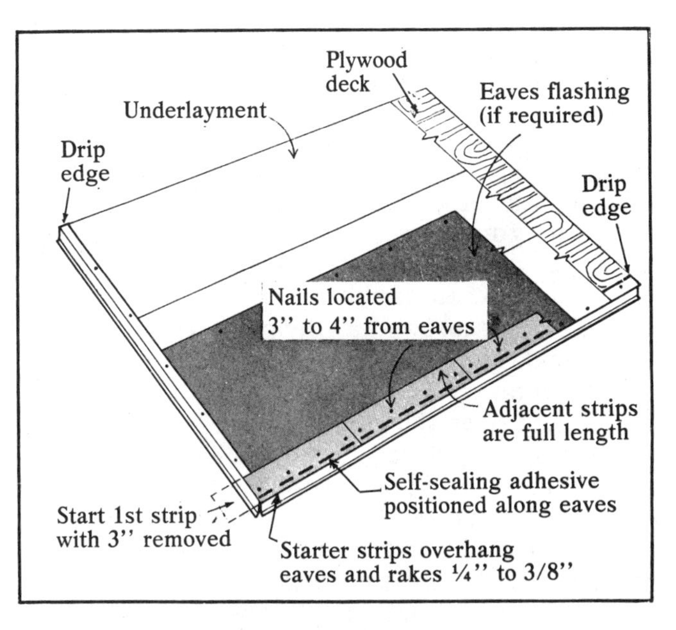

Starter strip application
Figure 13-20

First and Following Courses

The first course is the most critical. Be sure it's laid perfectly straight. Check it regularly against a horizontal chalk line. And to keep the cutouts in a straight line as they go up the roof, use a few vertical chalk lines aligned with the ends of the first-course shingles.

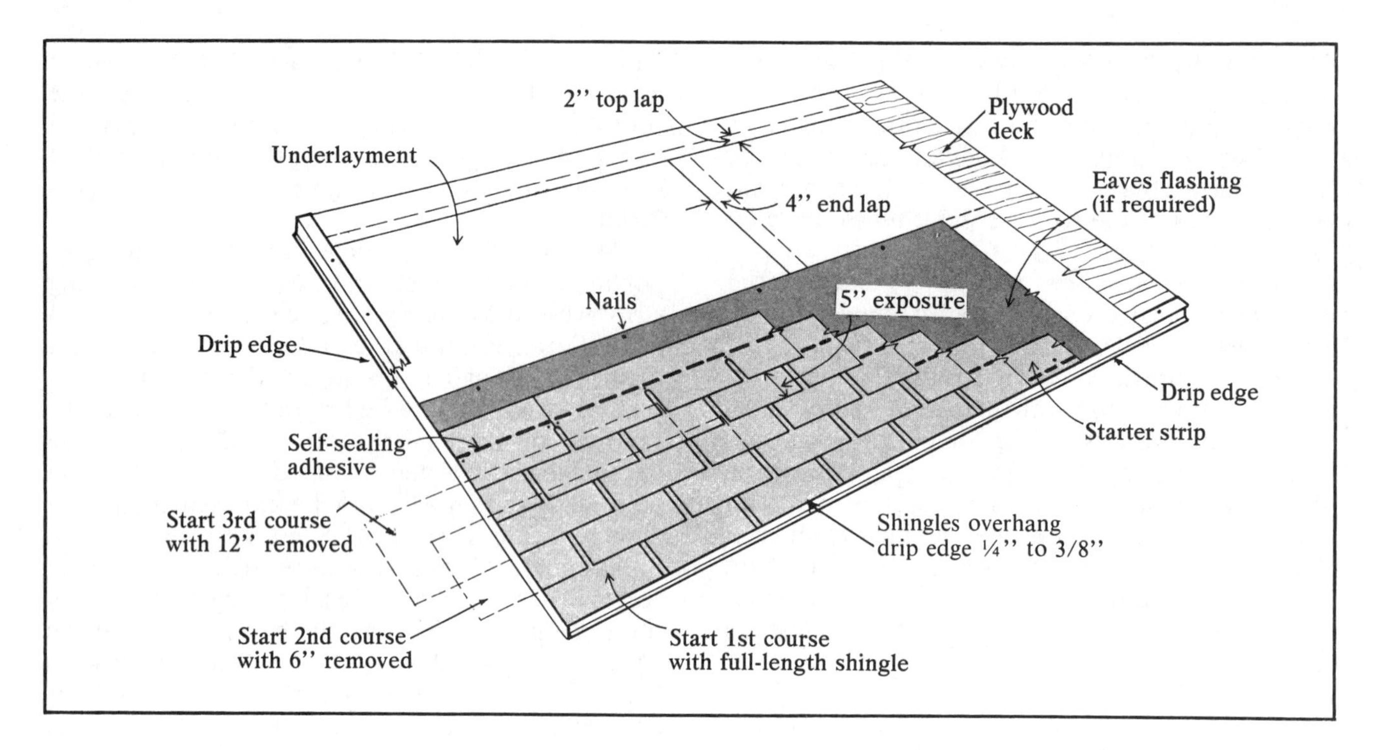

6-inch method of shingle application
Figure 13-21

If you use free-tab shingles or roll roofing for the starter strip, bond the tabs of each shingle in the first course to the starter strip below. Place a spot of asphalt plastic cement about the size of quarter on the starter strip under each tab. Then press the tab firmly into the cement. Don't use too much cement. It can cause blistering.

The first course begins with a full-length shingle. On most higher courses, you'll cut a little off the first shingle to conform to the method and pattern desired. Don't throw away the pieces cut from the first shingle in higher courses. You can probably use the pieces on the opposite end of the roof and for hip and ridge shingles.

There are three methods of applying three-tab strip shingles: the 6-, 5- and 4-inch methods. These methods are named for the amount of shingle removed from the first shingle in each course. Each produces a distinctive shingle pattern. By removing part of the first shingle in each row, the cutouts in that course of shingles will not line up directly with the cutouts in the course below.

The 6-inch method— Start the first course with a full-length shingle. The second through sixth courses start with a shingle from which a multiple of 6 inches has been removed. The second course starts with a shingle that has 6 inches removed, the third course with 12 inches removed, and so on through the sixth course, which starts with a shingle that has 30 inches removed. Adjacent shingles in each course are full-length shingles. The seventh course begins again with a full-length shingle. This pattern is repeated all the way up the roof. Figure 13-21 illustrates the 6-inch method.

The 5-inch method— Start the first course with a full-length shingle. The second through seventh courses start with a shingle from which a multiple of 5 inches has been removed. The second course starts with 5 inches removed from the first shingle, the third course with 10 inches removed, and so on through the seventh course, which has 30 inches removed from the first shingle. Adjacent shingles in each course are full-length shingles. The eighth course again begins with a full-length shingle, not with a shingle that has 35 inches removed. A 1-inch shingle won't stay on the roof. See Figure 13-22.

The 4-inch method— This method is illustrated in Figure 13-23. Start the first course with a full-length shingle. Start the second course with 4 inches removed from the first shingle, the third course with 8 inches removed, the fourth with 12 inches removed, and so on through the ninth course which has 32 inches removed from the first shingle. Adjacent shingles in each course are full-length shingles. The tenth course begins again with a full-length shingle, and the pattern is repeated.

It would be possible to remove multiples of 3 inches or less from each shingle. But this would make the shingle joints less than 3 inches apart. Be sure to avoid any alignment system where shingle joints are closer than 4 inches to each other.

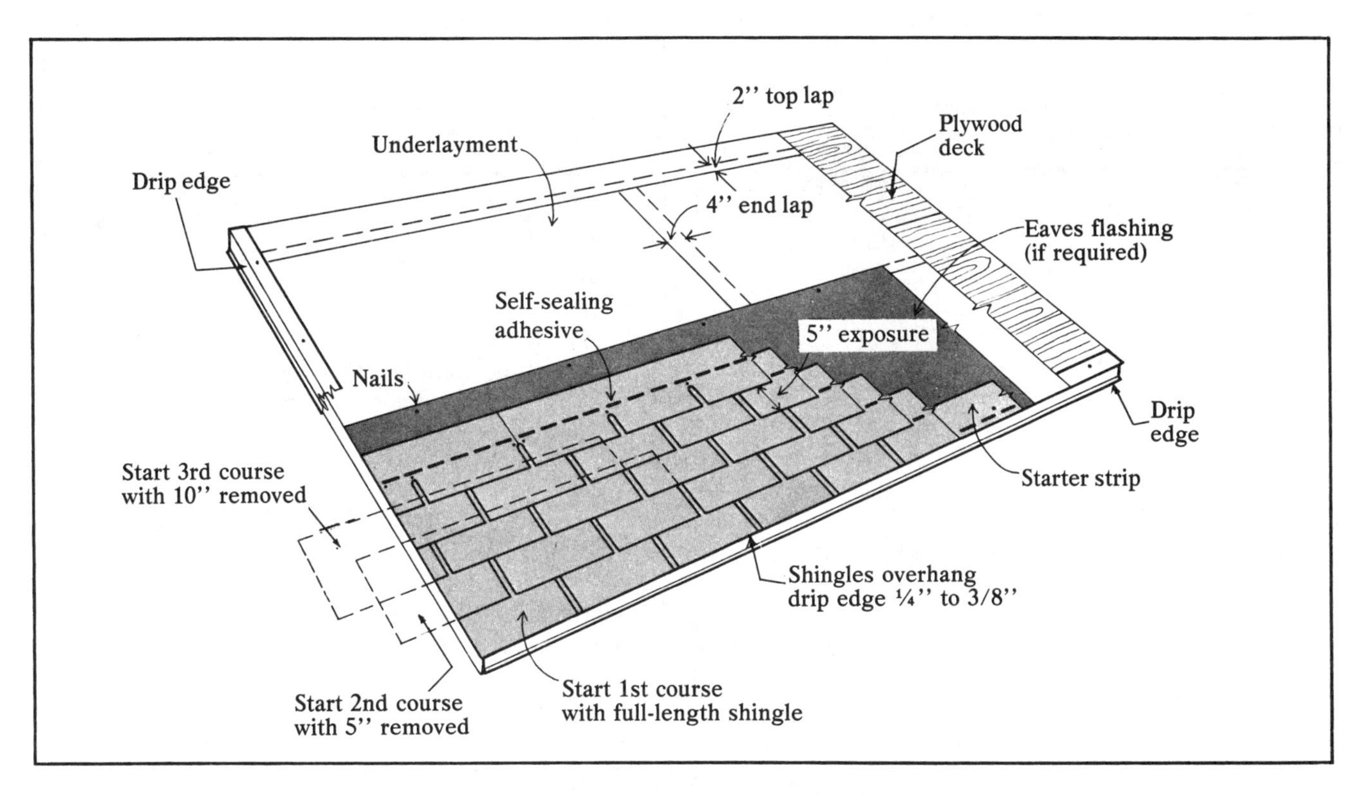

5-inch method of shingle application
Figure 13-22

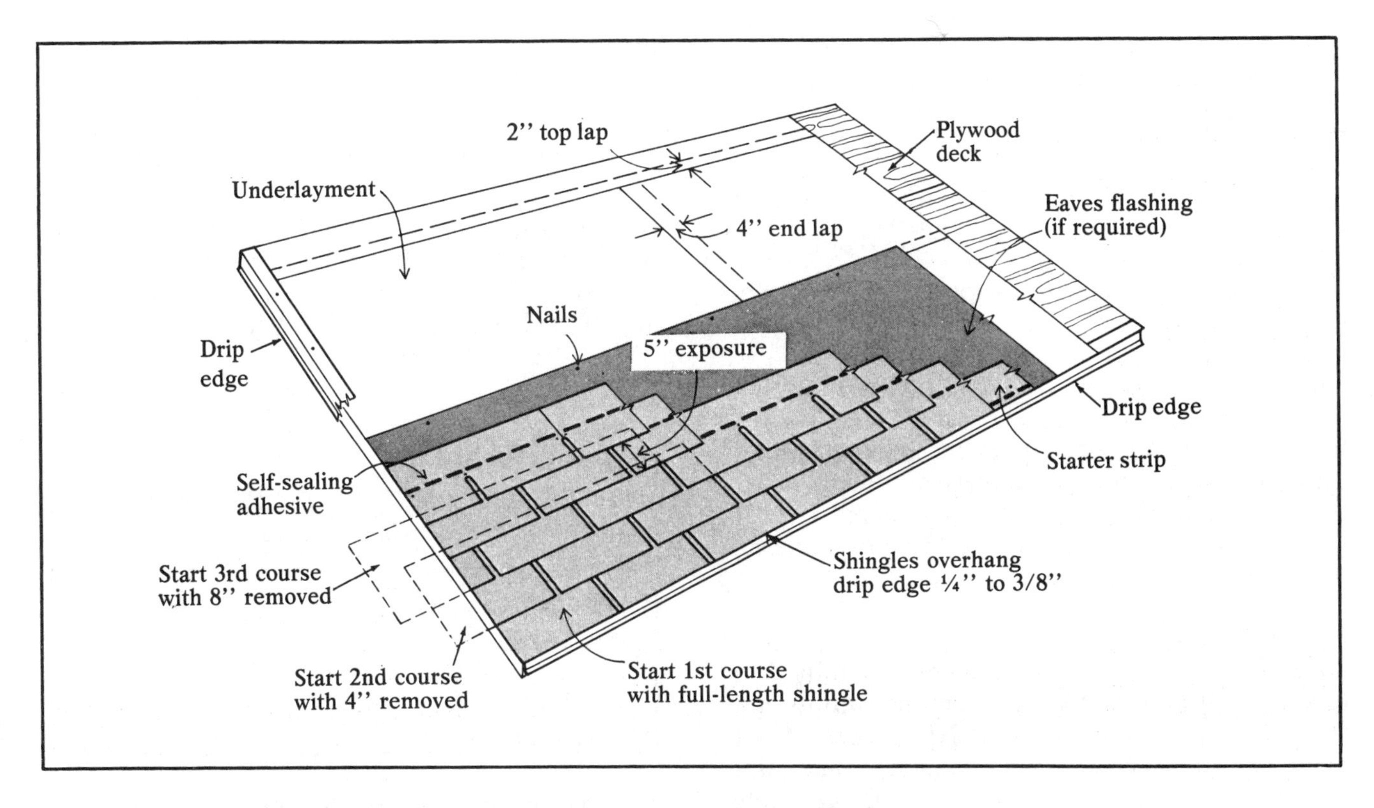

4-inch method of shingle application
Figure 13-23

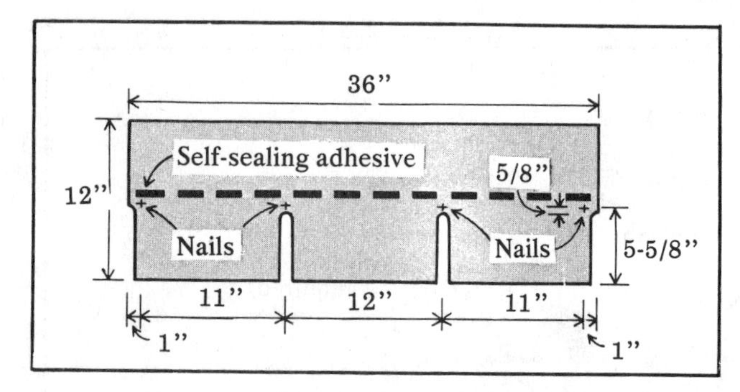

Nail locations for three-tab strip shingle
Figure 13-24

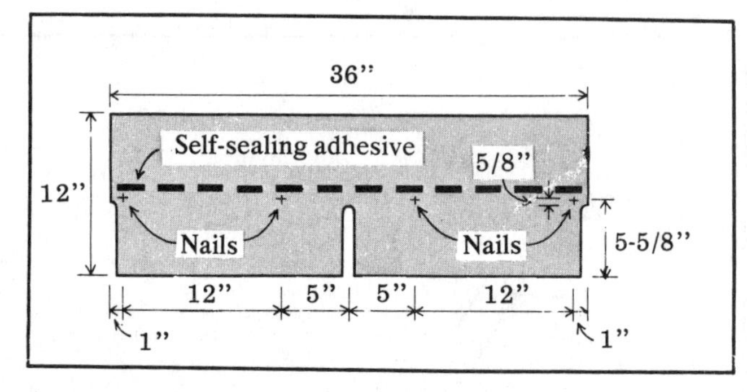

Nail locations for two-tab strip shingle
Figure 13-25

Fastening
Proper fastening is essential. Follow these points and you can't go wrong.

- Use the recommended size and grade of fasteners.

- Use zinc-coated fasteners for corrosion protection.

- Use the recommended number of fasteners per shingle.

- Drive fasteners according to the shingle manufacturer's recommendations.

- Align shingles so all fasteners are covered by the course above.

- Drive the fasteners straight, not at an angle.

- Don't break the shingle surface with the fastener head.

- Don't drive fasteners into knot holes or cracks in the roof deck.

- Repair faulty fastening immediately. If a fastener doesn't penetrate the deck properly, remove the fastener and use asphalt plastic cement to repair the hole in the shingle. Then place another fastener near but not directly on top of the repaired area. If it's not possible to repair the shingle properly, replace the entire shingle.

Nailing— Don't nail into or above factory-applied adhesives. Align each shingle carefully. Try to keep nails at least 2 inches from the cutouts and from the end joints of the underlying course. Start nailing from the end nearest the shingle just laid, and proceed across the shingle. This prevents buckling. Don't try to re-align a shingle by shifting the free end after two nails are in place. Drive nails straight so that the edge of the nail head is flush with the shingle surface and doesn't cut into it. Locate nails in three-tab, two-tab, and no-cutout strip shingles as described below.

1) Three-tab strip shingles: Each shingle requires four nails. When the shingles are applied with a 5-inch exposure, the nails should be placed on a line 5/8-inch above the top of the cutouts. The four nail-locations are; 1 inch from each end, and centered over each of the two cutouts. See Figure 13-24.

2) Two-tab strip shingles: Use four nails in each shingle. For a 5-inch exposure, the nails should be placed on a line 5/8-inch above the top of the cutouts. The four nail-locations are; 1 inch from each end, and 13 inches from each end. See Figure 13-25.

3) No-cutout strip shingles: Each shingle requires four nails. For a 5-inch exposure, the nails should be placed on a line 5⅝ inches above the butt edge. The four nail-locations are; 1 inch from each end, and 12 inches from each end. See Figure 13-26.

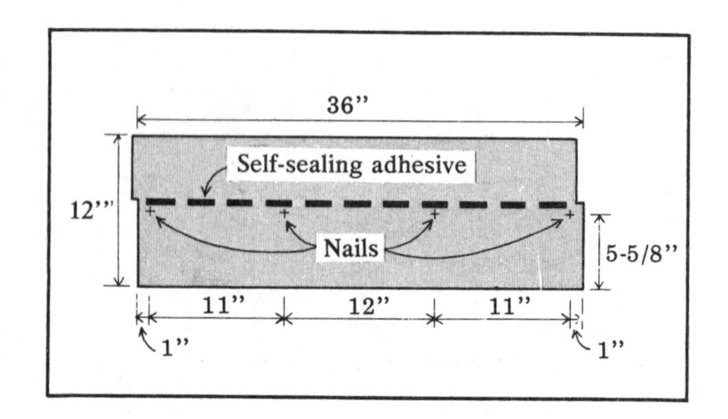

Nail locations for no-cutout strip shingle
Figure 13-26

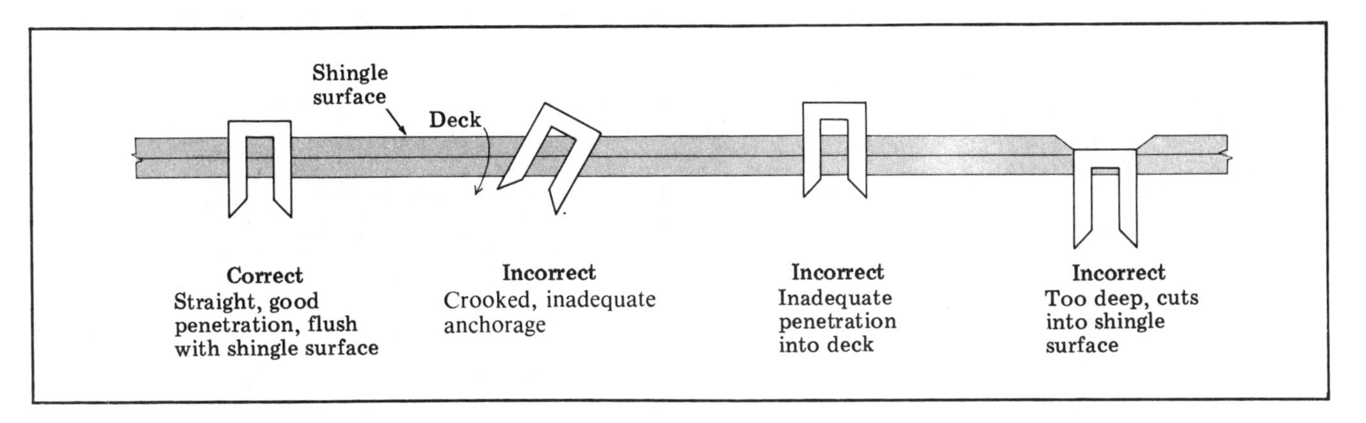

Staple application
Figure 13-27

Stapling— In new construction, staples can replace roofing nails on a one-for-one basis if you're using wind-resistant shingles with factory-applied adhesive. If there's any doubt about the acceptability of staples in a particular situation, call the shingle manufacturer or the factory representative in your area.

Drive staples into the same locations specified for nails. It's important that the staples be driven parallel to the length of the shingle. Be sure the stapler is set so that it will drive the staples all the way into the deck but will not over-penetrate. Figure 13-27 shows examples of good and bad stapling.

Wind protection— Wind-resistant shingles are a good choice in high wind areas. Several types of asphalt shingles offer this feature. They're made with either a factory-applied adhesive or an interlocking tab. Regardless of the type of shingle, look for a statement on the wrapper that says it complies with the Underwriters Laboratories "Standard 997 for Wind-Resistant Shingles."

The factory-applied adhesive on self-sealing shingles is hard when you apply the shingles. Once the shingles are installed, the sun softens the adhesive and it forms a secure bond after only a few days of exposure to warm weather. In winter, bonding can take longer. How much longer depends on temperature, sun conditions, roof slope, and which direction the roof slope faces.

You can give free-tab shingle roofs a degree of wind protection by cementing the tabs of each shingle to the course below. Place a spot of asphalt plastic cement the size of a quarter on the underlying shingle. Lift the tab of the overlying shingle just enough so that you can apply the cement. Press the tab into the cement to secure it. But don't squeeze the cement out beyond the edge of the tab. Be sure to cement every tab on the roof. But don't use too much cement. Excessive cement can cause blistering. See Figure 13-28.

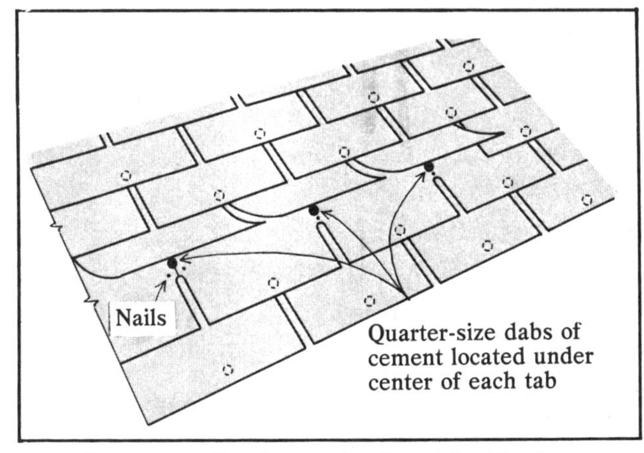

Cement application under free-tab shingles
Figure 13-28

Shingle Application in Valleys

Three methods of valley shingle application are: open, woven, and closed-cut. Woven and closed-cut valleys make the most attractive and most water-resistant roofs. For all methods, the valley flashing should be in place before you begin to apply the shingles, except for the open valleys around dormers where the valley flashing must overlap the top courses of shingles along the dormer side walls.

Open-valley shingle application for main roof— Assuming the valley flashing has been laid as described earlier in this chapter, here's how to shingle an open valley. Snap two chalk lines the full length of the valley flashing, one on each side of the valley centerline. The upper ends of the chalk lines should be 6 inches apart at the ridge (or 3 inches to each side of the valley centerline). Starting at the ridge and going down toward the eave, the chalk lines should diverge from one another 1/8-inch per foot. Thus, for an 8-foot-long valley, the chalk lines should be 7 inches apart at the eave. For a 16-foot-long valley, they should be 8 inches apart at the eave. See Figure 13-29.

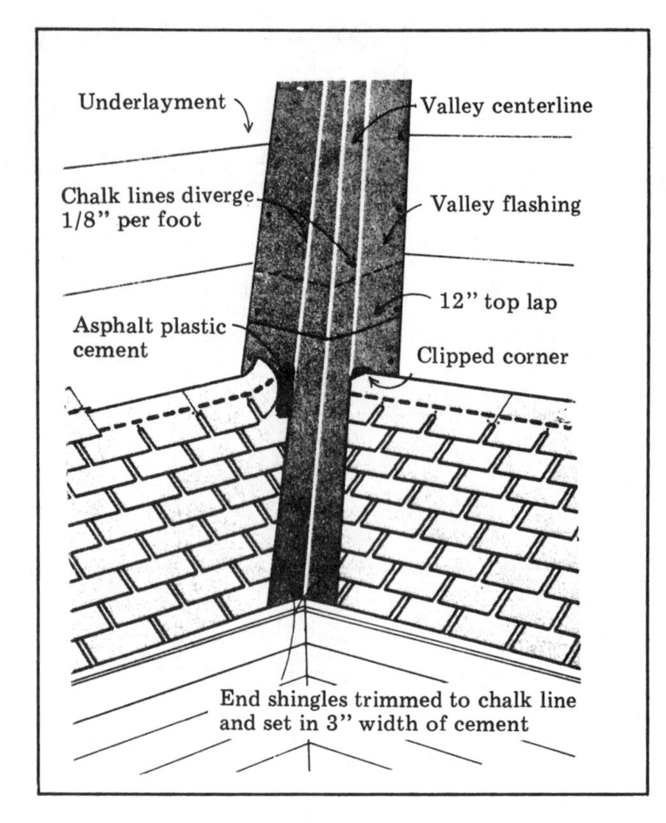

Open valley shingle application
Figure 13-29

As shingles are applied toward the valley, trim the last shingle in each course to fit the chalk line. At the upper corner of this shingle, clip 1 inch on a 45-degree angle to direct water into the valley and to prevent it from penetrating between the courses. Finally, to form a tight seal, use a 3-inch width of asphalt plastic cement to bond the shingle to the valley flashing. There should be no exposed nails along the valley flashing.

Open-valley shingle application for dormer roof— The valley between a dormer roof and main roof requires special treatment. Don't install the valley flashing until the main-roof shingles have been applied up to a point just above the lower end of the valley, as shown in Figure 13-30.

Apply the first or bottom layer of valley flashing (18-inch-wide, 90-lb. mineral surface roll roofing) the same way as for any open valley. Trim the lower section of the flashing so that it overhangs the dormer eave by 1/4-inch. The lower section of flashing in contact with the main roof deck should extend at least 2 inches below the point where the two roofs meet. The upper section of flashing on the main roof should extend 18 inches above the point where the dormer ridge intersects the main roof. Trim the portion on the dormer at the ridge. See Figure 13-31.

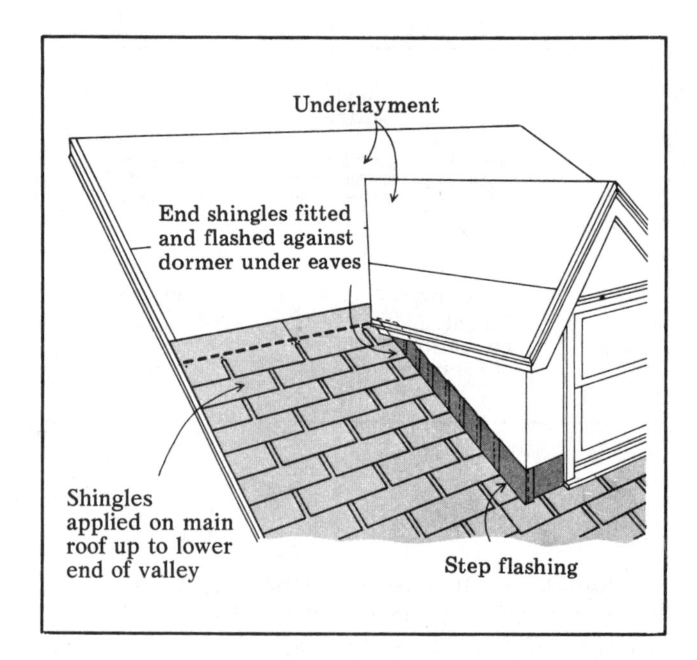

Starting point for open-valley flashing application at dormer roof
Figure 13-30

Apply valley flashing the same way on the other side of the dormer. Extend the main-roof portion of the flashing up and over that of the first valley. Cement and nail the overlap. The dormer-roof portion of flashing should be lapped over the ridge, and then cemented and nailed in place.

Apply the second or top layer of flashing (36-inch-wide, 90-lb. mineral surface roll roofing) in the following way. Trim the dormer-eave por-

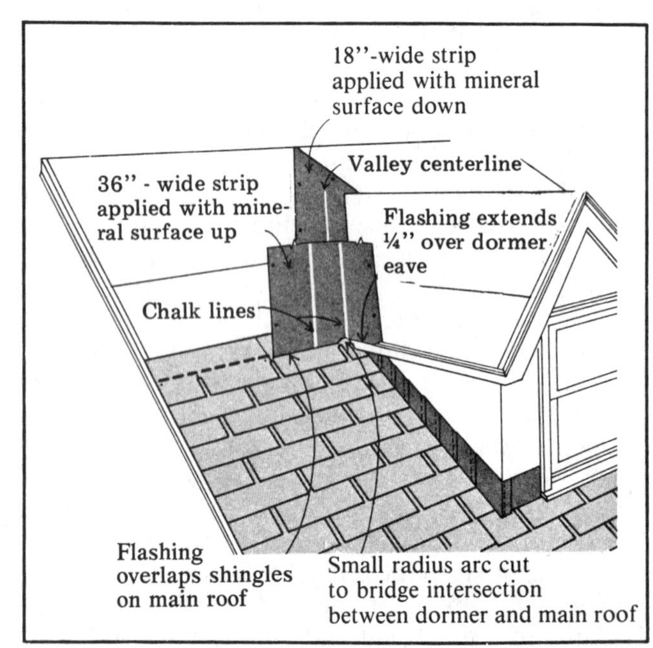

Open-valley flashing application at dormer roof
Figure 13-31

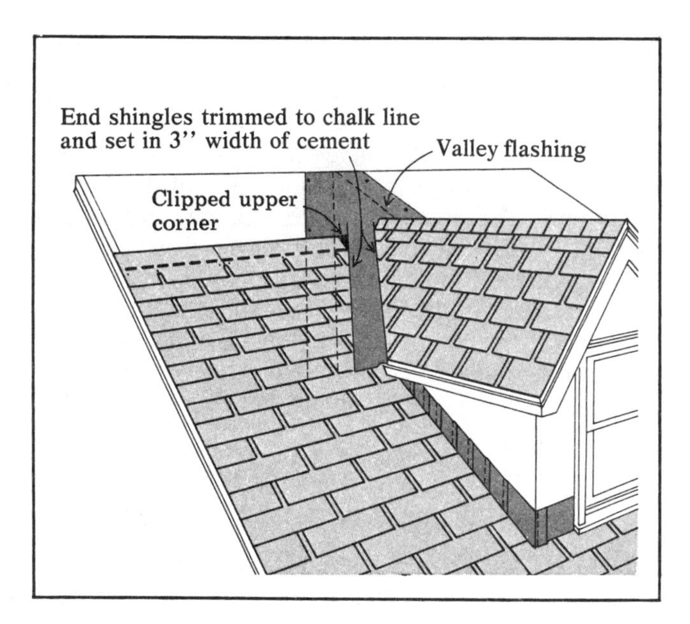

Open-valley shingle application at dormer roof
Figure 13-32

tion of the flashing so that it matches the underlying 18-inch strip. Trim the lower edge of the main-roof portion of the flashing so that it will overlap the nearest course of shingles. The amount of overlap is the same as the normal lap of one shingle over another. For example, for 12-inch-wide, three-tab strip shingles, the overlap extends to the top of the cutouts.

Nail the top layer of flashing over the bottom layer as in standard open-valley construction. Press the flashing into the valley joint so that it lies flat and smooth right up to the edge of the dormer eave. Trim the top part of the flashing horizontally along a line extended from the dormer ridge.

Apply the top layer of flashing in the valley on the other side of the dormer in the same manner, except at the dormer ridge where it overlaps the first valley flashing and is cemented and nailed to it.

At the point of intersection between the dormer eave and the main roof, trim the flashing in a small radius arc that bridges slightly over the point of intersection. This shape forms a small canopy over the joint between the two decks. See Figure 13-31.

Snap two chalk lines the full length of the valley flashing, one on each side of the valley centerline. The upper ends of the chalk lines are 6 inches apart at the ridge (or 3 inches to each side of the valley centerline). Starting at the ridge and going down toward the eave, the chalk lines should diverge 1/8-inch per foot. Resume the shingle application, trimming the end shingle in each course to fit the chalk lines. Clip the upper corner of the end shingle, and use a 3-inch-wide strip of asphalt plastic cement to seal it to the flashing. Complete the valley construction in the usual manner. See Figures 13-31 and 13-32.

After shingles have been applied to both sides of the dormer roof, apply the dormer ridge shingles. Start at the front of the dormer and work toward the main roof. Apply the shingles as described later in this chapter in the section, "Shingle Application at Hips and Ridges." Apply the final ridge shingle so that it extends at least 4 inches onto the main roof. Slit the center of the portion attached to the main roof and nail that portion into place. Then apply the main-roof shingle courses to cover that portion of the final ridge shingle. Snap vertical chalk lines so that the shingles on the main roof will continue the same alignment pattern on both sides of the dormer. See Figure 13-33.

Closed-cut valley shingle application— Look at Figure 13-34. With valley flashing already in place, apply the first course of shingles along the eave of one of the intersecting roof planes and across the valley. Extend the end shingle at least 12 inches onto the adjoining roof. Apply following courses the same way, extending them across the valley and onto the adjoining roof surface. Press the shingles tightly into the valley. Use normal shingle-fastening methods, except that no fastener should be within 6 inches of the valley centerline. And two fasteners should be placed at each shingle-end that crosses the valley.

Snap a chalk line the full length of the valley and on the valley centerline. The chalk line will be on top of the shingles you have just installed.

Then install the shingles on the adjoining roof surface. Apply the first course the same as before: along the eave and across the valley. The end shingle of the first course will overlap the end shingle of the first course installed on the other roof surface.

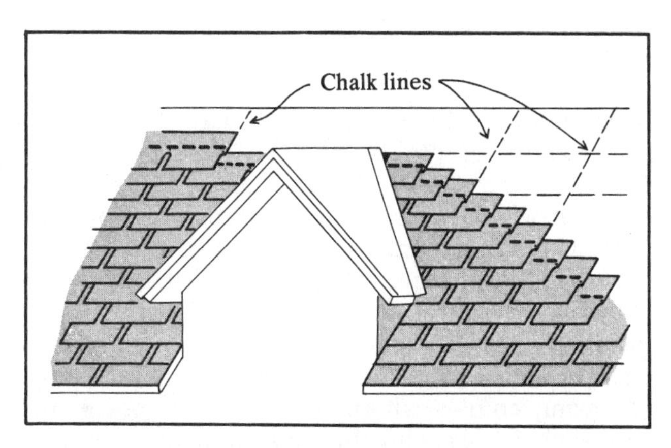

Chalk lines for proper alignment of shingles and cutouts around dormer
Figure 13-33

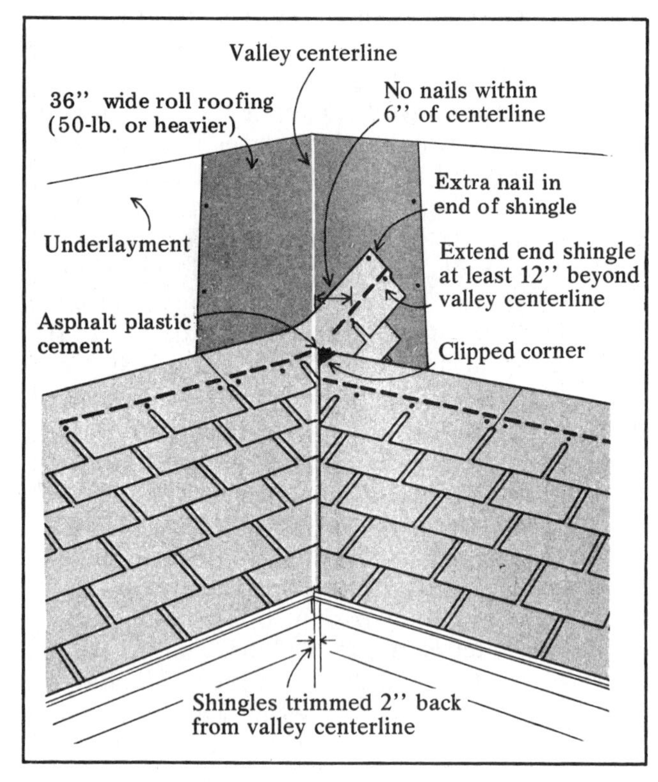

Closed-cut valley shingle application
Figure 13-34

Apply following courses in the same way, extending them across the valley and onto the previously installed shingles.

Using the chalk line as a guide, trim the end shingles so that they are no closer than 2 inches to the valley centerline. At the upper corner of each end shingle, trim 1 inch on a 45-degree angle to direct water into the valley. Embed each end shingle in a 3-inch-wide strip of asphalt plastic cement.

Woven-valley shingle application— Start a woven valley after the valley flashing is in place. On both of the intersecting roof surfaces, apply shingles toward the valley. Stop at a point about 3 feet from the center of the valley.

Apply the first course along the eave of one roof surface and across the valley. The last shingle should extend at least 12 inches onto the adjoining roof surface. Then go to the adjoining roof surface, and apply that first course (also along the eave and toward the valley). Extend it across the valley and over the top of the shingles previously installed. The end shingle should again extend at least 12 inches onto the other roof surface. Apply following courses alternately from the adjoining roof surfaces, weaving the valley shingles over each other, as shown in Figure 13-35. Press each shingle tightly into the valley and follow the same nailing procedures as for the closed-cut valley.

Flashing

All joints on a roof are natural leakage points. Any time roof surfaces join with one another or with vertical walls, you have a joint. Projections through the roof surface, such as soil stacks or chimneys, also create possible leakage points. Any material laid to give added protection at these points can be called flashing. Good flashing is essential for good roof performance.

Flashing against vertical side wall— Use metal flashing shingles to protect the point where a roof surface butts against a vertical wall. This is called step flashing.

The metal flashing shingles are rectangular, 10 inches long and should be 2 inches wider than the exposed face of the roofing shingles. For example, when used with roofing shingles that have a 5-inch exposure, the metal flashing shingles should be 10 x 7 inches. The 10-inch length is folded in half so that it can extend 5 inches across the roof deck and 5 inches up the wall surface. See Figure 13-36.

To install step flashing, place a metal flashing shingle over the end of the starter strip. Position the metal flashing shingle so that it will be completely covered by the tab of the end shingle in the first course of roofing shingles. Use two nails to secure the horizontal arm of the metal flashing shingle to the roof. Don't nail the flashing shingle to the wall. Settling of the roof could damage the seal.

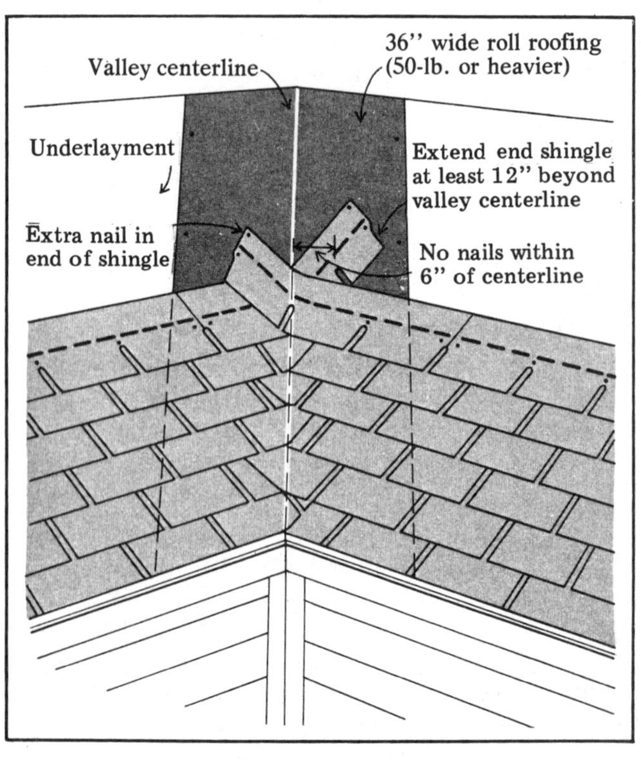

Woven-valley shingle application
Figure 13-35

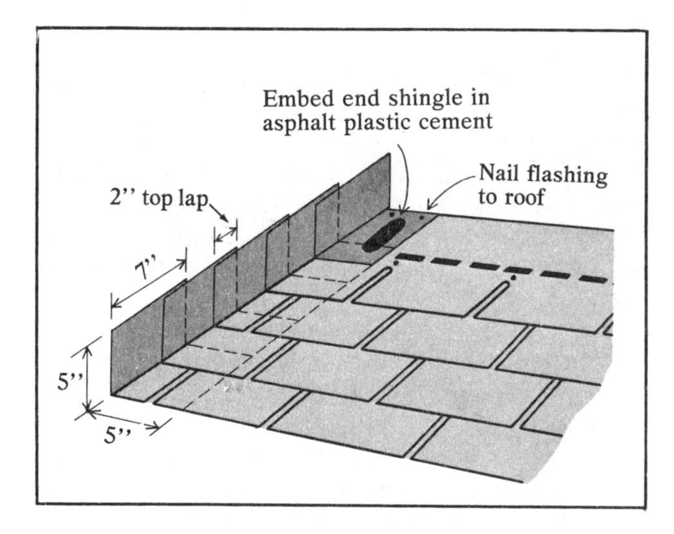

Step flashing application
Figure 13-36

Apply the first course of roofing shingles. Then position the second metal flashing shingle over the end shingle in the first course of roofing shingles. Place the flashing shingle 5 inches up from the butt so that the flashing shingle will be completely covered by the tab of the end shingle in the second course of roofing shingles. Again, fasten the horizontal arm of the flashing shingle to the roof. The second course of roofing shingles follows. The end is flashed as in preceding courses and so on to the top of the intersection. Because the metal flashing shingle is 7 inches wide and the roofing shingles are laid with a 5-inch exposure, each metal flashing shingle will overlap the one on the course below by 2 inches.

If the side wall has wood siding, bring the siding down (over the vertical portion of the step flashing) to serve as cap flashing. See Figure 13-37. Keep wood siding far enough away from the roofing shingles so that the siding can be painted without getting paint on the shingles.

Flashing against vertical front wall— Apply shingles up the roof until a course must be trimmed to fit the base of the vertical wall.

Plan ahead and adjust the exposure slightly in the two courses before the last course so that the last course is at least 8 inches wide. Apply a continuous piece of 26-gauge metal flashing over the last course of shingles. Fold the metal flashing strip so that it will extend at least 5 inches up the vertical wall and at least 4 inches onto the last course of shingles. Embed the flashing strip in asphalt plastic cement and then nail it to the roof. But don't nail the strip to the wall. Finish the job by applying an additional row of shingles over the metal flashing strip. You'll have to trim this course to the width of the strip. See Figure 13-38.

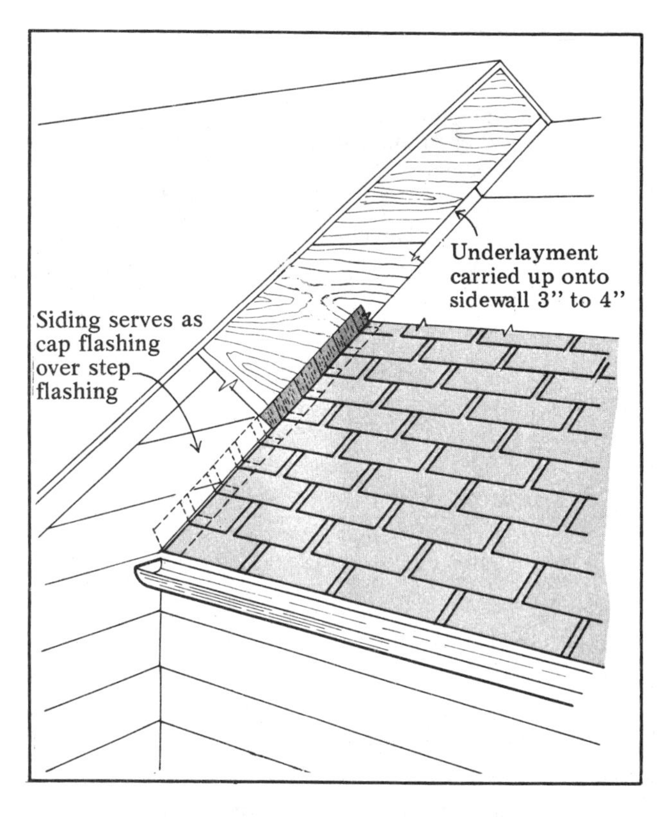

Step flashing against vertical wall
Figure 13-37

Bring the siding down (over the vertical portion of the flashing strip) to serve as cap flashing. But don't nail the siding to the vertical flashing. Keep wood siding far enough away from the roof shingles so that painting is easy.

If the vertical front wall meets a side wall, as in dormer construction, cut the flashing so that it extends at least 7 inches around the corner. Then continue up the side wall with step flashing as described earlier.

Flashing around soil stacks and vent pipes— Practically all dwellings have vent pipes or ventilators projecting through the roof. Most of these are circular and require special flashing.

Here's how to flash a vent pipe. Apply shingles up to the pipe. Then cut a hole in a shingle so that the shingle fits around the pipe, and set the shingle in asphalt plastic cement. See Figure 13-39. Then put a preformed flashing flange over the pipe and the shingle. Set this flange in asphalt plastic cement. Be sure the flange fits flush against the roof. See Figure 13-40.

When the flashing is in place, continue shingling. Cut shingles on the next higher course so they fit around the pipe. Embed these shingles in asphalt plastic cement where they overlap the flange. Don't apply the plastic cement too thick. Too much cement can cause blistering. And don't drive fasteners close to the pipe.

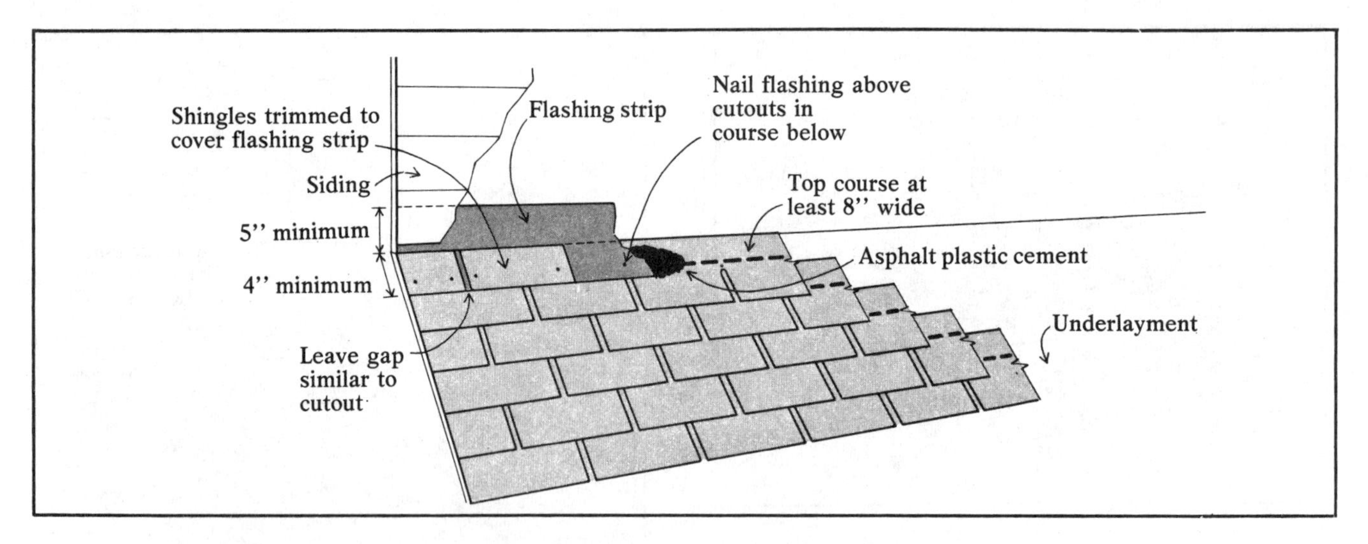

Flashing against vertical front wall
Figure 13-38

The finished job should look like Figure 13-41. Notice that the lower part of the flange overlaps the lower shingles, and the side and upper shingles overlap the flange.

The procedure is nearly the same when a ventilator or exhaust stack comes up through the ridge. Bring the shingles up to the pipe from both sides. Bend the flange over the ridge so it's flush against the roof on both sides of the ridge and overlaps the roof shingles. Lay ridge shingles so they cover the flange. Embed the ridge shingles in asphalt plastic cement where they overlap the flange.

Flashing around chimneys— Uneven settling can be a serious problem around chimneys. The chimney is usually placed on a separate foundation and may separate from the roof slightly as the building ages.

Flashing around the chimney has to allow movement and still keep rain out of the home. This is accomplished with a base flashing that is secured to the roof deck and a cap flashing that is secured to the chimney. So if movement does occur, the cap flashing slides over the base flashing without opening up a hole where water can enter the house.

If the chimney comes through a sloped roof surface, you should install a cricket on the roof deck at the high side of the chimney. The cricket (sometimes called a "saddle") keeps water, ice and snow from accumulating on the high side of the chimney where it could drain into the attic. See Figure 13-42.

The cricket should be in place before roofing begins, because all roofing materials, from the felt underlayment to the roofing shingles, are carried over the cricket.

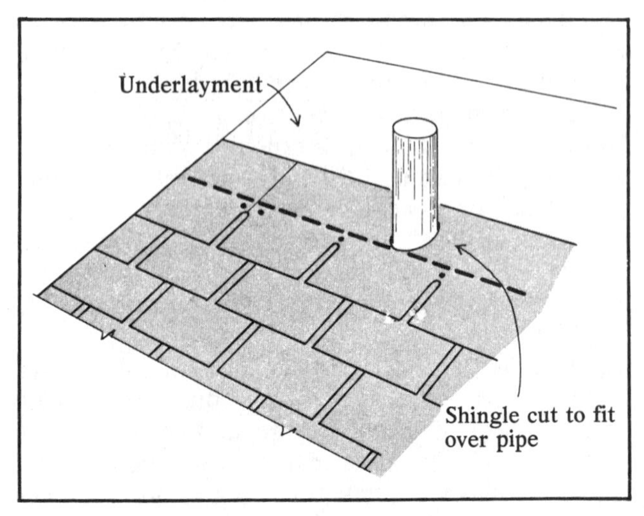

Shingle over vent pipe
Figure 13-39

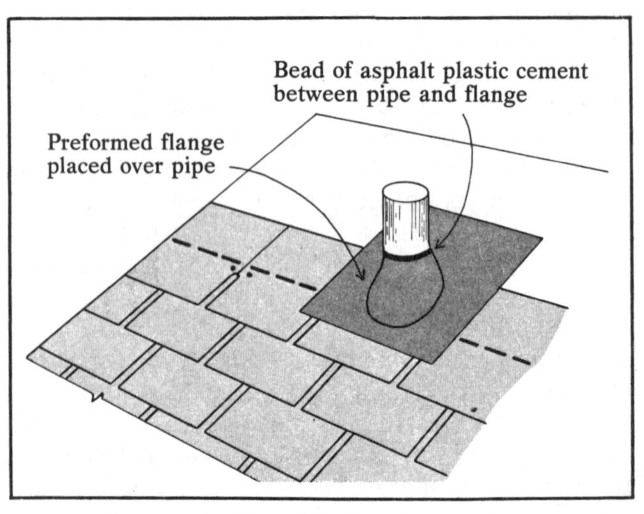

Flashing flange over vent pipe
Figure 13-40

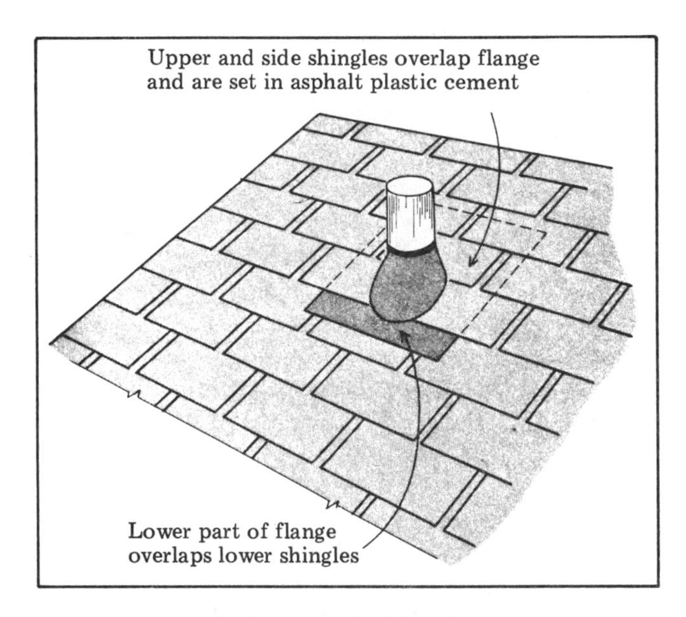

Shingles over flashing flange
Figure 13-41

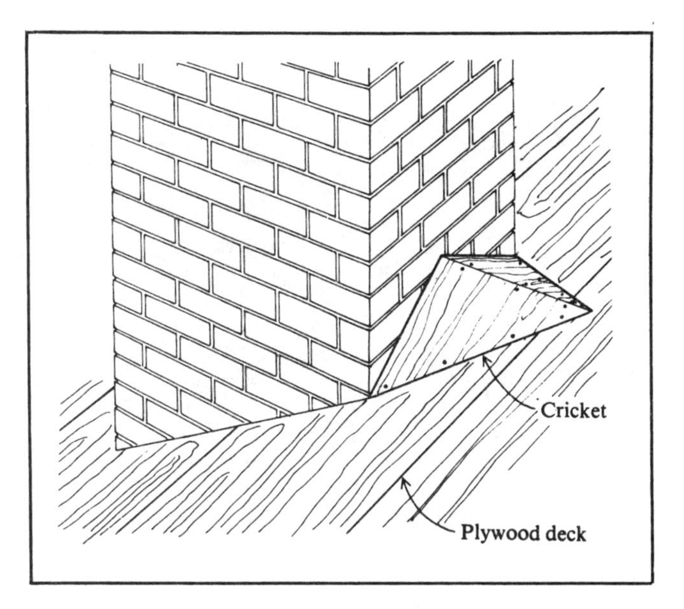

Chimney cricket
Figure 13-42

Make a simple cricket from two triangular sections of plywood joined to form a level ridge. The ridge should extend from the centerline of the chimney directly back to the roof deck. Nail the triangular sections to each other along their common edge. Fasten the cricket to the deck by nailing along the outside edges of the triangles.

Apply shingles up to the front edge (the low side) of the chimney before installing any flashing. Then apply a coat of asphalt primer to the entire chimney area where asphalt plastic cement will be applied. This seals the surface and provides good adhesion at the points where the cement is used.

Begin flashing the chimney by installing 26-gauge corrosion-resistant metal base flashing between the chimney and the roof deck on all sides of the chimney.

Apply the base flashing to the front first. See Figures 13-43 and 13-44. Fold the base flashing so that the lower section extends at least 4 inches over the shingles. The upper section should extend at least 12 inches up the vertical face of the chimney.

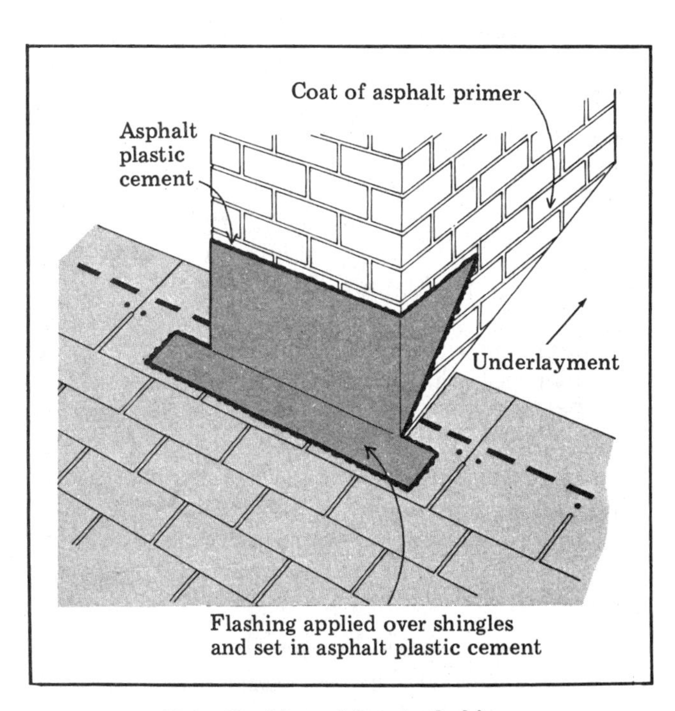

Base flashing at front of chimney
Figure 13-44

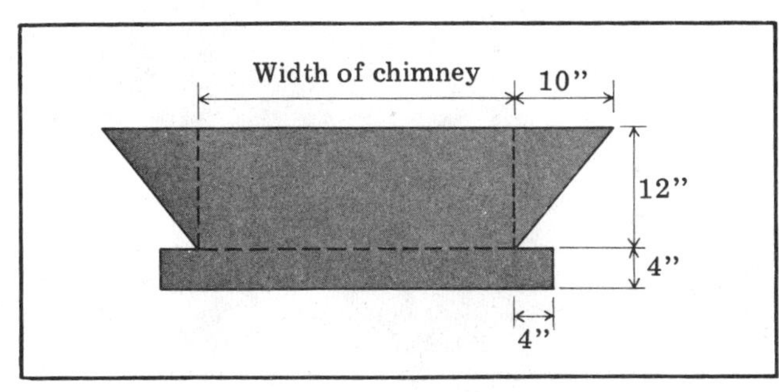

Pattern for cutting front base flashing
Figure 13-43

Work the flashing firmly and smoothly into the joint between the shingles and chimney.

The roof portion of the flashing should be set in asphalt plastic cement on the roof shingles. The chimney portion of the flashing should be set in asphalt plastic cement on the chimney face. While the cement dries, hold the flashing snug against the chimney face with one or two nails driven into the mortar joints.

Use metal step flashing for the sides of the chimney. Lay the metal step flashing the same way

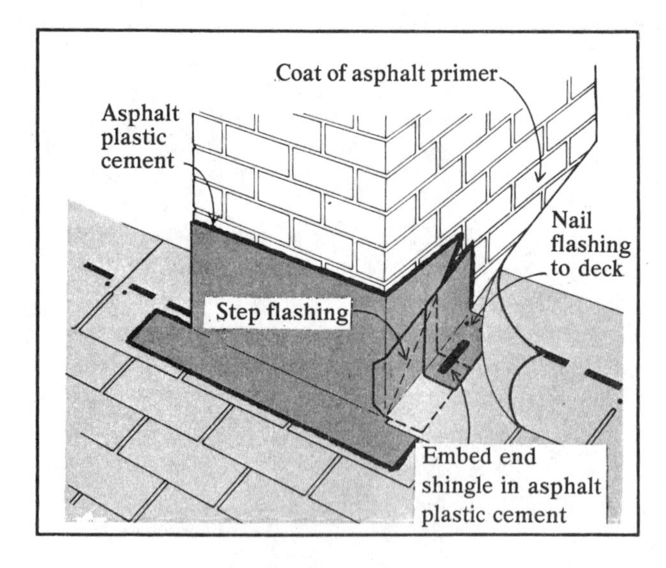

Base flashing at side of chimney
Figure 13-45

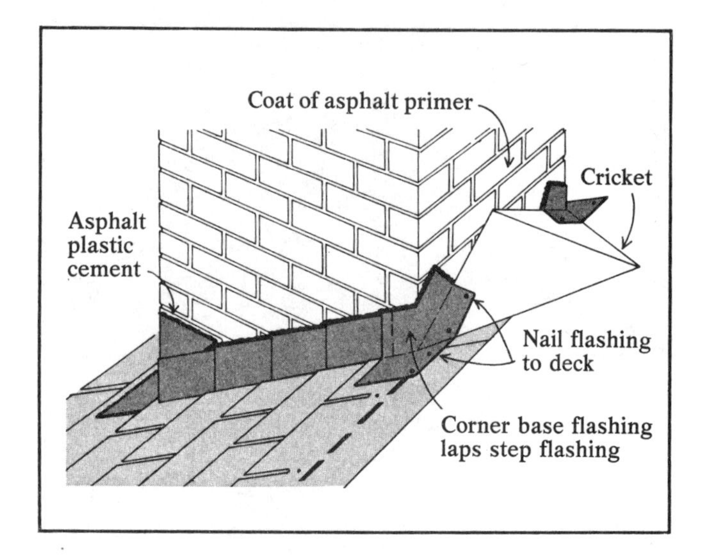

Corner base flashing at rear of chimney
Figure 13-46

as if you were flashing a vertical side wall. Cut, fold and apply the step flashing as shown in Figure 13-45. Secure each metal shingle to the masonry with asphalt plastic cement and to the deck with nails. Embed overlapping end shingles in asphalt plastic cement. See Figure 13-45.

The corner base flashing at the rear of the chimney should lap the step flashing just applied at the sides of the chimney. See Figure 13-46.

Place the rear base flashing over the corner base flashing, the cricket and the back of the chimney, as shown in Figures 13-47 and 13-48. Cut and fold the flashing to cover the cricket. The flashing should extend at least 6 inches onto the roof surface. It should also extend at least 6 inches up the back of the chimney.

If the cricket is large enough, consider covering it with shingles. Otherwise, apply the rear base flashing, then bring the end shingles in each course up to the edge of the cricket and cement them in place.

Then install the metal cap flashing. Begin by setting the cap flashing into the masonry, as shown in Figure 13-49. Rake out a mortar joint to a depth of 1½ inches, and insert the folded edge of the flashing into the cleared joint. Refill the joint with portland cement mortar. Finally, fold the flashing down so it covers the base flashing and lies snug against it.

Use one continuous piece of cap flashing for the front of the chimney. See Figure 13-50. On the sides and back of the chimney, use several pieces of flashing. All of these side pieces and back pieces are approximately the same size. Trim each so the upper edge fits along the line of a brick joint, and the lower edge fits along the line of the roof pitch.

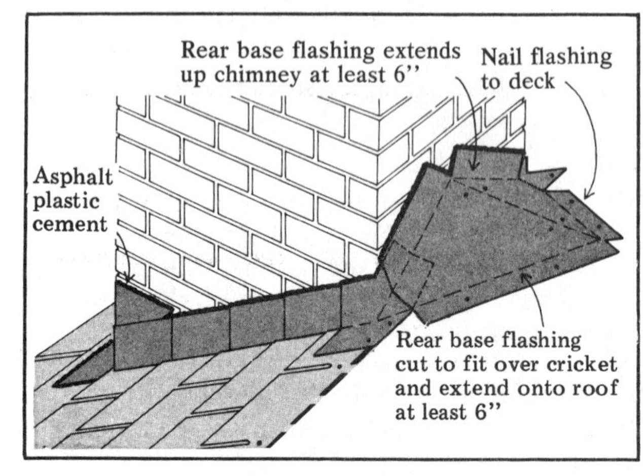

Base flashing over cricket
Figure 13-47

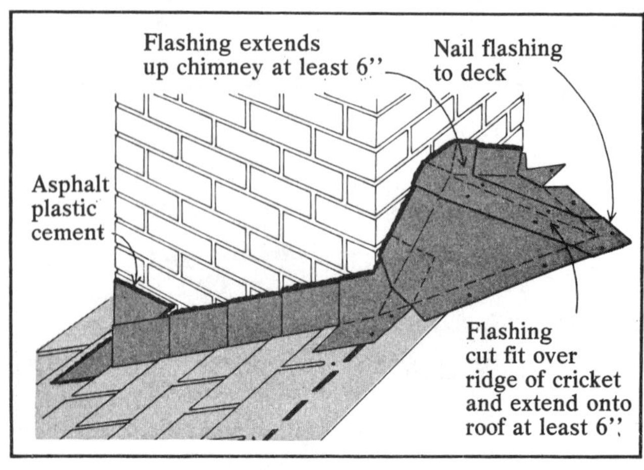

Base flashing over ridge of cricket
Figure 13-48

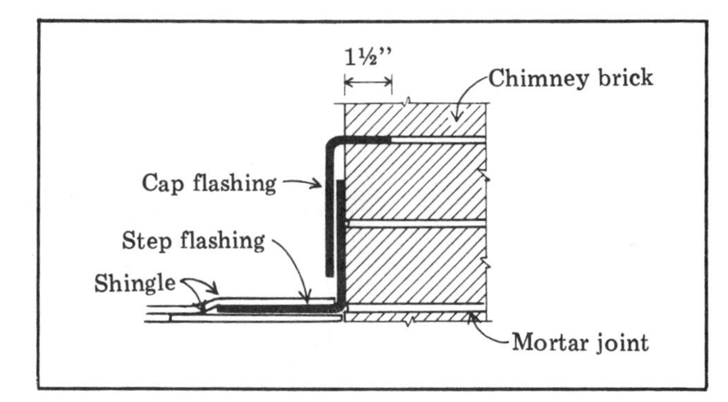

Cap-flashing application
Figure 13-49

See Figure 13-51. Start the side units at the lowest point and overlap each unit at least 3 inches.

Cap flashing isn't sealed to the base flashing. Any movement between the chimney and the roof is accommodated by the movement between the cap and base flashings.

Shingle Application at Hips and Ridges

Apply shingles up to a hip or ridge from both sides of the roof. Adjust the last few courses so that the ridge cap will cover the top course of shingles equally on both sides of the ridge.

Some manufacturers offer special hip and ridge shingles and give instructions on how they should be used. But hip and ridge shingles can be made from standard 12 x 36-inch strip shingles.

Start by cutting the strip shingle down as follows: Cut a three-tab shingle down to 12 x 12 inches. Cut a two-tab or no-cutout shingle down to a minimum of 9 x 12 inches. Then taper the top portion of the shingle slightly so that it's narrower than the exposed portion. This makes a neater job. See Figure 13-52.

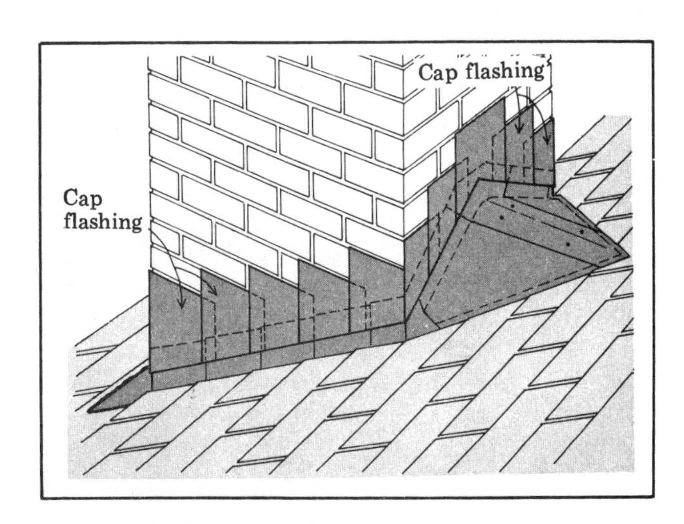

Cap flashing at side and rear of chimney
Figure 13-51

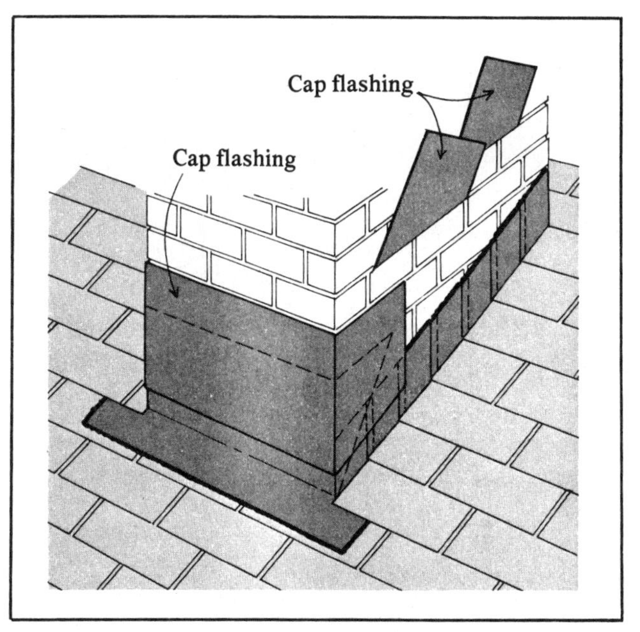

Cap flashing at front and side of chimney
Figure 13-50

Next, fold each shingle along the centerline of the longer (or tapered) dimension. In cold weather, you'll probably want to warm the shingle until it's pliable. Apply the shingles with a 5-inch exposure. On a hip, start at the bottom of the hip. On a ridge, start at the ridge end that is opposite the prevailing wind direction. See Figure 13-53. Drive fasteners as shown in Figure 13-54. Use one fastener on each side, placing it at a point 5½ inches from the exposed end of the shingle, and 1 inch up from its side edge.

Shingle Application on Low Slopes and Steep Slopes

Low slopes— Asphalt strip shingles can be used on slopes between 2 inches and 4 inches per foot if you

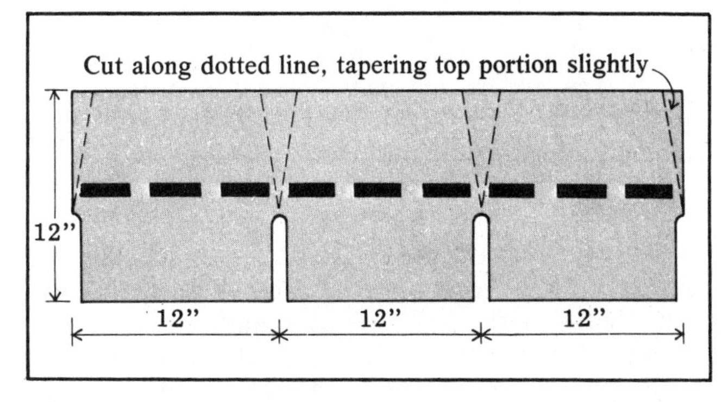

Make hip and ridge shingles from three-tab strip shingle
Figure 13-52

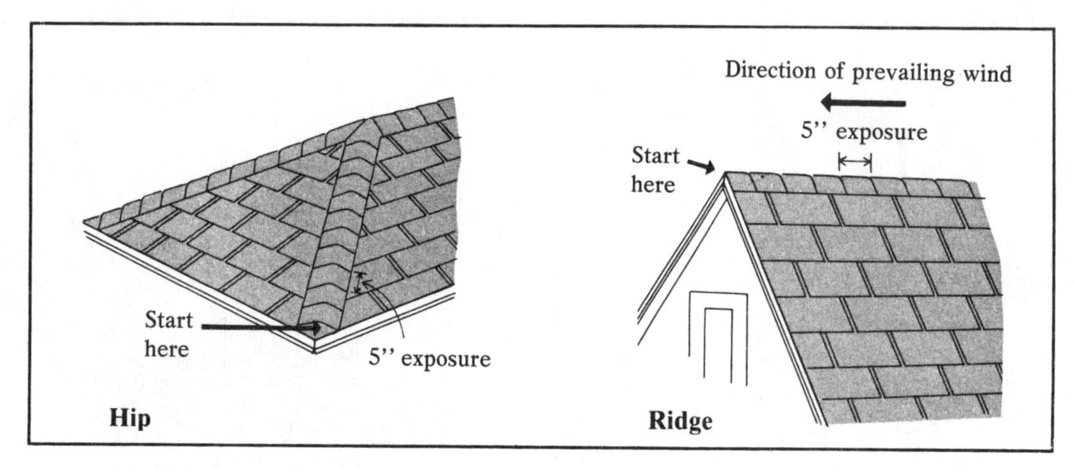

Hip and ridge shingle application
Figure 13-53

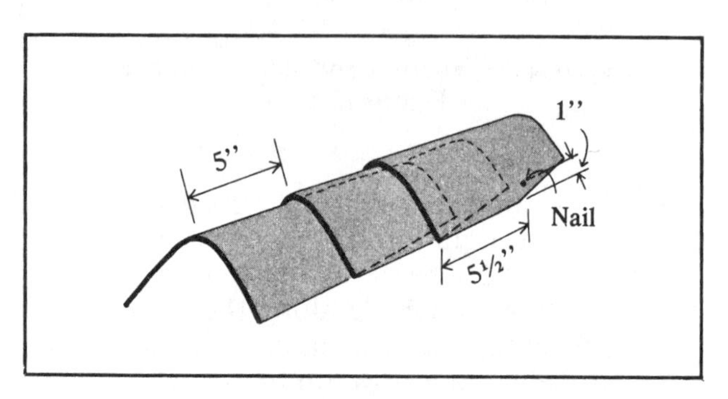

Nail location for hip and ridge shingles
Figure 13-54

observe special precautions. But don't ever use shingles on slopes less than 2 inches per foot.

Water drains slowly from low slopes, making it more likely that moisture will back up into the attic. The shingling procedure described below should insure a weathertight roof.

1) Underlayment. Cover the deck with two layers of 15-lb. nonperforated asphalt-saturated felt. Begin by nailing a 19-inch-wide strip of underlayment along the eave. It should overhang the drip edge by 1/4-inch to 3/8-inch. Then place a 36-inch-wide sheet of underlayment over the starter strip. The 36-inch width should completely cover the 19-inch strip.

Following courses are all 36-inch-wide sheets laid so they lap the preceding course by 19 inches. Nail each course with the minimum number of nails necessary to hold it in place until the shingles are down. End laps should be 12 inches wide and located at least 6 feet from the end laps in courses above or below.

2) Eave flashing. Wherever you expect icing along the eaves, cement all laps in the underlayment courses. Start the cement at the eave and continue to at least 24 inches beyond the interior wall line. This cemented double-ply underlayment serves as eave flashing. See Figure 13-55.

To install eave flashing, cover the entire surface of the starter strip with a continuous layer of asphalt plastic cement applied at a rate of 2 gallons per 100 square feet. Place the first course over the starter strip, pressing it firmly into the cement, as shown in Figure 13-55.

After the first course is down, coat the upper 19 inches with cement. Position the second course. Press it into the cement. Repeat the process for each course that lies within the eave-flashing distance. Apply the cement uniformly so that the overlapping felt floats on the cement but does not touch the felt in the underlying course. Avoid using too much cement.

After the eave flashing is finished, secure each following course of felt with the minimum number of nails necessary to hold it in place until the shingles are applied.

3) Shingle application. On a low slope in a high wind area, either use self-sealing shingles or put asphalt plastic cement under the tabs of each shingle. You can use any of the shingle-application methods discussed previously in this chapter.

Steep slopes— Many new homes are being built with steep-slope roofs. A mansard roof is a good example.

Application procedures for underlayment, drip edges, and flashing on a steep slope are the same as on a normal slope. But shingle application requires special treatment.

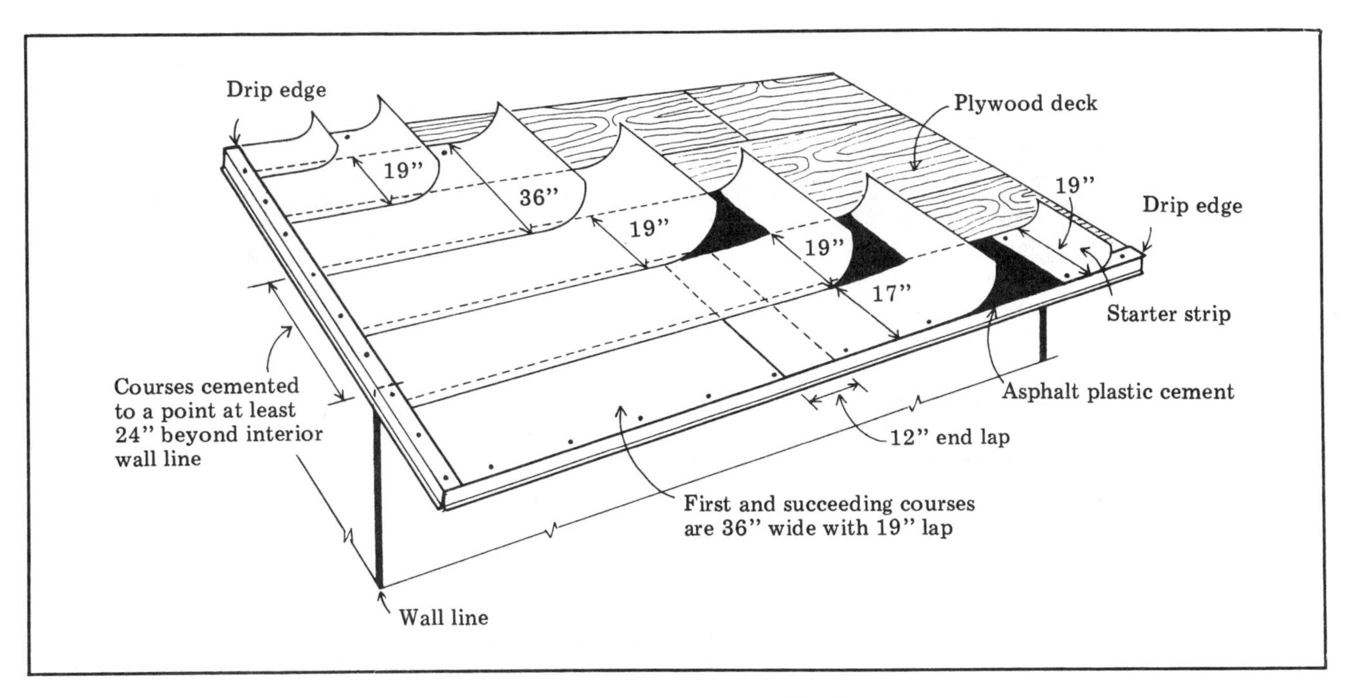

Low-slope underlayment application
Figure 13-55

Steep slopes tend to reduce the effectiveness of factory-applied, self-sealing adhesives, especially on colder or shaded areas of the roof. And don't even consider using asphalt shingles on a vertical side wall. The shingles can't get enough ventilation in that location.

If the roof slope is more than 21 inches per foot, use the steep-slope shingle application method described below. Details are shown in Figure 13-56.

Use the nails or staples recommended by the roofing manufacturer. And follow the manufacturer's directions on the number and position of fasteners for each shingle. If you use self-sealing shingles, don't drive fasteners into or above the factory-applied adhesive.

If you plan to cement the tabs down, do it right away. Use the quick-setting asphalt cement recommended by the manufacturer. Apply the adhesive in spots about the size of a quarter. For shingles with three or more tabs, place one spot of cement under each tab. For two-tab shingles, put two spots under each tab. For no-cutout shingles, place three spots of cement under the exposed portion of the shingle.

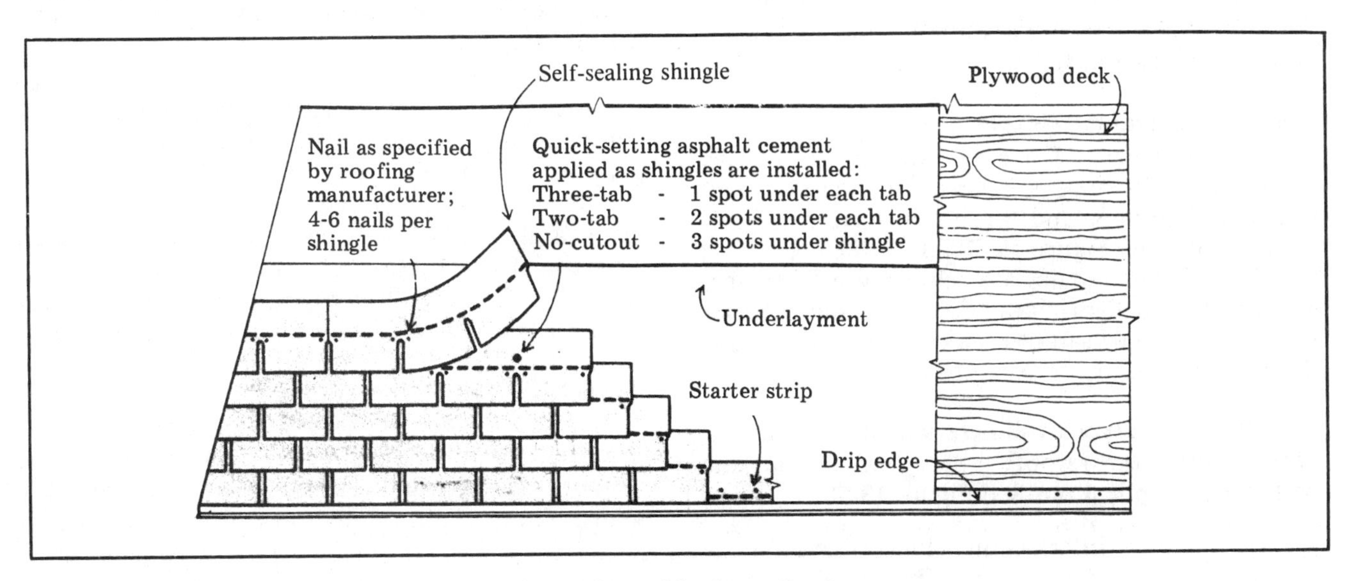

Steep-slope shingle application
Figure 13-56

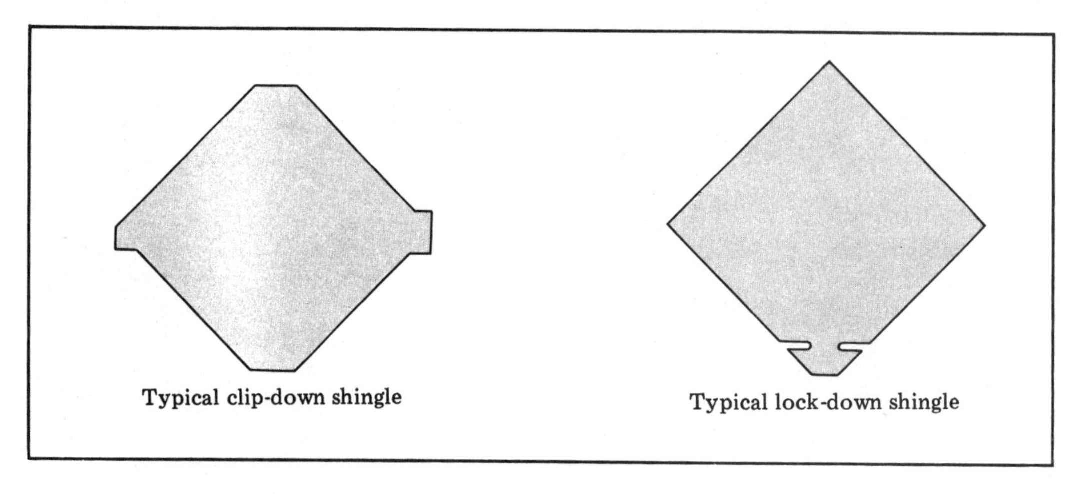

Individual "hex" shingles
Figure 13-57

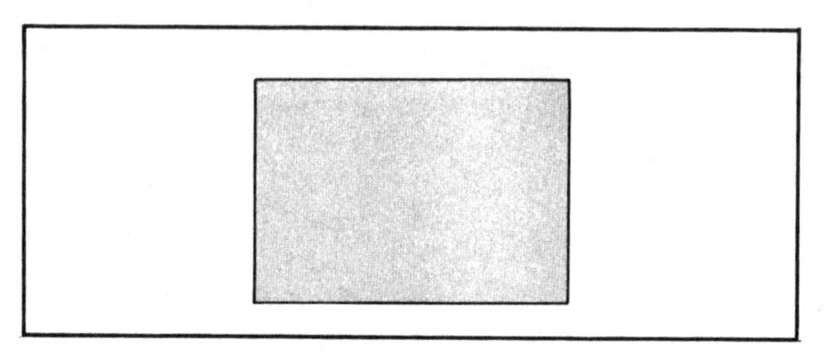

Individual giant shingle
Figure 13-58

Individual Shingles

Up to now, we've been talking about strip shingles. But individual shingles are available and are a good choice for some applications. Individual shingles come in three basic types: hexagonal, giant, and interlocking. Which type to use depends on slope, wind resistance, coverage, esthetics and cost. Each type has its own installation procedure.

Types of Individual Shingles

Hexagonal (Hex)— There are two common types of hexagonal shingles: those that are locked together by a clip, and those that have a built-in locking tab. See Figure 13-57. Both the clip-down and lock-down shingles are relatively lightweight, and are intended primarily for reroofing over an old roof. But they can be used on new construction if the slope is 4 inches or more per foot.

Giant— This type of shingle can be used for new construction or reroofing, depending on the method of application. See Figure 13-58.

Interlocking— Interlocking shingles have a locking device for wind resistance. The shingles can be used for reroofing over existing roofing, or on new construction when double coverage is required. Check your local building code before installing them on new roofs.

Placement of nails or staples is critical on interlocking shingles. Follow the shingle manufacturer's recommendations on nailing and on the use of starter material.

Although interlocking shingles are self-aligning, they're flexible enough to allow limited adjustment. Be sure to use horizontal and vertical chalk lines.

The integral locking tabs are manufactured to close tolerances. Engage the locking devices carefully and correctly. Figure 13-59 illustrates two common locking devices.

During installation, you may have to remove the locking tabs on the shingles along eaves and rakes. To prevent wind damage, cement these shingles down or nail them in place.

Preparing for Installation of Individual Shingles

Prepare the roof for individual shingles the same way you would prepare it for strip shingles. Be sure the roof is adequately ventilated. Sweep the roof deck and then cover it with an asphalt-saturated felt underlayment. Snap horizontal and vertical chalk lines. Install drip edges, eave flashing (if required) and valley flashing.

Shingle Application Methods

Dutch lap method— Used for reroofing over old roofing, provided that the old roofing has a smooth surface and offers good anchorage for nails. The Dutch lap method can also be used to cover new decks where single coverage gives enough protection. For either application, the slope should be at least 4 inches per foot.

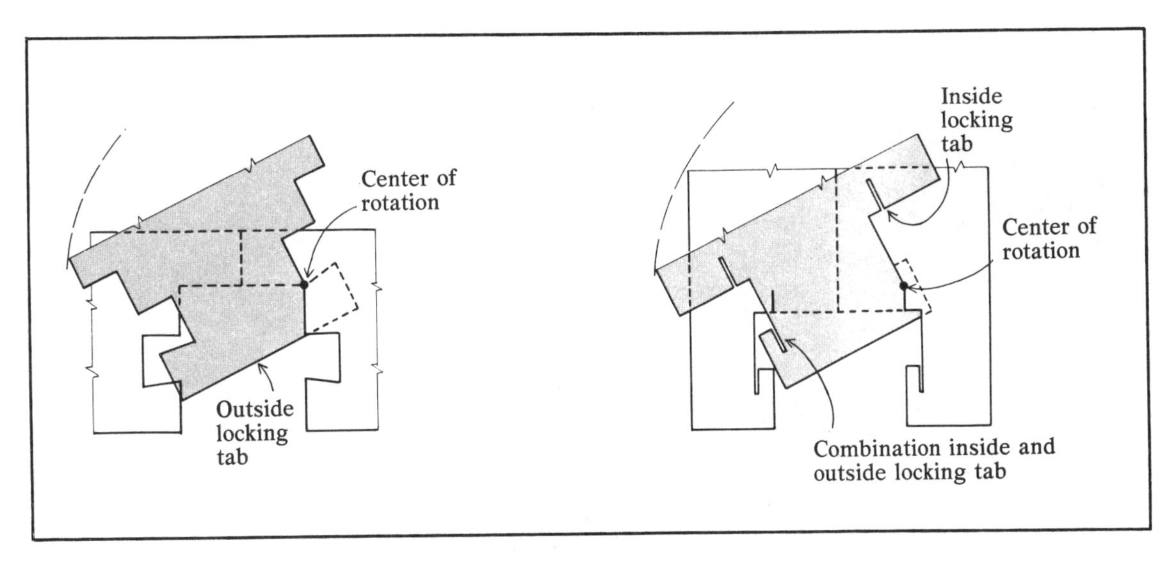

Common interlocking shingles and their methods of locking
Figure 13-59

American method— For either reroofing or new construction. In either case, the slope should be at least 4 inches per foot.

Roll Roofing

Asphalt roll roofing comes in 36-inch-wide sheets, in many weights and surface colors. It's used both as a final roof covering and as flashing under other roofing.

As the primary roof cover, roll roofing can be used on slopes down to 2 inches per foot when applied by the exposed-nail method. If you want to extend the life of roll roofing, use the concealed-nail method. With both methods, the roofing can be applied either parallel to the eave or parallel to the rake. Be sure to use the nails recommended by the manufacturer.

In temperatures below 45 degrees, store roll roofing in a warm place until ready for use. It's not good practice to apply roll roofing when the temperature is below 45 degrees. But if it's absolutely necessary to work with roll roofing in cold temperatures, be sure to warm the rolls before unrolling them. This will prevent the coating from cracking. Then cut the rolls into 12- to 18-foot lengths and spread them on a smooth surface until they flatten out.

Prepare the deck and install flashing the same way as for strip shingles. Valleys will be the open type.

All roll roofing is applied with a certain amount of top- and side-lapping. It is critical that you seal the laps properly. Use only the lap cement or asphalt plastic cement recommended by the roofing manufacturer. Store the cement in a warm place until ready for use. If you have to warm the cement an hour or two before application, put the unopened container in hot water. *Never* heat asphalt plastic cement directly over a flame. And don't try to thin the cement by diluting it with solvents.

Exposed-Nail Application Method

Figure 13-60 shows the exposed-nail method of applying roll roofing, including lapping, cementing and nailing.

First course— Lay a full-width sheet so that its lower edge and ends overhang the eave and rake 1/4-inch to 3/8-inch. Nail along a line 1/2- to 3/4-inch from and parallel to the top edge of the sheet, spacing the nails 18 to 20 inches apart. This preliminary top-nailing holds the sheet in place until the second course is placed over it and fastened. Nail along the eave and rake on a line 1 inch from and parallel to the edges of the roofing. Space the nails 3 inches on center and stagger them along the eave to avoid splitting the deck.

If two or more sheets must be used to complete the course, end-lap them 6 inches. Apply lap cement (the full width of the lap) on the underlying edge. Embed the overlapping sheet into it. Fasten the overlap with two rows of nails 4 inches apart, and 4 inches on center within each row. Stagger the rows so that the spacing is 2 inches between nails from row to row.

Second and following courses— Position the second course so that it overlaps the first course by 2 inches. Fasten the second course along the *top* edge following the same nailing directions as for the first course. Then lift the lower edge of the overlapping sheet, and apply lap cement evenly over the upper 2 inches of the first course. Embed the overlapping

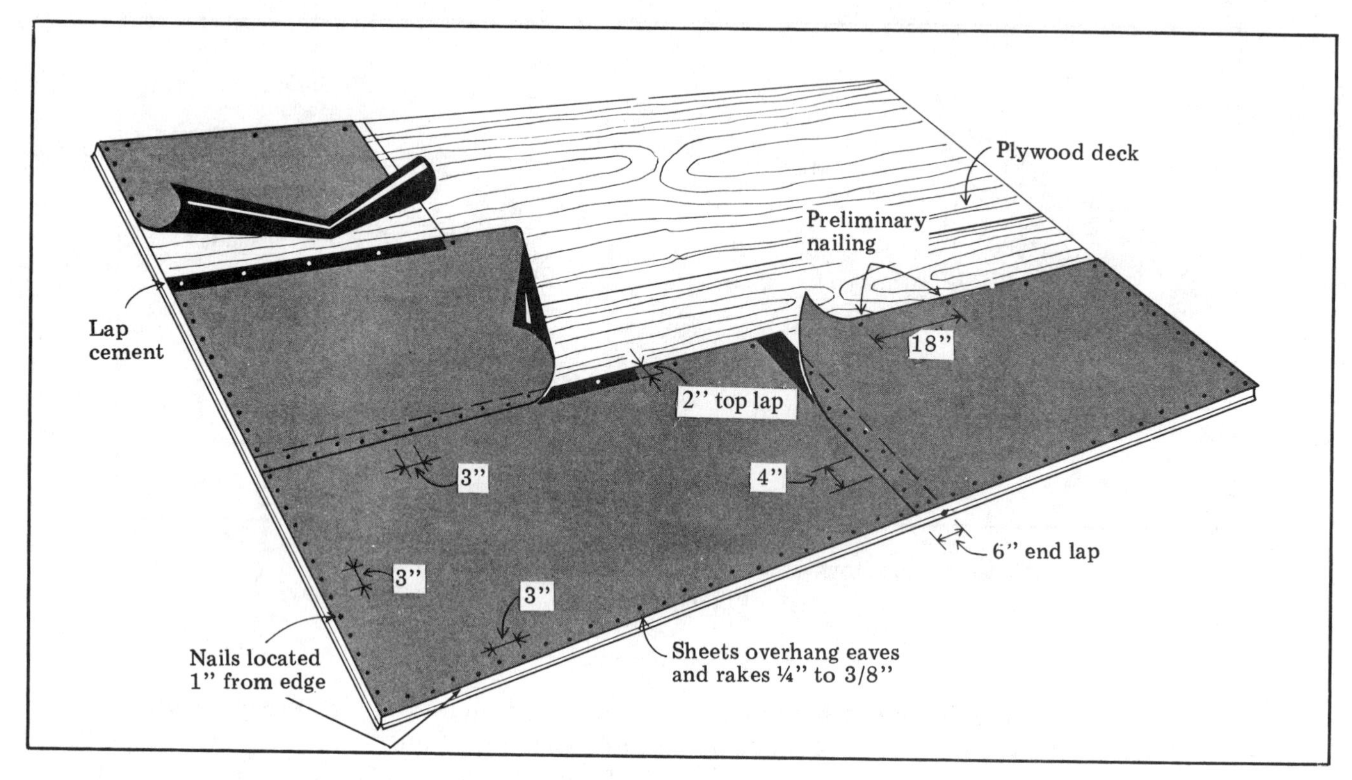

Exposed-nail method
Figure 13-60

sheet in the cement. Fasten the lap with nails spaced 3 inches on center and staggered slightly. Place the nails not less than 3/4-inch from the edge of the sheet. Nail the rake edges the same way as for the first course.

Follow the same procedure for each following course. End laps should be 6 inches wide, cemented and nailed the same way as for the first course. Stagger end laps so that an end lap in one course is never positioned over the end lap of a preceding course.

Hips and ridges— Trim, butt and nail the roofing as it meets a hip or ridge. On each side of the hip or ridge, snap a chalk line 5½ inches from the joint and parallel to it. Starting at the chalk lines and working toward the joint, spread a 2-inch-wide band of asphalt lap cement on each side of the hip or ridge. See Figure 13-61.

Cut strips of roll roofing 12 inches wide and fold them lengthwise along the centerline so that they will lie 6 inches on each side of the hip or ridge. In cold weather, warm the roofing before folding it. Lay the folded strip over the joint and embed it in the cement. Fasten the strip to the deck with two rows of nails, one row on each side of the hip or ridge. Each row should be located 3/4-inch from the edge of the strip. Space nails 3 inches on center. Be sure the nails penetrate the cemented zone underneath. This will seal the nail hole with cement. End laps should be 6 inches and cemented the full width of the lap. Don't apply the cement too thick.

Concealed-Nail Application Method

In the concealed-nail method, narrow edging strips are placed along the eaves and rakes before applying the roofing. Figure 13-62 shows the general installation procedure, including lapping, cementing and nailing.

Edging strips— Place 9-inch-wide strips of roll roofing along the eaves and rakes, positioning them to overhang the deck 1/4-inch to 3/8-inch. Fasten the strips with two rows of nails, located 1 inch and 8 inches from the roof edge. Nails are spaced 4 inches on center in each row.

First course— Position a full-width strip of roll roofing so that its lower edge and ends are flush with the edging strips at the eaves and rakes. Fasten the upper edge with nails spaced 4 inches on center and slightly staggered. Locate the nails so that the next course will overlap them a minimum of 1 inch. Lift the lower edge of the first course, and cover the edging strips with cement according to the manufacturer's specifications.

In cold weather, turn the course back carefully to avoid damaging the roofing material. Press the eave edges and rake ends of the first course firmly into the cement-covered edging strips. Work from one side of the sheet to the other to avoid wrinkling or bubbling.

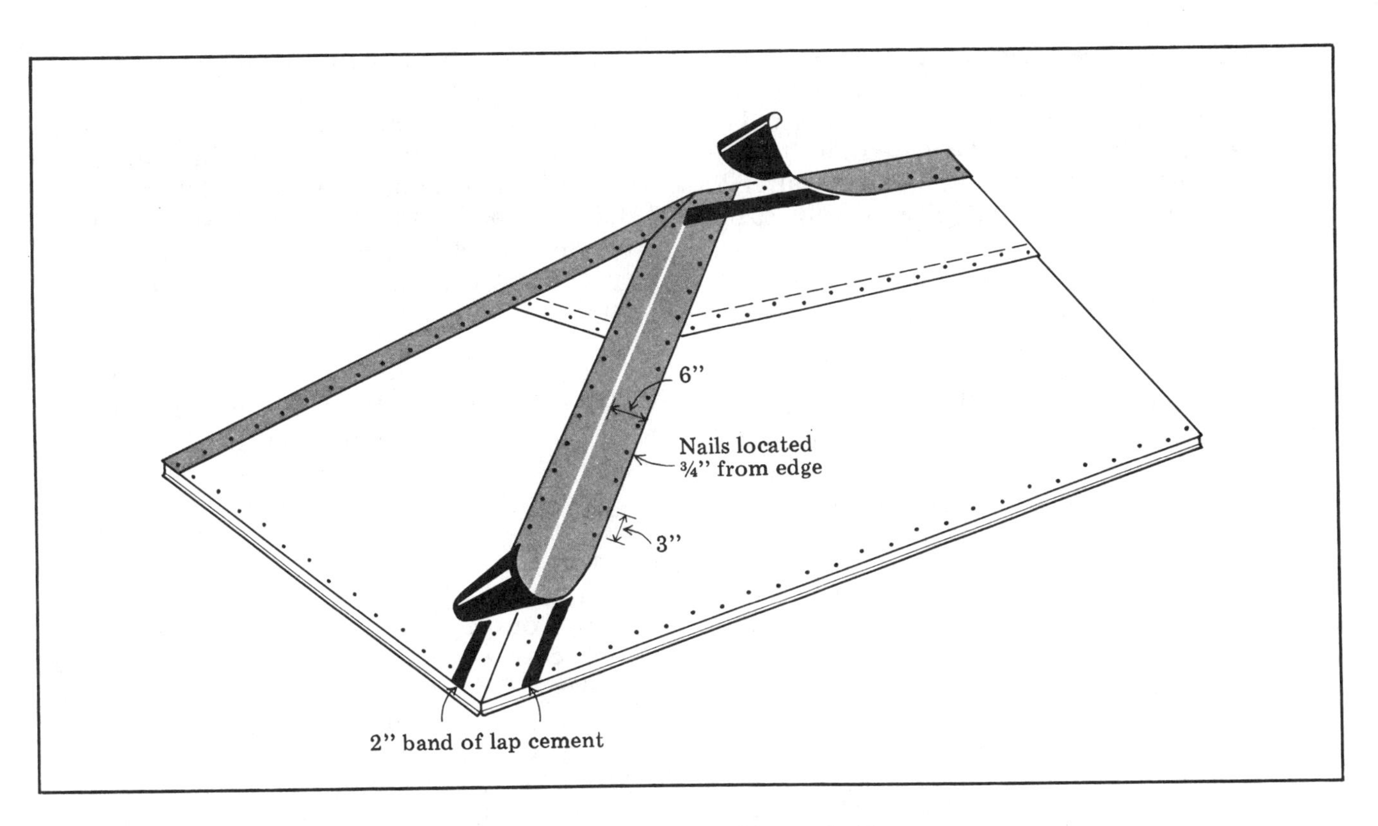

Exposed-nail method at hips and ridges
Figure 13-61

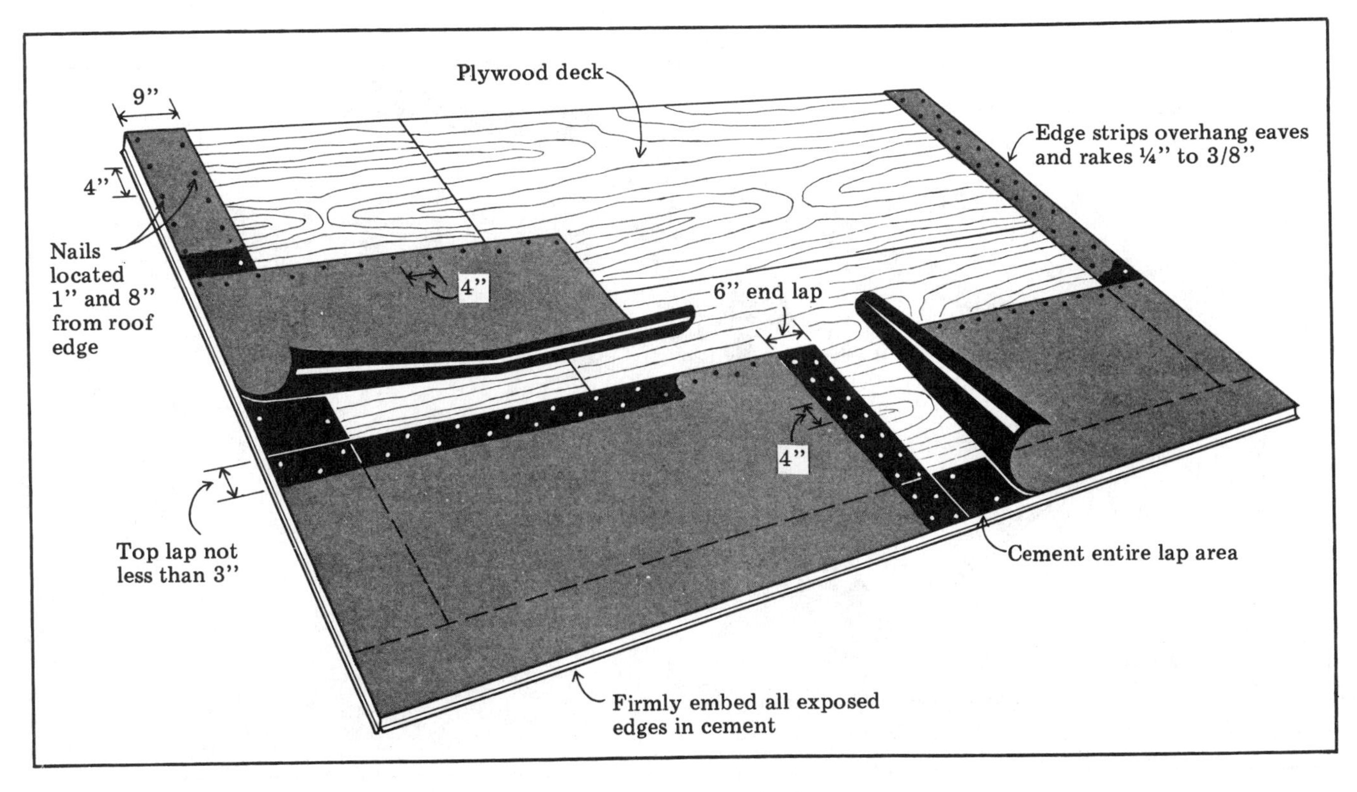

Concealed-nail method
Figure 13-62

End laps should be 6 inches wide and cemented the full width of the lap. Nail the underlying sheet with two rows of nails, 1 inch and 5 inches from the end of the sheet. Nails should be spaced 4 inches on center and slightly staggered. Make sure that end laps in adjacent courses don't line up with each other.

Second and following courses— Position the second course so that it overlaps the first course by at least 3 inches. Fasten the top edge to the deck, cement the laps and finish installing the sheet in the same way you did the first course. Follow the same procedure for each following course. Don't apply nails within 18 inches of the rake until the cement has been applied to the edging strip and the overlying strip has been pressed down.

When cementing the laps, apply the cement in a continuous layer across the full width of the lap. Press the lower edge of the upper course firmly into the cement until a small bead appears along the edge of the sheet. Use a roller to apply uniform pressure over the entire cemented area.

Hips and ridges— Trim, butt and nail the sheets where they meet at a hip or ridge. Next, cut 12 x 36-inch strips from the roll roofing and fold them lengthwise to lay 6 inches on each side of the joint. In cold weather, warm the strips before folding them. These strips will be used as "shingles" to cover the joint. Each one overlaps the next by 6 inches, as shown in Figure 13-63.

Start hips at the bottom and ridges at the end opposite the direction of the prevailing wind. To guide the installation, snap a chalk line 5½ inches from, and parallel to, the joint on each side of it. Apply asphalt plastic cement evenly over the entire area between the chalk lines. Fit the first folded strip over the joint and press it firmly into the cement.

Next, drive two nails 5½ inches from the end that will be overlapped by the next strip. Apply lap cement evenly over the 6-inch lap area. Then place the second strip over the first one. Nail and cement the second strip the same way as the first one. Continue the same procedure until the hip or ridge is finished.

Double-Coverage Roll Roofing

Double-coverage roll roofing comes in 36-inch-wide sheets. There are 17 inches of exposure and 19 inches of selvage edge. You can use double-coverage rolls on slopes down to 1 inch per foot.

The 17-inch exposed portion is covered with mineral granules. The 19-inch selvage portion can be saturated with asphalt or saturated and coated.

The selvage edge and all end laps should be cemented according to the manufacturer's recommendations. It's important to follow the directions carefully.

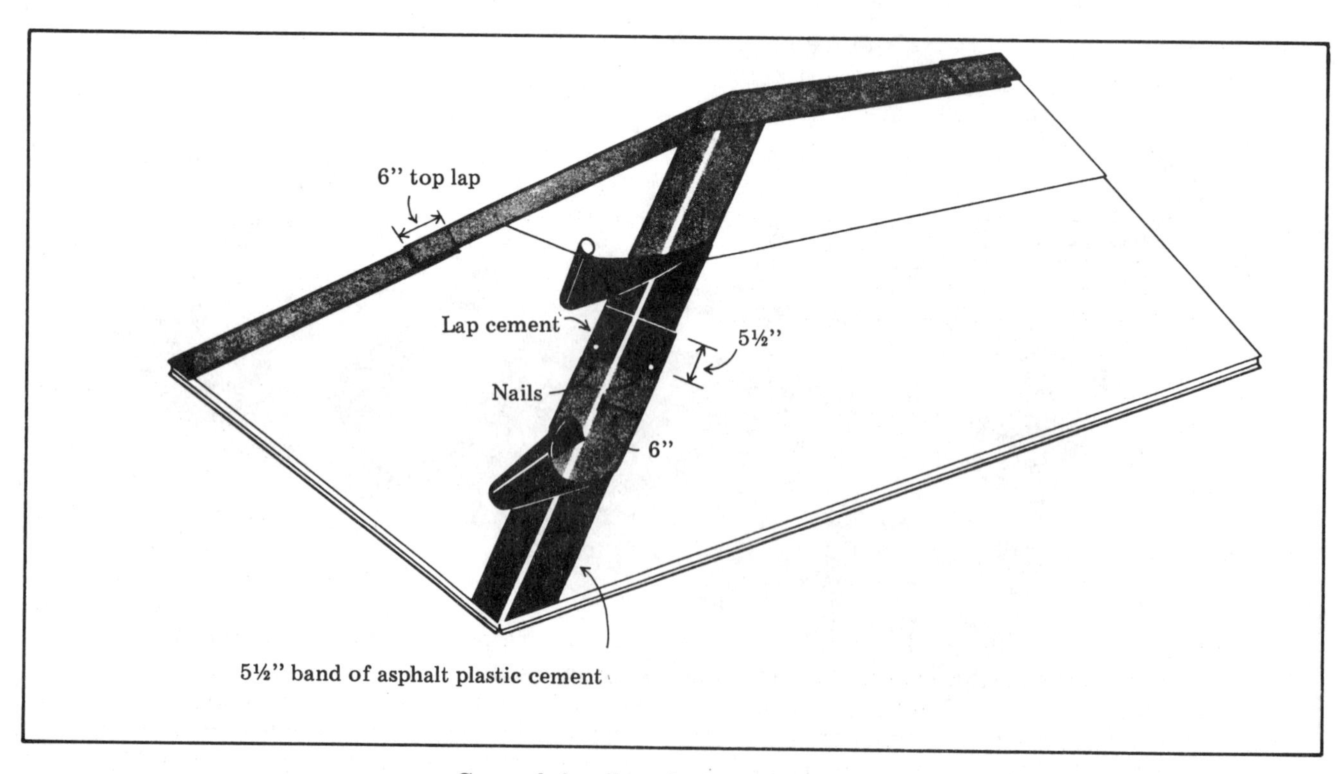

Concealed-nail method at hips and ridges
Figure 13-63

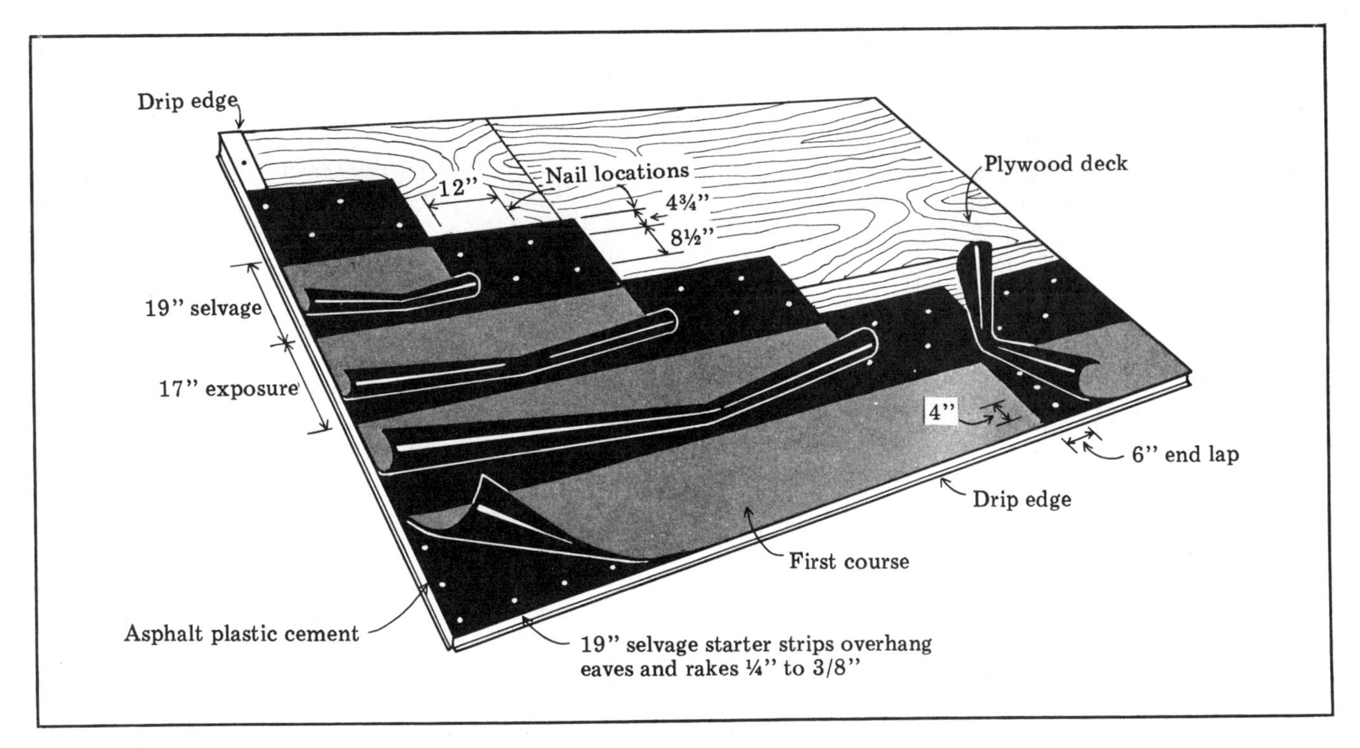

Double-coverage roll roofing parallel to the eave
Figure 13-64

Make certain there is adequate roof drainage. This is especially important on the low slopes where double-coverage roofing is commonly used. Choose the correct type and length of nail to fit the application.

Observe the same precautions concerning storage, application temperature and warming of the rolls as for single-coverage roll roofing. Store asphalt cement in a warm place until ready for use.

You can install double-coverage roll roofing either parallel to the eave or parallel to the rake. Although 19-inch-selvage roll roofing is discussed here, any roll roofing can be used if the selvage portion of the sheet is 2 inches wider than the exposed portion.

Before applying double-coverage roll roofing, prepare the deck and install flashing the same way as described for strip shingles. Valleys will be the open type.

Installation Parallel to the Eave

Starter strip— First, remove the 17-inch granule-surface portion from a sheet of double-coverage roll roofing. Place the remaining 19-inch selvage portion parallel to the eave so that it overhangs the drip edge 1/4-inch to 3/8-inch at the eave and rake. Fasten it to the deck with two rows of nails, one row on a line 4¾ inches from the top edge of the strip, the other on a line 1 inch above the lower edge of the strip. Space the nails 12 inches on center, staggering them slightly in each row. See Figure 13-64.

First course— Cover the entire starter strip with asphalt plastic cement. Then position a full-width sheet over the cement. Place the sheet so that the end and lower edge of the granule-surface portion are flush with the rake and eave edges of the starter strip. Fasten the sheet to the deck with two rows of nails on the selvage portion of the strip. Locate the first row 4¾ inches below the upper edge, and the second row 8½ inches below the first row. Space the nails 12 inches on center, staggering them slightly in each row.

Second and following courses— Position each following course so that it overlaps the full 19-inch selvage width of the course below. Nail the selvage portion the same way as for the first course. Turn the sheet back and apply cement to the full selvage width of the underlying sheet. In cold weather, turn the sheet back carefully to avoid damaging it.

Spread the cement to within 1/4-inch of the edge of the exposed portion. Press the overlying sheet firmly into the cement. To insure complete adhesion between sheets, apply pressure over the entire lap with a broom or light roller. It's important to apply the cement so that it flows to the edge of the overlying sheet. But don't use too much cement.

End laps— All end laps should be 6 inches wide. Fasten the underlying granule-surface portion of the lap to the deck with a row of nails 1 inch from the edge. Space the nails 4 inches on center. Then spread asphalt plastic cement evenly over the lap area. Fasten the overlying sheet to the deck with a row of nails on a line 1 inch from the edge of the lap. Space the nails 4 inches on center. Stagger all end laps so that those in following courses don't line up with each other.

Caution: Never cement roll roofing directly to the deck. This would cause the sheets to split when the deck moves and shrinks.

Installation Parallel to the Rake

With this method, the sheets are applied vertically from the ridge to the eaves. Begin by applying starter strips on both rakes. Use the same procedure as you used for horizontal application. Cover the starter strip with asphalt plastic cement and apply a full-width sheet over it for the first course. Position all top laps so that the overlapping sheet is the closest to the ridge. This carries drainage over the joint rather than into it. The remainder of the installation is then the same as if you were applying rolls parallel to the eave. Figure 13-65 shows installation parallel to the rake.

Hips and ridges— Trim, butt nail the roofing sheets as they meet at a hip or ridge. To guide the installation, snap chalk lines on each side of the joint. The chalk lines should be 5½ inches from, and parallel to, the joint.

Next, cut 12 x 36-inch strips of roll roofing that include the selvage portion. Fold the strips lengthwise to lie with 6 inches on either side of the joint. In cold weather, be sure to warm the strips before folding them. Start applying the strips at the lower end of the hip, or at the end of the ridge opposite the direction of prevailing wind.

Cut the selvage portion from one strip to use as a starter. Fasten this strip in place by driving nails 1 inch from each edge along the full length of the strip. Space the nails 4 inches on center. Cover the entire strip with asphalt plastic cement. Fit the next folded strip over the starter strip, and press it firmly into the cement. Nail it the same way as the starter, but only on the selvage portion. Cover the selvage portion with cement. Press the next strip over it, and nail and cement that selvage portion. Continue the process until the hip or ridge is completed. Figure 13-66 illustrates the procedure.

Double-coverage roll roofing is common on shed roofs that have no hips or ridges. To finish this type of roof, trim and nail the selvage portion of the last course to the edge of the roof. Then trim the exposed, granule-surface portion that was cut from the starter strip and fit it over the final selvage

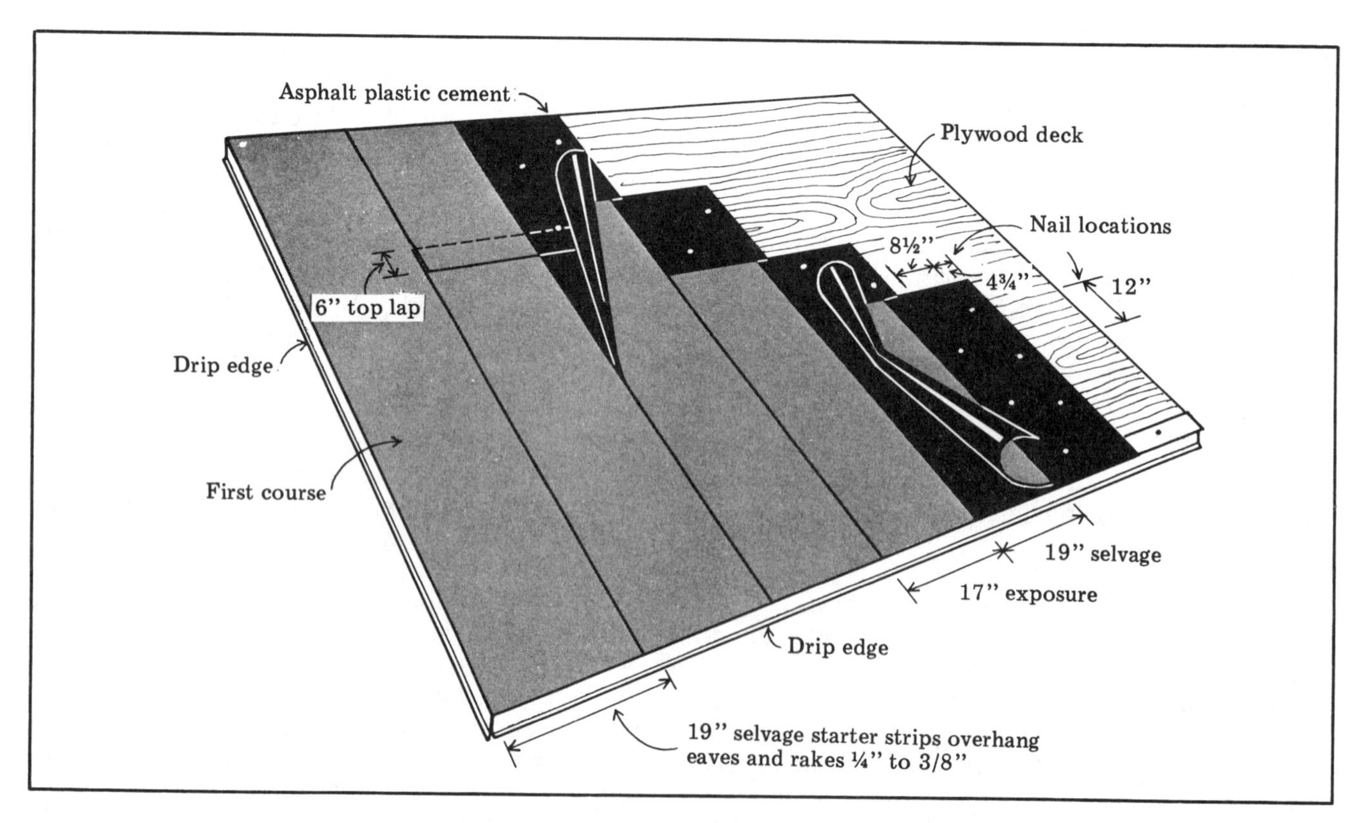

Double-coverage roll roofing parallel to the rake
Figure 13-65

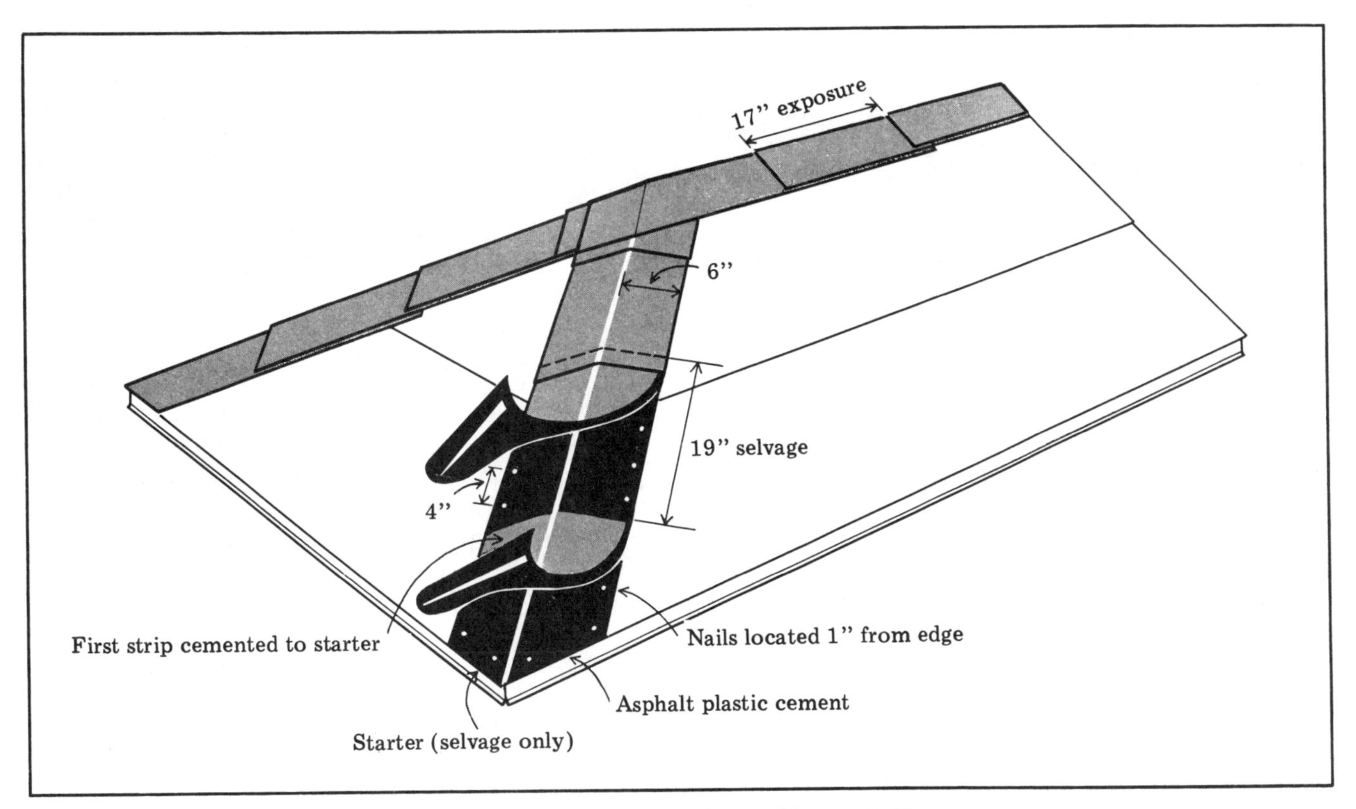

Double-coverage roll roofing at hips and ridges
Figure 13-66

portion. Cement it into place. Finally, overlay the entire edge with metal flashing and cement the flashing into place.

Shading

When a completed asphalt shingle roof is viewed from different angles, certain areas may appear darker or lighter. This difference in appearance is called *shading*. Shading varies with the position of the sun and the overall intensity of light. For example, slanting sun rays emphasize shading. Direct, overhead rays cause color shading to disappear.

Shading is caused by slight variations in the surface texture of shingles. It's a result of the normal manufacturing process, and does not affect the performance of the shingles.

Shading is usually more noticeable on black or dark-colored shingles. They reflect only a small part of the light shining on them, magnifying the slight differences in surface texture. White and light-colored shingles reflect a great amount of light, reducing noticeable shading. Blends made from a variety of colors actually tend to camouflage shading.

"Racking" or straight-up application accentuates shading. Be sure that shingles are applied across and diagonally up the roof. This blends shingles from one bundle into the next, minimizing the variations in texture.

Wood Shingles

Use No. 1 grade wood shingles for house roofs. They're all-heartwood, all-edge-grain, and tapered. For side walls, use No. 2 grade shingles. Most shingles are made from either Western red cedar or redwood. The heartwood of these trees is decay-resistant, and shrinkage is minimal.

Four bundles of 16-inch shingles laid 5 inches "to the weather" will cover 100 square feet. Wood shingles always come in random widths. The recommended exposures for standard shingle sizes are shown in Figure 13-67.

Figure 13-68 illustrates the right way to apply wood shingles. Underlayment or roofing felt is not required for wood shingles, except for protection of possible ice-dam areas. In areas with damp climates, spaced roof boards will help dry the shingles. Spaced or solid sheathing can be used in dry climates.

Here are some general rules to follow when applying wood shingles:

1) Shingles should extend about 1½ inches beyond the eave line and about 3/4-inch beyond the rake edge.

2) Use two rust-resistant nails in each shingle. Space them about 3/4-inch from the shingle edge and 1½ inches above the butt line of the next

Shingle Length (Inches)	Shingle Thickness (Green)	Maximum Exposure	
		Slope Less[1] than 4 in 12 inch	Slope 4 in 12 and over inch
16	5 butts in 2"	3¾	5
18	5 butts in 2¼"	4¼	5½
24	4 butts in 2"	5¾	7½

[1] Minimum slope for main roofs — 4 in 12. Minimum slope for porch roofs — 3 in 12.

Recommended exposures for wood shingles
Figure 13-67

course. Use 3d nails for 16- and 18-inch shingles, and 4d for 24-inch shingles on new construction. When applying shingles over plywood roof sheathing less than 1/2-inch thick, a ring-shank (threaded) nail is often recommended.

3) The first course of shingles is doubled. In all courses, allow a 1/8- to 1/4-inch space between shingles so that the shingles will have room to expand when wet. The joints between shingles should be offset at least 1½ inches from the joints in the course below. And the joints in following courses should be spaced so that they do not line up directly with joints in the second course below.

4) Shingle away from the valleys. And set aside the wide shingles as you work. Save these for use in the valleys.

5) Install metal edging along the gable end to help guide water away from the side walls.

6) When laying No. 1 all-heartwood, edge-grain shingles, it is not necessary to split the wide shingles.

Figure 13-69 shows a wood shingle job in progress.

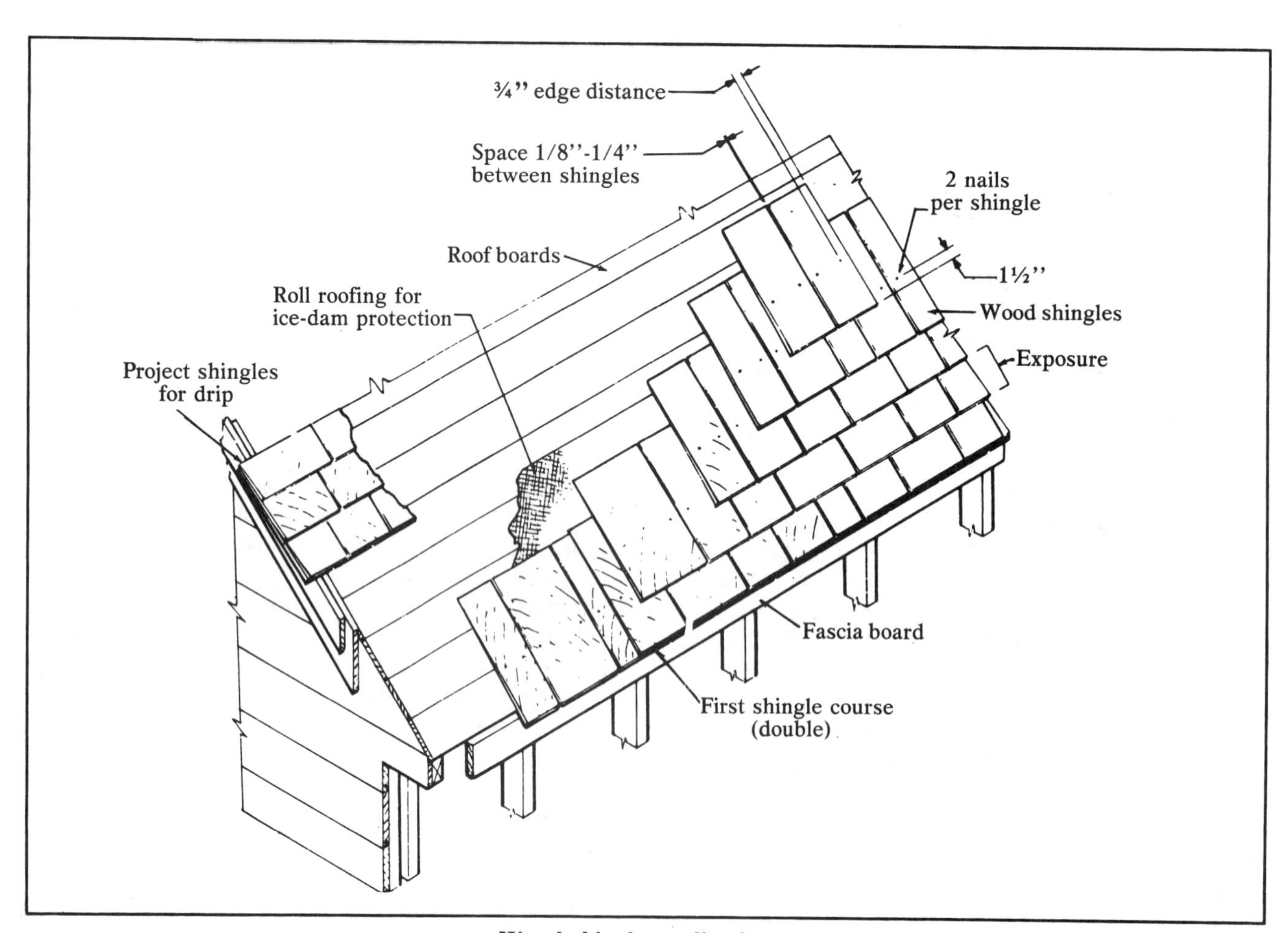

Wood shingle application
Figure 13-68

Wood shingle application on solid deck
Figure 13-69

Wood shakes are applied about the same as wood shingles. But shakes are much thicker (longer shakes have the thicker butts), so you have to use long galvanized nails. To create a rustic appearance, the butts can be laid unevenly.

Because shakes are longer than shingles, they have a greater exposure. Exposure distance is usually 7½ inches for 18-inch shakes, 10 inches for 24-inch shakes, and 13 inches for 32-inch shakes.

Shakes are not smooth on both faces. This makes it possible for wind-driven snow or rain to enter the attic between the shakes. That's why it's essential to use underlayment between courses. Use an 18-inch-wide layer of 30-lb. asphalt felt. Position the felt so that the bottom edge is located above the butt edge of the shakes a distance equal to double the weather exposure.

Use a 36-inch-wide starting strip of asphalt felt at the eave line. Use solid sheathing for the roof deck if wood shakes are used in areas where wind-driven snow is common.

Other Roof Coverings

Galvanized metal, built-up roofing, asbestos, slate, and tile make attractive and durable roofs. But these materials are less common than asphalt and wood shingles. The cost is almost always more than asphalt or wood shingles, and special tools and knowledge are needed for installation. Leave this work to knowledgeable, experienced subcontractors.

Estimating Manhours

Manhours for applying roofing materials are based on the square (100 square feet of roof). An experienced roofer will lay 3-tab asphalt shingles at a rate of about 1.5 squares per hour. A cut-up roof will take longer. The following chart shows a breakdown on labor and materials for asphalt roofing:

Type of roofing	Shingles per square	Manhours per square
Roll roofing	----	1.25
3-tab sq. butt	80	1.5
Hex strip	86	1.5
Giant American	226	2.5
Giant Dutch Lap	113	1.5
Individual Hex	82	1.5

The next chart gives a breakdown of labor for wood shingles and shakes per square:

Type of Roofing	Manhours per square
Wood shingles	
16" perfections, 5½" exposure	2.3
18" perfections, 6" exposure	2.0
Split or sawn cedar shakes	
10" exposure, 24" long, ½" to ¾" thick	3.0
White cedar	
16" long, 5" exposure	2.5
Fire-retardant cedar shingles	
16", 5" exposure	2.6
18" perfections, 5½" exposure	2.2

Roofing is a job I prefer to sub out. Most builders find that a good sub can do the work cheaper than they could do it themselves. Carpenters are generally slow roofers. Applying shingles or roll roofing is something you have to do every day to become proficient. The job requires leg muscles that non-roofers aren't accustomed to using. For anyone over 30, the second day on the roof will result in slow production due to aching muscles.

But every builder needs to know the difference between a good roof and a bad one. There are good roofers and careless roofers. Roofing is generally subcontracted on a per square basis. You're buying pure production from the lowest bidder. It's up to you, the general contractor, to control quality. And what you don't know *can* hurt you. Use this chapter as your reference guide to building a good roof. If your sub isn't following the procedures outlined here, he may not be doing a professional job.

Summary

A good-looking roof helps sell a house. This chapter is your guide to a neat, watertight, durable roof — whether your own crew does the work or you subcontract the job.

Figure 13-70 is your Roof Covering Checklist and Data Reference Sheet.

Roof Covering Checklist and Data Reference Sheet

1) The house is ________ feet in length and ________ feet wide, for a total of ________ square feet.

2) The house is ☐1-story ☐1½-story ☐2-story.

3) The roof is:

☐Plain gable, ______/12 pitch

☐Gable with dormers, ______/12 pitch

☐Gable dormers. Number: ____________

☐Shed-type dormers. Number: ____________

☐Variable gable

☐Hips, valleys, cut-up

☐Other (specify) ____

☐Hip, ______/12 pitch

☐Flat

☐Other (specify) ____

4) The roof covering is:

Asphalt shingles

☐Self-sealing random tab strip (multi-thickness)

☐Self-sealing random tab strip (single-thickness)

☐Self-sealing square tap strip (three-tab)

☐Self-sealing square tab strip (no-cutout)

☐Nonself-sealing square tab

☐Individual interlocking

Asphalt roll (weight: ____________)

☐Mineral-surface roll

☐Single coverage

☐Double coverage

☐Smooth-surface roll

5) Number 15-pound saturated felt (non-perforated) underlayment ☐was ☐was not used.

Figure 13-70

6) Metal drip edge ☐was ☐was not used.

7) Color of roofing material is ☐white ☐black ☐grey ☐red ☐brown ☐green.

8) Shingle exposure is ☐4" ☐4½"☐5" ☐5½" ☐5⅝" ☐6".

9) Roll roofing exposure is ☐17" ☐32" ☐33" ☐34".

10) Roll roofing nailing method: ☐exposed ☐concealed

11) Total roof area is ________ square feet.

12) Fasteners used:

☐Staples, size ____________

☐Roofing nails, size ________ ☐aluminum ☐steel.

13) Eave flashing (for ice dams) ☐was ☐was not installed.

14) Cements and coatings used:

☐Asphalt plastic cement

☐Lap cement

☐Quick-setting asphalt cement

☐Masonry primer

15) Number of valleys flashed: ____________

Materials used:

☐15-pound felt

☐50-pound roll roofing

☐90-pound mineral-surface roll roofing

16) Valleys are ☐open ☐closed-cut ☐woven.

17) Starter strips used:

☐Trimmed shingles

☐Strip of mineral-surface roll roofing

☐Manufactured starter strips

18) Horizontal and vertical chalk lines ☐were ☐were not used as a guide for roofing application.

19) Shingle application method was ☐6-inch method ☐5-inch method ☐4-inch method.

20) Step flashing with metal "flashing shingles" ☐was ☐was not installed at vertical side walls.

21) Continuous metal flashing strip ☐was ☐was not used against vertical front wall.

22) Material costs:

Asphalt shingles	______sqs	@	$______sq	=	$______
Roll roofing	______rolls	@	$______ea	=	$______
Felt underlayment	______rolls	@	$______ea	=	$______
Metal drip edge	______LF	@	$______LF	=	$______
Nails (size ____)	______lbs	@	$______lb	=	$______
Staples (size ____)	______lbs	@	$______lb	=	$______
Eave flashing	______rolls	@	$______ea	=	$______
Asphalt plastic cement	______gals	@	$______gal	=	$______
Lap cement	______gals	@	$______gal	=	$______
Quick-setting cement	______gals	@	$______gal	=	$______
Masonry primer	______gals	@	$______gal	=	$______
Valley flashing:					
15-lb.	______rolls	@	$______ea	=	$______
50-lb.	______rolls	@	$______ea	=	$______
90-lb.	______rolls	@	$______ea	=	$______
Starter strips	______pcs	@	$______ea	=	$______
Step flashing shingles	______pcs	@	$______ea	=	$______
Soil stack and vent flashing	______pcs	@	$______ea	=	$______
Chimney flashing:					
Crickets	______pcs	@	$______ea	=	$______
Metal step flashing	______pcs	@	$______ea	=	$______
Rear base flashing	______pcs	@	$______ea	=	$______
Cap flashing	______pcs	@	$______ea	=	$______

Other materials, not listed above:

________________	__________	@	$__________	=	$__________
________________	__________	@	$__________	=	$__________
________________	__________	@	$__________	=	$__________
________________	__________	@	$__________	=	$__________

Total cost of all materials $__________

23) Labor costs:

☐Roofing job subcontracted to:

Name:____

Address:____

Telephone:____

Cost (per square) $__________

☐Roofing job handled by my crew:

Skilled	______hours	@	$______hour	=	$________
Unskilled	______hours	@	$______hour	=	$________

Total labor costs:

Skilled	______hours	@	$______hour	=	$________
Unskilled	______hours	@	$______hour	=	$________
Other (specify)					
________________	______hours	@	$______hour	=	$________

Total $________

24) Total cost:

Materials $__________

Labor $__________

Other $__________

Total $__________

25) Problems that could have been avoided ____________________

____

____

____

____

26) Comments ____________________

____

____

____

____

27) A copy of this Checklist and Data Reference Sheet has been provided:

☐ Management

☐ Accounting

☐ Estimator

☐ Foreman

☐ File

☐ Other: ____________

Chapter 14

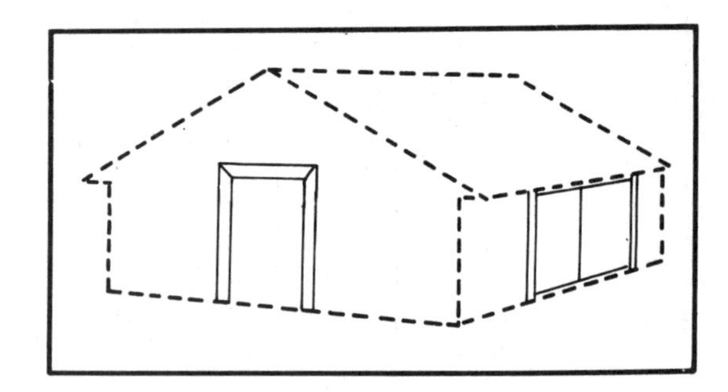

Doors

Prehung exterior and interior doors can save money and increase your profit on every job. These units are completely framed and ready to install. The holes for the lockset are bored and ready for installation. Prehung doors are the best thing to happen to home builders since the development of sheet rock in the late 1940's. If you're not using these units, you're probably spending too much time and money on carpentry.

Exterior Doors and Frames

Exterior doors are 1¾ inches thick. The standard height is 6 feet 8 inches. Most main entrance doors are 3 feet wide, and side or service doors are 2 feet 8 inches wide.

The frames for exterior doors are made of 1⅛-inch or thicker material. Rabbeting of the side and head jambs provides a stop for the door. See Figure 14-1.

The wood sill is usually oak. Oak is wear-resistant. When softer species are used, a metal nosing and wear strips are included. The outside casing provides space for the 1⅛-inch combination or screen door.

The frame is nailed (through the outside casing) to the studs and headers of the rough opening. The sill of most prehung units can rest on the subfloor. Underlayment, such as 5/8-inch-thick particleboard, is then butted to the sill. The finish floor covering is also butted to the sill.

In some prehung units, the sill must rest firmly on the header or stringer joists of the floor framing, and the joists usually must be trimmed. When the finish flooring is in place, a hardwood or metal threshold with a plastic weather stop covers the joint between the floor and sill. In on-slab construction, the sill can be omitted.

The exterior trim around the main entrance door can vary from a simple casing to a molded pilaster with a decorative head casing. Keep decorative designs compatible with the architectural style. Contemporary homes can use a wide variety of door and entry designs. And manufacturers produce a variety of millwork compatible with these designs.

If there's an entry hall, consider having glass in the main door (or a side light) to admit natural light during the daytime.

Types of Exterior Doors

Doors are classified by their size, system of paneling, and construction. The stiles are either solid or built-up. Built-up stiles are made from several pieces of lumber that are glued together and covered with veneer on the exposed surfaces. Stiles and rails can be molded in any of several forms. See Figure 14-2.

Exterior doors are available to fit almost any style of house. The traditional pattern is the common panel type. See Figure 14-3. These have stiles (solid vertical members), rails (solid cross members), and filler panels in a number of designs. Glazed upper panels can be combined with raised wood or plywood lower panels.

Exterior flush doors should be solid-core rather than hollow-core. A solid core tends to minimize

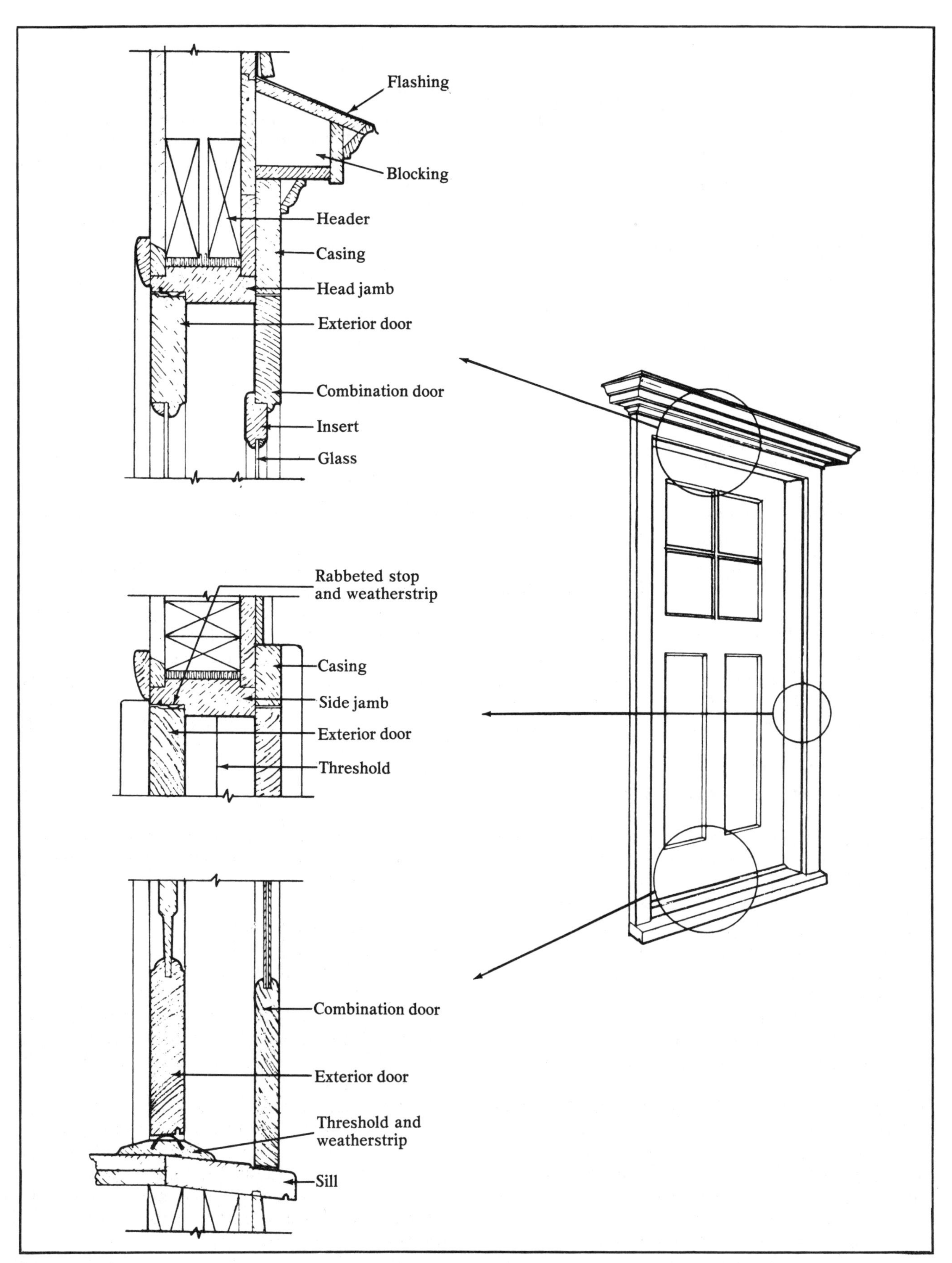

Exterior door and combination door
Figure 14-1

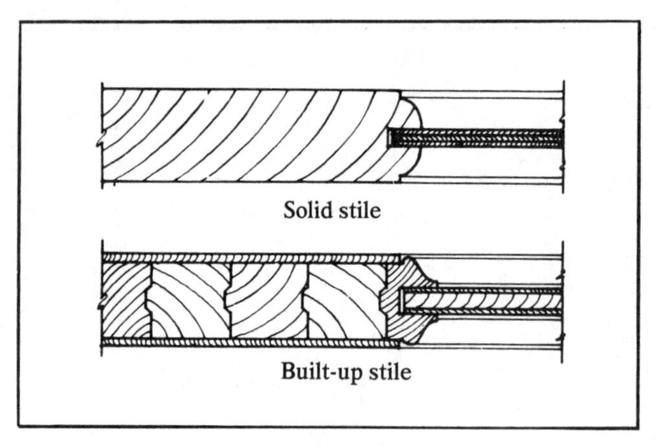

Door stiles
Figure 14-2

warping during the heating season. Warping is caused by the difference in the moisture content on the exterior and interior faces.

Flush doors have a thin plywood face applied over a framework of wood. The core can be wood block, composition mineral or particleboard. Common designs range from plain flush doors to a variety of panels and glazed openings. See Figure 14-3.

Wood combination doors have screened sections that can be opened in warm weather, and storm closures that give added insulation in cold weather. The screen and storm inserts are normally located in the upper portion of the door. Some have self-storing features so that the storm door is stowed internally like in a combination window.

Be aware that heat loss through metal combination doors is greater than through wood doors. However, insulated metal doors have excellent heat-retention properties.

Prehung exterior doors nearly always come equipped with aluminum or bronze weatherstripping. When installing the door, apply a bead of caulk on the sheathing where the door casing will make contact. This seals the joint between the casing and sheathing.

Interior Doors

Interior door installation is one of the last items in the construction sequence. Be sure that all sheet rock and paneling are finished *before* installing the interior doors. Don't install interior doors until the finish floor is in place. Carpeting and resilient flooring are put down after all carpentry and painting are done.

Types of Interior Doors

Two common types of interior doors are flush and panel doors. Novelty doors, such as the folding door unit, may be flush or louvered. Most standard interior doors are 1⅜ inches thick.

Flush interior doors usually have a hollow core covered with a thin veneer of plywood or hardboard. See Figure 14-4. Plywood-faced flush doors are available in gum, birch, oak and mahogany. All of these are suitable for a natural finish. Nonselect grades and hardboard-faced doors are usually painted.

The panel door has solid stiles, rails, and panel fillers. The five-cross panel and the colonial panel doors are perhaps the most common. See Figure 14-4. The louvered door is popular for closets because it provides ventilation even when closed. Large openings for wardrobe closets are finished with sliding or folding doors, or with flush or louvered doors. Wardrobe closet doors are usually 1⅛ inches thick.

Hinged doors should open or swing in the direction of natural entry, and against a blank wall whenever possible. Don't install them in a location where they will obstruct other swinging doors.

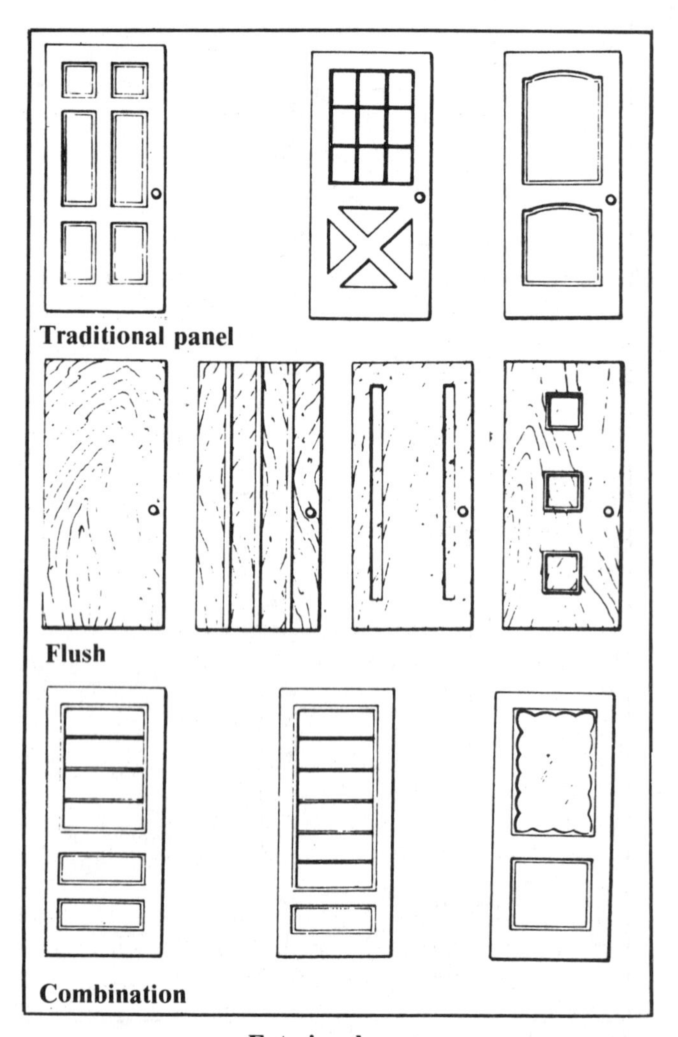

Exterior doors
Figure 14-3

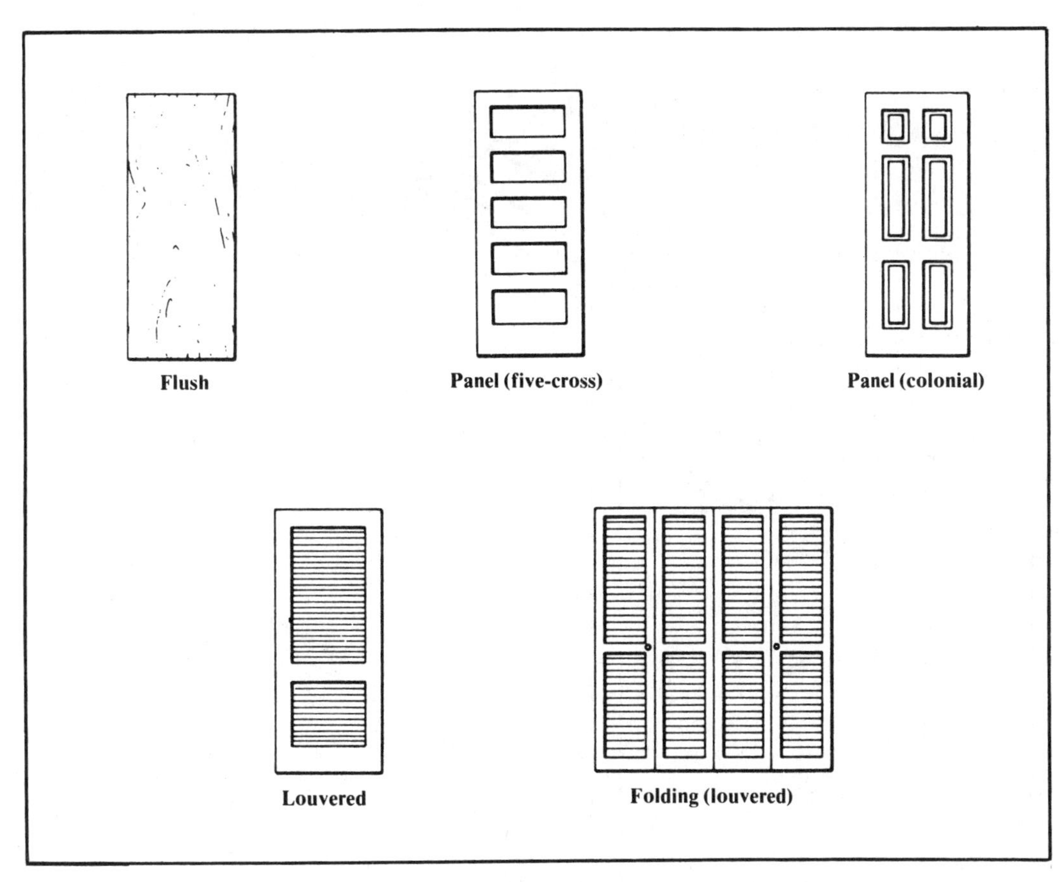

Interior doors
Figure 14-4

With the exception of closet doors, don't let doors swing into a hallway. And keep light switches and wall receptacles away from a location blocked by an open door.

Don't stockpile doors on the job. Doors shouldn't be delivered to the job until you're ready to install them. They'll be better protected against moisture, damage and theft if left at the lumber yard until you're ready for them. When they are delivered, set them upright on end.

Installing Prehung Doors

Prehung doors go directly into the rough opening. Even prehung double-door units are installed this way. So it's important to know the exact rough opening dimension. Check the rough opening door size *before* framing the wall. Also, be careful to specify the exact wall thickness. Once you've got the right size door to fit the opening, half the job is done.

Figure 14-5 shows how a prehung exterior door unit fits into the rough opening. Remember that a prehung exterior or interior door is installed as a complete unit. *Never remove the door from the hinges during installation.*

Figure 14-6 shows the two-piece and three-piece adjustable split jambs commonly found in prehung units. Another type of split jamb uses machine-fitted steel dowels to connect the front and reverse jambs. The dowels fit into countersunk receiving holes. This makes it easy to separate and reassemble the front and reverse jamb parts.

Install the prehung door unit in the following way:

1) Remove the packing nails connecting the keeper jamb to the door. See Figure 14-7. Separate the halves of the door frame.

2) Place the *front* half of the frame into the opening, as shown in Figure 14-8 A. (The front half of the frame has the hinged door attached to it.) Plumb the jamb on the hinge-side of the frame. Nail the hinge-side trim to the wall. Use 8d casing nails spaced 24 inches on center. Don't drive the nails all the way into the wall.

3) Press the keeper-side of the jamb against the spacer tabs. Be sure the head jamb fits uniformly at the top of the opening. Nail the keeper-side trim to the wall. Use 8d nails spaced 24 inches on center. Don't drive the nails all the way into the wall.

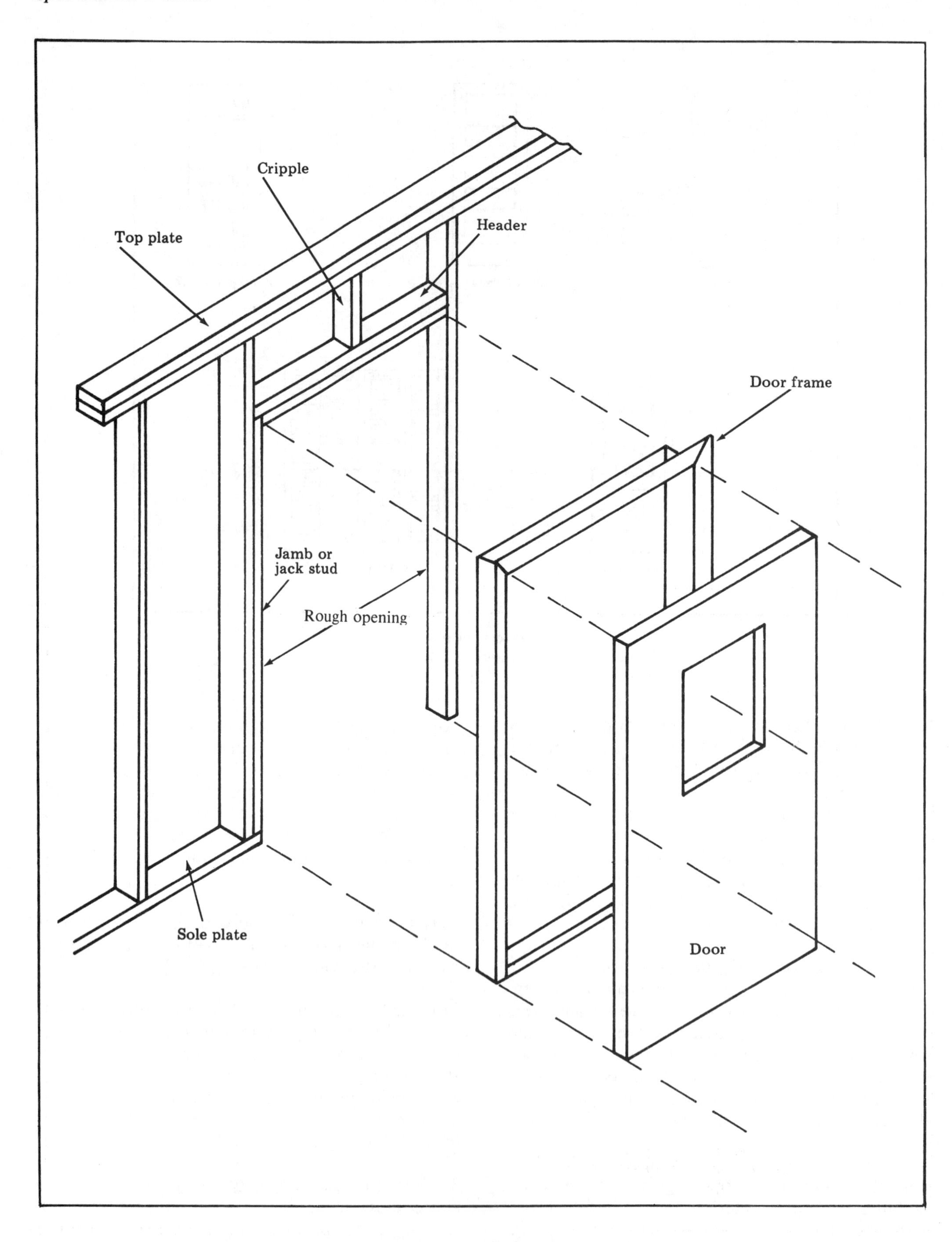

Fitting arrangement—exterior prehung door unit
Figure 14-5

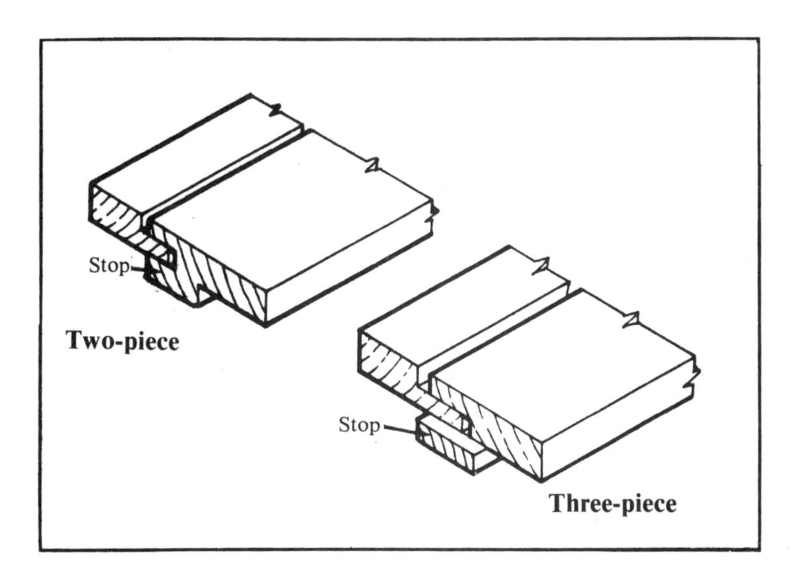

Two-and three-piece adjustable split jambs
Figure 14-6

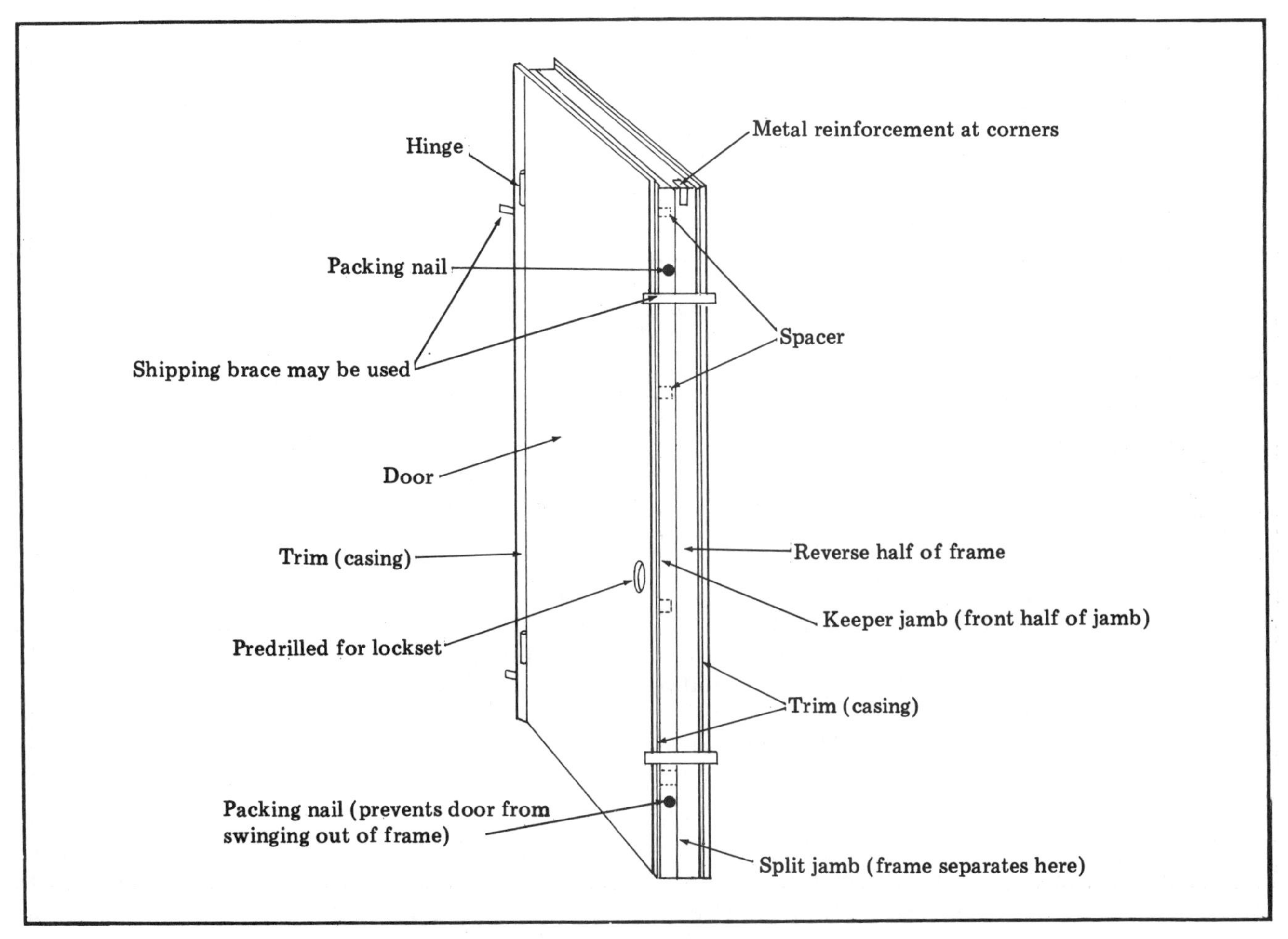

Prehung door unit ready to be installed in opening
Figure 14-7

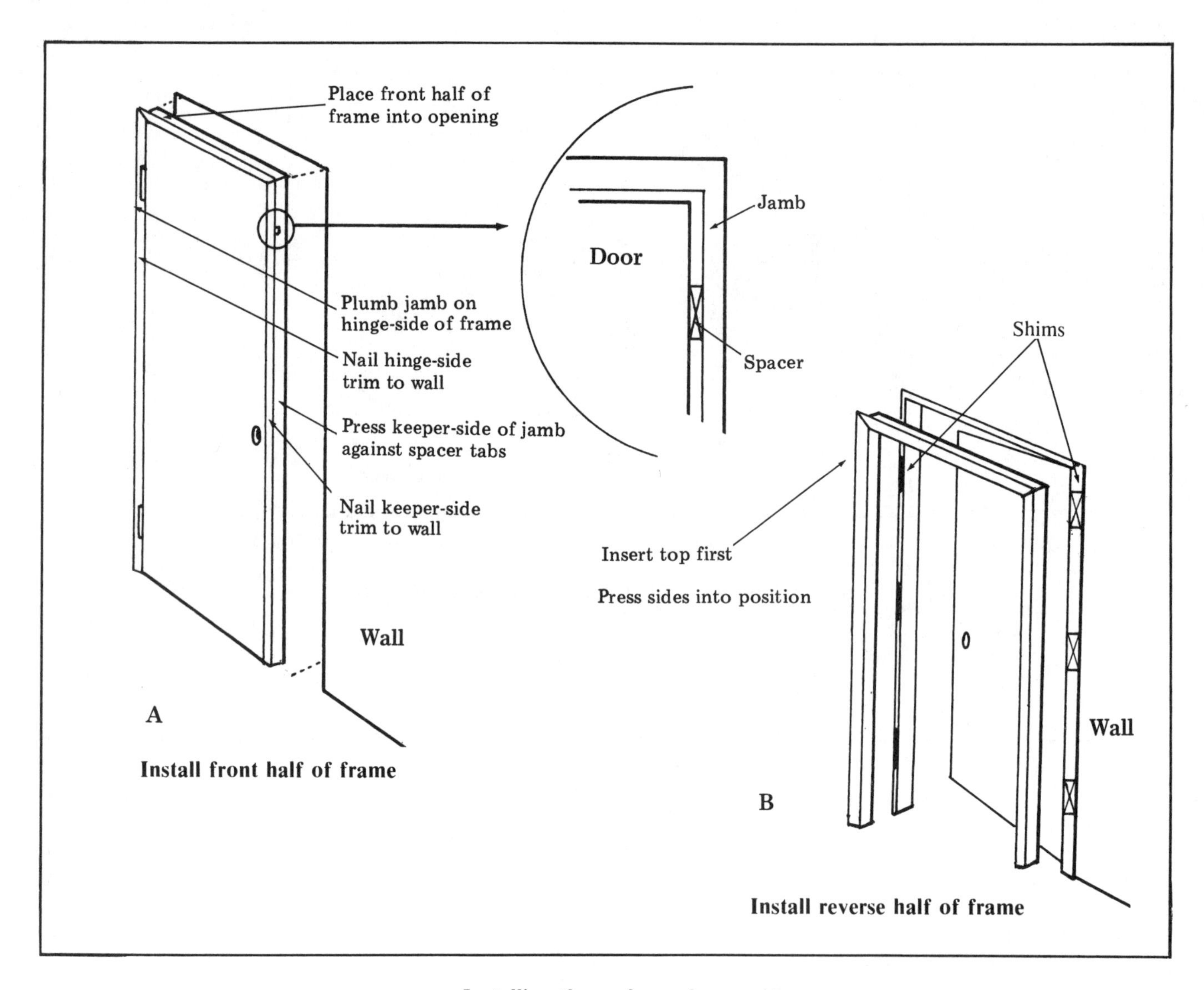

Installing the prehung door unit
Figure 14-8

4) With the door closed and spacers still attached, shim behind the jambs.

5) Open the door and nail through the jambs and shims. Use 8d casing nails. Remove all spacers.

6) Take the *reverse* half of the frame and slide it into position, as shown in Figure 14-8 B. Insert the top first. Then press the sides firmly into position against the wall. Nail through the trim into the wall. Use 8d casing nails spaced 24 inches on center. Don't drive the nails all the way into the wall.

7) Recheck the plumb on the hinge-side of the frame. Check the level of the head jamb. Install the lockset and keeper, and check the door for proper functioning.

8) Drive and set all casing nails. Nail side stops and header stops. Use 3d finishing nails, spaced 12 inches on center.

Sliding Glass Doors

There's a sliding glass door available to fit nearly every opening width. Door dimensions are based on unit panel sizes and panel combinations. Standard sizes are 6 feet 8 inches high and 3, 4, 5, or 6 feet wide per section. Assemblies are available in groups of two, three or four sections, some operable and others stationary. Generally, a single-piece metal track is used for both door and screen.

The most common sliding glass door has bottom rollers with adjustable sealed-bearing sheaves on either aluminum or stainless steel tracks. Doors with a stainless steel track usually have stainless steel inserts in the aluminum sill "cap" of the sill track. The movable panel or panels slide inside the fixed panel.

Most of these doors are reversible, so the door can be mounted to open either left or right. That's an important advantage.

Screens come with either glass fiber or aluminum mesh. Glass fiber is cheaper and lighter, but aluminum mesh lasts longer. The screen panel slides on a roller track in the sill and outside the stationary panel.

Sliding glass doors require exact fitting. Never force a sliding glass door into place. A glass door with a bent frame will never open or close smoothly. And it will leak cold air in winter and water when it rains.

Sliding glass doors first became popular in California, Florida and Hawaii where insulation is less important. Glass doors are now available with insulating glass and thermal break frames. These doors are more practical for colder climates.

Estimating Manhours

A good rule of thumb is to allow 1.5 skilled manhours for the installation of each standard-size prehung exterior unit and 1 skilled manhour for each standard prehung interior unit. This includes installation of the lockset.

Prehung exterior double-door units will require about 3 skilled hours for installation.

Closet by-pass doors (with pulls and tracks) require about 2 skilled hours.

Sliding glass doors measuring 6'x 6'8'' will require about 2.5 skilled hours for installation. Heavier doors or doors with insulating glass panels will take about an hour longer.

Summary

The house plan will indicate the door sizes and swing. Be economical in selecting prehung doors for your spec houses. Paint-grade units are less expensive than stain-grade units.

For efficient operation of forced-air heating and cooling equipment, leave about 3/4-inch space between the bottom of interior doors and the finish floor.

Improperly installed doors can be a source of constant aggravation for the occupant. That reflects on your professionalism as a builder.

Complete Figure 14-9, Prehung Door Checklist and Data Reference Sheet. It will be a valuable record of the doors installed on your jobs.

Prehung Door Checklist and Data Reference Sheet

1) Exterior Doors:

Main entrance doors:

☐Size: ☐3'-0'' x 6'-8'' ☐other (specify) ____________________

☐Wood (species: ____________________)

☐Metal

☐Combination

☐Insulated

☐Flush: ☐solid-core ☐hollow-core

☐Panel

☐Glazed

☐Paint grade

☐Stain grade

☐Threshold: ☐aluminum ☐wood

☐Lockset style ________ and number __________

Rear or side doors:

☐Size: ☐2'-8'' x 6'-8'' ☐other (specify) ____________________

☐Wood (species:________)

☐Metal

☐Combination

☐Insulated

☐Flush: ☐solid-core ☐hollow-core

☐Panel

☐Glazed

☐Paint grade

☐Stain grade

☐Threshold: ☐aluminum ☐wood

☐Lockset style _______ and number ________

Figure 14-9

Sliding glass doors:

☐Size 6'-0'' x 6'-8''☐other (specify) ___________

☐Lightweight ☐heavyweight

☐Insulated panels

☐Screen: ☐glass fiber ☐aluminum

☐Track: ☐aluminum ☐stainless steel

Storm doors:

☐Main entrance storm door size: ☐3'-0'' x 6'-8'' ☐other (specify) ___________

☐Rear or side storm door size: ☐2'-8'' x 6'-8'' ☐other (specify) ___________

☐Combination (screen and storm)

☐Aluminum

☐Wood

2) Interior doors:

☐2'-8'' x 6'-8'', number ___________

☐2'-6'' x 6'-8'', number ___________

☐2'-4'' x 6'-8'', number ___________

☐1'-6'' x 6'-8'', number ___________

☐Flush (wood species:___________) ☐hollow-core ☐solid-core

☐Panel (wood species:___________)

☐Paint grade

☐Stain grade

☐Lockset style _________ and number ___________

3) Special closet doors:

☐Sliding (by-pass) (wood species: ___________)

Size: ☐6'-0'' x 6'-8'' ☐other (specify) ___________

☐Flush ☐hollow-core ☐solid-core

☐Panel

☐Louver (size ___________) ☐sliding (by-pass) ☐bi-fold

4) When doors are opened, they ☐do ☐do not block light switches/receptacles.

5) Space between the bottoms of interior doors and the finish floor allows for adequate air circulation. The space is ☐ ½-inch ☐ ¾-inch.

6) Material costs:

Exterior doors	________pcs	@	$________ea	=	$__________
Exterior locksets	________pcs	@	$________ea	=	$__________
Storm doors	________pcs	@	$________ea	=	$__________
Glass sliding doors	________pcs	@	$________ea	=	$__________
Interior doors	________pcs	@	$________ea	=	$__________
Interior locksets	________pcs	@	$________ea	=	$__________
Special closet doors	________pcs	@	$________ea	=	$__________
Nails	________lbs	@	$________lb	=	$__________

Other materials, not listed above:

______________	________	@	$________	=	$__________
______________	________	@	$________	=	$__________
______________	________	@	$________	=	$__________
______________	________	@	$________	=	$__________

Total cost of all materials $__________

7) Labor costs:

Skilled	________hours	@	________hour	=	$__________
Unskilled	________hours	@	________hour	=	$__________

Other (specify)

______________	________hours	@	________hour	=	$__________

Total $__________

8) Total cost:

Materials	$__________
Labor	$__________
Other	$__________
Total	$__________

9) Problems that could have been avoided ________________________________

____

____

____

____

10) Comments ____

____

____

____

____

11) A copy of this Checklist and Data Reference Sheet has been provided:

☐ Management

☐ Accounting

☐ Estimator

☐ Foreman

☐ File

☐ Other: ____________

Chapter 15

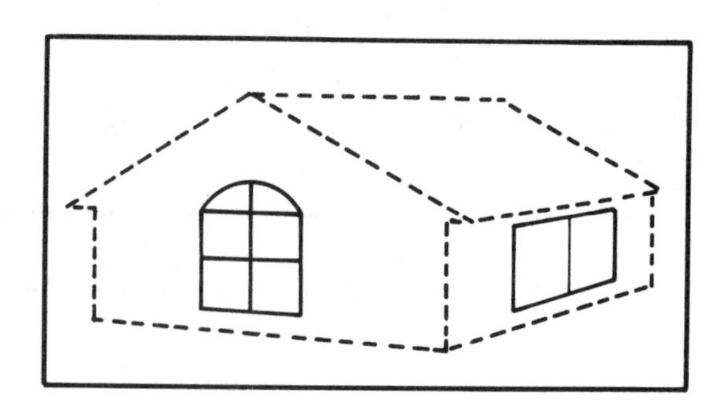

Windows

Any time you can order a standard building component already fabricated and ready to install, you're money ahead of the game. When building spec houses, use plans that require standard-size prehung doors and windows. Odd "outlaw" sizes rob you of legitimate profit, because both labor and material costs will usually be higher.

Like doors, windows have to be compatible with the style of the house. Don't use just any type of window. And consider the function of the window as well as esthetics. Type, size, material, and arrangement of windows all help establish the character of the building.

Windows are an integral part of building design. Make them an integral part of your planning process. When buying windows, consider daylight, privacy, ventilation, view, heat loss and weather-resistance.

Here are some generally accepted window standards that will apply on nearly every job:

• Habitable rooms should have windows with glass area of not less than 10% of the floor area.

• Natural ventilation should be not less than 4% of the floor area, unless a complete air-conditioning system is used. But avoid relying entirely on an air-conditioning system. Even in the hottest climates, there are days and evenings when an open window provides all the cooling you need — and at zero energy cost.

Window Dimensions

Experienced builders will tell you that many window units don't fit the rough opening (RO) that they're supposed to fit. The RO specified by one manufacturer may not be the same as the RO requirement for the same-size window made by a different manufacturer. The difference is often in the treatment of the jambs. Don't take the store clerk's word. Measure the window, or have sample sizes on the job. It can save you a lot of extra work (and lost profits).

Window dimensions are given by number of lights and size in inches. Thus, a "four-light 12/28 window" has four 12 x 28-inch panes of glass. When giving the size of a sash, the width of the glass is always given first, and then the height. If a double-hung window has a sash that is not divided by small wood members called *muntins,* or *sash bars*, it's called a "two-light" window.

On plans, window dimensions may also be shown by giving the exact outside dimensions of the sashes. Thus, a 2'8" x 4'6" window has a 2'8" x 4'6" sash. These figures also give the inside dimensions of the frame. See Figure 15-1.

A sash is usually rectangular in shape and consists of two side members called stiles, a top member called a top rail, and a bottom member called the bottom rail. A sash can carry one light (pane of glass), or a sash can carry several lights when it is subdivided by muntins. A sash is held in its frame by stops which are placed around the inside of the frame on each side of the sash.

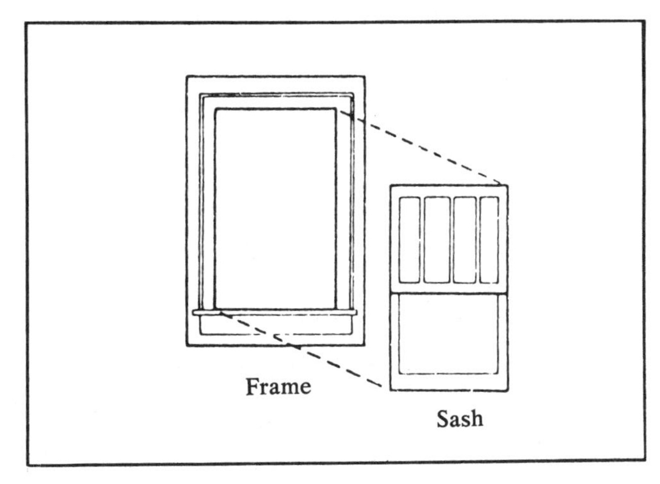

Window dimensions
Figure 15-1

When two sashes are set (one above the other) in a frame and they slide vertically past each other, the top rail of the lower sash and the bottom rail of the upper sash overlap and are called meeting rails or check rails. When a sash swings, the stile with the hinges is called the hanging stile.

The RO size for window frames varies slightly between manufacturers. Make the following allowances for the stiles and rails, thickness of jambs, and thickness and slope of the sill:

1) Double-hung window (single unit): Rough-opening width = glass width plus 6 inches. Rough-opening height = total glass height plus 10 inches.

For example, the following tabulation illustrates several glass and RO sizes for double-hung windows:

Window glass size (each sash)			Rough frame opening		
Width		Height	Width		Height
24"	x	16"	30"	x	42"
28"	x	20"	34"	x	50"
32"	x	24"	38"	x	58"
36"	x	24"	42"	x	58"

2) Casement window (one pair of sashes): Rough-opening width = total glass width plus 11¼ inches. Rough-opening height = total glass height plus 6⅜ inches.

Types of Windows

Windows can be classified by the type of sash-opening system. The most common types are: double-hung, single-hung (stationary), casement, awning, horizontal sliding, prefabricated metal, and storm windows. The sash can be wood, metal or a combination of wood and metal. Heat loss is greater through metal frame and sash units.

Insulated glass is used in both stationary and movable sashes. It consists of two or more glass sheets with hermetically-sealed edges. This type of glass has more resistance to heat loss than a single thickness. Insulated glass is often used without a storm sash.

Double-Hung Windows

The double-hung window is probably the most common. It has an upper sash and a lower sash that slide vertically in separate grooves in the side jambs or in full-width metal weatherstripping. See Figure 15-2. Notice that this type of window has a maximum face-opening of one-half the total window area. Springs, balances, or compression weatherstripping hold the sash in place. Compression weatherstripping is probably the best. It prevents air infiltration, provides tension, and acts as a counterbalance. When the window is hung, the sash should operate freely.

Double-hung windows can be arranged as a single unit, doubled (mullion), or in groups of three or more. To create the effect of a window wall, use one or two double-hung windows on each side of a large, stationary insulated window. When you frame an opening like that, be sure to use headers large enough to carry roof loads.

The jambs (sides and top of the frame) of a wood window are usually made of nominal 1-inch lumber. The width allows the use of either plaster or drywall interior finish. The sill is usually made of nominal 2-inch lumber, and is sloped at about 3 in 12 for good drainage toward the exterior. See Figure 15-2. The sash is normally 1⅜ inches thick. Wood combination storm and screen windows are usually 1⅛ inches thick.

A ranch-style house looks best with top and bottom sashes divided into horizontal lights. A colonial or Cape Cod house will usually have each sash divided into six or eight lights. A snap-in divider can be used on insulated windows. It can be removed for cleaning and painting.

Hardware for double-hung windows includes the sash lifts that are fastened to the bottom rail. These are not required where a finger groove is provided in the bottom rail. Other hardware includes sash locks or fasteners located at the meeting rails. They lock the window and draw the sashes together to provide an "airtight" fit.

Place strips of building paper, 30-lb. or heavier, around the RO perimeter. These minimize infiltration of air and moisture. See Figure 15-3. Then place the assembled window unit into the rough opening.

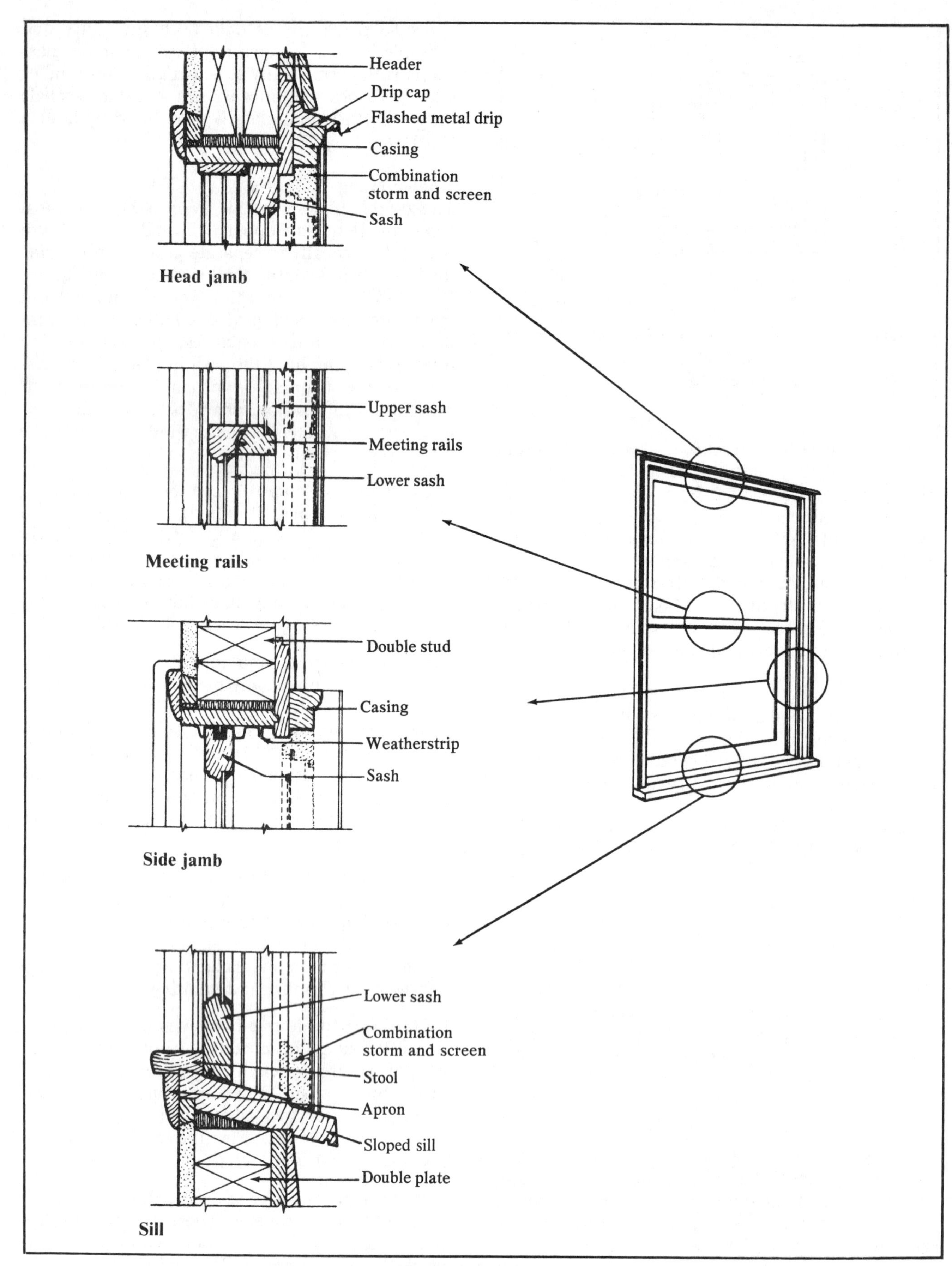

Double-hung window
Figure 15-2

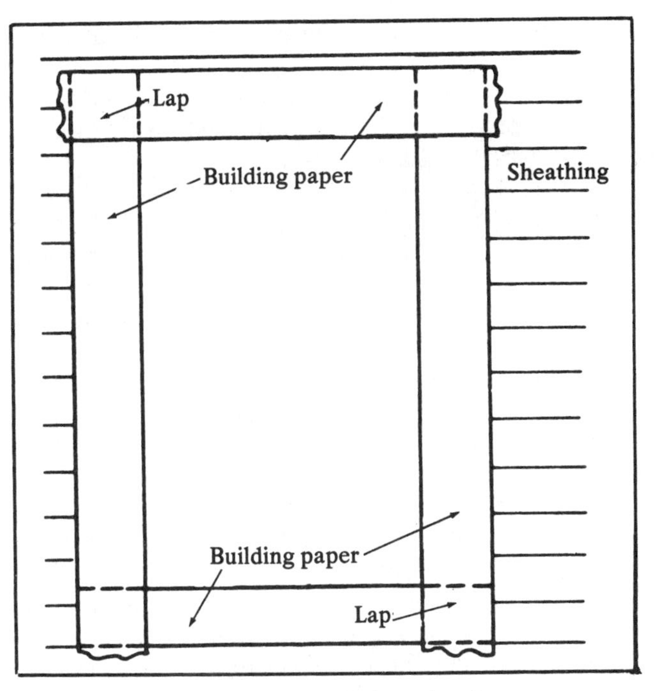

Building paper on RO perimeter
Figure 15-3

The frame is plumbed and fastened to the side studs and header by nailing through the casings or the blind stops at the sides. Where nails are exposed, such as on the casing, use corrosion-resistant nails.

Single-Hung (Stationary) Windows
The single-hung frame is made to hold a single sash. The sash may be fastened permanently in place, or it may swing in or out from either the side or head jamb.

Stationary window frames can be made of plain plank. See Figure 15-4. The sill in this case is a pine plank 1⅝ inches x 5⅝ inches. The plank is machined to receive a check strip and sash on the inside edge of the sill. The plank is also rabbeted to receive a storm sash or screen on the outside. The side- and head-jamb members are 1⅝-inch x 5⅝-inch stock, rabbeted to receive the sash on the inside and the storm sash on the outside.

Stationary windows used alone, or in combination with double-hung or casement windows, usually have a wood sash with a large single light of insulated glass. Stationary windows are fastened permanently into the frame. Because of their size, sometimes 6 to 8 feet wide, a 1¾-inch-thick sash may be needed to provide strength. Large lights of insulating glass will require a heavier sash, because insulating glass is relatively heavy.

Casement Windows
Casement windows have a side-hinged sash that swings outward. See Figure 15-5. The outward-swinging sash is more weathertight than one that swings inward. Screens are installed on the inside of the windows that swing outward. Some casement windows have a storm sash, but insulated glass is probably more practical in cooler climates. Unlike the double-hung unit, a casement window allows you to open the entire window for ventilation.

Casement windows usually arrive on the job fully assembled, weatherstripped, and with hardware in place. Closing hardware consists of a rotary operator and sash lock. Casement sashes can be used as a pair, or in a combination of two or more pairs with divided lights. Use snap-in muntins to provide a multiple-pane appearance in a home with more traditional styling.

Metal sash casements are popular in more moderate climates. But careful installation is necessary to help prevent condensation and frosting on the interior surfaces during cold weather. A full storm-window unit will usually eliminate this problem.

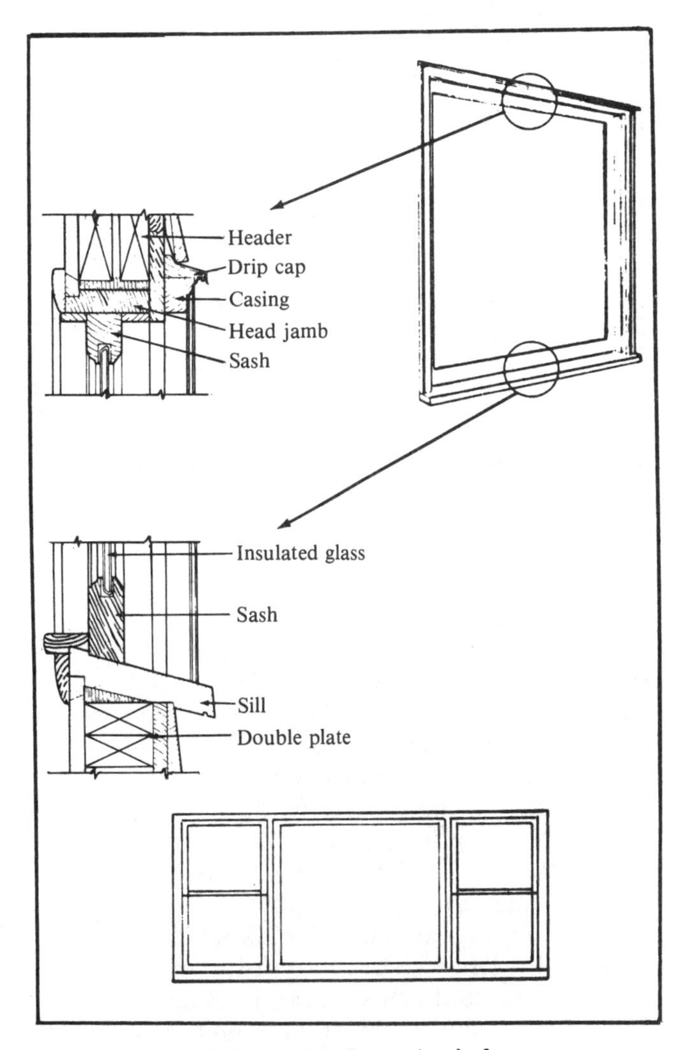

Single-hung (stationary) window
Figure 15-4

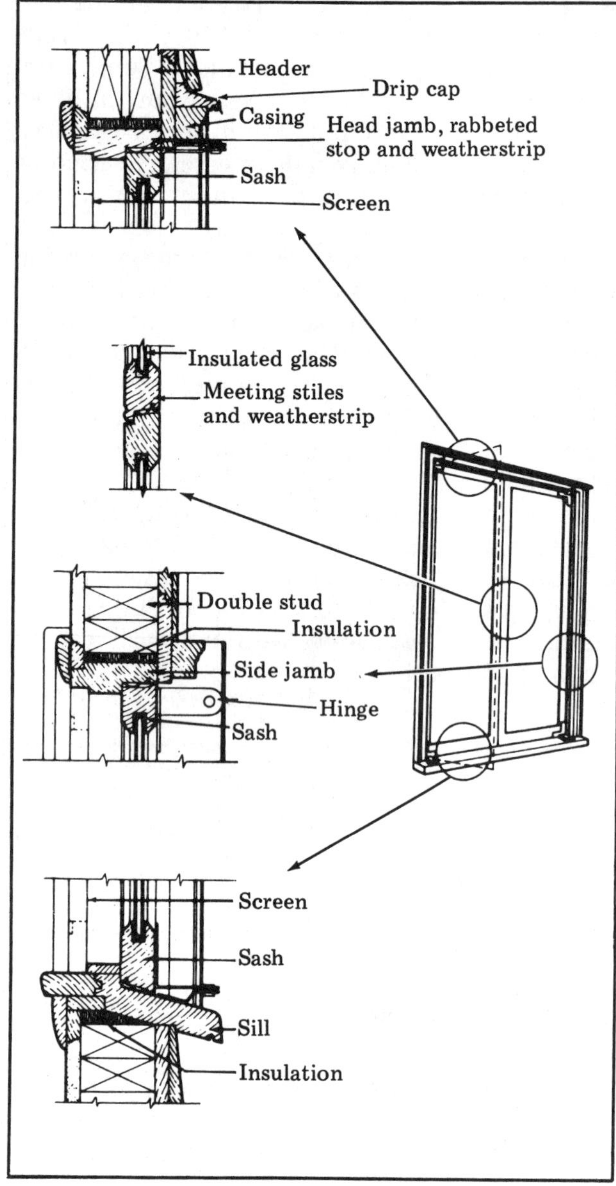

Casement window
Figure 15-5

Awning Windows

Awning windows have movable sashes that extend outward like an awning. See Figure 15-6. These windows are often grouped into multiple units to create a window wall.

The sashes swing outward at the bottom. This distinguishes awning windows from hopper windows that swing inward from the top. Both types provide protection from rain even when open.

Jambs are usually 1 1/16 inches (or more) thick because they are rabbeted. The sill is at least 1 5/16 inches thick when two or more sashes are used in a complete frame. Each sash can have an individual frame so that any combination of units can be joined together. Fixed sashes and operable sashes can be combined in the same multiple window units. Operable sashes have hinges, pivots, and sash supporting arms.

Weatherstripping, storm sash and screens are usually provided. The storm sash isn't needed when insulated glass is used.

Horizontal Sliding Windows

Horizontal sliders look similar to casement windows. But, the sash slides horizontally in a track or guides located on the sill and head jambs. Again, multiple window units can be joined to create a window wall. As in most modern window units, weatherstripping, water-repellent preservative treatments, and hardware, will be included in these factory-assembled units.

Sliding windows are somewhat less expensive, and tend to look good in pairs or groups. You can get them with or without muntins. Because of their vertical lines, they are well-adapted to long, low, one-story homes. The disadvantage of sliding windows is that they tend to allow more air infiltration than other types of windows.

Prefabricated Metal Windows

A wide variety of complete window units are available. Most come with frame, sash and trim already assembled. Often they have the appropriate screens, weatherstripping and hardware in place or on hand for quick assembly.

Several important points should be observed before and during installation. Regardless of the type of window used, it should be of the size, combination, and type specified. Most important, the unit should be quality-built. Some inexpensive units are not suitable for single-family homes. Here's how to distinguish poor quality from good quality in metal windows.

Welded joints should be solid, have excess metal removed, and be dressed smooth on exposed and contact surfaces. The dressing should be done so that no discoloration or roughness shows after finishing. Joints formed with mechanical fastenings should be closely fitted and made permanently watertight. Joints should be strong enough to resist deflection during normal construction and carry reasonable loads once the windows are installed.

Hardware should be of the right design and strength, and should be attached securely to the window with noncorrosive bolts or machine screws (not sheet metal screws). Where fixed screens are specified, the hardware should be specially adapted to permit satisfactory operation of ventilators.

For metal windows set in concrete or masonry walls, prepare openings very carefully. Unless otherwise specified, all windows should be built in as the work progresses. They should be installed without forcing them into the prepared openings. See Figure 15-7.

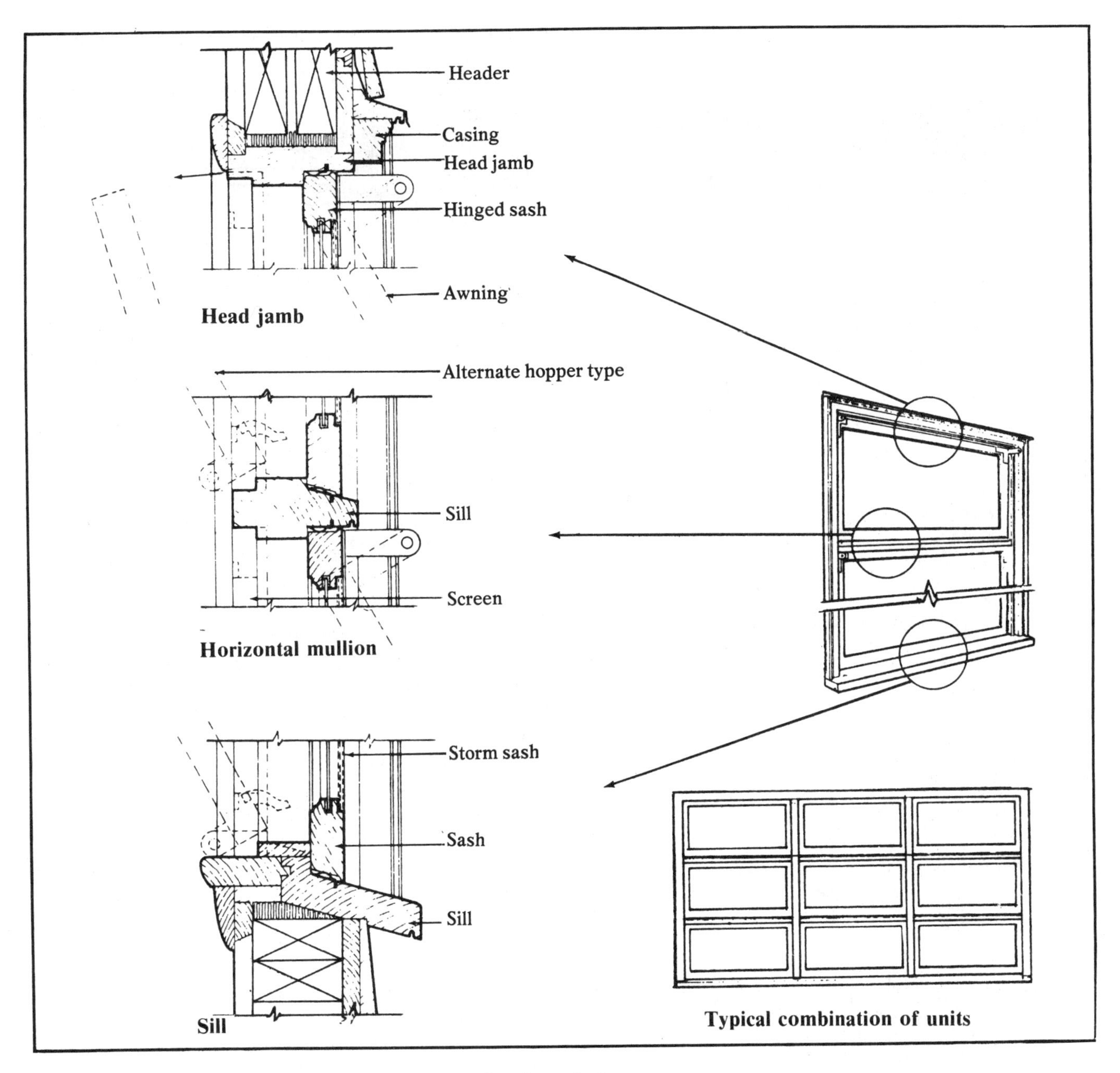

Awning window
Figure 15-6

Quality work requires that all windows be set plumb, square, level, and in alignment. Keep the window tightly closed during construction. This will protect movable parts against dirt and dust.

Any window installed in direct contact with masonry must have the head and jamb designed so that they enter not less than 7/16-inch into the masonry. Where windows are set in prepared masonry openings, the anchors should be placed during wall construction. They should be fastened securely to the windows or frames and to the adjoining construction. Unless otherwise indicated, anchors should be spaced not more than 18 inches apart on jambs and sills. Anchors and fastenings should be strong enough to hold the window in place against any reasonable attempt to dislodge it.

Storm Windows

Storm windows are the best and cheapest way to provide insulation. They not only keep wind and rain away from the sash edges and the frame, they also provide a double thickness of glass with a dead air space. This dead air space prevents heat loss and helps to avoid condensation on the glass surface in cold weather.

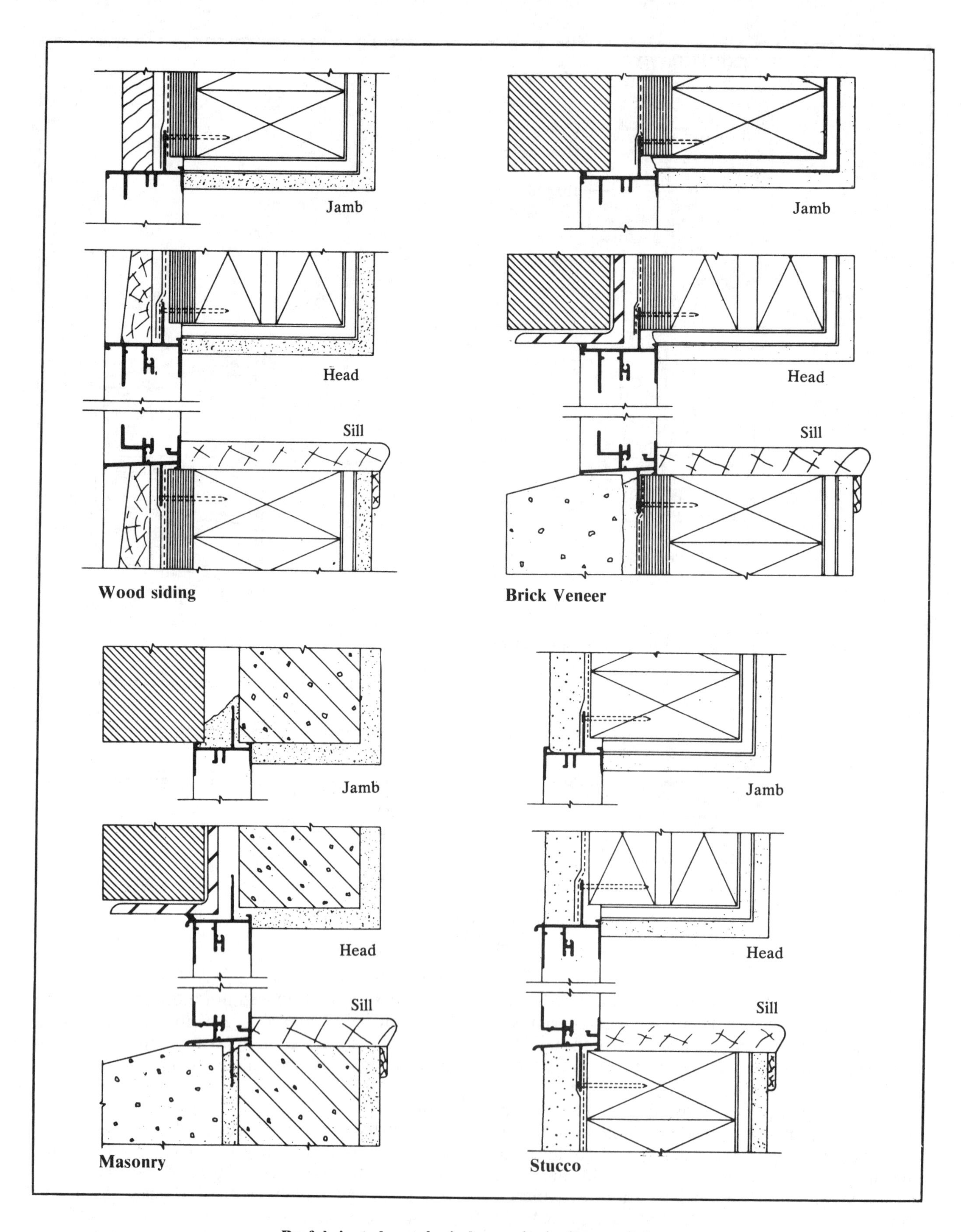

Prefabricated metal window units in four wall types
Figure 15-7

Estimating Manhours

Manhour requirements for installing prefabricated windows are shown below. The suggested crew is one carpenter and one laborer. Window trim is not included.

Work Element	Unit	Manhours per unit
Casement windows and screens		
1 leaf, 1'10'' x 3'2''	Each	1.4
2 leaves, 3'10'' x 4'2''	Each	1.9
3 leaves, 5'11'' x 5'2''	Each	2.4
Picture windows		
4'6'' x 4'6''	Each	3.0
5'8'' x 4'6''	Each	3.2
9' x 5'	Each	3.7
10' x 5'	Each	4.0
11' x 5'	Each	4.4
Double or single hung windows and screens		
2'0'' x 3'2''	Each	1.1
2'0'' x 4'6''	Each	1.6
2'8'' x 3'2''	Each	2.0
2'8'' x 5'2''	Each	2.1
Double or single hung windows and screens (cont.)		
3'4'' x 5'2''	Each	2.4
5'6'' x 5'2''	Each	3.4
8'4'' x 5'2''	Each	5.2

Summary

Windows set the tone of a house. The style, size, type, and location of windows are extremely important. Select windows that are suitable for the house plan and community where the house is built. In cold climates, include windows on the south side of the house where they can capture warmth from the winter sun. In warm climates, avoid windows on the west side where they will run up cooling costs.

With exterior doors and windows installed, the house is now ''dried-in.'' A rough-in of plumbing and wiring should be in progress. When the rough-in of plumbing and wiring is complete, you're ready to begin interior work. If the exterior finish is masonry veneer, that work should also be going forward.

Use Figure 15-8, the Window Checklist and Data Reference Sheet, to record *your* labor and material costs. This checklist will become a valuable reference for estimating future jobs.

Window Checklist and Data Reference Sheet

1) The house is ☐1-story ☐1½-story ☐2-story.

2) The exterior finish is:

☐Masonry: ☐brick ☐stone ☐stucco

☐Siding: ☐wood ☐hardboard ☐vinyl

☐Combination siding and masonry

3) Window style, type and placement were considered in relation to ☐orientation (north, south, east, west) ☐view ☐natural ventilation ☐esthetics ☐climate.

4) Rough-opening size was determined by:

☐Manufacturer's specifications

☐House plan

☐Measuring sample windows

☐Formula: On a double-hung window, the RO width = glass width plus 6 inches. RO height = glass height plus 10 inches. On a casement window, the RO width = glass width plus 11¼ inches. RO height = glass height plus 6⅜ inches.

5) Window types used:

☐Double-hung

☐Casement

☐Stationary

☐Awning

☐Horizontal sliding

☐Other (specify) ______________

6) Windows ☐are ☐are not insulated and have ☐wood sash and frame ☐metal sash and frame ☐other (specify) ______________.

7) Windows have ☐1 light ☐2 lights (horizontal) ☐4 lights ☐6 lights ☐8 lights or more and ☐conventional muntins ☐snap-in dividers.

8) To hold the sash in place, ☐compression weatherstripping ☐springs ☐balances were used.

9) Windows are equipped with:

☐Screens

☐Hardware ☐sash lock ☐sash lift

Figure 15-8

☐Rotary handle

☐Finger slot

10) Storm windows ☐were ☐were not installed. Types of storm windows installed: ☐metal frame ☐wood frame ☐combination storm and screen.

11) Building paper ☐was ☐was not used as flashing at each window.

12) Window combinations or pairs ☐were ☐were not used:

☐2 (or more) double-hung

☐Stationary and double-hung

☐Stationary and awning

☐Stationary and horizontal

☐2 (or more) awning

☐2 (or more) horizontal

☐Other (specify) ______________

13) Windows installed in each room:

☐Basement

Size(s): ___________

Number of units: ___________

☐Kitchen

Size(s): ___________

Number of units: ___________

☐Dining room

Size(s): ___________

Number of units: ___________

☐Family room or den

Size(s): ___________

Number of units: ___________

☐Living room

Size(s): ___________

Number of units: ___________

☐ Bedrooms

Size(s): ______________

Number of units: ______________

☐ Bathrooms

Size(s): ______________

Number of units: ______________

☐ Hall

Size(s): ______________

Number of units: ______________

☐ Porches

Size(s): ______________

Number of units: ______________

14) Material costs:

Windows:

Basement	________pcs	@	$________ea	=	$________
Kitchen	________pcs	@	$________ea	=	$________
Dining room	________pcs	@	$________ea	=	$________
Family room	________pcs	@	$________ea	=	$________
Living room	________pcs	@	$________ea	=	$________
Bedroom	________pcs	@	$________ea	=	$________
Bathroom	________pcs	@	$________ea	=	$________
Hall	________pcs	@	$________ea	=	$________
Porch	________pcs	@	$________ea	=	$________
Storm	________pcs	@	$________ea	=	$________
Screens	________pcs	@	$________ea	=	$________
Building paper	________ rolls	@	$________ea	=	$________
Nails	________lbs	@	$________ea	=	$________

Other materials, not listed above:

______________________ __________ @ $__________ = $__________

______________________ __________ @ $__________ = $__________

______________________ __________ @ $__________ = $__________

______________________ __________ @ $__________ = $__________

Total $__________

15) Labor costs:

Skilled ________hours @ $________hour = $__________

Unskilled ________hours @ $________hour = $__________

Other (specify)

______________________ ________hours @ $________hour = $__________

Total $__________

16) Total cost:

Materials $__________

Labor $__________

Other $__________

Total $__________

17) Average cost (materials and labor) per window: $__________

18) Problems that could have been avoided ______________________________

____

____

____

____

19) Comments ____

____

____

____

____

20) A copy of this Checklist and Data Reference Sheet has been provided:

☐Management

☐Accounting

☐Estimator

☐Foreman

☐File

☐Other: ____________

Chapter 16

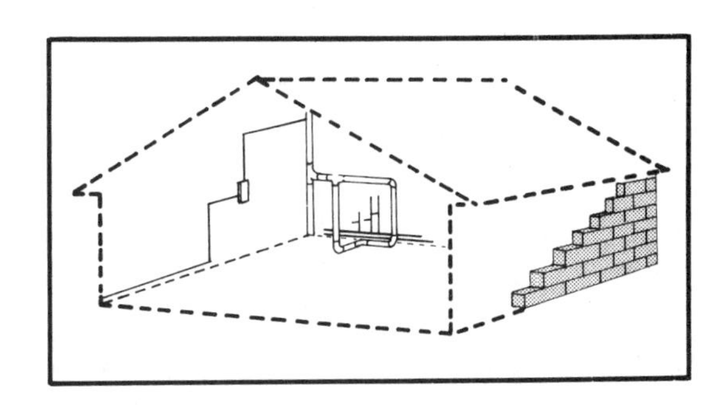

Subcontracted Work

Most home builders subcontract masonry, plumbing and electrical work. You may also want to sub out ceramic tile, sheet rock and plastering if there isn't a qualified tradesman on your payroll who can handle these specialties. Since you probably don't have a licensed plumber or electrician on staff, you'll sub out this work as a matter of necessity. Very few home builders use their own crews to do all the work. Some sub out nearly everything, including carpentry.

In this chapter, we'll take a close look at the three components you're most likely to subcontract at this stage of construction. Should you decide to tackle any of these on your own, or if you just want to know more about them, review the manuals in the order form at the back of this book.

Masonry

Experienced speculative builders often limit masonry work to a little decorative brick veneer. A combination of wood siding and brick gives the flavor of masonry to a home, but without the high cost usually paid for masonry construction.

A masonry contract is often for labor only. It may or may not include cleaning the brick wall afterwards. It probably excludes cleanup behind the mason. Most masons will contract at a price per 1,000 bricks or blocks. The mason furnishes the tools, scaffolding and labor. The builder furnishes the bricks, mortar, sand, wall ties, nails, lintels, and mortar color if required.

Begin planning a masonry job by finding the size of brick to be used. Dimensions will vary, because different factories use different-size dies. Even bricks from the same plant will vary in size. These size variations are caused by uneven distribution of heat in the kiln, and by chemical variations in the type of clay used. To allow for these differences, the size of the mortar joint is adjusted when the bricks are laid. This creates an even and level wall.

Standard or modular-size bricks simplify construction and reduce your costs. Modular bricks need less cutting, because the height and width of doors and windows are a multiple of the height and width of each brick (including the mortar joint). Modular bricks are based on 4-inch modules.

Joints have to be finished after the bricks are laid. Poor-quality masonry is most noticeable at the joints. Be sure the thickness and texture of the joints are uniform. Figure 16-1 shows the most common mortar-joint finishes. The struck joint is often used on standard brickwork.

Estimating Materials and Manhours

Here's a good rule of thumb for estimating the number of bricks required for veneer on a house: Allow 6.5 bricks per square foot of wall (without deducting window and door areas). This rule works remarkably well on houses with a normal number of window and door openings. It allows for average waste. It also allows for the mason's tendency to break a whole brick in half for a fit, even though he's standing on a pile of previously broken halves.

About five or six bags of mortar are required to lay 1,000 bricks. Practically all manufacturers of

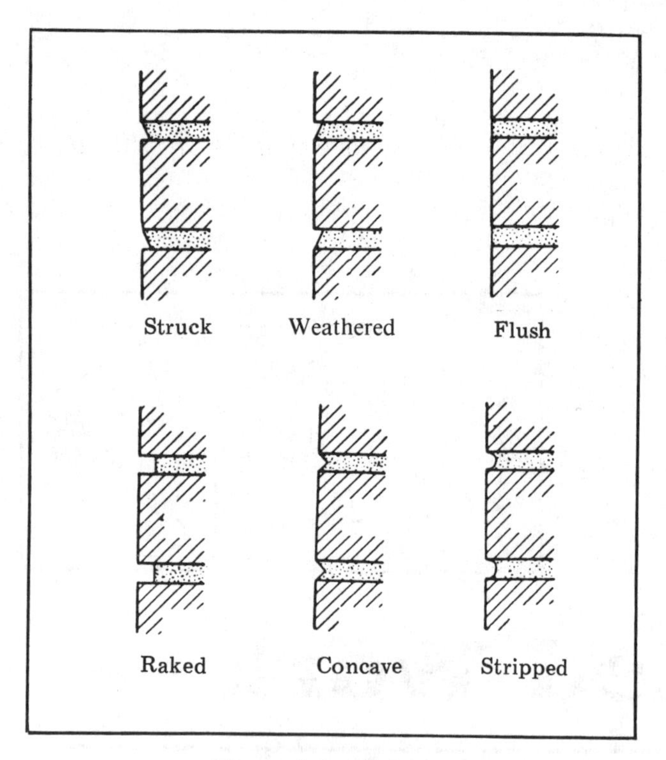

Mortar-joint finishes
Figure 16-1

masonry mortar recommend the following mix: One bag of masonry cement and three cubic feet of sand. Figure 16-2 shows typical brick-veneer applications.

The time required to lay 1,000 bricks varies according to weather conditions. Unless otherwise indicated, assume that 4-inch veneer on frame or masonry construction will require about 16 mason and 16 laborer hours per 1,000 bricks.

Some masons do quality work. Others do poor work. Before signing the subcontracting agreement, be sure to inspect some of the mason's previous work.

Plumbing

If you're building on a slab, part of the plumbing is done before the slab is poured. If it's a crawl-space house, now's the time to rough-in the plumbing.

At the rough-in stage, the pipes for the drainage system and water lines are run and "stubbed-out" so you can proceed with the construction. The bathtub will be set so the finish wall and floor can be completed. If you're installing a fiberglass shower-tub module, move it into the bathroom before the framing is completed. Some units won't pass through a door rough-opening.

The drainage system consists of all the piping, fittings and fixtures that carry waste water and sewage from a building to the street sewer or septic tank. It also includes the piping and fittings that ventilate the system. The drainage system may include both sanitary and storm sewers. A sanitary sewer conveys liquid or water-borne waste from plumbing fixtures. A storm sewer conveys surface or subsurface water, condensation, cooling water and similar discharges. In modern plumbing systems, storm sewers are separate from the sanitary sewer system.

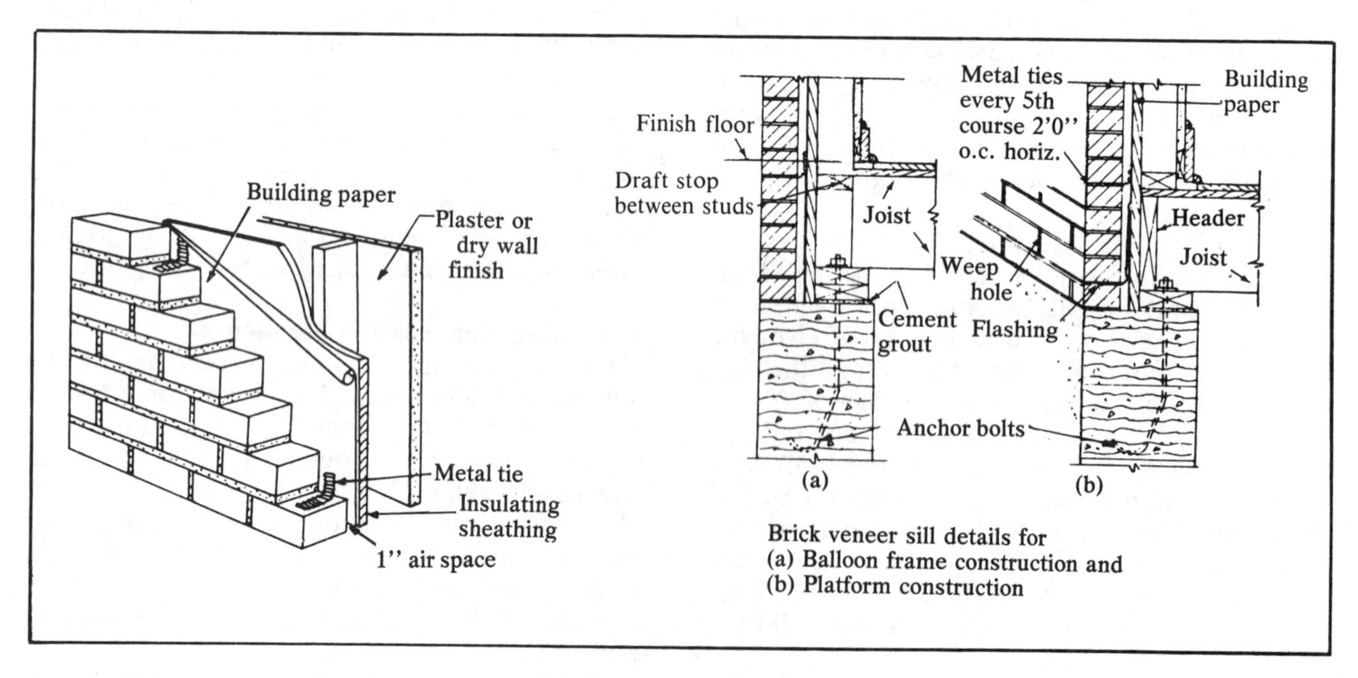

Brick veneer applications
Figure 16-2

Stacks, branch sewers, and fixture drains connect the waste outlets of the fixtures with the building drain.

Every building that has plumbing fixtures has to have soil or waste stacks that extend upward from the building drain. Usually these stacks exit through the roof, because they have to be as short, straight and direct as possible. Sharp bends, over 45 degrees, are not permitted. The stack must be carried (full-size) through the roof and project at least six inches above it.

If permitted by your code, use plastic pipe for the drainage system and water lines. But note that plastic water pipes tend to be noisy when they run through walls. Insulate the pipes so that vibrations are not transmitted to the building frame.

Some builders supply their own plumbing fixtures. They subcontract labor and materials for the rough-in, and labor only for the fixture hook-up.

Your plumbing sub should provide a list of the materials he proposes to use on the job and the total installed price. If he supplies the fixtures, make sure that they are identified clearly as to type, model number, manufacturer, capacity and color.

Electrical Work

Electrical rough-in should begin as soon as possible after the house is "dried-in." The rough-in includes all wiring, and the installation of the service panel, wall boxes, switch boxes and ceiling-fixture boxes. The electrician should install the wires for the door bell at this time.

This is also the time for the telephone company to run their wires.

The electrical code establishes the minimum number and the placement of wall outlets, switches and ceiling lights. Your electrician will be familiar with these requirements and plan the job accordingly. The electrician normally furnishes the rough-in materials, wall receptacles and switches. You may want to supply light fixtures, medicine cabinets and electrical appliances. His bid should include a list of the materials he proposes to use on the job and the total cost of the contract. The contract should include all final hook-ups of fixtures, switches and wall outlets.

Here's a rule of thumb for determining the amount of electrical wire required for general-purpose circuits: Allow 1 linear foot of wire for each square foot of floor space. Thus, a 2,000-square-foot house would require about 2,000 feet of wire. Separate 220-volt circuits for washer, dryer, range, and water heater should be measured to find the exact amount of wire needed.

Summary

Sloppy work by subcontractors can be your worst headache. Find good subs and deal fairly and honestly with them. Subcontractors will make or break you. Select them accordingly.

It's always worth your time to get several bids on every significant job. Subcontractors control their volume by adjusting their prices up or down. That's perfectly legitimate. It's really the only way an active subcontractor can keep a steady flow of work coming in the door. Save a few hundred dollars by accepting a low bid and you've added a few hundred dollars to your profit.

With the plumbing and wiring rough-in completed, you're ready to start insulating. But before going ahead with insulation, take a few minutes to record the costs of your subcontracted work. Figure 16-3 is your Checklist and Data Reference Sheet.

Checklist and Data Reference Sheet for Subcontracted Work

The house is _______ feet long and _______ feet wide, for a total of _______ square feet.

The house is ☐1-story ☐1½-story ☐2-story.

Masonry:

1) Brickmason contractor:

Name: ______________________________

Address: ______________________________

Telephone: ______________________________

Quality of work: ☐excellent ☐good ☐poor

2) Brickmason contracted to lay brick at $_______ per thousand.

3) Estimated number of bricks for job: _______ thousand.

4) Brick data:

Size: ______________________

Manufacturer: ______________________ Plant: ______________________

Brick name: ______________________ ID number: ______________________

Type: ______________________ Color: ______________________

5) Type of mortar joints: ☐struck ☐weathered ☐flush ☐raked ☐concave ☐stripped.

6) Brick washing was ☐included in contract ☐added to contract ($__________) ☐done by my crew.

7) Material costs:

Bricks	_________M	@	$_________M	=	$__________
Mortar	_______bags	@	$_________ea	=	$__________
Sand	_________CF	@	$_________CF	=	$__________
Wall ties	_________lbs	@	$_________lb	=	$__________
Steel lintels	_________pcs	@	$_________ea	=	$__________
Mortar color	_________lbs	@	$_________lb	=	$__________
Nails	_________lbs	@	$_________lb	=	$__________
Acid (wash)	_______gals	@	$_________ea	=	$__________
Brushes	_________pcs	@	$_________ea	=	$__________

Figure 16-3

Other materials, not listed above:

____________________ __________ @ $__________ = $__________

____________________ __________ @ $__________ = $__________

____________________ __________ @ $__________ = $__________

____________________ __________ @ $__________ = $__________

Total cost of all materials $__________

8) Labor costs:

Skilled ________hours @ $________hour = $__________

Unskilled ________hours @ $________hour = $__________

Other (specify)

____________________ ________hours @ $________hour = $__________

Total $__________

9) Total cost:

Materials $__________

Labor $__________

Other $__________

Total $__________

10) Cost of masonry per square foot of floor space $__________

Plumbing:

1) Plumbing contractor:

Name: ____

Address: ____

Telephone: ____

Quality of work: ☐excellent ☐good ☐poor

2) Contract price: $__________

3) The house has:

☐Basement bath with ☐tub ☐shower ☐commode ☐lavatory (☐one ☐two).

☐Basement sink.

☐First-floor bath with ☐tub ☐shower ☐lavatory (☐one ☐two) ☐commode ☐bidet ☐other (specify) ____________________.

☐Second-floor bath with ☐tub ☐shower ☐lavatory (☐one ☐two) ☐commode ☐bidet ☐other (specify) ____________________.

☐Kitchen with ☐sink ☐refrigerator-water hook-up ☐garbage disposal unit.

☐Laundry room with ☐washer ☐dryer (☐gas ☐electric) ☐water heater (☐gas ☐electric) ☐furnace (☐gas ☐electric).

4) Plumber furnished:

☐All materials for rough-in, including ☐drain, vent and waste lines ☐sewer line hook-up ☐septic tank hook-up ☐ water lines ☐meter hook-up ☐other (specify)____________________

☐All labor for ☐rough-in ☐fixture hook-up (final).

☐Water heater

☐Fixtures

5) Material costs:

Drain, vent, waste lines	________pcs	@	$________ea	=	$________
Sewer hook-up	________pcs	@	$________ea	=	$________
Septic hook-up	________pcs	@	$________ea	=	$________
Water lines	________pcs	@	$________ea	=	$________
Meter hook-up	________pcs	@	$________ea	=	$________
Water heater	________pcs	@	$________ea	=	$________
Fixtures	________pcs	@	$________ea	=	$________
Other materials, not listed above:					
____________________	________	@	$________	=	$________
____________________	________	@	$________	=	$________
____________________	________	@	$________	=	$________
____________________	________	@	$________	=	$________

Total cost of all materials $________

6) Total cost:

Contract $__________

Materials $__________

City or county fee $__________

Other $__________

Total $__________

7) Plumbing cost per square foot of floor space $__________

Electrical:

1) Electrical contractor:

Name: ____

Address: ____

Telephone: ____

Quality of work: ☐excellent ☐good ☐poor

2) Electrician furnished:

☐All materials for rough-in, including ☐service panel, stack and cable ☐wire ☐wall boxes ☐switch boxes ☐ceiling boxes ☐staples, nails and screws.

☐All labor for ☐rough-in ☐final connections.

☐Water heater.

☐Fixtures.

3) Material costs:

Wire	________	@	$________	=	$________
Wall boxes	________pcs	@	$________ea	=	$________
Switch boxes	________pcs	@	$________ea	=	$________
Ceiling boxes	________pcs	@	$________ea	=	$________
Staples, nails, screws	________lbs	@	$________lb	=	$________
Service panel	________pcs	@	$________ea	=	$________
Stack	________	@	$________	=	$________
Cable	________	@	$________	=	$________

Water heater ________pcs @ $________ea = $________

Fixtures ________pcs @ $________ea = $________

Other materials, not listed above:

________________ ________ @ $________ = $________

________________ ________ @ $________ = $________

________________ ________ @ $________ = $________

________________ ________ @ $________ = $________

Total cost of all materials $________

4) Total cost:

Contract $________

Materials $________

City or county fee $________

Other $________

Total $________

5) Electrical cost per square foot of floor space $________

Total cost for all subcontracted work:

Masonry $________

Plumbing $________

Electrical $________

Other $________

Total $________

Problems that could have been avoided:

Masonry: ____

____

____

Plumbing: ____

____

____

Electrical: ______________________________

____

____

Comments:

Masonry: ______________________________

____

____

Plumbing: ______________________________

____

____

Electrical: ______________________________

____

____

A copy of this Checklist and Data Reference Sheet has been provided:

☐ Management

☐ Accounting

☐ Estimator

☐ Foreman

☐ File

☐ Other: ____________

Chapter 17

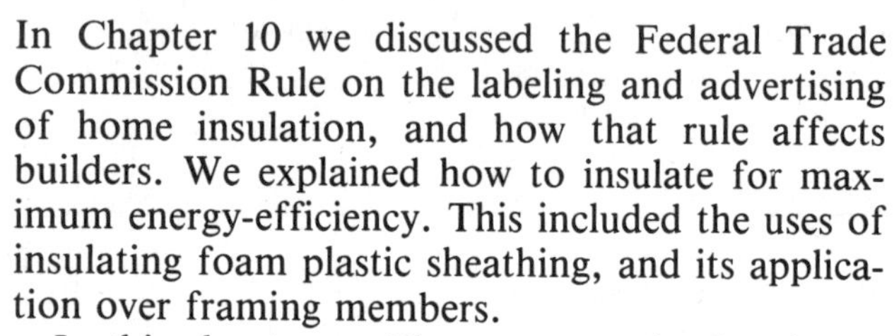

Insulation and Vapor Barriers

In Chapter 10 we discussed the Federal Trade Commission Rule on the labeling and advertising of home insulation, and how that rule affects builders. We explained how to insulate for maximum energy-efficiency. This included the uses of insulating foam plastic sheathing, and its application over framing members.

In this chapter, we'll cover several other forms of insulation, including: batts, blankets, reflective insulators, and loose fill. We'll explain how to install these forms of insulation *between* framing members.

Figure 17-1 shows the insulation R-values recommended for each climate zone. The figures are the recommendation of Owens-Corning Fiberglass. They take into account national weather data, energy costs and the cost of insulation.

Notice in Figure 17-1 that most areas in the United States require R-19 wall insulation. The colder the climate, the more insulation required. Insulation is also important in warmer climates, because it reduces cooling costs. Figure 17-2 shows the average winter-low-temperature zones in the United States.

Insulation Materials and Forms of Insulation

Insulation reduces the transmission of heat (and cold). Heat is transmitted three ways: by conduction, radiation, and convection.

Conduction— When heat travels through solid material and is transmitted to another solid body where they touch. Dense materials, like metal and glass, transmit heat (or cold) more easily than less dense materials, such as wood.

Radiation— When heat flows at a high speed in a direct line from a warmer body (such as the warm glow of a ceiling-mounted radiant heater) to a cooler one (such as the floor or carpet under the heater). Radiant heat passes through the air without warming it.

Convection— When heat is transmitted through a flow of air (by a forced-air heater, for example). Unlike radiated heat, convected heat involves the actual heating of the air.

A good insulating material blocks all three types of heat transmission. The material should be lightweight and may have a reflective surface. It should be moisture-resistant, fire-resistant, fungus-resistant, and provide no attraction for rats and mice. The material should not change volume, and should not deteriorate from aging or exposure to extreme temperatures. Fortunately, there are several good materials that qualify, and they're economical as well.

Insulation Materials

Common thermal insulating materials include: mineral fibers (made from glass, rock, slag, and asbestos), vegetable fibers (such as wood, cane, cotton, wool, and redwood bark), expanded mineral granules (such as perlite and vermiculite), vegetable granules (such as ground cork), foamed materials (both glass and synthetic resins, like styrene and urethane), and aluminum foil.

Insulating batts, blankets, and boards are made from fibers or granules. The fibers or granules are mixed with binders, and formed into useful widths, lengths, and thicknesses. The binder makes insula-

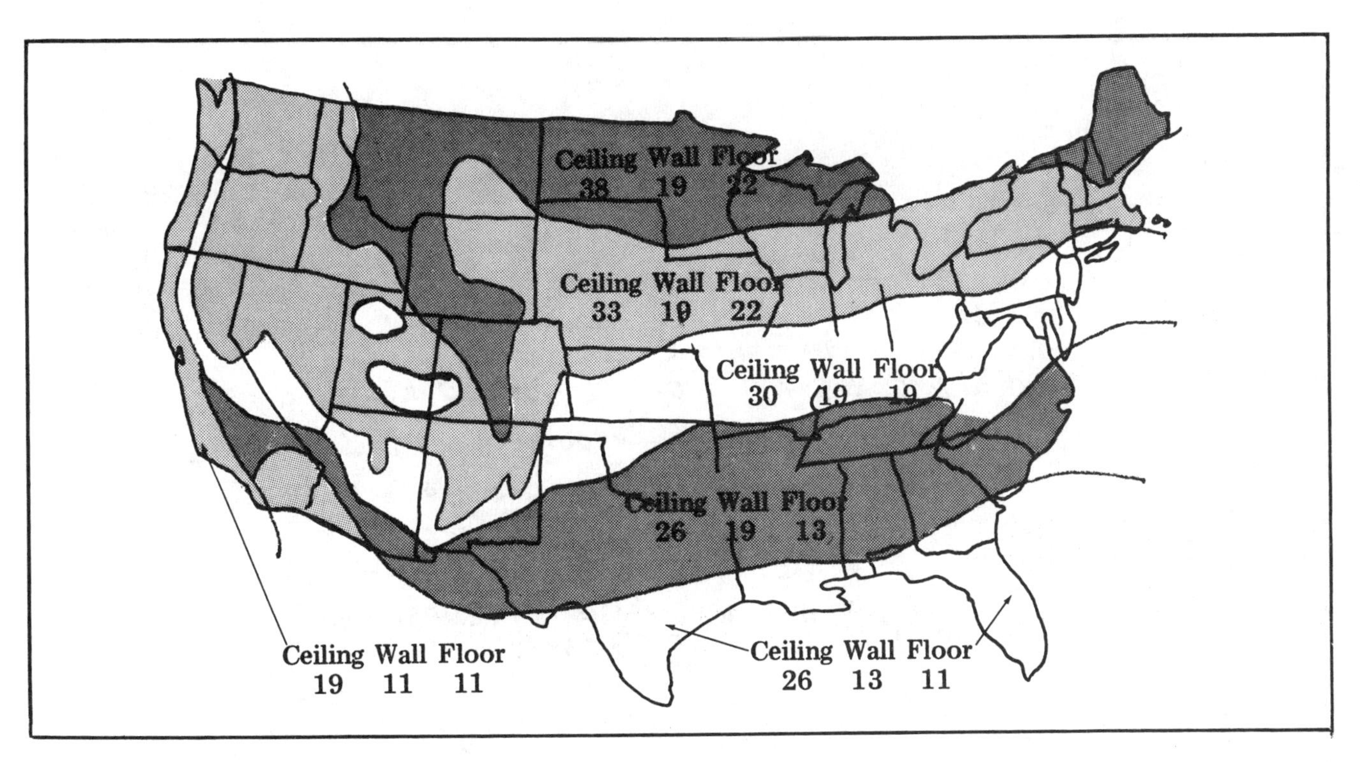

Recommended R-values
Figure 17-1

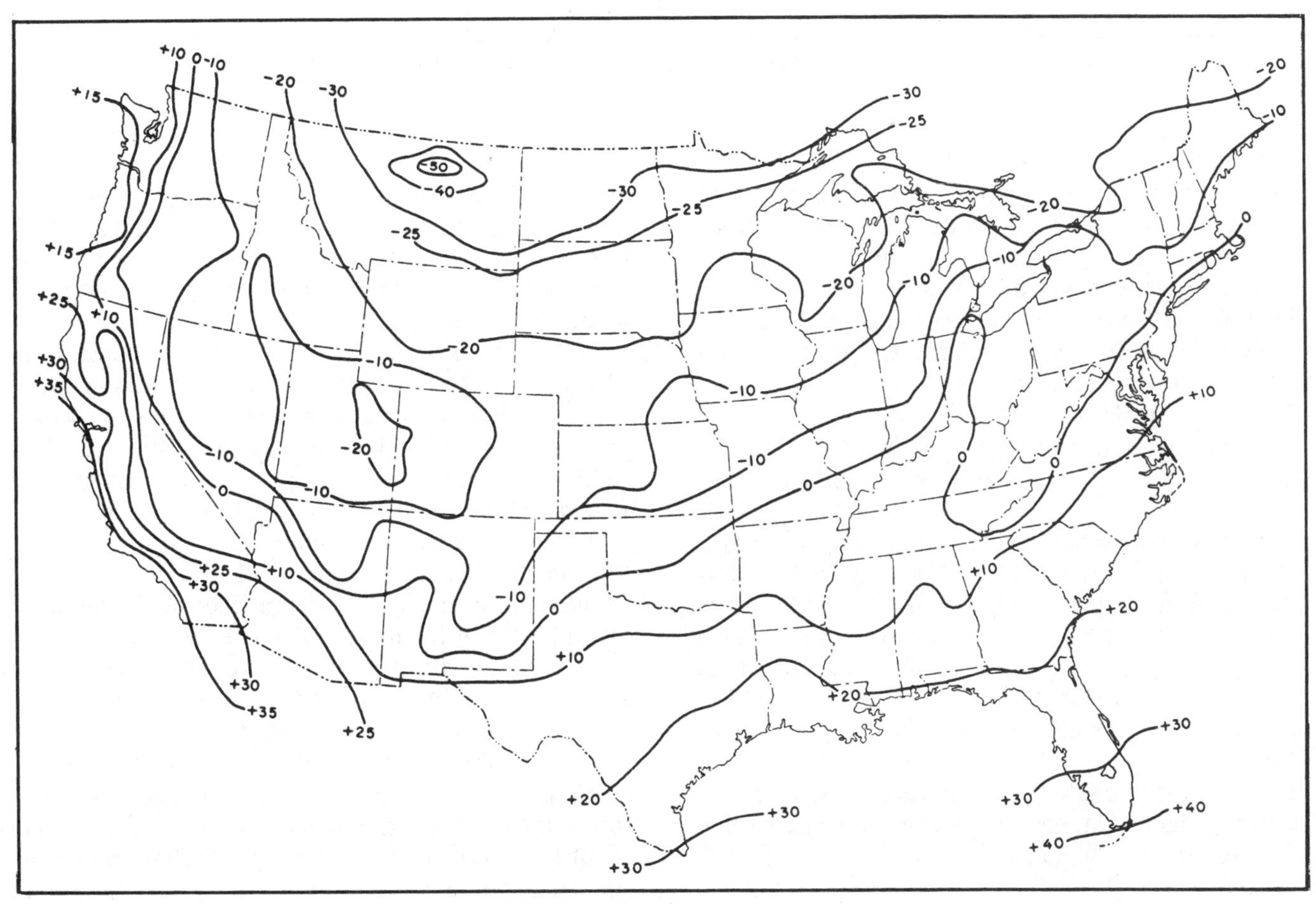

Average winter-low-temperature zones
Figure 17-2

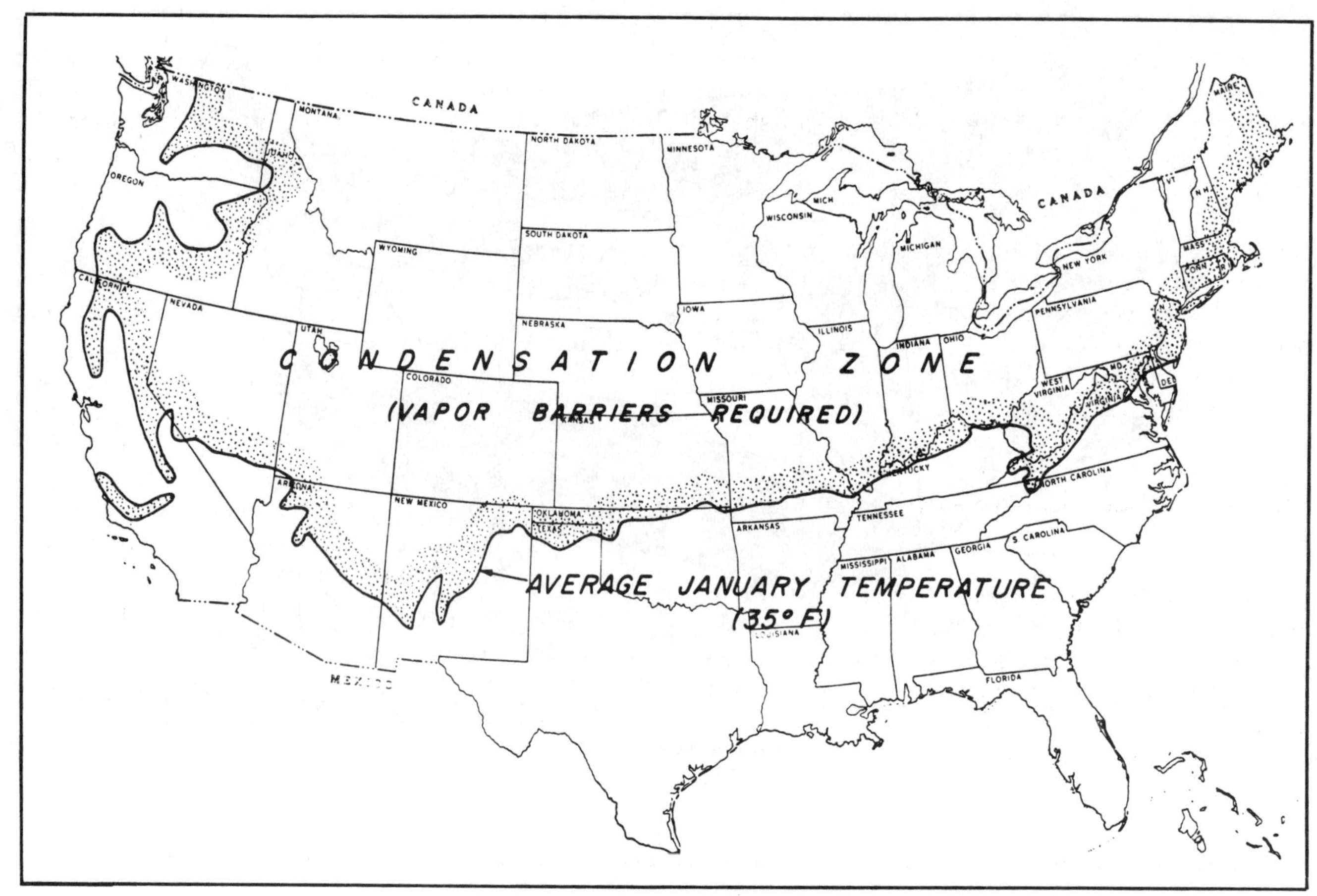

Condensation zone: Areas where average temperature for January is 35° F. or lower
Figure 17-3

tion more water- and mildew-resistant, and adds strength to the finished product. Vegetable fibers usually require chemical treatment to make them fire-resistant. Glass and synthetic resins are formed into blocks, sheets, or boards of various sizes. This is done by carefully controlling the foaming process to get the needed cell-size and density.

Forms of Insulation

Thermal insulation comes in a variety of forms. These include: batts, blankets, reflective insulators, and loose fill.

Batts or blankets are placed between the joists or studs. They're made in widths that fit standard joist and stud spacing. Batts or blankets may be unfaced, or they may be faced (on one or both sides) with paper, aluminum foil or plastic. If the batts or blankets are faced, the faced side has a lower vapor-permeability rate and acts as a vapor barrier.

Paper-faced batts or blankets are made with continuous paper flanges along the long edges. The flanges are easily nailed or stapled to studs and joists.

Reflective insulators are made with aluminum foil. The foil is usually reinforced with paper backing. The insulators are generally made to provide two or more reflective surfaces with air space between the layers. Like batts or blankets, they're used between studs or joists and also have flanges for nailing or stapling.

Loose fill is made from mineral or vegetable fibers. Or it can be made from mineral or vegetable granules. It is applied either by blowing or hand packing. You'll see it primarily in masonry cavity walls, and over ceilings in attic spaces. But it can also be blown into cavities between wall studs.

Blowing insulation is usually done by a specialist who has the necessary equipment. But it's your job to make sure that all areas are filled with the correct amount of insulation. Also be sure that the insulation is evenly distributed over the entire area and does not block any vents.

Vapor Barriers

Vapor barriers prevent condensation of moisture in insulated spaces. When warm, moisture-laden air comes in contact with a cold surface, little droplets of moisture collect on that surface. When moisture collects inside of insulating material, the insulation value of the material is soon lost. That's why vapor barriers must be installed on the warm side of the insulation. The warm side is almost always the side facing the interior of the home.

The vapor barrier should have a "perm" rating of 1 perm or less. Aluminum foil, plastic sheet material, and coated or laminated paper all qualify. Ordinary 15- or 30-lb. asphalt-saturated felts are not acceptable. The greater the temperature difference from one side of the wall to the other, and the greater the relative humidity of the air on the warm side, the more effective the vapor barrier must be.

Many batts and blankets have vapor barriers already attached to one side. Other forms of insulation will require a separate membrane of polyethylene. A vapor barrier is essential if you build in the condensation zone shown on the map in Figure 17-3.

Installing Insulation and Vapor Barriers

Figure 17-4 shows how insulation should form a blanket around the heated areas of the house. Insulation and vapor barriers are commonly installed in ceilings, walls, floors, second stories, and basement rooms. Here's how.

Ceilings

Most heat is lost upward through the ceiling. So you'll need to give the attic special attention. Provide weatherstripping for the attic-access door. A piece of insulation board (the size of the attic door) should be tacked to the attic side of the door. If there is an attic scuttle hole, weatherstrip it, and insulate the back of the scuttle-closure panel. Pack insulation around pipes, flues, or chimneys that penetrate the attic space. This is especially important in cold-climate areas.

When insulating a ceiling, extend the insulation over the top plate, as shown in Figure 17-5. Install mineral-fiber blankets by stapling the vapor-barrier flanges from below. Or, after the ceiling is in place, you can lay blankets in from above. If you use unfaced (no vapor barrier), pressure-fit blankets or batts, no fasteners are needed.

Place insulation on the cold (outside wall) side of pipes and ducts.

Repair rips or tears in the vapor barrier with cloth-backed tape or by stapling poly film over the tear.

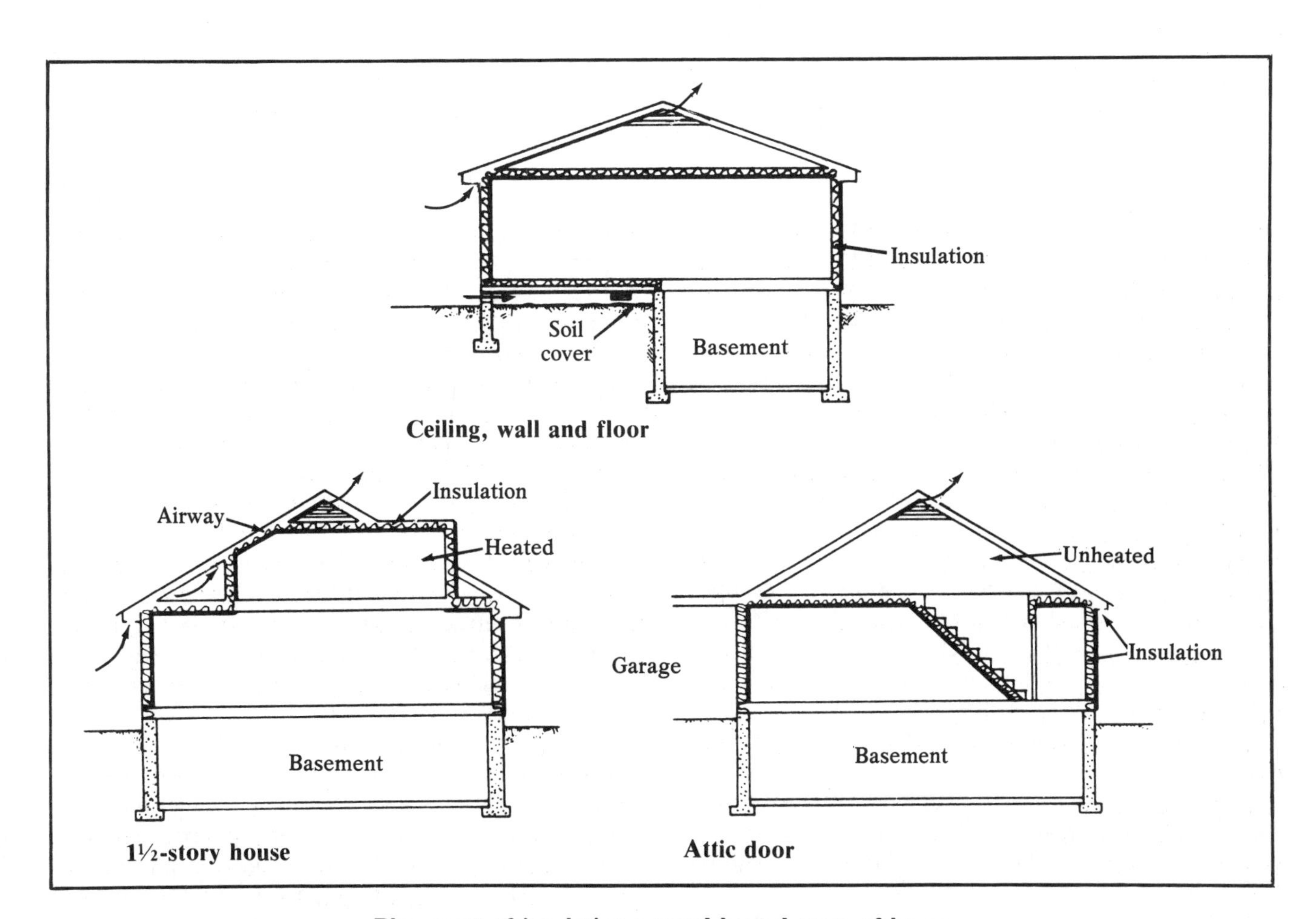

Placement of insulation around heated areas of house
Figure 17-4

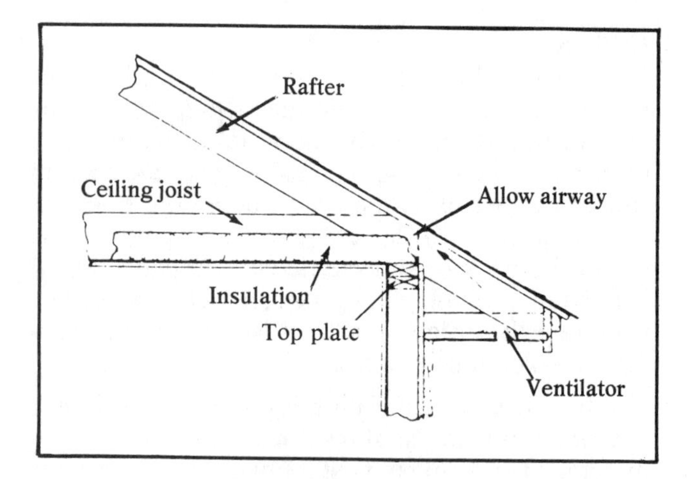

Ceiling insulation
Figure 17-5

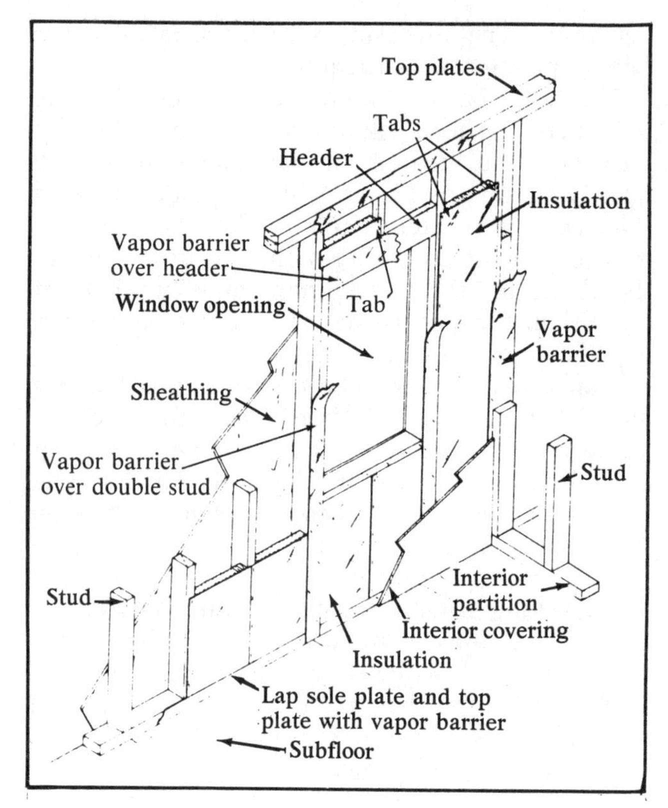

Wall insulation
Figure 17-6

In unfinished attic areas, pouring wool is often used. Empty the bags evenly between ceiling joists. Pay particular attention to the manufacturer's recommendations about proper thickness of the layer and the coverage per bag. Level the wool with a wood slat or garden rake. Small openings, such as those around a chimney, should be hand packed with mineral wool. Be sure that eave-ventilation openings aren't blocked. And don't cover recessed lighting fixtures or exhaust-fan motors that protrude into the ceiling. These fixtures run too hot when covered up by a good insulator.

Walls

Blankets and batts usually come with vapor barriers slightly wider than the insulation. The vapor barrier contains nailing tabs at each side of the blanket or batt. Fasten the tabs to the frame members as described below.

To minimize vapor loss and reduce condensation problems, you can staple the tabs over the edge of the studs, as shown in Figure 17-6. But many builders don't follow this procedure, because it makes drywall or lath application more difficult. Instead, the tabs are fastened to the inner faces of the studs. This usually leaves openings along the edges of the vapor barrier. Moisture can escape through these openings and cause problems. If you fasten the flanges to the interior of studs, plan to apply a vapor barrier over the entire wall, as described later in this section.

To insulate nonstandard-width spaces, cut insulation and vapor barriers about an inch wider than the space to be filled. Staple the uncut flange as usual. Then as you pull the cut side of the vapor barrier over to the other stud, compress the insulation behind the vapor barrier. (Wedge the slightly oversize blanket into place.) Staple through the vapor barrier to the stud.

Be especially careful to cover the area around window and door openings. Where the vapor barrier doesn't cover double studs and header areas, apply a strip of poly film for extra protection. See Figure 17-6.

Where interior partitions meet exterior walls, cover the intersection with some type of vapor barrier. Insulating the space between the double exterior wall studs and applying the vapor barrier should both be done before the corner post is assembled. See Figure 17-6.

There's an easy way to eliminate nearly all condensation problems in walls. It's called *enveloping*. This is a process of installing a vapor barrier over the entire wall, and it's done in the following way.

Start by installing an unfaced friction-type insulation batt without a vapor barrier. See Figure 17-7. The batt is made to fit tightly between framing members spaced 16 or 24 inches on center. Once the rough-in of wiring, plumbing, and duct work is complete, install the vapor barrier over the entire wall. Use 4-mil (or thicker) polyethylene in 8-foot-wide rolls. Be sure to completely cover window and door headers, top and bottom plates, and corners. Don't trim the plastic around window openings until after lath or drywall finish has been installed.

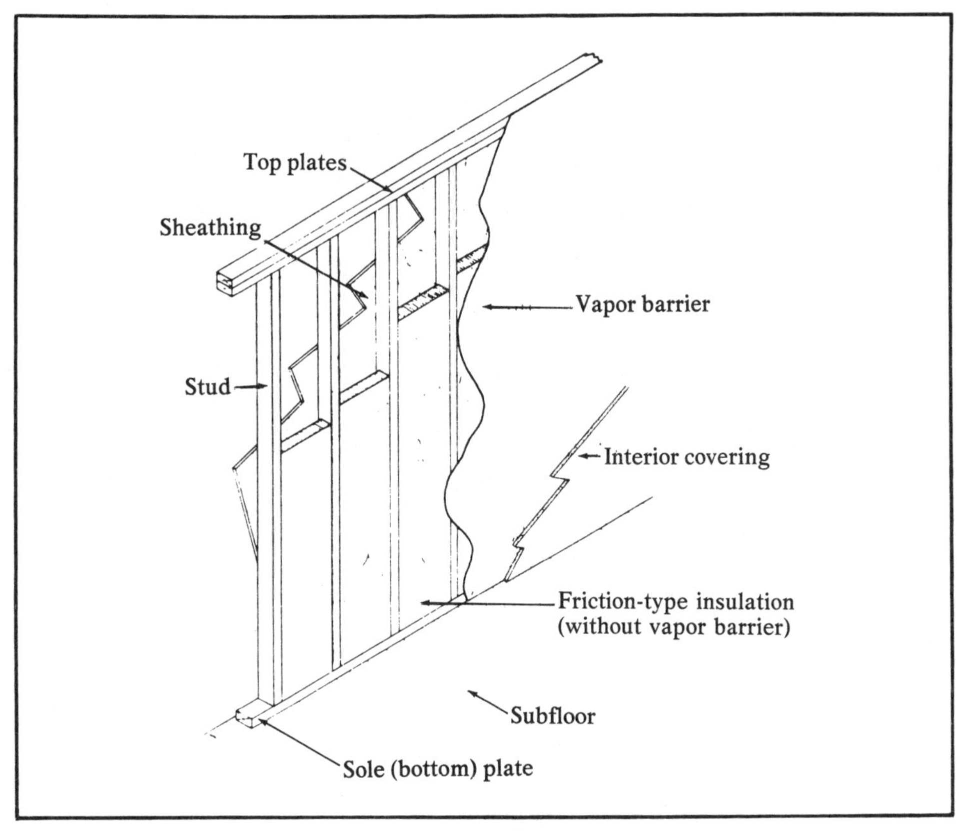

"Enveloping"
Figure 17-7

Be sure to pack insulation into areas behind pipes, ducts, and electrical boxes. These spaces can be filled with loose insulation or fitted with pieces of batt. Pack small pieces of insulation into spaces between the rough framing and door and window headers, jambs, and sills. Staple vapor-barrier paper or polyethylene over these small spaces.

Floors

During the winter months in cold-climate areas, an unheated crawl space is not enough protection for pipes. Without additional insulation, the pipes will freeze. If you build in colder climates, consider installing your supply and disposal pipes inside of a large-diameter vitrified clay pipe. Pack insulation into the void between the two pipes.

In many climates, it pays to insulate the floor above the crawl space. Before the subfloor is installed, place insulating batts or blankets between the floor joists. The batts or blankets should have an attached vapor barrier. If the vapor barrier is strong enough to support the insulation, fasten the tabs over the edges of the joists. It's good practice to use small dabs of mastic adhesive to insure that the batts remain in place against the subfloor. If you use unfaced batts, lay a separate vapor barrier between the subfloor and the underlayment.

If it's likely that the insulation will become wet before the subfloor is installed, delay insulating until framing and sheathing are complete.

Second Stories

Don't overlook the perimeter area of second-floor joists. To protect the spaces between the joists (at the header and along the stringer joists), use insulation that has a vapor barrier. See Figure 17-8. Both the vapor barrier and the insulation should fill the joist spaces. Install insulation and vapor barriers in exposed second-floor walls the same way as in first-floor walls.

A two-story house is sometimes designed so that part of the second floor projects beyond the first floor. The projection is usually about 12 inches. Insulate the projection and install a vapor barrier, as shown in Figure 17-9.

Many 1½-story houses have knee walls. These are short walls which extend from the floor to the rafters. See Figure 17-10. The wall height is between 4 and 6 feet. If a house has knee walls, install insulation and vapor barriers in the first-floor ceiling area, at the knee walls, and between the rafters. Use either blanket or batt insulation with integral vapor barriers, or use a separate vapor barrier as described for first- and second-floor walls.

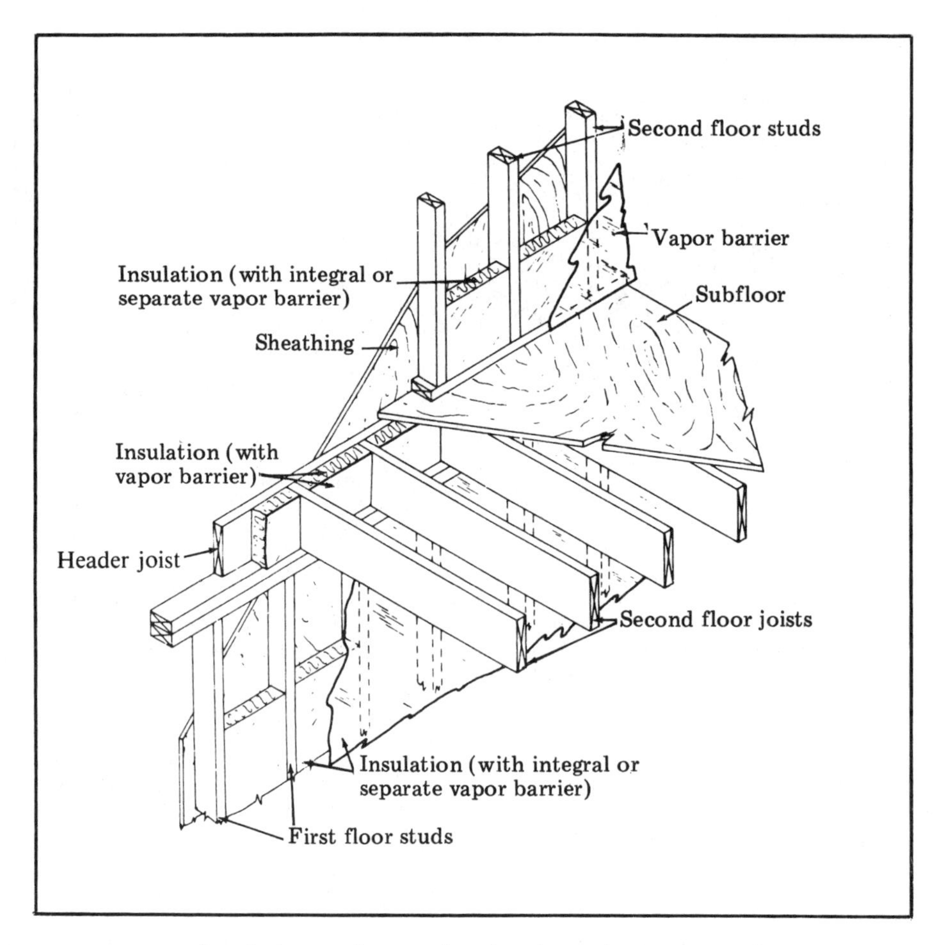

Insulation and vapor barriers in perimeter area of second-floor joists
Figure 17-8

Begin by placing insulation batts or blankets between the joists that run from the outside wall plate to the knee wall. Be sure to install the insulation with the vapor barrier facing down. The insulation should also fill the entire joist space directly under the knee wall, as shown in Figure 17-10. Make sure you leave an airway at the junction of the rafter and the exterior wall. This airway will ventilate the attic.

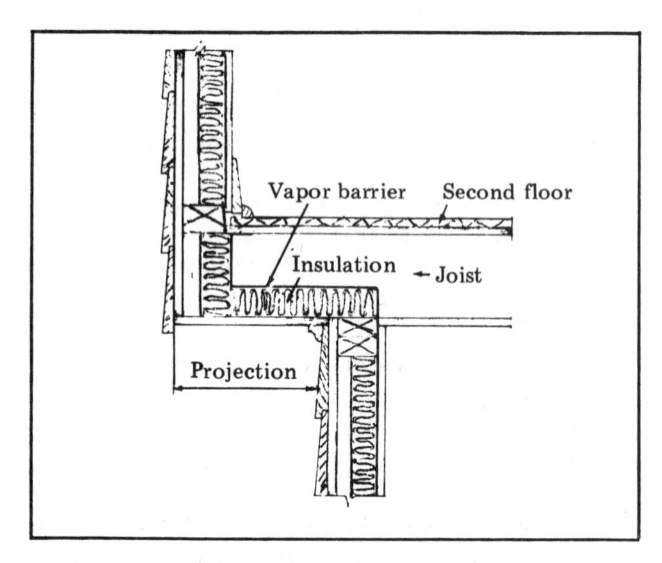

Insulation and vapor barrier at second floor projection
Figure 17-9

Then install batt or blanket insulation between the rafters of the sloping portion the room. The vapor barrier goes on the inner (warm) side of the roof or wall. Leave at least a 1-inch airway between the top of the insulation and the roof sheathing. This allows for the movement of air from the area behind the knee wall to the attic area above the second-floor rooms.

Finished Basement Rooms

Finished basement rooms often have fully or partly exposed masonry walls. When located on property that has poor drainage, drain tiles are needed on the outside of the footing. These are required by code in many areas. You must also provide a waterproof coating on both the exterior and interior surfaces of the wall.

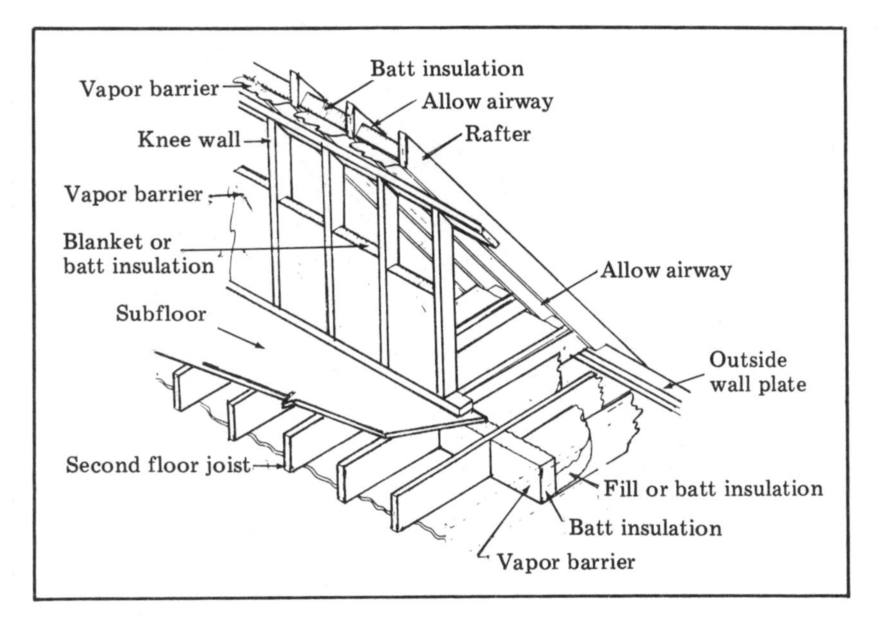

Insulation and vapor barriers in knee-wall area of 1½ story house
Figure 17-10

To provide space for blanket insulation on a masonry wall, apply furring strips (2 x 2- or 2 x 3-inch members) to the wall, as shown in Figure 17-11.

You can also use expanded polystyrene over a masonry wall. Apply it with a slurry of cement mortar. The wall is completed with a plaster finish. Expanded polystyrene has good resistance to vapor movement, and requires no other barrier.

If there is no vapor barrier under a concrete slab, lay a sheet of poly film over the slab before applying the sleepers. Another system uses treated 1 x 4-inch sleepers fastened to the slab before the vapor barrier is applied. Lay a vapor barrier on top of the sleepers. Then nail a second set of 1 x 4-inch sleepers over the first set. The subfloor and finish floor then go down over the sleepers.

To prevent heat loss and minimize the escape of water vapor at the top of the foundation wall, install blanket or batt insulation between the floor joists and along stringer joists. See Figure 17-11. The vapor barrier faces the interior of the basement, and should fit tightly against the joists and subfloor.

Insulation Values

To get the insulation values recommended in Figure 17-1, you may have to use double layers of insulation. You may need to combine layers having different insulation values. Here are some suggested combinations:

Ceilings, double layers of batts or blankets:
R-38, two layers of R-19 (6") mineral fiber
R-33, one layer of R-22 (6½") and one layer of R-11 (3½") mineral fiber
R-30, one layer of R-19 (6") and one layer of R-11 (3½") mineral fiber
R-26, two layers of R-13 (3⅝") mineral fiber

Ceilings, loose fill mineral wool and batts:
R-38, R-19 (6") mineral fiber and 20 bags of wool per 1,000 SF (8¾")
R-33, R-22 (6½") mineral fiber and 11 bags of wool per 1,000 SF (5")
R-30, R-19 (6") mineral fiber and 11 bags of wool per 1,000 SF (5")
R-26, R-19 (6") mineral fiber and 8 bags of wool per 1,000 SF (3¼")

Walls, 2 x 4-inch framing:
R-19, R-13 (3⅝") mineral fiber batts and 1" foam plastic sheathing
R-11, R-11 (3½") mineral fiber batts

Floors:
R-22, R-22 (6½") mineral fiber
R-19, R-19 (6") mineral fiber
R-13, R-13 (3⅝") mineral fiber
R-11, R-11 (3½") mineral fiber

Estimating Materials

When estimating batt or blanket insulation requirements, first calculate the total area of the ceiling, floor or wall to be covered. Then take 95% of that total area. (About 5% of the total area is already taken up by the thickness of the framing members.) Thus, a 1,000 SF floor area requires 950 SF of insulation. Allow 160 staples per 100 SF of insulation.

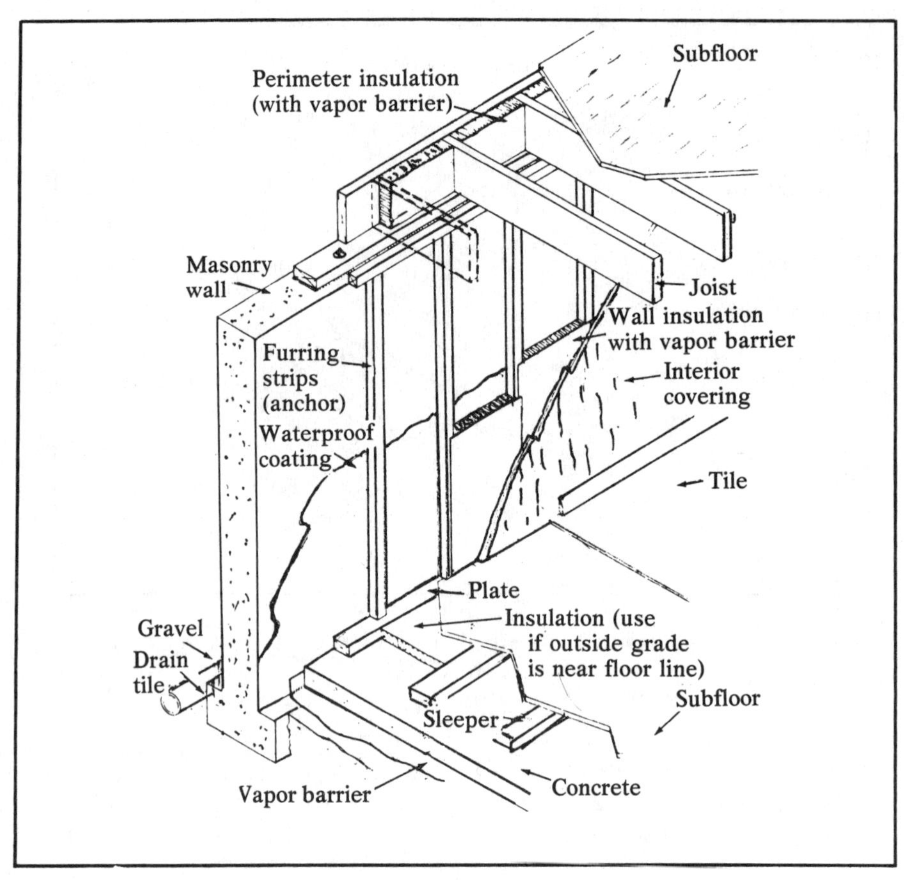

Insulation and vapor barriers in basement
Figure 17-11

Estimating Manhours

Here's a rule of thumb for estimating manhours required to install insulation batts or blankets in the ceilings, walls and floors of a one-story house with framing 16 inches on center:

The floor space (in square feet) divided by 30 equals the number of manhours required.

A 24' x 40' 1-story house requires insulating:

960 SF floor area
960 SF ceiling area
1,020 SF wall area (8' ceiling)

2,940 SF total area

(960 SF floor space divided by 30 = 32 manhours)

For a two-story house:

Floor space (in square feet) divided by 45 equals the number of manhours required.

A 24' x 40' 2-story house requires insulating:

960 SF floor area
960 SF ceiling area
2,040 SF wall area (8' ceiling)

3,960 SF total area

(1,920 SF floor space divided by 45 = 42.7 manhours)

Where a second layer of insulation is required in the ceiling, add one manhour for each 100 SF of ceiling area.

Summary

Home buyers are energy-conscious. They want a home that's easy to heat and cool. Your spec house has a major selling advantage if you have been conscientious about meeting recommended R-values and installing energy-efficient windows.

There are so many good insulation materials available today that you can hardly go wrong, no matter which one you select. I'm partial to the glass-fiber blanket insulation that comes in rolls. It can be cut to any length and goes up fast. But small glass particles fill the air when you're installing it. Wear a mask over your nose and mouth to keep these particles out of your lungs. Wear a long-sleeve shirt (or jacket) and gloves. Don't wash these clothes with other garments. Run the clothes through the washer twice before wearing them again.

Make Figure 17-12, your Insulation and Vapor Barrier Checklist and Data Reference Sheet, a permanent record of your insulation work.

Insulation and Vapor Barrier Checklist and Data Reference Sheet

1) The house is _______ feet in length and _______ feet wide, for a total of _______ square feet.

2) The house is ☐1-story`☐1½-story ☐2-story.

3) The house ☐does ☐does not have a heated basement.

4) R-values for this region

Recommended:	**Installed:**
Ceiling ______________________________	Ceiling ______________________________
Wall ______________________________	Wall ______________________________
Floor ______________________________	Floor ______________________________

5) Insulation used:

	Ceiling	**Wall**	**Floor**
☐Glass fiber:			
☐Batt	☐	☐	☐
☐Blanket	☐	☐	☐
☐Loose fill	☐	☐	☐
☐Rock wool:			
☐Batt	☐	☐	☐
☐Blanket	☐	☐	☐
☐Loose fill	☐	☐	☐
☐Reflective	☐	☐	☐
☐Pressure-fit batts	☐	☐	☐
☐Friction-type batts	☐	☐	☐
☐Vermiculite	☐	☐	☐
☐Perlite	☐	☐	☐
☐Other (specify)			
______________________________	☐	☐	☐

☐Unfaced

☐Faced (☐one side ☐ two sides ☐paper ☐aluminum foil ☐plastic)

Figure 17-12

6) Number of square feet in:

☐Ceiling ________________ SF

☐Wall ________________ SF

☐Floor ________________ SF

7) Thickness of insulation in:

☐Ceiling ______________ inches

☐Wall ______________ inches

☐Floor ______________ inches

8) Flanges of faced insulation are stapled ☐over edge of structural members ☐to inner face of structural members.

9) Vapor barrier is ☐attached to one side of the insulation ☐a separate membrane made of ☐4-mil polyethylene ☐other (specify)________________________________
(Note: A vapor barrier is essential in the condensation zone.)

10) Supply and disposal pipes in crawl space ☐are ☐are not insulated. They are ☐wrapped ☐enclosed in vitrified or similar tile and tile is filled with insulation ☐other (specify) ____________________

11) The builder ☐is ☐is not familiar with the Federal Trade Commission Rule concerning labeling and advertising of home insulation and how it affects the builder.

12) Structural members are ☐16 inches o.c. ☐24 inches o.c.

13) Material costs:

Insulation:

Rolls	________ rolls	@	$________	ea	=	$________
Bundles	________ bdls	@	$________	ea	=	$________
Batts	________ pcs	@	$________	ea	=	$________
Bags	________ bags	@	$________	ea	=	$________
Blown	________ SF	@	$________	SF	=	$________
4-mil polyethylene	________ rolls	@	$________	ea	=	$________
Staples	________ boxes	@	$________	ea	=	$________

Other materials, not listed above:

____________________	__________	@	$__________	=	$________
____________________	__________	@	$__________	=	$________

__________________ ________ @ $________ = $________

__________________ ________ @ $________ = $________

Total cost of all materials $________

14) Labor costs:

Skilled ______hours @ $______hour = $________

Unskilled ______hours @ $______hour = $________

Other (specify)

__________________ ______hours @ $______hour = $________

Total $________

(Note: Rule of thumb for estimating manhours required to install insulation batts or blankets in floor, wall, and ceiling of a one-story house with framing 16 inches o.c.)

The floor space (in square feet) divided by 30 equals the number of manhours required.

For a two-story house:

The floor space (in square feet) divided by 45 equals number of manhours required.

15) Total cost:

Materials $________

Labor $________

Other $________

Total $________

16) Insulation cost per square foot of floor space: $________

17) Problems that could have been avoided ______________________________

____

____

____

____

18) Comments ____

____

____

____

19) This copy of Checklist and Data Reference Sheet has been provided:

☐Management

☐Accounting

☐Estimator

☐Foreman

☐File

☐Other: ___________

Chapter 18

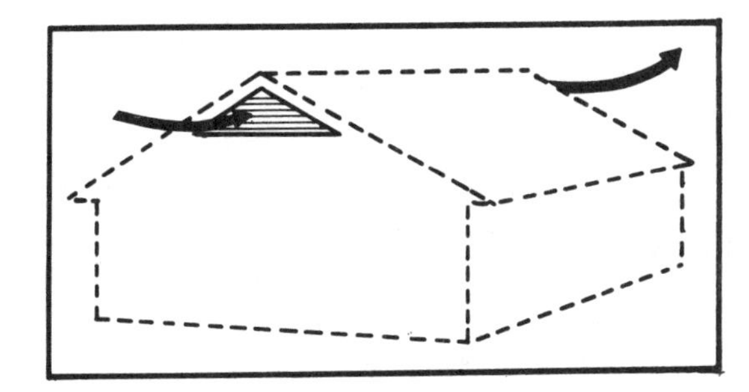

Ventilation

Earlier in this manual, we learned about the importance of ventilation. In this chapter, we'll look at the accepted standards for ventilation in homes, and how to meet these standards.

The Minimum Property Standards of the U.S. Department of Housing and Urban Development (HUD) state:

1) Attics and spaces between roof and top floor ceiling may have a free ventilating area of 1/300 of the horizontal area when (a) a vapor barrier having a transmission rate not exceeding one perm is installed on the warm side of the ceiling or (b) at least 50% of the required ventilating area is provided with fixed louvers located in the upper portion of the space to be ventilated (at least 3 feet above eave or cornice vents) with the remainder of the required ventilation provided by eave or cornice vents. Structural spaces shall include porch roofs, canopies and any enclosed structural space where condensation may occur. All spaces shall be cross-ventilated.

2) Basementless (crawl) spaces may have a free ventilating area of 1/150 of the ground area when ground surface is covered with a vapor barrier. These spaces shall be cross-ventilated.

During cold weather, condensation is common in attic spaces and under flat roofs. Even when there is a good vapor barrier, some moisture will collect around pipes and on exposed metal surfaces. This isn't a problem if the moisture is evenly distributed. But if the moisture is concentrated in cold spots, it will cause damage.

Wood shingle and wood shake roofs don't resist vapor movement. But asphalt shingles and built-up roofs are moisture-resistant. Natural ventilation (through space ventilators) is the most practical and cheapest method of moisture removal.

If a poorly-insulated attic does not have good ventilation, ice dams will form at the cornice in cold weather. This happens after a heavy snowfall, when heat trapped in the attic causes the snow on the roof to melt. Water runs down the roof and freezes on the surface of the cornice. This forms an ice dam at the gutter, and causes water to back up at the eaves and under the shingles. Similar dams can form in roof valleys.

Good ventilation generally solves the problem. With a well-insulated ceiling and adequate ventilation, attic temperatures are lower. Snow melting is greatly reduced.

In hot weather, attic ventilation removes hot air and lowers the temperature in the rooms below. Insulation installed between attic joists reduces heat flow even further. This means greater comfort for homes without mechanical cooling, and lower energy-costs in air-conditioned homes.

In a crawl-space home, ventilation is needed to remove moisture rising from the soil. Otherwise, vapor condenses on the wood under the floor. This causes the wood to decay. To reduce the amount of ventilation required, install a permanent vapor barrier over the soil of the crawl space.

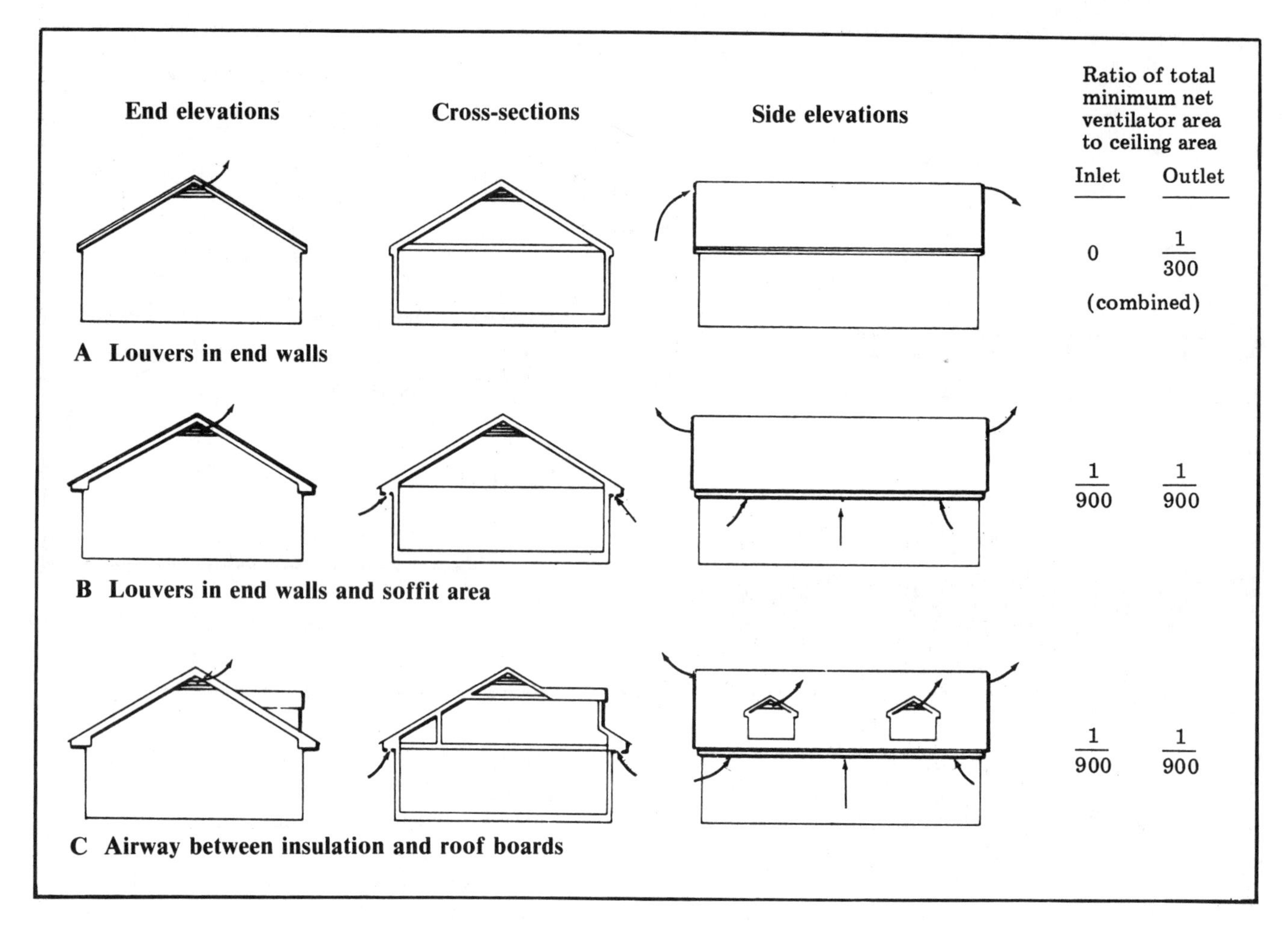

Gable-roof ventilators
Figure 18-1

Modern homes are often airtight. Tight construction (including storm windows and storm doors) and the use of humidifiers have created moisture problems for many homeowners. It takes only a little planning and knowledge to overcome moisture problems — before the house is built. After construction is complete, it's a much more difficult job. And repairing moisture damage is expensive. Take the time to plan for moisture control. It will pay big dividends for many years.

Sizing and Locating Ventilators

Each of the common roof styles requires a particular size and type of vent. Crawl-space construction also has specific ventilation requirements. Ventilator sizing standards are expressed as the ratio of minimum net vent area to the projected ceiling area of the rooms below. See Figure 18-1. "Net vent area" means the actual size of the ventilator opening. Usually, you must increase the net vent area if you plan to include any restrictions, such as louvers or screens. If you restrict the air flow with screens, you must double the net vent areas shown in Figures 18-1, 18-3, and 18-4.

Let's look at the standard sizing and locations for ventilators in homes with gable, hip, and flat roofs, and in homes with crawl-space construction.

Gable Roofs

For houses with gable roofs, it's common practice to install louvered openings in the end walls. Unfortunately, air moves through these openings only when there's enough wind blowing in the right direction. You'll get better air circulation if there are additional openings in the soffit area. You can also use a continuous ridge vent if the house is located in a moderate-climate area.

Locate end-wall vents as close to the ridge as possible. See Figure 18-1 A. The net vent area for the openings should be 1/300 of the ceiling area. For example, where the ceiling area is 1,200 square feet, the minimum total net vent area is 4 square feet.

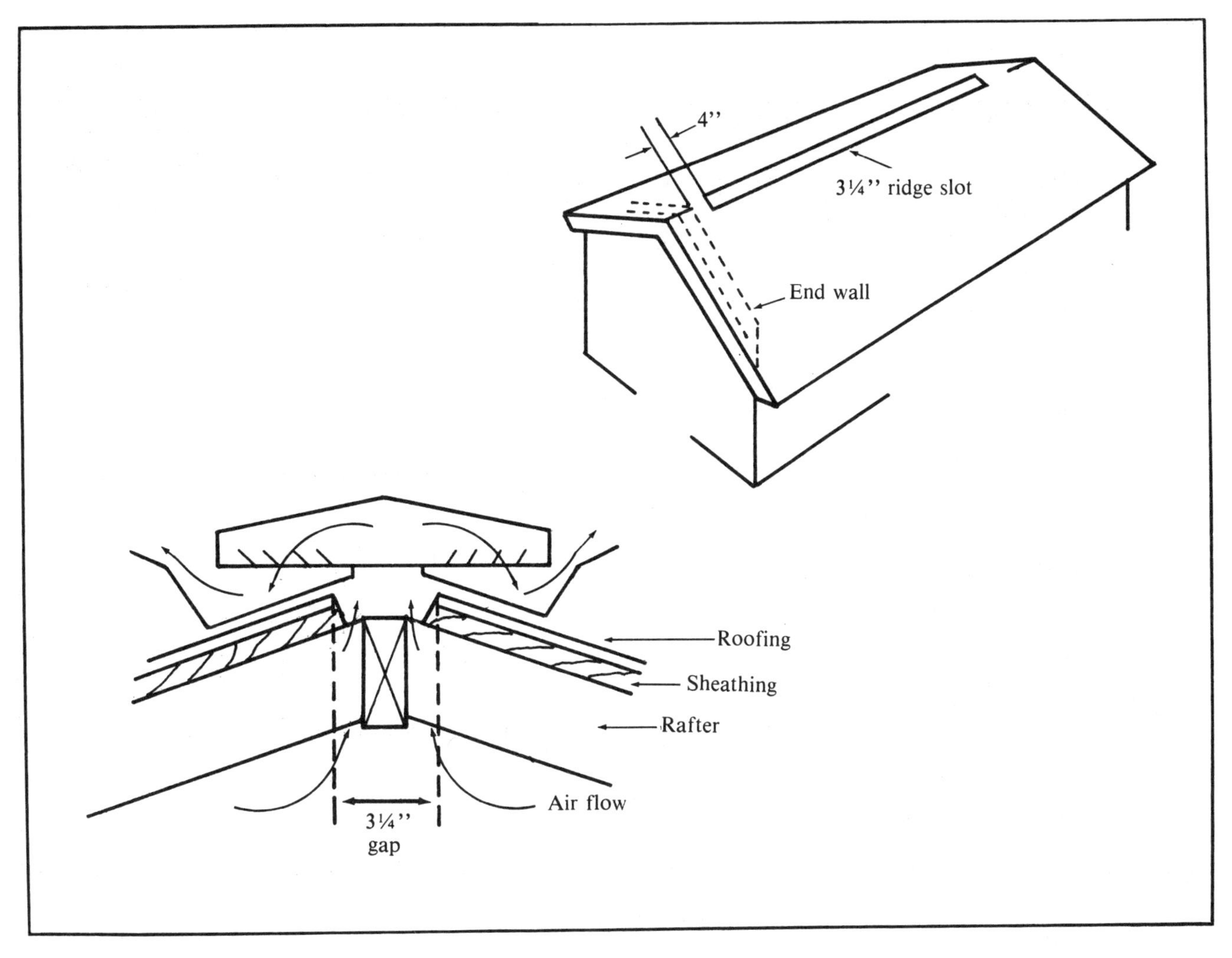

Continuous ridge vent
Figure 18-2

On a house with a wide overhang at the gable end, you can use a series of small vents or a continuous slot. Locate the small vents or slot on the underside of the gable-end soffit. When you use this system, be sure to drill 3/4-inch holes in the end rafters of the gable. These allow air to move from the soffit area to the attic area.

End-wall vents are combined with soffit ventilators in Figure 18-1 B. The net vent area for these openings should be 1/900 of the ceiling area.

If the attic area has rooms with sloped ceilings, the insulation should follow the roof slope. Be sure to install the insulation so that there is an open space of 1½ inches between the insulation and the roof boards. This provides for good air movement, as shown in Figure 18-1 C.

Continuous ridge vents are efficient outlet vents in climates not subject to heavy snowfall. The vent fits over a 3¼-inch-wide slot that runs along the length of the ridge. See Figure 18-2. Don't run the slot over onto the gable overhang. The slot should end at a point 4 inches short of the end wall. The continuous ridge vent eliminates the need for ridge cap shingles. Follow the manufacturer's instructions for installing the vent.

Hip Roofs

Hip roofs need an air-inlet in the soffit area and an air-outlet near the peak or ridge. The difference in temperature between the attic air and the outside air creates a chimney effect that sucks in the outside air and blows out the attic air. This air movement is completely independent of the wind outside the home.

The most efficient type of inlet opening is a continuous slot that is at least 3/4-inch wide. The air-outlet near the peak can be a gable-type metal ventilator. The minimum net vent areas are shown in Figure 18-3.

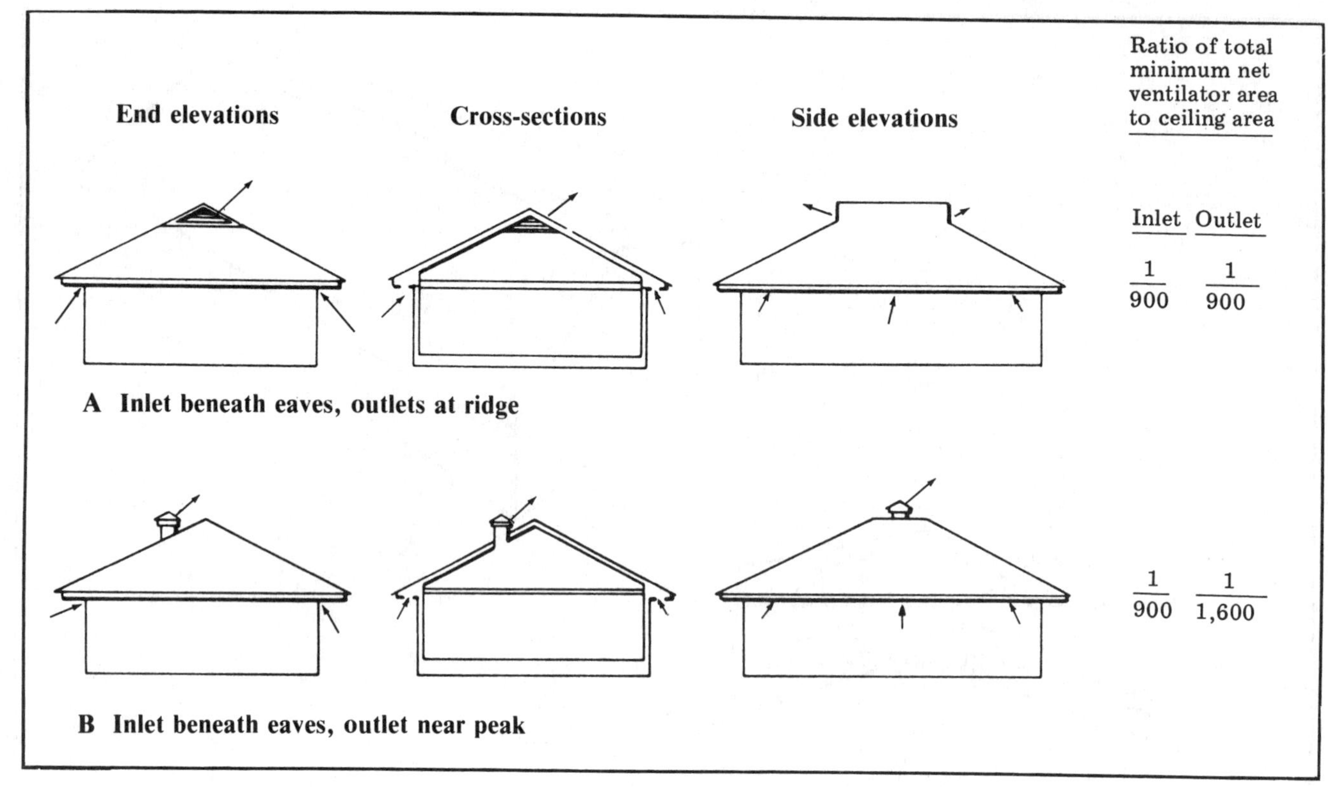

Hip-roof ventilators
Figure 18-3

The air-outlet can also be a series of small roof ventilators located near the ridge. See Figure 18-3 B. Locate the small ventilators on the rear slope of the roof so they don't clutter the front of the house.

A continuous ridge vent can also be used on a hip roof to provide efficient outlet vents.

Flat Roofs

A flat roof requires a greater ratio of net vent area, because air movement depends on wind, rather than on natural circulation of warm air. It's important to have a clear, open space above the ceiling insulation and below the roof sheathing. This allows air to move freely from inlet to outlet openings.

Avoid using solid blocking for bridging and for bracing over bearing partitions. Solid blocking stops air circulation in the attic area.

The most common type of flat roof has rafters that extend to form an overhang beyond the exterior wall. See Figure 18-4 A. When soffits are used, locate the inlet-outlet ventilators in the soffit area. Use continuous-slot vents that are at least 3/4-inch wide.

When a flat roof is combined with a parapet-type wall, the ceiling joists can be either separate from or combined with the roof joists. When they are separate, use the space between as an airway. See Figure 18-4 B. Place inlet and outlet vents as shown, or use a series of stack-outlet vents in combination with the inlet vents. Place the stack-outlet vents along the centerline of the roof.

If the ceiling joists are combined with the roof joists, locate the vents as shown in Figure 18-4 C. Use wall inlet-ventilators and center stack-outlet vents.

Crawl-Space Construction

The crawl space below the house and under porches should be ventilated. It may also be protected from ground moisture with a vapor barrier, as shown in Figure 18-5. Use a vapor barrier with a perm value of less than 1. Plastic films, roll roofing and asphalt-laminated paper all qualify. But make sure, before you begin construction, that you know what the local code requirements are for a vapor barrier in a crawl space.

A crawl space may open into a partial basement. If there's a basement window that can be opened, no wall vents are needed. But you must still use the vapor barrier over the soil if one is required.

In a crawl space that doesn't adjoin a basement, provide at least four foundation-wall vents near the corners of the building. The total net vent area should be equal to 1/160 of the ground area when

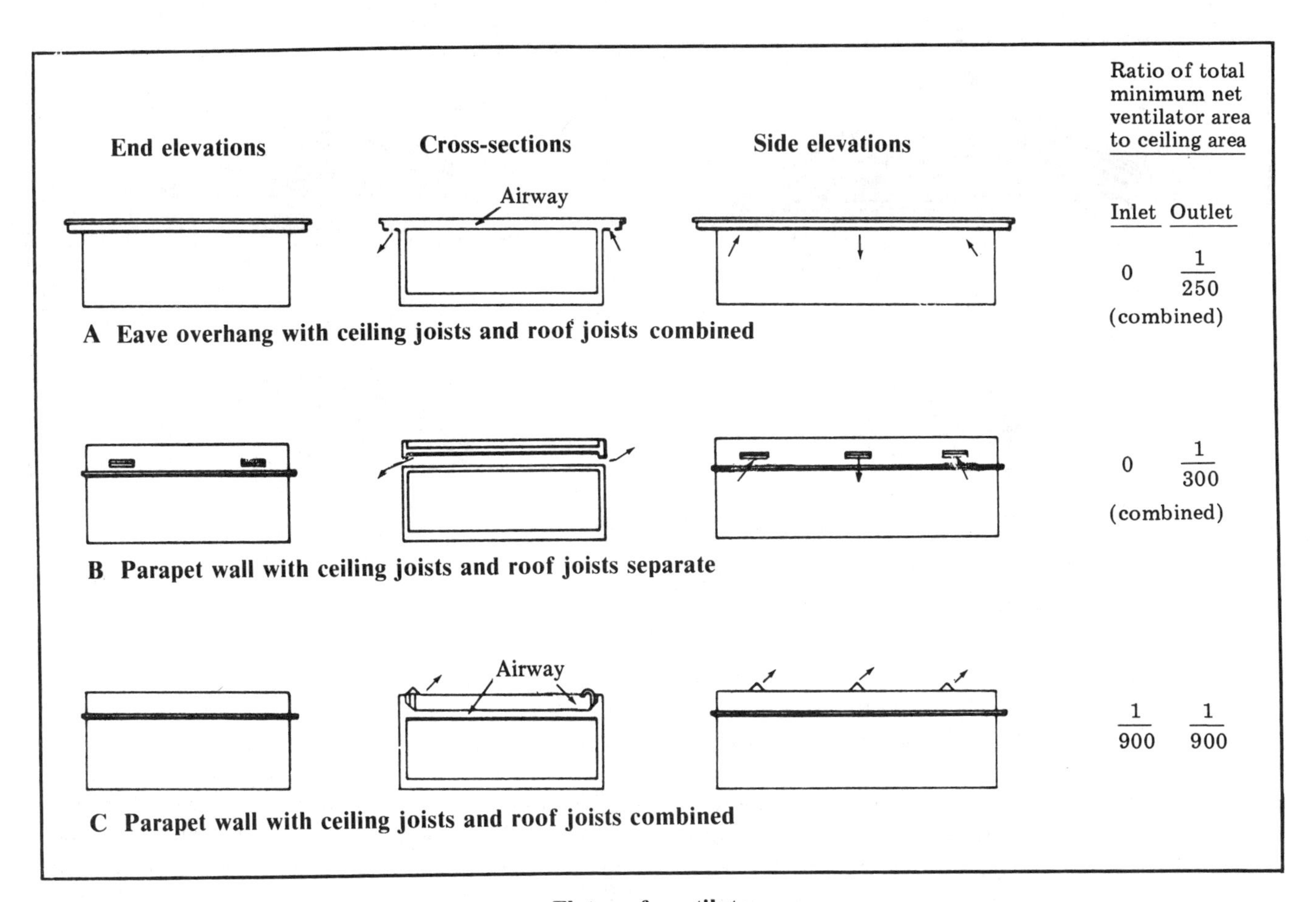

Flat-roof ventilators
Figure 18-4

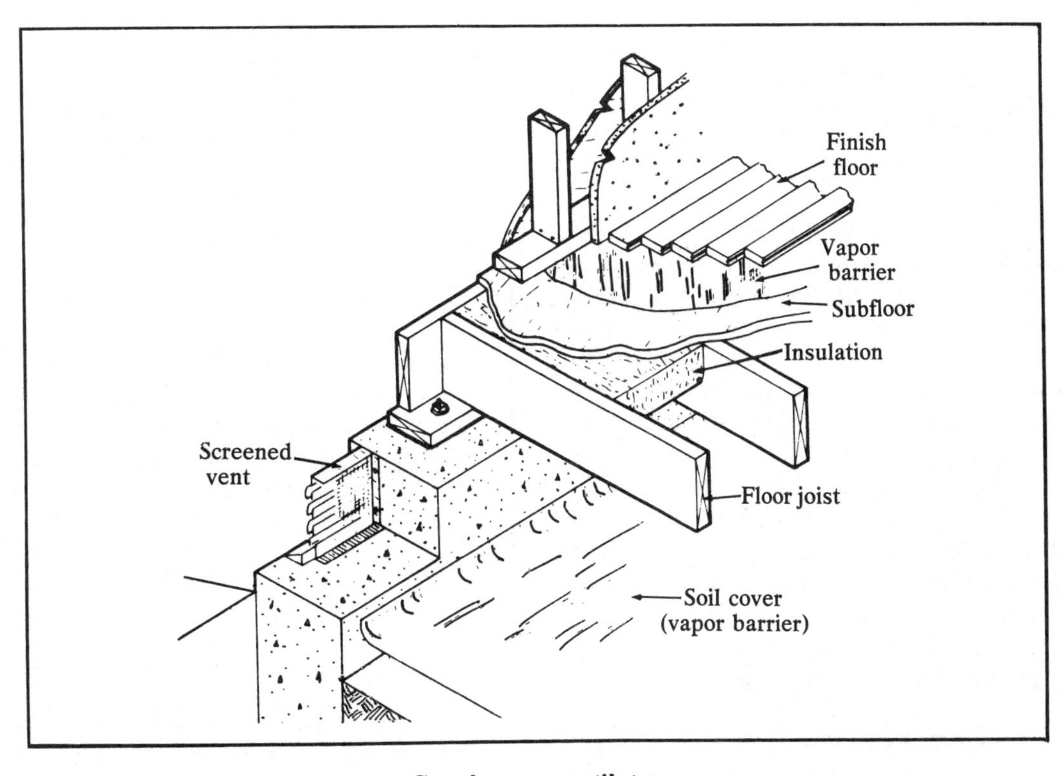

Crawl-space ventilator
Figure 18-5

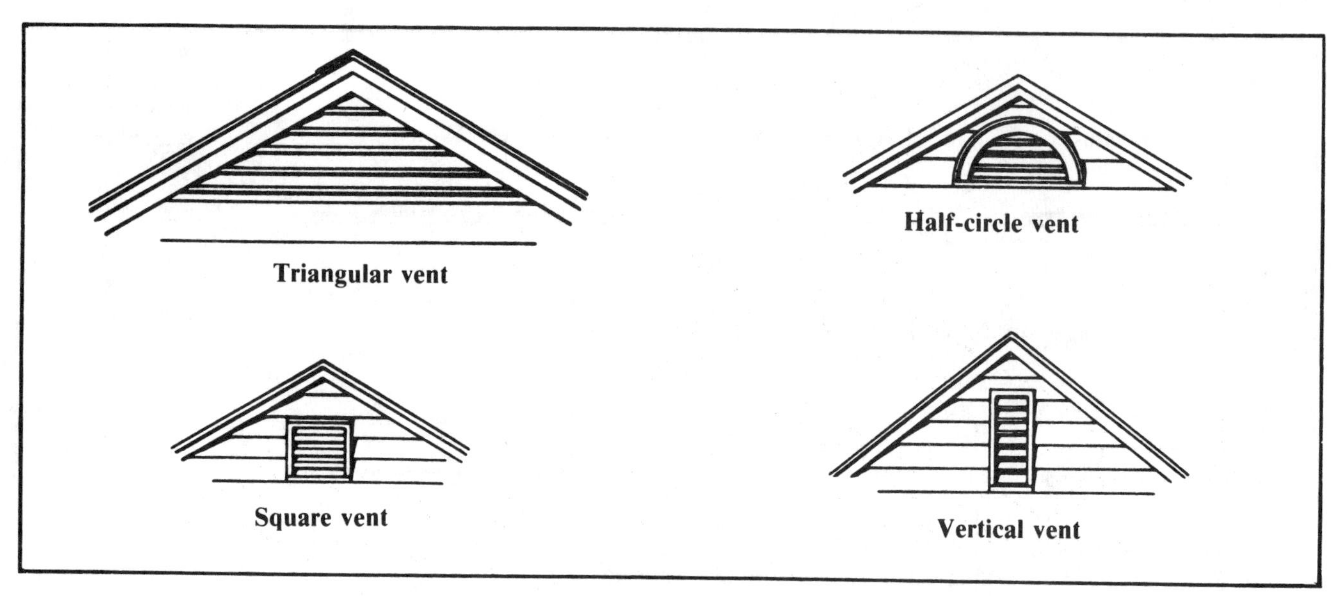

Gable-end ventilators
Figure 18-6

no soil cover is used, and 1/150 of the ground area when the ground surface is covered with a vapor barrier.

For example, when no soil cover is used on a ground area of 1,200 square feet, a total net ventilating area of about 8 square feet is needed. Each of the four vents should have 2 square feet of net vent area. If that makes the vents too large, use 8 one-foot vents. The best vents for crawl spaces have corrosion-resistant screens of No. 8 mesh.

In termite-prone areas, vents are doubly important. Moisture under the house attracts termites. Be sure to use enough vents. Many builders install a vent at each corner, plus additional vents spaced 10 feet on center between the corners.

Outlet Ventilators

Prefabricated gable-end ventilators are available in sizes and shapes that fit all common roof styles. Most of these ventilators have metal louvers and frames. You can make custom vents out of wood if needed for a particular house design. But the most important considerations are still the net vent area and the location of the vent.

Common types of gable-end vents are shown in Figure 18-6. The triangular vent shown in the figure fits the slope of the roof and is located near the ridge. It can be made of wood or metal. Metal vents can usually be adjusted to fit the roof slope. Wood vents are enclosed in a frame that fits into the rough opening, as shown in Figure 18-7.

Inlet Ventilators

Two commonly-used inlet ventilators are the continuous screened slot and small, louvered and screened vents. Both are located in the soffit area, as shown in Figure 18-8.

The small, louvered and screened vents (shown in Figure 18-8 A) are available from many building-supply dealers and are easy to install. You can cut out the necessary sections of soffit before the soffit is installed. These smaller vents provide efficient ventilation when properly spaced.

If you use a continuous screened slot, place it near the outer edge of the soffit and close to the fascia. See Figure 18-8 B. During the winter, snow is less likely to enter through a vent located in this area. This type of vent can also be used on a flat roof.

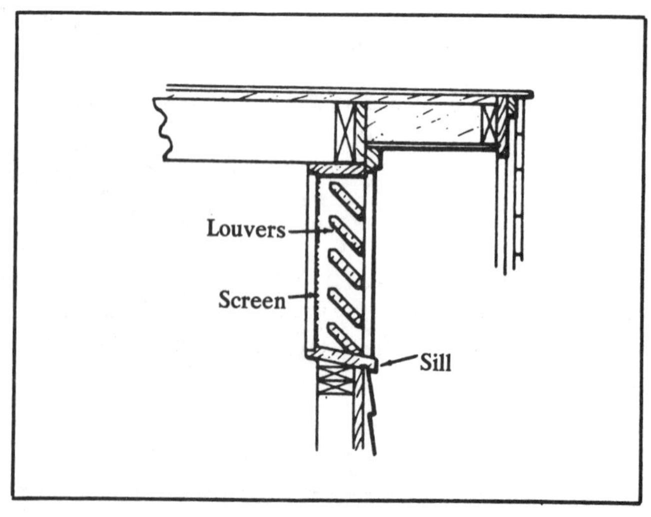

Cross-section of enclosed wood vent
Figure 18-7

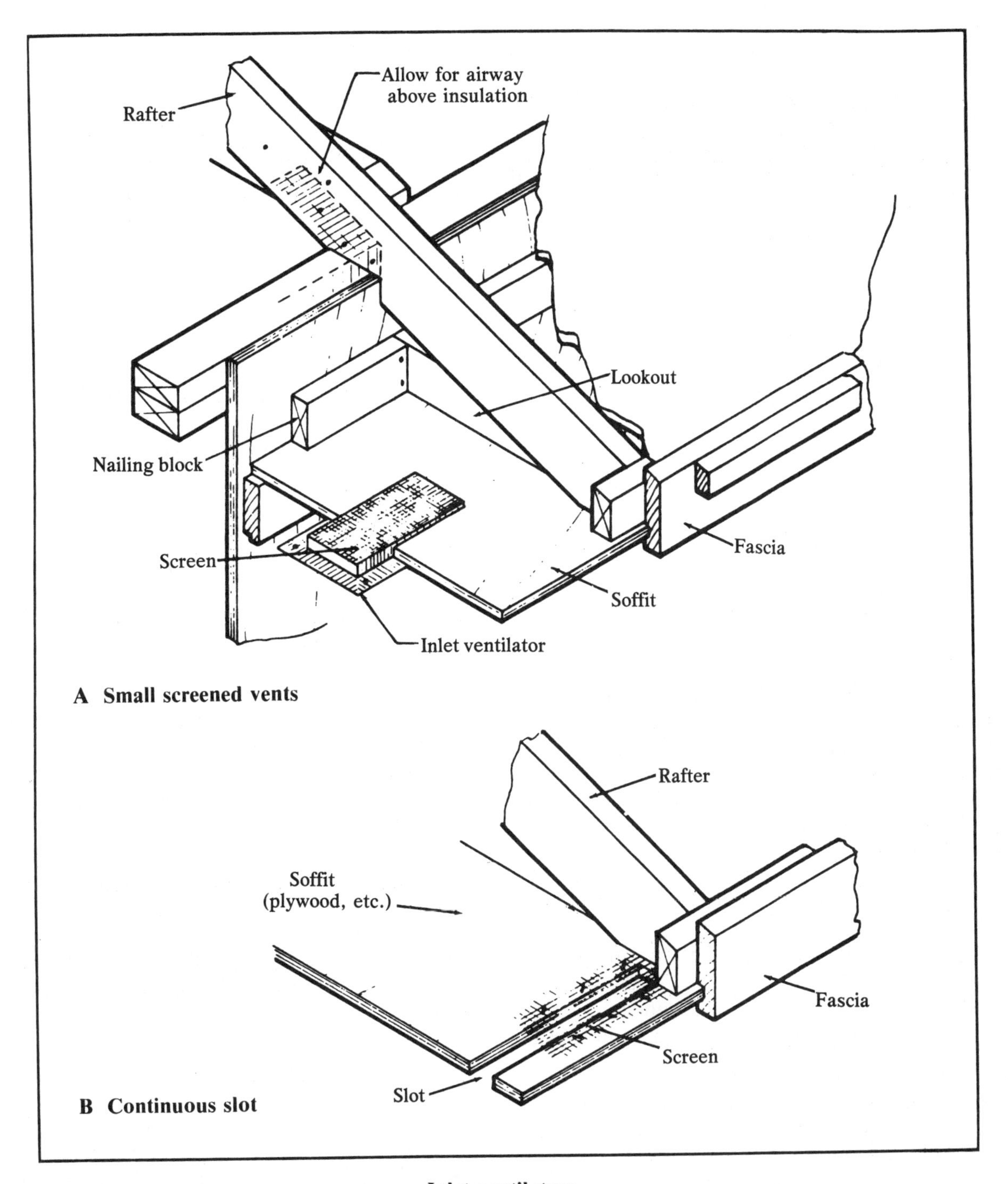

Inlet ventilators
Figure 18-8

Summary

Take a few minutes to give ventilation the attention it deserves. A well-ventilated home identifies you as a professional, conscientious, quality-minded builder.

No Checklist and Data Reference Sheet is provided for this chapter since ventilation costs are part of foundations and roof framing.

Chapter 19

Siding

As the builder of a spec house, you decide on the exterior finish. The site, the surrounding houses, and the cost will influence your decision. You might decide on brick veneer and subcontract the brickwork, as discussed in Chapter 16. Or you might elect to use one of the manufactured wood sidings now available in many patterns and textures. APA Sturd-I-Wall, plywood paneling, plywood lap siding, hardboard lap siding, hardboard paneling, hardboard shakes and wood siding all make excellent exterior coverings.

This chapter will cover the common siding materials and their application. We'll emphasize siding that can be installed by your own crew, with little or no specialized training and no special tools.

Plywood Siding

Plywood siding is a popular exterior covering. In many cases, it can be applied directly to studs. See Figure 19-1. If you apply the siding directly to studs, install the insulation after the siding is up.

Plywood siding systems include APA Sturd-I-Wall, plywood panel siding and plywood lap siding. Let's look at each system in detail.

APA Sturd-I-Wall

The APA Sturd-I-Wall system uses APA 303 plywood panel siding. The siding can be applied directly to studs or over nonstructural sheathing materials, such as fiberboard, gypsum or rigid foam insulation sheathing. "Nonstructural" means that the sheathing material does not meet the bending and racking strength requirements specified by building codes.

Sturd-I-Wall is accepted by HUD and most building codes. A single layer of Sturd-I-Wall is strong and rack-resistant enough to eliminate the need for separate structural sheathing and diagonal wall bracing.

Sturd-I-Wall siding is normally installed vertically, as shown in Figure 19-2. It can also be applied horizontally (Figure 19-3) if there is 2 x 4-inch blocking at the horizontal joints. Figure 19-4 shows maximum stud spacing for both vertical and horizontal applications.

Applying APA Sturd-I-Wall over nonstructural sheathing— When installing panel siding over rigid foam sheathing, drive nails flush with the siding surface. Avoid overdriving the nails. Overdriven nails will dimple the siding and compress the foam sheathing.

Plywood siding can be treated with a water repellent or wood preservative. This treatment will make the siding more durable and give it a nicer finish. If the siding has been treated, be sure it's dry when you hang it. Otherwise, the solvent may react chemically with the foam sheathing.

Foam sheathing is an excellent vapor barrier. Unfortunately, if you install foam sheathing on a wall exterior, you're placing a vapor barrier on the wrong side of the wall. You'll remember from Chapter 17 that the vapor barrier goes on the warm (in winter) side of the wall. If you use foam sheathing under plywood siding, install either a polyethylene vapor barrier or foil-backed gypsum wallboard on the warm side of the wall. This should prevent condensation problems in the wall cavity. Ordinary foil or paper-faced insulation

Plywood siding panels applied directly to studs
Figure 19-1

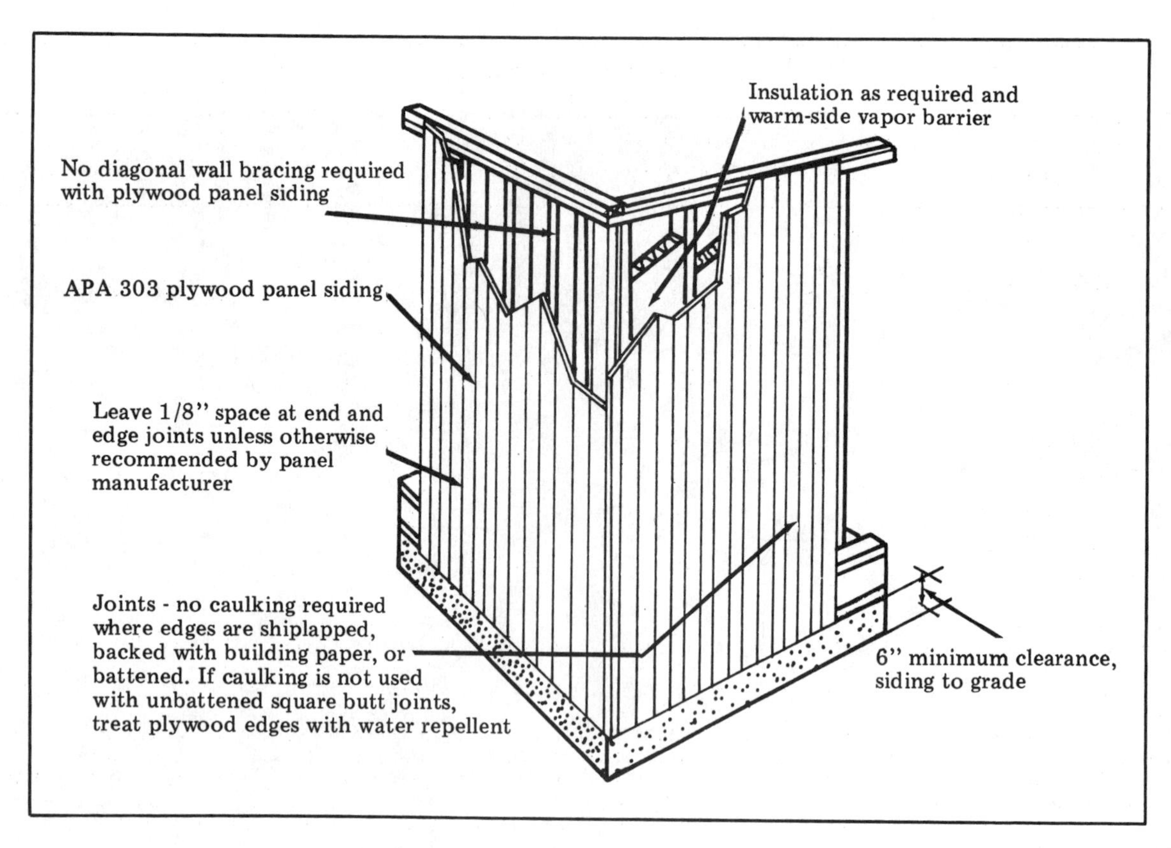

APA Sturd-I-Wall (vertical application)
Figure 19-2

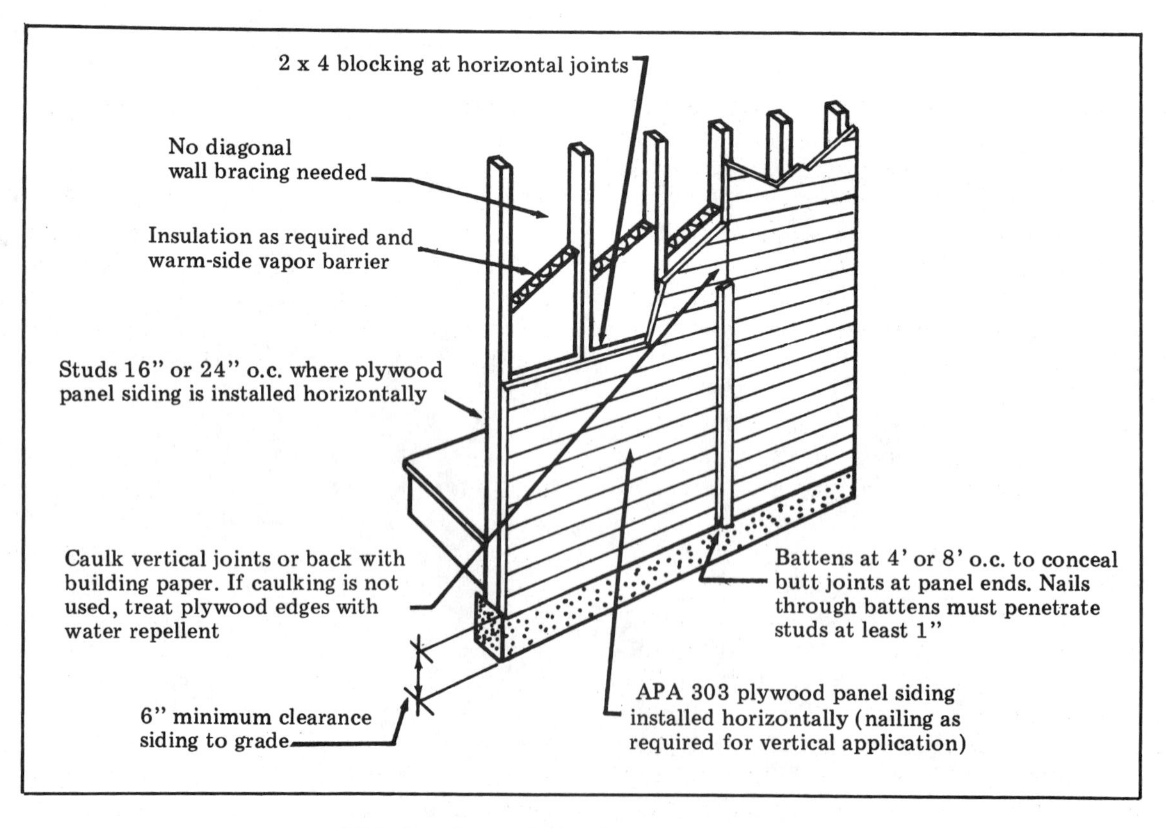

APA Sturd-I-Wall (horizontal application)
Figure 19-3

Plywood panel siding description (All species groups)	Nominal thickness (inches)	Maximum stud spacing (inches)		Nail size*	Nail spacing (inches)	
		Face grain vertical	Face grain horizontal		Panel edges	Intermediate
APA MDO EXT	11/32 & 3/8 ½ & thicker	16 24	24 24	6d for panels ½" thick or less, 8d for thicker panels	6***	12
APA 303 siding— 16 oc EXT (including T1-11)	11/32 & thicker	16	24			
APA 303 siding — 24 oc EXT	15/32 & thicker**	24	24			

Recommendations apply to siding direct to studs and over sheathing other than structural panels or lumber.

*If siding is applied over sheathing thicker than ½-inch, use next regular nail size. Use nonstaining box nails for siding installed over foam insulation sheathing.

Hot-dipped or hot-tumbled galvanized steel nails are recommended for most siding applications. For best performance, stainless steel nails or aluminum nails should be considered. APA tests also show that electrically or mechanically galvanized steel nails appear satisfactory when plating meets or exceeds thickness requirements of ASTM A641 Class 2 coatings, and is further protected by yellow chromate coating. *Note:* Galvanized fasteners may react under wet conditions with the natural extractives of some wood species and may cause staining if left unfinished. Such staining can be minimized if the siding is finished in accordance with APA recommendations,or if the roof overhang protects the siding from direct exposure to moisture and weathering.

**Only panels 15/32" and thicker which have certain groove depths and spacings qualify for 24 oc Span Rating

***For braced wall section with 11/32" or 3/8" siding applied horizontally over studs 24" oc, space nails 3" oc along panel edges.

APA Sturd-I-Wall construction
Figure 19-4

batts are *not* recommended. When rigid foam insulation sheathing is used, building codes generally require 1/2-inch gypsum wallboard on the inside surface of the wall for fire protection.

Applying APA Sturd-I-Wall directly to studs— Sturd-I-Wall applied directly to studs goes up about the same way as panels installed over nailable sheathing. Only the joint details and maximum stud spacing differ. See Figures 19-2, 19-3, 19-6 and 19-7.

In Sturd-I-Wall construction, all panel edges should be backed with framing or blocking. To prevent staining the siding, use corrosion-resistant nails.

Plywood Panel Siding and Lap Siding

Most plywood siding is installed over sheathing. Figure 19-5 gives installation recommendations for plywood panel siding and lap siding. Figures 19-6 and 19-7 show how these materials are installed.

Lap siding must be applied with the face grain across the supports. Siding joints should be staggered and don't have to fall on a stud if nailable panel or lumber sheathing is used under the siding. Nailable sheathing includes:

1) Nominal 1-inch boards with studs spaced 16 or 24 inches o.c.

2) Rated sheathing panels with a span rating of 24/0 or greater, and the long dimension either parallel to or perpendicular to studs spaced 16 or 24 inches o.c. But note that 3-ply plywood panels must be applied with the long dimension across studs when studs are spaced 24 inches o.c. Check your local code on this point.

When installing plywood siding, it's sometimes hard to maintain the uniformly flat appearance required for a finished wall. Using straight studs makes siding application easier, of course. And be sure to allow the recommended spacing between panel edges. This leaves room for small adjustments if the panels get out of plumb. It also lets the panels expand slightly without causing the siding to buckle.

The nailing sequence also affects the appearance of the finished wall. Here's an easy way to install plywood siding without compressing the panels. Install the first panel flush at a corner. Use a level to plumb it. Then position the second panel.

	Plywood siding description (All species groups)	Nominal thickness (inches)	Maximum stud spacing		Nail size*	Nail spacing (inches)	
			Face grain vertical	Face grain horizontal		Panel edges	Intermediate
Panel Siding	APA MDO EXT	11/32 & 3/8	16	24	6d for panels ½" thick or less, 8d for thicker panels	6	12
		1/2 & thicker	24	24			
	APA 303 Siding - 16 oc EXT (including T1-11)	11/32 & thicker	16	24			
	APA 303 SIDING - 24 oc EXT	15/32 & thicker**	24	24			
Lap Siding	APA MDO EXT	11/32 & thicker	--	24	6d for siding 3/8" thick or less; 8d for thicker siding	4d @ vertical butt joints; 6d along bottom edge	8d (if siding wider than 12")
	APA 303 SIDING	11/32 & thicker	--	24			

*If siding is applied over sheathing thicker than ½-inch, use next regular nail size.

Hot-dipped or hot-tumbled galvanized steel nails are recommended for most siding applications. For best performance, stainless steel nails or aluminum nails should be considered. APA tests also show that electrically or mechanically galvanized steel nails appear satisfactory when plating meets or exceeds thickness requirements of ASTM A641 Class 2 coatings and is further protected by yellow chromate coating. *Note:* Galvanized fasteners may react under wet conditions with the natural extractives of some wood species and may cause staining if left unfinished. Such staining can be minimized if the siding is finished in accordance with APA recommendations, or if the roof overhang protects the siding from direct exposure to moisture and weathering.

**Only panels 15/32" and thicker which have certain groove depths and spacings quality for 24 oc Span Rating.

APA 303 plywood siding over nailable panel or lumber sheathing
Figure 19-5

(Remember to observe the recommended edge-spacing.) Lightly tack the panel at each corner. Install the first row of nails along the edge nearest the preceding panel. Nail from top to bottom. Remove tacking nails. Then nail along the first intermediate stud. Continue by nailing at the second intermediate stud, and then along the outside edge. Complete the installation by nailing the siding to the top and bottom plates.

Hardboard Siding

Hardboard is made from wood chips and sawmill by-products. It's a popular exterior siding because it comes in many designs and patterns and is one of the least expensive siding materials. It's fairly easy to install and requires no special tools. Hardboard is free of knots and is uniform in thickness, density and appearance. It has to be maintained like all wood and kept free of termites. But there's no grain to rise or cause splitting.

Proper storage is important. Until hardboard is installed and primed, it must be protected from the elements. Store hardboard outside but off the ground. Protect the top and sides with a waterproof cover. But don't seal the bundle. Adequate ventilation is essential. Protect the siding from dirt, grease and rough handling. Be sure the siding is stacked flat so that it doesn't warp.

Cut hardboard siding with a fine-tooth hand saw. Or use a power saw with a combination blade. Be sure that the cutting action is toward (or into) the finish side. If you use a hand saw, put the finish side face up. For a power saw, the finish side goes face down.

Never apply hardboard siding to drying concrete, drying plaster, wet studs or wet sheathing

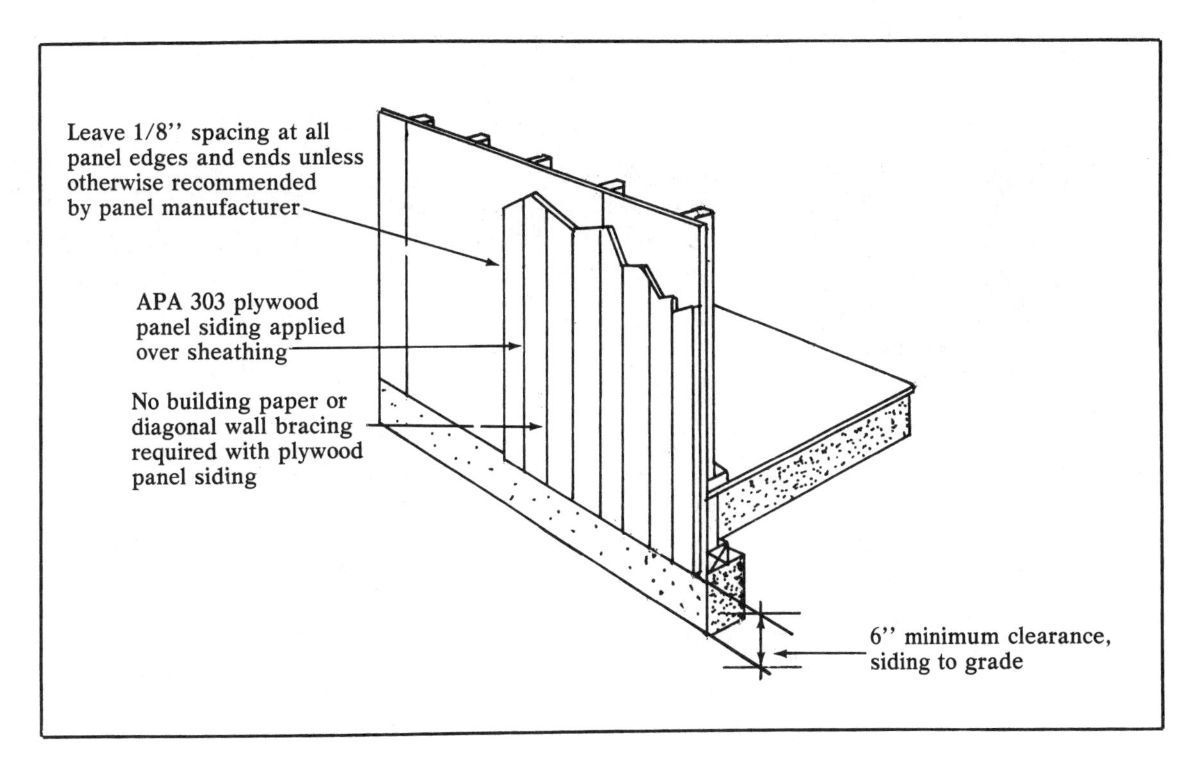

APA 303 plywood panel siding over nailable panel or lumber sheathing
Figure 19-6

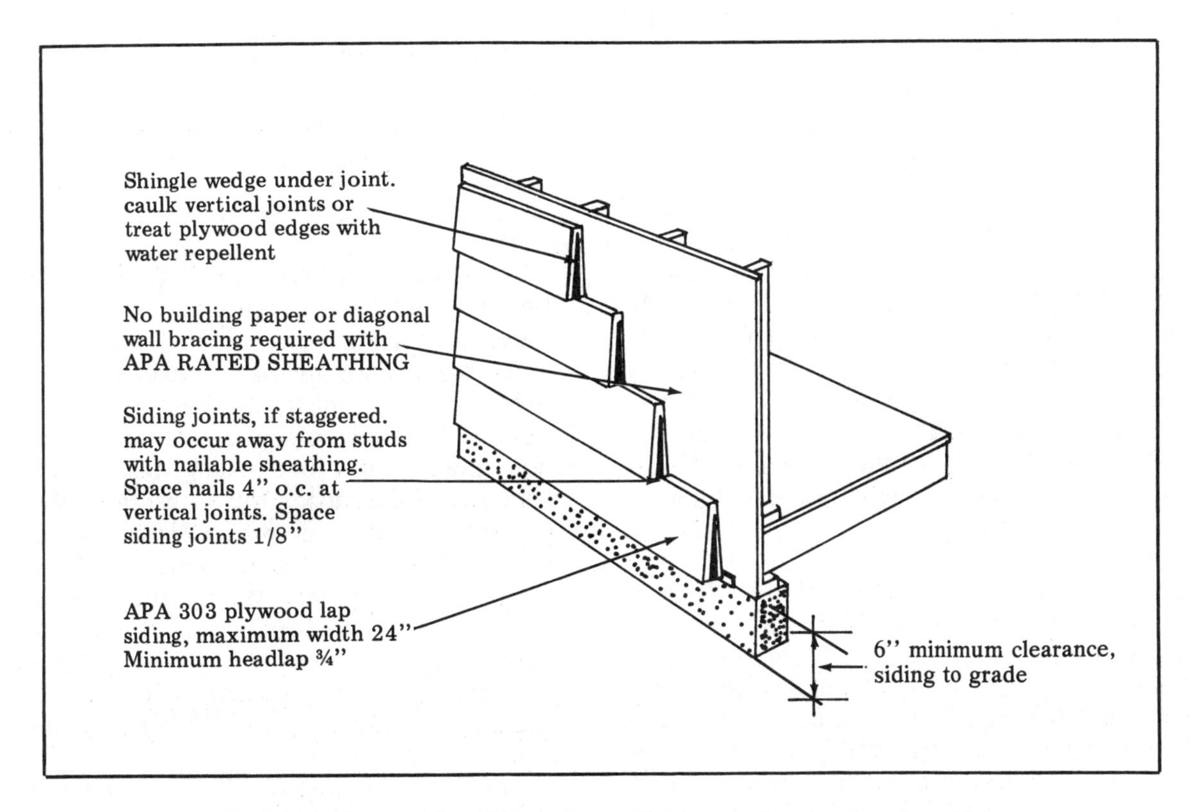

APA 303 plywood lap siding over nailable panel or lumber sheathing
Figure 19-7

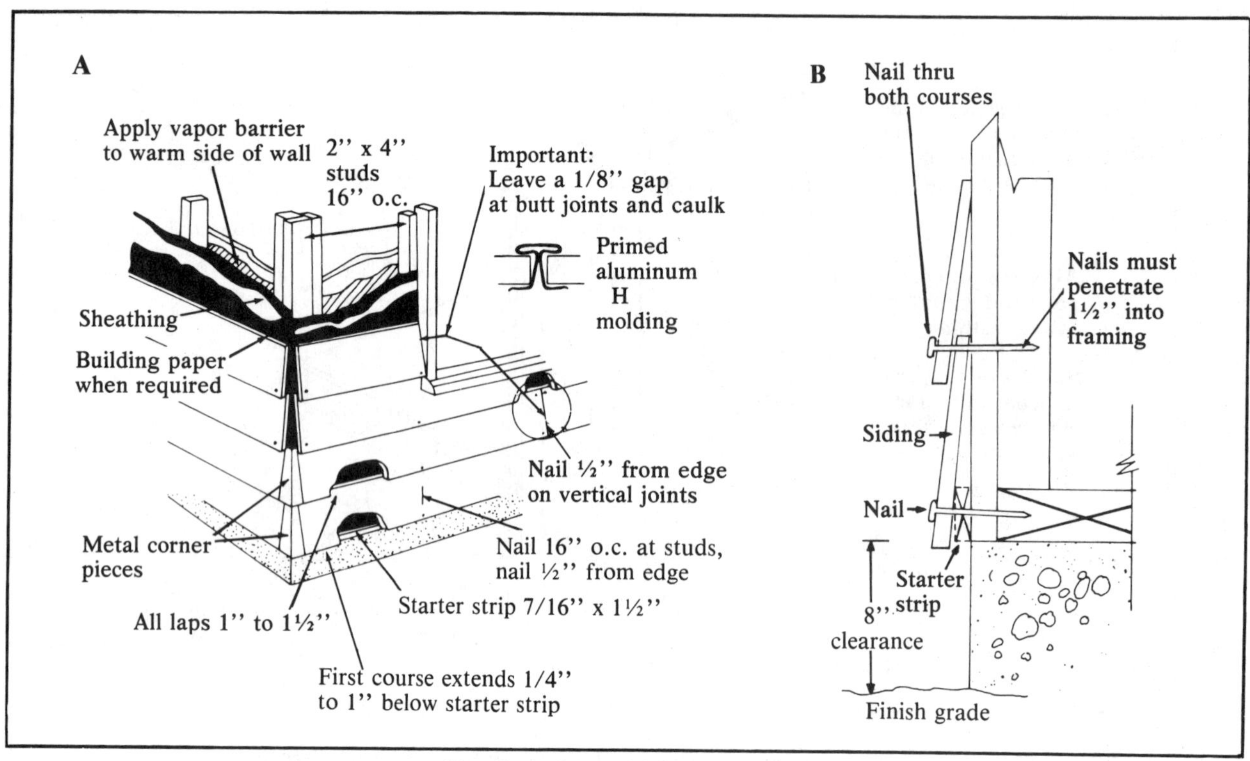

Hardboard lap siding application
Figure 19-8

materials. Let the concrete or framing dry before beginning application.

The two most common types of hardboard siding are hardboard lap siding and hardboard panel siding. Let's look at the installation procedures for both types.

Hardboard Lap Siding Application

Your building code may allow you to apply hardboard lap siding directly to studs. But this application often results in wavy siding. To minimize waviness, install the siding over plywood or board sheathing. Whether the siding is applied directly to studs or over sheathing, the maximum stud spacing is 16 inches o.c. Wall corners must be adequately braced.

Use a layer of building paper under the siding, even when the siding is applied over plywood or board sheathing. The building paper acts as a wind barrier.

Begin installation with a wood starter strip measuring 7/16-inch x 1½-inches, or use a strip of lap siding cut down to 2 inches wide. Level and nail it along the bottom edge of the sill plate. See Figure 19-8 A. Nail the first course of siding so that it extends at least 1/4-inch (but not more than 1 inch) below the starter strip. Drive a nail every 16 inches o.c. There should be a nail at each stud location. Allow at least an 8-inch clearance between the siding and the finish grade. See Figure 19-8 B.

Lap following courses at least 1 inch but not more than 1½ inches. Nails must go through both siding courses every 16 inches o.c. at stud locations. Always use corrosion-resistant nails with a head diameter of at least 3/16-inch. Shim behind the siding over doors or windows, or apply a drip cap. Leave a 1/8-inch space between the siding and the window or door frame. Also leave a 1/8-inch space between the siding and the corner boards. Caulk these spaces after the siding is completely installed.

At inside and outside corners, use wood corner boards, as shown in Figure 19-9 A and D. Or use formed metal corners.

All butt joints must be staggered and centered over a stud. See Figure 19-9 C. Nail the siding at the top and bottom of the butt joint, and also on both sides of the joint, using nails 1" to 1⅛" in length. Consider using the primed aluminum "H" expansion molding shown in Figure 19-8 A, or use the 1-inch butt-joint cover shown in Figure 19-9. "H" molding permits the natural expansion and contraction caused by changes in temperature and humidity. It also hides the gaps that often occur at butt joints. If "H" molding or joint covers aren't used, then caulk the gaps. A 1/8-inch gap is essential at all butt joints.

Applying hardboard lap siding over foam sheathing— If you apply hardboard lap siding over foam sheathing, set the nails flush with the surface

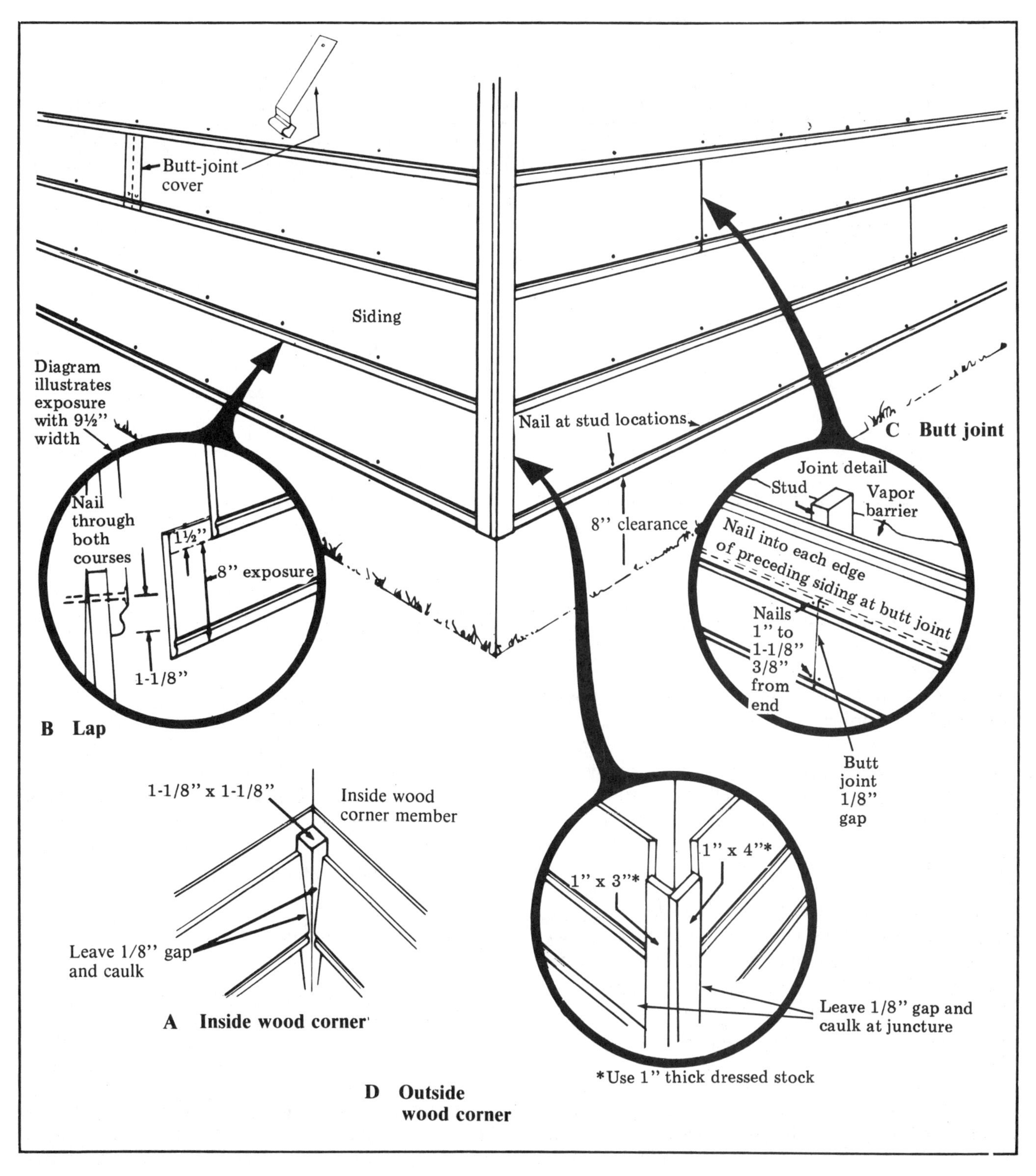

Hardboard lap siding application details
Figure 19-9

of the hardboard. If you overdrive the nails, it will dimple the hardboard and give the siding a wavy appearance.

The siding nails should penetrate at least 1½ inches into each stud. Otherwise, foam sheathing may cause the hardboard lap siding to hang out from the wall as much as 1 inch. With time, the siding may begin to sag. For 1/2-inch lap siding and 1-inch foam sheathing, use a 4-inch (20d) nail.

As with plywood, use a vapor barrier on the interior side of the wall when hardboard siding is applied over foam sheathing.

Edge detail	Grooves	Substrate	2 x 4 framing. Maximum spacing inches o.c.	Nail size	Nail Spacing: Siding only*	Nail Spacing: Racking strength**	Joint gap	Gap around openings
Shiplap or square edge	No	Direct to studs	24	6d	6" o.c. edges 12" o.c. intermediate	4" o.c. edges 8" o.c. intermediate	1/16"	1/8"
Shiplap or square edge	No	Over sheathing	24	8d	6" o.c. edges 12" o.c. intermediate	4" o.c. edges 8" o.c. intermediate	1/16"	1/8"
Shiplap	Yes	Direct to studs	16	6d	6" o.c. edges 12" o.c. intermediate	4" o.c. edges 8" o.c. intermediate	1/8"	1/8"
Shiplap	Yes	Over sheathing	16	8d	6" o.c. edges 12" o.c. intermediate	4" o.c. edges 8" o.c. intermediate	1/8"	1/8"

*Racking resistance per FHA Circular No. 12 provided by sheathing or corner braces
**Racking resistance provided by siding

Hardboard panel siding nail size and spacing requirements
Figure 19-10

Hardboard Panel Siding Application
Grooved panel siding with shiplap joints can be applied over sheathed or unsheathed walls with studs spaced (maximum) 16 inches o.c. Nongrooved panels can be applied over sheathed or unsheathed walls with studs spaced up to 24 inches o.c. Hardboard panel siding applied (without corner bracing) directly to studs, and nailed as recommended in Figure 19-10, has good racking resistance and meets racking strength requirements in FHA Technical Circular No. 12.

Applying nongrooved panel siding (square edges) — Application details for nongrooved panel siding are shown in Figure 9-11 A. Here's the installation procedure.

1) Apply panels parallel to framing. Cover the sill and the top plates.

2) All panel edges must fall on studs or other nailing members. Don't force panels into place. Leave a 1/8-inch space where siding butts against trim, Caulk the gap.

3) Square-edge panels can be applied directly to studs (24 inches o.c. or less) without corner bracing. Nail as recommended in Figure 19-10.

4) Begin nailing at the middle of the panel and work toward the edges.

5) Apply sealant to vertical joints, at windows, and at doors.

6) Install battens over all vertical joints. Make battens from wood or strips of siding cut to the desired width. Install battens only where there's an adequate nailing base. Nail through the panels and into the framing. Space nails 12 inches o.c.

7) Intermediate battens can also be applied if there's an adequate nailing base.

8) Nail outside wood corner boards 12 inches o.c.

9) Apply sealant to any butt joints at inside corners. Install a 1 x 1-inch inside wood corner post. Nail every 12 inches o.c.

Applying grooved panel siding (shiplap edges)— This type of panel does not require battens. And you can apply the panels directly to studs without using sheathing paper or corner bracing. If you apply the panels without using corner bracing, the maximum stud spacing is 16 inches o.c. See Figure 19-11 B.

Follow the same application instructions as for nongrooved panels. But remember to omit the battens, and don't caulk the vertical shiplap joints.

When field-cut butt joints are required, butter the edges with sealant before making the joints. Then remove excess sealant.

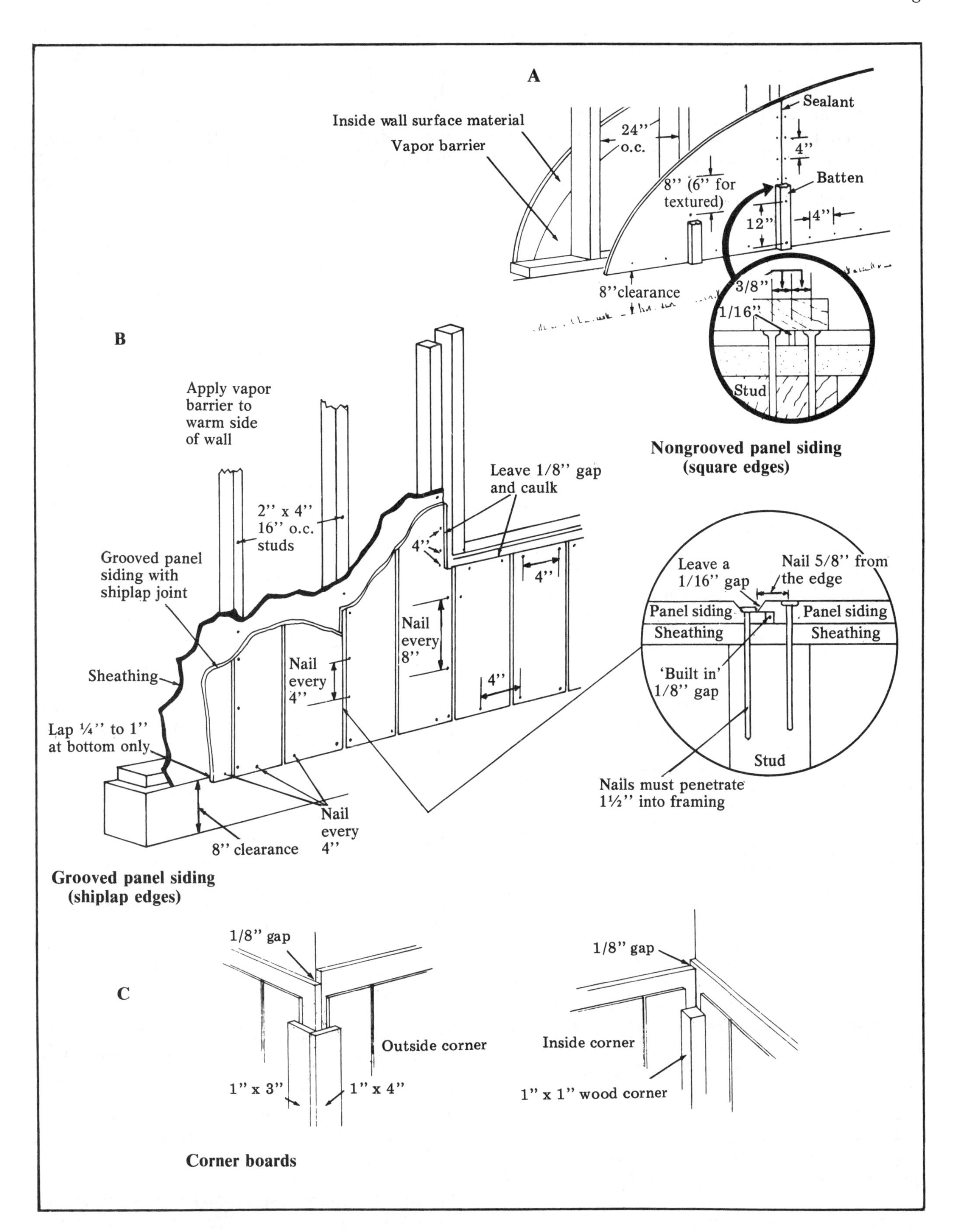

Hardboard panel siding application details
Figure 19-11

Finishing Plywood Siding and Hardboard Siding

Siding is available both factory-primed and unprimed. The two most common finishes are paint and stain. Let's look at the finishing details for plywood and hardboard siding.

Factory-Primed Siding

Primed siding should be painted within 60 days after installation. If exposed for a longer period of time, reprime the siding with a quality exterior-grade primer compatible with the existing finish.

Unprimed Siding

This siding should be primed within 3 days after installation and finished within 30 days after installation. If you're going to paint the siding, use a quality exterior-grade primer first.

Painting

The number of coats depends on the type of paint, how it's applied, and the finish you want. Use a good quality exterior acrylic, acrylic latex, latex, or oil-base paint. Follow the paint manufacturer's recommended application procedure. Avoid flat alkyd-type house paints. They give a poor appearance and don't last. Be sure that all cut edges are also primed and painted.

Staining

Only textured siding should be stained. Use a quality, opaque exterior acrylic latex stain on primed or unprimed textured sidings. This will provide a finish with good color and durability.

For top-grade appearance, use two coats of stain. Follow the application procedure recommended by the stain manufacturer. Properly stained (or painted) hardboard siding will last as long as the more expensive siding materials. But hardboard siding isn't intended to be left unpainted or to "weather."

Hardboard Shakes

Like plywood siding and hardboard siding, hardboard shakes are also a popular exterior covering. You can usually apply shakes either over sheathing or directly to studs. No building paper is needed over vertical sheathing unless required by your code.

Shakes can be applied to walls that slope as much as 15 degrees from vertical. But be sure to apply sheathing to the framing, and cover the sheathing with building paper. Then apply the shakes.

Corner bracing is required when hardboard shakes are installed directly over studs, or when nonstructural sheathing is used. When shakes are applied over sheathing, corner bracing requirements are determined by the type of sheathing.

Use quality, nonhardening caulk where siding meets windows, doors, and trim. Shiplap edges are not sealed. If the siding rests on a wood or concrete sill, there must be enough slope for drainage. Use a nonhardening, permanent sealant at the joint between siding and sill.

Always leave a slight gap at joints and openings. Never spring panels into place. Keep the siding at least 8 inches above the finish grade.

Hardboard Shake Application

Figure 19-12 shows hardboard shake application details. Install hardboard shakes in the following way:

1) Apply shakes horizontally over framing members spaced no more than 16 inches o.c.

2) On a horizontal run of 50 feet or more, use expansion joints to break the continuity of the wall.

3) Install a starter strip (3/8-inch x 1½-inch wood lath) at the bottom plate. Mark the location of this strip with a chalk line run parallel to the top plate. Nail the starter strip to the framing members at each stud. Use 6d (2-inch) galvanized siding nails.

4) Install the inside and outside wood corner boards. Leave a 1/16-inch gap between the corner boards and the shakes. Caulk this gap after the shakes are installed.

5) Level and install the first course of siding. The bottom edge of the first course should extend at least 1/2-inch below the starter strip.

Begin the first course at the lower left corner with the first shake. You may need to trim the *left* edge of the shake so that the *right* edge will fall over a stud. Use full pieces of shake for the rest of the first course until you reach the next corner. When you reach the corner, trim the shake to leave a 1/16-inch gap.

6) The second course begins with a shake that has 16 inches trimmed from it. See Figure 19-12. Overlap the first course at least 1½ inches.

7) The third course begins with a shake that has 32 inches trimmed from it. Overlap the preceding course at least 1½ inches.

8) Begin the fourth course the same as you did the first course. You may need to trim the left edge of the first shake so that the right edge will fall over a stud.

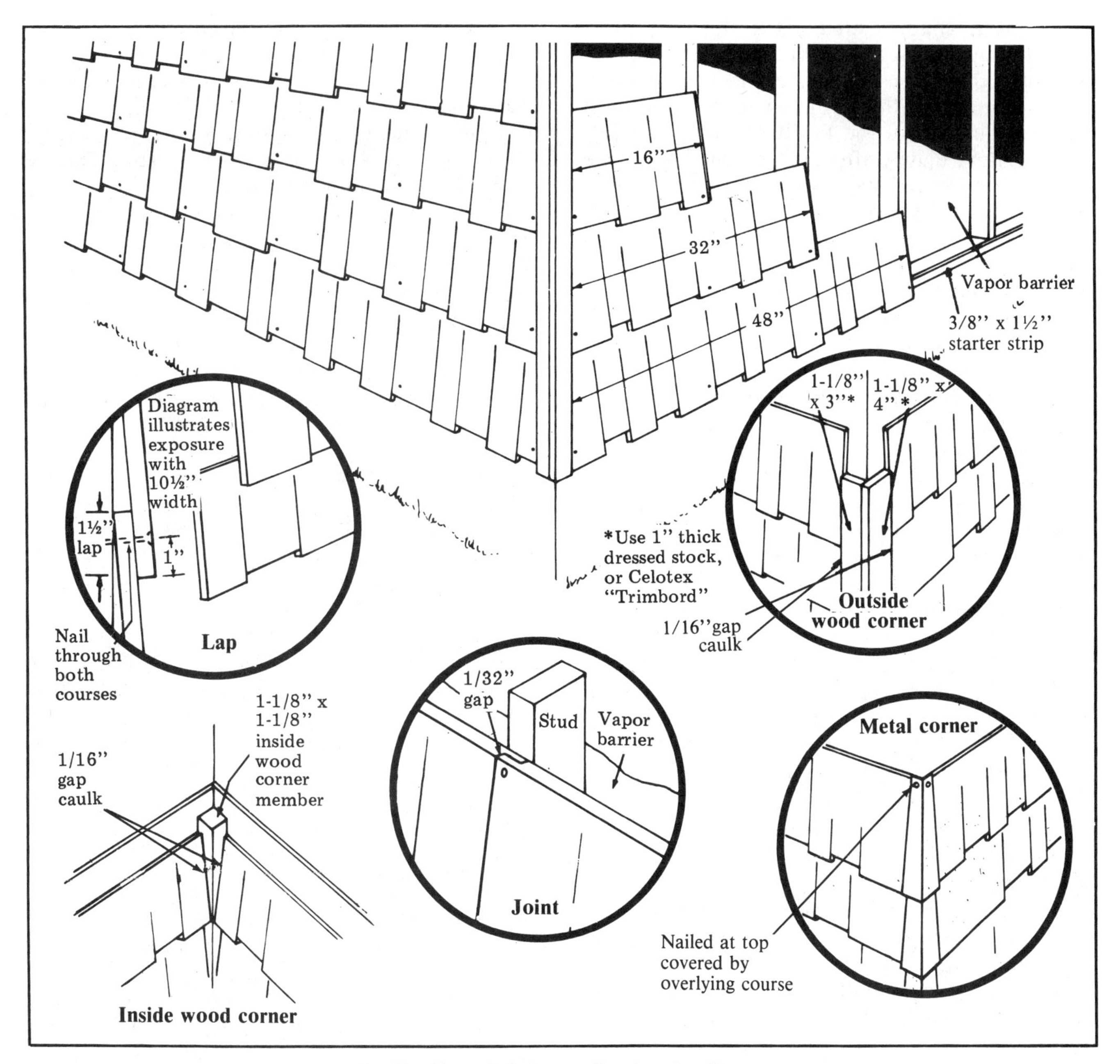

Hardboard shake application details
Figure 19-12

9) Start the fifth course with a piece trimmed 16 inches. Begin the sixth course with a piece trimmed 32 inches. Then start the sequence again, and continue for all following courses.

10) Nail along a line 1 inch from the bottom edge and 3/8-inch from the side edge. Space nails 16 inches o.c. All nails should penetrate a minimum of 1 inch into the framing. Use at least two nails on any small, trimmed pieces. Use 8d (2½-inch) or 10d (3-inch) corrosion-resistant nails. Be careful you don't overdrive nails. Shim to avoid deforming the shakes around windows and doors. Shim also when the studs are not straight and even.

Finishing Hardboard Shakes

All exposed surfaces and bottom edges of unprimed, textured shakes must be painted or stained within 30 days after installation.

Exterior acrylic latex stains or paints are recommended. Use a brush or roller to apply the coating. This insures good penetration into the textured siding. Apply at least two coats and follow the manufacturer's instructions for mixing and coverage.

Airless spray application is O.K. But after spraying, you must use a dry brush to work the paint or stain into the crevices. Two coats are needed. Apply the first coat as thick as possible, but stop short of the point where the finish runs or sags. Dry brush it immediately after application. The second coat can be sprayed and dry brushed when the first coat is dry to the touch.

Wood Siding

Wood siding is more expensive than hardboard siding. But wood siding is still used on quality homes in many parts of the country.

Wood siding should meet the following requirements: It should be easy to work with and should not warp. If you plan to paint the siding, the wood should be reasonably free from knots, pitch pockets and tapered edges.

These properties are present in cedar, Eastern white pine, sugar pine, Western white pine, cypress, and redwood. Other species used for siding include Western hemlock, ponderosa pine, spruce, yellow poplar, Douglas fir, Western larch, and Southern pine.

Vertical-grain lumber makes the best siding, because it doesn't expand and contract much with changes in temperature and humidity. Redwood and Western red cedar are usually available in vertical-grain and mixed-grain grades.

At the time of installation, the moisture content of the siding should be 10% to 12% except in dry Southwestern states, where it should be closer to 8% or 9%.

Before installation, give the wood a 3-minute dip in a water-repellent preservative. This will prolong paint life, and help the wood resist moisture-entry and decay. Some manufacturers supply pretreated siding. Freshly cut ends should be brush-treated on the job.

Nailing Wood Siding

When installing wood siding, use plenty of corrosion-resistant galvanized nails. Aluminum and stainless steel nails cost more, but they prevent the rust-spotting that can occur when the paint surface begins to deteriorate.

Two types of nails commonly used with wood siding are finishing nails and siding nails. Finishing nails have a small head. Siding nails have a larger, flat head. Small-head finishing nails are set about 1/16-inch below the face of the siding. Fill the nail hole with putty after the prime coat has been applied. Flat-head siding nails are driven flush with the face of the siding and then painted. Don't overdrive flat-head siding nails. Overdriving leaves hammer marks and can split or crush the wood. This is especially important on wood siding with a prefinished surface or overlay.

Common wire nails will rust in a short time and leave an ugly stain on the face of the siding. Small-head wire nails may even show rust spots through the putty and paint. Use galvanized nails. They're available at all lumber yards.

Siding that is going to have a "natural finish" (a water-repellent preservative or stain) should be fastened with stainless steel or aluminum nails. Some types of prefinished sidings come with nails that are color-matched to the siding.

Nails with modified shanks tend to stay in the siding better. This means you can use a shorter nail. Both annular-threaded and helical-threaded shank nails can be used on wood siding.

Horizontal Wood Siding

Some wood siding patterns are used only horizontally. Others are applied only vertically. Some can be used either way if enough nailing area is available. Three common horizontal siding patterns are bevel siding, drop siding and matched paneling.

Bevel siding— Three types of horizontal bevel siding are plain bevel, "Anzac" bevel, and Dolly Varden. See Figure 19-13 A.

Plain bevel siding comes in 1/2-inch thicknesses that are 4 to 8 inches wide and in 3/4-inch thicknesses that are 8 to 10 inches wide.

"Anzac" bevel siding is 3/4-inch thick by 12 inches wide. The actual width is usually about 1/2-inch less than the nominal width.

Dolly Varden siding is similar to plain bevel siding. But Dolly Varden siding has shiplap edges that permit uniform exposure distance. Because this type of siding lies flat against the studs, it's sometimes used for garages and other nonhabitable buildings without sheathing. But diagonal bracing is needed to provide racking resistance for the wall.

One side of bevel siding is a smooth, planed surface. The other side is a rough, resawn surface. If you plan to apply a stain finish to bevel siding, install the siding so that the rough side is exposed. Stain looks better and lasts longer on rough surfaces.

Installed bevel siding is shown in Figure 19-14. The lower edge of one board overlaps the upper edge of the board below it. This lap should never be less than 3/4-inch.

Narrow siding of this type has a tendency to pull apart (where upper and lower courses overlap) in the area between the studs. But if you use wood sheathing on the side walls, the siding can be nailed to the sheathing between the studs. If the siding is 3/4-inch-thick, only composition sheathing is needed.

A

Type	Nominal Sizes
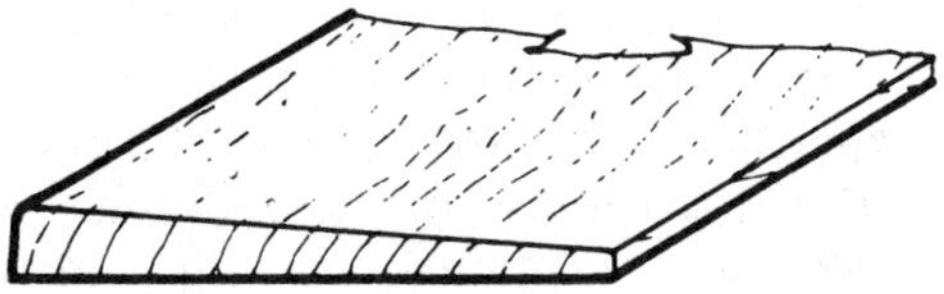 Bevel	½ x 4 to ¾ x 10
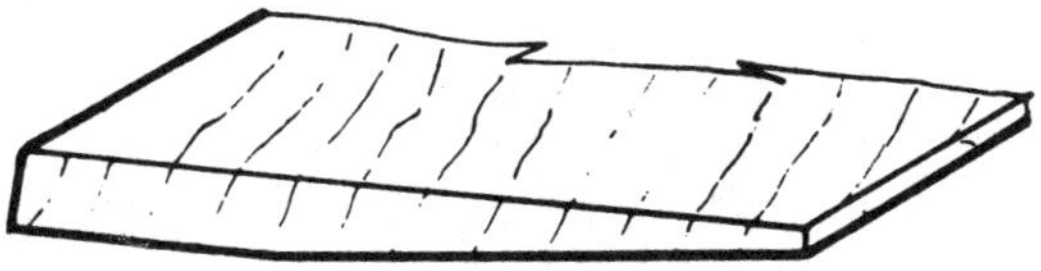 "Anzac" (bevel)	¾ x 12
 Dolly Varden (bevel)	¾ x 6 to ¾ x 10
 Drop (pattern 106)	1 x 6 to 1 x 8
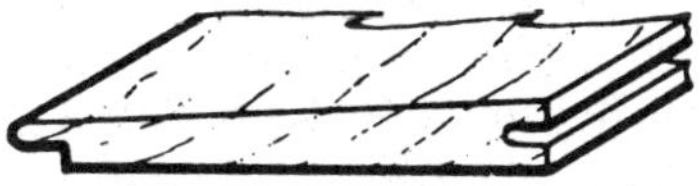 Drop (pattern 124)	1 x 6 to 1 x 8

Horizontal application

B

Type	Nominal Sizes
 Paneling (wc 130)	1 x 4 to 1 x 12
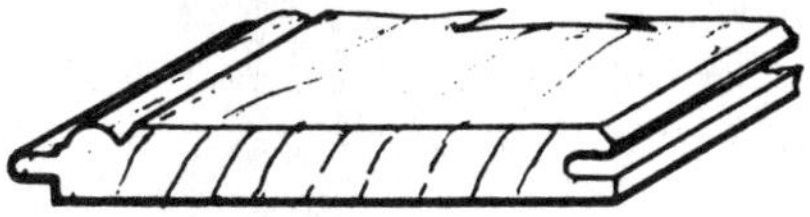 Paneling (wc 140)	1 x 4 to 1 x 12

Horizontal or vertical application

Horizontal and vertical wood siding
Figure 19-13

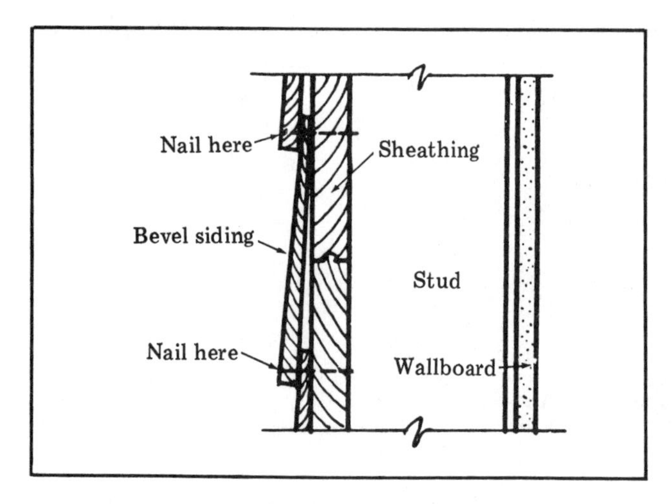

Bevel siding installed
Figure 19-14

If wide Colonial siding is used, install it as shown in Figure 19-15. Colonial siding is sometimes beveled on the back so the board can lie close to the sheathing. This provides a solid nailing base.

To calculate the average exposure distance for bevel siding, first measure the distance from the underside of the window sill to the top of the drip cap. Then divide that distance by the number of courses planned for that area. This gives you the exposure distance for each piece of bevel siding. See Figure 19-16.

Appearance and weather-resistance are better if the butt edge of a siding course coincides with the top of the window drip cap. In many one-story houses with an eave overhang, this course is replaced by a frieze board.

The butt edge of a siding course should also be flush with the underside of the window sill. But varying window heights may make this impractical.

Another method for setting exposure distance is to use an equal exposure distance for the entire wall and then notch the siding at the window sill. In either case, be sure there's a good fit between sill and siding, and remember to caulk the joint between the sill and siding.

Avoid butt joints whenever possible. Use the longer sections of siding under windows and for other long runs. Use shorter pieces between windows and doors. If absolutely necessary to make a butt joint, locate it over a stud. Stagger butt joints between courses as much as possible.

Use a miter box when cutting bevel siding. Square cuts help provide a good joint at window and door casings and at butt joints. Open joints allow moisture to enter. Moisture causes the wood to deteriorate. Before siding boards are nailed in place, brush or dip freshly cut ends of siding in a water-repellent preservative.

Figure 19-17 shows several corner treatments for horizontal siding. The miter-corner method, shown in Figure 19-17 A, requires precise work and takes longer than other methods. Actually, it's a poor system because the miter leaves thin wood exposed at the corners of the building. This thin wood deteriorates rapidly when exposed to weather. The metal corner and the outside corner boards, shown in Figure 19-17 B and C, are better methods.

To make the corner shown in Figure 19-17 C, determine the size and type of corner board based on the type of siding you're using. Outside corner boards are usually 1⅛ or 1⅜ inches thick. They extend from the bottom of the siding to the top of the drip cap or to the bottom of the frieze. The corner boards are edge-butted and nailed together before being nailed in place. This guarantees a tight joint.

Be sure that a strip of building paper is tacked over the corner before the corner board is nailed into position. Allow an overlap of paper to cover the point where the ends of the siding butt against the corner board.

Here's how to install horizontal bevel siding with corner boards:

1) Once the corner boards are in place, apply a full-width strip of building paper flush with the bottom of the wall. Apply the next higher strip of building paper so it laps over the first strip about 4 inches. Continue applying strips of building paper in this way. But don't cover too much wall area with paper all at once. Instead, cover part of the wall with paper, and then apply siding to that part of the wall. Continue applying more paper as the application of siding progresses. Tuck the paper tightly under all window frames.

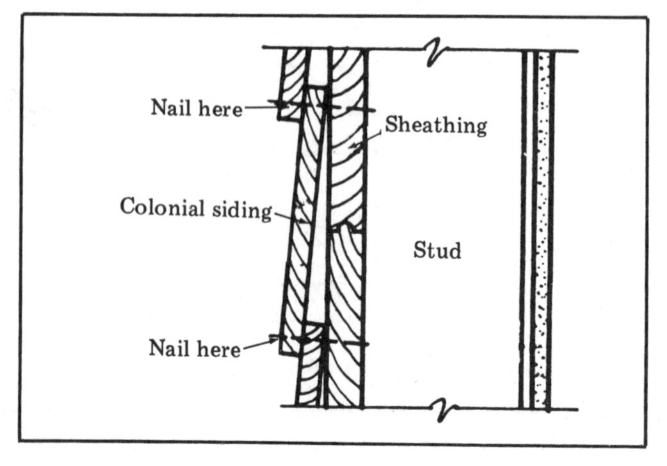

Colonial siding
Figure 19-15

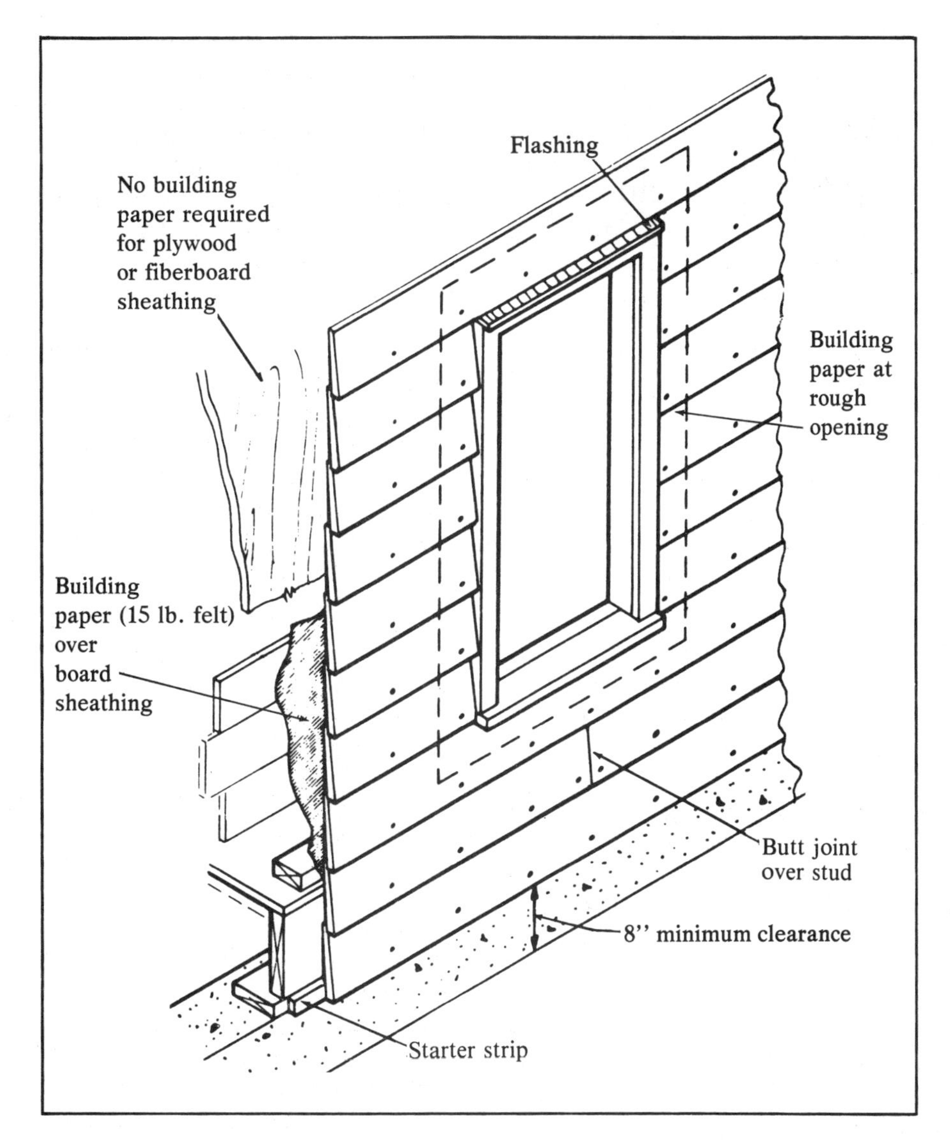

Bevel siding exposure
Figure 19-16

2) Get a board about 3/4-inch x 3 inches and long enough to reach from the bottom of the sheathing to the top of the first-floor windows. This board will be used as a "spacing rod."

3) With one end of the rod 1/2-inch below the bottom edge of the sheathing, mark the location of the bottom of the window sill and the top of the drip cap. See Figure 19-18.

4) Use dividers (small finishing nails) on the rod to mark the spacing of the courses of siding. The width of the spaces will be determined by the type of siding and the exposure distance. The dividers should be adjusted so the spaces are equal, and the butt edge of a siding course coincides with the top of the window. The butt edge of a siding course should also coincide with the underside of the window sill. If the spacing below and at the sides of the windows does not come out equal, adjust the spacing slightly so it appears equal. Mark the spacing on the rod and remove the finishing nails.

The position of the courses can be stepped off directly on the building without using a rod. However, this process would then have to be repeated for each side of the building. The rod saves time and makes errors less likely.

5) To start the first course, make a mark on each end of the building (at all corners) where the top edge of the first course should be.

6) Snap a chalk line between these marks to show the top edge of the first course.

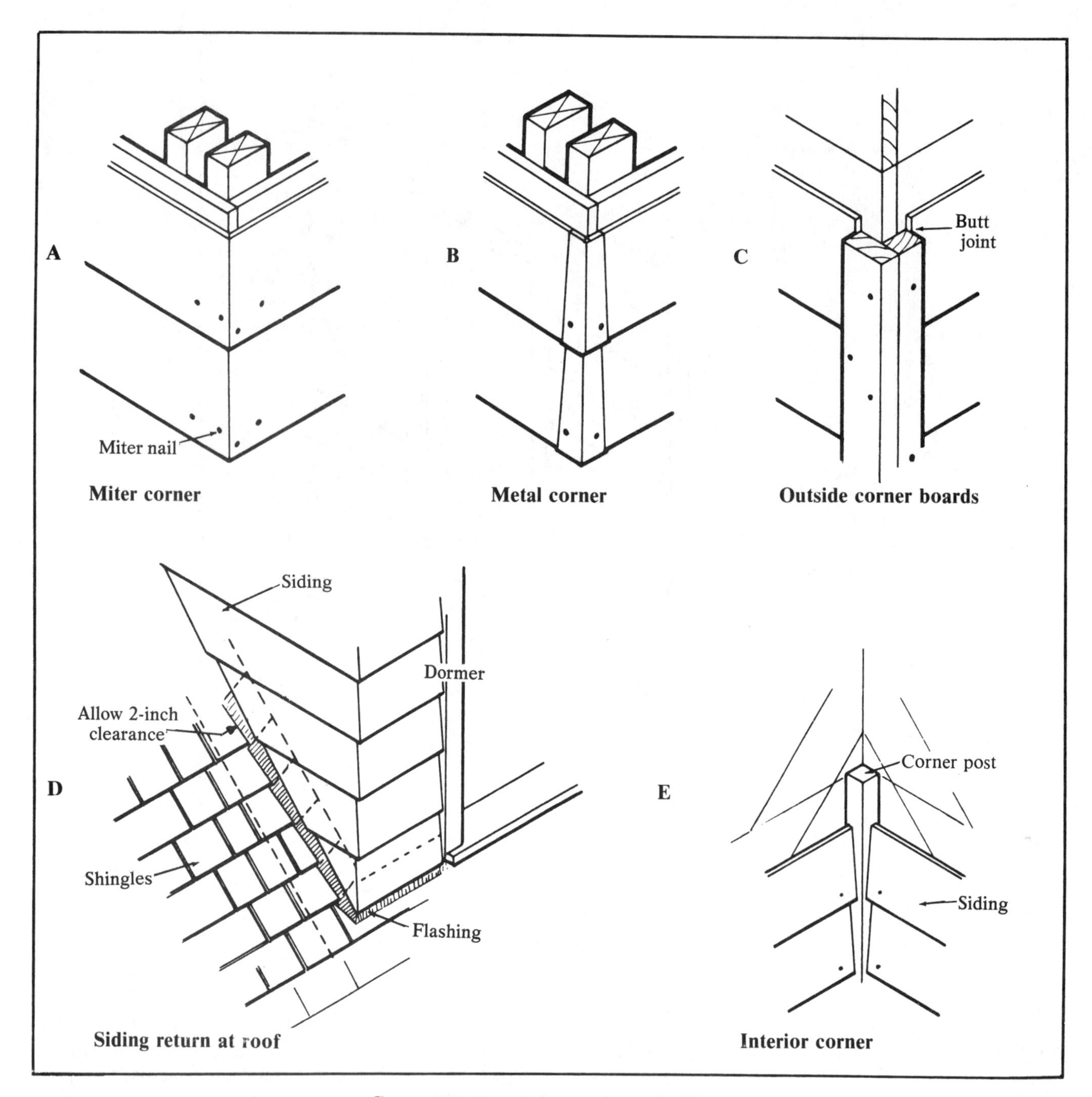

Corner treatments for horizontal siding
Figure 19-17

7) Nail furring strips along the bottom edge of the sheathing. See Figure 19-19.

8) Square the ends of the first piece of siding, and put one end against the left corner board. The top edge of the siding should be on the chalk line. Tack the top edge with small nails driven part-way in.

9) Continue applying the siding across the wall all the way to the right corner board.

10) Sight across the bottom of the first course. If it looks level, nail it securely into place. Nail through the thick part of the board and into the furring strip at the bottom. You may want to begin with a water table and drip cap rather than the furring strip. See Figure 19-20.

The siding should be nailed to each stud or 16 inches o.c. If you use plywood sheathing, wood sheathing, or spaced wood-nailing strips over non-wood sheathing, use 7d or 8d nails for 3/4-inch-thick siding. But if you use gypsum or fiberboard

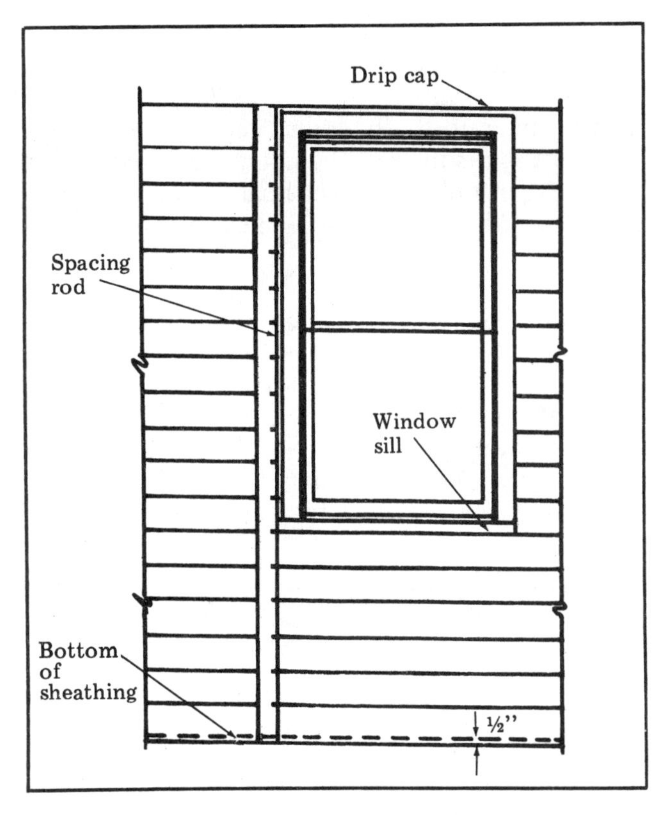

Spacing bevel siding
Figure 19-18

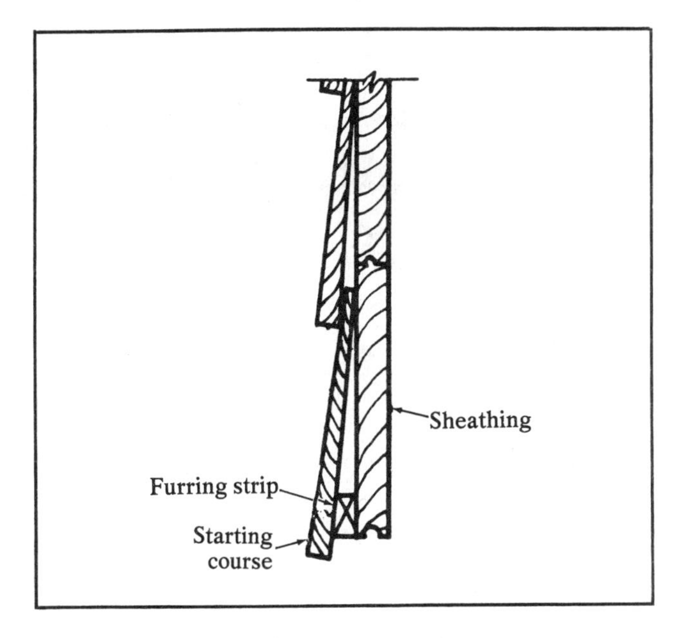

Bevel siding first course over furring strip
Figure 19-19

sheathing, 10d nails are recommended because they penetrate deeper into the stud.

The nails should be located far enough up from the butt edge so they miss the top of the siding course below. The clearance distance is usually 1/8-inch. See Figure 19-21. This allows the siding to move slightly without splitting. This is especially important for siding 8 inches wide or wider.

11) Use the spacing rod to mark the second course. Snap a chalk line and continue applying siding until all siding is installed.

12) Caulk butt joints at corner boards, windows and doors. Interior corners are butted against a square corner post 1¼ or 1⅜ inches thick, depending on the thickness of the siding. See Figure 19-17 E.

Drop siding— Regular drop siding comes in several patterns, two of which are shown in Figure 19-13 A. This siding has either matched or shiplap edges and is 1-inch-thick by 6 or 8 inches wide. It's often used for lower-cost homes, garages and other buildings without sheathing.

Drop siding is installed the same way as bevel siding except that spacing and nailing are different. Drop, Dolly Varden, and similar sidings have a set exposure distance. The face width is normally 5¼ inches for 1 x 6-inch siding and 7¼ inches for 1 x 8-inch siding. Depending on the width of the siding, one or two 8d or 9d nails should be used at each stud crossing. See Figure 19-22 A. The nails should be long enough to penetrate into the stud or wood sheathing at least 1½ inches.

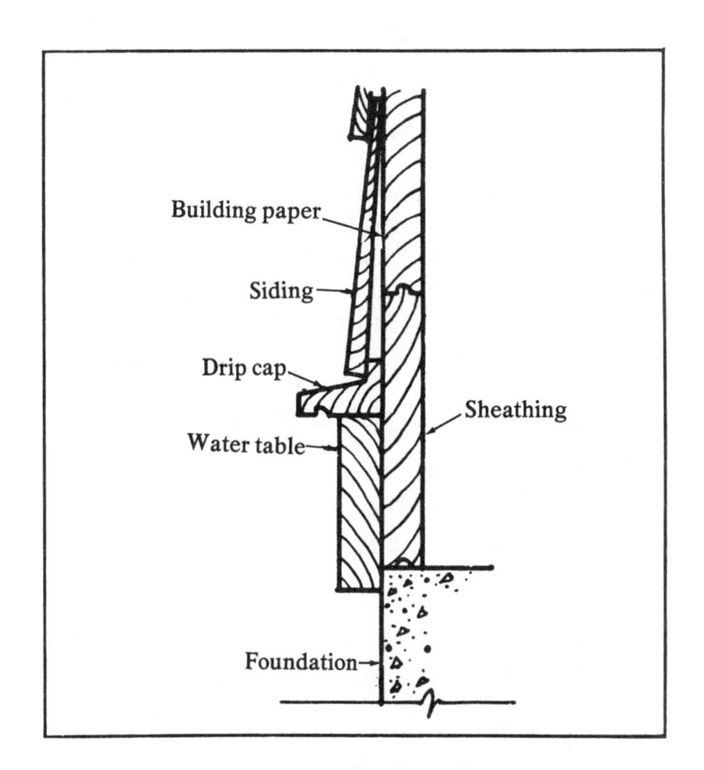

Bevel siding first course over drip cap and water table
Figure 19-20

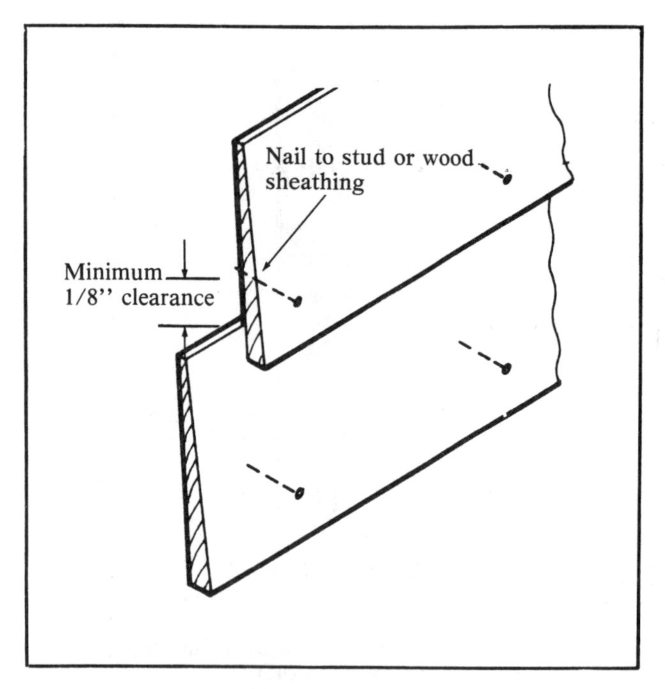

Nailing bevel siding
Figure 19-21

Metal corners, as shown in Figure 19-17 B, are more commonly used on drop siding than are miter corners. It's easy to put the metal corners in place as the siding is installed. The metal corners should fit tightly without openings and should be nailed (on each side) to the sheathing or corner stud. Corners made of galvanized metal should be cleaned with a mild acid wash and primed with a metal primer before the house is painted. This prevents early peeling of paint.

Matched paneling— Two common types of matched paneling are shown in Figure 19-13 B. These patterns can be applied horizontally or vertically.

Horizontal matched paneling in narrow widths should be blind-nailed with a corrosion-resistant finishing nail at the tongue. See Figure 19-22 B. For panel widths greater than 6 inches, use one extra nail as shown.

Here's how to install horizontal matched paneling:

1) Select a straight piece of siding. Start at the bottom of the sheathing and flush with the corner. The tongue faces up.

2) Cut the piece so that the end falls on the center of a stud. Tack it in place with nails driven partway in.

3) Continue across the building, staying level with the first board. Cut the last piece flush with the corner, and tack it in place.

4) Sight this course for straightness. If it is straight, nail it firmly in place at each stud.

5) Continue up the building, keeping all matched and butt joints tight. If the siding is applied directly to studs, install the corner boards *over* the siding and finish the corner boards the same as for bevel siding.

Figure 19-17 D shows a corner siding return at the roof.

Vertical Wood Siding

Vertical siding is appropriate for some architectural styles. Vertical siding is usually made of rough-sawn boards and battens. The boards and

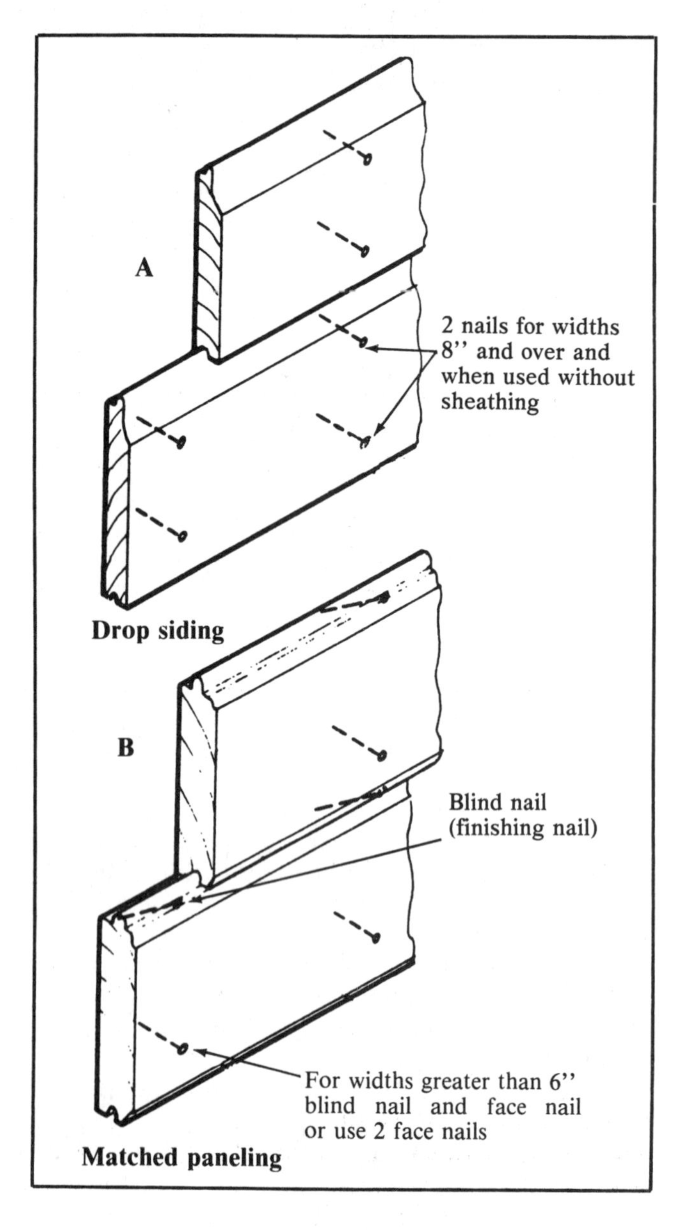

Nailing drop siding and matched paneling
Figure 19-22

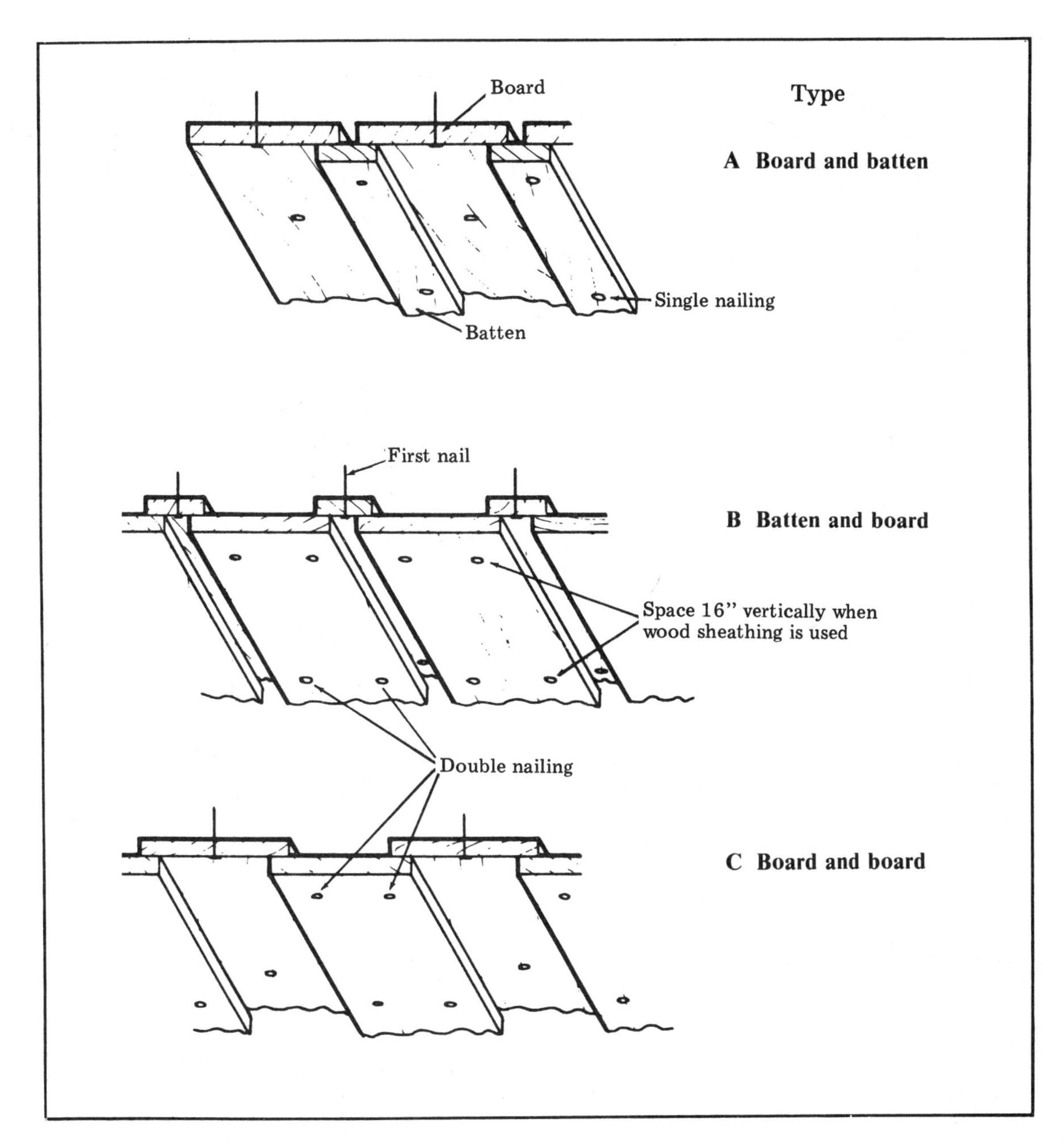

Vertical board and batten siding
Figure 19-23

battens can be arranged in several ways: board and batten; batten and board; and board and board. See Figure 19-23.

Vertical siding should be applied over 1-inch sheathing boards or over plywood sheathing that is ½-inch or 5/8-inch thick. Sheathing provides the required nailing surface.

If you use composition or foam sheathing, or thinner plywood sheathing, install 1 x 4-inch nailing blocks between the studs. Apply the nailers horizontally. Allow vertical spacing of 16 to 24 inches between nailing blocks. This system of sheathing plus nailers requires thicker casing around door and window frames. And it's good practice to apply building paper over the sheathing before applying vertical siding.

The first board or batten should be fastened with one 8d or 9d nail at each nailing block. Provide at least 1½-inch penetration. When wide boards are used, drive two nails spaced about 2 inches apart, rather than a single row along the center. This double nailing is shown in Figure 19-23 B and C.

The second or top board or batten should be nailed with 12d nails. Don't nail through both board and batten. Nails in the top board or batten should always miss the underboard. This permits normal expansion without risk of splitting.

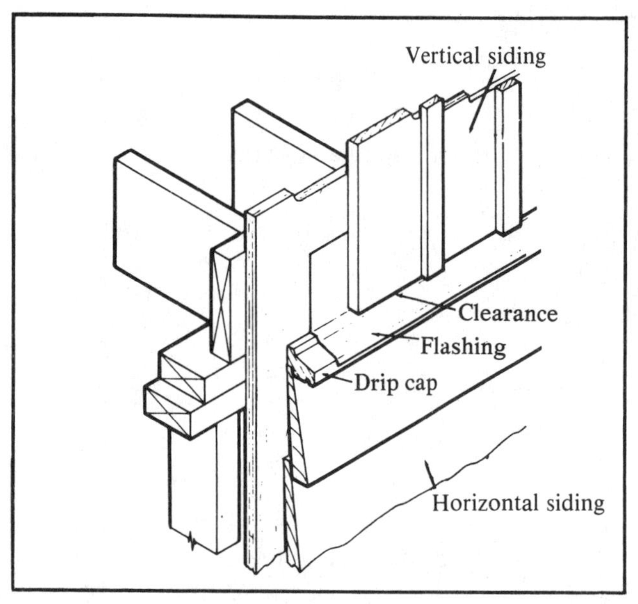

Drip cap at material transition
Figure 19-24

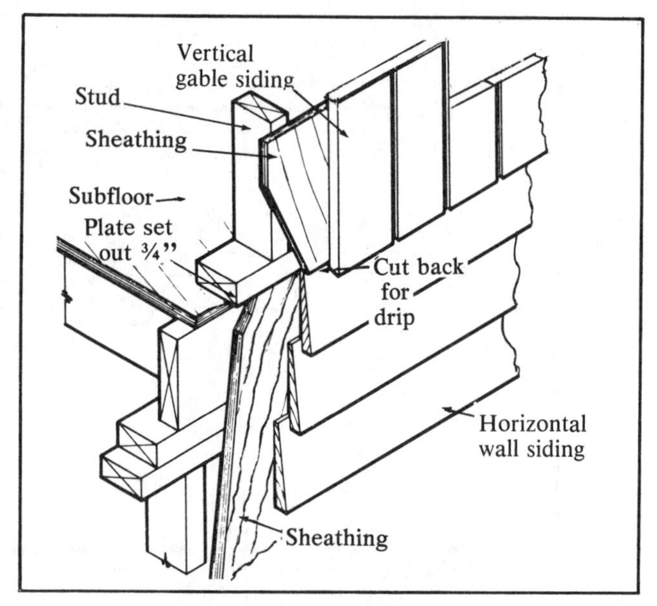

Extended plate at material transition
Figure 19-25

Material Transition

Some houses use two types of siding. Others use both vertical and horizontal siding of the same type. The siding used for gable ends may be different than the siding used for walls.

The joint between the two types of siding should provide good drainage. For example, if vertical board-and-batten siding is used at the gable ends, and horizontal siding is used on the walls below, a drip cap or similar molding is required. See Figure 19-24. Use flashing over and above the drip cap so that moisture will fall clear of the joint between the two types of siding.

Here's another way to make the material transition. Extend the plate and studs of the gable end out from the wall a short distance. The gable siding will project beyond the wall siding and provide good drainage. See Figure 19-25.

Summary

We haven't covered all types of siding — just those commonly used by builders who want to use their own construction crews to do as much of the work as possible. Sheet panel materials, such as plywood and hardboard, go up quickly and with minimal labor. They're an excellent choice for your spec houses. If low cost is the prime consideration, use hardboard siding. When building spec houses, you can always put too much money into the job. You'll seldom put in too little.

The difference between a profit and loss on most jobs is the ability to estimate with accuracy the labor and materials required to do the work. The estimating tables shown below will help you arrive at reasonable estimates. But the best estimating table is the one you compile yourself — from your own job records. Use Figure 19-26, Siding Checklist and Data Reference Sheet, to compile your cost references for future siding jobs.

Labor Installing Siding

Work Element	Unit	Man-hours Per Unit
Bevel siding		
½" x 6", 3' to 7' long	1000 SF	19.0
½" x 6", 6' to 18' long	1000 SF	17.0
½" x 8", 3' to 7' long	1000 SF	17.0
¾" x 8", 3' to 7' long	1000 SF	18.0
¾" x 10", 6' to 18' long	1000 SF	18.5
Tongue and groove siding		
1" x 4"	1000 SF	21.5
1" x 6"	1000 SF	20.5
Board and batten siding		
1" x 12"	1000 SF	29.0
Coved channel siding, 1" x 8"	1000 SF	24.5
Plywood siding, 4' x 8' panels		
3/8"	1000 SF	14.0
5/8"	1000 SF	16.0
Trim pieces		
Edging, 1" x 3"	10 LF	0.4
Corner, 1" x 3" x 3"	10 LF	0.5
Base trim, 1" x 4"	10 LF	0.35

No waste or allowance for coverage is included.
Suggested Crew: 1 carpenter and 1 laborer.

Labor and Materials for Bevel Siding

Size	Material: Exposed to Weather	Material: Add for Lap	Material: SF of Siding per 100 SF of Wall	Lbs. Nails Per 100 SF	Man-Hours Per 100 SF
1/2" x 4"	2¾	46%	151	1½	3.2
1/2" x 5"	3¾	33%	138	1½	2.5
1/2" x 6"	4¾	26%	131	1	1.9
1/2" x 8"	6¾	18%	123	¾	1.7
5/8" x 8"	6¾	18%	123	¾	1.8
3/4" x 8"	6¾	18%	123	¾	1.8
5/8" x 10"	8¾	14%	119	½	2.1
3/4" x 10"	8¾	14%	119	½	2.2
3/4" x 12"	10¾	12%	117	½	2.1

Labor and Materials for Drop Siding

Size	Exposed to Weather	Add for Lap	SF of Siding per 100 SF of Wall	Lbs. Nails Per 100 SF	Man-Hours Per 100 SF
1" x 6"	5¼"	14%	119	2½	2.0
1" x 8"	7¼"	10%	115	2	1.9

Quantities include 5% for end cutting and waste. Deduct for all openings over ten square feet.

Coverage of T&G and Shiplap Boards

Measured Size, Inches	Finished Width, Inches	Add For Shrinkage, Percent	Quantity Required, Multiply Area by	S.F. of Lumber Required, Per 100 SF
1 x 2	1-3/8	50	1.50	150
1 x 2¾	2	42½	1.425	142½
1 x 3	2½	38-1/3	1.383	138
1 x 4	3¼	28	1.28	128
1 x 6	5¼	20	1.20	120
1 x 8	7¼	16	1.15	115
1¼ x 3	2¼	38-1/3	1.73	173
1¼ x 4	3¼	28	1.60	160
1¼ x 6	5¼	20	1.50	150
1½ x 3	2¼	38-1/3	2.08	208
1½ x 4	3¼	28	1.92	192
1½ x 6	5¼	20	1.80	180
2 x 4	3¼	28	2.60	260
2 x 6	5¼	20	2.40	240
2 x 8	7¼	16	2.32	232
2 x 10	9¼	13	2.25	225
2 x 12	11¼	12	2.24	224
3 x 6	5¼	20	3.60	360
3 x 8	7¼	16	3.48	348
3 x 10	9¼	13	3.39	339
3 x 12	11¼	12	3.36	336

This data applies to most dressed and matched lumber. Waste allowance shown includes width lost in dressing and lapping. Add 5% for end-cutting and matching.

Siding Checklist and Data Reference Sheet

1) The house is _______ feet in length and _______ feet wide, for a total of _______ square feet.

2) The house is ☐1-story ☐1½-story ☐2-story.

3) The wall sheathing is ☐board (size: ___________) ☐plywood (thickness: ___________ inch) ☐foam plastic (thickness: ___________ inch) ☐fiberboard (☐structural ☐nonstructural) ☐gypsum board ☐other (specify) ___________

4) Studs are spaced ☐16 inches o.c. ☐24 inches o.c.

5) Building paper ☐was ☐was not used.

6) Siding ☐was ☐was not applied directly to studs.

7) Siding is:

☐APA Sturd-I-Wall:

☐Panel siding (size and thickness: ___________)

☐Lap siding (size and thickness: ___________)

☐Plywood (pattern and size: ___________)

☐Hardboard:

☐Panel (size: ___________)

☐Lap siding (pattern and size: ___________)

☐Shakes

☐Textured

☐Plain (smooth)

☐Primed

☐Unprimed

☐Square edge

☐Shiplap edge

Wood:

☐Horizontal (size: ___________)

☐Bevel

☐"Anzac" bevel

☐Dolly Varden

Figure 19-26

☐Drop

☐Other (specify) ____________

☐Vertical (size: ____________)

☐Tongue and groove panel

☐Board and batten

☐Batten and board

☐Board and board

8) Corner treatment is ☐corner boards ☐metal ☐miter.

9) Gable-end treatment is ☐same as wall ☐material transition (☐drip cap belt course ☐projected) ☐horizontal ☐vertical.

10) Nails are:

☐Flathead siding ☐small head finish ☐threaded shank.

☐Stainless steel ☐galvanized ☐aluminum.

☐6d ☐8d ☐10d ☐12d.

☐Flush ☐countersunk.

☐Color-matched.

☐Other (specify) ____________

11) Finish is ☐paint (☐one coat ☐two coats) ☐stain (☐one coat ☐two coats) ☐natural.

12) Finish application method was ☐brush ☐roller ☐spray (☐brushed ☐unbrushed).

13) Material costs:

Sturd-I-Wall:

Panels	________pcs	@	$________ea	=	$__________
Lap siding	________BF	@	$______M BF	=	$__________
Plywood panels	________pcs	@	$________ea	=	$__________
Hardboard:					
Panels	________pcs	@	$________ea	=	$__________
Lap siding	________BF	@	$______M BF	=	$__________
Shakes	________SF	@	$________SF	=	$__________
Wood siding	________BF	@	$______M BF	=	$__________

Furring strips	________LF	@	$________LF	=	$__________
Drip cap	________LF	@	$________LF	=	$__________
Corner boards	________LF	@	$________LF	=	$__________
Metal corners	________pcs	@	$________ea	=	$__________
"H" molding	________pcs	@	$________ea	=	$__________
Butt joint cover	________pcs	@	$________ea	=	$__________
Building paper	________rolls	@	$________ea	=	$__________
Caulk	________tubes	@	$________ea	=	$__________
Nails	________lbs	@	$________lb	=	$__________
Primer paint	________gals	@	$________ea	=	$__________
Paint	________gals	@	$________ea	=	$__________
Stain	________gals	@	$________ea	=	$__________
Other materials, not listed above:					
______________	________	@	$________	=	$__________
______________	________	@	$________	=	$__________
______________	________	@	$________	=	$__________
______________	________	@	$________	=	$__________
			Total cost of all materials		$__________

14) Labor costs:

Skilled	________hours	@	$________hour	=	$__________
Unskilled	________hours	@	$________hour	=	$__________
Other (specify)					
______________	________hours	@	$________hour	=	$__________
			Total		$__________

15) Total cost:

Materials	$__________
Labor	$__________
Other	$__________
Total	$__________

16) Siding cost per SF of floor space $__________

17) Problems that could have been avoided ______________________________

18) Comments ______________________________

19) A copy of this Checklist and Data Reference Sheet has been provided:

☐ Management

☐ Accounting

☐ Estimator

☐ Foreman

☐ File

☐ Other __________

Chapter 20

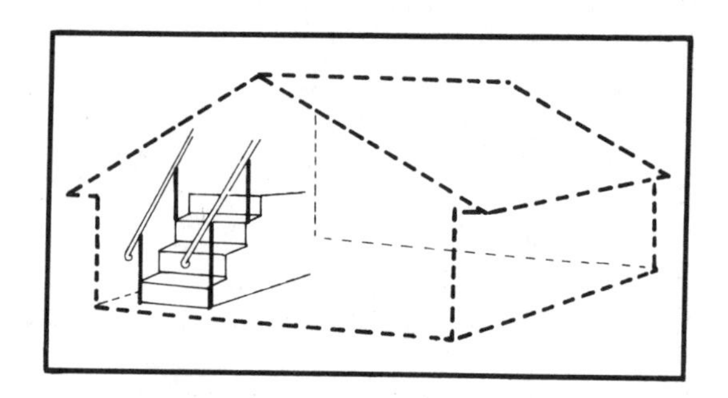

Stair Construction

Stair construction is considered specialty work and is usually done only by experienced carpenters. More intricate and decorative stairs are often subcontracted out to stair specialty shops located in mills.

The stairway is a critical part of the house. Poor stair design or faulty construction can result in dangerous falls. The stairs you build must have adequate headroom, as well as enough space for the passage of furniture from one floor to another.

The two most common types of stairs are main stairs and basement (or service) stairs. Main stairs lead up to the second floor. They must provide easy ascent and descent and are a feature of the interior design of the home. Basement stairs lead to a basement or garage area. These stairs can be somewhat steeper and are usually built with less expensive materials.

Most main stairs and basement stairs are constructed in place. The main stairs can be assembled from prefabricated parts, including housed stringers, treads, and risers. Basement stairs are usually made of plank treads and 2 x 12-inch carriages.

Both hardwood and softwood lumber species are used in stair construction. Use oak, maple or birch for main-stairway components, such as treads, risers, handrails and balusters. Basement-stairway treads and risers are usually made of Douglas fir or Southern pine. For an economical, wear-resistant stairway, combine a hardwood tread with a softwood riser.

Stairway Runs

The three types of stairway runs most commonly used in house construction are the straight run (Figure 20-1 A), the long "L" (Figure 20-1 B), and the narrow "U" (Figure 20-1 C).

A fourth type of run, the winder, is shown in Figure 20-1 D. It's similar to the long "L" except that winders are substituted for the landing. Avoid winders whenever possible. They're not as convenient or as safe as the long "L," and they're restricted by many codes. Use winders only when the available space is too short for more conventional stairs with a landing.

Stairway Code Restrictions

Know the code requirements governing stair construction. There are restrictions on winders, landings, headroom, stairway width and handrails, isolated flights of interior stairs, tread width and riser height.

Winders

If you must use winders, make sure the winder tread width meets code requirements. (See Figure 20-2 A.) Check the building code that applies *before* you start. Measure along a line 18 inches from the narrow end of the winder tread. At the 18-inch mark, the winder tread width must be no less than the minimum tread width on the straight run of the stairs. If the straight-run tread is 10 inches wide, the winder tread should be at least 10 inches wide at the 18-inch mark.

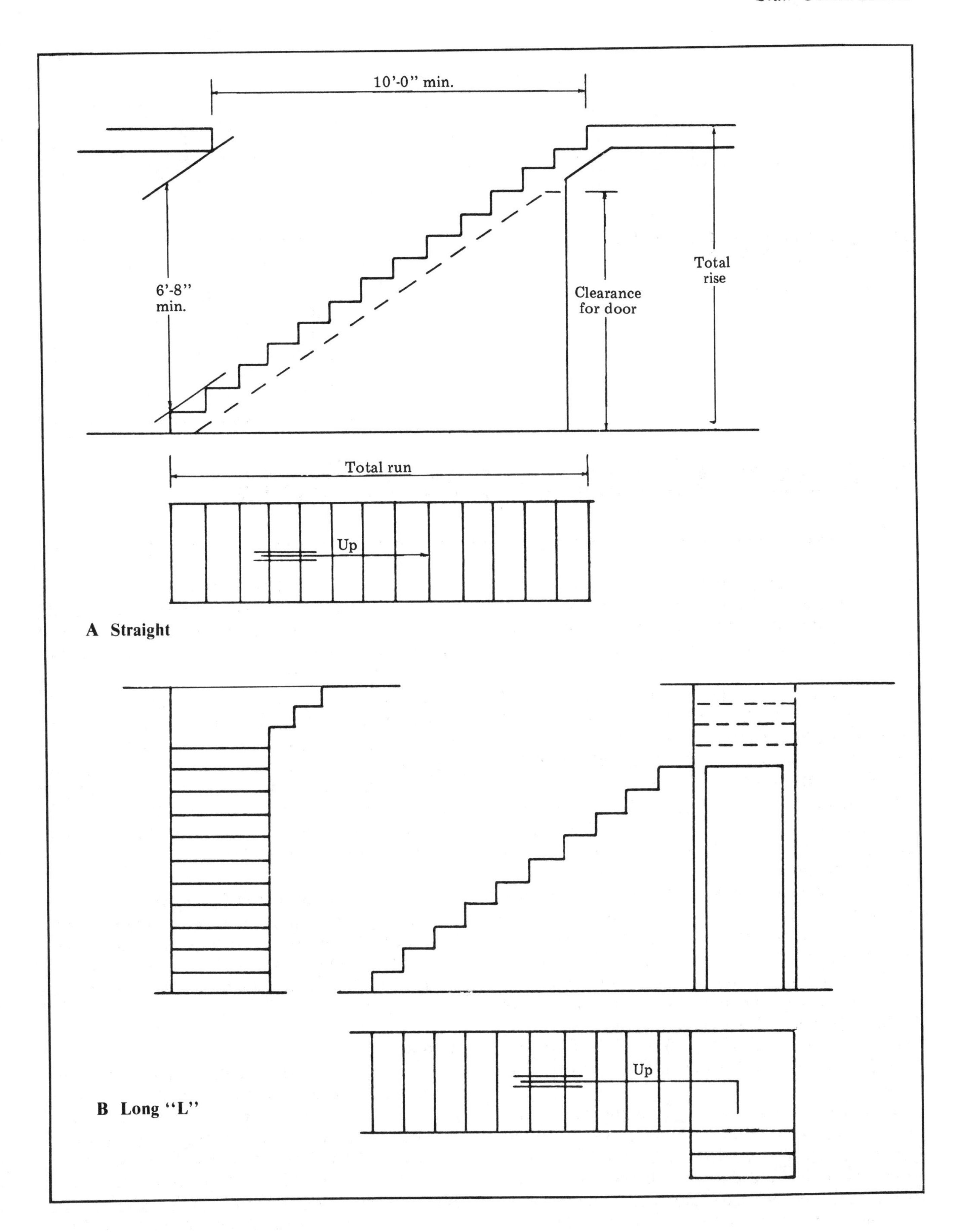

Common stairway runs
Figure 20-1

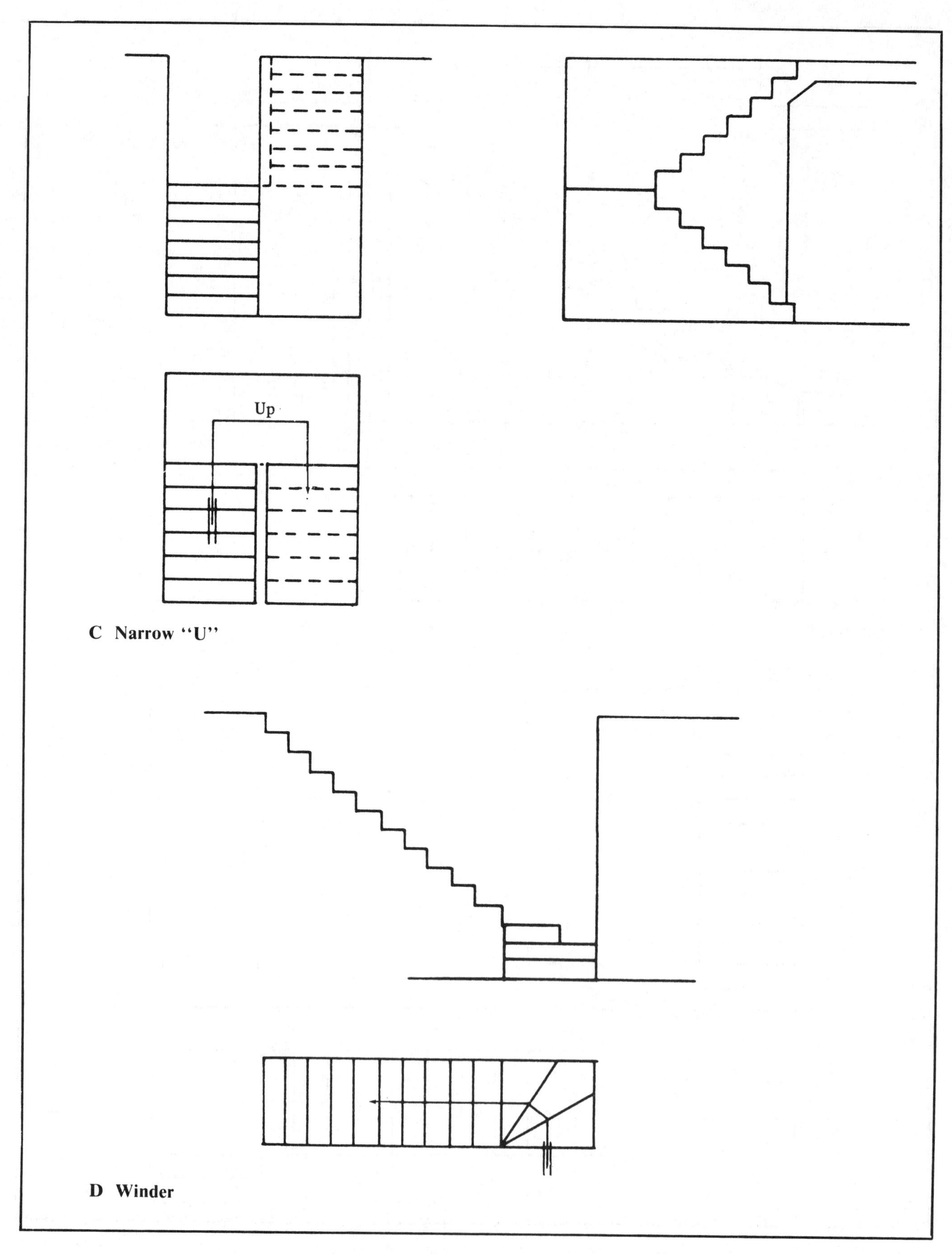

Common stairway runs (continued)
Figure 20-1

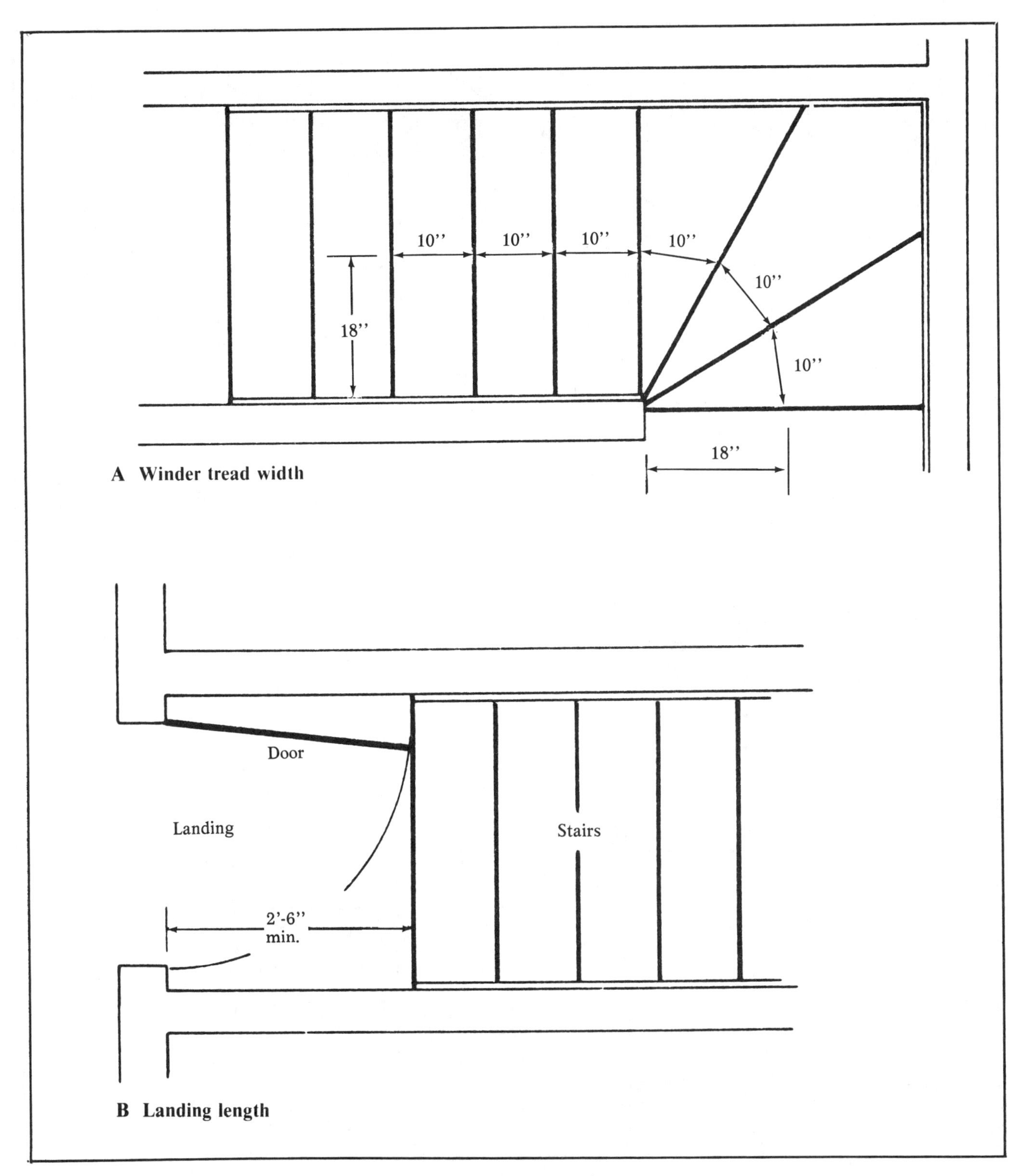

Stairway code restrictions
Figure 20-2

Landings

If the landing is at the top of a stair run and the door opens into the stairway, the landing must be at least 2'6'' long. Middle landings must also be at least 2'6'' long. See Figure 20-2 B.

Headroom

For main stairways, the clear vertical distance must be at least 6'8'', as shown in Figure 20-2 C. Basement stairs must provide not less than 6'4'' clearance.

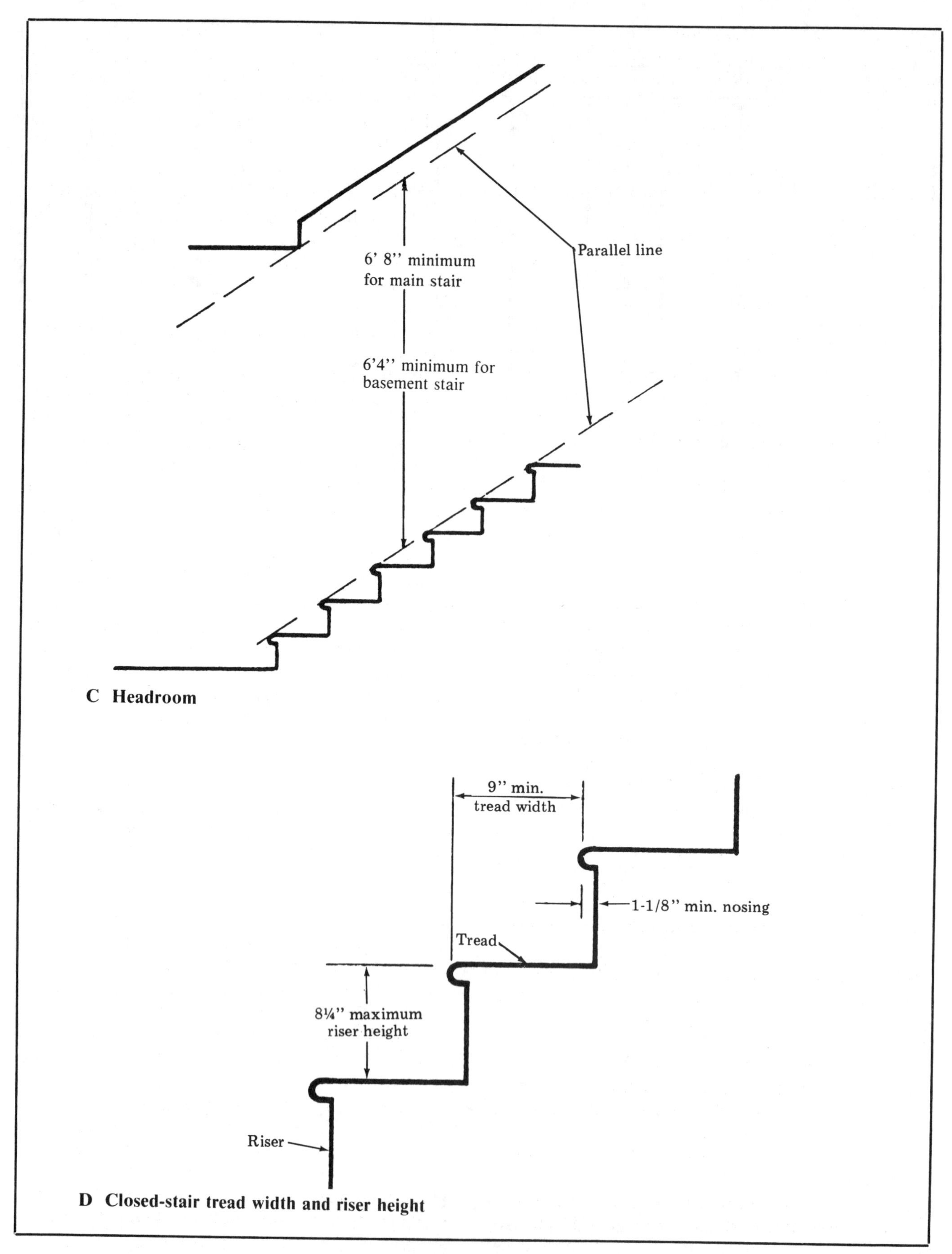

Stairway code restrictions (continued)
Figure 20-2

Stairway Width and Handrails

The width of main stairs (clear of the handrail) should not be less than 2'8''. In better homes, main stairways are usually designed with a distance of 3'6'' between the centerlines of the enclosing side walls. This results in a stairway width of about 3 feet. Split-level entrance stairs should be even wider. For basement stairs, the minimum clear width is 2'6''.

Have a continuous handrail on at least one side of the stairway if there are more than three risers. When stairs are open on two sides, provide protective railings on *both* sides.

Isolated Interior Stairs

An isolated flight of interior stairs should have at least three and no more than 18 risers. A flight with two risers is acceptable only if it is separated from another flight of stairs by a landing.

Tread Width and Riser Height

On closed (interior) stairs, the minimum tread width is 9'', and the maximum riser height is 8¼''. See Figure 20-2 D. (On exterior stairs, minimum tread is 11'', and maximum riser is 7½''.)

Ratio of riser to tread— Many studies have been done to determine the ideal ratio between the height of a riser and the width of a tread. The ratio should produce a step that is comfortable for most people. A riser 7½ to 8¼ inches high seems to meet this requirement.

A rule of thumb for the ratio between riser and tread is: *The tread width in inches times the riser height in inches should equal 72 to 75.*

The stairs shown in Figure 20-2 D conform to this rule — 9 times 8¼ equals 74¼. If the tread is 10 inches, however, the riser will be 7½ inches, which is more desirable for common stairways.

Another rule sometimes used is: *The tread width plus twice the riser height should equal about 25.*

Use these rules and the ideal riser height when calculating the number of steps between floors. For example, 14 risers are normally used for main stairs between the first and second floors. The 8-foot ceiling height of the first floor plus the upper-story floor joists, subfloor, and finish floor result in a floor-to-floor distance of about 105 inches. Thus, 105 divided by 14 is exactly 7½ inches, the height of each riser. Fifteen risers used for this height would mean a 7-inch riser height.

No matter how many risers you use, make sure that each riser in the flight is the same height (to within 1/4-inch) as all the other risers. Your inspector probably won't pass the stairs if the riser height varies more than a fraction of an inch.

Framing the Stairway

Frame stairway floor openings at the same time you frame the rest of the floor. See Figure 6-22 in Chapter 6. For a main stairway, the rough opening in the second floor is usually at least 10' long x 36'' (or more) wide. For a basement stairway, the rough opening should be about 9'6'' long x 32'' wide.

The long dimension of the stair opening can be either parallel to or at right angles to the joists. But it's easier to frame a stairway opening when the length runs parallel to the joists. See Figure 20-3 A. A short header may be required for one or both ends.

When the length of the stair opening runs perpendicular to the joists, a long double-header is required. See Figure 20-3 B. Simplify construction by using a load-bearing wall under part or all of the opening. The joists can then bear on the top plate of the wall and won't need to be supported by joist hangers at the header. If you don't use a supporting wall, the maximum allowable header length is 10 feet. Follow the nailing schedule in Figure 6-4 in Chapter 6.

Framing for a long ''L'' stairway is usually supported in the basement. Use a load-bearing wall, or use a post at the corner of the stair opening. When a similar stairway leads from the first to the second floor, frame the landing as shown in Figure 20-4. The landing frame is nailed to the wall studs. The frame provides a nailing area for the subfloor, as well as a support for the stair carriages.

Stair Construction Details

Let's look at construction details for the two most common types of stairs: basement (or service) stairs and finished main stairs. We'll also briefly discuss construction details for attic stairs.

Basement Stairs

Stair carriages anchor the treads and support the load on the stairs. There are two common types of stair carriages: notched and unnotched.

Rough stair carriages, commonly used for basement stairs, are made from notched, 2 x 12-inch planks. The effective depth below the tread and riser notches must be at least 3½ inches. See Figure 20-5 A.

These carriages are usually placed only at each side of the stairs. But an intermediate carriage is required at the center of the stairs when the treads are 1$\frac{1}{16}$ inches thick and the stairs are wider than 2'6''. Three carriages are also required when treads are 1⅝ inches thick and the stairs are wider than 3 feet.

The carriage is fastened to the joist headers at the top of the stairway. Or it can rest on a supporting ledger nailed to the header. See Figure 20-5 B.

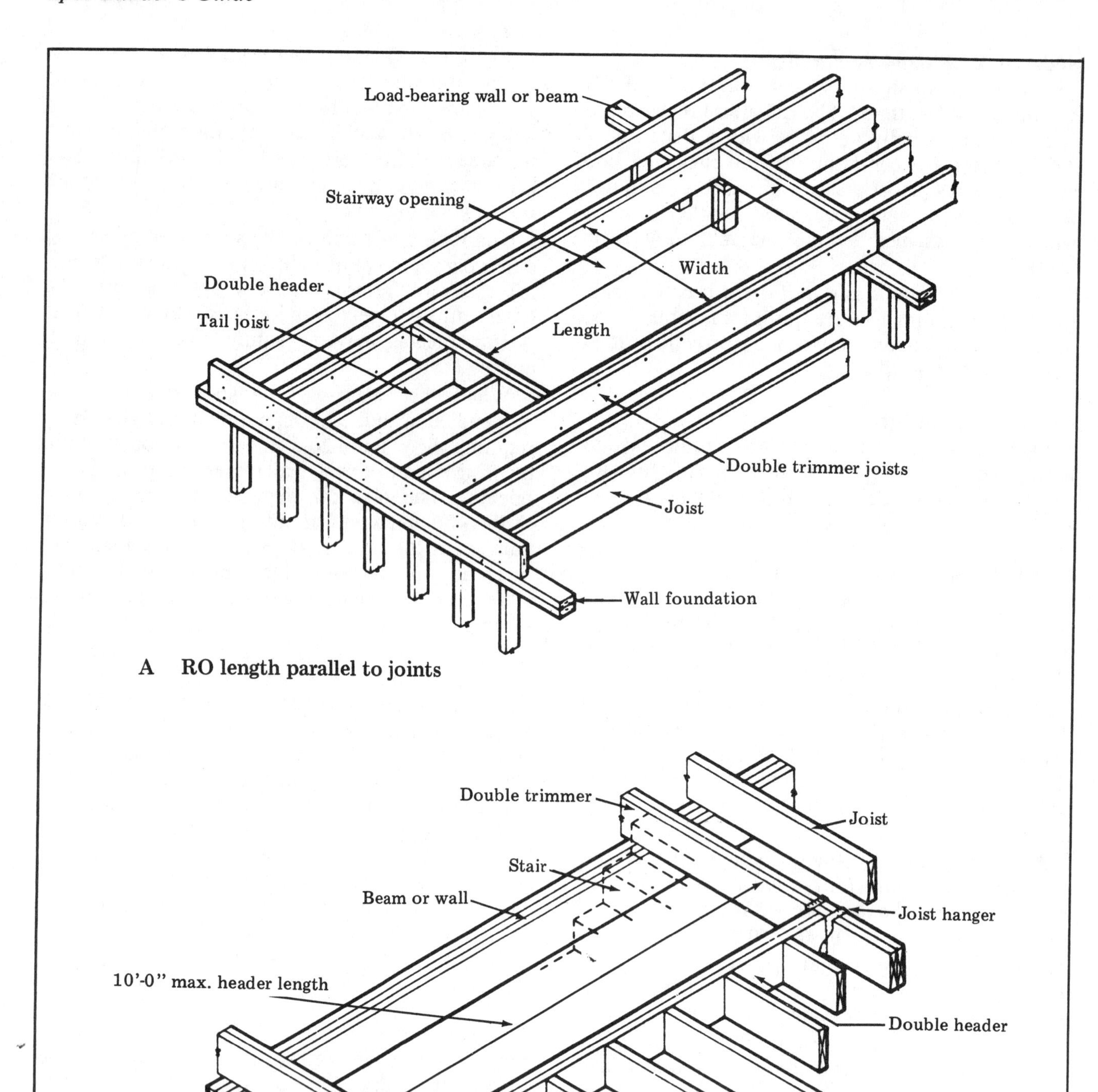

Framing the stairway
Figure 20-3

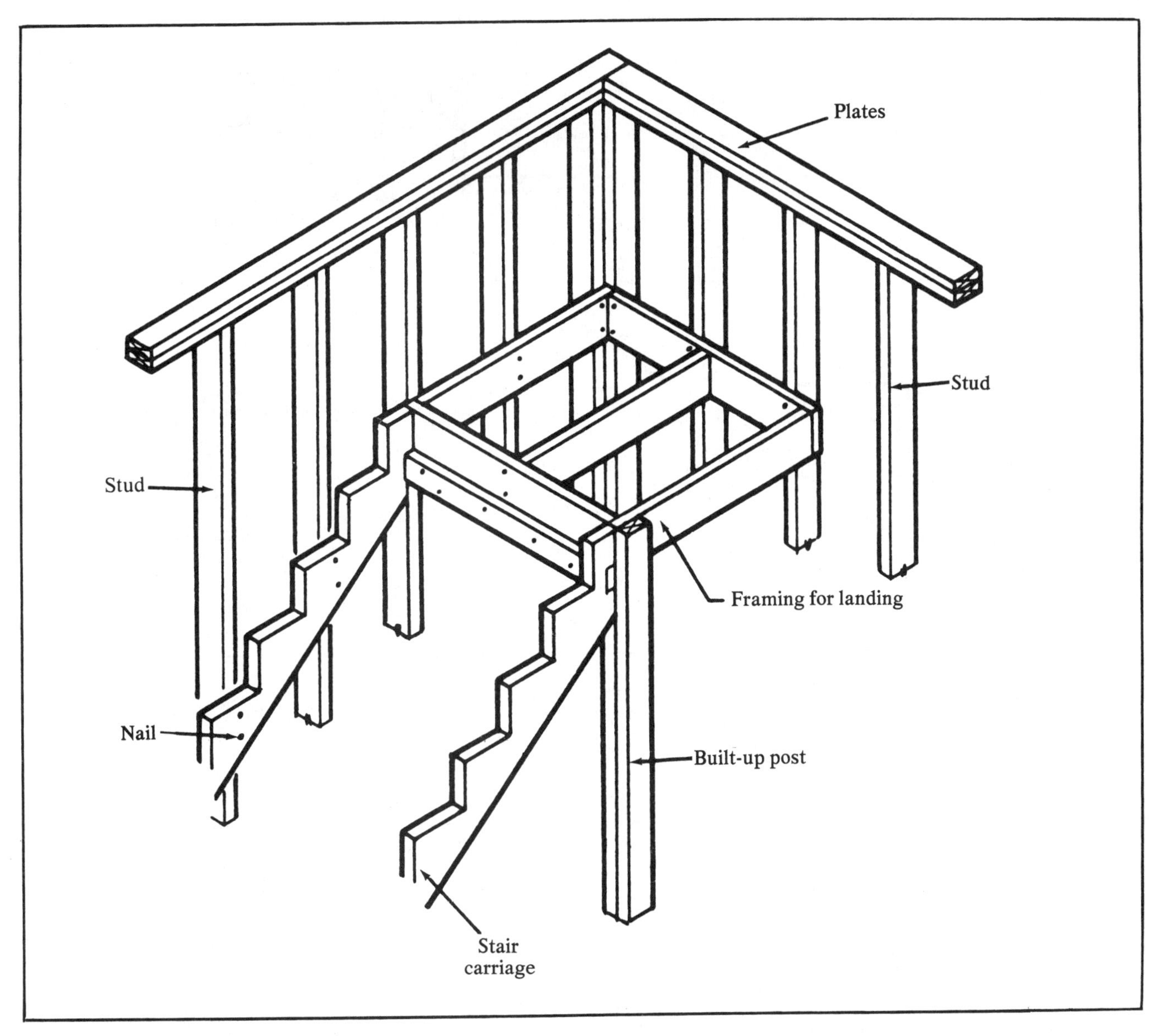

Framing the landing
Figure 20-4

Carriages can also be supported by walls located below them.

The bottom of the stair carriage can rest on the basement floor. But a better method is to use an anchored 2 x 4-inch or 2 x 6-inch treated kicker plate, as in Figure 20-5 A. Be sure to provide fire stops at the top and bottom of all stairs, as shown.

The simplest stair system has carriages that are not notched for the treads and risers. Instead, cleats are nailed to the side of the unnotched carriage, and treads are nailed to the cleats. When walls are present, however, it's more practical to use a notched carriage.

You can make basement stairs out of 1½-inch-thick plank treads without risers. However, the stairway looks better and is more durable if you use 1⅛-inch finished tread material and nominal 1-inch boards for risers. Use finishing nails to secure the risers to the treads.

Main Stairs

In a two-story traditional home, it's nearly essential to have an open main stairway with railings and balusters ending in newel posts. Many contemporary homes also have dramatic stairways with bold handrails.

On main stairways, the supporting members are housed stringers. The stringers are routed to fit both the tread and the riser. Fasten the housed stringers directly to the finish wall, as shown in Figure 20-6 A.

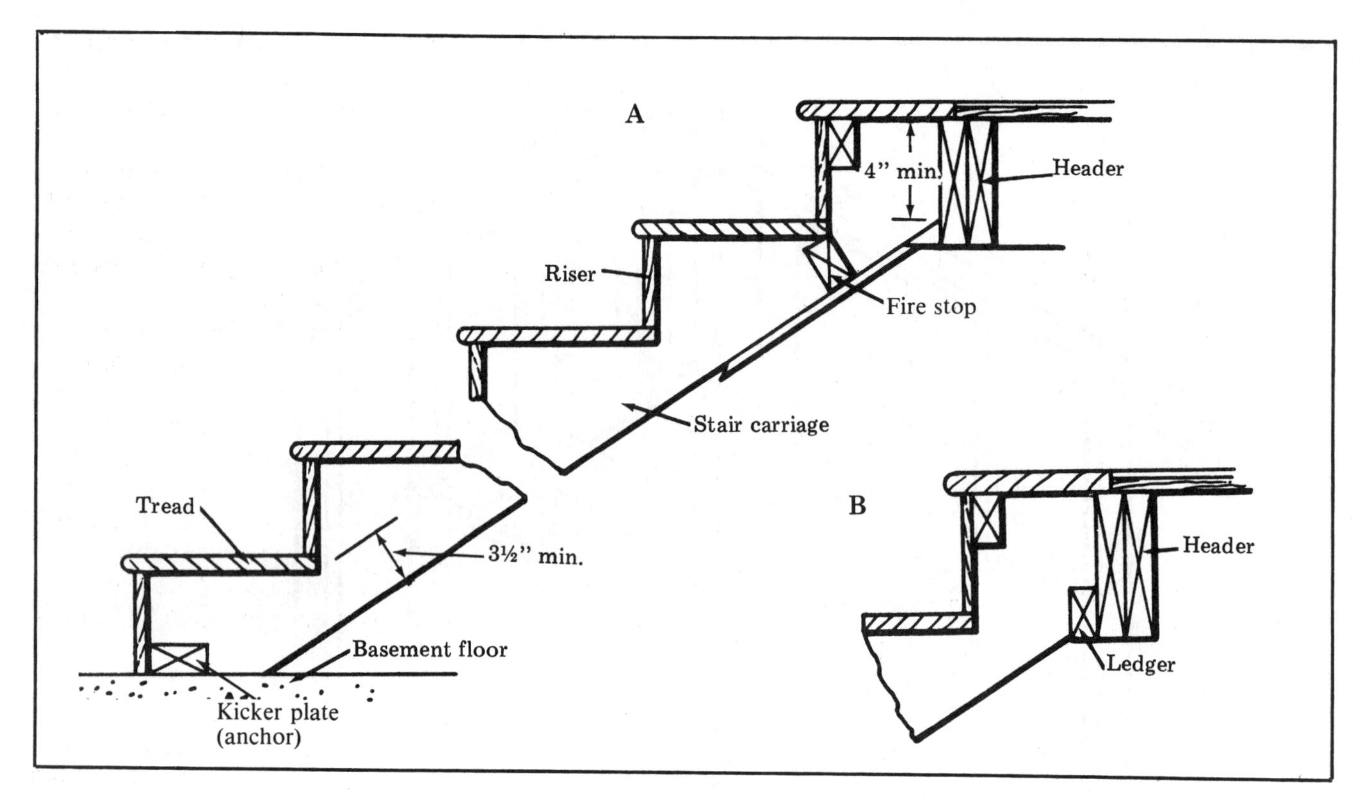

Basement stairs
Figure 20-5

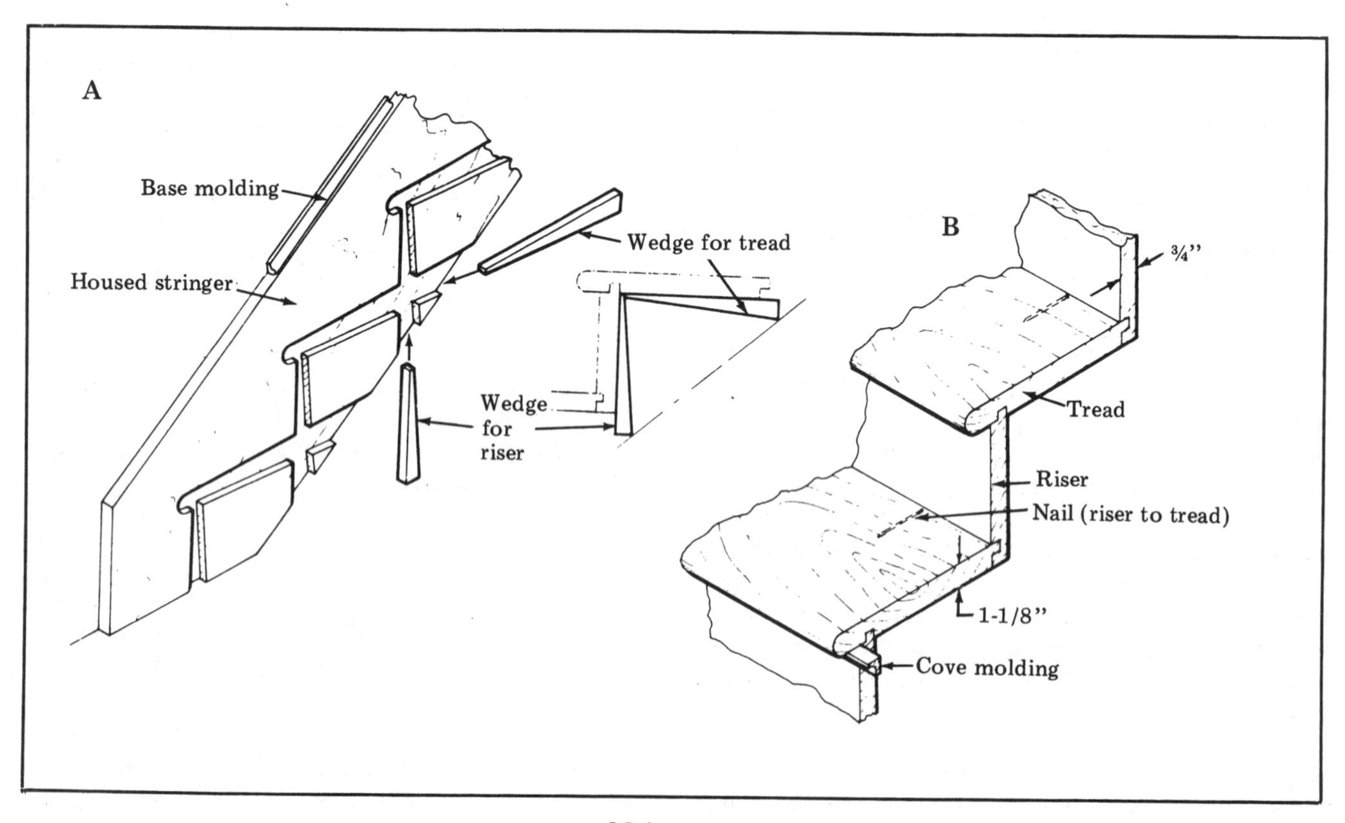

Main stairs
Figure 20-6

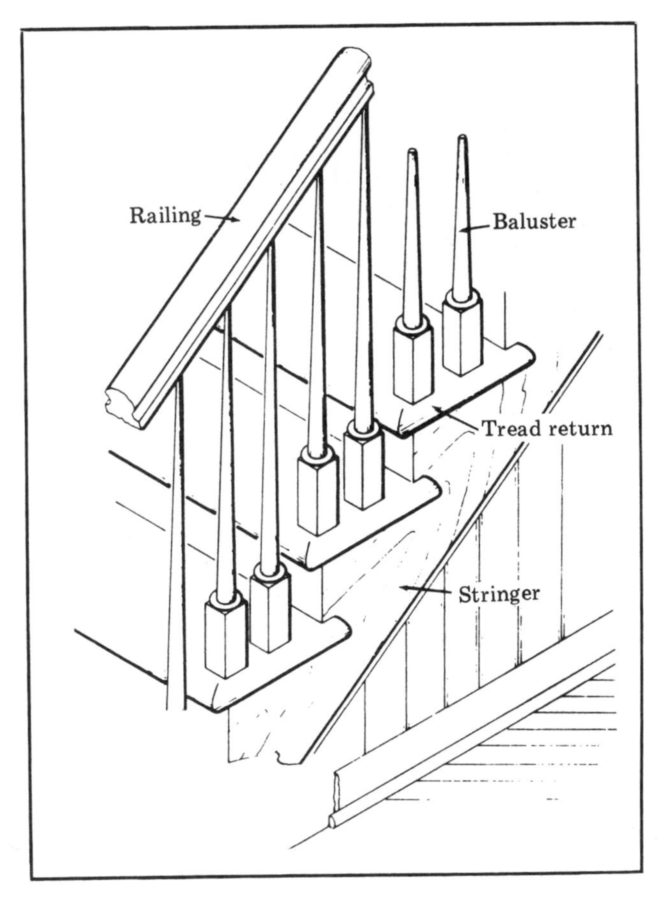

Open main stairs
Figure 20-7

The stairway is assembled with hardwood wedges coated with glue. Assembly is usually done from the underside of the stairway. Drive the wedges under the ends of the treads. Drive additional wedges behind the risers, as shown in Figure 20-6 A. Use nails to fasten the risers to the treads. See Figure 20-6 B.

When treads and risers are wedged and glued into housed stringers, the maximum allowable stair width is 3'6''. For wider stairs, provide a notched carriage between the housed stringers.

When stairs are open on one side, a railing and balusters are needed. See Figure 20-7. Balusters are made with doweled ends that fit into drilled holes in the railing and in the treads. Balusters should be fastened to tread ends that have finished returns. The railing usually ends at a newel post. A stringer and appropriate moldings complete the stairway trim.

Attic Stairs

A quality, fully enclosed stairway is appropriate between the main floor and the attic. Use a rough, notched carriage with a finish stringer, as shown in Figure 20-8 A. Fasten the stringers to the wall. Then install the carriages. Cut treads and risers to fit snugly between the stringers. Use finishing nails to secure the treads and risers to the carriages. See Figure 20-8 A.

Figure 20-8 B shows an alternate method of installing this fully enclosed stairway. Begin by nailing the rough carriages directly to the wall. Then notch the stringers to fit the carriages. Treads and risers are installed as described above.

If the attic is intended primarily for storage, or if space is not available for a fixed stairway, consider installing hinged or folding stairs. This type of stairway provides attic access through an opening in a hall ceiling. When the stairs aren't in use, they swing up into the attic space. If you install folding stairs, be sure the attic floor joists are designed for the loads expected. Use the same framing procedure as for other stair openings. One common size of folding stairs requires only a 26 x 54-inch rough opening.

Laying Out a Stair Carriage

Figure 20-9 shows dimensions for various heights of straight stairs. Notice that the riser height and tread width are given. Let's assume that you have already calculated the tread and riser dimensions. The next step is to establish the carriage dimensions.

Establishing carriage dimensions takes a lot of time. So when you select a piece of stock for the layout work, make sure the stock is long enough to do the job.

Figure 20-10 shows the difference between the mathematical carriage length and the actual carriage length. You can see why the actual length isn't known until the layout is completed.

The mathematical carriage length can be found easily with a steel square. On the body of the steel square, locate the total run (in feet). (Refer to the dimensions given in Figure 20-9.) On the tongue of the steel square, locate the floor-to-floor height (in feet). Measure the distance between these two points on the square. This distance will be the approximate carriage length.

Here's an example. Suppose the total run is 12'6'' and the floor-to-floor height is 9'6''. The distance between 12'6'' (on the body of the steel square) and 9'6'' (on the tongue of the square) measures about 14'3''. So we'll use a 16-foot board for the carriage. (The stock should be 18 to 20 inches longer than the carriage length.)

Start your layout by placing the carriage board on a pair of saw horses. The right-angle portion of the square should be toward you. On the body of the square, locate the number representing the tread width (10 inches in our example). On the tongue of the square, locate the number represen-

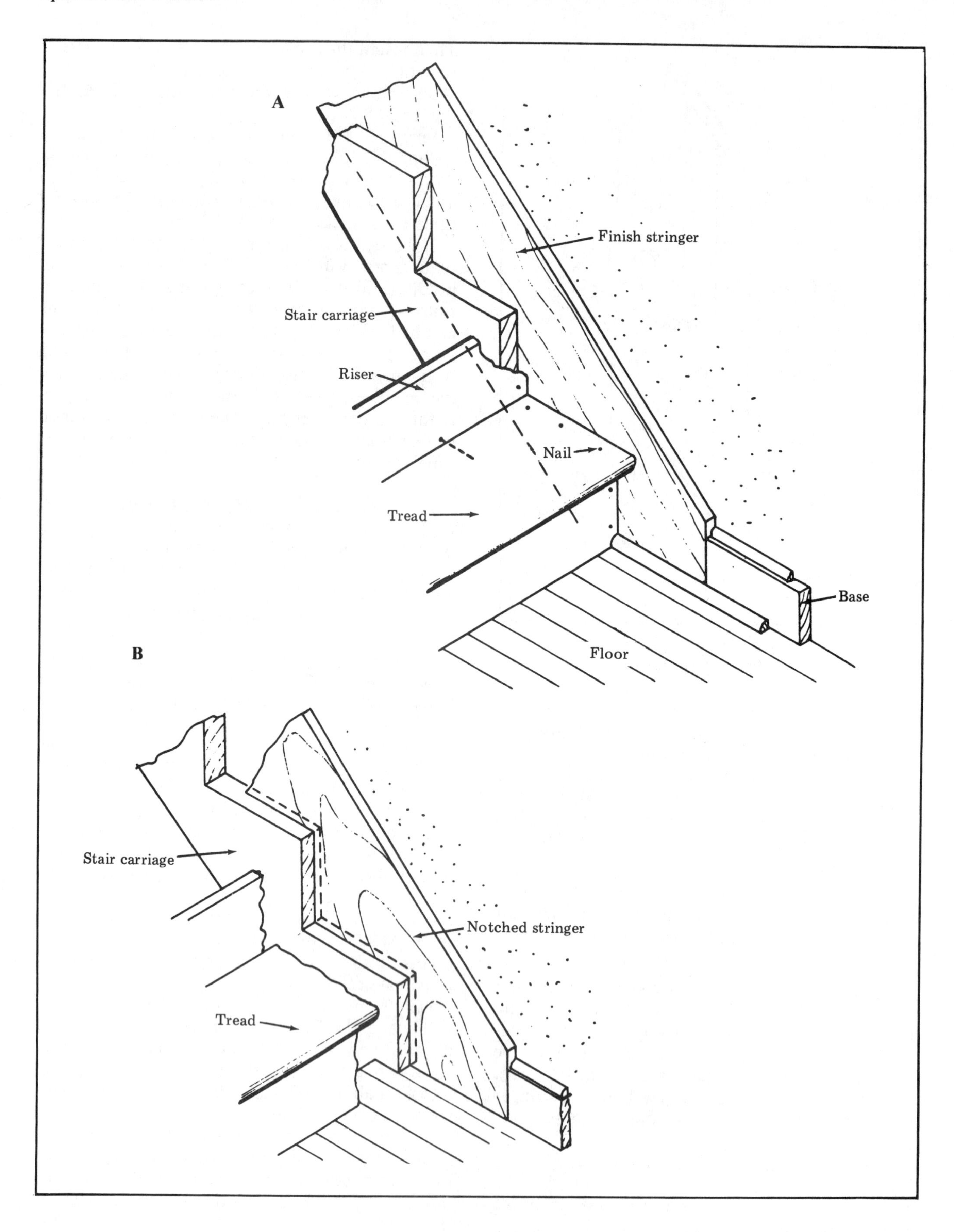

Attic stairs
Figure 20-8

Dimensions For Straight Stairs

Height Floor-to-Floor H	Number of Risers	Height of Risers R	Width of Treads T	Total Run L	Minimum Headroom Y	Well Opening U
8'-0"	12	8"	9"	8'- 3"	6'-6"	8'- 1"
8'-0"	13	7 3/8" +	9 1/2"	9'- 6"	6'-6"	9'- 2 1/2"
8'-0"	13	7 3/8" +	10"	10'- 0"	6'-6"	9'- 8 1/2"
8'-6"	13	7 7/8"—	9"	9'- 0"	6'-6"	8'- 3"
8'-6"	14	7 5/16"—	9 1/2"	10'- 3 1/2"	6'-6"	9'- 4"
8'-6"	14	7 5/16"—	10"	10'-10"	6'-6"	9'-10"
9'-0"	14	7 11/16" +	9"	9'- 9"	6'-6"	8'- 5"
9'-0"	15	7 3/16" +	9 1/2"	11'- 1"	6'-6"	9'- 6 1/2"
9'-0"	15	7 3/16" +	10"	11'- 8"	6'-6"	9'-11 1/2"
9'-6"	15	7 5/8"—	9"	10'- 6"	6'-6"	8'- 6 1/2"
9'-6"	16	7 1/8"	9 1/2"	11'-10 1/2"	6'-6"	9'- 7"
9'-6"	16	7 1/8"	10"	12'- 6"	6'-6"	10'- 1"

Dimensions shown under well opening "U" are based on 6'-6" minimum headroom. If headroom is increased well opening also increases.

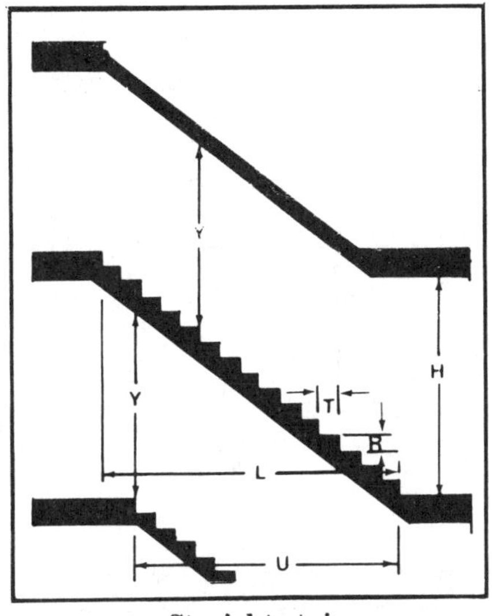

Straight stair

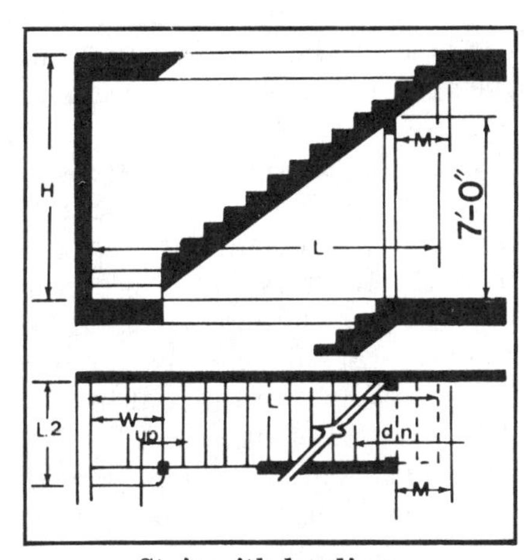

Stair with landing

Dimensions For Stairs With Landings

Height Floor to Floor H	Number of Risers	Height of Risers R	Width of Tread T	Run: Number of Risers	Run: L	Run: Number of Risers	Run: L2
8'-0"	13	7 3/8" +	10"	11	8'- 4" + W	2	0'-10" + W
8'-6"	14	7 5/16"—	10"	12	9'- 2" + W	2	0'-10" + W
9'-0"	15	7 3/16" +	10"	13	10'- 0" + W	2	0'-10" + W
9'-6"	16	7 1/8"	10"	14	10'-10" + W	2	0'-10" + W

Stairs with landings are safer and reduce the required stair space. The landing provides a resting point and a logical place for a right angle turn.

Stair dimensions
Figure 20-9

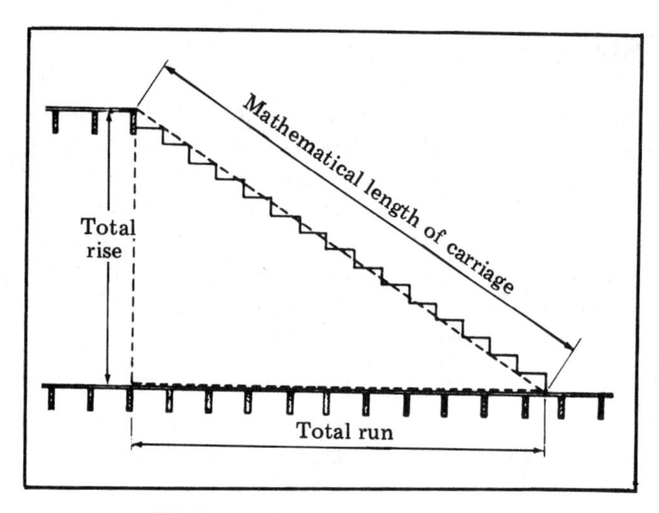

Mathematical length of carriage
Figure 20-10

ting the riser height (7⅛ inches for our example). Beginning at the end of the board, place the square so that these two numbers intersect the top edge of the board. Mark a line along the outside edges of the square. Then move the square to the next step. See Figure 20-11.

Continue laying out the steps until all treads and risers are marked. The number of treads is always one less than the number of risers. In our example, there are 16 risers. Number each riser so you know that the right number of risers have been laid out.

Next, cut the carriage board along the lines marked. When cutting out the first step, allow for the thickness of one tread. If the treads are 1½ inches thick, cut 1½ inches from the bottom of the carriage so the entire stairway is 1½ inches lower. See Figure 20-12.

Figure 20-12 also shows alternate methods of fastening the carriage board to the floor framing above. See Figure 20-12 A, B and C.

Estimating Materials

Once you know the stair dimensions, you can easily estimate the necessary materials. Figure 20-13, Stair Checklist and Data Reference Sheet, will help you create an organized list of materials.

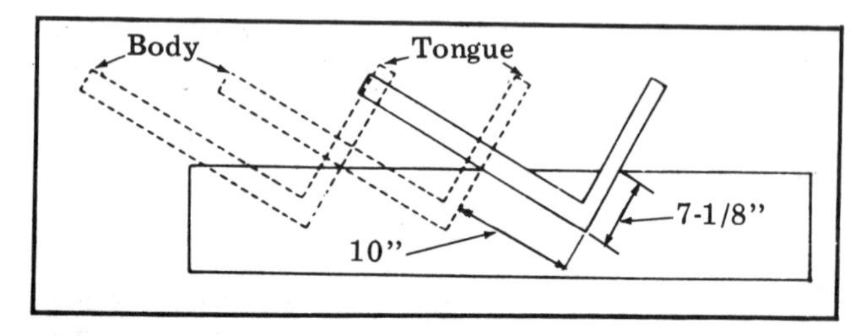

Laying out a carriage
Figure 20-11

Estimating Labor

Here are some guidelines for estimating labor on stair jobs:

Work	Unit	Manhours Per Unit
Erecting stairs, hours per 9' rise		
Building ordinary plain box stairs on the job	Ea.	8 to 16
Rails, balusters and newel posts for above	Ea.	4 to 8
Erecting plain flight of stairs built-up in shop	Ea.	6 to 8
Erecting two short flights	Ea.	10 to 12
Erecting open stairs	Ea.	10 to 12
Erecting open stairs with two flights	Ea.	12 to 16
Newels, balusters and handrails for the above	Ea.	6 to 8
Erecting prefabricated wood stairs, hours per 9' rise		
Circular, 6' diameter, oak	Ea.	23.0
Circular, 9' diameter, oak	Ea.	31.0
Straight, 3' wide, assembled	Ea.	3.0
Straight, 4' wide, assembled	Ea.	3.2

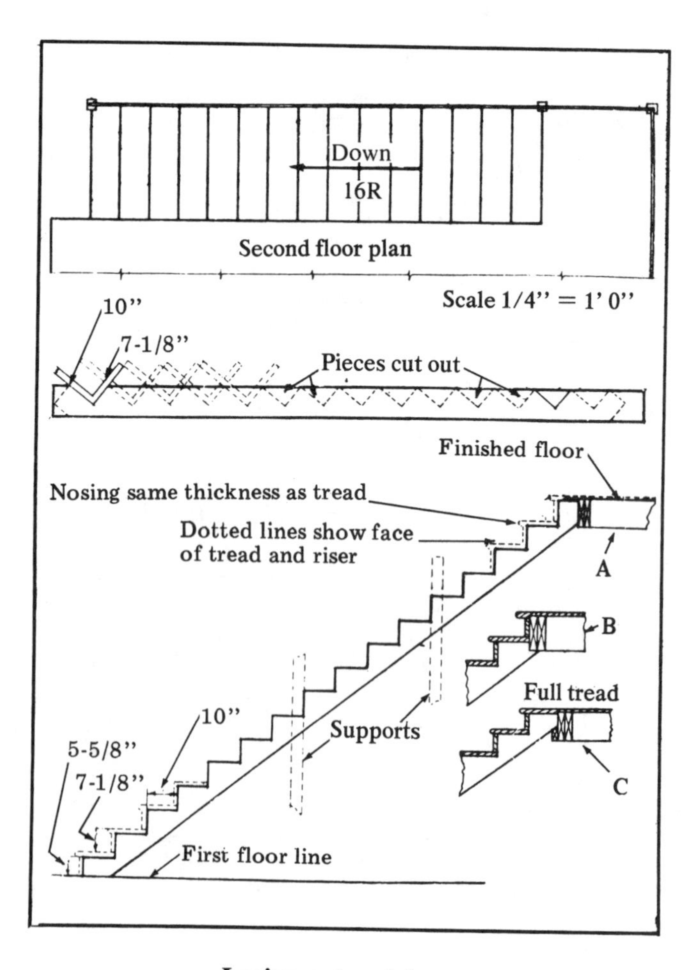

Laying out a stairway
Figure 20-12

Summary

Stair construction is often left to the most experienced carpenter on the job. Yet, most journeymen carpenters would agree that more labor and materials have been wasted on tearing down poorly designed or poorly built stairways than on any other framing job. No matter how experienced the craftsman, each stairway presents its own design and construction problems.

Even in highly repetitive jobs where many similar stairways are constructed on one site, each stairway must be measured accurately. The craftsman must select the right tread and riser combination so that all the rises and all the runs are exactly (to within 1/4-inch) the same size. This isn't easy when plans are inadequate or when the actual floor-to-floor dimensions aren't what the plans indicate. On many jobs, the craftsman who builds the stairway has to design it first.

Stair Checklist and Data Reference Sheet

1) The house is: ☐1-story ☐1½-story ☐2-story.

2) Stairs required: ☐basement ☐main ☐attic ☐attic folding ☐exterior

3) Stair type:

☐Long "L" ☐main ☐basement ☐exterior

☐Narrow "U" ☐main ☐basement ☐exterior

☐Winders ☐main ☐basement ☐exterior

☐Landing ☐main ☐basement ☐exterior

4) Stair dimensions:

☐Floor-to-floor: ☐main ____________ ☐basement ____________

☐Number of risers: ☐main ____________ ☐basement ____________

☐Height of risers: ☐main ____________ ☐basement ____________

☐Width of treads: ☐main ____________ ☐basement ____________

☐Total run: ☐main ____________ ☐basement ____________

☐Minimum headroom: ☐main ____________ ☐basement ____________

☐Well opening: ☐main ____________ ☐basement ____________

5) Number of carriages or stringers:

☐Main

☐Two ☐three

☐Basement

☐Two ☐three

☐Exterior

☐Two ☐three

6) Number of handrails:

☐Main

☐One ☐two

☐Basement

☐One ☐two

Figure 20-13

☐Exterior

☐One ☐two

7) Main stairway is:

☐Open

☐Railing

☐Balusters

☐Newel post

☐Closed

☐Handrail

8) Material costs:

Carriages:					
2 x 10	________LF	@	$______LF	=	$________
2 x 12	________LF	@	$______LF	=	$________
Stringers:					
25/32"	________LF	@	$______LF	=	$________
1 1/16"	________LF	@	$______LF	=	$________
Treads:					
1 1/16"	________LF	@	$______LF	=	$________
1 1/2"	________LF	@	$______LF	=	$________
1 5/8"	________LF	@	$______LF	=	$________
Risers:					
25/32"	________LF	@	$______LF	=	$________
1"	________LF	@	$______LF	=	$________
Handrail	________LF	@	$______LF	=	$________
Newel posts	________ea	@	$______ea	=	$________
Balusters	________ea	@	$______ea	=	$________
Molding	________LF	@	$______LF	=	$________
Nails	________lbs	@	$______lb	=	$________

Framing and bracing:

2 x 8 ________LF @ $________LF = $________

2 x 6 ________LF @ $________LF = $________

2 x 4 ________LF @ $________LF = $________

Prefabricated stairs $________

Other materials, not listed above:

________________ ________ @ $________ = $________

________________ ________ @ $________ = $________

________________ ________ @ $________ = $________

________________ ________ @ $________ = $________

Total cost of all materials $________

9) Labor costs:

Skilled ________hours @ $________hour = $________

Unskilled ________hours @ $________hour = $________

Other (specify)

________________ ________hours @ $________hour = $________

Total $________

10) Total cost:

Materials $________

Labor $________

Other $________

Total $________

11) Problems that could have been avoided ________________

____

____

____

____

12) Comments ____

____

____

____

____

13) A copy of this Checklist and Data Reference Sheet has been provided:

☐ Management

☐ Accounting

☐ Estimator

☐ Foreman

☐ File

☐ Other: ____________

Chapter 21

Fireplaces and Chimneys

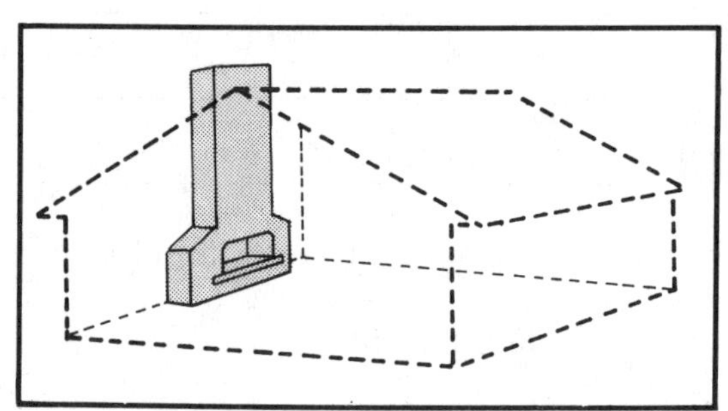

The prefabricated fireplace is becoming increasingly popular. It's energy-efficient, easy to install, and can be finished with the same mantel, hearth and chimney as a masonry fireplace. Best of all, the cost may be just a fraction of the cost of a conventional fireplace.

Let's look at fireplace and chimney construction details.

Fireplace Construction

Figure 21-1 shows a masonry fireplace from footing to top cap. The design of the fireplace is critical. A poorly-designed fireplace won't "draw" properly. It will spew smoke into the room, and it won't give off much heat.

Figure 21-2 shows the features of a well-designed fireplace and provides a table giving the dimensions. The left columns of the table give the more common sizes of fireplace openings. Select any size opening, and then read from left to right to find the correct dimensions.

In addition to fireplace design and dimensions, also pay careful attention to the fireplace opening, the throat, the smoke shelf, the smoke chamber, the hearth and the ash drop.

Fireplace opening— The fireplace opening should be rectangular and wider than it is high. Most fireplaces are from 36 to 42 inches wide. A typical height is 30 inches. A room with 300 square feet of floor space is served well by a fireplace 30 to 36 inches wide, 28 to 30 inches high, and 16 inches deep. For larger rooms, increase the width, but add very little to the height or depth.

There's an important relationship between the size of the fireplace opening and the size of the flue. A flue that is too small can't carry away all the smoke and gases from a large fireplace. It's better to have a flue that's too large than one that's too small. Use the table of dimensions shown in Figure 21-2 to select the right size fireplace opening and flue.

Throat— The throat is the narrow opening between the fireplace and the smoke chamber, as shown in Figure 21-2. Proper draft and heat production depend on the size relationship between the fireplace, throat and flue. If an oversize flue is brought straight down (without a longitudinal narrowing at the throat) to the fireplace opening, the fireplace will have poor draft and little heat production.

The area of the throat should not be less than the cross-section area of the flue. The throat should not be more than 8 to 12 inches wide. The length of the bottom of the throat should be equal to the width of the fireplace opening. The sides of the fireplace opening are vertical up to the throat. Above the throat, the sides taper to the size of the flue. The throat contains the damper and should be as near the front of the fireplace as possible. Locate the throat two to four brick courses above the fireplace.

Smoke shelf— The smoke shelf is formed by extending the brickwork at the top of the throat back to the line of the flue wall. The depth of the smoke shelf depends on the depth of the fireplace, but the shelf should not be less than 4 inches. The purpose

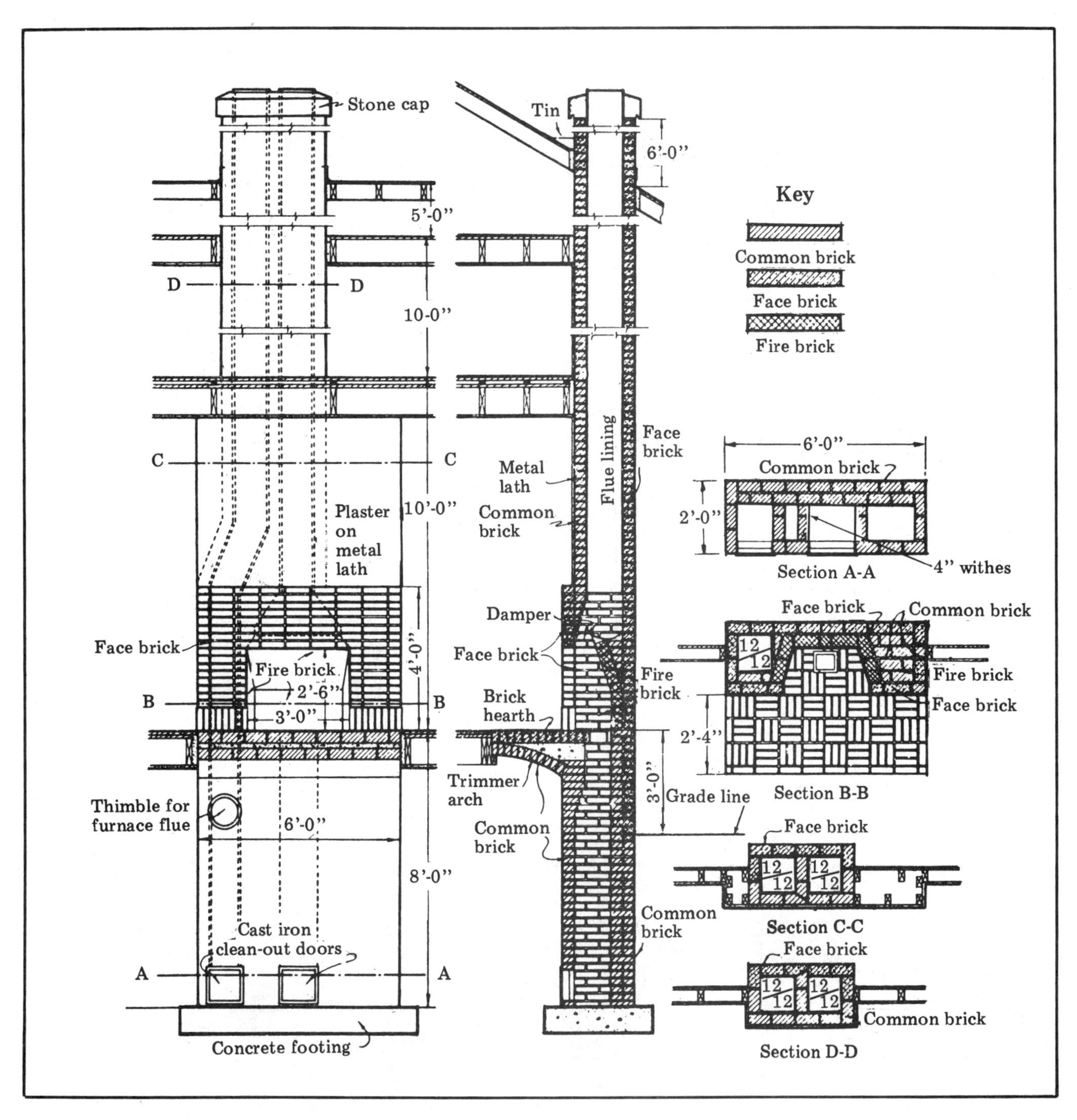

Masonry fireplace
Figure 21-1

of the smoke shelf is to change the direction of the down draft so that smoke and hot gases can rise without obstruction.

In Figure 21-3, the arrows marked "A" designate the path of smoke and hot gases through the fireplace, through the throat and smoke chamber, and on up the flue. The arrows marked "B" indicate down drafts, and show how the down drafts are turned upward by the smoke shelf.

Smoke chamber— Figures 21-2 and 21-3 show that the smoke chamber is the space between the top of the throat and the bottom of the flue. The chamber is larger at the bottom than at the top. This shape forms a reservoir which will hold the smoke in case the draft is cut off by a gust of wind. Chamber sides should have a slope of about 7 inches per foot of rise. If the sides are sloped too abruptly, the smoke can't rise properly.

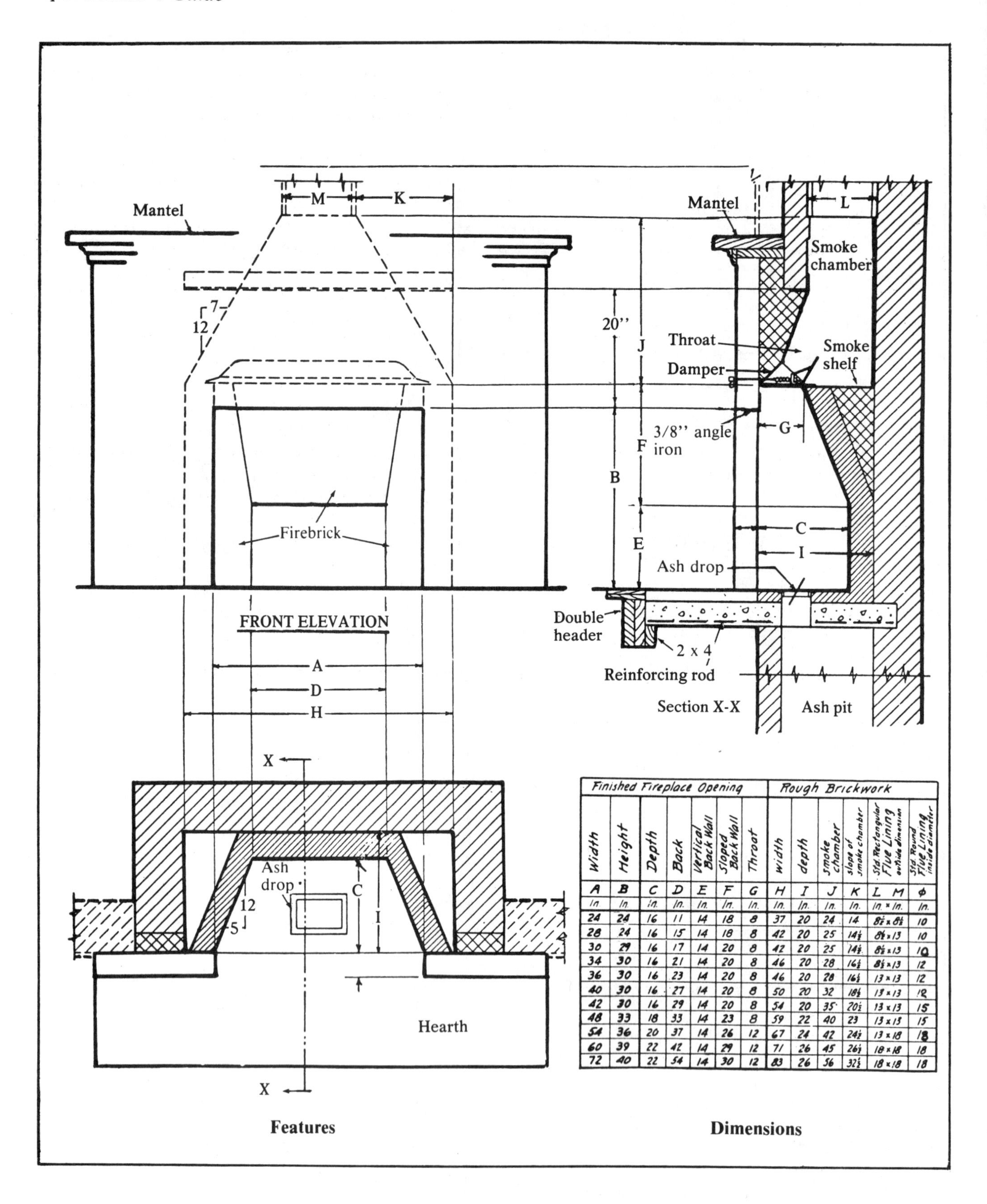

Finished Fireplace Opening							Rough Brickwork					
Width	Height	Depth	Back	Vertical Back Wall	Sloped Back Wall	Throat	width	depth	smoke chamber	slope of smoke chamber	Std. Rectangular Flue Lining outside dimension	Std. Round Flue Lining inside diameter
A	B	C	D	E	F	G	H	I	J	K	L M	ϕ
In.	In.	In.	In.	In.	In.	In.	In.	In.	In.	In.	In. × In.	In.
24	24	16	11	14	18	8	37	20	24	14	8½ × 8½	10
28	24	16	15	14	18	8	42	20	25	14½	8½ × 13	10
30	29	16	17	14	20	8	42	20	25	14½	8½ × 13	10
34	30	16	21	14	20	8	46	20	28	16½	8½ × 13	12
36	30	16	23	14	20	8	46	20	28	16½	13 × 13	12
40	30	16	27	14	20	8	50	20	32	18½	13 × 13	12
42	30	16	29	14	20	8	54	20	35	20½	13 × 13	15
48	33	18	33	14	23	8	59	22	40	23	13 × 13	15
54	36	20	37	14	26	12	67	24	42	24½	13 × 18	18
60	39	22	42	14	29	12	71	26	45	26½	18 × 18	18
72	40	22	54	14	30	12	83	26	56	32½	18 × 18	18

Fireplace features and dimensions
Figure 21-2

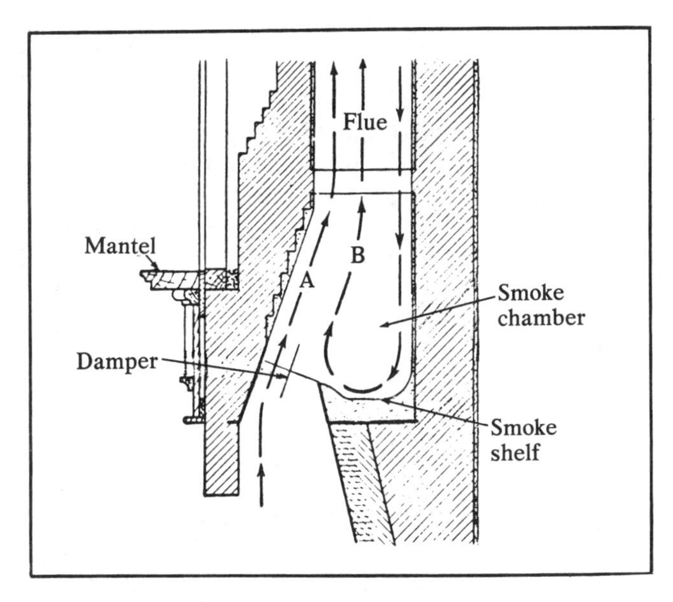

Down drafts and smoke shelf
Figure 21-3

Hearth— The hearth is the part of the fireplace in front of the opening. It's usually at the level of the floor and is built of fireproof material, such as tile, stone, brick or concrete. Some modern hearths are raised 10 inches or more above the floor line. It should extend 16 or 20 inches in front of the fireplace to protect the floor from sparks and embers.

Ash drop— Most owners appreciate having an ash drop at the back of the fireplace. It consists of a cast-iron metal frame with a pivoting cover. Ashes are brushed through the ash drop and into an ash pit below. The ash pit should be made of tight masonry and fitted with a clean-out door. The floor should be concrete.

Chimney Construction

A chimney's main purpose is to create a draft. Draft is caused by the difference in weight between the heated air in the chimney and the relatively cooler air outside. As the heat increases, air expands and becomes lighter. The pressure at the base of the flue becomes unequal and the lighter gases rise. Suction draws air into the fireplace and promotes combustion.

Important chimney construction details include: chimney height, flue construction, the foundation and the chimney cap.

Chimney height— The chimney must extend 3 feet above flat roofs. It must extend at least 2 feet above the roof ridge or any raised part of the roof within 10 feet of the chimney. See Figure 21-4.

Flue construction— The flue should be as vertical as possible and have no internal obstructions. Chimney flues work best when lined with a clay flue lining. Most codes require liners. Fire clay flue lining is made of a special mixture of clays that have been thoroughly burned to withstand excessive changes in temperature.

Don't use the same flue for both a fireplace and a furnace. If more than one heating unit is connected to the same flue, the draft for each unit is reduced to half.

The size of the flue lining depends on the draft needed. As a rule, a fireplace flue lining should be 1/10 to 1/12 the area of the fireplace opening.

For example, a fireplace opening 42 inches wide and 30 inches high has an area of 1,260 square inches. A 12-inch circular flue lining has an actual inside area of 113 square inches so it is satisfactory for the 42 x 30-inch fireplace opening.

A 13 x 13-inch flue lining (outside measurement) has a gross inside area of 127 square inches. But the effective inside area would be only 100 square inches. That's not enough. A square or oblong flue isn't effective over its entire cross-sectional area, because the column of smoke doesn't fill the corners. Gases and smoke rise with a circular swirling motion. A round flue is the most efficient because it offers less resistance to the rising column of smoke.

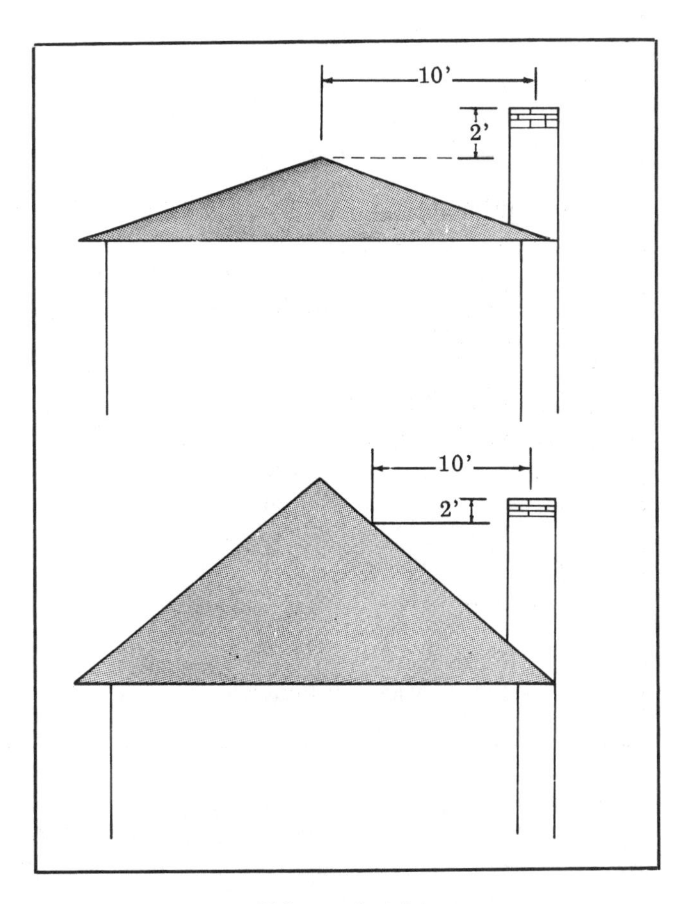

Chimney height
Figure 21-4

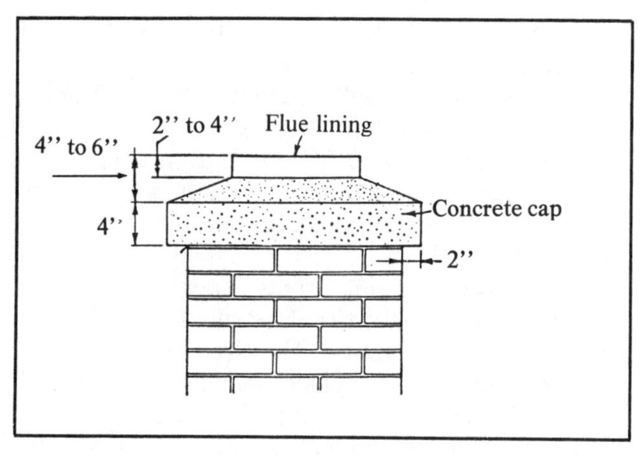

Chimney cap
Figure 21-5

Foundation— Every chimney needs a good foundation. The chimney weight should not be supported by any part of the building (except by properly corbeled walls). The foundation and footings must be proportioned to the weight of the chimney and to the safe bearing capacity of the soil. The footing is usually 6 inches wider than the chimney (on all sides of the chimney) and 8 to 12 inches deep. But note that the footing must always go below the frost line.

Chimney cap— To prevent weathering caused by rain and changing temperatures, cap the top of the chimney with stone or concrete. Figure 21-5 shows a typical cap for a one-flue chimney. Note that the flue lining extends above the top of the concrete cap by 2 to 4 inches. The flue lining is surrounded with mortar about 2 inches thick. Slope this mortar from the sides of the flue lining to the edges of the concrete.

Estimating Materials

Figure 21-6 shows different sizes of chimneys and the number of bricks required per foot of height for each size. This table assumes a 1/2-inch mortar joint. For example, if your chimney is the size shown in "h" and is 20 feet high, you need 1160 bricks.

Standard-size bricks for the fireplace hearth are estimated at 6.5 bricks per square foot when laid on edge. Use 4 bricks per square foot when bricks are laid flat.

It normally takes 5 or 6 bags of mortar to lay 1,000 bricks. Three cubic feet of sand are required per bag of mortar. Allowing for waste, usually one cubic yard (approximately 2,550 lbs.) of sand is required for 10 sacks of mortar.

Rectangular flue lining comes in 2-foot lengths and in sizes from 8 x 8 inches to 24 x 24 inches. Some codes do not require flue lining when the chimney is made of masonry at least 8 inches thick. But the cost of the extra brick and the labor involved is probably greater than the cost of the flue lining. And flue lining makes a safer chimney.

Fireplace mantels are available in prefabricated units or can be constructed from various sizes of stock material.

Angle iron is used to support brick or masonry over the fireplace opening. The bottom of the inner hearth and the sides and back of the fireplace are built of firebrick. Mix fire clay with mortar when laying firebrick. Follow the fire clay manufacturer's instructions for the mix ratio.

Use Figure 21-7, Fireplace and Chimney Checklist and Data Reference Sheet, to compile your total list of materials.

Estimating Manhours

Not every mason is a chimney master. Fireplace and chimney construction is exacting work. Be sure the mason is qualified in this craft.

The mason will often contract to furnish the labor, scaffolding, and tools. The mason will determine his cost based on the number and type of bricks, the style of fireplace and chimney, the height of the chimney and any special work or conditions. You may furnish only the materials. The approximate manhours required are:

Per 1,000 Bricks	Mason	Labor
Common brick chimneys	16	16
Common brick fireplaces	20	20
Brick veneer fireplaces	24	24
Washing (cleaning) per 100 SF	1	1

Labor for installing fireplace mantels is:

Work Element	Unit	Manhours Per Unit
Prefabricated milled decorative unit, 42" high, 6' wide	Ea	3.2
Bracket mounted 10" wide, 3"-thick hardwood beam		
6' long	Ea	1.7
8' long	Ea	2.0
Rough sawn oak or pine beam		
4" x 8"	LF	0.25
4" x 10"	LF	0.27
4" x 12"	LF	0.31

Time includes layout, cutting, drilling and placing of shields where required, repairs and clean-up.
Suggested Crew: 1 carpenter

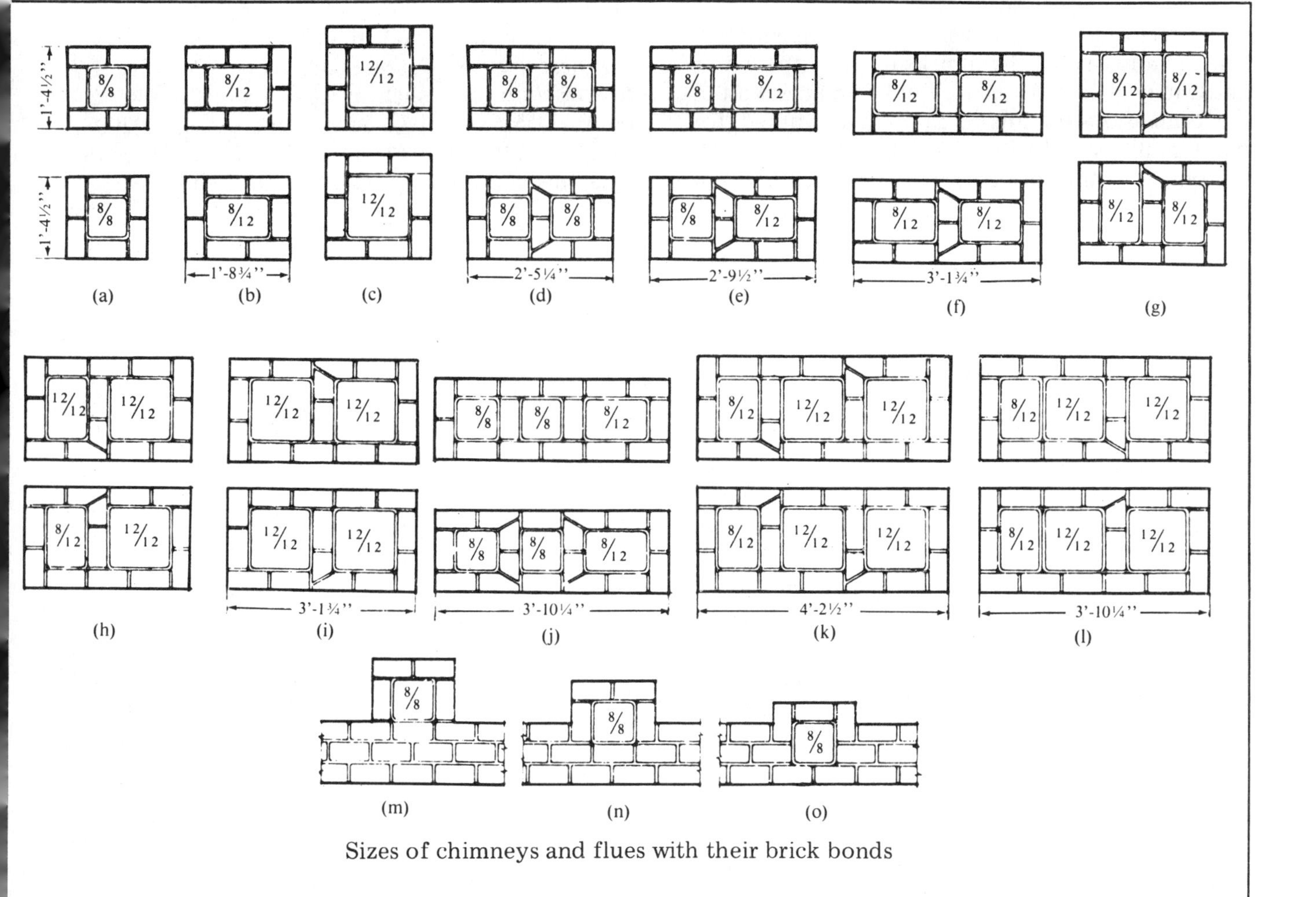

Sizes of chimneys and flues with their brick bonds

Size and Number of Flues	Number of Brick	Cubic Feet Mortar
(a) 1 - 8" x 8" flue	27	0.5
(b) 1 - 8" x 12" flue	31	0.5
(c) 1 - 12" x 12" flue	35	0.6
(d) 2 - 8" x 8" flue	46	0.8
(e) 1 - 8" x 8" and 1 - 8" x 12" flue	51	0.9
(f) 2 - 8" x 12" flue	55	0.10
(g) 2 - 8" x 12" flue	53	0.9
(h) 1 - 8" x 12" and 12" x 12" flue	58	1.0
(i) 2 - 12" x 12" flue	62	1.1
(j) 2 - 8" x 8" and 1 - 8" x 12" flue	70	1.2
(k) 1 - 8" x 12" and 2 - 12" x 12" flue	83	1.4
(l) 1 - 8" x 12" and 2 - 12" x 12" flue	70	1.2
(m) 1 - 8" x 8" extending 12" from face of wall	18	0.4
(n) 1 - 8" x 8" extending 8" from face of wall	9	0.3
(o) 1 - 8" x 8" extending 4" from face of wall	0	0.0

Number of brick required in chimneys per foot in height

Chimney sizes and brick requirements
Figure 21-6

Summary

The secret of a good fireplace is in the dimensions. Change any one of the dimensions and the whole battle may be lost. Fireplaces and chimneys built to the dimensions in Figure 21-2 have worked for many builders and will work for you.

Consider using prefabricated fireplace units in your spec houses. The units are economical and highly efficient. Detailed installation instructions for these units are found in *Manual of Professional Remodeling,* published by Craftsman Book Company, 6058 Corte del Cedro, Carlsbad, CA 92008. Most manufacturers of prefabricated fireplaces include installation instructions with their units.

Fireplace and Chimney Checklist and Data Reference Sheet

1) The house is ☐1-story ☐1½-story ☐2-story.

2) The fireplace is ☐masonry ☐prefabricated.

3) Fireplace-opening dimensions (refer to Figure 21-2):

(Inches)	Typical Size (Inches)
☐Width (A): ____________	36
☐Height (B): ____________	30
☐Depth (C): ____________	16
☐Back (D): ____________	23
☐Vertical back wall height (E): ____________	14
☐Sloped back wall height (F): ____________	20
☐Throat (G): ____________	8

4) The chimney has:

☐1 flue ☐2 flues ☐3 flues

☐Flue lining or ☐8-inch-thick masonry without linings

☐Ash pit and ☐ash clean-out door

☐Cap

5) The chimney height is ______ feet ______ inches.

(The top of the flue liner should not be less than 2 feet above the ridge or raised part of the roof within 10 feet of the chimney. The flue liner must extend 3 feet above flat roofs.)

6) Material costs:

Prefab pipes	__________pcs	@	$__________ea	=	$____________
Bricks	__________pcs	@	$______Per M	=	$____________
Firebricks	__________pcs	@	$__________ea	=	$____________
Mortar	________bags	@	$________bag	=	$____________
Sand	______cu. yds	@	$______cu. yd.	=	$____________
Angle iron (size: ______)	__________pcs	@	$__________ea	=	$____________
Dampers (size: ______)	__________pcs	@	$__________ea	=	$____________

Figure 21-7

Ash drops ________pcs @ $________ea = $________

Clean-out doors ________pcs @ $________ea = $________

Flue linings (size: ______) ________pcs @ $________ea = $________

Muriatic acid ________gals @ $________gal = $________

Brushes ________pcs @ $________ea = $________

Mantels:

Prefab (size: ______) ________pcs @ $________ea = $________

Pine (size: ______) ________pcs @ $________ea = $________

Hardwood (size: ______) ________pcs @ $________ea = $________

Other materials, not listed above:

____________________ ________ @ $________ = $________

____________________ ________ @ $________ = $________

____________________ ________ @ $________ = $________

____________________ ________ @ $________ = $________

Total cost of all materials $________

7) Masonry work performed by:

☐Contract: $________

Name: ____________________

Address: ____________________

Telephone: ____________________

My crew:

Skilled ________hours @ $________hour = $________

Unskilled ________hours @ $________hour = $________

8) Labor for installing prefab fireplace unit:

Skilled ________hours @ $________hour = $________

Unskilled ________hours @ $________hour = $________

9) Labor costs:

Skilled ________hours @ $________hour = $________

Unskilled ________hours @ $________hour = $________

Other (specify)

____________________ ________hours @ $________hour = $________

Total $________

10) Total cost:

Materials $____________________

Labor $____________________

Other $____________________

Total $____________________

11) Problems that could have been avoided ____

12) Comments ____

13) A copy of this Checklist and Data Reference Sheet has been provided:

☐ Management

☐ Accounting

☐ Estimator

☐ Foreman

☐ File

☐ Other: ____________________

Chapter 22

Interior Walls, Ceilings and Trim

For your spec houses, you'll want to use interior finish materials that are economical and easy to install. Sheet rock (gypsum wallboard) meets these requirements. Like B/B board (bath/backer board), plywood, hardboard and fiberboard, sheet rock is a drywall finish. At least 80% of all new homes use drywall construction.

In this chapter, we'll look at each of these interior finish materials in detail. We'll also cover interior trim (molding).

Sheet Rock

Sheet rock is an engineered building panel made of gypsum (hydrated calcium sulfate) and other materials. It's finished on both sides with special paper to provide a smooth and attractive wall finish. Sheet rock is noncombustible, nontoxic, dimensionally stable and resists sound transmission. And it's the least expensive wall-surfacing material offering these advantages.

Sheet rock comes in several thicknesses and sizes. Use the 1/4-inch thickness over old walls and ceilings or in sound-rated systems, where the 1/4-inch thickness is used in combination with other gypsum board to reduce through-the-wall sound transmission. Use the 3/8-inch thickness for single-layer application over framing members spaced no more than 16 inches o.c. Framing-member spacing is also 16 inches o.c. when 1/2-inch and 5/8-inch thicknesses are installed vertically (with the long dimension parallel to the framing). When installed horizontally (with the long dimension at right angles to the framing), maximum framing-member spacing is 24 inches o.c.

When a ceiling has to support insulation or is going to be sprayed with a water-base textured coating, use either 1/2-inch or 5/8-inch sheet rock. The 1/2-inch thickness should be applied horizontally over framing members spaced 16 inches o.c. The 5/8-inch thickness can be applied horizontally over framing members spaced 16 or 24 inches o.c., or vertically over framing members spaced 16 inches o.c. Be sure to coat the sheet rock with a pigmented primer-sealer before the textured coating is applied.

Sheet rock requires special protection when exposed to continuous moisture or high humidity. In bathrooms and laundry rooms, for example, it's good practice to use B/B board.

Apply sheet rock finishes only when the air temperature is above 55 degrees. And the temperature must have been above 55 degrees for at least 24 hours beforehand. This may require special heating equipment. If the humidity is high, make sure there is adequate ventilation.

Installing Sheet Rock

Here's how to install sheet rock in your spec houses.

Framing members— Any 2 x 4 or larger framing can support sheet rock. But note that alignment is important. The wall will only be as straight as the studs behind it. Use only dry lumber for framing

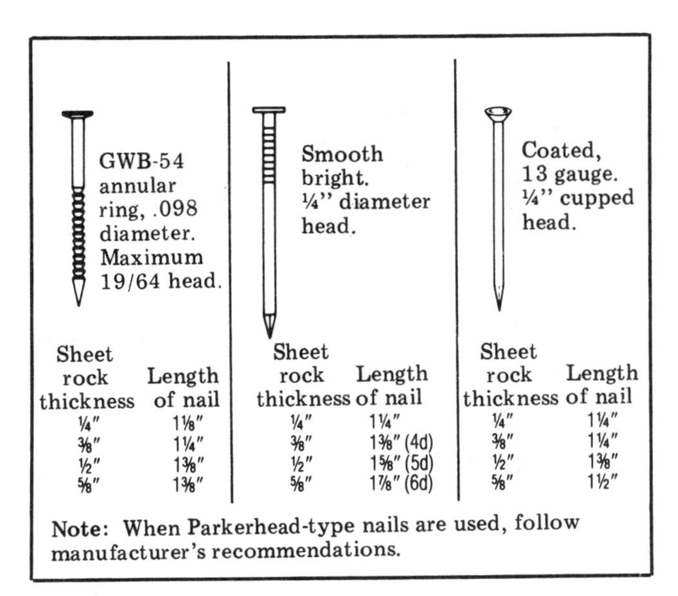

Nails recommended for sheet rock installation
Figure 22-1

members. Green lumber shrinks as it ages and will usually cause nails to pop out of the sheet rock.

Application methods— The four most common methods of applying sheet rock are: nailing, glue-nailing, "floating angles," and wood furring. Let's look at all four methods.

1) Nailing: Figure 22-1 shows nails recommended for sheet rock installation. Single nailing should be done as shown in Figure 22-2. Double nailing, shown in Figure 22-3, minimizes nail pops. Starting from the center of the panel and working toward the edge, nail the panel at intermediate studs 12 inches o.c. Each pair of nails should have about two inches between them, as shown in the figure. Nail as usual around the panel edges.

2) Glue-nailing: You can save time by using adhesive and nails to apply sheet rock. This method requires less nailing and less finishing. Apply the adhesive to studs and joists and allow it to set for a few seconds. Then place the panel into position. Drive the minimum number of nails

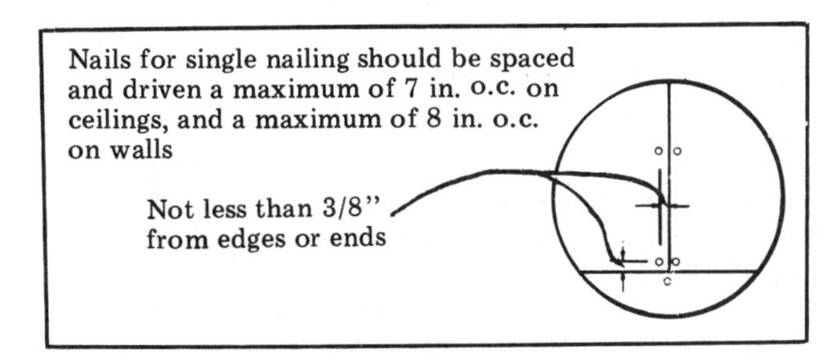

Single nailing
Figure 22-2

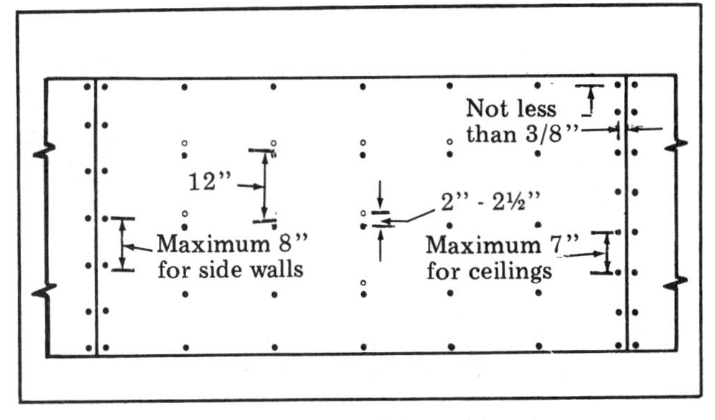

Double nailing
Figure 22-3

necessary to hold the panel in place until the glue dries. For side-wall applications, space nails 24 inches o.c. For ceiling applications, space nails 16 inches o.c. Adhesive application patterns are shown in Figure 22-4.

3) "Floating angles:" This method is designed to reduce the panel stress that often occurs when the framing settles. See Figures 22-5 and 22-6. Omit some of the nails at interior corners (where ceiling and side walls meet and also where side walls intersect). Then nail as usual for the rest of the ceiling and wall area. Always install the ceiling first, and be sure to fit the sheet rock snugly into all corners.

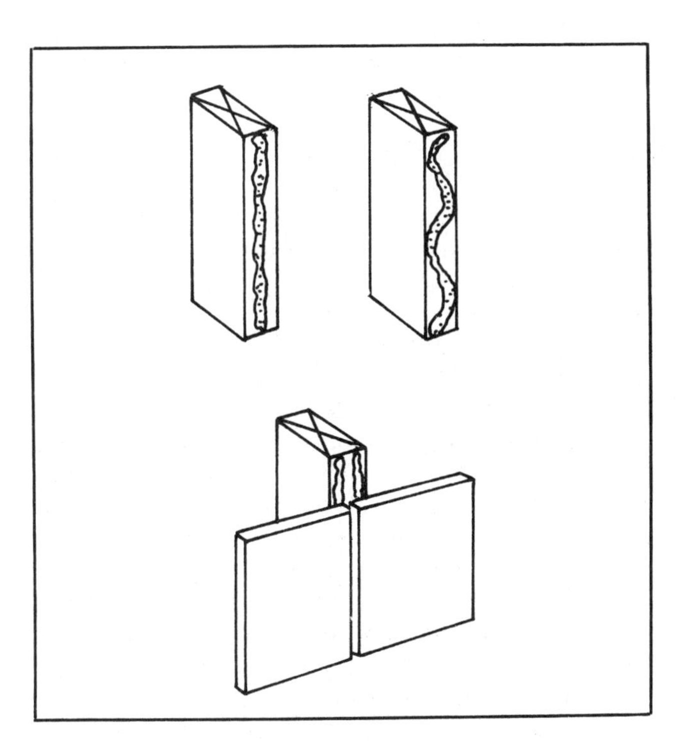

Adhesive application patterns
Figure 22-4

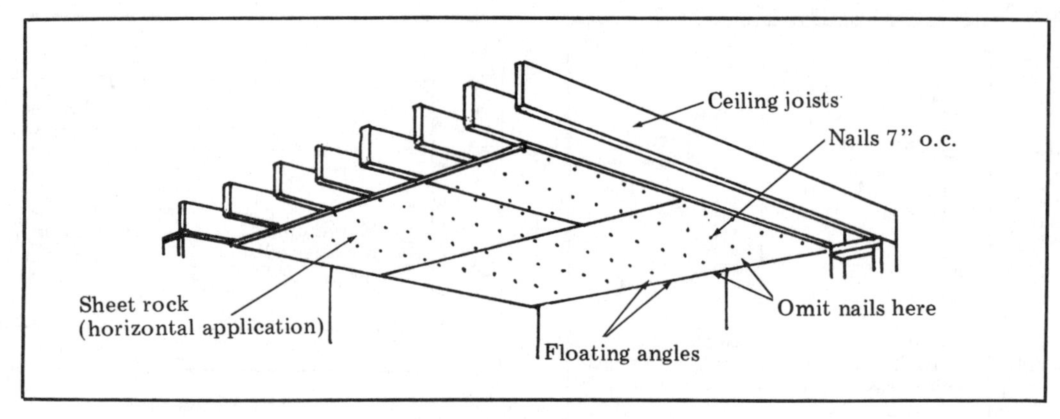

Ceiling application details
Figure 22-5

When ceiling sheet rock is applied horizontally, use ordinary nailing procedure where the panel ends abut the wall intersection. But along the long edges of the panel, the first row of nails should be set back about 7 inches from the wall intersection.

When ceiling sheet rock is applied with long edges parallel to the joists, again use ordinary nailing procedure where the long edges abut the wall intersection, but where panel ends meet the wall intersection, set the first row of nails back about 7 inches from the intersection.

The 7-inch border on each panel is allowed to "float" freely against the side walls. Any expansion or contraction is absorbed at the joint between wall and ceiling.

Side-wall sheet rock has to be in firm contact with the ceiling panels just applied. Wall panels actually help support the ceiling panels. Along the wall-ceiling intersection, omit nails directly below the ceiling angle. See Figure 22-6. The highest nail should be about 8 inches from the ceiling intersection.

Where side walls intersect, omit corner-nailing on panels applied to the first of the two intersecting walls. Don't drive any nails within 7 inches of the corner. Then hang the panels on the adjoining wall so that they overlap the corner panels just applied. Nail the adjoining-wall panels in the usual way, spacing nails 8 inches o.c. See Figure 22-6.

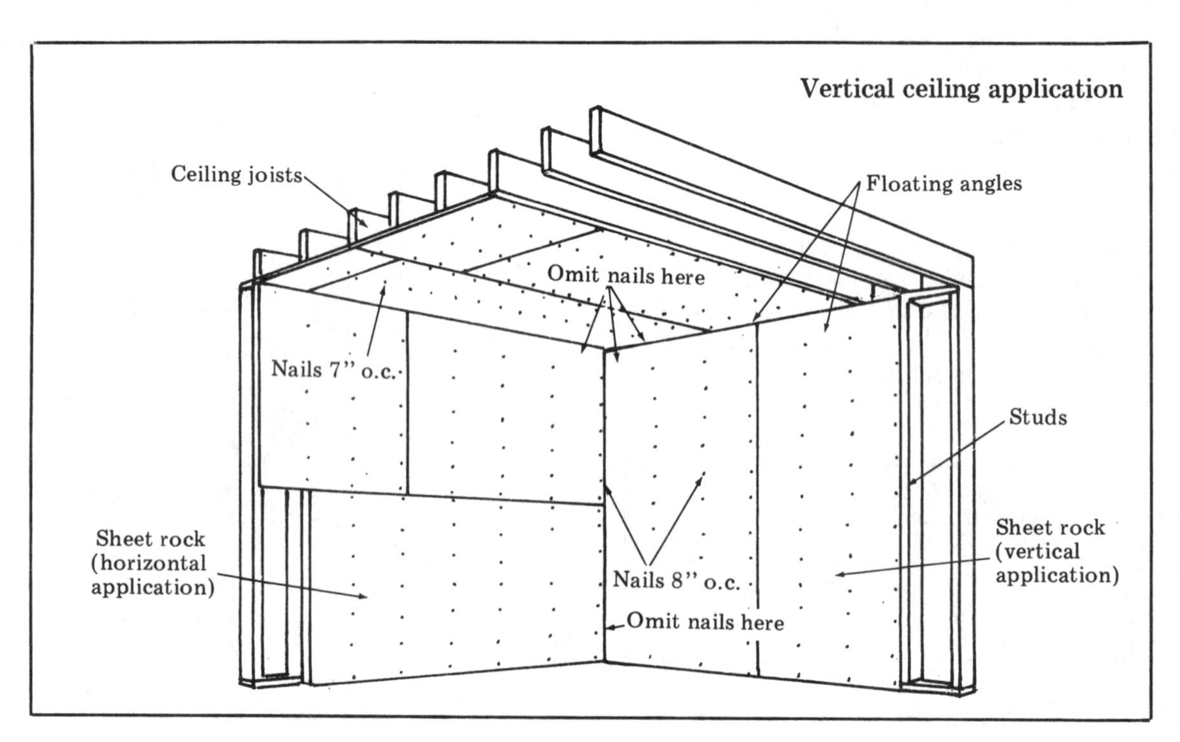

Side-wall application details
Figure 22-6

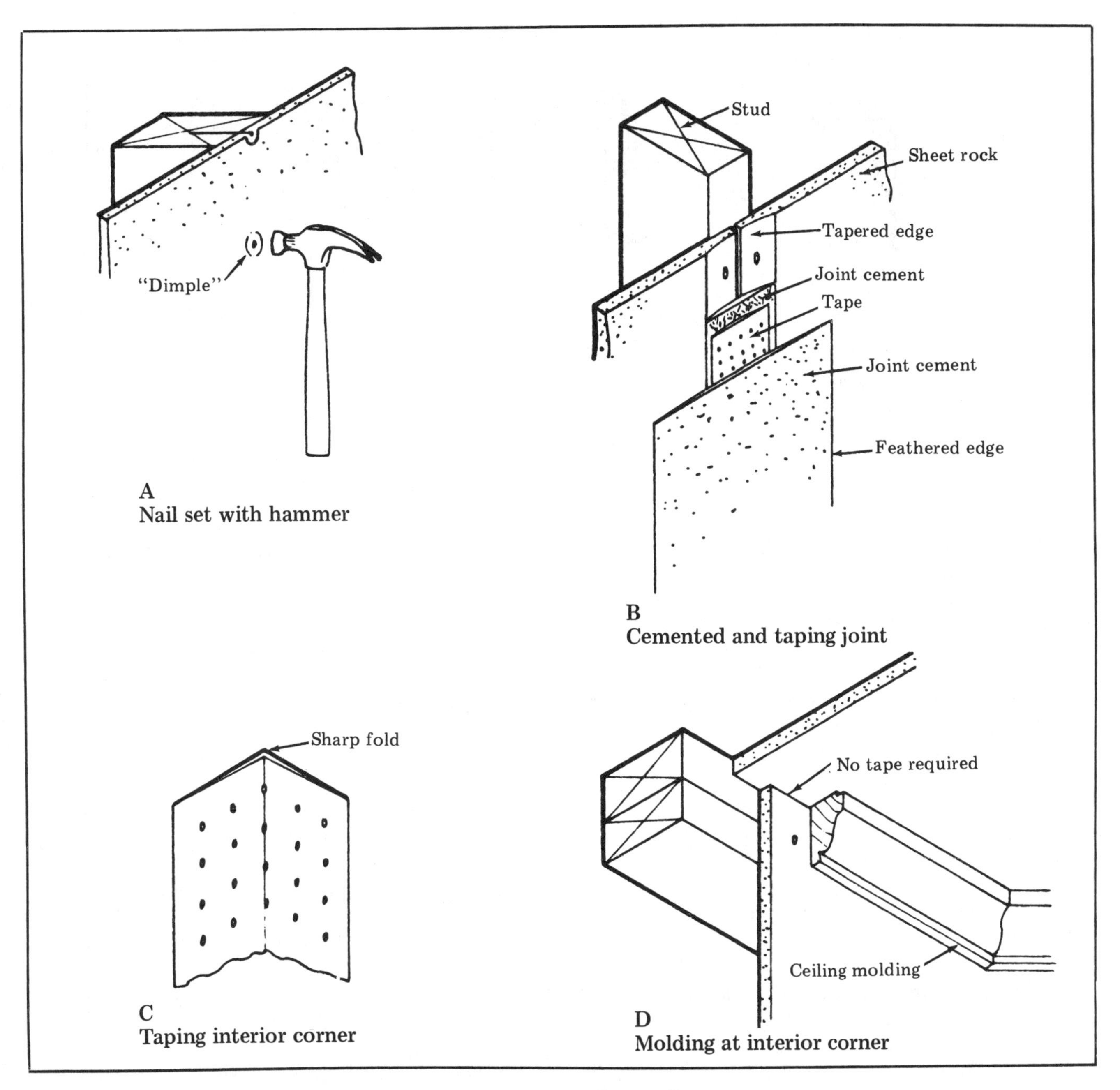

Sheet rock finishing details
Figure 22-7

4) Wood furring: When sheet rock is fastened to wood furring on ceilings, the furring should be at least nominal 1 x 3-inch boards spaced no more than 24 inches o.c. If insulation is going to rest on the ceiling panels, space framing members no more than 16 inches o.c. Apply the sheet rock panels at right angles to the framing.

Here's a tip that can reduce manhours spent on sheet rock application. Use the longest panels possible. This can cut your labor by about 10% on many jobs. And use horizontal application whenever possible.

Cementing and taping joints— When installing sheet rock, drive nails so that heads are set slightly below the panel surface. But don't use a nail set to drive the nails. Use a hammer, and note that the head of the hammer will make a small "dimple," as shown in Figure 22-7 A. Be careful you don't break the paper face on the sheet rock. Fill nail dimples with joint cement. When the cement is dry, sand it smooth. Repeat with a second coat of cement, if necessary.

Joints between panels are covered with joint cement and tape. This helps smooth and level the wall surface. The cement comes in powder form

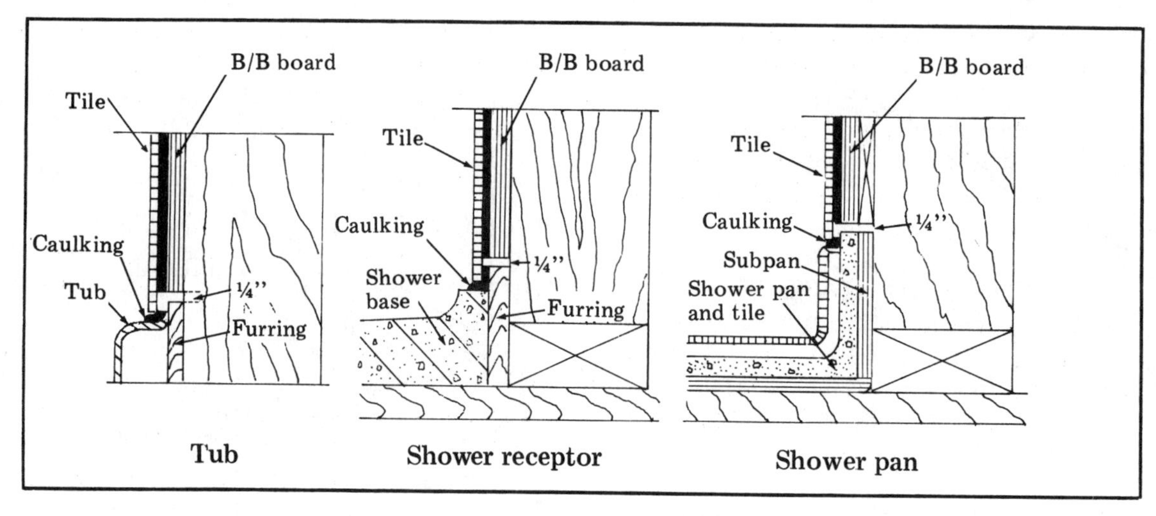

B/B board application details
Figure 22-8

that you mix with water to make a soft putty that can be spread with a putty knife. Joint cement is also available in premixed form. Here's the procedure for cementing and taping, as shown in Figure 22-7 B.

1) Starting at the top of the wall, use a 5-inch-wide spackling knife and spread the cement into the tapered edges at the joint.

2) Using the same knife, press tape into the recess until the joint cement is forced through the perforations.

3) Cover the tape with additional cement. Feather the edges.

4) Allow the cement to dry. Then apply a second coat of cement, and feather the edges so they extend beyond the edges of the first coat. A steel trowel works best for the second coat. On special jobs, use a third coat, and feather it beyond the second coat.

5) After the joint is dry, sand it smooth.

Interior corner joints can be finished with either tape or molding. If you use tape, fold a strip of tape down the center so it forms a sharp angle, as shown in Figure 22-7 C. Apply cement at the corner. Press the tape into place, and finish the corner with joint cement. When the cement is dry, sand it smooth. Then repeat the procedure with a second coat.

Figure 22-7 D shows an interior corner finished with molding. If you use molding, tape isn't necessary. At outside corners, use corner beads to prevent damage to the corners. Metal corner beads are fastened in place over the sheet rock and then covered with joint cement.

B/B Board

B/B board is used as a base for ceramic, metal and plastic wall tiles used in high-moisture areas, such as shower stalls, tub enclosures, bathrooms and laundry rooms.

B/B board has an asphalt-treated core and a heavy, water-repellent paper face. It doesn't need surface-sealing before the tile is applied, and joints don't have to be taped. If tile isn't applied over B/B board, the surface can be taped, finished and decorated like ordinary sheet rock. Hang B/B board with 1⅜-inch annular-ring nails. Don't use B/B board for ceilings or soffits.

Installing B/B Board

Before hanging the B/B board, you should install the tub, shower receptor or shower pan. Each of these should have a lip or flange that is 1 inch higher than the water drain or threshold of the shower. Let's look at the B/B board installation details shown in Figure 22-8.

1) To eliminate butt joints, apply B/B board horizontally. Leave a 1/4-inch gap between the lip of the fixture and the paper-bound edge of the board. Space nails 8 inches o.c. If using ceramic tile thicker than 5/16-inch, space nails 4 inches o.c. Dimple the nail heads, but don't tear the paper face.

2) In areas that are to be tiled, use a waterproof tile adhesive to treat joints and angles. Do *not* use regular joint cement and tape.

3) Use a waterproof nonsetting caulking compound to caulk openings around pipes and fixtures.

4) Apply tile down to the top edge of the shower floor surface, or to the return of the shower pan, or so that it extends over the tub lip. Make sure that all joints are completely grouted.

Plywood Paneling

Sheet rock is an excellent interior wall covering. But occasionally you'll want to use one or more of the other interior finish materials. Plywood paneling is a good choice for dens and family rooms, especially if you need to impress prospective buyers.

For 16-inch stud spacing, you'll need (minimum) 1/4-inch-thick panels. Use color-matched-panel nails spaced 8 inches o.c. The edge-nailing distance should be at least 3/8-inch. Butt panels together lightly. Before installation, plywood paneling should be exposed to the room conditions. Spot the panels around the room for at least 24 hours before installing them.

Hardboard and Fiberboard Paneling

Hardboard and fiberboard paneling are generally less expensive than plywood paneling.

You can use 1/4-inch-thick hardboard over studs spaced 16 inches o.c. But fiberboard (in tongue-and-groove plank or sheet form) must be 1/2-inch-thick when studs are spaced 16 inches o.c. For studs spaced 24 inches o.c., use 3/4-inch-thick fiberboard. Use casing or finishing nails slightly longer than you would use for 1/4-inch plywood or hardboard panels.

Fiberboard is also used as acoustic tile on ceilings and can be nailed to furring strips fastened to the ceiling joists. This system is similar to the suspended ceiling grids often found in commercial buildings.

Trim (Molding)

Carefully mitered corners on trim and molding make a favorable impression on many buyers. Some will judge the quality of your home by the appearance of the interior trim. If you're building quality homes for buyers who expect attention to detail, spend some time and a few dollars to make sure your trim measures up to standard.

On most jobs, you'll use unfinished softwood moldings of pine, fir, cedar or hemlock. Installing these moldings is called "trimming out" the house. It's usually the final step in construction. It may take one man a few hours or several days, depending on the amount and type of trim needed.

Moldings are strips of wood ripped from kiln-dried boards up to 16 feet long. There are about 30 different stock patterns or profiles. Each profile is designed for a specific purpose, but most have many secondary uses as well. Standard molding patterns and places where they can be used are shown in Figure 22-9.

Cutting and Fitting the Molding

Figure 22-10 shows how to make ceiling moldings. Here's how to miter, splice, cope, and install the molding.

Mitering— Most molding joints are cut on a 45-degree angle, as shown in A in the figure. Set your miter box accordingly, and trim (at opposite angles) each of the two pieces to be mated. For tight joints, use both glue and nails. When measuring moldings to be mitered, add the width of the molding to the length of each miter. If your molding is 3 inches wide and you have two miters, add 6 inches.

When mitering cove molding, put the molding in the miter box so that the molding is upside down and end-reversed to the way it will be installed.

Splicing— When splicing two lengths of molding, as shown in B, place the molding flat on its backside in the miter box. Miter (at identical 45-degree angles) the ends to be joined. When the ends are butted together in their correct position, you get a neat match without unevenness.

Coping— Place the molding in the miter box and position it upright against the backplate. Cut at a 45-degree angle. The cut exposes the profile of the molding. With the profile as a template, use the coping saw and follow along the profile. Trim away a wedge at another 45-degree angle. This duplicates the pattern so that it fits over the face of the adjoining molding. Coping is shown in C in Figure 22-10.

Installation— Figure 22-11 shows installation of base molding. Figure 22-12 shows installation of ceiling molding.

Finishing the Molding

Unfinished softwood molding will accept any wood finish: paint, lacquer, enamel, stain, shellac or varnish. But make sure the wood is smooth and clean. Fill nail holes with wood putty, and use #00 or finer sandpaper to sand with the grain. Primers are a must if you use paint or enamel. Prefinished moldings are available in many wood-grain color tones.

Estimating Materials

Estimate sheet rock, plywood and hardboard panel requirements by the room size. Multiply the wall length by the wall height. Then deduct for door and window openings, unless the paneling has to be cut out to maintain panel continuity. For ceiling re-

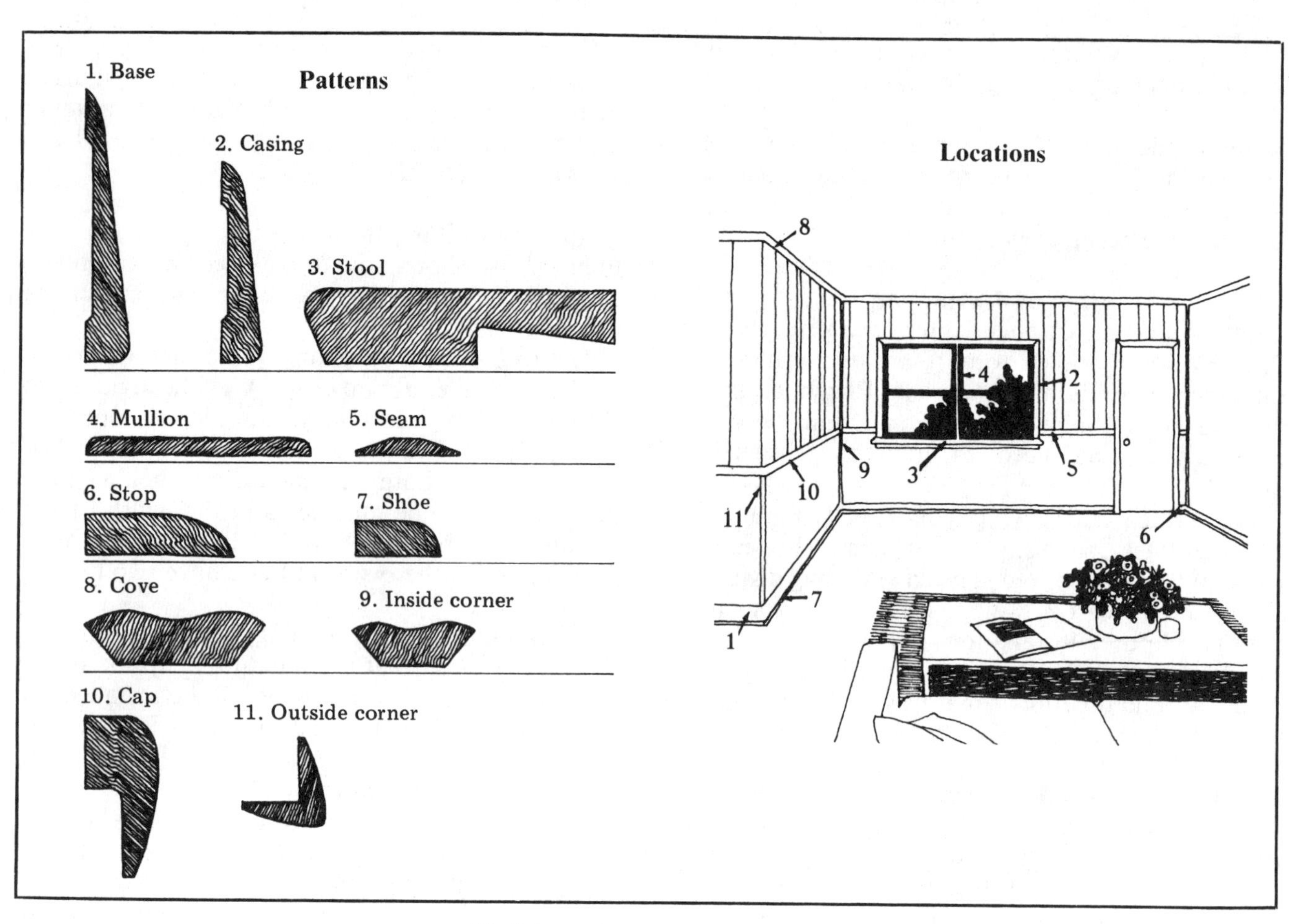

Molding patterns and locations
Figure 22-9

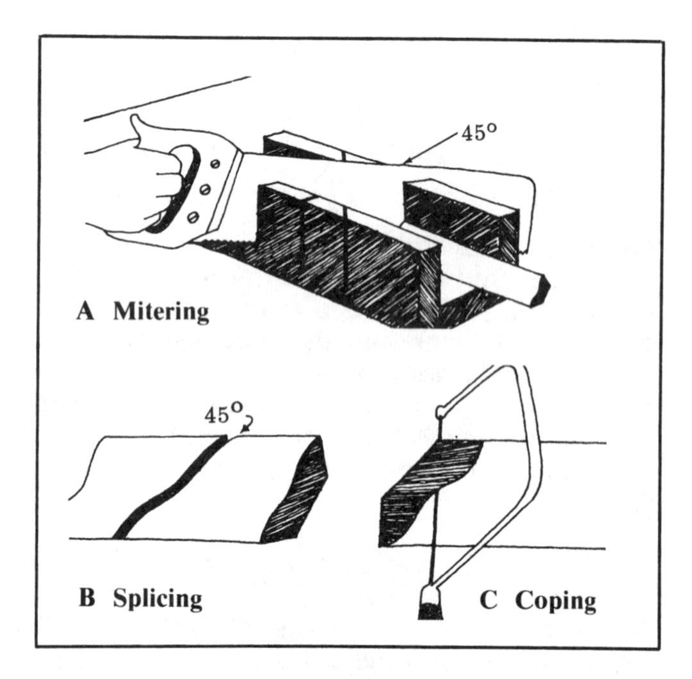

Cutting molding
Figure 22-10

quirements, multiply ceiling length by ceiling width.

Here's how to figure the material required for furring:

Size of Strips	Spacing of Furring	B.F. per S.F. of Wall/Ceiling	Lbs. Nail Per 1,000 B.F.
1" x 3"	16" o.c.	.21	37
1" x 3"	24" o.c.	.14	37
1" x 4"	16" o.c.	.28	30
1" x 4"	24" o.c.	.20	30

Use the floor plan to calculate the material needed for interior trim. Allow 10% for waste and end-cutting.

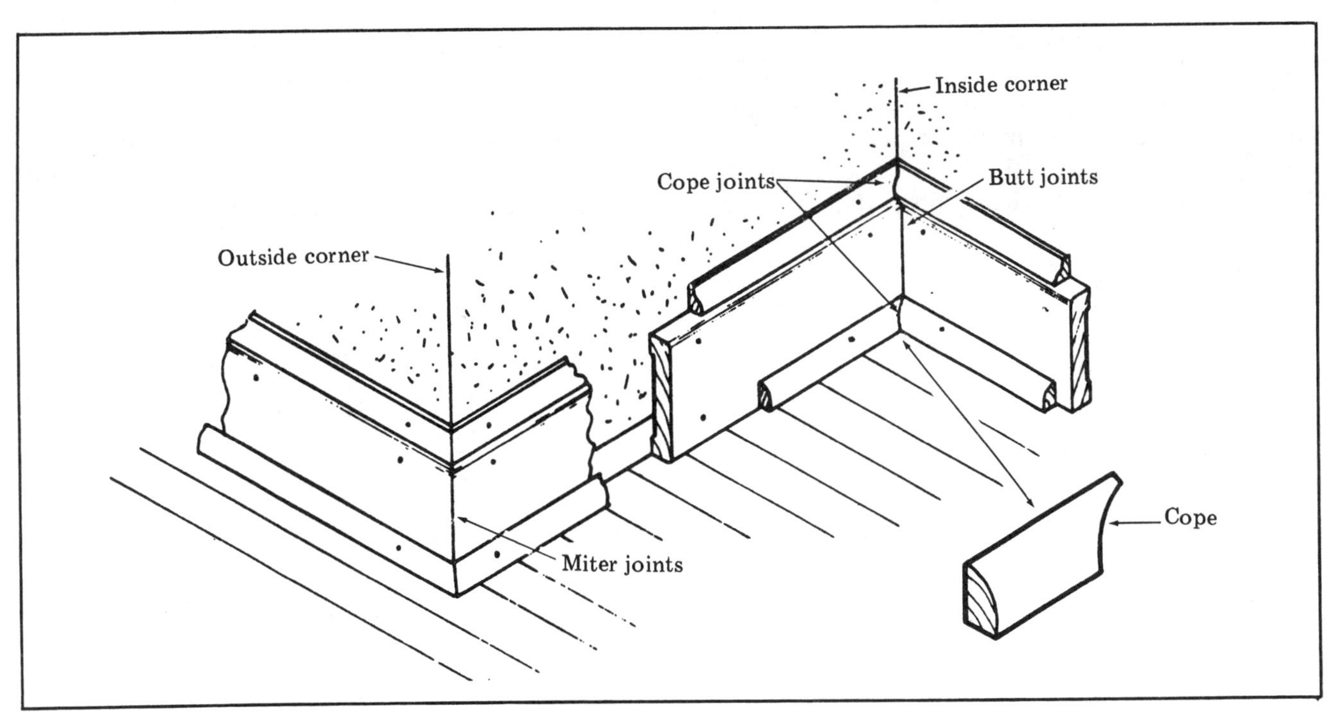

Base molding installation
Figure 22-11

Estimating Manhours

Here's how to estimate labor for walls, ceilings, trim and closets.

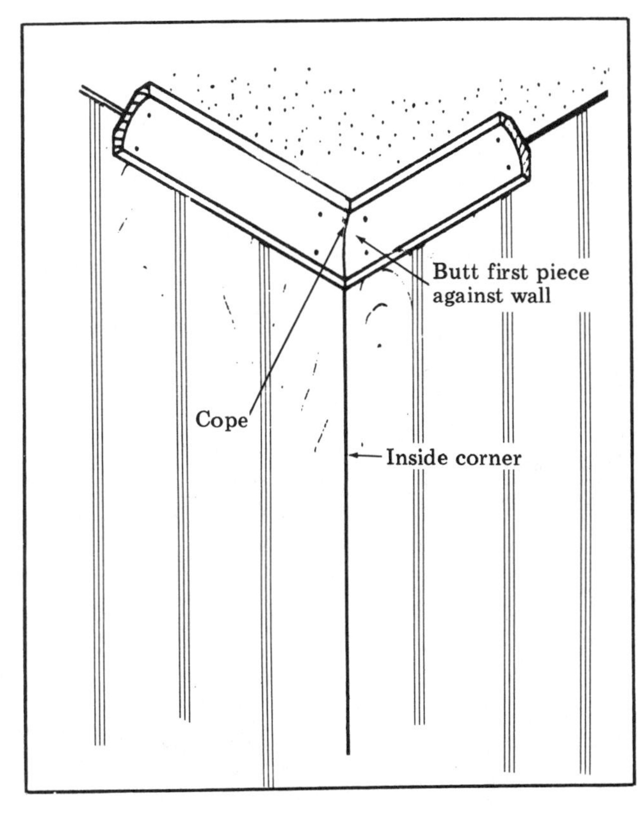

Ceiling molding installation
Figure 22-12

Sheet rock— For installation, taping, joint-finishing and sanding, allow 2½ skilled hours per 100 SF of wall and ceiling area.

You'll probably find that wallboard contractors can hang and finish sheet rock cheaper than you can do it yourself. But consider hanging it yourself and then bringing in a sub to tape and texture the walls and ceilings. Acoustical finishes require specialized equipment.

Trim— Estimate labor for trim as follows:

Work Element	Unit	Manhours Per Unit
Baseboard		
Two member, ordinary work	100 LF	4 to 6
Two member, hardwood, first class difficult work	100 LF	6 to 8
Three member, ordinary work	100 LF	5 to 7
Three member, hardwood, first class or difficult work	100 LF	7 to 9
Cove molding		
Ordinary work	100 LF	2.7 to 3
Hardwood, first class or difficult work	100 LF	4 to 5

Closets— Shelving, hookstrips, hooks and poles will average about 2½ manhours per closet.

Summary

Spec building is a good business when costs are held to a minimum. Sheet rock saves time and money and is an excellent finish material. It will help you turn a nice profit if the work is done to professional standards. Remember that a buyer may never see the framing, but he'll have to live with the interior wall covering and trim as long as he owns the house.

Record your material and labor costs on Figure 22-13, your Interior Wall, Ceiling and Trim Checklist and Data Reference Sheet.

Interior Walls, Ceilings and Trim Checklist and Data Reference Sheet

1) The house is ________ feet long ________ feet wide, for a total of ________ square feet.

2) The house is ☐1-story ☐1½-story ☐2-story.

3) The house has ________ square feet of finish wall area and ________ square feet of finish ceiling area.

4) Type of finish material for:

	Thickness (inches)	Number of rooms
Walls:		
☐Sheet rock	______	______
☐Hardboard paneling	______	______
☐Plywood paneling	______	______
☐Other (specify)		
________________	______	______
Ceiling		
☐Sheet rock	______	______
☐Other (specify)		
________________	______	______

5) Ceiling ☐is ☐is not furred ☐16 inches o.c. ☐24 inches o.c.

6) Ceiling finish is ☐paint ☐spray-applied texture ☐other (specify) ________________

7) B/B board was used in ☐bathroom ☐laundry room ☐other (specify) ________________

8) Ceiling/wall joint treatment:

☐Tape and cement (number of rooms ________)

☐Cove molding (number of rooms ________)

9) Sheet rock installed by:

☐Contract

☐Hang, tape and finish @ $________SF = $________

☐Tape and finish only @ $________SF = $________

☐Spray-applied texture @ $________SF = $________

Figure 22-13

Name: ______________________________

Address: ______________________________

Telephone: ______________________________

☐My crew

☐Hang, tape and finish	______hours	@	$______hour	=	$________
☐Tape and finish only	______hours	@	$______hour	=	$________
☐Spray-applied texture	______hours	@	$______hour	=	$________

Total $________

10) Material costs:

Sheet rock	__________SF	@	$_______ MSF	=	$__________
B/B board panels	__________pcs	@	$__________ea	=	$__________
Plywood panels	__________pcs	@	$__________ea	=	$__________
Hardboard panels	__________pcs	@	$__________ea	=	$__________
Tape	________ rolls	@	$__________ea	=	$__________
Joint cement	________ gals	@	$________ gal	=	$__________
Caulk	________tubes	@	$__________ea	=	$__________
Nails:					
Sheet rock	__________lbs	@	$__________lb	=	$__________
Paneling	__________lbs	@	$__________lb	=	$__________
Moldings:					
Base	__________LF	@	$________LF	=	$__________
Casing	__________LF	@	$________LF	=	$__________
Stool	__________LF	@	$________LF	=	$__________
Mullion	__________LF	@	$________LF	=	$__________
Seam	__________LF	@	$________LF	=	$__________
Stop	__________LF	@	$________LF	=	$__________
Shoe	__________LF	@	$________LF	=	$__________
Cove	__________LF	@	$________LF	=	$__________

Inside corner	________LF	@	$________LF	=	$__________
Outside corner	________LF	@	$________LF	=	$__________
Cap	________LF	@	$________LF	=	$__________
Nails	________lbs	@	$________lb	=	$__________
Furring:					
1 x 3	________LF	@	$________LF	=	$__________
1 x 4	________LF	@	$________LF	=	$__________
Nails (8d)	________lbs	@	$________lb	=	$__________
Closets:					
Shelving	________LF	@	$________LF	=	$__________
Hookstrip	________LF	@	$________LF	=	$__________
Hooks	________pcs	@	$________ea	=	$__________
Poles	________pcs	@	$________ea	=	$__________
Cleats	________LF	@	$________LF	=	$__________
Nails	________lbs	@	$________lb	=	$__________
Other materials, not listed above:					
____________________	________	@	$__________	=	$__________
____________________	________	@	$__________	=	$__________
____________________	________	@	$__________	=	$__________
____________________	________	@	$__________	=	$__________
			Total cost of all materials		$__________
11) Labor costs:					
Sheet rock:					
Skilled	________hours	@	$________hour	=	$__________
Unskilled	________hours	@	$________hour	=	$__________
Plywood/hardboard paneling:					
Skilled	________hours	@	$________hour	=	$__________
Unskilled	________hours	@	$________hour	=	$__________

B/B board:

Skilled	______hours	@	$______hour	=	$______
Unskilled	______hours	@	$______hour	=	$______

Trim:

Skilled	______hours	@	$______hour	=	$______
Unskilled	______hours	@	$______hour	=	$______

Total labor costs:

Skilled	______hours	@	$______hour	=	$______
Unskilled	______hours	@	$______hour	=	$______

Other (specify)

______________	______hours	@	$______hour	=	$______
				Total	$______

12) Total cost:

Materials $ ______

Labor $ ______

Other $ ______

Total $ ______

13) Cost per square foot of area:

Ceilings $ ______

Walls $ ______

Trim $ ______

Total $ ______

14) Problems that could have been avoided ______________________________

____

____

____

____

15) Comments ______________________________

____

____

____

____

____

16) A copy of this Checklist and Data Reference Sheet has been provided:

☐Management

☐Accounting

☐Estimator

☐Foreman

☐File

☐Other: ____________

Chapter 23

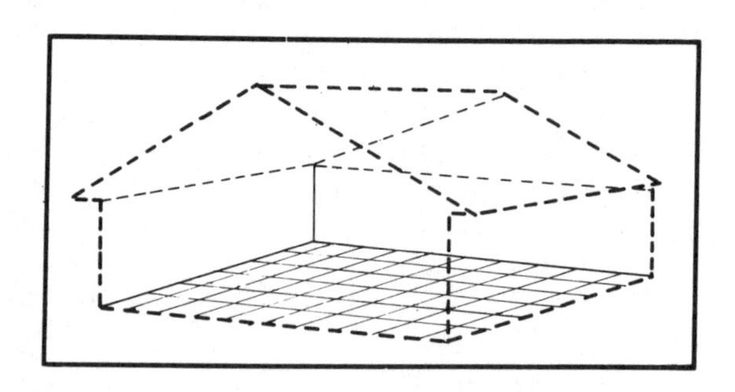

Floor Coverings

Interior finish materials have changed a lot in the last 30 years, but ceramic tile is still the best finish material for the high-moisture areas in a home. And, it's still popular as a floor covering. Other popular floor coverings include: carpeting, sheet vinyl, resilient vinyl tiles, wood strip and wood block flooring.

In this chapter, we'll look again at floor underlayment. Then we'll discuss each of the popular floor coverings mentioned above.

Floor Underlayment

We have already briefly discussed floor underlayment. It's installed after the interior wall finish is applied. Interior doors and baseboards can't be installed until the underlayment is down.

Particleboard makes the best underlayment for carpeting and resilient vinyl flooring. But if the carpeting or resilient vinyl is in a bathroom, use exterior plywood for underlayment. For non-bathroom areas, use 5/8-inch particleboard over 1/2-inch plywood subflooring. Some codes require a 5/8-inch plywood subfloor when using 5/8-inch particleboard underlayment.

Particleboard is made of wood particles combined with resin binders and then heat-pressed into panels. Store these panels flat and in a dry place. Don't have them delivered to the job until you're ready to install them.

Here's an important rule for installing underlayment over a plywood subfloor:

If the plywood is laid at right angles to the joists, the underlayment panel joints and plywood panel joints must be offset by at least 2 inches. If the plywood is laid parallel to the joists, the underlayment panel joints and plywood panel joints must be offset by at least one joist.

Use ring-grooved underlayment nails. Start nailing in the center of each panel, and work toward the edges. Drive the nails perpendicular to the panel surface, and set them flush. Edge-nailing should be done at least 1/2-inch but no more than 3/4-inch from panel edges. Nail a complete panel before starting the next panel. Leave a 1/16-inch space between panels.

Floor Coverings

The most commonly used floor coverings are: ceramic tile, carpeting, sheet vinyl, resilient vinyl tiles, wood-strip and wood-block flooring. Let's look at each of these interior floor finish materials.

Ceramic Tile

The bathroom in the home where you grew up probably had ceramic tile around the tub or shower — and may even have had a tile floor. If you're using ceramic tile for finish flooring in your spec houses, note that the color and design of the floor tile must be carefully coordinated with the color and design of the wall tile and bathroom fixtures.

The most common wall tile measures 4½ inches by 4½ inches and usually has a bright, glazed finish. There's a full spectrum of colors available, but the prevailing trend is toward pastel tones and solid colors, rather than contrasting border prints.

But watch your colors. It's hard to get an exact match between the color of the tile and the color of the bathroom fixtures. You're dealing with products from two different manufacturers. It's better to use a contrasting color scheme. Let the wall color be lighter or darker than the fixtures, or use a completely different color that harmonizes.

If you do use border-print tile, remember that it usually becomes the most important design element in the room. Fixtures and interior decor will have to be coordinated with the border print. Solid-color, pastel tile is better for your spec houses.

Unglazed mosaic tile makes a highly durable floor. It's expensive but adds a distinctive flavor to the home. Floor tile comes in hexagons, octagons, rectangles, small squares, large squares, Spanish, and Moorish shapes. Most tile companies manufacture floor tile to harmonize with their wall tile.

You'll probably want to sub out ceramic tile work to a tradesman who specializes in it and has the necessary tools.

Carpeting, Sheet Vinyl and Vinyl Tiles

Carpeting is the cheapest way to cover a floor. That's why wall-to-wall carpeting has become the rule for most new homes. But most buyers prefer to have sheet vinyl or resilient vinyl tiles in the kitchen.

Indoor-outdoor carpeting and sheet vinyl can also be used in bathrooms. But vinyl tiles are not recommended for bathroom floors since water can seep between the tiles, causing them to lift at the corners.

If your spec house has ceramic tile bathroom floors, a sheet vinyl kitchen floor, and the remainder of the floors are carpeted, it will appeal to the vast majority of home buyers. Be sure that carpeting and vinyl flooring meet HUD and local code requirements.

Wood-strip Flooring

Softwood finish flooring costs less than most hardwood flooring. It's less dense, less wear-resistant, and shows surface abrasions more readily than hardwood. But it's acceptable for use in bedroom and closet areas where traffic is light.

Softwoods most commonly used for flooring are Douglas fir, Western hemlock, Southern pine, and redwood. Grades and descriptions of softwood flooring are shown in Figure 23-1. Softwood flooring has tongue-and-groove edges and is either hollow-backed or groove-backed. Some types are end-matched. Vertical-grain flooring is more wear-resistant than flat-grain flooring.

Hardwoods most commonly used for flooring are red and white oak, beech, birch, maple, and pecan. Figure 23-1 lists grades, types, and sizes. Manufacturers supply both prefinished and unfinished hardwood flooring.

The most common hardwood flooring size is 25/32-inch x 2¼-inch. The strips are laid along the length of the room and at right angles to the floor joists. Use a subfloor of diagonal boards or plywood under the finish floor.

Hardwood strip flooring has tongue-and-groove edges and is end-matched. See Figure 23-2. The

Species	Grain	Thickness (in.)	Width (in.)	First grade	Second grade	Third grade
			Softwoods			
Douglas fir and	Vertical grain	25/32	2⅜-5³⁄₁₆	B and better	C	D
Western hemlock	Flat grain	25/32	2⅜-5³⁄₁₆	C and better	D	---
Southern pine	Vertical grain and flat grain	5/16-1 5/16	1¾-5 7/16	B and better	C and better	D (and No. 2)
			Hardwoods			
Oak	Vertical grain	25/32	1½-3¼	Clear	Select	---
	Flat grain	⅜, ½	1½, 2	Clear	Select	No. 1 Common
Beech, birch maple, and pecan*		25/32	1½-3¼	First grade	Second grade	---
		⅜, ½	1½, 2			

*Special grades are available in which uniformity of color is a requirement.

Wood-strip flooring description and grades
Figure 23-1

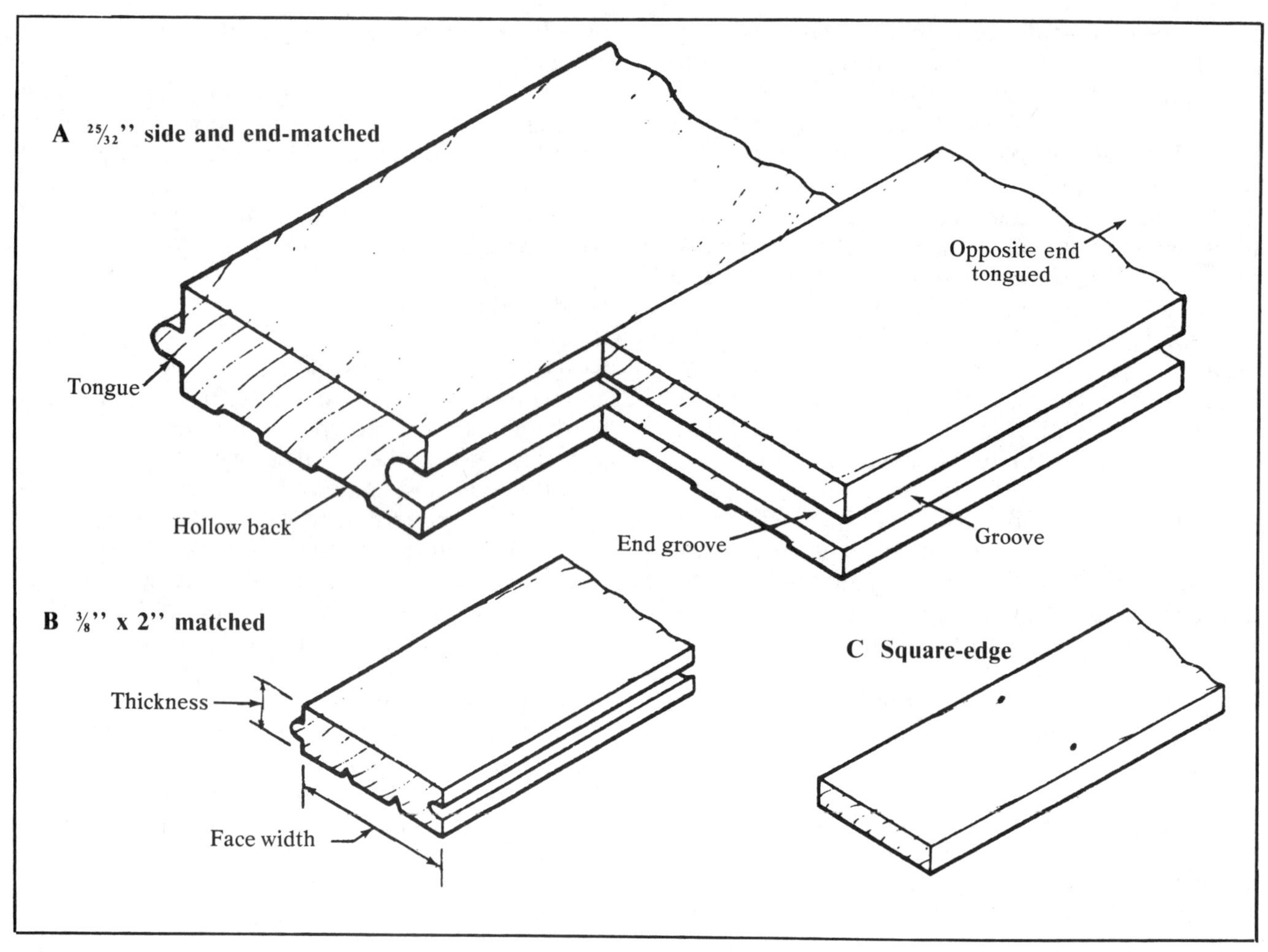

Types of wood-strip flooring
Figure 23-2

strips come in random lengths and may vary from 2 to 16 feet or more.

End-matched strip flooring in the 25/32-inch thickness is generally hollow-backed, as shown in Figure 23-2 A. The face is slightly wider than the back so that the joints are tight when the flooring is laid. To prevent movement and floor "squeaks," the tongue fits tightly into the groove.

Another matched pattern uses 3/8-inch x 2-inch strips. See Figure 23-2 B. This is commonly used for remodeling work or when the subfloor is edge-blocked or thick enough to provide very little deflection under loads.

Square-edge strip flooring (Figure 23-2 C) is also occasionally used. It's usually 3/8-inch x 2-inch and is laid over a firm subfloor. Face-nailing is required.

Wood-block Flooring

Wood-block flooring (Figure 23-3) is made in a number of patterns. Blocks vary in size from 4 x 4-inch to 9 x 9-inch and larger. Thickness varies from 1/8-inch (for stabilized veneer) to 25/32-inch (for laminated blocking or for the plywood block tile shown in the figure). The thicker tile has tongue-and-groove edges. Also illustrated is a solid wood tile. This is often made up of narrow strips of wood splined or keyed together. Many block floors are prefinished and require only waxing after installation.

Estimating Materials

Most home builders subcontract ceramic tile, carpeting and sheet vinyl work. Get three bids from contractors you know and trust.

Some builders hire an interior decorator to plan the floor covering and select the wall colors. A few dollars spent here may yield major dividends.

Estimate underlayment material requirements the same way as plywood subflooring requirements. Allow for 5% waste due to cutting and fitting.

For wood-strip flooring material, add 33% to the area to be covered. And for 25/32-inch x

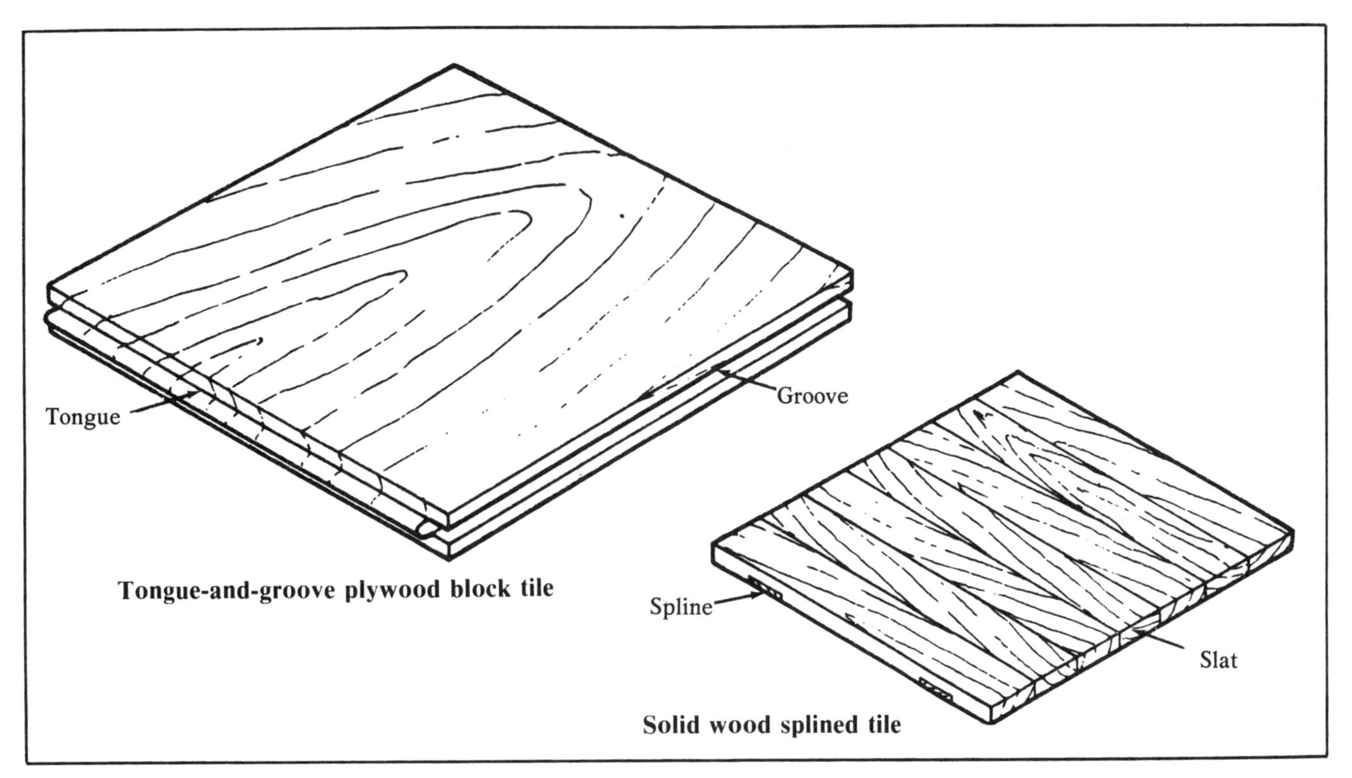

Wood-block flooring
Figure 23-3

2¼-inch strip flooring, allow about 40 lbs. of cut nails per 1,000 SF of flooring.

Estimating Manhours

Particleboard underlayment requires about 12 manhours per 1,000 SF.

Strip flooring requires about 25 manhours per 1,000 BF. 1,000 BF of 25/32-inch x 2¼-inch strip flooring will cover about 750 square feet.

Record your material and labor costs on Figure 23-4, your Floor Covering Checklist and Data Reference Sheet.

Summary

To increase your chances of building a house that sells quickly, use finish materials that are popular, such as ceramic tile, carpeting and resilient vinyl.

Most builders need help when it comes to color schemes, floor-covering patterns, and other finish-material choices. Either get expert advice on these points, or see what other spec builders are doing on their houses. The colors, textures and designs must appeal to the majority of people shopping for a house in the price range of the house you're building.

Floor Covering Checklist and Data Reference Sheet

1) The house is ________ feet in length and ________ feet in width, for a total of ________ square feet.

2) The house is ☐1-story ☐1½-story ☐2-story

3) Ceramic tile installed:

	Floor area	**Wall area**
Bathroom #1	____________SF	____________SF
Bathroom #2	____________SF	____________SF
Other (specify) ______________________	____________SF	____________SF

Contractor: ______________________________

Address: ______________________________

Telephone: ______________________________

Cost: ____________SF @ $____________SF = $____________

4) Floor covering:

	Carpeting	**Sheet vinyl**	**Other (specify)**
Living room	☐	☐	________________
Dining room	☐	☐	________________
Kitchen	☐	☐	________________
Family room	☐	☐	________________
Bedroom #1	☐	☐	________________
Bedroom #2	☐	☐	________________
Bedroom #3	☐	☐	________________
Bedroom #4	☐	☐	________________
Hall(s)	☐	☐	________________
Powder room	☐	☐	________________
Bathroom #1	☐	☐	________________
Bathroom #2	☐	☐	________________
Laundry room	☐	☐	________________

Figure 23-4

Floor covering contractor: ______________________________

Address: ______________________________

Telephone: ______________________________

5) Material costs:

Underlayment:

⅝'' particleboard	______pcs	@	$______ea	=	$______
Other (specify)					
______________	______	@	$______	=	$______
Nails	______lbs	@	$______lb	=	$______

Floor covering:

Ceramic tile	______SF	@	$______ SF	=	$______
Carpeting	______SY	@	$______ SY	=	$______
Sheet vinyl	______SY	@	$______ SY	=	$______
Vinyl tile	______SF	@	$______ SF	=	$______
Wood-strip	______BF	@	$______ BF	=	$______
Wood-block	______BF	@	$______ BF	=	$______
Adhesive	______ gals	@	$______ gal	=	$______
Nails	______lbs	@	$______lb	=	$______

Other materials, not listed above:

______________	______	@	$______	=	$______
______________	______	@	$______	=	$______
______________	______	@	$______	=	$______
______________	______	@	$______	=	$______

Total cost of all materials $______

6) Labor costs:

Underlayment:

Skilled	______hours	@	$______hour	=	$______
Unskilled	______hours	@	$______hour	=	$______

Ceramic tile:						
Skilled	_______hours	@	$_______hour	=	$__________
Unskilled	_______hours	@	$_______hour	=	$__________
Carpeting:					
Skilled	_______hours	@	$_______hour	=	$__________
Unskilled	_______hours	@	$_______hour	=	$__________
Sheet vinyl:					
Skilled	_______hours	@	$_______hour	=	$__________
Unskilled	_______hours	@	$_______hour	=	$__________
Vinyl tile:					
Skilled	_______hours	@	$_______hour	=	$__________
Unskilled	_______hours	@	$_______hour	=	$__________
Wood-strip:					
Skilled	_______hours	@	$_______hour	=	$__________
Unskilled	_______hours	@	$_______hour	=	$__________
Wood-block:					
Skilled	_______hours	@	$_______hour	=	$__________
Unskilled	_______hours	@	$_______hour	=	$__________
Other (specify)					
____________________	_______hours	@	$_______hour	=	$__________
Skilled	_______hours	@	$_______hour	=	$__________
Unskilled	_______hours	@	$_______hour	=	$__________
Total labor costs:					
Skilled	_______hours	@	$_______hour	=	$__________
Unskilled	_______hours	@	$_______hour	=	$__________
Other (specify)					
____________________	_______hours	@	$_______hour	=	$__________
				Total	$__________

7) Total cost:

Materials $____________

Labor $____________

Other $____________

Total $____________

8) Cost of finish floor covering per square foot of floor area: $____________.

9) Problems that could have been avoided ________________________

10) Comments ________________________

11) A copy of this Checklist and Data Reference Sheet has been provided:

☐ Management

☐ Accounting

☐ Estimator

☐ Foreman

☐ File

☐ Other: ____________

Chapter 24

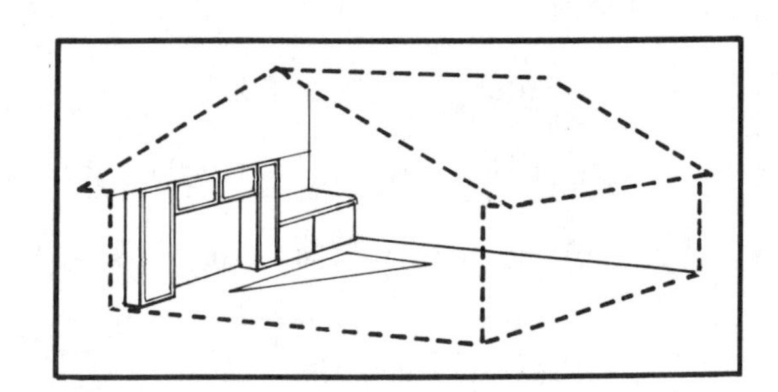

Kitchen Layout

Most homes built today have enough kitchen cabinet space to comply with the Minimum Property Standards published by the U.S. Department of Housing and Urban Development (HUD). Many local codes are based on these standards.

In this chapter, we'll discuss the standard dimensions for kitchen cabinets, fixtures, countertops and storage space. We'll also look at the four most popular kitchen layouts, and touch briefly on special requirements for wood cabinets.

Standard Dimensions

Here are the standard dimensions for kitchen cabinets, fixtures, countertops and storage space.

Cabinets

Figure 24-1 shows minimum and maximum dimensions for kitchen cabinets.

Fixtures and Countertops

Figure 24-2 shows minimum frontages for kitchen fixtures and countertops. Frontages vary, depending upon the number of bedrooms in the house. Be sure to note the following restrictions on frontage dimensions for sink, range, range countertop, refrigerator and refrigerator countertop:

Sink— When a dishwasher is provided, a 24-inch sink is acceptable even for three- and four-bedroom homes.

Range— Don't locate range burners under a window or within 12 inches of a window. If there's a cabinet above a range, there must be a 30-inch clearance from the range to the bottom of an unprotected cabinet (or a 24-inch clearance to the bottom of a protected cabinet). Where no range is provided, you must leave a 30-inch-wide space.

Range countertop— There should be at least 9 inches between the edge of the range and an adjacent corner cabinet. Where there's a built-in wall oven, provide an 18-inch-wide countertop adjacent to the oven.

Refrigerator— If a refrigerator is provided and the door opens within its own width, the refrigerator space can be 33 inches wide.

Refrigerator countertop— There should be at least 15 inches between the side of a refrigerator and an adjacent corner cabinet.

Apart from these restrictions, countertops should be approximately 24 inches deep and 36 inches high. Allow a (minimum) 40-inch clearance between base cabinet fronts in the food preparation area.

Required countertops can be combined when they are located between fixtures, such as stove, refrigerator and sink. The combined countertops should have a minimum frontage equal to the larger of the countertops being combined. The combined counter can also serve as the mixing counter if the combined counter is long enough. Countertop frontages can continue around corners.

Storage Area

Figure 24-3 suggests minimum dimensions for shelf and drawer storage area. At least one-third of the required area should be located at the base of wall

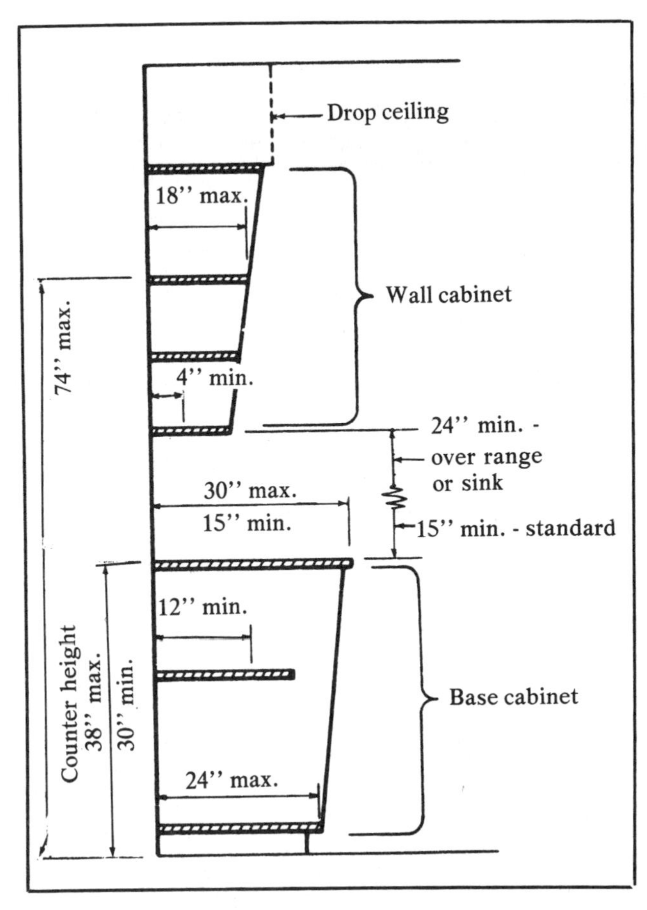

Kitchen cabinet dimensions
Figure 24-1

cabinets. At least 60% of the required area should be enclosed by cabinet doors. Note the following restrictions:

Shelf area— A dishwasher can be counted as 4 SF of base cabinet storage. But required shelf area doesn't include wall cabinets located over refrigerators, or any shelf area located 74 inches (or higher) above the finish floor. You can include 50% of the shelf area in inside corner cabinets unless revolving shelves are used. When revolving shelves are used, count the entire shelf area.

Drawer area— If drawers exceed the 6-inch depth requirement, any excess drawer area can be counted as shelf area.

Kitchen Layouts

The efficient kitchen has a "work triangle" of no more than 22 feet. The triangle is the distance between the storage and mixing center, the cleaning and preparation center, and the cooking and serving center. In other words, consider the distance from the refrigerator to the sink to the range as more or less a triangle. The sum of the sides should be no more than 22 feet. Plan an efficient work triangle in the kitchen, and make that plan a selling point of the house.

	Number of Bedrooms				
	0	1	2	3	4
	Minimum Frontages (Linear Inches)				
Sink	18	24	24	32	32
Countertop, one side	15	18	21	24	30
Range space	21	21	24	30	30
Countertop, one side	15	18	21	24	30
Refrigerator space	30	30	36	36	36
Countertop, one side	15	15	15	15	18
Mixing countertop	21	30	36	36	42

Kitchen countertop and fixture dimensions
Figure 24-2

Layout Designs

Figure 24-4 shows the four most popular kitchen layouts: U-type, L-type, "parallel-wall" and "side-wall."

U-type— This layout has the sink at the bottom of the U. The range and refrigerator are on opposite sides of the sink. This tends to be a very efficient layout.

L-type— This layout has the sink and range on one leg and the refrigerator on the other. There's sometimes a dining space in the opposite corner.

"Parallel-wall"— Also called a corridor kitchen, this layout is good for narrow kitchens. It can be quite efficient with proper arrangement of the sink, range and refrigerator.

"Side-wall"— Also called a one-wall kitchen, this layout is used only in small apartments. The sink, range, refrigerator and all cabinets are located along one wall. Counter space is usually quite limited in a side-wall kitchen.

	Number of Bedrooms				
	0	1	2	3	4
Minimum shelf area (square feet)	24	30	38	44	50
Minimum drawer area (square feet)	4	6	8	10	12

Kitchen storage area dimensions
Figure 24-3

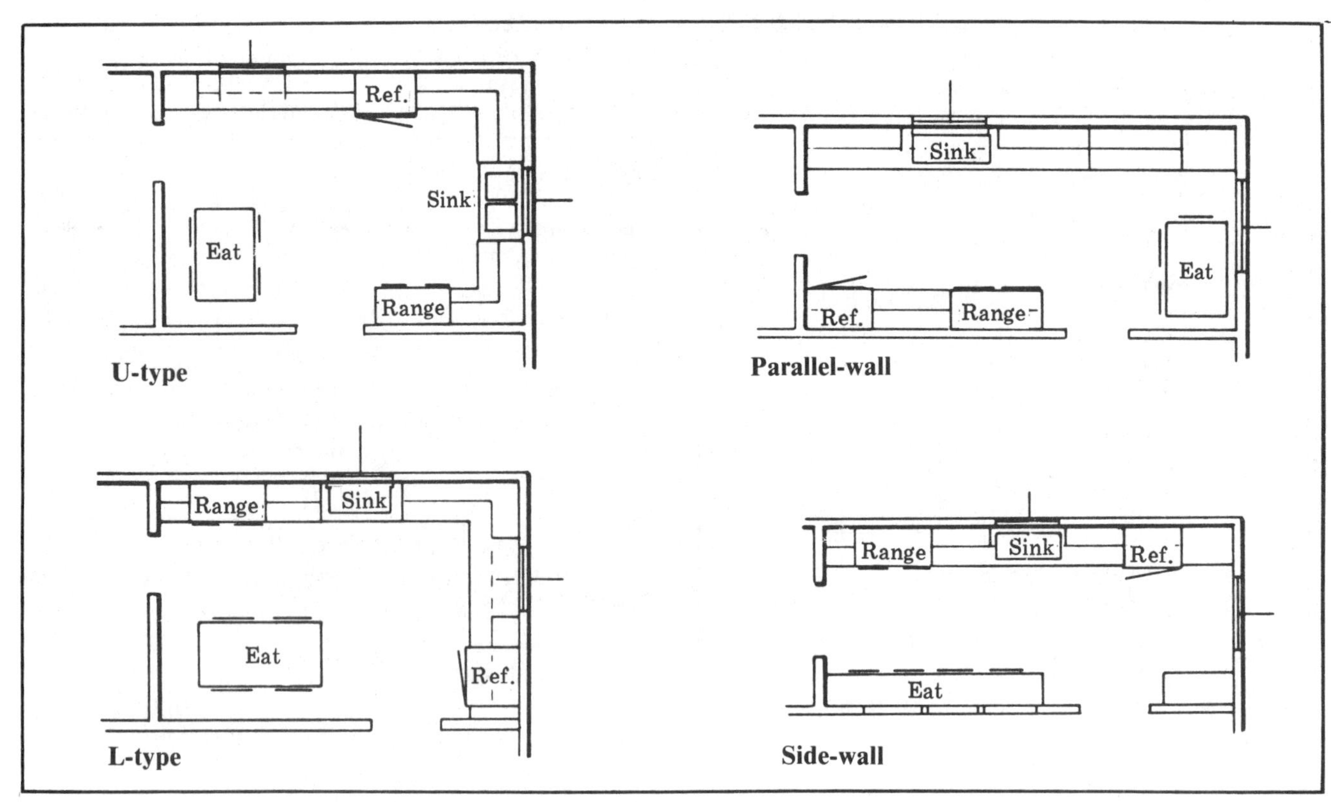

Kitchen layouts
Figure 24-4

Island cabinets
Figure 24-5

A kitchen that will sell the house
Figure 24-6

Layout Dimensions
We've already looked at minimum-dimension requirements for kitchen cabinets, fixtures, countertops and storage space. And we've discussed the four most popular layout designs. Now let's look at the dimensions most commonly used with these popular kitchen layouts.

Use not less than 4½ feet, nor more than 5½ feet, of countertop length between the refrigerator and sink. Use no less than 3 feet, nor more than 4 feet, between the sink and range. Of course, there are many reasons to vary these dimensions. But if you do, make sure that your reasons are valid.

Snack bars are often extended out at right angles from a kitchen wall or located at the end of a wall. In large kitchens, consider using island cabinets to save steps, as shown in Figure 24-5. An island cabinet is really two base cabinets set back-to-back and covered with a one-piece top.

Doors and windows should be coordinated with cabinets, appliances and interior corners. Most appliances and cabinets are 24 to 28 inches deep. A door opening should be at least 30 inches from a corner to allow space for installing cabinets or placing appliances. In cases where a range is next to a door, there should be a clearance of 48 inches between range and door. There should be at least 15 inches between a window opening and a corner, because standard, flat wall cabinets are about 13 inches deep. If this clearance is allowed, wall cabinets can be installed all the way to the corner.

The backsplash at the back of countertops is usually 3 to 8 inches high. This means that the underside of window sills should be a minimum of 44 inches from the floor.

Good kitchen-planning requires more than a layout of work areas and equipment. People spend a lot of time in the kitchen. Do what you can to make the kitchen bright and cheerful, as well as useful. Include open shelves for pottery or plants. Provide large window areas. A good kitchen can sell a home all by itself. And a large window, especially over the sink, as in Figure 24-6, can sell the kitchen.

Skylights and wallpaper can add much to a kitchen
Figure 24-7

A basic L-type kitchen
Figure 24-8

You can brighten and add charm to a kitchen with a skylight and wallpaper, as shown in Figure 24-7.

Figure 24-8 shows a basic L-shape kitchen. Note the attractive corner placement of window and sink.

Wood Cabinets

Figure 24-9 shows sizes and types of wood kitchen cabinets. Your cabinets should include a bread board or chopping board (that slides into a slot under the counter) and at least one special compartment. Notice that the bottom of the base cabinets is recessed about 3 inches. The bottom shelf is also raised about 4 inches off the floor. This toe space makes it easier to work when standing at the counter.

Make the countertop the same height as the top of the stove and kitchen sink. The least expensive tops are made from laminated plastic. The top can be custom-formed with an integral front edge and backsplash.

Stock cabinets are available either as assembled units or knockdown units. But many builders find that they can get custom cabinets made in a local cabinet shop for about the same price as assembled stock units. Your local cabinet shop may quote an installed price that can save you from the uncertainty of estimating the installation labor.

Stock cabinets come in widths from 12 to 48 inches (in 3-inch increments). Narrow filler strips are available to fill in spaces smaller than 3 inches wide.

If you decide to buy stock units and install them yourself, see the cabinet installation details shown in Figure 24-10.

Estimating Materials

If you subcontract cabinet work, your cabinet sub will probably not take any measurements until after the finish wall and ceiling are installed. He'll probably use the house plans and cabinet specs to quote prices.

Use Figure 24-11 to compile a complete list of kitchen layout materials.

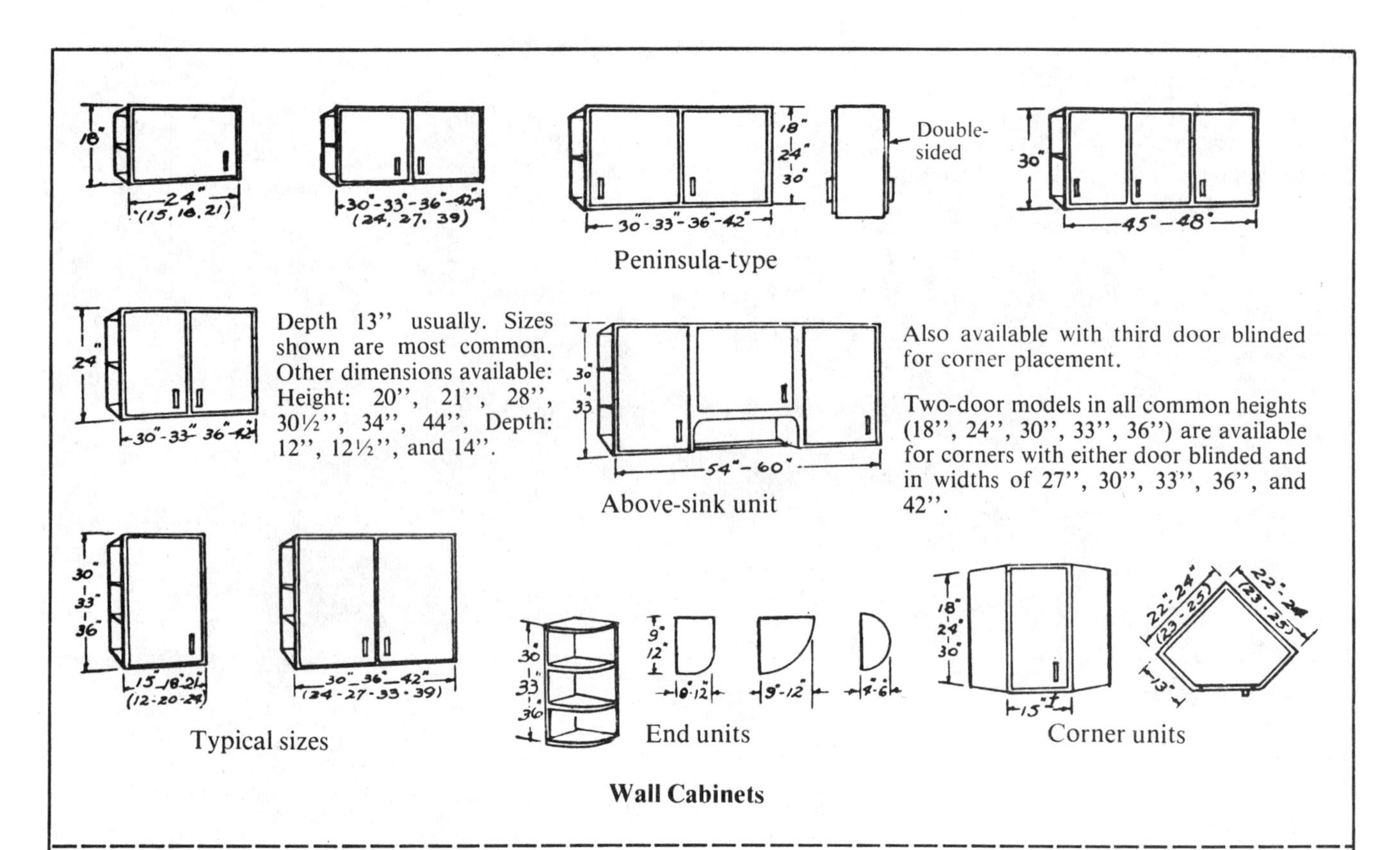

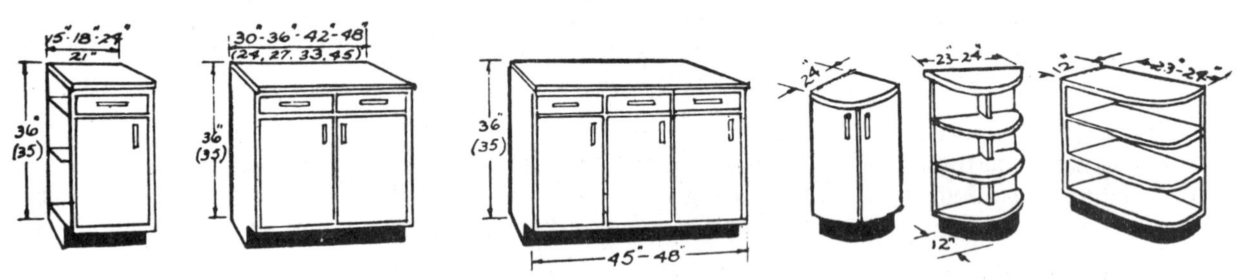

Also available with third door blinded for corner placement

End units

Depth 23", 24", 24½". Units available without top drawer, with 3 or 4 drawers (no cupboard), and in many uncommon combinations of drawers and cupboards. Also with special accessories: towel rack; dryer; utensil rack; cutting board; vegetable, sugar, and flour bins; tray storage; pan rack; bread box; flour sifter; soap tray; sliding table top; etc. Two-door models available with one door blinded for corner placement.

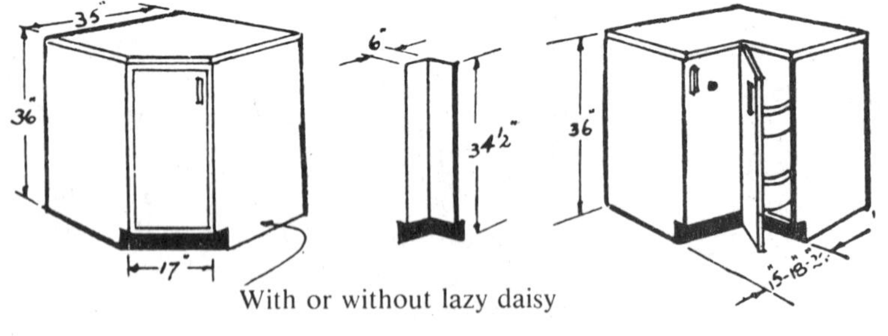

Typical sizes

Corner units

Base Cabinets

Dimensions given in inches. Sizes vary among manufacturers. Numbers in parentheses ( ) are less common dimensions. Countertops available in stainless steel, laminated plastic.

Wood kitchen cabinets
Figure 24-9

Tips on Cabinet Installation

Be sure to plan carefully the location of your kitchen cabinets, fixtures, countertops and storage space. The efficient kitchen has a "work triangle" of no more than 22 feet. Notice that the work triangle is a step-saver for all of the popular kitchen layouts, including U-type, L-type and parallel-wall type kitchens.

After choosing the best layout for your kitchen, draw up a complete floor plan of the kitchen. As you draw up your plan, remember these basics:

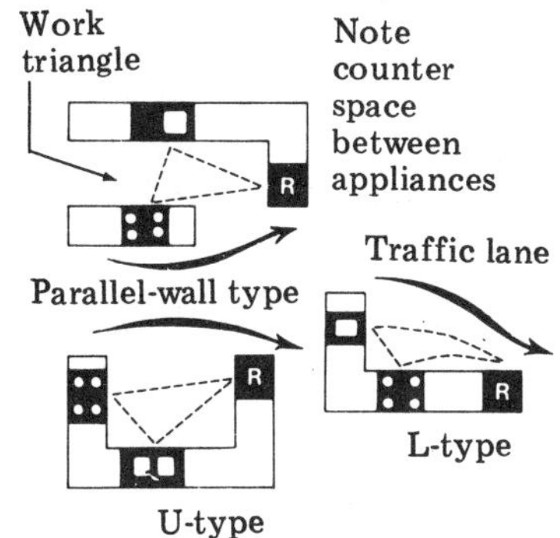

1) Allow 2' of counter space on each side of the range and sink.

2) Allow 15" of counter space next to the refrigerator.

3) Refrigerator doors should swing away from the adjoining work surface. Be sure there's enough space to open the refrigerator door and pull out the shelves without the shelves hitting the door.

4) Use a sink cabinet that's at least 3" wider than the actual sink dimension.

5) Locate tall utility cabinets at the end of a row of lower cabinets.

Tools You'll Need

- Chalkline
- Countersink
- Electric drill with several bits
- 4' level
- 10' or 12' wood extension rule
- 2' carpenter's square
- Stepladder
- Screwdriver
- Wood screws
- Wood shims (made from shingles or scrap wood)
- Two 4" C-clamps

1) Prepare Walls

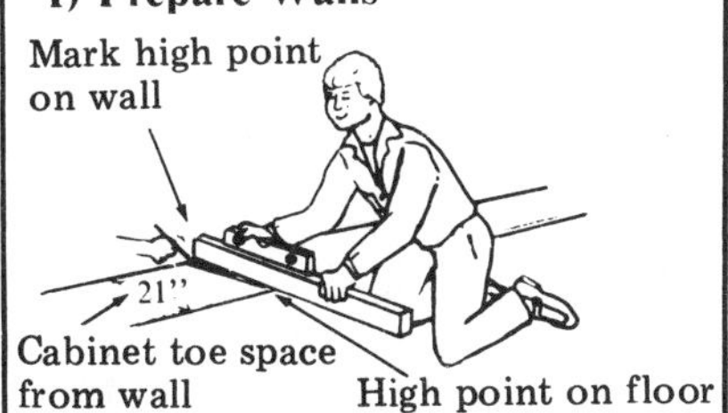

Cabinets must be mounted perfectly level and square. To prepare walls for cabinet installation, first remove all moldings, base boards and other protruding objects. Use a level to locate the high point on the floor. Then mark this point on the wall.

2) Measure and Mark Walls

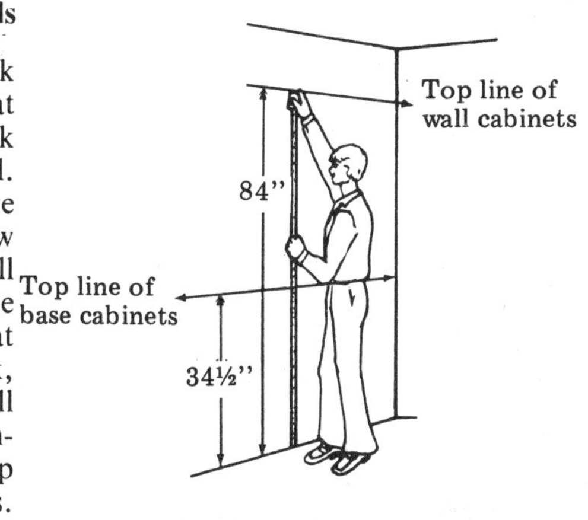

To measure and mark cabinet locations, begin at the floor high point mark you just made on the wall. From this point, measure 34½" up the wall and draw a horizontal line. This will be the top line of your base cabinets. Starting again at the floor high point mark, measure 84" up the wall and draw another horizontal line. This will be the top line of your wall cabinets. These measurements are based on standard 8-foot-high ceilings.

3) Locate Wall Studs

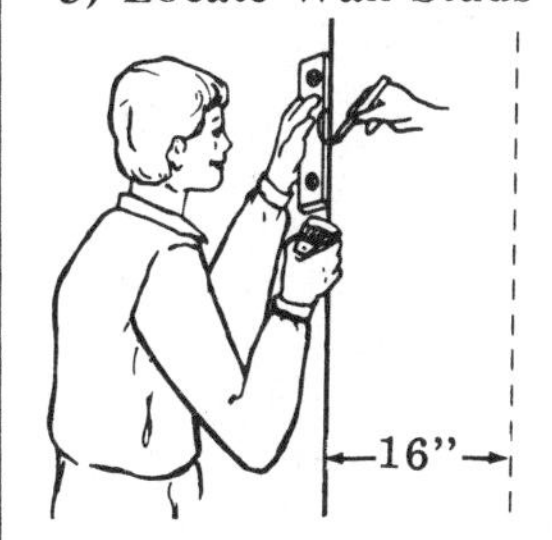

Cabinets must be attached to studs. To locate studs, you can use a magnetic stud finder, tap the wall with a hammer, or test the wall with a nail that will later be hidden from view. Studs are often spaced 16" o.c.

4) Wall Cabinet Installation

Always start with the corner wall cabinets. After locating the wall studs, place the top of your wall cabinet along the 84" line. Use a 3/16" twist bit to drill installation holes through the hanging strips built into the back of the cabinet. Then use a 3/32" twist bit to drill (through the hanging-strip holes) 1¾" into the studs. Use two 2½" flat-head wood screws to fasten the cabinet to the wall. To make stronger fasteners that have a nicer appearance, install chrome-head grommets behind the screw heads. All screws must penetrate studs at least 1". Do not tighten down screws until all wall cabinets have been installed. When all wall cabinets are installed, then the final leveling and shim adjustments can be made, and all screws can be tightened down. *Warning: If you don't use shims when tightening the back rails against a crooked wall, you can break the joint between the rail and the cabinet. This will cause the cabinet to fall off the wall.*

Step-by-step cabinet installation details
Figure 24-10

5) Blind Corner Wall Cabinets

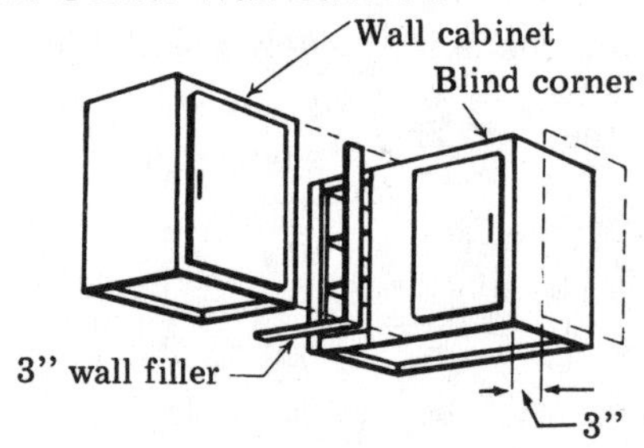

Blind corner cabinets can be pulled up to 4½" out from the wall. Use normal cabinet-hanging procedure, but add a 3" wall filler. To attach the filler to the cabinet, use the same procedure as for joining front frames. See Step 7.

6) Diagonal Corner Wall Cabinets

Diagonal cabinets extend out from the corner 24" along each wall. Be sure the cabinets are perfectly plumb on both sides and front. Use your carpenter's level to make sure the cabinets are perfectly level, both horizontally and vertically. It's especially important that the corner cabinets are plumb and level — the correct positioning of the other cabinets depends on it!

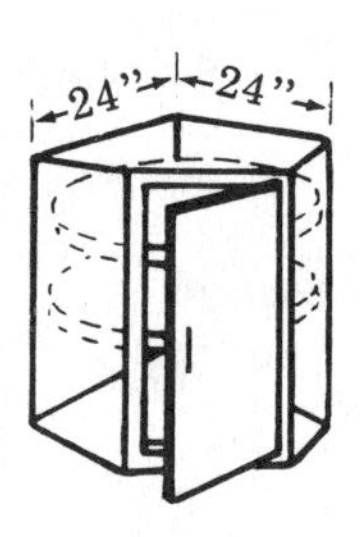

7) Join Front Frames

Frames must be flush at top and bottom. Use C-clamps to secure cabinets together while drilling and fastening. Use an 11/64" twist bit to drill a hole in the cabinet stile. Then use a 7/64" bit to drill a pilot hole approximately 1" deep in the adjoining cabinet stile. Fasten cabinets together with No. 8 x 2½" flat-head wood screws. Countersink for screw heads. For fillers that are to be trimmed, you may need to use shorter screws.

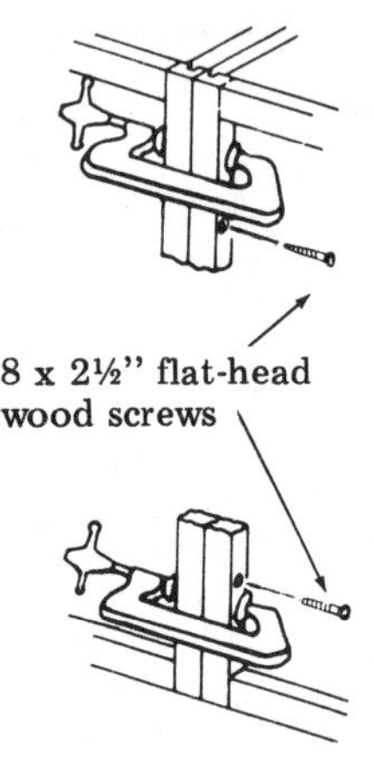

8) Ceiling-hung Wall Cabinets

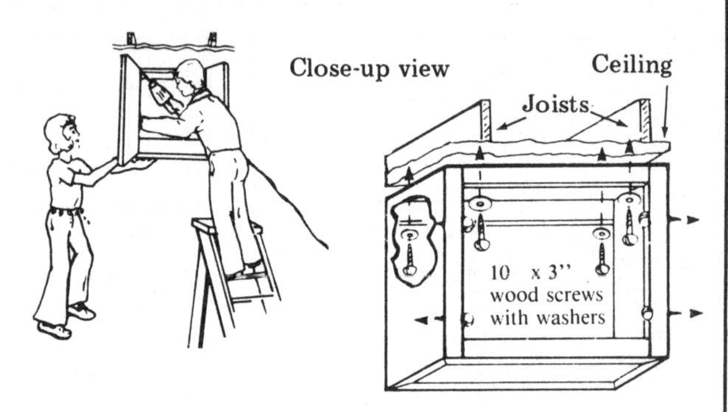

Locate ceiling joists above wall cabinets. Use a 3/16" twist bit to drill four holes through the top frame of the cabinet. Use a 3/32" twist bit to drill (through the existing holes) about 1½" into ceiling and joists. Use No. 10 x 3" round-head wood screws with washers. Attach cabinet to front frame as described in Step 7.

9) Level Cabinets

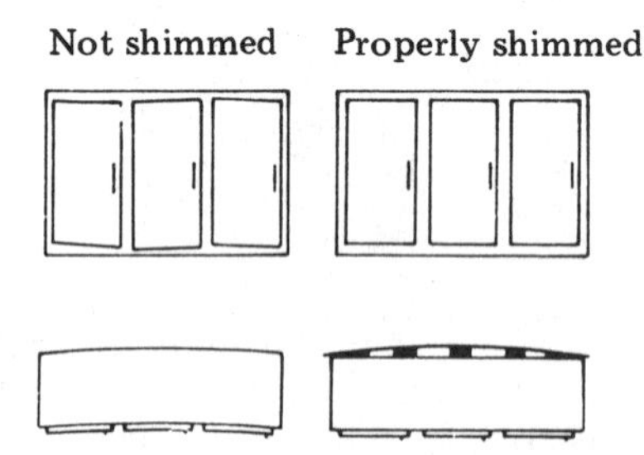

After all cabinets have been hung, use your carpenter's level to make sure they are plumb and level on front, sides and bottom. To correct uneven walls, shim behind cabinets. After cabinets and doors are perfectly aligned, tighten all screws.

10) Base Cabinet Installation

After all wall cabinets have been installed, begin your base cabinet installation. *Again, start with the corner base cabinets.* Drill small (1/8") holes through the top rail at the back of the cabinet and into the studs. Insert No. 10 x 2½" screws. Assemble the base corner unit by adding one cabinet on each side of the corner cabinet. Before fastening cabinets together, remove cabinet doors and hinges. Between the hinge holes, drill holes for the screws you're using to fasten the cabinets together. Insert screws. When door is replaced, these screws will be hidden. Join front frames as described in Step 7.

Cabinet installation details
Figure 24-10 (continued)

11) Void Corner Base Cabinets

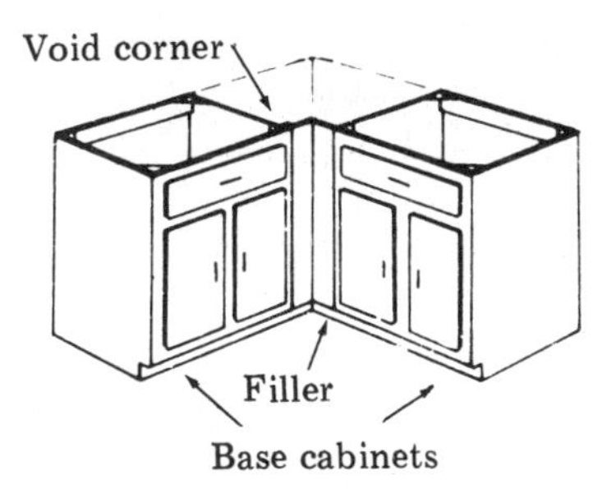

Use a filler when installing base cabinets with a void corner. A properly used filler can give you a void corner of 27'' x 27''. Attach filler as described in Step 7

12) Blind Corner Base Cabinets

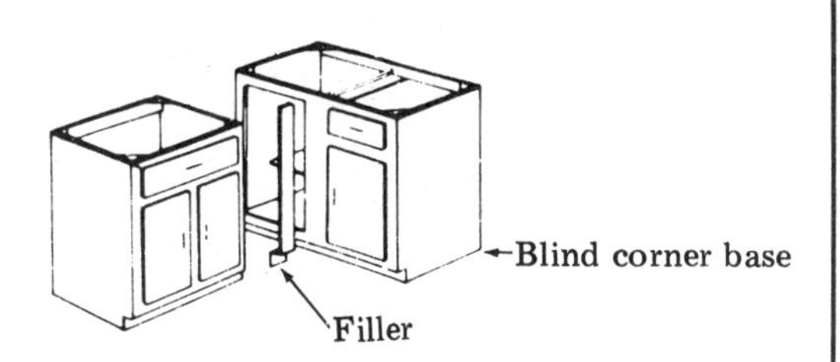

To determine the proper distance between wall and cabinet, check the layout drawing. The distance will vary, depending on the size of each unit. Position and fasten cabinets in the usual way, but always use a filler. Attach filler as described in Step 7.

13) Corner Carousel Base Cabinets

This unit must be perfectly plumb, or the carousel won't operate properly. To provide support for the countertop, secure two 36'' wood strips (1'' x 2'') to the wall. Make sure the strips are level with the front rails.

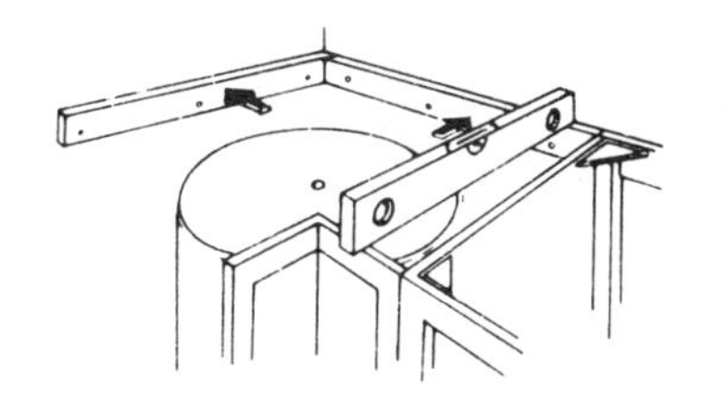

14) Sink Front

Align sink front flush with adjacent cabinets. Attach front frames as described in Step 7.

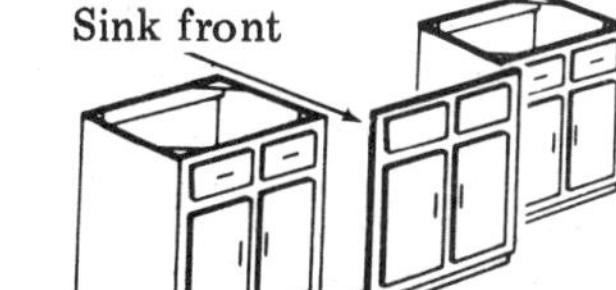

15) Level Cabinets

As you install each cabinet, use a level to check it across the front edge and from front to back. And make sure the front frame is plumb. Use shims, if necessary.

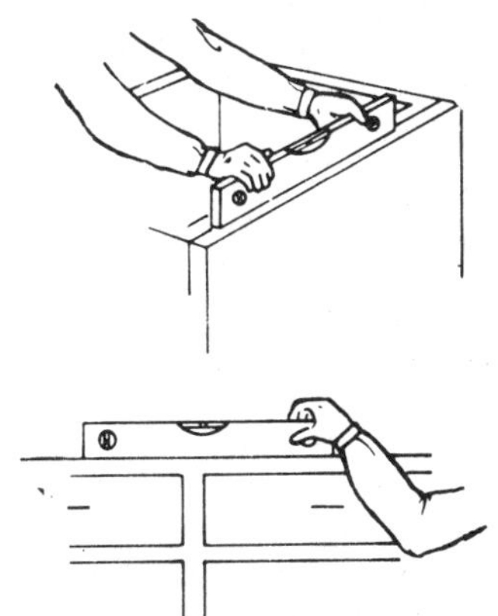

Not shimmed

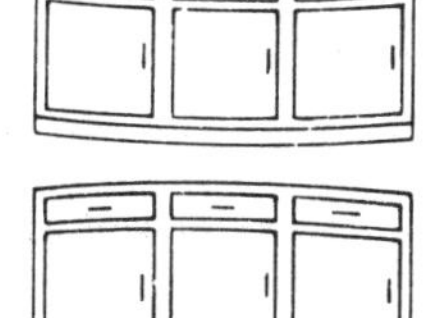

Properly shimmed

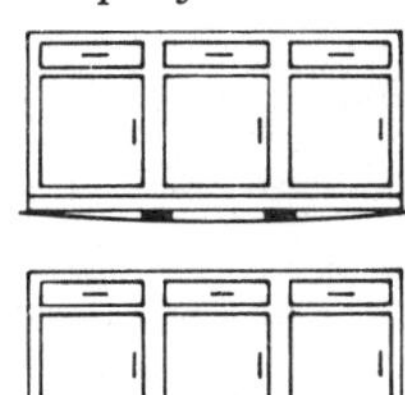

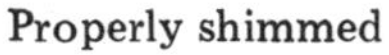

16) Countertops

Attach the countertop to the corner blocks of the cabinet. To insure proper installation, follow the countertop manufacturer's instructions.

17) Adjust Cabinet Doors

Loosen the adjustment screw on each hinge. Align door in frame. Then tighten screws. If hinge is bent, use a screwdriver to ease hinge into the correct position.

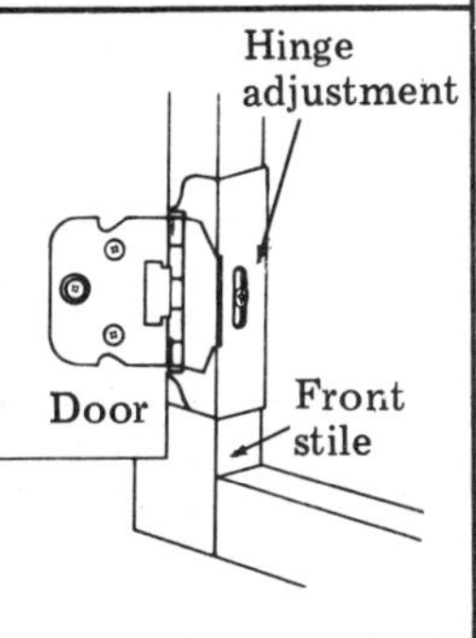

Cabinet installation details
Figure 24- 10 (continued)

Estimating Manhours

Labor for installing cabinets and countertops is estimated as follows:

Work	Unit	Manhours Per Unit
Base cabinets, 36" high		
24" wide	Each	1.0
36" wide	Each	1.2
Base corner cabinets, 36" wide	Each	3.2
Wall cabinets, 12" deep		
12" x 18"	Each	0.8
18" x 18"	Each	0.9
18" x 36"	Each	1.0
24" x 36"	Each	1.3
Cabinet stain finish	100 SF	1.0
Cabinet paint finish	100 SF	1.2
Cabinet vinyl finish	100 SF	1.4
Factory-formed tops, 4" backsplash		
24" wide	LF	0.20
32" wide	LF	0.21
Backsplash only, 4" high	LF	0.08
Cutting blocks, custom sizes	SF	0.42
Broom closets, 7' high	Each	1.8

Suggested Crew: 1 carpenter and 1 laborer, 1 painter for finishing.

Estimated manhours include: layout, unloading, all necessary trim work, cleanup and repairs. Estimate does not include: demolition work or any cleanup required prior to installation.

Summary

The kitchen is probably the key to most home sales. Your buyers will scrutinize the kitchen. Take extra care to develop eye-catching appeal, and you'll find a responsive buyer.

Keep an accurate record of costs. The record you create on this job will become your reference manual for future jobs. Figure 24-11, Kitchen Layout Checklist and Data Reference Sheet, will help you accumulate accurate kitchen layout data.

Kitchen Layout Checklist and Data Reference Sheet

1) Kitchen is ☐L-type ☐U-type ☐parallel-wall ☐side-wall.

2) The work triangle is:

☐Sink to refrigerator: ___________ feet

☐Refrigerator to range: ___________ feet

☐Range to sink: ___________ feet

☐Total:___________ feet

3) Cabinets meet or exceed ☐HUD minimum standards ☐local code minimum standards

4) Cabinets are:

☐Stock cabinets (☐knockdown ☐assembled)

☐Subcontracted

5) Cabinets were manufactured and installed on contract by:

Name: ____

Address: ____

Telephone: _______________________________________

Cost:

Base cabinets	_________LF	@	$_______LF	=	$__________
Wall cabinets	_________LF	@	$_______LF	=	$__________
Countertops	_________LF	@	$_______LF	=	$__________
Other (specify)					
____________________	_________LF	@	$_______LF	=	$__________
				Total	$__________

6) Material costs:

Base cabinets:

Corner unit	_________pcs	@	$_________ea	=	$__________
Sink unit	_________pcs	@	$_________ea	=	$__________
1-door unit	_________pcs	@	$_________ea	=	$__________
2-door unit	_________pcs	@	$_________ea	=	$__________
3-door unit	_________pcs	@	$_________ea	=	$__________

Figure 24-11

End unit	______pcs	@	$______ea	=	$______
Island	______pcs	@	$______ea	=	$______
Wall cabinets:					
Above sink	______pcs	@	$______ea	=	$______
Peninsula-type	______pcs	@	$______ea	=	$______
1-door unit	______pcs	@	$______ea	=	$______
2-door unit	______pcs	@	$______ea	=	$______
3-door unit	______pcs	@	$______ea	=	$______
Corner unit	______pcs	@	$______ea	=	$______
End unit	______pcs	@	$______ea	=	$______
Refrigerator enclosure	______pcs	@	$______ea	=	$______
Broom closet	______pcs	@	$______ea	=	$______
Linen closet	______pcs	@	$______ea	=	$______
Sealer	______gals	@	$______gal	=	$______
Stain	______gals	@	$______gal	=	$______
Paint	______gals	@	$______gal	=	$______
Countertops:	______LF	@	$______LF	=	$______

☐ Laminated plastic

☐ Factory-formed edges and integral backsplash

☐ Metal-trimmed edges and separate backsplash

☐ Other (specify)

____

Other materials, not listed above:

______________	______	@	$______	=	$______
______________	______	@	$______	=	$______
______________	______	@	$______	=	$______
______________	______	@	$______	=	$______

Total cost of all materials $______

7) Labor costs:

Install cabinets:

Skilled	________hours	@	$________hour	=	$____________
Unskilled	________hours	@	$________hour	=	$____________

Install countertops:

Skilled	________hours	@	$________hour	=	$____________
Unskilled	________hours	@	$________hour	=	$____________

Finish cabinets:

Skilled	________hours	@	$________hour	=	$____________
Unskilled	________hours	@	$________hour	=	$____________

Total labor costs:

Skilled	________hours	@	$________hour	=	$____________
Unskilled	________hours	@	$________hour	=	$____________
Other (specify)					
__________________________	________hours	@	$________hour	=	$____________
				Total	**$**____________

8)Total cost:

Materials	$____________
Labor	$____________
Other	$____________
Total	$____________

9) Cabinet cost per room:

☐Bathroom #1	$____________
☐Bathroom #2	$____________
☐Laundry room	$____________
☐Other (specify)	
____________________	$____________

10) Problems that could have been avoided ________________________________

____

____

____

11) Comments ____

____

____

____

12) A copy of this Checklist and Data Reference Sheet has been provided:

- ☐ Management
- ☐ Accounting
- ☐ Estimator
- ☐ Foreman
- ☐ File
- ☐ Other: ____________

Chapter 25

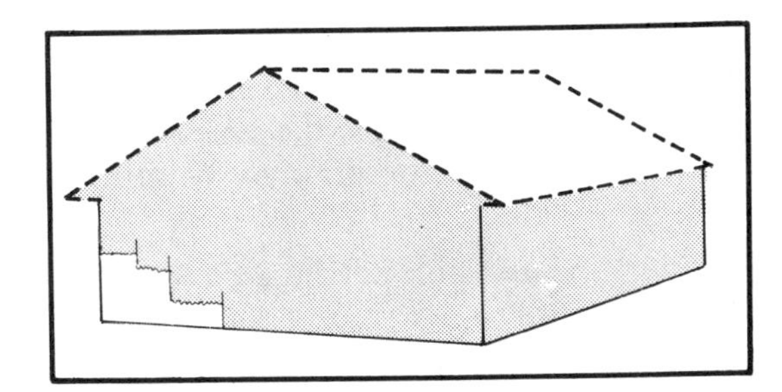

Painting

Paint provides more than just color. It also forms a protective coating that resists weather damage. Different paints are designed for different uses. As a builder, your challenge is to find the paint or coating appropriate for each situation, and to see that it's applied correctly.

This chapter explains most of what you need to know. The rest of the necessary information is printed on the side of the paint can. That information will identify the type of exposure (exterior or interior), the type of surface to be covered (wood, metal, masonry, concrete, plaster) and the weathering characteristics of the paint.

Let's begin with some basic principles that apply to nearly every painting job. Then we'll look at painting procedure details, including: color selection; primer and finish coat; paint application tools; paint application; pigmented penetrating stains, and interior trim finishes.

Basic Principles

Previously unpainted exterior wood surfaces nearly always need a primer coat and at least a 0.004-inch finish coat. Two finish coats are usually required. All exterior wood doors and all wood windows need two coats of paint or sealer. Don't forget to cover the top and bottom edges of the doors. These areas are seldom seen, but are ideal entry points for moisture. Give them the same weather protection as exterior surfaces.

On masonry surfaces, use low-chalking paint. This will avoid streaks and stains.

Interior sheet rock can be painted or papered. Paint is the easiest and cheapest way to decorate interior walls. Latex paint in a flat, semigloss or gloss finish is popular. Latex dries fast, has little odor and can be cleaned up with water. On sheet rock, one coat of sealer and one finish coat may be enough. Finish coats in kitchens and bathrooms should have a semigloss or gloss finish so that the surface is washable.

Interior wood surfaces can be painted, stained or left with a natural finish. A fill material is usually applied to keep the wood grain from rising. Interior wood surfaces need three coats of paint or stain (including primer). Be sure to sand between coats. Give the top and bottom edges of all interior doors one coat of paint or sealer.

Some builders use wallpaper or wall fabrics throughout the house. Others use wallpaper only as a decorative accent. Wall coverings are available in a variety of patterns and materials. Kitchens and bathrooms should have waterproof wall coverings.

Now let's look at painting procedure details.

Color Selection

Paint color plays a vital role in winning or losing the approval of a prospective buyer. Few new owners are anxious to repaint as soon as they move in. As a spec builder, you must find colors that are appropriate and blend well with the surroundings.

Lighter, less intense colors are more likely to win a buyer's approval. But there's more to selecting colors than just using whites and off-whites. You can use color to hide the shortcomings of a house, bring out the strong points and add a feeling of warmth. Here are the most popular exterior and interior color schemes.

Exterior Colors

A combination of light and dark colors gives the most interesting visual effect. Lighter colors command more attention. Use them to highlight the attractive features of the house. To emphasize an appealing entry, use a light color on the area around the door. Paint the rest of the entry a shade darker. The lighter color brings the entry forward. Then draw attention to the door by painting it a vivid color.

The same principle can be used on larger sections of the house. Give more definition to a long, straight ranch-style home by highlighting a section of the long front. Paint the front entrance and window areas a lighter shade than the rest of the house front. The lighter area comes forward, lifts upward and is framed by the darker, subdued colors.

Lighter shades give an impression of height. Darker shades minimize size. If a two- or three-level house is too tall for its width, correct this by painting the top level slightly darker than the bottom. But when mixing colors like this, keep the colors in the same color family.

An unattractive or poorly-placed roof extension or dormer window is less obvious when it's painted the same color as the roof.

Houses with large porches and overhangs need special attention. For a pleasing, shadow-box frame effect at the front entrance, try a dark trim color on the overhang and supporting columns.

Choosing a color scheme that blends with the surroundings doesn't mean that, on a wooded site, every house should be painted green. There are many other tones suitable in wooded areas. Greys, browns, beiges and golds blend well with the bark, rock and undergrowth in wooded areas.

Keep in mind that the surroundings will intensify or weaken the color you select for the house. If a green house is surrounded by shrubs and grass, the house will look greener.

There should be just enough contrast between surroundings and house color so that house definition is not lost and house color is not overwhelmed. Also, be sure to note the color of neighboring houses.

Northern climates need warm colors to compensate for intermittent sunlight. Cooler pastel tones are better for warmer climates. The cooler tones should, however, be bright enough so the color will not look weak in intense sunlight.

Interior Colors

A small room looks larger if the walls, ceiling and woodwork are all the same color. A pale, bright color works best.

A square room is more interesting when one wall is painted a dramatic, contrasting color. The contrasting wall is usually the wall directly opposite the entrance to the room.

Square up a long, narrow room by painting the two end walls a deeper color.

A room that's too large seems better proportioned if painted in darker shades.

Create a feeling of more space and continuity between rooms by using similar colors in the rooms. The total living area looks larger if all rooms are done in coordinated colors.

Primer and Finish Coat

The primer and finish coat work hand-in-hand on painted surfaces. Together they become the paint system.

Primer

The primer's job is to seal the surface material. Primer prevents absorption of the finish coat of paint. Primer also protects the finish coat from damage caused by chemicals in the surface material.

There are two types of primer: general-purpose, sealing primer (for wood, metal and masonry) and anti-corrosive primer (to protect metal against rust and corrosion).

Primers are available in many colors, but white is easiest to use. White can be tinted to the approximate color of the finish coat. This avoids color contrasts that are difficult to cover.

All unpainted wood, metal and masonry surfaces need a coat of either general-purpose primer or anti-corrosive primer.

It's essential to use caulking around window and door trim areas (and other joints) after the wood has been primed. Use quality caulking. It has greater elasticity and will last longer.

Finish Coat

The finish coat protects the surface from weather damage. It also provides the finish (flat, low-lustre, soft-gloss or full gloss) and color.

Flat and low-lustre paints can be used on most wood, metal or masonry surfaces. If you use a gloss finish on a textured surface, the gloss finish will emphasize the rough, uneven nature of the surface. Use soft-gloss and gloss paints only on smooth surfaces.

There are two finish-coat paint formulas: latex-base (water-thinned) and oil-base (solvent-thinned). Oil-base paint is still the choice of many professional painters. But latex-base paint is the most popular finish coat.

Latex-base paint— Most painters agree that latex paint is easier to apply. It dries rapidly and cleans up quickly with soap and water. If the surface you are painting is new, the "breathing film" of latex paint discourages blistering.

Latex resists fading, making it ideal for sunny climates and locations. Latex paint is alkali-resistant. It's a natural choice for masonry surfaces, which are alkaline by nature. On new masonry, latex serves as both primer and finish coat. In this case, use two coats of latex.

Latex is available in both low-lustre and soft-gloss finishes.

Oil-base paint— This paint provides better penetration and adhesion. It's probably the best choice for surfaces that have been painted many times.

Although they require more drying time and cleanup than latex paints, oil-base paints brush out easily and provide a level finish. The paint dries to a somewhat thicker film than latex paint and is available in both flat and gloss finishes.

Pigmented Penetrating Stains

Many stains are nearly transparent so the grain pattern shows through. They penetrate into the wood without forming a continuous film on the surface of the wood. This penetration keeps the surface from blistering, cracking or peeling, even when there's excessive moisture in the wood.

Penetrating stains are used on both rough and smooth surfaces. But they look best when applied on rough-sawn or weathered wood. Stain is also the best coating on lumber that doesn't hold paint well, such as flat-grained, dense species.

One coat of penetrating stain applied to a smooth surface may last 2 to 4 years. But the second application, after the surface has been roughened by weathering, will last 8 to 10 years.

If you apply two coats of stain on a rough-sawn surface, you'll get close to ten years out of the first application. Two-coat staining also has the advantage of reducing lapping and uneven application. The second coat should always be applied the same day as the first and before the first coat dries. That way, both coats penetrate into the wood.

Paint Application Tools

The most commonly used paint application tools are: brushes, rollers, paint pads and spray applicators.

Brushes

Use quality brushes that don't shed bristles. A quality brush will be springy and resilient. Cheap bristles become limp and bunch together in clumps while you paint, making it difficult to apply a smooth coat. Better bristles are shaped to a tapered edge so you can feather the paint at the end of each stroke and avoid "lap marks."

Brushes come in two types: natural bristle and synthetic filament. Natural bristle brushes are made from animal fibers. Quality varieties, such as hog bristle, have flagged ends that fan out like branches on a tree. Flagged bristles carry more paint so you get the coverage you want with fewer dips of the brush. Don't use natural bristle brushes with latex-base paints. The bristles swell and distort when washed with water.

Synthetic brushes are made of nylon or polyester filament. Better synthetic types are artificially flagged so they hold paint like hog bristle. Synthetic brushes can be used with oil-base paint, but they're better suited to latex paint.

A 3½-inch to 4-inch brush is about right for exterior siding. For trim areas, a 2½-inch angular brush makes precise edging easier. A round, sash brush is ideal for narrow window moldings. For masonry surfaces, use a stiff-fiber 6-inch brush.

Rollers

Rollers differ in the type of fabric and the length of pile. Choose a roller fabric that's compatible with the type of paint you're using. Lamb's wool is good with oil-base paints but will mat with latex-base paints. Mohair or synthetic fabrics are better with latex.

Choose a pile length that's right for the surface being painted. For common stucco, a 1/2-inch-pile roller will do the job. Rougher surfaces, such as concrete block, textured stucco, brick or poured concrete, need the extra penetration of a 3/4-inch-pile roller.

Rollers also vary in quality. A "bargain" roller sleeve doesn't carry as much paint as a quality sleeve. The pile will mat quickly and begin leaving a blotchy appearance on the surface being painted. As you paint, the paper core may be softened by water or solvent, making the roller virtually useless.

Paint Pads

Flat, rectangular pad applicators are sometimes used on shake and shingle surfaces. They have a rug-like texture. One or two downward strokes force paint into the surface grooves. Pad applicators don't spatter, and they let you paint right up to the trim.

The trick with pad applicators is loading the pad with the right amount of paint — not too little and not too much.

Spray Applicators
Modern spray equipment is easy to operate. It's versatile enough to use with nearly all types of paint and on most exterior surfaces. Spraying cuts application time in half on many jobs and leaves a smooth, uniform surface. But an inexperienced spray-painter can do a lot of damage in a very short period of time. If you allow any spray-painting on your jobs, supervision is critical.

Exterior Paint Application

Avoid painting when the air or surface temperature is below 50 degrees. And don't paint when it's raining or when rain is expected. A shower can ruin a fresh paint job.

Don't ever use oil-base paint on a damp surface, in foggy weather, or before the dew has evaporated. This moisture will be trapped by the paint and will cause blistering and peeling. Latex paint can be used on damp surfaces.

Plan the job so it begins with a side of the house that will be in the shade while you're painting it. Then move to the side that's becoming more shady, and paint that side. Continue in this way until the entire house is painted. Painting in the shade is important. Direct sun heats the surface and accelerates the drying rate. Premature drying keeps the paint from leveling properly and leaves lap marks, especially with latex paint.

When oil-base paint is applied in the sun, dry blisters will develop on the surface. This is especially true of darker colors. A strong wind will also dry oil-base paint too fast. Reschedule painting for a calmer day.

Begin painting at the roof peak or at the eave. Steadily move down (in sections) and across the face. Paint cornice and overhang areas as you go, if they're the same color as the siding.

On clapboard siding, the first surface to paint is the area where boards overlap. Use the narrow side of the brush and force paint up under the lap. Coat as much board as you can comfortably reach, then flip to the wider portion of the brush and coat the face of the board. Paint down four or five boards, then move the ladder horizontally to a new position, a comfortable arm's-length away from the painted area. Continue painting across the house in this manner. Repeat this procedure until the entire side of the house is painted.

For a better look on horizontal siding, don't try to paint one section of the house completely from top to bottom. By the time you're ready to work in the new area, edges of the paint will have set and will resist any additional brushing or blending. Lap marks will form where the wet and dry paint layers overlap. This disturbs the uniform appearance of the siding surface.

Paint the trim last. Coat window sash and door paneling, then do the remainder of the window frames, sills, and door trim. Don't worry about getting paint along the putty line of the window glass. It will help protect the puttied surface from water seepage. After paint has dried, use a razor blade scraper to remove excess paint. This creates a neat edge.

Interior Trim Finishes

Unlike exterior surfaces, interior trim does not need to be protected from weathering. But interior painting is more difficult because attention to detail is more important. There's a wider variety of interior paints than exterior paints. A good interior finish should last much longer than exterior paint.

There are two types of interior trim finishes: opaque and transparent. Opaque finishes include enamel, semigloss enamel and flat paint. Transparent finishes include stain, varnish, lacquer, sealer and wax.

Opaque Finishes
Interior surfaces can be painted with the same materials, tools and procedures as recommended for exterior surfaces. But interior woodwork, and especially trim, demands a smoother finish, better color, and a more lasting sheen. Enamels or semigloss enamels are often used.

When using enamel, start with a wood surface that's extremely smooth. Imperfections, such as planer marks, hammer marks and raised grain, will stand out when the enamel finish is dry.

Raised grain can be a real problem on the flat-grain surfaces of the heavier softwoods. During planing, the hard bands of summerwood get crushed into the softer springwood. And the hard bands are pushed up again later when the wood dries. For a quality job, moisten the wood surface with water, and then let it dry thoroughly. That raises any grain that is likely to come up during painting. Sand lightly with sharp sandpaper before applying primer or enamel.

Some hardwoods, such as oak and ash, have large pores. These pores must be filled with wood filler before primer can be applied. The primer coat for interior wood surfaces can be the same as for exterior woodwork, or you can use special interior wood primer. After the primer is dry, use a knot-sealer to shellac or seal the knots in white pine, ponderosa pine and Southern yellow pine. This reduces discoloration and pitch run.

Next, apply one or two coats of enamel undercoat. The undercoat should completely cover the wood and also create a surface that can easily be sanded smooth. For best results, sand the surface before applying the finish enamel. The finish surface can be left with its natural gloss or rubbed to a dull finish.

Using a flat paint finish is easier and quicker than using enamel. Surface preparation is faster, and you don't need to hand-rub the final coat. This saves time and money — an important consideration if you plan to make a profit on a spec house.

Transparent Finishes

Transparent finishes are used on some softwood trim, most hardwood and doors. Before finishing, remove any cutting marks or other blemishes in the wood.

Both softwoods and hardwoods can be finished without staining, especially if the wood has an attractive natural color and grain. But most builders prefer to use stain, because it provides more than just color. Stain is absorbed unequally by different parts of the wood. This brings out the natural variations in the grain of the wood. The most commonly used stains are water stains, non-grain-raising stains and pigmented-oil stains.

Water stains give the best results but also raise the grain of the wood. An extra sanding is needed after the stain is dry.

Many non-grain-raising stains are available. They dry quickly and, like water stains, have clarity and uniformity of color. But when you apply this type of stain on softwood, the stain colors the springwood more than the summerwood. This reverses the natural gradation in color and will look strange to anyone who knows painting and wood.

Pigmented-oil stains are essentially thin paints. They're a better choice for softwood because they don't color the springwood as strongly as the non-grain-raising stains do. To get a nearly uniform color on softwood, coat the wood with a clear sealer before applying the pigmented-oil stain.

Varnish and lacquer are also common transparent finishes. When applying them to hardwood that has large pores, be sure to fill the pores first. The filler can be transparent so it doesn't change the color of the finish. Or the filler can be a color that contrasts with the surrounding wood.

Sealer is no more than thinned-out varnish or lacquer. It keeps the wood from absorbing later coats and prevents the bleeding of stains and fillers into the surface coatings. Lacquer sealers are especially fast-drying.

Use a transparent surface coating over the sealer. This coating can be semigloss varnish, gloss varnish, nitrocellulose lacquer, or wax. Varnish and lacquer coatings bring out the natural luster of some hardwoods. They seem to let you look down into the wood. And sometimes you want a shine that looks a foot deep. But for most homes, a gloss like that shows wear and fingerprints too easily. Wax gives a sheen without forming a thick coating and without adding too much luster to the wood. There are also some less-glossy or "satin finish" lacquer and varnish products available.

Estimating Materials

Many builders subcontract all of their painting. Paint subcontractors will give you a bid for both labor and materials or for labor only. Most painters prefer to see the house before submitting a bid.

If you sub out the painting, have a clear set of specs, and get the paint sub's signature on those specs. There are slapdash painters in most communities, and you don't want them working on your job. The specs should cover the type of paint to be used, the number of coats, the caulking and sanding required, any unusual treatments, and cleanup of adjacent surfaces. A sloppy paint job or a painter who leaves paint on walkways, driveways (or in other unexpected places), can ruin an attractive home.

You don't need a painting sub if you have someone on the payroll who can handle a paint brush. Painting doesn't require much equipment. Consider doing the painting yourself, especially if it seems impossible to find a conscientious and reasonable subcontractor.

Let's look at material requirements for exterior surfaces, interior surfaces and special areas.

Exterior Surfaces

To calculate the amount of paint needed, first calculate the square footage of the surface to be painted. Here's a simple formula for determining the square footage of exterior surface:

A. (Length of house + width of house) x 2 = total distance around house.

B. Distance around house x height of house = sidewall area.

C. (½ gable height x width of gable base) x 2 = total gable area.

D. Total sidewall area + total gable area = total surface to be painted.

Note: Do not subtract window and door areas from the total surface figure if each opening is less than 100 square feet.

Example (See Figure 25-1):

A. (50' + 42') x 2 = 184'

B. 184' x 14' = 2,576 square feet

C. ½ (14') x 42' x 2 = 588 square feet

D. 2,576 square feet + 588 square feet = 3,164 square feet.

Divide the total surface area by the spreading rate (coverage rate) of a gallon of paint. The coverage rate will almost always be listed on the paint can. Figure primer and finish coats separately.

In Figure 25-1, the total exterior surface to be painted is 3,164 SF. If the approximate spreading rate of a gallon of paint is 400 SF, divide 3,164 by 400. The result is 7.9 gallons. Figure 8 gallons for one-coat coverage.

Interior Surfaces

Figure 25-2 gives the square footage of interior wall and ceiling surfaces for various room sizes.

Figure 25-3 gives single-roll wallpaper requirements for various room sizes. The table assumes the use of a standard roll of wallpaper, 8 yards long by 18 inches wide. Deduct one roll of wallpaper for every two doors (or windows) or for each 50-square-foot opening. The labor is shown in Figure 25-4, which includes time for moving on and off site, for limited surface and material preparation, and for any necessary cleanup and repair. An appropriate crew would be one paperhanger and one laborer.

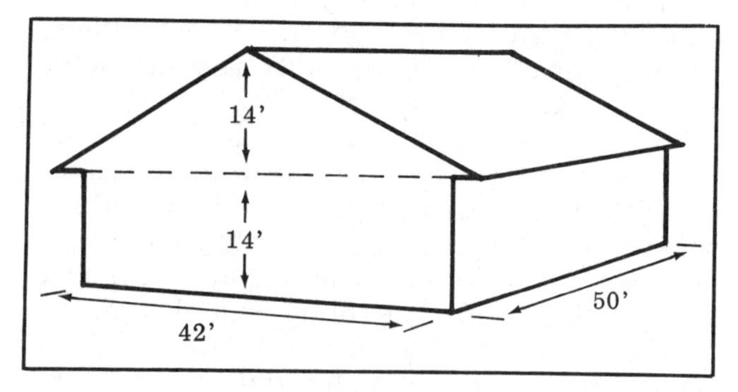

Square footage of exterior surface
Figure 25-1

Wall Length	6'	8'	10'	12'	14'	16'	18'	20'	22'	24'	26'	28'	30'
6 Feet	C 36	C 48	C 60	C 72	C 84	C 96	C 108	C 120	C 132	C 144	C 156	C 168	C 180
	W 192	W 224	W 256	W 288	W 320	W 352	W 384	W 416	W 448	W 480	W 512	W 544	W 576
8 Feet	C 48	C 64	C 80	C 96	C 112	C 128	C 144	C 160	C 176	C 182	C 198	C 224	C 240
	W 224	W 256	W 288	W 320	W 352	W 384	W 416	W 448	W 480	W 512	W 544	W 576	W 608
10 Feet	C 60	C 80	C 100	C 120	C 140	C 160	C 180	C 200	C 220	C 240	C 260	C 280	C 300
	W 256	W 288	W 320	W 352	W 384	W 416	W 448	W 480	W 512	W 544	W 576	W 608	W 640
12 Feet	C 72	C 96	C 120	C 144	C 168	C 192	C 216	C 240	C 264	C 288	C 312	C 336	C 360
	W 288	W 320	W 352	W 384	W 416	W 488	W 480	W 512	W 544	W 576	W 608	W 640	W 672
14 Feet	C 84	C 112	C 140	C 168	C 196	C 224	C 252	C 280	C 308	C 336	C 364	C 392	C 420
	W 320	W 352	W 384	W 416	W 448	W 480	W 512	W 544	W 576	W 608	W 640	W 672	W 704
16 Feet	C 96	C 128	C 160	C 192	C 224	C 256	C 288	C 320	C 352	C 384	C 416	C 448	C 480
	W 352	W 384	W 416	W 448	W 480	W 512	W 544	W 576	W 608	W 640	W 672	W 704	W 736
18 Feet	C 108	C 144	C 180	C 216	C 252	C 288	C 324	C 360	C 396	C 432	C 468	C 504	C 540
	W 384	W 416	W 448	W 480	W 512	W 544	W 576	W 608	W 640	W 672	W 704	W 736	W 768
20 Feet	C 120	C 160	C 200	C 240	C 280	C 320	C 360	C 400	C 440	C 480	C 520	C 560	C 600
	W 416	W 448	W 480	W 512	W 544	W 576	W 608	W 640	W 672	W 704	W 736	W 768	W 800
22 Feet	C 132	C 176	C 220	C 264	C 308	C 352	C 396	C 440	C 484	C 528	C 572	C 616	C 660
	W 448	W 480	W 512	W 544	W 576	W 608	W 640	W 672	W 704	W 736	W 768	W 800	W 832
24 Feet	C 144	C 182	C 240	C 288	C 336	C 384	C 432	C 480	C 528	C 576	C 624	C 672	C 720
	W 480	W 512	W 544	W 576	W 608	W 640	W 672	W 704	W 736	W 768	W 800	W 832	W 864
26 Feet	C 156	C 198	C 260	C 312	C 364	C 416	C 468	C 520	C 572	C 624	C 676	C 728	C 780
	W 512	W 544	W 576	W 608	W 640	W 672	W 704	W 736	W 768	W 800	W 832	W 864	W 896
28 Feet	C 168	C 224	C 280	C 336	C 392	C 448	C 504	C 560	C 616	C 672	C 728	C 784	C 840
	W 544	W 576	W 608	W 640	W 672	W 704	W 736	W 768	W 800	W 832	W 864	W 896	W 928
30 Feet	C 180	C 240	C 300	C 360	C 420	C 480	C 540	C 600	C 660	C 720	C 780	C 840	C 900
	W 576	W 608	W 640	W 672	W 704	W 736	W 768	W 800	W 832	W 864	W 896	W 928	W 960

Example: A 14' x 20' room has 280 square feet of ceiling and 544 square feet of wall area.

Square footage of interior wall and ceiling surfaces
Figure 25-2

Size of Room	Height of Ceiling 8'	9'	10'	Yards of Border	Rolls for Ceiling
4 x 8	6	7	8	9	2
4 x 10	7	8	9	11	2
4 x 12	8	9	10	12	2
6 x 10	8	9	10	12	2
6 x 12	9	10	11	13	3
8 x 12	10	11	13	15	4
8 x 14	11	12	14	16	4
10 x 14	12	14	15	18	5
10 x 16	13	15	16	19	6
12 x 16	14	16	17	20	7
12 x 18	15	17	19	22	8
14 x 18	16	18	20	23	8
14 x 22	18	20	22	26	10
15 x 16	15	17	19	23	8
15 x 18	16	18	20	24	9
15 x 20	17	20	22	25	10
15 x 23	19	21	23	28	11
16 x 18	17	19	21	25	10
16 x 20	18	20	22	26	10
16 x 22	19	21	23	28	11
16 x 24	20	22	25	29	12
16 x 26	21	23	26	31	13
17 x 22	19	22	24	23	12
17 x 25	21	23	26	31	13
17 x 28	22	25	28	32	15
17 x 32	24	27	30	35	17
17 x 35	26	29	32	37	18
18 x 22	20	22	25	29	12
18 x 25	21	24	27	31	14
18 x 28	23	26	28	33	16

Single-roll wallpaper requirements
Figure 25-3

Work	Unit	Manhours Per Unit
Wallpaper		
Light to medium weight, butt joint	100 SF	1.4
Heavy weight, butt joint	100 SF	1.5
Vinyl wall covering		
Light to medium weight, butt joint	100 SF	1.9
Heavy weight, butt joint	100 SF	2.1
Special wall coatings	100 SF	3.4
Flexwood	100 SF	7.1
Flexi-wall	100 SF	8.1

Labor for installing wall coverings
Figure 25-4

Special Areas
Here's how to estimate paint requirements for special areas:

Balustrades— Measure the front area and multiply by 4.

Lattice work— Measure the front area and multiply by 2.

Cornices— Measure the front area and multiply by 2.

Gutters and downspouts— Measure the front area and multiply by 2.

Stairs— Count the risers and multiply by 8.

Porches— Multiply the length by the width.

Eaves— Measure the area and multiply by 2.

Eaves with rafters— Measure the area and multiply by 3.

Estimating Manhours

Use the following tables to estimate manhour requirements for exterior painting, interior painting and millwork.

Exterior Painting
Before painting begins, the surface must be properly prepared. Preparation for exterior painting includes wirebrushing or sandblasting mill scale from metal surfaces, removing dust, oil and grease, masking adjacent surfaces and removing the masking, any light sanding needed between coats, and any necessary scraping, spot-puttying, and sizing and filling. Figure 25-5 gives typical manhours for exterior painting, including the above preparation work. If other surface preparation is necessary, this will add to the manhours.

A good crew would be one or two men spraying, and one or two men tending (one man mixes and prepares paint for larger crews).

Interior Painting
See Figure 25-6 for manhour requirements for interior painting. This table includes labor for minimum surface preparation, mixing, and application of paint.

Millwork
Figure 25-7 gives the labor requirements for painting millwork. Manhours include moving on and off site, set-up, surface preparation, masking, light

Work	Unit	Manhours Per Unit
Brush, per coat		
Wood siding	1,000 SF	7.5
Wood doors and windows, area of opening	1,000 SF	9.5
Trim	1,000 SF	8.5
Steel sash, area of opening	1,000 SF	5.0
Flat metal	1,000 SF	7.0
Metal roofing and siding	1,000 SF	7.5
Masonry	1,000 SF	7.5
Roller, per coat		
Masonry	1,000 SF	5.5
Flat metal	1,000 SF	4.5
Doors	1,000 SF	7.0
Spray, per coat		
Wood siding	1,000 SF	4.0
Doors	1,000 SF	5.0
Masonry	1,000 SF	6.0
Flat metal	1,000 SF	5.0
Metal roofing and siding	1,000 SF	6.0
Highway or airfield lines and symbols, including glass beads	1,000 SF	8.5
Cementitious paint, including curing	1,000 SF	10.0
Sandblasting steel	1,000 SF	55.0
Wire brush cleaning of steel	1,000 SF	17.5
Clean and spray waterproofing on masonry	1,000 SF	10.0

Labor for exterior painting
Figure 25-5

Work	Unit	Manhours Per Unit
Exterior wood trim, 3 coats	100 LF	1.7
Interior wood trim, 3 coats	100 LF	1.7
Kitchen cabinets, 3 coats	100 SF	2.7
Wood casework, 3 coats	100 SF	2.7
Metal casework, 2 coats	100 SF	1.7
Wardrobes, 3 coats	100 SF	2.7
Bookcases, 3 coats	100 SF	2.7

Labor for painting millwork
Figure 25-7

sanding between coats, removing masking, cleanup, and touch-up as needed.

Work	Unit	Manhours Per Unit
Brush, per coat		
Wood flat work	1,000 SF	8.5
Doors and windows, area of opening	1,000 SF	9.0
Trim	1,000 SF	8.0
Plaster, sand finish	1,000 SF	7.0
Plaster, smooth finish	1,000 SF	6.0
Wallboard	1,000 SF	5.5
Metal	1,000 SF	8.5
Masonry	1,000 SF	7.0
Varnish flat work	1,000 SF	8.5
Enamel flat work	1,000 SF	6.5
Enamel trim	1,000 SF	8.0
Roller, per coat		
Wood flat work	1,000 SF	6.0
Doors	1,000 SF	8.5
Plaster, sand finish	1,000 SF	2.5
Plaster, smooth finish	1,000 SF	3.0
Wallboard	1,000 SF	3.0
Metal	1,000 SF	5.5
Masonry	1,000 SF	3.0
Spray, per coat		
Wood flat work	1,000 SF	2.0
Doors	1,000 SF	3.0
Plaster, wallboard	1,000 SF	2.5
Metal	1,000 SF	3.5

Labor for interior painting
Figure 25-6

In using this table, remember that labor can vary greatly in painting trim. Location, height, amount of masking, surface conditions, extensive preparation, specifications and types of materials used can increase manhours by as much as 300%. Figure 25-7 assumes no abnormal or difficult conditions. Adjust your manhours accordingly when you see problems.

Summary

Many spec builders handle their own painting, and most do good work. If you follow the instructions in this chapter, there's no reason why you can't do a professional job using your regular crew.

Figure 25-8 is your Painting Checklist and Data Reference Sheet.

Painting Checklist and Data Reference Sheet

1) The house is ________ feet in length and ________ feet in width, for a total of ________ square feet.

2) The house is ☐1-story ☐1½-story ☐2-story.

3) The exterior finish is:

☐Brick veneer

☐Siding ☐wood ☐hardboard ☐plywood ☐other (specify) ________________________________

☐Brick veneer and siding (combination)

4) Exterior surfaces received ☐primer plus ☐1 coat ☐2 coats

☐Walls: ____________ SF

☐Gables: ____________ SF

☐Balustrades: ____________ SF

☐Lattice work: ____________ SF

☐Cornices: ____________ SF

☐Gutters and downspouts: ____________ SF

☐Stairs: ____________ SF

☐Porches: ____________ SF

☐Eaves: ____________ SF

☐Eaves with rafters: ____________ SF

Total: ____________ SF

5) Interior surfaces received ☐primer plus ☐1 coat ☐2 coats

☐Walls: ____________ SF

☐Ceilings: ____________ SF

☐Balustrades: ____________ SF

☐Stairs: ____________ SF

☐Shelves: ____________ SF

☐Doors: ____________ SF

Total: ____________ SF

Figure 25-8

6) Interior trim (stain or varnish) ☐sealer plus ☐1 coat ☐2 coats

☐Moldings

☐Base: ____________ LF

☐Cove: ____________ LF

☐Other (specify)

________________ LF

Total: ____________ LF

☐Doors

☐Flush (number: ____________)

☐Panel (number: ____________)

☐Louver (number: ____________)

☐Other (specify)

________________ SF

Total: ____________ SF

7) Painting method: ☐brush ☐roller ☐pad

☐Other (specify) ____

8) Material costs:

Exterior:

Primer paint	________gals	@	$________gal	=	$___________
Latex paint	________gals	@	$________gal	=	$___________
Oil-base paint	________gals	@	$________gal	=	$___________
Stain	________gals	@	$________gal	=	$___________
Enamel	________gals	@	$________gal	=	$___________
Solvent	________gals	@	$________gal	=	$___________
Caulk	______tubes	@	$________ea	=	$___________
Sandpaper	________pcs	@	$________ea	=	$___________

Interior:

Primer paint	________gals	@	$________gal	=	$___________
Latex paint	________gals	@	$________gal	=	$___________

Oil-base paint	________gals	@	$________gal	=	$__________
Solvent	________gals	@	$________gal	=	$__________
Varnish	________gals	@	$________gal	=	$__________
Stain	________gals	@	$________gal	=	$__________
Sandpaper	________pcs	@	$________ea	=	$__________
Brushes	________pcs	@	$________ea	=	$__________
Rollers	________pcs	@	$________ea	=	$__________

Other materials, not listed above:

____________________	__________	@	$__________	=	$__________
____________________	__________	@	$__________	=	$__________
____________________	__________	@	$__________	=	$__________
____________________	__________	@	$__________	=	$__________

9) Painting done by subcontractor:

Name:____

Address:____

Telephone:____

Cost:

☐Materials and labor

☐Primer plus ☐1 coat ☐2 coats $__________

☐Labor only

☐Primer plus ☐1 coat ☐2 coats $__________

10) Painting done by my crew:

☐Primer plus ☐1 coat ☐2 coats

Exterior:

Skilled	________hours	@	$________hour	=	$__________
Unskilled	________hours	@	$________hour	=	$__________

Interior:

Skilled	________hours	@	$________hour	=	$__________
Unskilled	________hours	@	$________hour	=	$__________

Total labor costs:

Skilled ________hours @ $________hour = $__________

Unskilled ________hours @ $________hour = $__________

Other (specify)

__________________________ ________hours @ $________hour = $__________

Total $__________

11) Total cost:

Materials $________________

Labor $________________

Other $________________

Total $________________

12) Painting cost per square foot of floor space: $____________

13) Problems that could have been avoided ______________________________________

____

____

____

____

14) Comments ____

____

____

____

____

15) A copy of this Checklist and Data Reference Sheet has been provided:

☐ Management

☐ Accounting

☐ Estimator

☐ Foreman

☐ File

☐ Other: ____________

Chapter 26

Heating and Cooling Systems

Every heating and cooling system has to be designed for the building it's going to heat or cool. On most jobs, you'll have an HVAC (heating, ventilating, air conditioning) subcontractor select and install the equipment. This isn't the type of work you'll want to do yourself unless heating and cooling requirements are minimal. Most residential builders subcontract both system design and installation to a qualified subcontractor.

But that doesn't mean you can ignore what your HVAC sub does. You inspect his work the same way you inspect the work of your own carpenters.

The only way to be sure your HVAC subcontractor does his job right is to know the difference between quality work and slipshod work. That's the purpose of this chapter.

You'll find more information here than you're likely to need unless you're building apartments on spec. But the concepts in this chapter are basic to all heating and cooling work. We'll cover HVAC bids, system capacity, the most common heating and cooling systems, and ducting.

HVAC Bids

The final plans for your spec house probably won't include a very detailed description of the HVAC work. Every HVAC subcontractor who submits a bid will prepare an HVAC plan and give you a price for that plan.

Each subcontractor may have a different idea about what HVAC equipment is needed. Compare the bids and select the one that seems most competitive. You aren't expected to find a technical flaw in any of the plans. But satisfy yourself that the bidder you select has an adequate plan and is recommending the equipment that is right for your house.

System Capacity

Avoid buying too much or too little heating and cooling capacity. If the equipment is too small, it won't give off enough warm or cool air to handle extreme weather conditions. If the system is too large, it will cycle off before reaching peak efficiency. This makes the system expensive to operate. The correct system capacity depends on the size of the house and the calculated heating and cooling loads.

Heating Systems

Gas, oil and electric systems are the most popular, but heat pumps and solar heating are common in many areas. In some communities, coal heating is still used. Steam and hot water heating are the best choice for larger buildings and for colder climates where a large heating and cooling system is essential.

The type of system to use depends on the climate and what form of energy is available. If your local utility company offers gas hookups, a gas system will probably be the best choice. In many homes, forced-air furnaces are the most economical.

Let's look at each of the common heating systems in detail.

Heat Pumps

Heat pumps have become more popular and more reliable in the last ten years. A heat pump is both a heating system and a cooling system. Think of it as cooling equipment that can run backward. Like any air conditioner, the pump provides cool air during hot weather. In winter, the cycle is reversed so that hot air moves into the home and cold air is expelled into the atmosphere.

But if the weather is extremely cold, the heat pump can't operate efficiently enough to generate adequate heat. So most heat pumps have built-in auxiliary resistance heaters that automatically switch on when the outside temperature drops too low.

Heat pumps cost less to run than resistance-type heaters. But in most cases, a heat pump will still cost more than gas heat.

The big advantage of the heat pump is that the initial cost is little more than the cost of a cooling system alone. And you save the entire cost of gas or oil hookups and piping.

A rule of thumb for estimating the size of heat pump required for your spec house is: *Use 1 ton of refrigeration (12,000 Btu's) for each 600 SF of living area.*

For example, a home with 1,800 SF of living area on one floor will probably need a 3-ton heat pump. For a two-story house with 1,200 SF of living area on the first floor and 800 SF on the second floor, use a 2-ton heat pump for the first floor and a 1½-ton heat pump for the second floor.

Warm-air Heating Systems

This type of system operates by circulating warm air through ducts to each room. A warm-air heating system has three main components:

1) A supply duct system that includes diffusers or registers to carry the warm air from the furnace to the rooms, and a return air system to carry the cool air back to the furnace

2) A source of warm air (usually a furnace or heat exchanger)

3) A control system (thermostat)

The two most common warm-air heating systems are the gravity system and the forced-air system. The most commonly-used warm-air furnaces are coal-fired, oil-fired and gas-fired. Let's look at each of these systems and furnaces.

Gravity system— This system uses natural convection to circulate warm air through the ducts. There's a difference in weight between the heated air leaving the furnace and the cooler air entering the furnace. This difference creates circulation.

Since this force is small, the heat moves slowly and requires large ducts. Also, every heated room must be at least a few feet higher than the bonnet of the furnace. This makes the gravity system impractical for homes without basements.

The amount of heat available depends on the temperature and weight of the air as it leaves the registers. So if you're thinking of installing a gravity system, be sure to consider:

- Size and location of air ducts
- Heat loss from the building
- Heat available from the furnace

Figure 26-1 shows a simple gravity warm-air heating system. Warm air passes from the furnace bonnet (top section of furnace casing), through metal ducts, and to the rooms to be heated.

The horizontal ducts are called *leaders.* The vertical ducts are called *stacks* or *risers.* Stacks connect the registers and are usually located inside partition walls so that duct air is separated from cooler exterior walls. Registers are usually installed in baseboards, in floors, or in the side walls (just above the baseboards).

The cooler air returns to the furnace through return registers and ducts, usually located in the floor. Gravity systems usually have only one or two centrally located return registers, all on the first floor. Return registers can be placed against partition walls. But the best place is against cold exterior walls as long as it doesn't cause the duct run to be too long.

Before connecting with the furnace casing, return ducts usually join a single larger duct which enters the casing near the floor or furnace foundation. This connection must be located below all furnace parts that might radiate enough heat to cause a countercurrent of warm air. A countercurrent would defeat the natural flow required by this type of heating system.

Gravity heating was once much more popular than it is today. Its appeal was that there were no moving parts in the system. That made it both quiet and maintenance-free.

But the disadvantages outweigh the advantages. It's slower than forced-air heating. There's no rush of warm air each time the furnace kicks on. Some rooms always seem to be colder than others. This is because it's difficult to balance low-velocity, high-diameter ducting. Ducts and registers have to be much larger than forced-air ducts and registers. Finally, every heating system that includes cooling capacity needs forced-air circulation.

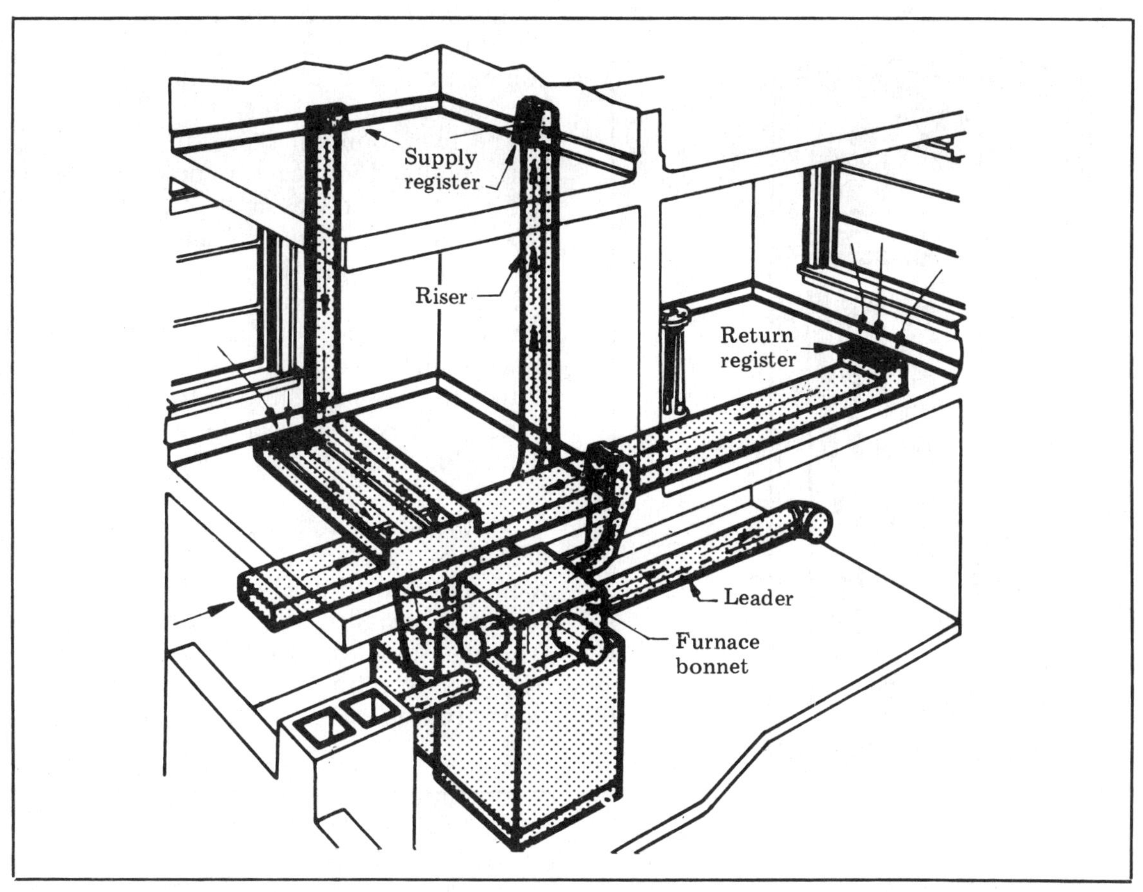

Gravity warm-air heating system
Figure 26-1

For all practical purposes, gravity heat has been replaced by forced-air heating systems. In your spec home, do what other builders do. Install the furnace that buyers expect to find.

Forced-air system— This system uses a centrifugal fan or blower, not natural convection, to circulate the air. Figure 26-2 shows a forced-air furnace unit that circulates air with a motor-driven blower.

For good distribution of heated air, registers must be in the right locations. Many designers locate supply registers near the floor, and set them so they direct air upward along the wall. You can also place them high in the wall and use horizontal vanes to deflect air downward. In either case, the supply registers should blanket a cold wall. That way, warm air mixes with cold air descending from an exterior wall or glass areas.

Locate return registers near the greatest outside exposure and as close to the furnace as possible. This will insure that the coldest air goes directly to the furnace for heating. It also helps eliminate cold drafts across the floor.

To control warm-air distribution, dampers are installed at the registers or in the warm-air ducts. Locate dampers either at the branch take-off or at the warm-air outlet. Use arrows to indicate the damper position, and be sure to label the proper summer and winter settings.

Ducts can be either rectangular or round. They're usually made of sheet steel covered with insulation and a synthetic fabric. Duct size will vary depending on system capacity.

Warm-air ducts that pass through cold spaces, such as attics or crawl spaces, are normally insulated with at least 2-inch-thick duct insulation. On the special connections to and from the furnace casing and fan casings, use strips of fire-resistant fabric to reduce noise and vibration.

Warm-air furnaces— These furnaces can be made of either cast iron or steel. The minimum thickness for cast-iron furnace sections is 1/2-inch. The cast-iron furnace resists corrosion and high temperature. Because of its relatively large mass, it has a large heat-storage capacity. However, the cast iron furnace is slow to respond to changes in heat requirements.

The metal parts of a steel furnace are joined by riveting, welding or both. Because of their relatively small mass, they deliver heat quickly and can

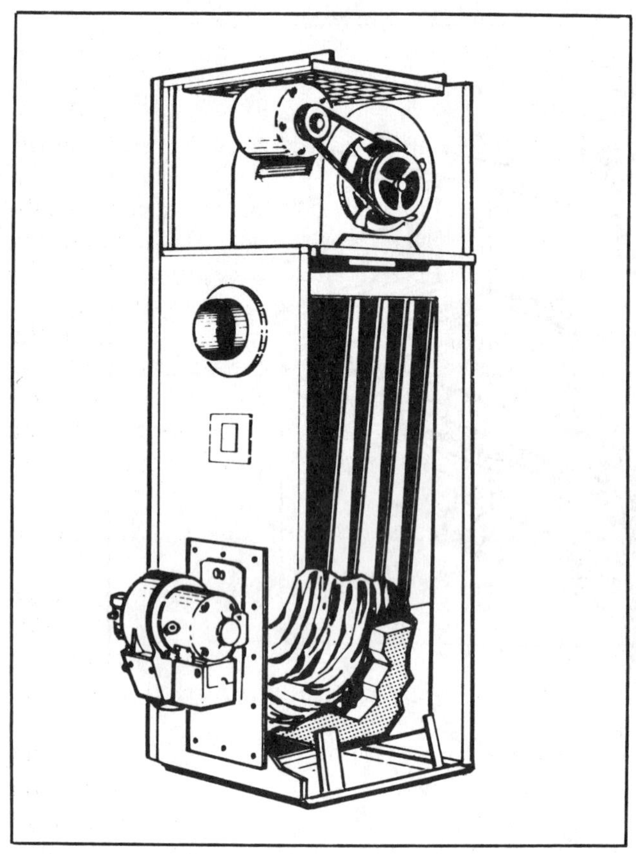

Forced-air furnace unit
Figure 26-2

adapt to changes in heat requirements. However, their heat-storage capacity is limited.

The three most common types of warm-air furnaces are: coal-fired, oil-fired and gas-fired.

Coal-fired furnaces are made of cast iron or steel. Cast-iron coal-fired furnaces are built in sections and then made gastight with furnace cement, asbestos rope, or both. The radiator (where room air is heated) is usually located on top of the combustion chamber.

Steel coal-fired furnaces are made of heavy-gauge steel. They are riveted and caulked or welded at the joints to make them gastight. The fronts, which include the fire doors, ash pit doors and draft doors, are usually cast iron. Small steel furnaces have a single radiator attached to the rear of the combustion chamber. Large steel furnaces may have two radiators installed on the furnace sides.

Oil-fired furnaces need greater combustion-chamber volume and larger heating surfaces. Oil furnaces are usually the blow-through type. The air-space pressure is kept higher than the gas-space (combustion chamber or flue) pressure.

Most oil furnaces are larger than gas furnaces. But there are compact fan-furnace-burner units available for basements, closets and attics.

Gas-fired furnaces are direct-fired. This means that the heat is transferred directly from the hot combustion gases to the air that circulates around the furnace and radiator.

The two common types of gas-fired furnaces, horizontal and vertical, are described below. Each has a gas burner, gas controls, fan (on forced-air types), filters, heat exchangers, casing, and sometimes a humidifier.

Every gas-fired furnace also has a draft diverter. On some models, it's built into the furnace. On other models, it's installed separately on the smoke outlet to the breeching. Most gas burners are Bunsen types and operate with a nonluminous flame. Figure 26-3 shows a gas-fired warm-air furnace.

On the horizontal furnace, the fan, filters, and heat exchanger are side-by-side. This means that the horizontal furnace can be suspended from the ceiling or installed in attics or crawl spaces. This type of installation won't take up any floor space. The only objection to attic or crawl-space installation is that it makes service difficult.

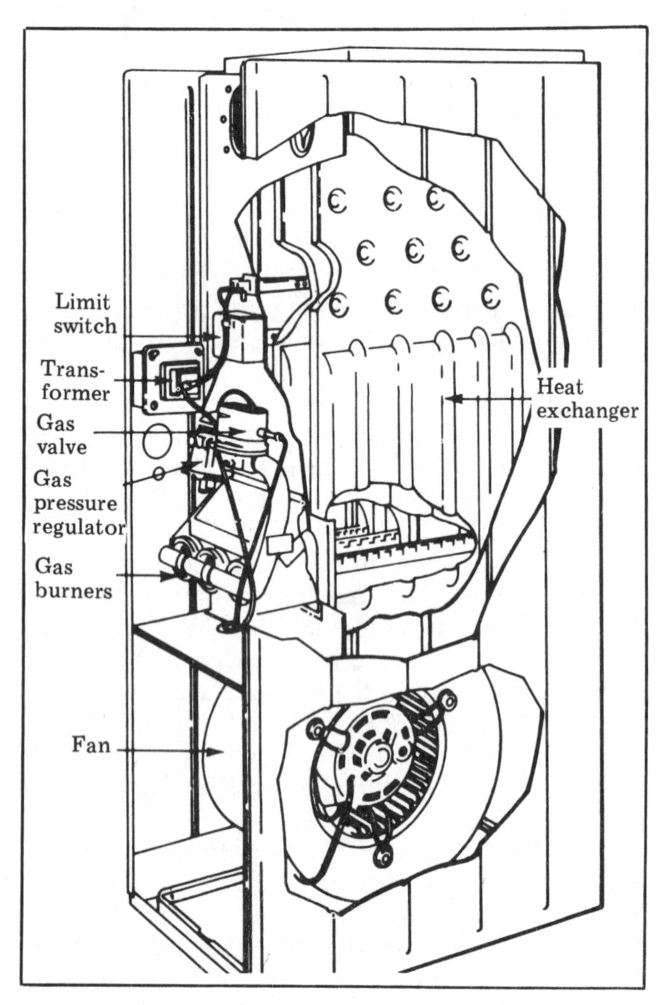

Vertical gas-fired warm-air furnace
Figure 26-3

The vertical furnace has the same components as the horizontal furnace, but the vertical furnace is designed for floor installation. There are two types: the counterflow-type (discharges air downward) and the duct-type (for basement installation).

Counterflow furnaces are used when heated air is to be ducted through a crawl-space, and the furnace can't be placed in a basement.

The duct-type furnace is designed especially for installation in a duct system with a central air supply. It's built exactly like a gas-fired unit heater except that it has no fan. Flanges are added to allow for mounting in the duct system.

Be careful to allow enough clearance between a gas-fired duct-type furnace and the nearest combustible material. Clearances are specified by the furnace manufacturer. Include a limit control to prevent overheating in case of an air-supply failure.

If the duct (in which the furnace is installed) also carries refrigerated air for an air-conditioning system, install a bypass duct with dampers to shut off and isolate the furnace unit when the system is used for cooling.

Gas-fired furnace manufacturers use several types of manifold controls. They usually consist of the following items:

1) A main valve (or stop cock) at the upstream side of the manifold controls. The pilot takeoff valve is usually installed just upstream from this stop cock so that the pilot can be lighted without turning on the main gas supply.

2) A low-pressure gas regulator to reduce gas pressure from the system or from the building gas pressure regulator. The low-pressure gas regulator delivers gas to the burners at a uniform pressure, regardless of the fluctuation in upstream pressure.

3) An automatic shut-off safety valve to stop gas flow to the burner if the flame fails or the pilot goes out.

4) A gas burner primary control to regulate the gas flow to the burners by opening or closing in response to signals from the thermostat.

Here are some points to remember when installing the gas line:

Use a short gas line and as few fittings as possible. Use pipe joint compound sparingly and on male threads only. Place piping where it will be accessible, and support it adequately. Avoid installing gas pipe through air ducts, supporting beams, or under floor slabs. Install a ground joint union close to the control assembly of the unit, and provide a drip leg on the vertical riser that supplies the unit.

The diaphragm-operated gas regulator is installed in the gas line and upstream from the operating valve. It provides constant gas pressure at the burners. A bleeder pipe is installed in the chamber (above the diaphragm) and is vented into the burner chamber, or into an opening in the flue gas vent above the downdraft diverter. A constant escape of gas from the bleeder pipe warns the homeowner that the diaphragm is ruptured and must be replaced.

Here are some important considerations when venting any gas furnace:

Flue pipe diameter must be at least as large as the flow connection and never less than 3 inches. For horizontal runs, maintain a minimum upward slope of 1 inch per linear foot. A pitch of at least 45 degrees is advisable in cold attics.

The flue pipe must run as directly as possible and with a minimum number of turns. Chimneys and flues running through roofs must be extended at least two feet above the roof peak (or other object) within a 10-foot radius. All flue pipe that extends through the roof must be equipped with a hood.

Every gas furnace vent should develop a positive draft that sucks up all the products of combustion and expels them outside the building. These products are mainly carbon dioxide, nitrogen and water vapor.

As you might expect, condensation can be a problem. The vent should be designed to prevent condensation. Burning 100,000 Btu's of fuel gas will produce about 11.5 gallons of water. Your subcontractor should be aware of this potential problem and have a plan for avoiding it.

The vent shouldn't leak flue gas at the joints or at the draft hood. Discharge of flue gas within a building can be deadly. And, of course, the vent system must protect the building from fire.

Your building code probably requires that furnace rooms be ventilated. A common requirement is for an air inlet opening of at least one square inch of free vent area for each 1,000 Btu's per hour of furnace input rating. Two hundred square inches of free vent area is considered the minimum.

Locate the inlet opening at or near the floor line whenever possible. Another louvered outlet-opening is needed at or near the ceiling. It should have free vent area of at least one square inch per 1,000 Btu's per hour of furnace input rating. Again, 200 square inches is considered the minimum allowable free vent opening.

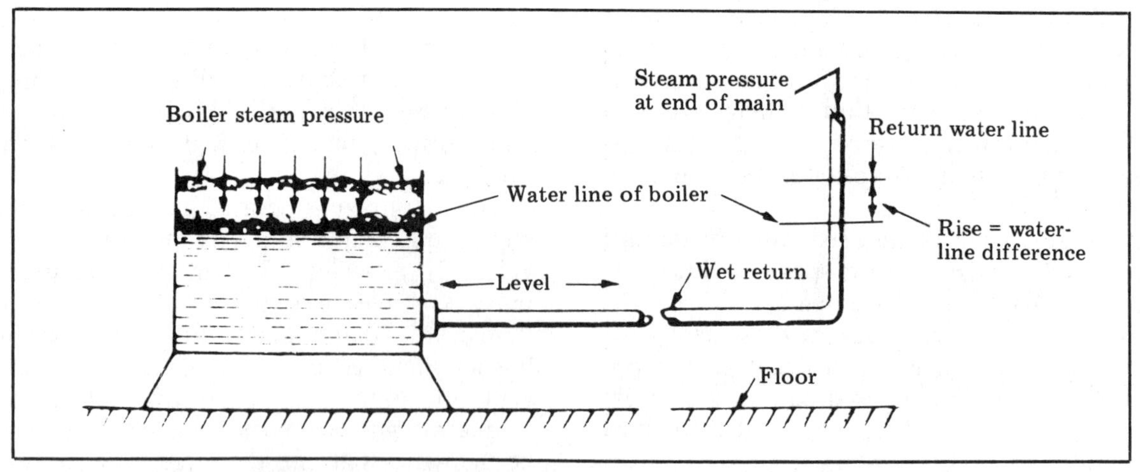

Boiler steam pressure and steam pressure at end of main
Figure 26-4

Steam Heating Systems

When water is heated to the boiling point, it evaporates and produces steam. This steam is a very effective heat source. The quantity of heat in each pound of steam depends on the pressure and temperature. The higher the pressure and temperature, the more heat available.

Steam can be generated and used either as saturated steam (which is dry) or as wet steam. Wet steam has moisture in the form of mist or spray. Saturated steam is commonly used for space heating.

When steam has a temperature higher than its corresponding saturation pressure, it's called "superheated" steam. This is often used in central heating plants when it's important to avoid condensation in the steam lines.

Here are the most common classifications of steam heating systems:

One-pipe and two-pipe systems— A steam heating system is known as a one-pipe system when a single main line serves the dual purpose of supplying steam to the heating unit and returning condensate back to the boiler. The two-pipe system has two piping connections. Steam and condensate flow in separate mains and branches.

Up-feed and down-feed systems— This describes the direction of steam flow in the risers.

Dry-return and wet-return systems— This tells you whether the condensate mains are located above or below the water line of the boiler (or condensate receiver).

Gravity-return systems— When condensate is returned to the boiler by gravity, the system is known as a gravity-return system. In this system, all heating units must be located higher than the water line of the boiler.

Note in Figure 26-4 that the boiler and wet return form a U-shaped container. The boiler steam pressure is on the top of the water at one end. The steam main pressure is on the top of the water at the other end. The difference between these two pressures is the pressure drop in the system. The water in the steam main pressure end will rise to balance the two pressures. It will rise far enough to produce a flow through the return pipe and overcome the resistance of any check valves.

Mechanical-return systems— When steam condensate can't be returned to the boiler by gravity, traps or pumps must be used. This is called a mechanical-return system.

There are three general types of mechanical condensate return devices: the alternating return trap, the condensate return pump, and the vacuum return pump.

Alternating-return systems— In systems where pressure conditions vary between that of a gravity-return system and a mechanical-return system, a boiler return trap or alternating receiver is used. This system is known as an alternating-return system.

Condensate-pump return systems— Where condensate is pumped (under atmospheric pressure) to the boiler.

Vacuum-pump return systems— Where condensate is pumped (under vacuum conditions) to the boiler.

Steam heating systems are also classified as high-pressure, low-pressure, vapor, and vacuum systems, depending on the operating pressure.

Now that we know the terminology used to classify steam heating systems, let's look at each of the most popular systems in detail. We'll also touch on three important steam heating system

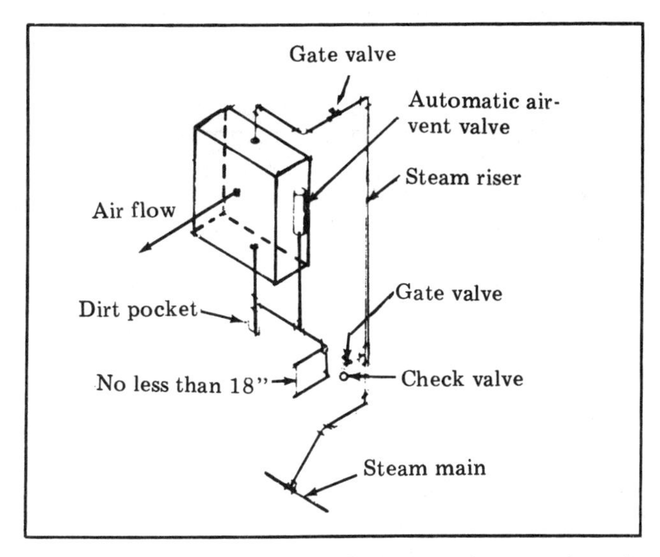

Two-pipe connections to unit heaters in one-pipe air vent systems
Figure 26-5

components: boilers, pumps and traps.

One-pipe systems— Radiators and other heating units generally have only one piping connection from the steam main to the unit heater as shown in Figure 26-5. But it's possible to use two connections to the same main.

There are several variations in the piping arrangement of a one-pipe system:

Up-feed one-pipe systems have radiators and other heating units above the supply mains. The mains carry both steam and condensate. See Figure 26-6. Figure 26-7 shows a typical connection to radiators or risers. Figure 26-8 shows a piping detail when the size of the main is changed.

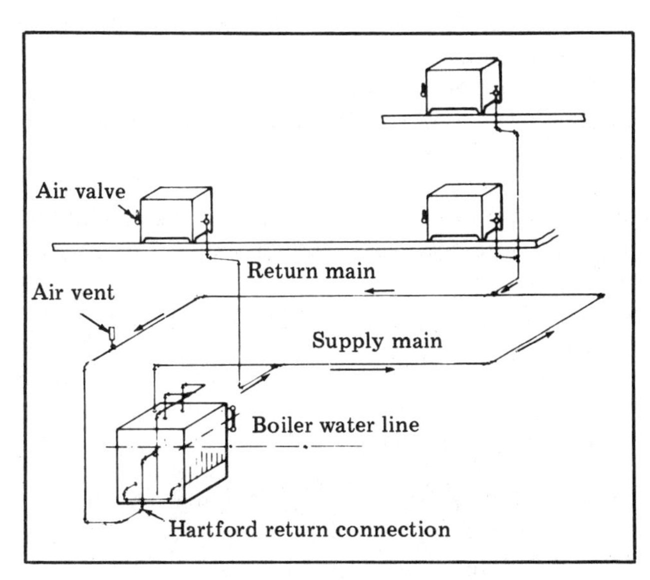

Up-feed gravity one-pipe system
Figure 26-6

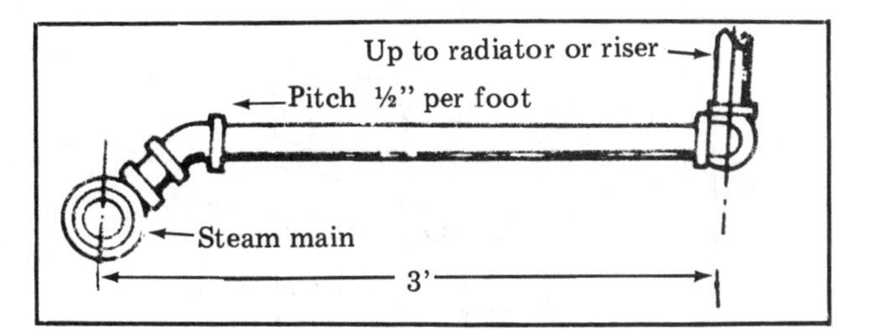

Connection to radiator or riser
Figure 26-7

In larger up-feed one-pipe systems, the mains are dripped at each radiator connection to a wet return. This way, the steam main carries a minimum of condensate. See Figure 26-9. Typical connections to radiators and risers are illustrated in Figure 26-10. Up-feed systems are not recommended for use in buildings higher than four stories.

Down-feed one-pipe systems have the radiators and other heating units located below the supply main. Only risers and connections to heating units carry both steam and condensate, and both flow in the same direction. The steam main is kept relatively free of condensate by dripping through the drop risers.

Each radiator or heating unit in a one-pipe system needs a thermostatic air valve. This valve opens to release air from the heating unit, and then closes when steam heats the thermostatic element of the valve.

Steam circulates better in one-pipe systems if there are quick-vent air valves at the ends of each line and at intermediate points where the steam main is brought to a higher elevation.

There are two types of air valves: pressure valves and vacuum valves. The pressure valve opens to admit air into the system when steam pressure falls below normal air pressure. The vacuum valve has a small check valve that keeps air from flowing back into the system. This helps maintain a vacuum in the piping. Systems that use vacuum valves are known as vapor or vacuum one-pipe systems. Vapor or vacuum systems usually maintain more uniform temperatures than pressure systems.

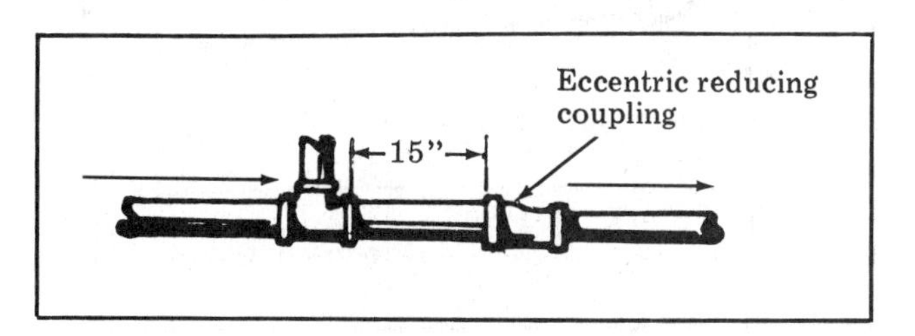

Changing sizes of steam main
Figure 26-8

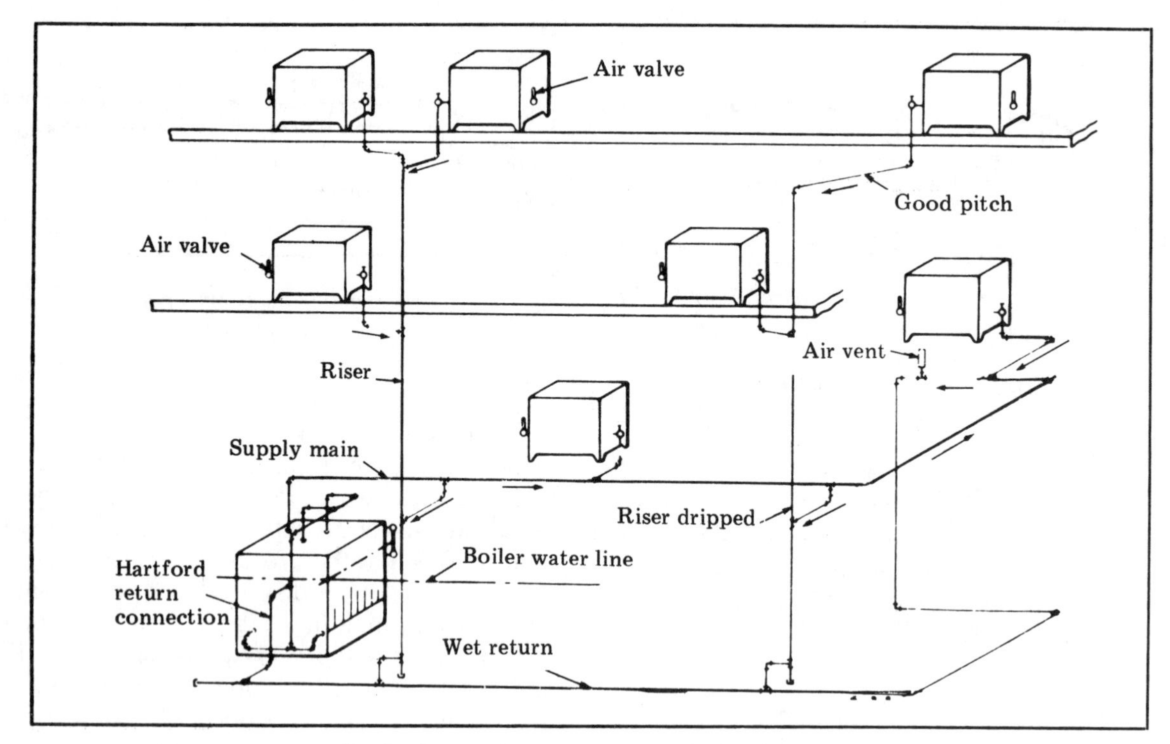

Up-feed one-pipe system with dripped risers
Figure 26-9

You'll probably want to provide a valve at each heating unit in a one-pipe system. The valve gives the room occupant a way to reduce the heat if necessary. It's not essential to provide a valve at each unit, but it helps save energy when the room isn't occupied. Valves on one-pipe systems have to be either fully opened or fully closed.

Two-pipe systems— The two-pipe high-pressure system operates at pressures above 15 psi, usually from 30 to 150 psi. They're common in large in-

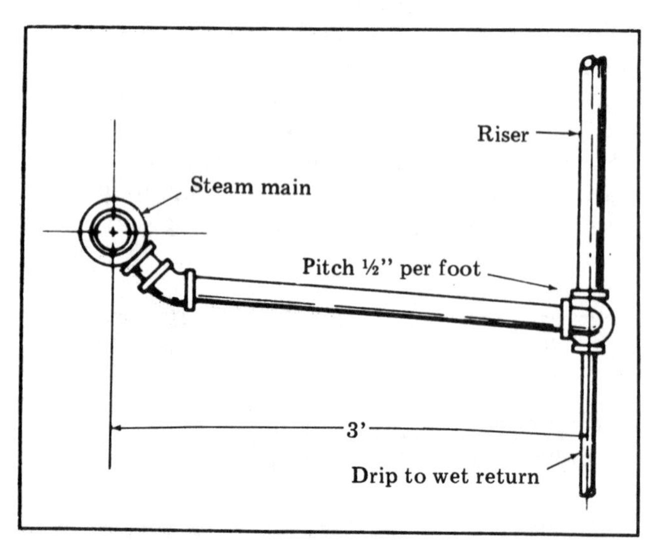

Steam runout where risers are dripped
Figure 26-10

dustrial and commercial buildings where steam is already available. Heat is transferred to the rooms by fan units.

Figure 26-11 shows a typical high-pressure system. Because of the difference in steam pressure between steam and return mains, the returns can be above the heating units. Pressure lifts the condensate to these returns.

The condensate can be flashed into steam in low-pressure mains if any are available, or passed through an economizer heater before being discharged to a vented receiver.

Return traps in high-pressure systems will be either the bucket, inverted bucket, float or impulse type.

Low-pressure systems— These operate at pressures between 0 and 15 psi. The piping layout for both up-feed and down-feed low-pressure systems is the same as piping in a two-pipe vapor system. The only real difference between the two systems is in the type of air valve used. The air valves in low-pressure systems usually don't have check discs. Without check discs, operation under a vacuum is impossible.

Low-pressure systems aren't as popular as vapor systems. The disadvantage is that they don't hold heat when the rate of steam generation is falling. They also tend to corrode faster than vapor systems, because new air keeps coming into the system.

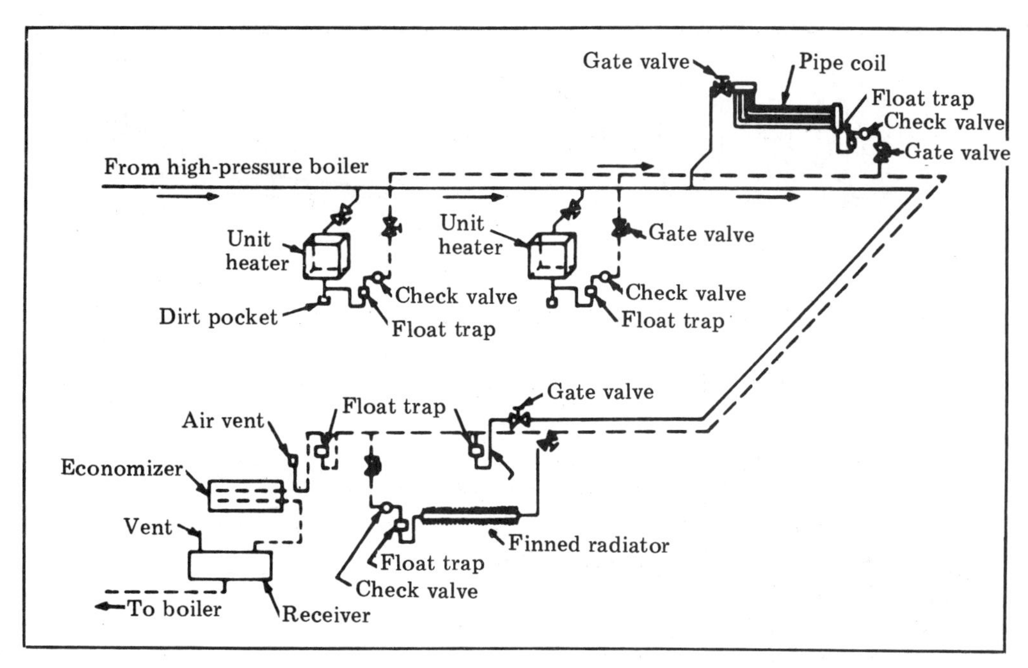

High-pressure steam heating system
Figure 26-11

Low-pressure systems do have some advantages though. They return condensate to the boiler without retaining it in the piping. Figure 26-12 shows a typical low-pressure system with condensate pump.

Vapor systems— Two-pipe vapor systems operate at pressures from 20 inches of vacuum to more than 15 psi without the use of a vacuum pump. Figure 26-13 shows a typical two-pipe up-feed vapor system. Figure 26-14 shows a common two-pipe down-feed system. The way to drip drop-risers in a down-feed system is illustrated in Figure 26-15. Radiators discharge their condensate and air through thermostatic traps to the dry-return main.

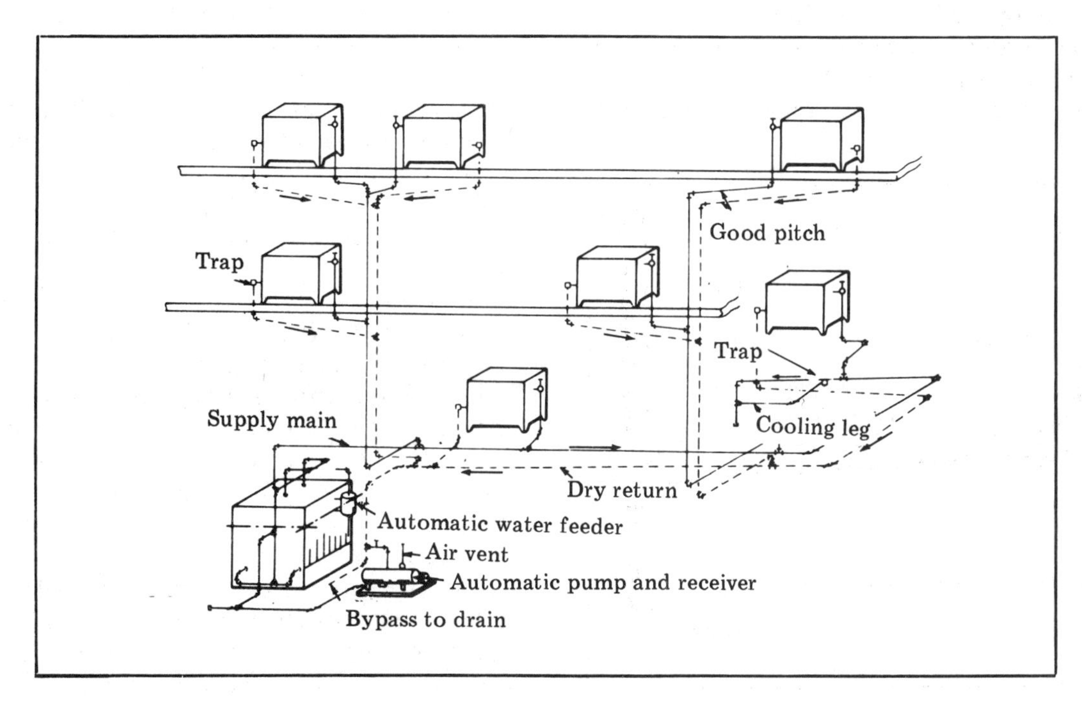

Low-pressure system with condensate pump
Figure 26-12

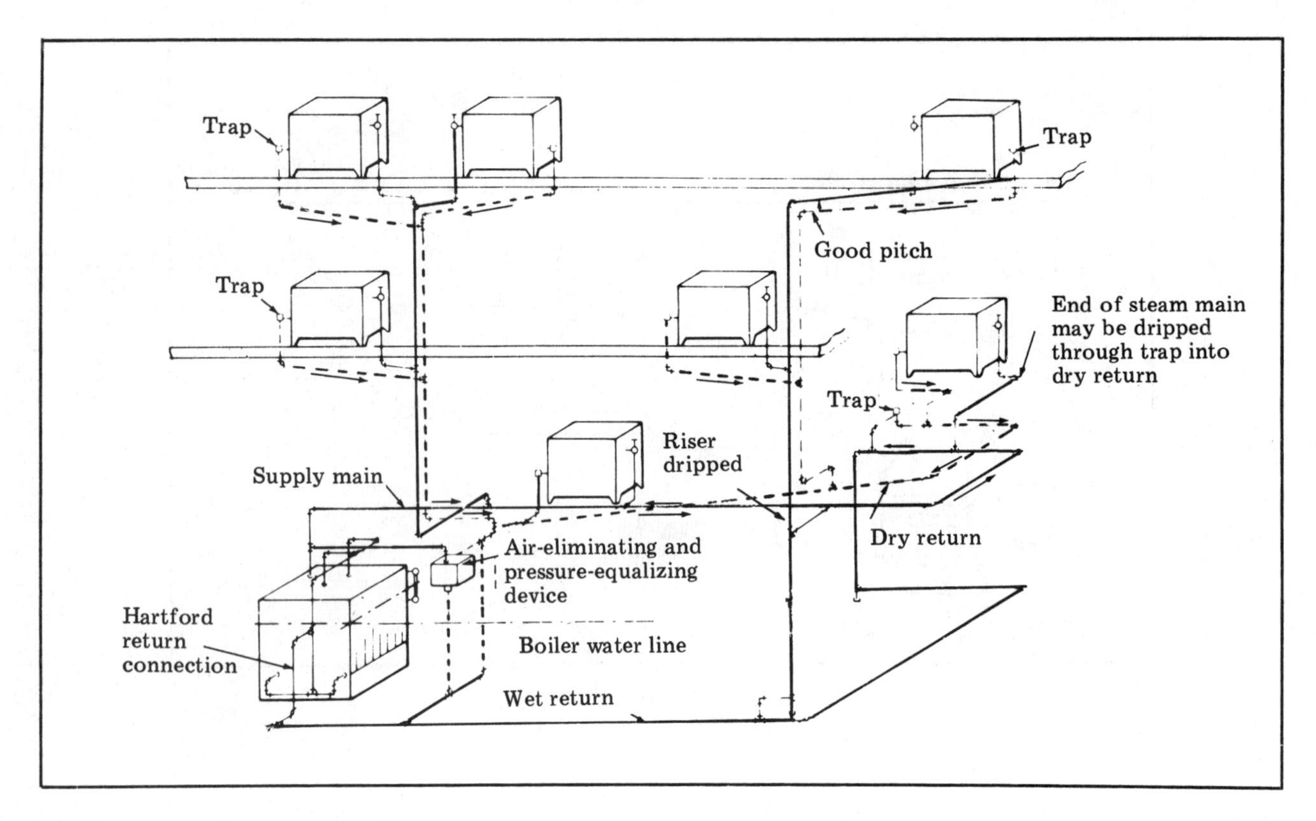

Two-pipe up-feed vapor system
Figure 26-13

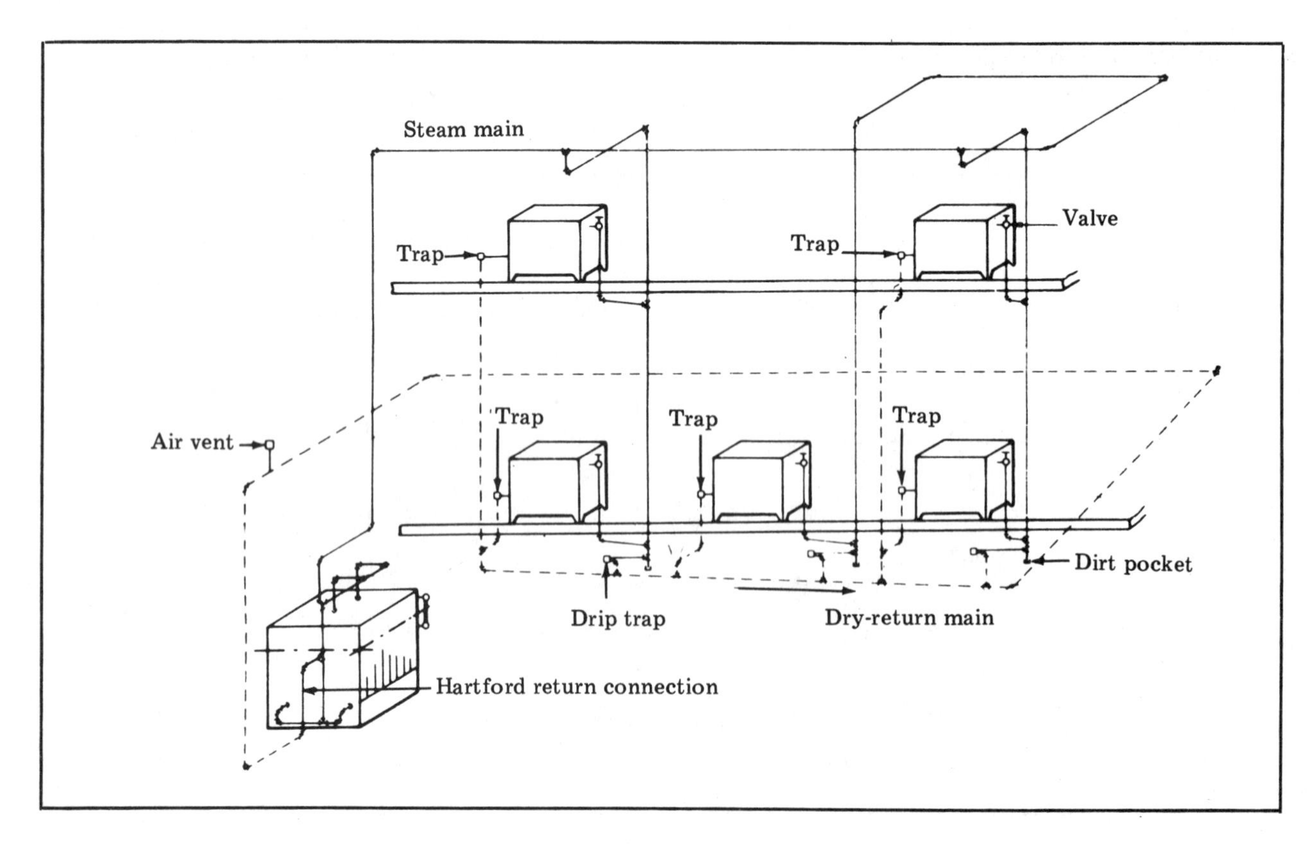

Two-pipe down-feed system
Figure 26-14

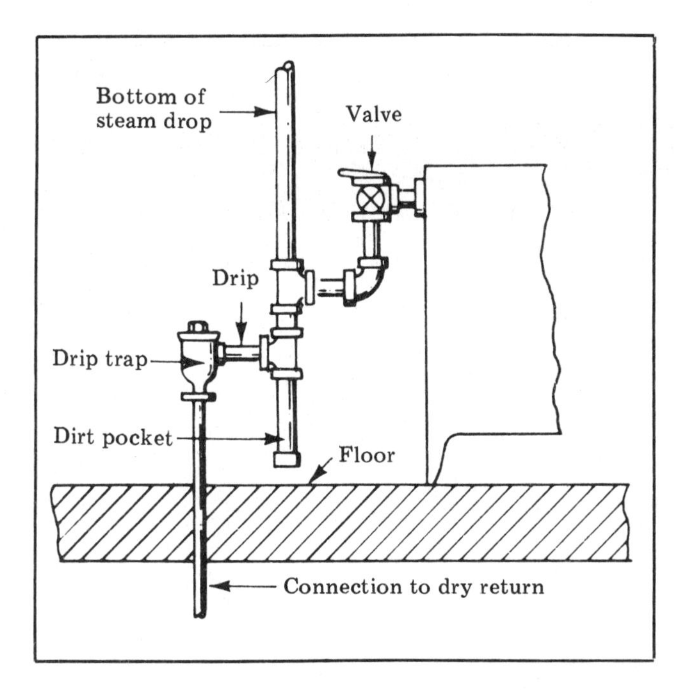

Drip connections at bottom of down-feed steam drop
Figure 26-15

When the system is pressurized, air is eliminated at the ends of the supply and return mains just before they drop to the wet return.

Vapor systems can also have an automatic return trap or alternating receiver. This automatically returns condensate to the boiler when boiler pressure prevents the return of condensate by gravity. Typical connections for an automatic return trap are illustrated in Figure 26-16.

Each heating unit in a vapor system, as in all two-pipe systems, must have a graduated or modulating valve. The valve controls heat in the radiator by controlling the passage of steam.

Vacuum systems— These can operate either under low pressure or under a vacuum. But there will always be a vacuum pump on the return piping to guarantee a vacuum whenever the system is working.

Figure 26-17 shows a typical two-pipe up-feed vacuum system. A down-feed arrangement is shown in Figure 26-18.

Return risers are connected in the basement to a common return main which slopes downward toward the vacuum pump. The vacuum pump sucks air and water out of the return pipe and separates the air from the water. Air is expelled, and water goes back to the boiler. It's important that there's no connection from the supply side to the return side at any point except through a trap.

Subatmospheric systems— These are similar to vacuum systems. The difference is that the temperature can be controlled by varying the heat output from each radiator. Radiator heating varies with the pressure, temperature and volume of steam in circulation.

Subatmospheric systems have a controllable partial vacuum on both the supply and return sides. Ordinary vacuum systems control only the return side. In a vacuum system, steam pressure will usually be higher than air pressure in the supply mains and radiator. In a subatmospheric system, steam supply-line pressure will be above air pressure only in the coldest weather. Under most conditions, steam will be under partial vacuum.

We've looked at the common ways a steam heating system can be laid out. You'll agree that most of what we've covered is better left to the subcontractor who designs and installs the heating system for your building. But what you've learned in this chapter will provide important background knowledge when you have to select among competing bids or evaluate the performance of your HVAC subcontractor.

Steam heating system components— Before we leave steam heating systems and go on to cooling, we should touch on three important system components: boilers, pumps and traps.

Boilers: Cast-iron sectional heating boilers usually have several outlets at the top. It's good practice to use two or more of these outlets. They help reduce steam velocity in the vertical uptakes. They also help keep water out of the steam main.

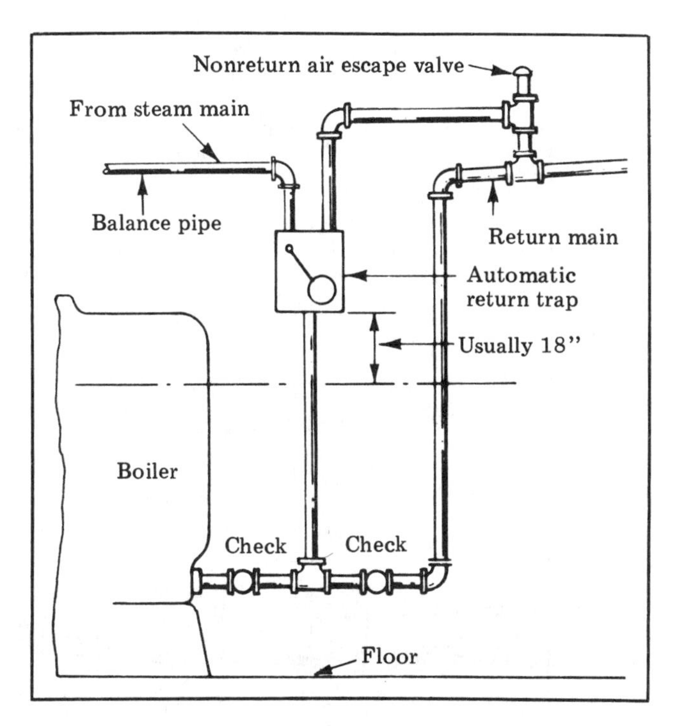

Connections for automatic return trap
Figure 26-16

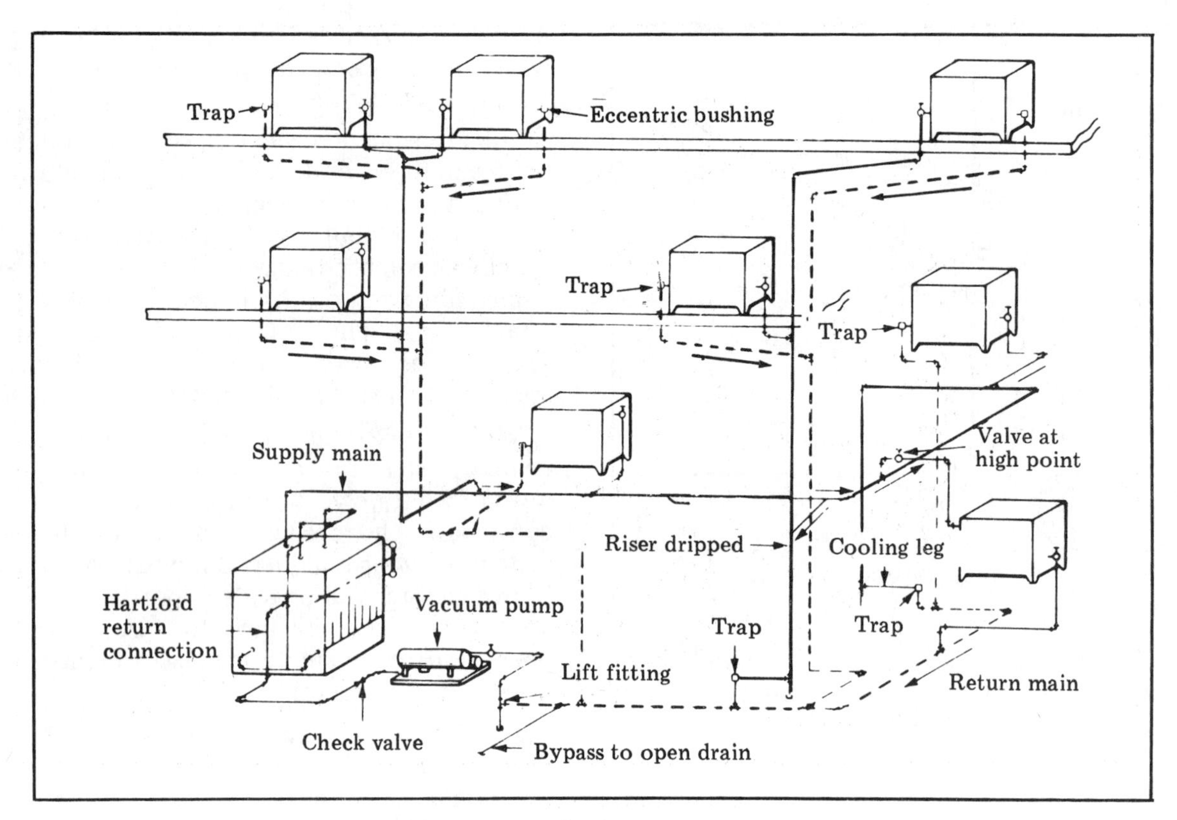

Two-pipe up-feed vacuum system
Figure 26-17

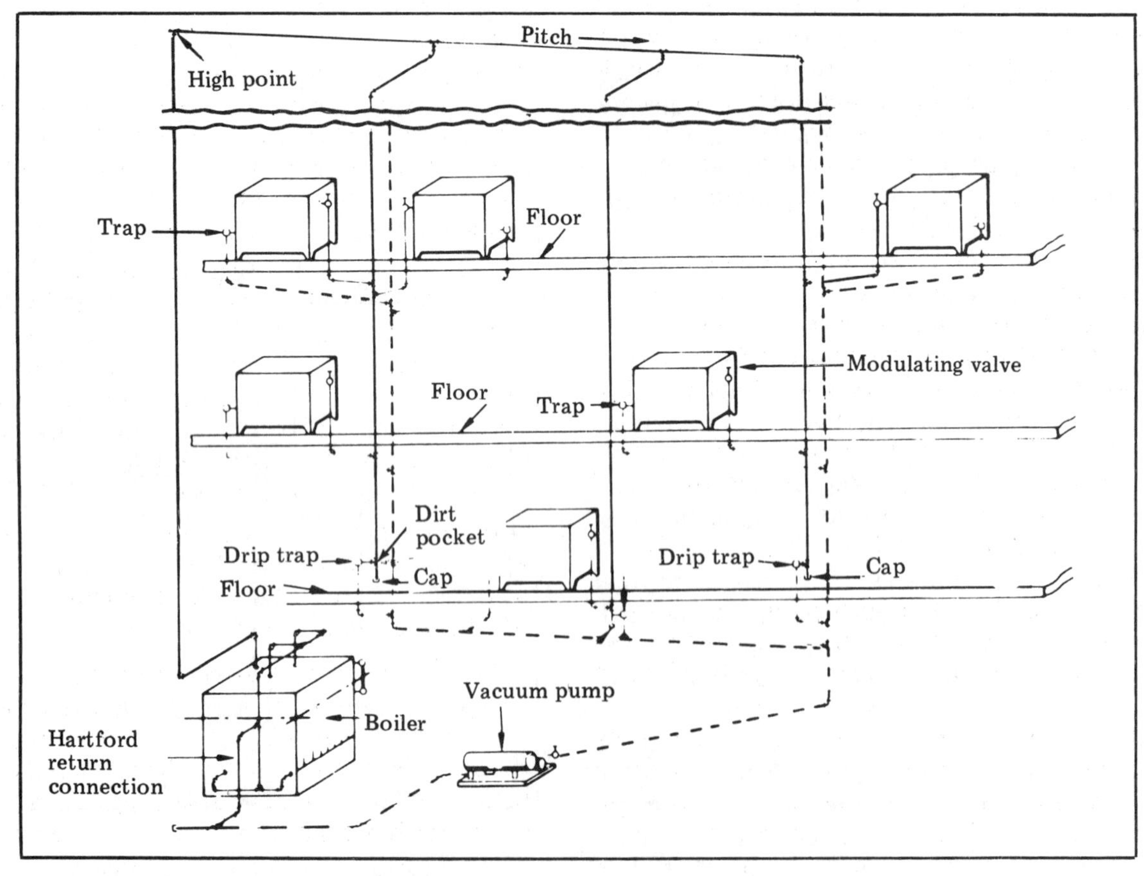

Two-pipe down-feed vacuum system
Figure 26-18

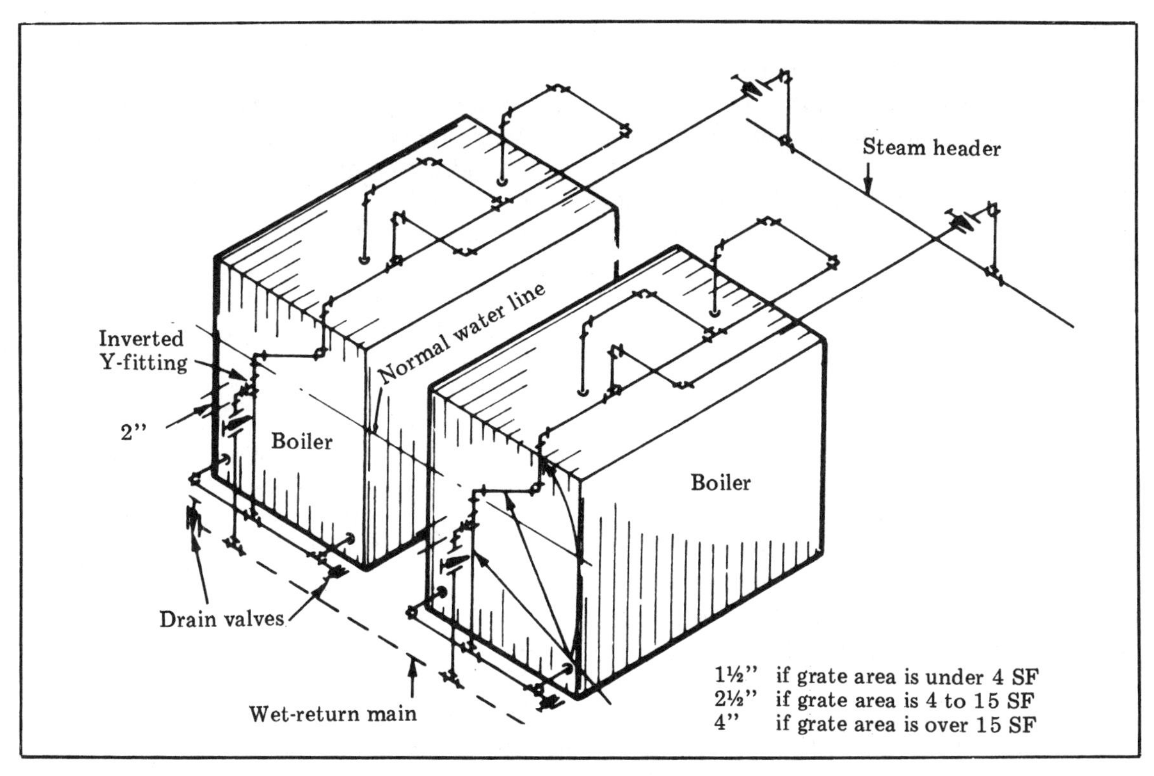

Hartford return connection
Figure 26-19

Cast-iron boilers usually have return tappings on two sides. Steel boilers usually have only one return tapping. If the boiler your subcontractor installs has two tappings, be sure he uses both. This insures better circulation through the boiler.

The return needs either a Hartford return connection or a check valve to prevent accidental loss of boiler water. Whenever possible, use the Hartford return connection. Check valves sometimes stick slightly open. Also, check valves tend to restrict the flow of condensate coming back to the boiler. The Hartford return connection for a one- or two-boiler installation is shown in Figure 26-19.

The Hartford return connection provides a direct connection (made without valves) between the steam side of the boiler and the return side of the boiler. It includes an inverted Y-fitting (or close nipple) about 2-inches below the normal boiler water line.

Pumps: Condensate return pumps are used in gravity systems when the building shape won't permit condensate to return under system pressure to the boiler.

A condensate pump on a low-pressure heating system will have a motor-driven centrifugal pump with a receiver and an automatic float control. Rotary, screw, turbine and reciprocating pumps are also available with either steam turbine or motor drive. There are also direct-acting steam reciprocating pumps.

Vacuum pumps are needed on vacuum systems where the returns are under a vacuum. They are also necessary on subatmospheric systems where the supply piping, radiation and returns are under a vacuum. The pump discharges air and non-condensable gases into the atmosphere and returns condensate to the boiler.

The common type of vacuum pump comes complete with receiver, separating tank and automatic controls mounted as a single unit on one base.

In most systems, the vacuum pump has a regulator that cuts in when the vacuum drops to a certain point. It then cuts out when vacuum reaches the high limit point. These points can be set to suit operating conditions.

In addition to this vacuum control, the pump will have a float control which starts the pump whenever there's enough condensate in the receiver, regardless of the vacuum in the system.

Traps: These will trap and hold steam until it has given up its heat. The trap allows condensate and air to pass as soon as they accumulate. Traps have a bucket to accumulate the condensate, a hole through which the condensate passes, a valve to close that hole, a device to operate the valve, and inlet and outlet openings so condensate can get in and out.

Cooling Systems

In many parts of the country, buyers will expect to find some type of cooling system in their prospective new home. This usually means central air conditioning.

Cooling capacity is measured in tons of refrigeration. One ton of refrigeration equals 12,000 Btu's per hour.

There are four basic parts in every mechanical refrigeration unit: the evaporator, the expansion valve, the condenser and the compressor. We'll look at each of these components and also touch briefly on cooling towers and air handlers.

Evaporators

The evaporator is the coil where the cooling system captures unwanted heat. Pressurized liquid refrigerant is released into the evaporator by an expansion valve. The refrigerant expands into gas, becoming very cold as the pressure drops. This cold refrigerant gas then absorbs heat from the air or water that surrounds the coil.

You can feel the same thing happen at a gas station if you hold your hand in front of a tire inflator and open the valve. Pressure drops suddenly as air leaves the hose. The air turns cold as the pressure drops. Your hand gives up heat as the air passes over it.

This process is reversed when air (or refrigerant) is compressed. The air gives off heat, warming whatever holds the compressed air (or refrigerant). The refrigerant leaves the evaporator loaded with heat and in the form of a gas.

Two common types of evaporators are direct-expansion evaporators and shell-and-tube evaporators.

Direct-expansion evaporators— These use a fin-and-tube type coil similar to a car radiator. The coil cools any air that comes in direct contact with the cold exterior surface of the coil. The ducts can then move this cool air to where it's needed.

Some direct-expansion (or "DX") systems don't use any ducts. The evaporator coil is installed right where the cooling is needed. A window-type air conditioner is an example.

Shell-and-tube evaporators— These use direct-expansion principles, but have a heavy shell that holds pressurized water against the outside of the refrigerant tubes.

The liquid refrigerant enters the evaporator and passes back and forth through the inside of the tubes. By the time the refrigerant reaches the discharge point, it has absorbed enough heat to evaporate. As it evaporates, it cools the water surrounding the tubes.

Water being chilled moves from one end of the shell to the other, flowing up and down over the refrigerant-carrying tubes. Baffles direct the flow. Water temperature at the discharge point can be precisely regulated.

Expansion Valves

The expansion valve controls the flow of the liquid refrigerant into the evaporator. The expansion valve also maintains high pressure in the condenser.

Condensers

The condenser pressurizes refrigerant gas into liquid, causing it to give off the heat it absorbed in the evaporator. There are three types of condensers: air-cooled, water-cooled and evaporative.

Air-cooled condensers— These are usually a fin-and-tube coil and look about like the direct-expansion evaporator coil. The refrigerant vapor is pumped into the tubes, giving up heat to air circulating across the outer surface. This air is usually discharged outdoors.

Air-cooled condensers are common on small cooling systems and when it would be impractical to circulate water across the condenser coils. Cars, buses, trucks, and window air-conditioners use air-cooled condensers.

There are two disadvantages to using air-cooled condensers. They need power to operate the fan that moves the air. And the condenser's capacity is lowest on hot days when maximum capacity is required. Still, air-cooled condensers are popular in both residential and commercial buildings. In many areas, they're still the most economical and convenient cooling system.

Water-cooled condensers— If the water supply is adequate and economical, a water-cooled condenser is the best choice, especially in larger commercial buildings. Most water-cooled condensers are the shell-and-tube type. They're more compact than either the air-cooled condenser or the evaporative condenser, and no fan is required.

At first glance, a water-cooled condenser looks like a direct-expansion water chiller. But there's one big difference. Most water-cooled condensers have cooling water flowing inside the shell but outside the tubes. This is exactly the opposite of the direct-expansion evaporator.

In a shell-and-tube water-cooled condenser, water is brought in one end, runs back and forth inside the tubes, and is finally removed. Most shell-and-tube water-cooled condensers have an even number of passes so the outlet water connection is at the same end of the shell as the inlet connection.

Water picks up heat as it flows through the condenser and will leave the condenser several degrees warmer than it was when it entered.

Evaporative condensers— This is a different type of condenser than either the air-cooled or water-cooled condenser. It uses an entirely different principle, evaporation, to cool and condense the refrigerant.

As the refrigerant is passed through the finless coil tubes, water is sprayed downward over the outside of the tubes. At the same time, air from either outdoors or indoors is blown or drawn upward through the water spray and over the outside of the tubes. Some of this water evaporates as the air moves over it. The remainder is collected in a tank under the tubes and eventually gets pumped back to the spray nozzles again.

Compressors

Compressors fall into two categories: reciprocating and centrifugal.

Reciprocating compressors— In this type of compressor, pistons pressurize the refrigerant. The pistons are driven by a connecting rod and crankshaft which is turned by an electric motor.

Small reciprocating compressors are usually sealed in one case with the motor that drives the compressor. That's called a hermetically sealed unit.

Centrifugal compressors— If you fill a bucket with water and then swing it fast enough in a circular plane, all the water will stay in the bucket. Centrifugal compression uses this principle.

The pressure develops as the refrigerant turns at high speed. Pressure is lowest at the center of the circle. That's where the vapor enters. Gas is thrown to the edge of the wheel by centrifugal force. This provides a continuous flow of pressurized refrigerant.

Centrifugal compressors run at high speeds and pump a large volume of gas at relatively low compression ratios. They can be single- or multi-stage, depending on how much cooling density is needed.

For maximum efficiency, compression ratios up to 2.5 are used for multi-stage units and up to 4.5 for single-stage units. Most centrifugal compressors use one or two stages of compression.

Centrifugal compressors are common in large buildings. A single centrifugal compressor can replace many reciprocating units. Reliability tends to be good, because there are relatively few moving parts.

In general, the capacity of the largest reciprocating compressors is about 100 tons. If you need more than that, the best choice is a centrifugal compressor or an absorption refrigeration system. Absorption units are available for small loads. But they need a heat source (like steam or high-temperature water) to be practical.

HVAC subcontractors occasionally talk about the "high side" or "low side" of air-conditioning equipment. These terms refer to the high-pressure section and low-pressure section of air conditioners. From the expansion valve through the evaporator to the compressor is the low-pressure section. The high-pressure section extends from the compressor through the condenser and back to the expansion valve.

Cooling Towers

Cooling towers are used in larger systems to cool the water that's circulated through the condensers. If the humidity isn't high, a cooling tower will cool water below air temperature.

Cooling towers are classified by the method of moving air through the tower: natural-draft, induced-draft and forced-draft.

Natural-draft cooling towers— This tower doesn't have a fan. The air just circulates naturally and at a low velocity through the tower.

Induced-draft cooling towers— A top-mounted fan sucks air up through the tower as warm water falls downward.

Forced-draft cooling towers— A fan forces air through the tower.

Air Handlers

Air handler fans are needed in nearly all mechanical refrigeration systems to move the cold air to where it's needed. They fall into two broad classifications: centrifugal or radial flow (squirrel cage) and axial flow (like a desk fan).

The air distribution system for a central air-conditioning system has a supply fan or blower with capacity to change the air often enough in the occupied space. The number of air changes per hour depends on the type of activity expected in each room. The air supply to the room should include both fresh air and return air, both of which pass through an air filter before returning to the building.

Air from the supply fan enters a discharge plenum and then goes through either a heating coil or a cooling coil. After passing through these coils, the air enters the duct system and is transmitted to the various rooms.

Ducts

Ducting distributes cooled or heated air to where it's needed, and returns air from rooms to the heating or cooling equipment. Ducts can be located underground, in basements, crawl spaces, attics or walls.

Wherever you decide to locate your duct system, make sure that no vibration is transmitted from the supply fan to the ducts.

Thermal insulation is needed on supply and return air ducts and plenums of heating and cooling systems. If you don't see insulation on the outside of the ducts, it may be inside the ducts. To control moisture and humidity, a vapor barrier may also be installed over the ducts.

Most ducting is made of galvanized steel. But other materials are acceptable. Asbestos-cement, concrete, clay and ceramic ducts can be used underground. Ducts made of gypsum are acceptable but have certain temperature and humidity restrictions. Tin-plated steel, aluminum, copper, wood and reinforced fiberglass are other approved materials.

The thickness of the metal and the use of combustible materials are controlled by the *Uniform Mechanical Code.*

Duct size is determined by the heat loss of the space to be heated, the cooling load requirements, the air-delivery requirements and the static pressure of the duct system.

Volume dampers can be used to control the direction and volume of air that comes through the ducts. These air-flow adjustment controls are important, because more air is needed to cool a building than to heat it. In larger buildings, the dampers will probably be controlled actuators that respond to temperature changes in the rooms.

A well-designed duct system keeps duct runs as short as possible and has outlets located for good air distribution. Larger ducts minimize both air resistance and duct noise. But they also cost more. Your subcontractor has to find the compromise that gives the best value for your HVAC dollar.

Duct Systems

There are several ways to run ducting. Your subcontractor will probably prefer one of the following systems:

Perimeter ducting— This system uses perimeter ducting that extends around the outside of the building. Conditioned air enters the rooms through floor registers.

Radial ducting— In this system, ducts are run from the central furnace to the perimeter walls of the building. This is similar to the old gravity warm-air distribution system.

Extended plenum ducting— In this system, a large duct is run lengthwise through the building. As the duct gets farther away from the furnace, the diameter of the duct gets smaller. Branches are run at right angles to the perimeter and end at registers. This system is the most popular ducting layout.

Summary

Before you decide on the best heating and cooling system for your spec house, get bids from several reliable HVAC subcontractors. Compare their prices and the equipment listed in their bids. Don't accept a bid just because it's the lowest — the lower bid may be based on inadequate equipment.

If one contractor suggests smaller equipment, ask the other bidders whether or not the suggested unit is large enough to do the job. Then select the subcontractor who offers the most attractive price for an adequate system.

Figure 26-20, your Heating and Cooling System Checklist and Data Reference Sheet, will help you evaluate bids and organize information both for this job and for future reference.

Heating and Cooling System Checklist and Data Reference Sheet

1) The house is ________ feet in length and ________ feet in width, for a total of ________ square feet.

2) The house is ☐1-story ☐1½-story ☐2-story and has ________ SF of living space.

3) Heating and cooling requirements:

First floor

☐Heating: ____________ Btu's

☐Cooling: ____________ Btu's

Second floor

☐Heating: ____________ Btu's

☐Cooling: ____________ Btu's

4) System identification:

First floor

Manufacturer: ____________

Serial number ____________ Model number ____________ Catalog number ____________

Capacity: ________________________

Type: ☐gas ☐oil ☐electric

Duct size: ____________ Room register size: ____________

Cooling condenser size: ____________ Serial number ____________

Model number ____________ Catalog number ____________

Second floor

Manufacturer: ____________

Serial number ____________ Model number ____________ Catalog number ____________

Capacity: ________________________

Type: ☐gas ☐oil ☐electric

Duct size: ____________ Room register size: ____________

Cooling condenser size: ____________ Serial number ______________

Model number ____________ Catalog number ____________

Figure 26-20

5) Heating and cooling system installed by subcontractor:

Name: ______________________________

Address: ______________________________

Telephone: ______________________________

Cost:

First-floor system $__________

Second-floor system $__________

Total $__________

6) Heating and cooling system:

☐ Installed (date): __________

☐ Tested (date): __________

☐ Approved (date): __________

7) Problems that could have been avoided ______________________________

____

____

____

____

8) Comments ______________________________

____

____

____

____

9) A copy of this Checklist and Data Reference Sheet has been provided:

☐ Management

☐ Accounting

☐ Estimator

☐ Foreman

☐ File

☐ Other: __________

Chapter 27

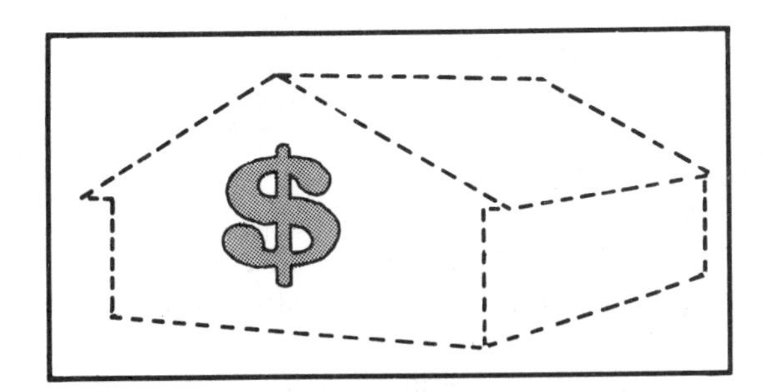

Estimates and Profit

Every successful builder learns the importance of accurate records — records that show actual material, labor and overhead costs. He learns to use these records to make sound estimates on every job and to guarantee a reasonable profit for his business.

In this chapter, we'll lay out the details of a sound estimating system. It will include material and labor costs, overhead, and a profitable return on the money you've invested in your business.

Material and Labor Costs

All good estimating systems have one thing in common: an accurate listing of every significant cost item required for the job. This list is called a "take-off" sheet.

List the materials you expect to use and in the order you expect to use them. Work from the footing up to the roof. List materials by category: footings, foundations, floor framing, studding, etc. Include nails and all accessory materials associated with each major component.

Once every material has been listed, start estimating the labor. Labor estimates are more difficult. But the guidelines offered in this book should be helpful. If you've been keeping track of the labor costs on your jobs, use your own figures to create an accurate labor estimating standard.

Here's a rule of thumb for estimating labor:

For a house without a basement, allow 1 manhour (2/3 skilled, 1/3 unskilled) per square foot of floor space.

Thus, a 1,000-square-foot house would require 1,000 manhours (667 skilled hours and 333 unskilled hours). This means that you and a crew of two can build the house in 1,000 manhours. If you sub out the electrical, plumbing and HVAC work, your manhours are reduced by that much. If your labor costs $20 per hour, the cost for 1,000 square feet is $20,000.

A rule of thumb for estimating the total cost of a house, including all materials and labor, is:

Multiply the square footage of the house times the hourly cost for three union carpenters.

Thus, if a union carpenter's hourly wage is $30, the hourly cost of three union carpenters would be $90. A 1,500-square-foot house would cost $135,000, or $90 per square foot.

Remember that these rules are just guidelines. And they won't work well in areas where materials and labor are out of proportion to one another. When you finish a job, check the ratio of manhours to square feet and the ratio of construction cost to hourly wage cost. Then adjust the rules to fit your situation.

Overhead

Your cost for every job is more than just what you pay for materials and labor. You carry an overhead that includes several kinds of taxes, insurance and miscellaneous expenses. Don't forget to include these costs in your estimate! They're just as real as the cost of lumber and concrete.

Taxes and Insurance

Most states levy an unemployment insurance tax on employers. This tax is based on the total payroll for each calendar quarter. The actual tax percentage is usually based on the employer's history of unemployment claims and may be from 1% to 4% or more.

The federal government also levies an unemployment insurance tax based on payroll (F.U.T.A.). This tax runs about 0.7%. And the federal government collects social security taxes (F.I.C.A.) and medicare taxes. These two taxes come to about 7% of payroll, and are collected from the employer each calendar quarter.

States generally require employers to maintain worker's compensation insurance to cover their employees in the event of job-related injuries. The cost of the insurance is a percentage of payroll and is based on the type of work each employee performs. Most light construction trades have a rate between 5% and 15%. The cost varies from one period to the next, depending on the history of injuries for the previous period.

You should also carry liability insurance to protect yourself in the event of an accident. Liability insurance usually costs about 1% to 2% of payroll. The higher the liability limits, the higher your cost.

Here are your approximate total costs for taxes and insurance:

State Unemployment Insurance	4.0%
F.I.C.A. and Medicare	7.0%
F.U.T.A.	0.7%
Worker's Compensation Insurance	5.0 to 15.0%
Liability Insurance	2.0%
Total	18.7 to 28.7%

Miscellaneous Expenses

Other important "cost-of-doing-business" expenses include:

- Supervision
- Payroll and reports
- Interest on borrowed money
- Car and truck expenses
- Office expenses
- Licenses

If you are building on spec, you can add the following expenses to your overhead:

- Fire and liability insurance
- Real estate taxes
- Advertising
- Sales commission
- Legal fees

These additional expenses will vary from job to job. To cover these expenses, some builders add 22% to 28% to the estimated payroll. Others prefer to list each item separately. Just be sure that you account for these expenses, or your profit will be reduced accordingly.

You'll build few houses where the estimated cost and the actual cost are identical. Labor is always an estimate. Inflation can increase material costs weekly. And some costs can't be foreseen, even with the most careful estimate. But your actual cost should be within 5% of your estimate.

To allow for the difference between the estimated cost and the actual cost, add a contingency percentage to your estimated cost of the house (before profit). Your experience from previous jobs is your best guide to the appropriate percentage. From 2% to 3% should be an adequate allowance.

Profit

You are entitled to wages for your time on the job, plus a profit on the money you have invested in your business.

Profit often depends on the competition. If you include a high profit in your bid, some other builder may get the job. For most residential jobs, a profit of 8% to 15% is reasonable. Just make sure that the profit percentage you add is really profit. It shouldn't cover any costs or expenses. It's your return on the money you have invested in your business.

Some states and cities have a business and occupation tax that builders must pay on gross receipts. Where this tax is imposed, the builder is exempt from paying sales tax on the materials he uses. Add the business and occupation tax to the estimate after you have added your profit percentage.

Summary

The Master Checklist and Data Reference Sheet, Figure 27-1, brings together costs from the other checklists in this book. Use it as your cost summary for the job at hand.

Master Checklist and Data Reference Sheet

Project: ____

____

☐Custom building ☐Spec building

Date completed: ____

Date sold: ____

Material and Labor Costs:

	Materials	Labor
1) The building site	$________	$________
2) Staking out the house	$________	$________
3) Footings and foundations	$________	$________
4) Slab-on-grade construction	$________	$________
5) Floor framing	$________	$________
6) Wall framing	$________	$________
7) Ceiling and roof framing	$________	$________
8) Modular construction	$________	$________
9) Insulation sheathing	$________	$________
10) Sheathing	$________	$________
11) Cornices and rakes	$________	$________
12) Roof covering	$________	$________
13) Doors	$________	$________
14) Windows	$________	$________
15) Subcontracted work:		
☐Electrical ☐plumbing ☐masonry	$________	$________
16) Insulation and vapor barriers	$________	$________
17) Siding	$________	$________

Figure 27-1

	Materials	Labor
18) Stair construction	$________	$________
19) Chimney and fireplace construction	$________	$________
20) Interior walls, ceilings and trim	$________	$________
21) Floor coverings	$________	$________
22) Kitchen cabinets	$________	$________
23) Painting	$________	$________
24) Heating and cooling system	$________	$________
25) Appliances:		
☐Range	$________	$________
☐Range hood	$________	$________
☐Oven	$________	$________
☐Microwave oven	$________	$________
☐Refrigerator	$________	$________
☐Dishwasher	$________	$________
☐Other (specify)		
____________________	$________	$________
____________________	$________	$________
26) Work not covered by checklists:		
☐Concrete driveway and walks	$________	$________
☐Gutters	$________	$________
☐Fences	$________	$________
☐Termite control	$________	$________
☐Other (specify)		
____________________	$________	$________
Total	**$________**	**$________**

Overhead:

1) Taxes and insurance:

☐State unemployment (____________%) $____________

☐F.U.T.A. (____________%) $____________

☐F.I.C.A. (____________%) $____________

☐Medicare (____________%) $____________

☐Worker's compensation (____________%) $____________

☐Personal liability (____________%) $____________

Total **$____________**

2) Miscellaneous expenses:

☐Supervision $____________

☐Payroll and reports $____________

☐Interest on borrowed money $____________

☐Car and truck expenses $____________

☐Office expenses $____________

☐Licenses $____________

☐Fire and liability insurance $____________

☐Real estate taxes $____________

☐Advertising $____________

☐Sales commission $____________

☐Legal fees $____________

Total **$____________**

Total Cost of House:

Total materials $____________

Total labor $____________

Total overhead $____________

Subtotal $____________

Contingency (______%) $__________

Subtotal $__________

Profit (______%) $__________

Business and occupation tax $__________

Total cost of house $__________

Cost per square foot $__________

A copy of this Checklist and Data Reference Sheet has been provided:

☐Management

☐Accounting

☐Estimator

☐Foreman

☐File

☐Other: ____________

Chapter 28

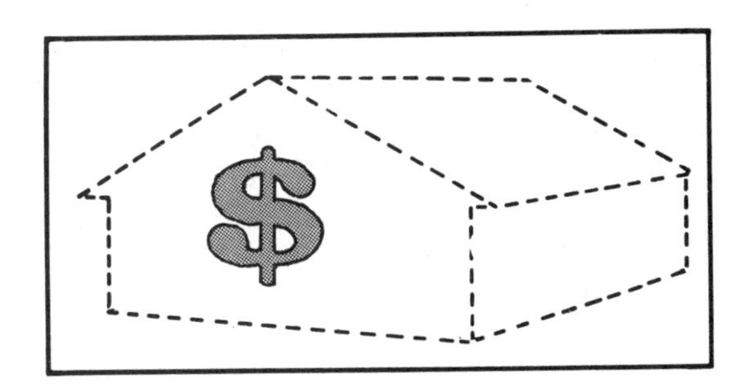

Construction Loans

Borrowed money makes the construction world turn. In fact, one of the most successful builders I know never uses a dime of his own money. His buyers and lenders advance all the cash he needs to pay suppliers, tradesmen and subs.

"I use credit and loans," he said. "If I have to put my own money behind a project, something's wrong with my planning."

On the other hand, another successful home builder I know pays nearly all of his expenses out of his own pocket right up to the time of sale. "Why should I pay the fees and interest rates they're quoting when I have the cash available myself?" he says. Of course, he has to be satisfied with building on a smaller scale than some of his competitors.

Most builders fall somewhere between these two extremes. If you're like most spec builders, you'll use loans and arrange for credit on most jobs. Sometimes that's the only way to continue building.

But there's a trap here that every builder should recognize. The bank isn't taking any risk when they lend you money to build a house. They'll get their money back even if you default and walk away from the project. But you can lose everything, both time and money, that you've invested.

Since you're taking the risk, you're entitled to the lion's share of the rewards. But if you guess wrong, there may not be any rewards. Or the profit may be so slim that you've wasted your time. The bank still gets their fees and interest. They've made a nice profit, and you're left with a big zero.

An economist would look at the problem this way. If you have to pay 15% interest on the money you borrow, be sure you're going to get more than a 15% return on your investment. If you can't earn more on the money you borrow than what the bank is charging, let the bank keep their money.

Here's how to test your return on an investment. Calculate the amount of money you have invested in your business. Let's say you begin the year with a truck, some office equipment and small tools. The total value is $40,000. You also own a lot worth $60,000. Your company has assets of $100,000 and no loans outstanding at the beginning of the year. The net worth is $100,000.

During that year you build a house on the lot and sell it at year-end for $30,000 more than your out-of-pocket costs. Have you made $30,000? Not quite! Here's why.

First, out-of-pocket costs don't include the value of your services. Deduct from that $30,000 what you could have earned if you had worked on some other contractor's payroll instead of planning and building the house you've just sold. Say that's $20,000. That reduces your profit to $10,000.

Now deduct what you could have earned by selling off all your assets at the beginning of the year and putting the proceeds in a bank. On $100,000 you could have earned close to $10,000 with no risk!

After deducting the $20,000 you could have earned working on another contractor's payroll and the $10,000 interest you could have earned, it looks like your building operation just broke even for the year.

Now assume that you began the year with the same assets. But this time your net worth was zero because the whole $100,000 was borrowed. Say your fees and interest expense for the loan would be $20,000 for the year. The $30,000 "profit" on the house sold is reduced by $20,000, leaving only $10,000 to go into your pocket. And remember, you could have earned $20,000 working for someone else! Instead of a $30,000 gain, you've suffered a $10,000 loss. The bank and your buyer come out O.K. But what you're running is an expensive hobby, not a construction business.

Does this sound farfetched? It's not. Some years, even very experienced builders with very deep pockets and big payrolls would be better off selling off their assets and depositing the proceeds in a bank. Of course, that isn't a real option for most contractors. But it's easy to underestimate the burden of debt payments and overestimate the return on borrowed funds. Don't fall into the debt trap. Don't borrow unless your return on that money will exceed its rental cost.

Now that I've explained why to avoid borrowing, I'll explain how to do it. The rest of this chapter will review the types of loans that institutional lenders make, outline loan procedures, describe how to find a bank that meets your needs, and suggest ways to avoid the cash crisis that can sink a speculative builder.

Credit

The best form of credit doesn't come from a bank. And every builder uses it. It's called *trade credit*. You take delivery of materials and have work done before making payment. Trade credit gives you operating room. And it's good credit, because there's no interest charge.

I've used trade credit to build and remodel houses without borrowing. Of course, you have to meet payroll out-of-pocket. But payment for the lot, materials and sub work can be delayed until construction is completed or until the house is sold. But make sure you clear this with your material suppliers, subcontractors and the landowner before work begins.

Abuse of trade credit is the quickest way to develop a reputation that will put you out of business. Subs and material suppliers won't do business with builders that don't pay their bills on time.

If your need for capital extends beyond trade credit, you're going to have to borrow from a savings and loan association, a bank, or an insurance company. Or you can use a loan guarantee program run by an agency like the FHA (Federal Housing Administration) or VA (Veterans Administration).

FHA-Insured Loans

The Federal Housing Administration provides mortgage insurance on houses that meet its construction standards, its eligibility requirements and its guidelines on site and location.

FHA insurance makes the mortgage more attractive to banks and other private lending institutions. If the lender defaults, the FHA guarantees payment to the lender.

The procedure is to submit a complete plan, specifications and information on the site and surrounding area to your local FHA office. The FHA will go over the plans and specs and inspect the site. Once approved, the FHA will issue a *commitment*. This commitment doesn't provide any mortgage funds. It does, however, become the basis for a loan to be made by a bank or savings institution.

Many lenders will help you arrange for the FHA guarantee. You submit your proposal to them first.

If you or your customers plan to get FHA mortgage insurance, check with the nearest FHA insuring office before you buy the land. You should also contact the FHA before spending money to develop complete plans and specifications for a customer who already has the lot.

Savings and Loan Associations

S&L's are major lenders in nearly every community. Most write long-term loans. A larger down payment may be required if the loan is not insured by the FHA. The interest rate may be lower but the term may be shorter on conventional loans. Familiarize yourself with the S&L's in your area and know the type of loans they offer to spec builders and potential buyers.

Here's how the loan procedure works at my local S&L. The explanation that follows is by Mrs. Carol Glasgow, Vice President, First Federal Savings and Loan Association, Warner Robins, Georgia. Procedures will be at least slightly different in your area.

"When a builder comes to us for a construction loan, he gets a pile of blank forms we call our loan package. It has a specification sheet, an application, an estimate form and our energy recommendations sheet. We ask him to return these, completed, with a full set of plans, his current financial statement, the last two years' Federal Income Tax returns, and either a sales contract or a deed on the land.

Once we've made a loan to a builder, we don't require new tax forms or a financial statement except to update the file (usually every six months). Also, once we know a contractor, we dispense with the estimate form and just ask for his total figures.

When the contractor comes back in, we complete the construction worksheet and discuss it with him. At this point, we get an appraisal on the lot, look carefully at his plans and specifications and order a credit report.

When all of the documentation is in order, we present his application to our loan committee for approval. A speculative construction loan is limited to 75% of the appraised value. If the house is presold, we can make the loan for up to 95% of the lesser of sales or appraised value. However, we want the buyer to come in and qualify for a take-out loan.

After approval, the loan is sent to our attorney for closing. At closing, the builder executes the construction loan agreement, the preconstruction affidavit, the note, and the deed to secure the debt.

At the start, we disburse from loan proceeds no more than 75% of the lot cost and closing fees. The balance of the funds are placed into an account called a loans-in-process account to be disbursed under the schedule in the loan agreement as work progresses. We disburse no more than 80% of the construction loan to a contractor until the house is completed.

We require that a contractor call us by noon on Thursday if he wants to draw against the loan that week. His check will be ready by noon on Friday after we have inspected the construction site. The final draw comes when the house is completed. We charge interest quarterly for funds used. Interest begins on the date of the draw."

Savings and loan associations are probably the most active lenders in most communities. But they aren't the only lenders for spec builders.

Other Sources of Funds

Insurance companies are a good source of construction funds, either indirectly through FHA-insured loans or through direct loans made without FHA insurance. In many communities, local cooperative groups have some type of financing available to members.

Never start building until financing has been arranged for the house. Some lending institutions require a certificate from you stating that no work has been started and no materials have been delivered to the site. Don't try to jump the gun!

The Elements of Borrowing

With few exceptions, the more successful you become, the more time you'll spend with loan officers. Many bankers and lenders that approve construction loans have a good knowledge of construction. Some worked in construction before joining the bank. I never met one who didn't know the difference between a 2 x 4 and a 2 x 6. You're a step ahead if you assume that bankers and lenders are as impressed by quality building as the building inspector.

Bankers also know how to evaluate prospective borrowers. They expect you to produce evidence that your loan will be secure and profitable to them. You have to convince the lender that the funds are required for sound reasons and that the loan can be paid back under the agreed terms. The lender wants profitable, trouble-free, secure loans just like you want to complete profitable, quality buildings.

The lender's business is renting money. Yours is building houses. Convince a loan officer that your business is solvent, profitable and well-managed and you'll get the loan you need. Borrowing will be easier if you have a good relationship with the bank before applying. Builders who are just getting started should open an account in the business name at the bank where they have their personal account. If you've had a loan with that bank or have one of their credit cards, all the better, provided your payment record is good.

Discuss your plans with the loan officer and find out what the bank can do for you. Banks have competitors just like you do. They offer a wide range of services to attract and keep good accounts that will grow into major revenue producers.

Services offered by larger banks include credit reports on your potential customers, financial, investment and estate advisory services, discounting customer's accounts and notes payable, check certification, night depositories, safe deposit boxes, payroll preparation services, tax filing services, and balance sheet services, among others.

Assume you're in the process of building two spec homes. You've already got loans on these jobs. But you have a chance to buy a choice lot for your next spec house. You want to act quickly before someone else buys the lot. Will your bank come to your aid?

Ask your banker about this before the problem comes up. What collateral is required? How much collateral? How long does it take to have a loan approved? Does the bank have limitations on the number or size of loans they can make to spec builders? What are the repayment terms? What supporting information do you have to provide? Will the bank give you an open line of credit? Do you have to maintain certain balances before the bank can consider a loan?

Of course, your bank should do more than provide loans when needed. Many banks collect information about construction in their area and will share this information with their customers. Even more important, they can refer business to you.

Some of your customers probably use the same bank that you do. This is particularly true in small towns. If they do, it's to your advantage for several reasons. First, the "float" is minimized. Checks you receive can be cashed immediately in the bank branch where the account is maintained. Second, your bank will usually cooperate in releasing credit information on their other customers.

When shopping for a bank, ask your suppliers and subcontractors what bank they use. Your trade association may also be a good source of information.

Getting Started

Here are some tips on opening a company account:

First, get a letter of introduction from your former bank . . . where you had your personal account. The reference will help to establish good relations right from the start.

Second, establish good relations *before* you need to borrow money. When you need money, you don't want to negotiate under pressure. This is true whether you're applying for a construction loan or for a quick loan to buy that choice lot.

Third, develop a personal contact with an officer or other responsible employee in the bank. Be on a first-name basis with one or two key people where you bank. Your aim isn't to have a fishing buddy. It's to have someone in the bank who associates your face and your reputation in the community with the loan applications and financial documents you submit.

If your loan officer isn't already familiar with construction, invite him (or her) to lunch or to one of your jobs. Demonstrate that you are a competent and progressive manager.

A banker who knows you and has a professional interest in your problems can save you time. For example, when you need a quick credit reference on a potential customer, the loan officer can give you the information over the telephone.

The foundation for good relations is mutual respect. Banks respect builders who show good character and conduct their affairs in a professional manner.

I remember the loan officer who came out to see the first spec house I ever built. All the rough framing was completed and we had just finished "drying" the house in. After walking through every room and asking questions for at least ten minutes, he complimented me on the quality job we were doing. The house sold before it was completed. When that deal was closed, the loan officer called to congratulate me on the sale and offered to finance any spec house I wanted to build thereafter.

Formula for Success in Borrowing

As I see it, there are four sides to a good relationship with a bank: (1) show good faith, (2) give the bank volume, (3) provide good financial data, and (4) keep the bank informed of your plans.

Good faith is the most important single factor in daily dealings with your bank. Always keep your word just as you expect them to keep theirs.

One way to show good faith is to observe the bank's policies. The bank has to comply with the law just like you have to follow the building code. In addition, bank employees are responsible to the bank's board of directors and stockholders. It isn't very likely that they'll ignore the law or written policy to help you. In fact, it's probably foolish even to ask for that kind of special treatment.

Another way of showing good faith is to keep them informed of important changes in your business . . . both good and bad. Some builders want the bank to be the last to find out when something important has gone wrong. But they create a much worse impression at the bank when a late payment comes as a surprise.

As one banker describes it, some contractors "use bad judgement the first time they can't meet their obligations. Instead of discussing it with their banker, they try to evade the issue." That creates ill will which erodes confidence far more than a simple admission that you're a little short of cash at the moment.

Give the Bank Volume

Give the bank enough volume so that they recognize you as a desirable customer. After all, banking is a business too. Banks watch average checking account balances the way you watch for vacant land to build on.

Volume helps the bank cover the costs of their services. They want to see their customers grow and prosper. That's how they expand their banking business.

Provide Financial Data

Your balance sheet and income statement help the bank evaluate your success as a manager and speculative builder. A good manager keeps his business in sound financial condition. But bankers worry more about not having current financial information than they do about having information that doesn't look too good. So right from the start, be ready to supply the financial information that they're going to request: profit and loss statement, balance sheet, estimate and budget.

These financial statements mean more to your banker if they're prepared by an independent auditor.

Supplying accurate financial statements adds to your reputation for integrity. Loan officers recognize a padded statement a mile away. Skimpy data or unexplained assets just invite them to question your professionalism.

Discuss Future Plans

Bankers hate surprises. When your banker knows your plans, he can estimate what his role is going to be. That lets him offer specific advice that could help you avoid a loss or escape what might become an embarrassing situation.

Interest rates change and the availability of money changes. You need to know how these changes affect your plans. Your banker can help if he knows what your plans are.

Loans to Creditworthy Borrowers

Suppose you have selected your bank carefully and have a good working relationship with one of the officers. You've worked up a projected cash budget that shows the loan needed on your next job. Can you get that loan?

All banks decline some loans. It's not that they don't trust the figures that the applicant put in the loan package. Nor is it a question of good intentions. It's just a matter of allocating the money available among the most creditworthy borrowers.

Get the money you need by putting yourself high on the list of creditworthy applicants. Take the application very seriously. Try to answer every question and provide the best copy available of every document requested. Bankers routinely decline applications that are incomplete or carelessly prepared. That shows a lack of professionalism in the banker's eyes. He wants to make loans only to competent professionals.

Here are two key points that the loan committee is sure to focus on. First, is the amount you need going to be enough? The bank doesn't want you coming back for more money to finish a project that was supposed to be covered by a current loan. Second, where's the money really going to go? Lenders have been burned many times by builders who "make a profit" on the loan . . . diverting part of the money to some other purpose.

Both of these are hard questions for the banker to analyze. Essentially, they have to trust your judgement and rely on your professionalism. A good construction estimate and written quotes by competent subcontractors are the best guarantee that the loan will cover all construction costs.

A builder who's having a hard time making ends meet is more likely to divert some of the loan proceeds. The bank will scrutinize cash flow to be sure there is enough cash to meet current obligations. Cash needed for working capital shouldn't be absorbed for other purposes.

Avoiding a Cash Crisis

Every experienced builder has had the feeling . . . the desperate moment when he discovers that there isn't enough cash to make an important payment. And all work is going to stop if the money isn't there!

The money in your checking account is the lifeblood of your business. When that money runs out, the heart of your business stops beating.

I'm not going to suggest how to raise cash when that desperate moment arrives. But there's plenty every builder should do to make sure it never comes.

Most builders who run out of cash tend to feel that their needs are purely financial. Just a few hundred or a few thousand dollars would solve the problem. But most financial consultants insist that the problem usually isn't money at all. Some extra cash might ease the pressure temporarily. But another crisis is likely before too long. In most cases, failure to manage debt and anticipate cash requirements brought on the cash crisis. Only better planning can prevent the next one.

A well-run construction company doesn't necessarily have any more money. It's just more likely to balance receipts and expenditures better and keep debt within manageable limits.

A poorly-managed building business goes through regular financial convulsions which tend to confuse, embarrass, and alarm everyone the company deals with. Employees, subcontractors and suppliers are the first to sense this problem. In every transaction they have with the company, they have to consider the chance that it may not survive.

And note this carefully. A cash shortage doesn't mean that the company does shoddy work or that sales are down. Even contractors who do first-rate work and have no trouble selling their homes at a good profit can run out of cash.

Cash Requirements

Avoid the cash crisis by being aware of how much cash your business needs. Then plan to meet that goal. As a rule of thumb, between 5 and 10% of your assets should be in cash. For a builder with $100,000 in assets, that translates into $5,000 to $10,000.

Some builders use a month's expenses as the yardstick. They feel that cash on hand is adequate when it would meet all normal expenses for a month, assuming there was no income and no draw against loans for that month.

What if you can't meet these tests? No problem. You're just doing business too fast for the cash

	Jan	Feb	Mar	Apr	May	Jun	Jul	Aug	Sept	Oct	Nov	Dec
Estimated cash available												
Cash balance on first day												
Expected receipts												
Job A												
Job B												
Job C												
Etc.												
Bank loans												
Total expected cash receipts												
Estimated cash requirements												
Job A												
Job B												
Job C												
Etc.												
Equipment rentals												
Taxes												
Insurance												
Overhead												
Loan Repayments												
Total cash required												
Cash balance at month end												

Estimated cash flow
Figure 28-1

available. It's like driving too fast on a rain-slick street. There is some speed that's safe, no matter how severe the weather. With every company, there's some level of business that's right for the size of its bank account, no matter how slim that account may be.

If you aren't satisfied with the level of business that your bank account warrants, there are many ways to build your cash balance: Rent tools and equipment rather than buying. Reduce your own salary. Buy materials in smaller volume. Delay major purchases such as land. Cut back or eliminate the credit you offer to buyers. Avoid unnecessary prestige items such as a new car or truck, expensive advertising or luxury office space.

But the first and most important step is to recognize your limitations. Find the level of business that you can handle with the cash available. Stay within that limit until some unusual opportunity comes up. When it does, draw up a cash budget to anticipate problems before the crunch comes. Your budget will show what's possible with existing cash or how much you need to borrow. If a loan is necessary, you can begin shopping for cash before the actual need arises.

Cash Budgeting

Drawing up a cash budget is like planning a football play. Your team is the income, cash arriving in the form of sales, and draws against loans. The other team is expenses, bills arriving and coming due for payment. For every move your team makes (each dollar of income), the opposing team will probably have a counter-move (dollar of expense). When income is exceeding expenses, you're marching downfield toward the goal line. When expenses exceed income, you're getting pushed back toward insolvency.

For every spec builder, the game starts on your own goal line. You have to put up a few dollars and make a few short gains before you have room to maneuver. Then you can go for plays that can make longer yardage. Your cash budget shows the gains (income) and losses (expense) after each play (month) of operation.

Figure 28-1 suggests what your cash budget form might look like. Begin by going one month into the future. Let's say that the next month is January. At the top of the January column opposite "Cash balance on first day," write in the bank balance you expect to have on the first day of January. Then, opposite each job you have going in January, list the receipts you expect from that job.

Add any loan proceeds you expect to receive. Then total January receipts.

Opposite "Estimated cash requirements" under the January column, enter all bills that you'll pay in January. Don't be too optimistic. Expect some extra expenses, a few delays due to bad weather and at least one minor emergency. Total the cash required. The difference between income and expenses is the net change in your cash account. Adjust your first day cash balance for that change and enter that result opposite "Cash balance at month end" for January and at the top of the February column. Fill in the February column and the following columns as far as you can foresee expenses and income with reasonable accuracy.

Of course, no one can predict income and expenses exactly, especially six or eight months into the future. But exact calculations aren't needed unless you expect to be down to your last dollar at some point. Just be accurate enough to anticipate critical points when careful management of cash is important.

It's harder to forecast income and expenses if you're just getting started. But a builder with several years of construction experience should be able to estimate with reasonable accuracy when money will come in and be paid out on each job under way.

If you need guidance, go back through your checkbook. Think of a job you've finished that's similar to a job you have now. Note the pattern of payments and receipts on the completed job. The day of the first major payment on that job is day zero. Every following payment or receipt is assigned a day number which shows the number of days that have passed since the first payment. When all receipts and payments have been assigned day numbers, you have a cash flow pattern that should apply more or less to jobs currently under way. Use that pattern when you make out the cash budget for each job.

A cash budget is the most effective way to plan cash requirements for your business. Your goal, of course, is to stay in the black. But it will help in other ways too. A properly prepared cash budget can:

1. Help you make payments coincide with receipts. This can reduce the need for loans.

2. Highlight periods when cash will be short and show how much cash will be needed, and at what times.

3. Help you estimate how long any cash shortage might last.

4. Show when it is or isn't wise to take discounts for prompt payment.

5. Show when surplus cash is available for investment or to buy inventory or equipment.

6. Emphasize the need for additional working capital.

The important thing to keep in mind in making a cash budget is the word *cash.* Be as factual as you can be. Try not to overestimate sales or underestimate expenses. Your sales forecast should be as accurate as possible because it's the basis for figuring your cash and expenses.

How Much Capital Do You Need?

If you haven't been running a construction business for several years, making up a cash budget will be much harder. The rest of this chapter will suggest the major items you should include in the cash budget for your first spec house.

How much capital you'll need depends on your situation and your plans. It also depends on the cost of supplies, materials, labor, and land in your area.

As a speculative builder, willing to take a chance, you may be able to start with only enough cash to carry you part way through the construction process. Of course, you'll be on a much sounder basis if you own the land free and clear or have the owner subordinate the land to your use.

If the land is clear and you have enough cash to begin construction, you'll be able to borrow enough to carry most of the construction expense. The real test of your staying power comes if the house is not sold when construction is complete. How long can you continue carrying the unsold house before your capital is exhausted? You need enough margin to keep up loan payments, and to pay insurance and taxes until the house is sold.

If you're starting as a custom contractor — building a house for someone on his land — you may need only enough money or credit to carry your expenses until you get the house underway. After you've brought it to a certain stage of completion, your customer will pay a percentage of the total price. How much, and at what stages, will depend on your contract with the owner, the loan he has and the policy of the lender. In most cases you can draw weekly to meet payroll costs and sub work. But the lender will usually limit these draws to 80% of the loan until work is 100% completed.

Here are the important expenses to include in your cash budget: (1) living expenses, (2) possibly an office, (3) tools and equipment, (4) materials, (5) payroll, (6) land expenses, and (7) incidentals.

Living expenses— These will probably be the same regardless of the type of building you do. Don't leave a secure job with a regular paycheck until there's enough money in the bank to tide you over until your new company can start paying a salary.

To be safe, you should probably have 6 to 12 months' living expenses saved up. Many builders work only on a part-time basis until the construction company starts paying off. Some have the type of job that leaves them several hours a day and lengthy vacation periods to use on a construction project. If you have a reliable supervisor, the job can go on while you keep your present position.

Office or place of business— At the start, you'll work out of your home or your pickup truck. That's all many small builders need. It keeps overhead to a minimum. I prefer to remain a small builder and have never had an office. To me, an office is an unnecessary expense I can do without. You can keep several houses going at the same time from your pickup!

Tools and equipment— You probably already own some of the tools you'll need. You may be planning to rent tools or to sub out work that requires expensive equipment like a bulldozer. Don't burden yourself owning and maintaining heavy equipment. Minor repairs on a small tractor or loader can cost hundreds of dollars. When you talk about major repairs on a bulldozer, you're talking about thousands. I know. I owned one. I bought it not for my construction business but to clear farmland. A D7 Cat can do a lot of work in a short time. But the cost of repairs can ruin you. Sub out the heavy stuff and let someone else pay the repair bills!

At this stage the important thing is that you be aware of what tools are actually needed and plan to have them available when the need arises. Fortunately, it doesn't take a wide assortment of expensive tools to build a house.

Materials— How much capital you need for building materials depends on at least two things; the kind of building you're doing, and the financing on that building. You can buy most materials on credit and squeeze by for 30 days without paying anything. But this credit costs you. When you can pay cash and take your discounts, your profit goes up by the discount taken.

Make a list of the materials needed, their cost on credit, and their cash cost. This list should help you decide how to handle material purchases.

Payroll— This item covers carpenters, brickmasons, plumbers, electricians, etc., who work for you. List the labor you need and what each will cost. For instance, list the labor by trade, hours, cost per hour and total.

Land expenses— You can tie up a lot of cash in land. But spec builders need good parcels to develop. If you can buy the land on credit, you may need only enough cash to pay for the title search. But make sure the land is clear of other liens before you buy it.

Incidental expenses— Broken tools, a ruined tire on your pickup or a $50 advance of salary are the type of incidental expenses that come up on every job. Have a few hundred dollars on hand to cover incidentals.

Few builders start out with enough money. That could be why many start on a part-time basis, taking on a room addition or a small spec house while remaining in their current jobs. There are many ways to get into the building or construction contracting business. Those who really want to always seem to find a way to do it. I know of one builder of spec houses who has been in business for ten years and has remained in his full-time job. He was lucky to find the right person to supervise the building of his houses.

Conclusion

The quality builder is a professional. He invests his money, time and know-how for the purpose of making a product at a fair profit. This makes him a businessman. Every businessman knows that a cash shortage can ruin him.

In my 25 years in the building business, I've seen several people with limited education, little training, and less cash, enter the building profession and become successful at it. They *knew* they could do it. If they didn't know how to do something, they studied it until they learned how. People who are willing to work hard and keep learning can overcome any problem a home builder can have.

Contractor's Estimate

____
Contractor

____
Address

Office ________________ Home ________________
Telephones

Estimate for: ________________________________

Address: ________________________ City ____________

Phone: ______________ Office ________________ Home

Street

Street

LOT INFORMATION
(Show location, size and house set back)
Distance to corner

Street

Street

BUILDING SITE

Street ________________

Subdivision ______________

Lot No. ______ Block ______

UTILITIES AVAILABLE

Water ______

Electricity ______

Sewer ______

Gas ______

Paving ______

Front ____________________ Rear
LOT GRADE

This Estimate Based on Current Cost and Supplies and is good not exceeding

a period of __________ days or until............................, 19....

Courtesy: Standard Homes Plan Service

ITEMS	COST ESTIMATES		Estimated Cost	Actual Cost
PRELIMINARY COST	Plans	$	$	$
	Survey and Laying off House			
	Insurance: Fire — Liability			
	Building Permit			
	Temporary Service Water & Lights			
PREPARATION OF LOT	Removing trees			
	Clearing site			
	Excavation			
	Hauling			
FOOTINGS	Excavation			
	Concrete footings			
	Backfilling			
MASONRY MATERIALS	Block			
	Common brick: Piers			
	Foundation			
	Chimney: Brick			
	Fire brick			
	Damper, ash dump,			
	cleanout			
	Tile for hearth			
	Flue lining			
	Face brick			
	Cleaning brick			
	Sand			
	Brixment			
	Waterproofing (where used)			
	Drain tile (where used)			
	Back filling (basement)			
	Scaffold			
STEEL	Lintels			
	Reinforcing rods and mesh			
	Steel sash or vents			
	Pipe piers and columns			
	Flitch plates, bolts, and anchors			
MASONRY LABOR	Skilled			
	Common			
CONCRETE WORK	Basement floor			
	Porches			
	Stoops			
	Steps			
	Walks .... Drives			
	Garage or carport floor			
	Forms			
ROUGH LUMBER	Joist, girders and wall plates			
	Stud, plates and purlins			
	Rafters, ridge and valleys			
	Sheathing and Sub-floors			
	Bracing and headers			
	Stringers, ties, and bridging			
ROOFING	Shingles			
	Building paper			
	Flashing chimney and eaves			
ROUGH HARDWARE	Nails			
	Building paper			
	Bolts, anchors and ties			
EXTERIOR MILLWORK	Front doorway			
	Doors and frames			
	Windows and frames			
	Porch columns and rails			
	Moulds			
	Louvres			
	Shutters			
	Garage Door			
FINISH LUMBER	Porch and carport ceiling			
	Porch and carport box			
	Soffit and fascias			
	Moulds			
	Siding or shingles			
SCREENS	Doors .... Windows			
	CARRY FORWARD ☞			

			Estimated	Actual
	BROUGHT FORWARD ☞		$	$
STORM SASH	Doors .... Windows			
INTERIOR MILLWORK	Doors and frames			
	Interior casings			
	Mantel			
	Flooring (Hardwood)			
	Moulds, base and shelving			
	Stairway (where shown)			
KITCHEN CABINETS	Shop built			
	Built on job			
CARPENTER LABOR	Rough			
	Finish			
SHEET METAL WORK	Doors and Window Flashing			
	Termite Shields			
	Gutters and Downspouts			
	Splash blocks			
INTERIOR WALL FINISH	Lath and Plaster yds........ @			
	Sheet-rock sq. feet.......... @			
	Special Dry Wall finish sq. feet			
	Corner bead			
FINISH HARDWARE	Locks .... Hinges			
	Cabinet hardware			
	Screen hardware			
WIRING	Electric openings			
	Fixtures and installation			
	Dishwasher ...... Dryer			
	Hot water heater .... Bell			
	Furnace .... Washer			
	Range ..... Air Conditioning			
PAINTING AND PAPERING	Caulking			
	Exterior woodwork			
	Interior woodwork			
	Interior walls and ceiling			
	Kitchen .... Bath			
FLOORS	Sanding and finishing (hardwood)			
SPECIAL WALLS AND FLOORS	Kitchen wall.......... floor.......... ...			
	Bath wall .... floor			
	Other			
PLUMBING	Cutting street and ditches			
	Septic tank....... Well & Pump.......			
	Fixtures .... and piping..........			
	Hot water heater			
	Medicine Cabinets			
	Towel, paper, and soap racks			
	Labor			
HEATING	Unit .... Tank......................			
	Ducts			
AIR CONDITIONING	Unit .... Installation..............			
ORNAMENTAL IRON	Columns .... Rails			
INSULATING	Walls .... ceilings .... floors			
WEATHER STRIPPING	Windows .... Doors			
CLEANING AND HAULING	Windows .... House................			
	....			
GRADING, SEEDING AND PLANTING	....			
	....			
SHADES OR BLINDS	....			
SPECIAL EQUIPMENT	Refrigerator......... Dishwasher.........			
	Range top & Oven ...... Washer........			
	Exhaust Fan.......... Dryer.............			
	Hood & duct .... Disposal...........			
OTHER ITEMS	....			
	Labor and Materials (Total)			
	Taxes and Insurance			
	Contractor's Fee			
	Total Cost			

Index

T

U

V

W

Other Practical References

National Construction Estimator
Current building costs in dollars and cents for residential, commercial and industrial construction. Prices for every commonly used building material, and the proper labor cost associated with installation of the material. Everything figured out to give you the "in place" cost in seconds. Many time-saving rules of thumb, waste and coverage factors and estimating tables are included. **544 pages, 8½ x 11, $19.50. Revised annually.**

Building Cost Manual
Square foot costs for residential, commercial, industrial, and farm buildings. In a few minutes you work up a reliable budget estimate based on the actual materials and design features, area, shape, wall height, number of floors and support requirements. Most important, you include all the important variables that can make any building unique from a cost standpoint. **240 pages, 8½ x 11, $14.00. Revised annually**

Estimating Home Building Costs
Estimate every phase of residential construction from site costs to the profit margin you should include in your bid. Shows how to keep track of manhours and make accurate labor cost estimates for footings, foundations, framing and sheathing finishes, electrical, plumbing and more. Explains the work being estimated and provides sample cost estimate worksheets with complete instructions for each job phase. **320 pages, 5½ x 8½, $17.00**

Estimating Painting Costs
Here is an accurate, step-by-step estimating system, based on a set of easy-to-use manhour tables and material tables that anyone can use for estimating painting: from simple residential repaints to complicated commercial jobs. Explains taking field measurements, doing take-offs from plans and specs, predicting labor productivity, figuring labor, material, equipment, and subcontract costs, factoring in miscellaneous and contingency costs, and modifying overhead and profit according to the variables of each job. **448 pages, 8½ x 11, $28.00**

Handbook of Construction Cont. Vol. 1 & 2
Volume 1: Everything you need to know to start and run your construction business; the pros and cons of each type of contracting, the records you'll need to keep, and how to read and understand house plans and specs to find any problems before the actual work begins. All aspects of construction are covered in detail, including all-weather wood foundations, practical math for the jobsite, and elementary surveying. **416 pages, 8½ x 11, $24.75**

Volume 2: Everything you need to know to keep your construction business profitable; different methods of estimating, keeping and controlling costs, estimating excavation, concrete, masonry, rough carpentry, roof covering, insulation, doors and windows, exterior finish, specialty finishes, scheduling work flow, managing workers, advertising and sales, spec building and land development and selecting the best legal structure for your business. **320 pages, 8½ x 11, $24.75**

Rough Carpentry
All rough carpentry is covered in detail: sills, girders, columns, joists, sheathing, ceiling, roof and wall framing, roof trusses, dormers, bay windows, furring and grounds, stairs and insulation. Many of the 24 chapters explain practical code approved methods for saving lumber and time without sacrificing quality. Chapters on columns, headers, rafters, joists and girders show how to use simple engineering principles to select the right lumber dimension for whatever species and grade you are using. **288 pages, 8½ x 11, $16.00**

Wood Frame House Construction
From the layout of the outer walls, excavation and formwork, to finish carpentry, and painting, every step of construction is covered in detail with clear illustrations and explanations. Everything the builder needs to know about framing, roofing, siding, insulation and vapor barrier, interior finishing, floor coverings, and stairs ... complete step by step "how to" information on what goes into building a frame house. **240 pages, 8½ x 11, $14.25. Revised edition.**

Manual of Professional Remodeling
This is the practical manual of professional remodeling written by an experienced and successful remodeling contractor. Shows how to evaluate a job and avoid 30-minute jobs that take all day, what to fix and what to leave alone, and what to watch for in dealing with subcontractors. Includes chapters on calculating space requirements, repairing structural defects, remodeling kitchens, baths, walls and ceilings, doors and windows, floors, roofs, installing fireplaces and chimneys (including built-ins), skylights, and exterior siding. Includes blank forms, checklists, sample contracts, and proposals you can copy and use. **400 pages, 8½ x 11, $19.75**

Carpentry Estimating
Simple, clear instructions show you how to take off quantities and figure costs for all rough and finish carpentry. Shows how much overhead and profit to include, how to convert piece prices to MBF prices or linear foot prices, and how to use the tables included to quickly estimate manhours. All carpentry is covered: floor joists, exterior and interior walls and finishes, ceiling joists and rafters, stairs, trim, windows, doors, and much more. Includes sample forms, checklists, and the author's factor worksheets to save you time and help prevent errors. **320 pages, 8½ x 11, $25.50**

Concrete Construction & Estimating
Explains how to estimate the quantity of labor and materials needed, plan the job, erect fiberglass, steel, or prefabricated forms, install shores and scaffolding, handle the concrete into place, set joints, finish and cure the concrete. Every builder who works with concrete should have the reference data, cost estimates, and examples in this practical reference. **571 pages, 5½ x 8½, $20.50**

Estimating Tables for Home Building
Produce accurate estimates in minutes for nearly any home or multi-family dwelling. This handy manual has the tables you need to find the quantity of materials and labor for most residential construction. Includes overhead and profit, how to develop unit costs for labor and materials and how to be sure you've considered every cost in the job. **336 pages, 8½ x 11, $21.50**

Remodeling Kitchens and Baths
This book is your guide to succeeding in a very lucrative area of the remodeling market: how to repair and replace damaged floors; how to redo walls, ceilings, and plumbing; and how to modernize the home wiring system to accommodate today's heavy electrical demands. Shows how to install new sinks and countertops, ceramic tile, sunken tubs, whirlpool baths, luminous ceilings, skylights, and even special lighting effects. Completely illustrated, with manhour tables for figuring your labor costs. **8½ x 11, 384 pages, $26.25**

Stair Builders Handbook
If you know the floor to floor rise, this handbook will give you everything else: the number and dimension of treads and risers, the total run, the correct well hole opening, the angle of incline, the quantity of materials and settings for your framing square for over 3,500 code approved rise and run combinations—several for every 1/8 inch interval from a 3 foot to a 12 foot floor to floor rise. **416 pages, 8½ x 5½, $15.50**

Builder's Guide to Accounting Revised

Step-by-step, easy to follow guidelines for setting up and maintaining an efficient record keeping system for your building business. Not a book of theory, this practical, newly-revised guide to all accounting methods shows how to meet state and federal accounting requirements, including new depreciation rules, and explains what the tax reform act of 1986 can mean to your business. Full of charts, diagrams, blank forms, simple directions and examples. **304 pages, 8½ x 11, $17.25**

Contractors Guide to the Building Code Rev.

This completely revised edition explains in plain English exactly what the Uniform Building Code requires and shows how to design and construct residential and light commercial buildings that will pass inspection the first time. Suggests how to work with the inspector to minimize construction costs, what common building short cuts are likely to be cited, and where exceptions are granted. **400 pages, 5½ x 8½, $24.25**

Video: Stair Framing

Shows how to use a calculator to figure the rise and run of each step, the height of each riser, the number of treads, and the tread depth. Then watch how to take these measurements to construct an actual set of stairs. You'll see how to mark and cut your carriages, treads, and risers, and install a stairway that fits your calculations for the perfect set of stairs. **60 minutes, VHS, $24.75**

Roof Framing

Frame any type of roof in common use today, even if you've never framed a roof before. Shows how to use a pocket calculator to figure any common, hip, valley, and jack rafter length in seconds. Over 400 illustrations take you through every measurement and every cut on each type of roof: gable, hip, Dutch, Tudor, gambrel, shed, gazebo and more. **480 pages, 5½ x 8½, $22.00**

Roofers Handbook

The journeyman roofer's complete guide to wood and asphalt shingle application on both new construction and reroofing jobs: How professional roofers make smooth tie-ins on any job, the right way to cover valleys and ridges, how to handle and prevent leaks, how to set up and run your own roofing business and sell your services as a professional roofer. Over 250 illustrations and hundreds of trade tips. **192 pages, 8½ x 11, $14.00**

Plumbers Handbook Revised

This new edition shows what will and what will not pass inspection in drainage, vent, and waste piping, septic tanks, water supply, fire protection, and gas piping systems. All tables, standards, and specifications are completely up-to-date with recent changes in the plumbing code. Covers common layouts for residential work, how to size piping, selecting and hanging fixtures, practical recommendations and trade tips. This book is the approved reference for the plumbing contractors exam in many states. **240 pages, 8½ x 11, $18.00**

Estimating Electrical Construction

A practical approach to estimating materials and labor for residential and commercial electrical construction. Written by the A.S.P.E. National Estimator of the Year, it explains how to use labor units, the plan take-off and the bid summary to establish an accurate estimate. Covers dealing with suppliers, pricing sheets, and how to modify labor units. Provides extensive labor unit tables, and blank forms for use in estimating your next electrical job. **272 pages, 8½ x 11, $19.00**

Audio Tape: Estimating Remodeling Work

Listen to these tapes, and use the workbook and forms to follow along as you learn estimating with the unit price method, preparing an operating budget, preparing an actual estimate from a set of plans and estimating with pre-printed project survey and estimating forms. If you've ever wanted to attend a "hands on" remodeling estimating seminar but couldn't take the time off work, you should have this complete estimating remodeling course. **Six-20 minute tapes, workbook & forms, $65.00**

Contractor's Survival Manual

How to survive hard times in construction and take full advantage of the profitable cycles. Shows what to do when the bills can't be paid, finding money and buying time, transferring debt, and all the alternatives to bankruptcy. Explains how to build profits, avoid problems in zoning and permits, taxes, time-keeping, and payroll. Unconventional advice includes how to invest in inflation, get high appraisals, trade and postpone income, and how to stay hip-deep in profitable work. **160 pages, 8½ x 11, $16.75**

Video: Roof Framing 1

A complete step-by step training video on the basics of roof cutting by Marshall Gross, the author of the book **Roof Framing**. Shows and explains calculating rise, run, and pitch, and laying out and cutting common rafters. **90 minutes, VHS, $80.00**

Video: Roof Framing 2

A complete training video on the more advanced techniques of roof framing by Marshall Gross, the author of **Roof Framing,** shows and explains layout and framing an irregular roof, and making tie-ins to an existing roof. **90 minutes, VHS, $80.00**

Contractor's Growth and Profit Guide

Step-by-step instructions for planning growth and prosperity in a construction contracting or subcontracting company. Explains how to prepare a business plan: selecting reasonable goals, drafting a market expansion plan, making income forecasts and expense budgets, and projecting cash flow. Here you will learn everything required by most lenders and investors, as well as solid knowledge for better organizing your business. **336 pages, 5½ x 8½, $19.00**

Masonry Estimating

Step-by-step instructions for estimating nearly any type of masonry work. Shows how to prepare material take-offs, how to figure labor and material costs, add a realistic allowance for contingency, calculate overhead correctly, and build competitive profit into your bids. **352 pages, 8½ x 11, $26.50**

Construction Surveying and Layout

A practical guide to simplified construction surveying: How land is divided, how to use a transit and tape to find a known point, how to draw an accurate survey map from your field notes, how to use topographic surveys, and the right way to level and set grade. You'll learn how to make a survey for any residential or commercial lot, driveway, road, or bridge — including how to figure cuts and fills and calculate excavation quantities. If you've been wanting to make your own surveys, or just read and verify the accuracy of surveys made by others, you should have this guide. **256 pages, 5½ x 8½, $19.25**

Paint Contractor's Manual

How to start and run a profitable paint contracting company: getting set up and organized to handle volume work, avoiding the mistakes most painters make, getting top production from your crews and the most value from your advertising dollar. Shows how to estimate all prep and painting. Loaded with manhour estimates, sample forms, contracts, charts, tables and examples you can use. **224 pages, 8½ x 11, $19.25**

Running Your Remodeling Business

Everything you need to know about operating a remodeling business, from making your first sale to insuring your profits: how to advertise, write up a contract, estimate, schedule your jobs, arrange financing (for both you and your customers), and when and how to expand your business. Explains what you need to know about insurance, bonds, and liens, and how to collect the money you've earned. Includes sample business forms for your use. **272 pages, 8½ x 11, $21.00**

Carpentry for Residential Construction

How to do professional quality carpentry work in homes and apartments. Illustrated instructions show you everything from setting batter boards to framing floors and walls, installing floor, wall and roof sheathing, and applying roofing. Covers finish carpentry, also: How to install each type of cornice, frieze, lookout, ledger, fascia and soffit; how to hang windows and doors; how to install siding, drywall and trim. Each job description includes the tools and materials needed, the estimated manhours required, and a step-by-step guide to each part of the task. **400 pages, 5½ x 8½, $19.75**

Builder's Office Manual, Revised

Explains how to create routine ways of doing all the things that must be done in every construction office — in the minimum time, at the lowest cost, and with the least supervision possible: Organizing the office space, establishing effective procedures and forms, setting priorities and goals, finding and keeping an effective staff, getting the most from your record-keeping system (whether manual or computerized). Loaded with practical tips, charts and sample forms for your use. **192 pages, 8½ x 11, $15.50**

Builder's Comprehensive Dictionary

Never let a construction term stump you. Here you'll find almost 10,000 construction term definitions, over 1,000 detailed illustrations of tools, techniques and systems, and a separate section containing most often used legal, real estate, and management terms. **532 pages, 8½ x 11, $24.95**

Video: Contracting a Home Vol. 1

How to take plans from a plan agency and customize them for your use. How to budget and keep track of your expenses throughout the project. Gives tips on choosing, hiring, and scheduling subcontractors. Shows what goes into excavation and foundation construction, framing, roof trusses, and rough-in of HVAC. Shows how to solve most of the major problems in building — before they come up. **1 hr. 30 min., VHS, $15.50**

Video: Contracting a Home Vol. 2

How to arrange for windows, doors, siding, and brick. Shows how to schedule for insulation and drywall contractors, coordinate tiling, cabinets and countertops. You'll see how to supervise final trim-outs and exterior finishing, and calculate final costs and savings for the home. **1 hr. 30 min., VHS, $15.50**

Carpentry in Commercial Construction

Covers forming, framing, exteriors, interior finish and cabinet installation in commercial buildings: designing and building concrete forms, selecting lumber dimensions, grades and species for the design load, what you should know when installing materials selected for their fire rating or sound transmission characteristics, and how to plan and organize the job to improve production. Loaded with illustrations, tables, charts and diagrams. **272 pages, 5½ x 8½, $19.00**

Craftsman Book Company
6058 Corte del Cedro
P.O. Box 6500
Carlsbad, CA 92008

Phone Orders

For charge card orders call *1-800-829-8123*
Your order will be shipped within 48 hours of your call.

Mail Orders

We pay shipping when you use your charge card or when your check covers your order in full.

Name (Please print clearly) ____________________

Company ____________________

Address ____________________

City ________ State ________ Zip ________

Total Enclosed ____________ (In California add 6% tax)

If you prefer, use your ☐ Visa ☐ MasterCard or ☐ AmEx

Card Number ____________________

Expiration date ____________ Initials ________

10 Day Money Back GUARANTEE

- ☐ 65.00 Audio: Estimating Remodeling Work
- ☐ 24.95 Builder's Comprehensive Dictionary
- ☐ 17.25 Builder's Guide to Accounting Revised
- ☐ 15.50 Builder's Office Manual Revised
- ☐ 14.00 Building Cost Manual
- ☐ 25.50 Carpentry Estimating
- ☐ 19.75 Carpentry for Residential Construction
- ☐ 19.00 Carpentry in Commercial Construction
- ☐ 20.50 Concrete Const. & Estimating
- ☐ 19.25 Construction Surveying & Layout
- ☐ 19.00 Contractor's Growth & Profit Guide
- ☐ 24.25 Contractor's Guide to the Building Code Rev.
- ☐ 16.75 Contractor's Survival Manual
- ☐ 19.00 Estimating Electrical Construction
- ☐ 17.00 Estimating Home Building Costs
- ☐ 28.00 Estimating Painting Costs
- ☐ 21.50 Estimating Tables for Home Building
- ☐ 24.75 Handbook of Const. Contracting Vol. 1
- ☐ 24.75 Handbook of Const. Contracting Vol. 2
- ☐ 19.75 Manual of Professional Remodeling
- ☐ 26.50 Masonry Estimating
- ☐ 19.50 National Construction Estimator
- ☐ 19.25 Paint Contractor's Manual
- ☐ 18.00 Plumber's Handbook Revised
- ☐ 26.25 Remodeling Kitchens and Baths
- ☐ 22.00 Roof Framing
- ☐ 14.00 Roofers Handbook
- ☐ 16.00 Rough Carpentry
- ☐ 21.00 Running Your Remodeling Business
- ☐ 15.50 Stair Builder's Handbook
- ☐ 15.50 Video: Contracting a Home Vol. 1
- ☐ 15.50 Video: Contracting a Home Vol. 2
- ☐ 80.00 Video: Roof Framing 1
- ☐ 80.00 Video: Roof Framing 2
- ☐ 24.75 Video: Stair Framing
- ☐ 14.25 Wood-Frame House Construction
- ☐ 27.00 Spec Builder's Guide

These books are tax deductible when used to improve or maintain your professional skill.

10 Day Money Back GUARANTEE

spk card

Craftsman Book Company
6058 Corte del Cedro
P. O. Box 6500
Carlsbad, CA 92008

Craftsman

In a hurry?
We accept phone orders charged to your MasterCard, Visa or Am. Ex.
Call 1-800-829-8123

Name (Please print clearly) ______________________

Company ______________________

Address ______________________

City/State/Zip ______________________

Total Enclosed ______________________ (In California add 6% tax)

Use your ☐Visa ☐MasterCard ☐Am. Ex.

Card # ______________________

Exp. date ____________ Initials ______

☐ 95.00 Audio: Construction Field Sup.
☐ 65.00 Audio: Estimating Remodeling
☐ 19.95 Audio: Plumbers Exam
☐ 22.00 Basic Plumbing with Illust.
☐ 30.00 Berger Building Cost File
☐ 11.25 Blprt Read. for Blding Trades
☐ 19.75 Bookkeeping for Builders
☐ 24.95 Builder's Comprehensive Dictionary
☐ 17.25 Builder's Guide to Accting. Rev.
☐ 15.25 Blder's Guide to Const. Fin.
☐ 15.50 Builder's Office Manual Revised
☐ 14.00 Building Cost Manual
☐ 11.75 Building Layout
☐ 25.50 Carpentry Estimating
☐ 19.75 Carp. for Residential Const.
☐ 19.00 Carp. in Commercial Const.
☐ 16.25 Carpentry Layout
☐ 17.75 Computers: Blder's New Tool
☐ 14.50 Concrete and Formwork
☐ 20.50 Concrete Const. & Estimating
☐ 26.00 Const. Estimating Ref. Data
☐ 22.00 Construction Superintending
☐ 19.25 Const. Surveying & Layout
☐ 19.00 Cont. Growth & Profit Guide
☐ 24.25 Cont. Guide to the Blding Code Rev.
☐ 16.75 Contractor's Survival Manual
☐ 16.50 Cont.Year-Round Tax Guide
☐ 15.75 Cost Rec. for Const. Est.
☐ 9.50 Dial-A-Length Rafterule
☐ 18.25 Drywall Contracting
☐ 13.75 Electrical Blueprint Reading
☐ 25.00 Electrical Const. Estimator
☐ 19.00 Estimating Electrical Const.
☐ 17.00 Estimating Home Blding Costs
☐ 28.00 Estimating Painting Costs
☐ 17.25 Estimating Plumbing Costs
☐ 21.50 Esti. Tables for Home Building
☐ 22.75 Exca. & Grading Handbook, Rev.
☐ 9.25 E-Z Square
☐ 15.25 Finish Carpentry
☐ 23.00 Kitchen Designer
☐ 24.75 Hdbk of Const. Cont. Vol. 1
☐ 24.75 Hdbk of Const. Cont. Vol. 2
☐ 14.75 Hdbk of Modern Elec. Wiring
☐ 15.00 Home Wiring: Imp., Ext., Repairs
☐ 17.50 How to Sell Remodeling
☐ 24.50 HVAC Contracting
☐ 20.25 Manual of Elect. Contracting
☐ 19.75 Manual of Prof. Remodeling
☐ 13.50 Masonry & Concrete Const.
☐ 26.50 Masonry Estimating
☐ 19.50 National Const. Estimator
☐ 23.75 Op. the Tractor-Loader-Backhoe
☐ 19.25 Paint Contractor's Manual
☐ 21.25 Painter's Handbook
☐ 23.50 Pipe & Excavation Contracting
☐ 13.00 Plan. and Design. Plumbing Sys.
☐ 19.95 Plumber's Exam Audiotape
☐ 21.00 Plumber's Exam Prep. Guide
☐ 18.00 Plumber's Handbook Revised
☐ 14.25 Rafter Length Manual
☐ 23.00 Remodeler's Handbook
☐ 26.25 Remodeling Kitchens & Baths
☐ 11.50 Residential Electrical Design
☐ 16.75 Residential Electrician's Hdbk.
☐ 18.25 Residential Wiring
☐ 22.00 Roof Framing
☐ 14.00 Roofers Handbook
☐ 16.00 Rough Carpentry
☐ 21.00 Running Your Remodeling Bus.
☐ 27.00 Spec Builder's Guide
☐ 15.50 Stair Builder's Handbook
☐ 15.50 Video: Asphalt Shingle Roofing
☐ 15.50 Video: Bathroom Tile
☐ 15.50 Video: Contracting a Home 1
☐ 15.50 Video: Contracting a Home 2
☐ 32.00 Video: Designing Your Kitchen
☐ 24.75 Video: Drywall Contracting 1
☐ 24.75 Video: Drywall Contracting 2
☐ 15.50 Video: Electrical Wiring
☐ 15.50 Video: Exterior Painting
☐ 15.50 Video: Finish Carpentry
☐ 15.50 Video: Hanging An Exterior Door
☐ 15.50 Video: Int. Paint & Wallpaper
☐ 15.50 Video: Kitchen Renovation
☐ 24.75 Video: Paint Contractor's 1
☐ 24.75 Video: Paint Contractor's 2
☐ 15.50 Video: Plumbing
☐ 80.00 Video: Roof Framing 1
☐ 80.00 Video: Roof Framing 2
☐ 15.50 Video: Rough Carpentry
☐ 24.75 Video: Stair Framing
☐ 15.50 Video: Windows & Doors
☐ 15.50 Video: Wood Siding
☐ 7.50 Visual Stairule
☐ 14.25 Wood-Frame House Const.

10 Day Money Back GUARANTEE

spk card

Craftsman Book Company
6058 Corte del Cedro
P. O. Box 6500
Carlsbad, CA 92008

Craftsman

In a hurry?
We accept phone orders charged to your MasterCard, Visa or Am. Ex.
Call 1-800-829-8123

Name (Please print clearly) ______________________

Company ______________________

Address ______________________

City/State/Zip ______________________

Total Enclosed ______________________ (In California add 6% tax)

Use your ☐Visa ☐MasterCard ☐Am. Ex.

Card # ______________________

Exp. date ____________ Initials ______

☐ 95.00 Audio: Construction Field Sup.
☐ 65.00 Audio: Estimating Remodeling
☐ 19.95 Audio: Plumbers Exam
☐ 22.00 Basic Plumbing with Illust.
☐ 30.00 Berger Building Cost File
☐ 11.25 Blprt Read. for Blding Trades
☐ 19.75 Bookkeeping for Builders
☐ 24.95 Builder's Comprehensive Dictionary
☐ 17.25 Builder's Guide to Accting. Rev.
☐ 15.25 Blder's Guide to Const. Fin.
☐ 15.50 Builder's Office Manual Revised
☐ 14.00 Building Cost Manual
☐ 11.75 Building Layout
☐ 25.50 Carpentry Estimating
☐ 19.75 Carp. for Residential Const.
☐ 19.00 Carp. in Commercial Const.
☐ 16.25 Carpentry Layout
☐ 17.75 Computers: Blder's New Tool
☐ 14.50 Concrete and Formwork
☐ 20.50 Concrete Const. & Estimating
☐ 26.00 Const. Estimating Ref. Data
☐ 22.00 Construction Superintending
☐ 19.25 Const. Surveying & Layout
☐ 19.00 Cont. Growth & Profit Guide
☐ 24.25 Cont. Guide to the Blding Code Rev.
☐ 16.75 Contractor's Survival Manual
☐ 16.50 Cont.Year-Round Tax Guide
☐ 15.75 Cost Rec. for Const. Est.
☐ 9.50 Dial-A-Length Rafterule
☐ 18.25 Drywall Contracting
☐ 13.75 Electrical Blueprint Reading
☐ 25.00 Electrical Const. Estimator
☐ 19.00 Estimating Electrical Const.
☐ 17.00 Estimating Home Blding Costs
☐ 28.00 Estimating Painting Costs
☐ 17.25 Estimating Plumbing Costs
☐ 21.50 Esti. Tables for Home Building
☐ 22.75 Exca. & Grading Handbook, Rev.
☐ 9.25 E-Z Square
☐ 15.25 Finish Carpentry
☐ 23.00 Kitchen Designer
☐ 24.75 Hdbk of Const. Cont. Vol. 1
☐ 24.75 Hdbk of Const. Cont. Vol. 2
☐ 14.75 Hdbk of Modern Elec. Wiring
☐ 15.00 Home Wiring: Imp., Ext., Repairs
☐ 17.50 How to Sell Remodeling
☐ 24.50 HVAC Contracting
☐ 20.25 Manual of Elect. Contracting
☐ 19.75 Manual of Prof. Remodeling
☐ 13.50 Masonry & Concrete Const.
☐ 26.50 Masonry Estimating
☐ 19.50 National Const. Estimator
☐ 23.75 Op. the Tractor-Loader-Backhoe
☐ 19.25 Paint Contractor's Manual
☐ 21.25 Painter's Handbook
☐ 23.50 Pipe & Excavation Contracting
☐ 13.00 Plan. and Design. Plumbing Sys.
☐ 19.95 Plumber's Exam Audiotape
☐ 21.00 Plumber's Exam Prep. Guide
☐ 18.00 Plumber's Handbook Revised
☐ 14.25 Rafter Length Manual
☐ 23.00 Remodeler's Handbook
☐ 26.25 Remodeling Kitchens & Baths
☐ 11.50 Residential Electrical Design
☐ 16.75 Residential Electrician's Hdbk.
☐ 18.25 Residential Wiring
☐ 22.00 Roof Framing
☐ 14.00 Roofers Handbook
☐ 16.00 Rough Carpentry
☐ 21.00 Running Your Remodeling Bus.
☐ 27.00 Spec Builder's Guide
☐ 15.50 Stair Builder's Handbook
☐ 15.50 Video: Asphalt Shingle Roofing
☐ 15.50 Video: Bathroom Tile
☐ 15.50 Video: Contracting a Home 1
☐ 15.50 Video: Contracting a Home 2
☐ 32.00 Video: Designing Your Kitchen
☐ 24.75 Video: Drywall Contracting 1
☐ 24.75 Video: Drywall Contracting 2
☐ 15.50 Video: Electrical Wiring
☐ 15.50 Video: Exterior Painting
☐ 15.50 Video: Finish Carpentry
☐ 15.50 Video: Hanging An Exterior Door
☐ 15.50 Video: Int. Paint & Wallpaper
☐ 15.50 Video: Kitchen Renovation
☐ 24.75 Video: Paint Contractor's 1
☐ 24.75 Video: Paint Contractor's 2
☐ 15.50 Video: Plumbing
☐ 80.00 Video: Roof Framing 1
☐ 80.00 Video: Roof Framing 2
☐ 15.50 Video: Rough Carpentry
☐ 24.75 Video: Stair Framing
☐ 15.50 Video: Windows & Doors
☐ 15.50 Video: Wood Siding
☐ 7.50 Visual Stairule
☐ 14.25 Wood-Frame House Const.

Charge Card Phone Orders — Call 1-800-829-8123

10 Day Money Back GUARANTEE

Craftsman Book Company
6058 Corte del Cedro
P. O. Box 6500
Carlsbad, CA 92008

Craftsman

In a hurry?
We accept phone orders charged to your MasterCard, Visa or Am. Ex.
Call 1-800-829-8123

Name (Please print clearly) ______________________

Company ______________________

Address ______________________

City/State/Zip ______________________

Total Enclosed ______________________ (In California add 6% tax)

Use your ☐Visa ☐MasterCard ☐Am. Ex.

Card # ______________________

Exp. date ____________ Initials ______

☐ 95.00 Audio: Construction Field Sup.
☐ 65.00 Audio: Estimating Remodeling
☐ 19.95 Audio: Plumbers Exam
☐ 22.00 Basic Plumbing with Illust.
☐ 30.00 Berger Building Cost File
☐ 11.25 Blprt Read. for Blding Trades
☐ 19.75 Bookkeeping for Builders
☐ 24.95 Builder's Comprehensive Dictionary
☐ 17.25 Builder's Guide to Accting. Rev.
☐ 15.25 Blder's Guide to Const. Fin.
☐ 15.50 Builder's Office Manual Revised
☐ 14.00 Building Cost Manual
☐ 11.75 Building Layout
☐ 25.50 Carpentry Estimating
☐ 19.75 Carp. for Residential Const.
☐ 19.00 Carp. in Commercial Const.
☐ 16.25 Carpentry Layout
☐ 17.75 Computers: Blder's New Tool
☐ 14.50 Concrete and Formwork
☐ 20.50 Concrete Const. & Estimating
☐ 26.00 Const. Estimating Ref. Data
☐ 22.00 Construction Superintending
☐ 19.25 Const. Surveying & Layout
☐ 19.00 Cont. Growth & Profit Guide
☐ 24.25 Cont. Guide to the Blding Code Rev.
☐ 16.75 Contractor's Survival Manual
☐ 16.50 Cont.Year-Round Tax Guide
☐ 15.75 Cost Rec. for Const. Est.
☐ 9.50 Dial-A-Length Rafterule
☐ 18.25 Drywall Contracting
☐ 13.75 Electrical Blueprint Reading
☐ 25.00 Electrical Const. Estimator
☐ 19.00 Estimating Electrical Const.
☐ 17.00 Estimating Home Blding Costs
☐ 28.00 Estimating Painting Costs
☐ 17.25 Estimating Plumbing Costs
☐ 21.50 Esti. Tables for Home Building
☐ 22.75 Exca. & Grading Handbook, Rev.
☐ 9.25 E-Z Square
☐ 15.25 Finish Carpentry
☐ 23.00 Kitchen Designer
☐ 24.75 Hdbk of Const. Cont. Vol. 1
☐ 24.75 Hdbk of Const. Cont. Vol. 2
☐ 14.75 Hdbk of Modern Elec. Wiring
☐ 15.00 Home Wiring: Imp., Ext., Repairs
☐ 17.50 How to Sell Remodeling
☐ 24.50 HVAC Contracting
☐ 20.25 Manual of Elect. Contracting
☐ 19.75 Manual of Prof. Remodeling
☐ 13.50 Masonry & Concrete Const.
☐ 26.50 Masonry Estimating
☐ 19.50 National Const. Estimator
☐ 23.75 Op. the Tractor-Loader-Backhoe
☐ 19.25 Paint Contractor's Manual
☐ 21.25 Painter's Handbook
☐ 23.50 Pipe & Excavation Contracting
☐ 13.00 Plan. and Design. Plumbing Sys.
☐ 19.95 Plumber's Exam Audiotape
☐ 21.00 Plumber's Exam Prep. Guide
☐ 18.00 Plumber's Handbook Revised
☐ 14.25 Rafter Length Manual
☐ 23.00 Remodeler's Handbook
☐ 26.25 Remodeling Kitchens & Baths
☐ 11.50 Residential Electrical Design
☐ 16.75 Residential Electrician's Hdbk.
☐ 18.25 Residential Wiring
☐ 22.00 Roof Framing
☐ 14.00 Roofers Handbook
☐ 16.00 Rough Carpentry
☐ 21.00 Running Your Remodeling Bus.
☐ 27.00 Spec Builder's Guide
☐ 15.50 Stair Builder's Handbook
☐ 15.50 Video: Asphalt Shingle Roofing
☐ 15.50 Video: Bathroom Tile
☐ 15.50 Video: Contracting a Home 1
☐ 15.50 Video: Contracting a Home 2
☐ 32.00 Video: Designing Your Kitchen
☐ 24.75 Video: Drywall Contracting 1
☐ 24.75 Video: Drywall Contracting 2
☐ 15.50 Video: Electrical Wiring
☐ 15.50 Video: Exterior Painting
☐ 15.50 Video: Finish Carpentry
☐ 15.50 Video: Hanging An Exterior Door
☐ 15.50 Video: Int. Paint & Wallpaper
☐ 15.50 Video: Kitchen Renovation
☐ 24.75 Video: Paint Contractor's 1
☐ 24.75 Video: Paint Contractor's 2
☐ 15.50 Video: Plumbing
☐ 80.00 Video: Roof Framing 1
☐ 80.00 Video: Roof Framing 2
☐ 15.50 Video: Rough Carpentry
☐ 24.75 Video: Stair Framing
☐ 15.50 Video: Windows & Doors
☐ 15.50 Video: Wood Siding
☐ 7.50 Visual Stairule
☐ 14.25 Wood-Frame House Const.

spk card

These books are tax deductible when used to improve or maintain your professional skill.

NO POSTAGE NECESSARY IF MAILED IN THE UNITED STATES

BUSINESS REPLY MAIL

FIRST CLASS PERMIT NO. 271 CARLSBAD, CA

POSTAGE WILL BE PAID BY ADDRESSEE

Craftsman Book Company
6058 Corte Del Cedro
P. O. Box 6500
Carlsbad, CA 92008-0992

NO POSTAGE NECESSARY IF MAILED IN THE UNITED STATES

BUSINESS REPLY MAIL

FIRST CLASS PERMIT NO. 271 CARLSBAD, CA

POSTAGE WILL BE PAID BY ADDRESSEE

Craftsman Book Company
6058 Corte Del Cedro
P. O. Box 6500
Carlsbad, CA 92008-0992

NO POSTAGE NECESSARY IF MAILED IN THE UNITED STATES

BUSINESS REPLY MAIL

FIRST CLASS PERMIT NO. 271 CARLSBAD, CA

POSTAGE WILL BE PAID BY ADDRESSEE

Craftsman Book Company
6058 Corte Del Cedro
P. O. Box 6500
Carlsbad, CA 92008-0992